Handbook of Research on Inventive Bioremediation Techniques

Jatindra Nath Bhakta
University of Kalyani, India

A volume in the Advances in Environmental Engineering and Green Technologies (AEEGT) Book Series

IGI Global
DISSEMINATOR OF KNOWLEDGE

www.igi-global.com

Published in the United States of America by
IGI Global
Engineering Science Reference (an imprint of IGI Global)
701 E. Chocolate Avenue
Hershey PA, USA 17033
Tel: 717-533-8845
Fax: 717-533-8661
E-mail: cust@igi-global.com
Web site: http://www.igi-global.com

Library of Congress Cataloging-in-Publication Data

Names: Bhakta, Jatindra Nath, editor.
Title: Handbook of research on inventive bioremediation techniques / Jatindra Nath Bhakta, editor.
Description: Hershey, PA : Engineering Science Reference, 2017. | Includes bibliographical references.
Identifiers: LCCN 2016058807 | ISBN 9781522523253 (hardcover) | ISBN 9781522523260 (ebook)
Subjects: LCSH: Bioremediation--Handbooks, manuals, etc.
Classification: LCC TD192.5 .H357 2017 | DDC 628.5 --dc23 LC record available at https://lccn.loc.gov/2016058807

This book is published in the IGI Global book series Advances in Environmental Engineering and Green Technologies (AEEGT) (ISSN: 2326-9162; eISSN: 2326-9170)

British Cataloguing in Publication Data
A Cataloguing in Publication record for this book is available from the British Library.

For electronic access to this publication, please contact: eresources@igi-global.com.

Advances in Environmental Engineering and Green Technologies (AEEGT) Book Series

ISSN:2326-9162
EISSN:2326-9170

Mission

Growing awareness and an increased focus on environmental issues such as climate change, energy use, and loss of non-renewable resources have brought about a greater need for research that provides potential solutions to these problems. Research in environmental science and engineering continues to play a vital role in uncovering new opportunities for a "green" future.

The **Advances in Environmental Engineering and Green Technologies (AEEGT)** book series is a mouthpiece for research in all aspects of environmental science, earth science, and green initiatives. This series supports the ongoing research in this field through publishing books that discuss topics within environmental engineering or that deal with the interdisciplinary field of green technologies.

Coverage

- Water Supply and Treatment
- Electric Vehicles
- Alternative Power Sources
- Renewable Energy
- Policies Involving Green Technologies and Environmental Engineering
- Pollution Management
- Green Transportation
- Radioactive Waste Treatment
- Biofilters and Biofiltration
- Industrial Waste Management and Minimization

IGI Global is currently accepting manuscripts for publication within this series. To submit a proposal for a volume in this series, please contact our Acquisition Editors at Acquisitions@igi-global.com or visit: http://www.igi-global.com/publish/.

The Advances in Environmental Engineering and Green Technologies (AEEGT) Book Series (ISSN 2326-9162) is published by IGI Global, 701 E. Chocolate Avenue, Hershey, PA 17033-1240, USA, www.igi-global.com. This series is composed of titles available for purchase individually; each title is edited to be contextually exclusive from any other title within the series. For pricing and ordering information please visit http://www.igi-global.com/book-series/advances-environmental-engineering-green-technologies/73679. Postmaster: Send all address changes to above address.

Titles in this Series

For a list of additional titles in this series, please visit: www.igi-global.com

Applied Environmental Materials Science for Sustainability
Takaomi Kobayashi (Nagaoka University of Technology, Japan)
Information Science Reference • copyright 2017 • 416pp • H/C (ISBN: 9781522519713) • US $205.00 (our price)

Environmental Issues Surrounding Human Overpopulation
Rajeev Pratap Singh (Banaras Hindu University, India) Anita Singh (University of Allahabad, India) and Vaibhav Srivastava (Banaras Hindu University, India)
Information Science Reference • copyright 2017 • 325pp • H/C (ISBN: 9781522516835) • US $200.00 (our price)

Reconsidering the Impact of Climate Change on Global Water Supply, Use, and Management
Prakash Rao (Symbiosis International University, India) and Yogesh Patil (Symbiosis International University, India)
Information Science Reference • copyright 2017 • 430pp • H/C (ISBN: 9781522510468) • US $215.00 (our price)

Environmental Sustainability and Climate Change Adaptation Strategies
Wayne Ganpat (The University of the West Indies, Trinidad and Tobago) and Wendy-Ann Isaac (The University of the West Indies, Trinidad and Tobago)
Information Science Reference • copyright 2017 • 406pp • H/C (ISBN: 9781522516071) • US $200.00 (our price)

Technologies for the Treatment and Recovery of Nutrients from Industrial Wastewater
Ángeles Val del Río (University of Santiago de Compostela, Spain) José Luis Campos Gómez (University Adolfo Ibáñez, Chile) and Anuska Mosquera Corral (University of Santiago de Compostela, Spain)
Information Science Reference • copyright 2017 • 391pp • H/C (ISBN: 9781522510376) • US $200.00 (our price)

Agricultural Development and Food Security in Developing Nations
Wayne G. Ganpat (The University of the West Indies, Trinidad and Tobago) Ronald Dyer (Grenoble School of Management, France) and Wendy-Ann P. Isaac (The University of the West Indies, Trinidad and Tobago)
Information Science Reference • copyright 2017 • 370pp • H/C (ISBN: 9781522509424) • US $200.00 (our price)

Oceanographic and Marine Cross-Domain Data Management for Sustainable Development
Paolo Diviacco (Istituto Nazionale di Oceanografia e di Geofisica Sperimentale (OGS), Italy) Adam Leadbetter (Marine Institute, Ireland) and Helen Glaves (British Geological Survey, UK)
Information Science Reference • copyright 2017 • 425pp • H/C (ISBN: 9781522507000) • US $205.00 (our price)

Examining the Development, Regulation, and Consumption of Functional Foods
Sailas Benjamin (University of Calicut, India)
Medical Information Science Reference • copyright 2017 • 344pp • H/C (ISBN: 9781522506072) • US $185.00 (our price)

List of Contributors

Table of Contents

Detailed Table of Contents

Global modernization demands a huge energy for developing new technologies to meet the increasing human needs. The initial source of energy was fossil fuels that might release some harmful gases. Burning of fossil fuels was practiced since the 18th century, which consequently resulted in 40% increase in atmospheric CO_2 according researcher. Presently, the atmospheric level of C is increasing day by day. However, this level would reach ~1000 ppm at the end of this century. The consistent discharge of CO_2 into the atmosphere is a major threat to global warming; hence, it is important to minimize the atmospheric CO_2 by using innovative carbon sequestration technologies. Therefore, soil and water can act as a sink for C storage with the assistance of microorganisms. Here, an attempt is made to discuss these processes in details, interactions between plant and microbes, and the requirement of C sources for C sequestration in the context of climate change. We also conferred about the development of microbial inoculants for C sequestrations and their application for sustainability of agro-ecosystems.

Bioremediation is applied to eliminate various contaminants, such as organic, inorganic or other pollutants from the environment. Environment worldwide is under great stress due to industrialization and human interfering on the limited natural resources. The release of chemicals pollution needs several techniques to treat some of these chemicals, but due to their cost, new technologies should be developing in order to create cost-effective and eco-friendly bioremediation technologies for environmental conversions. Bioremediation is an increasingly popular using microbial and algae strains for degrading waste contaminants. It is using of microorganisms and its enzymes to protect the environment from severe pollution. Bioremediation may be employed in order to eliminate specific contaminants, such as chlorinated pesticides or other pollutants from the environment. Microorganisms degrade the different pollutants in a natural environment but some modifications can be done to enhance its degradation efficiency at a faster rate in a limited time frame by using the genetically engineered microorganisms and microalgae. In this chapter, the role of the bacteria, fungi and algae in bioremediation of different environmental pollutants was highlighted.

The diversity of agro industrial wastes makes them an attractive group of organic wastes for potential use in a wide variety of industrial and biotechnological applications. The new stimulating development in this current area of research approaches in combination with the technologies of large-scale production and biotechnology engineering, agro industrial wastes will be economically successful materials of the future. Increased public awareness of issues related to hydrocarbon pollution strongly influences the development of technologies that speed up cleaning hazardous contaminants. The cost of biodegradation technology and the low bioavailability including mass transfer limitations of hydrocarbons, especially those recalcitrant components, from various mediums into the aqueous phase for effective enzyme-based microbial biodegradation still constitute major challenges. Sustainable replacement of traditional microbiological media with agroindustrial wastes as substrates for biosurfactant production holds great

Metal toxification has remained one of the problems with the advent of industrial revolution. Plant based remediation are showing increasing promise for use in soils contaminated with organic and inorganic pollutants. A large number of plant families has been identified which has shown significant result in detoxification of heavy metals. Hyperaccumulator plant is capable of sequestering heavy metals in their shoot tissues. High tolerance to HM toxicity is dependent on a reduced metal uptake or increased internal sequestration, which depends on plant and environmental condition. Recent progresses on understanding cellular/molecular mechanisms of metal tolerance by plants are reviewed. This chapter aims to focus on molecular mechanism involved in heavy metal detoxification and tolerance by plants. A different method by which plant effectively converts toxic metal in less toxic compounds has been explained in this chapter. Further, mode of accumulation and sequestration of metals have been explained which are utilized by hyper accumulators.

The most abundant aromatic biopolymer on earth Lignin is extremely recalcitrant to degradation. It creates a barrier to solutions or enzymes by linking to both hemicellulose and cellulose preventing the penetration of lignocellulolytic enzymes into the interior lignocellulosic structure. Global attention has been gained by fungi owing to the potential use of their versatile enzymes for agriculture, medicines, industries and bioremediation. The combination of extracellular ligninolytic enzymes, mediators, organic acids and accessory enzymes make some of the basidiomycete white-rot fungi to be able to degrade lignin efficiently. This review describes remediation of lignocelluloses by fungi, properties of fungi, their spatial distribution and the mechanisms of action which render them attractive candidates in biotechnological applications like biopulping, animal feed, genetic engineering and space exploration.

The potential use of coal as source of carbon nano structure is seldom investigated. Herein we report a facile fungal solubilization method to extract mixed phase carbon structure from low grade coal. Coal had been used as a primary source for the production of carbon nanostructure with novel property, in addition to its main utility as a fuel. The major hurdle in its application is the inherent mineral embedded in it. An environmentally benign demineralization procedure make coal as a widely accepted precursor for the novel carbon materials. With Aspergiilus niger leaching, the randomly oriented preformed crystalline mixed phase nanocarbon in coal can be extracted. Raman studies revealed the presence of E2g scattering mode of graphite. The sp3 domains at ~1355 cm-1 (D band) is an indication of diamond like structure with disorder or defect. In the 2D region, multilayer stacking of graphene layers is noticed. The ratio of the defect to graphitic bands was found to be decreasing with increasing rank of coal. Bio leaching of

from different sources like Industrial, domestic and agricultural etc. having different kinds of chemical compound i.e. organic or inorganic. Traditional methods are not able to deal with some of these chemical compounds. Bioremediation process is good option in such environmental problems. Bioremediation provides a technique for cleaning up pollution by enhancing the natural biodegradation processes. It treats such waste with the help of microorganism. Number of microbes including aerobes, anaerobic and fungi are involved in bioremediation process. Specific types of microbes are used to treat specific type of chemical contaminant. The chapter include all the techniques of bioremediation used to treat different kinds of contaminant.

The astounding increase in the use of pharmaceuticals in the last decade has raised concerns over their occurrence in the soils and wastewaters posing potential dangers to the general public health and environment. Considering the limitations of chemical remediation measures in treatment of recalcitrant xenobiotics, several bioremediation measures are being currently investigated and proposed for removal of pharmaceutical contaminants from the environment. Several bacterial, fungal and plant species have shown promising bioremediation potential with regard to the removal of pharmaceuticals. Varying configurations of anaerobic and aerobic reactors have been utilized for pharmaceutical wastewater treatment. This chapter is intended to give a compilation and overview of the various types of bioremediation measures currently being employed or investigated to remove the pharmaceutical pollutants.

Heavy metal(loid)s are hazardous, biologically non-essential, non-biodegradable and persistent in nature, which can accumulate in plants and animals as well as in environment especially agri- and aqua- culture ecosystems. It is severely responsible for causing several health hazards problems in human, such as, cardiovascular, pulmonary, hepatic, nephrological, dermatological, neurological disorders as well as carcinogenic effects. Removal of these heavy metals from living systems is extensively expensive and also unsuccessful in sent percent removal. Therefore, in order to protect the environment, the removal of heavy metal(loid)s from polluted effluents is essential before discharging into environment. Besides various treatment technologies, sorption of metal(loid)s using bio-wastes are highly potent alternatives in recent years. The present chapter deals with the removal efficiencies of various bio-wastes, orange peels, waste tea leaves, rice husk, wheat stalk, sugar cane bagasse, coconut husk, sun flower stalk, corn cob, nut shell, water hyacinth, crab shell particle, activated carbons etc. The present discussion has also revealed that bio-waste could be a low-cost eco-friendly and green emerging alternative technology in treating the metal(loid)s contaminated environment without posing any further adverse environmental impacts.

Environmental pollution has been an irrefutable fact of life for many centuries; but it has become a real problem, since the start of the industrial revolution. Discharge of these toxic compounds without treatment results in serious health risks to humans and the marine ecosystem. Several physical, chemical and biological methods have been employed for the remediation of the phenolics. Bioremediation is identified as the most efficient, cost effective and eco-friendly ways for treatment of phenolic compounds. This article is a comprehensive review on the sources of phenolic compounds, their hazards, and their fate once released into the environment; the treatment technologies employed and bioremediation of these compounds using both non-extremophlic and extremophilic organisms. The review, throws light on the enzymes involved in the remediation of phenolic compounds, highlights the importance of extremophilic organisms and biological treatment of phenol containing industrial wastewaters. Such comprehensive information on the research work performed for the remediation of phenolic compounds provide ways to explore the role played by micro organisms in the remediation of phenolic compounds, which could be applied in the remediation of phenol /contaminated sites even under extreme conditions.

Pharmaceutical and personal care products (PPCPs) are regularly used by human being for their day to day life. These products contains various chemical compounds which are regularly added in our surface and ground water sources either through untreated or partially treated domestic and industrial wastes or through agricultural runoffs etc. These are present at the concentration of µg liter-1, and at such low concentration also these have shown harmful affect to living being. A lot of studies have been carried out for studying their effects on animal biodiversity, which can directly or indirectly affect human being also. So their removal from waste water is essential. Various technologies are being tested for removing these compounds from polluted water; bioremediation is also one of them. Present chapter will briefly give the description of various PPCPs present in waste water, their impact and removal technologies available for their removal with special emphasis on bio remediation.

Environmental pollution is one of the most acute global problems. Pollution is present at the every corners of the globe. Several methods are used for monitoring the pollutants in the environment. The present chapter includes the wide variety of testing such as physical, chemical, biological and as well molecular to determine the level of biodegradable contaminated substances in the environment and also check the feasibility of different bioremediation techniques. The monitoring and testing are essential and have an enormous impact on the cost of full-scale remediation. Successful remediation techniques completely degrade the contaminants. A successful bioremediation approach requires sufficient evidences for the non-toxicity of the contaminant. Current monitoring techniques require the disappearance of the contaminants. Different recombinant techniques also play great role to assess the pollution level. In the field and focus on interdisciplinary research, bioremediation technology will go a long way in cleaning our polluted environment in near future. Research on improved microbial strains and bioanalytical methods for measuring the level of contaminants should be strengthen.

Foreword

Modern and advanced technologies in all dimensions of human life are responsible for continuous releasing various hazardous pollutants, which seriously degrading the total environmental as well as human health since last few decades worldwide. Severe impact of environmental pollution increasing like "snowball effect" with time due to the phenomena of environmental and biological accumulation and magnification of hazardous and persistent pollutants, such as, heavy metal(loid)s, pesticides, insecticides, herbicides, polycyclic aromatic hydrocarbons (PAHs) and polychlorinated biphenyls (PCBs), dyes, phenolic compounds and other diversified xenobiotics. To overcome the tremendous and dangerous thread of environmental pollution, scientists continuously paying their much attention in various fields by developing ultra modern physico-chemical and biological technologies. However, very few technologies developed so far are environmental friendly in developing clean environment by treating precious soil, water and air compartments. Recently, bioremediation has been emerged as an important eco-friendly technology effectively treating, rescuing and conserving different polluted domains in various parts of the worlds employing potential pollutant detoxifying biological agents, microbes, bacteria, algae and fungi as well as various higher plants and animals, etc. in natural ecosystem and designed (or constructed) ecosystem using the approach of ecological engineering.

The present book idea has been conceived by editor, Dr. Jatindra Nath Bhakta who is expert in this field having long experiences in various international institutions and Universities with enough concerned quality research publications. The contributed chapters are of advance and current technologies presented critically dealing with current environmental pollutions and its bioremediation using ultra modern, low-cost and practically applicable technologies by a number of reputed researches, scientists and professionals from various countries. It is a very practical and field oriented book excellently explored and comprised the current and updated technologies in bioremediation. The advanced bioremediation technologies discussed in this book will be an eco-friendly solution for conserving the environments. Relevant researchers, scientists and professions can be benefited from various chapters of this book gaining knowledge in order to flourish new cutting edge research in the field of bioremediation by identifying research gaps. Moreover, I strongly believe that the book is a ready reckoner to the new researcher and also relevant students.

B. B. Jana
University of Kalyani, India

Preface

Twentieth and beginning of twenty-first centuries brought unprecedented developments in human history, with major breakthroughs in various scientific and technical fields and a fourfold increase in human population (from 1.6 billion people in 1900 to over 6 billion at present). An awful consequence of technology's recent fast progress, rapid and continuous industrialization, urbanization, population explosion, economic development and growth in developed, developing and underdeveloped countries around the world is the immense deterioration of total environment by polluting and contaminating the different compartments of environment especially soil, water and air through massive emission of various (such as - solid, liquid and gases) pollutants from different point and nonpoint sources i.e., anthropogenic and natural or geogenic sources. Recently, the tremendous pollution and contamination of total environment are the primary causes for global warming and climate change resulting in the frequently occurrence of sever natural disasters and calamities such as, cyclone, floods, tornadoes, volcanic eruptions,blizzards, draughts, wildfires, heat waves, etc., which is the alarming threats to all forms of lives on earth in near future. Therefore, protection and conservation of environment from pollution and contamination should be one of the important and primary objectives for sustaining the present civilization. These environmental problems are indeed critical and problematic challenges to scientists during last few decades, since it has been integrated to the economic development of the country.

To control the pollution, though varieties of technologies have been developed by researchers in order to remove the pollutants from environment throughout the world over time for the conservation of environment, such as – extraction, complexation, adsorption, absorption, sequestration, precipitation, membrane filtration, coagulation, ion exchange and diagenesis, etc., all are expensive, not feasible and ecofriendly in practical field and also have several further deleterious environmental impacts. Therefore, environmental friendly pollution controlling green technologies only can solve and overcome such problems in order to develop clean soil, water and air environments without posing further adverse impacts.

Recently, bioremediation, an environmental remediation technology using biological agents, evolved as an emerging eco-friendly and green biotechnology which is becoming inevitable and widely applied as promising and low-cost technology for controlling the pollution and sustainable conservation of environment. The main phenomenon of bioremediation process is to immobilize contaminants or to transform them to chemical products no longer hazardous to human and environment health. It is pronouncing an unimaginable performance in remediation of various pollutants from different domains of environment without exerting any adverse, deleterious and extraneous side effects in environment. Researches have evidenced that it has potential significance and merit in remediation of numerous categories of pollutants or contaminants those are recently playing devastating role in environment, such as metal, antibiotic, herbicides, pesticides, insecticides, phenolic, dyestuff compounds, hydrocarbons, etc. The processes broadly involved in the bioremediation are sorption and chemical transformation in order to detoxify and decontaminate the pollutants. Varieties of technologies are practicing globally in bioremediation approaches of different afore said pollutants using potential biological agents of the environment. The systematic presentation of very novel currently developed bioremediation technologies would be a great credential in respect to its practical application for sustainable environmental conservation from dangerous effects of hazardous pollutants. There is lack of literatures concerning the very current and updated novel bioremediation technologies developed so far. Therefore, idea of the present book has been conceived by the Editor.

The prime objectives of present book are to present and to draw a novel picture in current advancements of theoretical and practical frameworks of application from latest empirical research findings considering various aspects of bioremediation.

The book contains the following topics in twenty different chapters: basic concept and principle of process, modern biochemical and molecular approaches of contamination and toxicity, bioremediation, genomics, proteomics, remediation strategies of various hazardous pollutants, microbial remediation, phytoremediation, bioleaching, biosorption, upscaling of system, and merit and demerits. Various expert scientists, researchers, engineers and professionals have contributed updated and current research findings on afore said diversified topics of bioremediations.

The special characteristics of book are:

1. It includes basic concepts and principle as well as applied aspects of bioremediation,
2. Information jotted down herein are of depth concept, updated and current research findings
3. Each chapters are with plenty of relevant natural pictures and drawn pictorial illustrations to realize the presented facts,
4. Book considers current bioremediation approaches of some specific hazardous materials, metals, hydrocarbons, greenhouse gas carbon dioxide, pharmaceutical and personal care products, agro-industrial wastes, municipal wastes, phenolic compounds, various xenobiotics, etc.,
5. Inclusion of proteomics and genomics approaches of bioremediation, and
6. Vivid description of bioremediation characteristics of various biological agents.

The above fascinating criteria of present book will make it different from other similar book which is helpful for readers. To develop the basic and advance concepts on bioremediation as well as application of bioremediation technologies in practical field, therefore, readers should go through all chapters of book which help to develop overall and depth knowledge by gaining updated information concerning various current concepts of bioremediation. Due to having basic and applied aspects of bioremediation, it is a kind of reference text book to graduate and master students of relevant subjects along with a potential reference book of researchers and scientist.

Since, the book is a potentially rich source of different theoretical and practical cutting-edge research aspects of bioremediation of environmental pollution and its low-cost application technologies, which is principally conducive for practitioners of developed, developing and underdeveloped counties for practical applications, therefore, environmental professionals, engineers, researchers, scientist, teachers and students working in the fields of environmental science, bioremediation, pollution, engineering, management, microbiology, biotechnology, biochemistry, life sciences, proteomics, genomics and metagenomics, toxicity, waste treatment, industries, etc. will be benefited from this book.

All member of editorial advisory board, authors and reviewers of the book are carefully, sincerely and cordially supported and extended the hand of cooperation by spending their valuable times in every steps of reviewing process for the improvement of the chapters' quality. I express my heartiest thanks and profound gratitude to all authors and reviewers for their endless endeavours in compilation of the present book. I am also pleased to convey my sincere thanks towards all concern professionals, teachers, students, researchers and engineers involved and engaged to carryout work on present topic to save earth.

Jatindra Nath Bhakta
University of Kalyani, India

Chapter 1
Metal Toxicity in Microorganism

Jatindra Nath Bhakta
University of Kalyani, India

ABSTRACT

An awful consequence of metal contamination in environment is one of the global problems posing severe hazardous and toxic impacts in microorganisms. The objective of the present chapter is to elucidate how metals cause toxicity at biochemical and molecular levels in microorganisms. The excess concentration of metals is responsible for causing various toxicity reactions in microbial cell, such as, over production of reactive oxygen species; protein and enzyme dysfunction, destruction of thiol and iron-sulfide cluster, metal substitution and inhibition of nutrient assimilation; lipid peroxidation; and DNA damage. Consequently, toxicity causes mutagenicity effects and/or cell death that lead to immeasurable damage in microorganisms and microbial community. The biochemical and molecular mechanisms of metal toxicity may be helpful to depth metal toxicity study in microbes and other organisms for controlling and treating the metal toxicity in further. Moreover, metal-resistant microbes have potential significance in environmental and human health perspectives.

INTRODUCTION

Fast progress technologies are primarily responsible for causing tremendous environmental pollution. Metal pollution is one of the prime problems worldwide posing severe damage in environmental and human health. Metals[1] are naturally occurring major constituting elements in the earth's crust. Although, many of them play vital role in daily human life and various physiological process of organisms, it becomes potential contaminant in the environment under certain condition. Anthropogenic and geogenic activities generate massive amount of metals and its derivative chemicals and discharge as pollutants, which are awfully responsible for causing the environmental metal pollution. Therefore, metal pollution is a growing concern worldwide during last few decades due to posing severely hazardous and toxic impacts in all forms of life on earth especially microorganisms by bioconcentration, bioaccumulation and biomagnifications phenomena through food chain and food web (Jillian, Robert, & Rajakaruna, 2015, Bhakta, 2016). Microorganisms (such as bacteria, protists, fungi, yeast, algae, etc.) are omnipresent vital ecosystem components playing pivotal roles in various biogeochemical cycling process of environment (Gadd, 2010). Microbial communities are tremendously affected and intoxicated by environmental metal contamination (Bhakta, Ohnishi, Munekage, & Iwasaki, 2010; Bhakta, Munekage, Ohnishi, &

DOI: 10.4018/978-1-5225-2325-3.ch001

Jana, 2012a; Bhakta, Ohnishi, Munekage, Iwasaki, & Wei, 2012b; Olaniran, Balgobind, & Pillay, 2013; Lenart-Boroń & Boroń, 2014; Bhakta, Munekage, Ohnishi, Jana, & Balcazar, 2014; Kuperman, Siciliano, Römbke, & Oorts, 2014).

Certain metals are crucial for structure and function of cell, and hence indispensable for the biochemical process of life. Approximately half of all known proteins are predicted to be dependent on metal atoms for their structure and their participation in key cellular processes, such as electron transfer and catalysis (Waldron & Robinson, 2009; Andreini, Bertini, & Rosato, 2004). Several metal elements e.g. Na, Mg, K, Ca, Mn, Fe, Co, Ni, Zn, Mo, etc. are essential for functioning the life. These "essential metal" elements are present in a certain concentration range in biological systems and take part in various biochemical reactions in cell. They act as essential cofactors for structural and catalytic roles in enzymes and proteins such as, metalloenzyme and metalloproteins and for stabilizing biological molecules. They are also used in electron transfer and utilization of dioxygen and osmotic balance (Bruins, Kapil, & Oehme, 2000). The essential metal elements can alter (decrease/increase) metabolic activity and lead to develop an adverse situation at too low or insufficient/high concentrations. Besides, there are a number of metals, for example Al, Au, Ag, Bi, Cd, Cr, Hg, Pb, Sn, Ti etc. (metals) and As, Sb, Te (metalloids), those have no known positive biological role and showed hazardous toxic impacts on life are commonly known as "non-essential" metals. In addition, there are some other metals those have no known biological role and may be toxic to microbes with little known biological interactions, these are Be, Cs, Li and Sr. Non-essential metals are generally toxic at very low concentrations and inhibit metabolic activity at certain concentrations in organism. Some of them are referred to as "heavy metals". These metals are generally called as "toxic metals" under certain conditions. Such, different conditions of metals, (i) insufficient levels of essential metal ions, (ii) excess levels of essential metal ions and (iii) the presence of toxic metal ions, can cause cellular stress conditions.

Paradoxically, although the lists of essential, non-essential and toxic metals appear straightforward, many essential metals can be toxic to bacteria if their intracellular concentration becomes too high (Lippard & Berg, 1994; Nies, 1999).

However, it is apparent that all metals can accumulate and exert lethal toxicity impacts to cell when present in excess or above certain threshold concentrations. At high or specific concentrations, all metals can act in a deleterious manner by blocking essential functional groups, displacing other metal ions, or modifying the active conformation of biological molecules (Collins & Stotzky, 1989). These adverse effects of metals for both higher organisms and microorganisms are generally termed as metal toxicity. Such, the mechanism of adverse interactions between metals (and/or its derivatives) and microorganisms resulting in the development of deleterious impacts in microorganism can be defined as "metal toxicity" of microorganism. It alters various biochemical metabolic processes and may damage the microbial cell.

On the basis of different natures of metals available and exposed for experiencing a complex metal-microbial interactions and ultimately toxicity to microbes, the metal toxicity can be characterized into two following categories:

1. **Mono-Metal Toxicity:** One type of metal present over the threshold level in ambience is responsible for causing the metal toxicity to microbes. Microbes can easily tackle this toxicity by improvising responses of mono-metal resistance and
2. **Multi-Metal Toxicity:** When more than one type of metals are present in environment in higher concentrations compared to that of the threshold level, microbes are suffered by critical toxicity of numbers of metals.

Microbes are generally unable to control such critical situation of multi-metal toxicity. Occasionally, microbes may acquire multi-metal resistant properties in this condition.

From the above points of view, it has been evident that metals cause toxicity by altering metabolic process under certain concentrations in microorganisms. Although several studies have randomly been performed by various scientists to determine and elucidate the metal toxicity process in microbes during recent years, nonetheless there is indeed a lack of comprehensive and systematic information so far of a question what is the exact mechanism of metal toxicity in microorganism. Therefore, the present chapter has aimed to find out answer of the question, how metals cause toxicity in microbes by illustrating the biochemical and molecular reactions and influencing factors involved in metal toxicity mechanism of microorganism.

CHEMICAL PROPERTIES OF METALS RESPONSIBLE FOR TOXICITY

The important chemical properties of metal are specific and different for each type of metal, which play significant roles in living cell. This property also determines the degree of interaction between metals and biomolecules of cell, which leads to explain why some metals are toxic, and others are not. The following three most pertinent chemical properties of metals are found to be important for microbial toxicity, such as - Selectivity properties of atom, redox potential properties, chemical speciation properties (Lemire, Harrison,& Turner, 2013).

Selectivity Properties of Atom

In membrane and cytoplasm of cell, metal and biomolecule together form a metal-biomolecule complex for transport as well as for facilitating the vital biochemical reactions. According to principle of coordination chemistry, a central metal ion binds to the atoms of donor ligand (an ion or molecule or functional group, such as O, N and S) to form the coordination complex of metal-ligand. The bonding between metal and ligand generally involves formal donation of one or more of the ligand's electron pairs. The nature of metal-ligand bonding (one to three) can range from covalent to ionic. The selectivity criteria of atom of donor ligand are specific and strong to form metal-ligand coordination complex (Haas & Franz, 2009; Ma, Jacobsen, & Giedroc, 2009) which also underlies the ability of the donor groups of biomolecules to confine essential metals to specific roles. The selectivity criteria are governed by the structural specificity of ligand and metal atoms. The ionic size of a metal ion decreases as nuclear charge increases from left to right across the Periodic Table, and the polarizing power of metal ions increases. Metal ions of smaller ionic size have higher polarizing power. Therefore, a metal cation with a high polarizing power (charge/radius ratio) interacts strongly with a ligand than a metal cation with a lower polarizing power. It follows a universal order of binding preference of given donor ligands for divalent transition metal ions is called the Irving–Williams series (Ca(II) < Mg(II) < Mn(II) < Fe(II) < Co(II) < Ni(II) < Cu(II) > Zn(II)) (Lippard & Berg, 1994; Waldron & Robinson, 2009; Irving & Williams, 1953). Using this specific property ligand protein leads to select the correct metals required for protein folding and functions. Such, protein also can discriminate toxic, non-essential metal species from essential inorganic ions. The good example of this discrimination property, high-affinity periplasmic phosphate-binding protein PstS binds phosphate with a 500-fold greater affinity than it binds arsenate, in arsenate-rich environments, is experimentally found in *Pseudomonas fluorescens*. In some exception, proteins also unable to recognize the metals of interest due to mimic nature of metals that lead to cellular dysfunction. This misconduct phenomenon of proteins is termed as ionic mimicry.

Another important concept of metal and ligand interaction is the Hard–soft acid base theory (HSAB theory; also termed Pearson's acid base theory) (Pearson, 1963). According to the HSAB, the ligand preferences of hard, intermediate (borderline) and soft metal cations (acids) and bases are classified shown in Table 1 (Pearson, 1963; Lippard & Berg 1994; Lemire et al., 2013) which indicates the preferences of metals for different ligands. A different classification of metals by Nieboer and Richardson (1980) into class A, class B and intermediate is also shown in Table 2 (Duffus, 2002; Lemire et al., 2013). The harder acids and bases tend to have weaker polarizing power and higher oxidation state. In contract, softer acids and bases tend to have stronger polarizing power and a lower oxidation state. In addition, some acids and bases are classified as borderline due to having intermediate character between hard and soft. Hard acids react preferentially with hard bases, whereas soft acids react with soft bases. The polarizing power of metal ion is important to the strength of interaction with ligand as described above. The more polarizable atom that is "softer" metal ion has stronger interactions capacity with ligands. Many of the most toxic metals are soft Lewis acids (Pearson, 1963). The soft Lewis acids form complex with soft Lewis bases and have preference to coordinate to S and N groups (soft Lewis bases) such as, cysteine sulfydryl (SH) groups and histidine nitrogens.

Redox Potential Properties

Redox potential or reduction potential or oxidation/reduction potential (i.e., ORP or pE or ε, or E_h) is a measure (by volts (V) or millivolts (mV)) of the tendency of a chemical species to acquire electrons and thereby be reduced. Each species has its own intrinsic reduction potential; the more positive the

Table 1. Classification of hard, soft and intermediate acids and bases

Class	Hard	Intermediate	Soft
Acid (Metals)	Mn^{2+}, Cr^{3+}, Na^{+}, Al^{3+}, Co^{3+}, K^{+}, Ga^{3+}, Fe^{3+}, Mg^{2+}, Ca^{2+}, Ti^{3+}	Fe^{2+}, Co^{2+}, Ni^{2+}, Cu^{2+}, Zn^{2+}	Cu^{+}, Pb^{2+}, Cd^{2+}, Ti^{+}, Hg^{2+}, Au^{+}
Base (ligands)	H_2O, NH_3, RHO, RNH_2, CO_3^{2-}, Cl^-, PO_4^{3-}, $ROPO_3^{2-}$, $(RO)_2PO_2^-$, SO_4^{2-}, OH^-, NO_3^-, CO_2^-, RO^-	Pyramidine, NO_2^-, SO_3^{2-}, Br^-, N_3^-	CN^-, RS^-, RSH, RNC, R_2S, $S_2O_3^{2-}H^-$, CO, $(RS)_2PO_2^-$, $(RO)_2P(O)S^-$

Pearson, 1963; Lippard & Berg 1994; Lemire et al., 2013.

Table 2. Classification of metals as Class A, B and intermediate

Class	Metals
Class A (hard metals)	Li, Be, Na, Mg, Al, K, Ca, Sc, Ti, Fe(III), Rb, Sr, Y, Zr, Cs, Ba, La, Hf, Fr, Ra, Ac, Th
Borderline (intermediate metals)	V, Cr, Mn, Fe(III), Co, Ni, Cu(I), Zn, Rh, Pb(IV), Sn
Class A (soft metals)	Cu(I), Pb, Ag, Cd, Ir, Pt, Au, Hg, Ti, Pb(II)

Nieboer & Richardson, 1980; Duffus, 2002.

potential, the greater the species' affinity for electrons and tendency to be reduced. Therefore, reduction potential is an important character of metal for binding with ligand of biomolecules. Since, the donor ligand loses electrons to metal and becomes oxidized for the formation of bond between metal and ligand; thus, reduction and oxidation always occur simultaneously and each individual reaction is termed a half-reaction. In addition to the affinity of a metal for a donor ligand, reduction potential is a thermodynamic parameter that determines the tendency of a metal species to acquire electrons from a donor and become reduced.

On the basis of redox potential, metals can be grouped into redox-active such as, iron, copper and chromium; and redox-inactive such as, lead, cadmium, mercury etc. Redox-active metals have multiple oxidation states because they can accommodate a discrete but variable number of valence electrons in their outermost atomic orbital, undergo redox cycling. Redox-active essential metals function as catalytic cofactors in a wide range of enzymes. In addition, cellular donor ligands can make subtle changes in the relative reduction potential of a chelated metal. In contrast, redox-inactive metals and others have just one oxidation state because they can accommodate only a specific number of electrons, and hence deplete cells' major antioxidants, particularly thiol-containing antioxidants and enzymes. Both redox-active and redox-inactive metals can increase the production of reactive oxygen species (ROS) such as hydroxyl radical (HO.), superoxide radical (O_2.-) or hydrogen peroxide (H_2O_2). Enhanced generation of ROS can overwhelm cells' intrinsic antioxidant defenses, and results in a condition known as "oxidative stress". Cells under oxidative stress display various dysfunctions due to lesions caused by ROS to lipids, proteins and DNA. Consequently, it is suggested that metal-induced oxidative stress in cells can be partially responsible for the toxic effects of heavy metals.

Chemical Speciation Properties

The metal can exist as multiple different chemical species in any single milieu which is termed as metal speciation. For example, Fe(II) and Fe(III) species of iron (Fe), As(III) and As(V) species of arsenic (As), Hg(0) and Hg(II) species of mercury (Hg) etc. It is an important character of metals for solubility, bioavailability, reactivity and sequestration in the cell. Therefore, speciation of metals is the key determinants for metal toxicity. The speciation and abundance of metal are influenced by several factors, including pH, reduction potential, ionic strength, temperature etc. Essential metals are never free inside a cell, and an intricate network of transporters, metalloregulatory sensors and metallochaperones restricts metal speciation and guides metal atoms to specific enzymes, structural proteins and storage proteins (Finney & O'Halloran, 2003). The metal speciation influencing factors can also vary between subcellular compartments, such as, the cytoplasm is a strong reducing environment relative to the periplasm in Gram-negative bacteria (Stewart, Aslund, & Beckwith, 1998), and this can affect the oxidation state of a metal and thereby change its speciation. Besides, various metabolic processes are characteristically able to alter the chemical state of metals in microorganism belonged to same or different phyla (Gadd, 2010). The metal speciation is changed by geological time and habitats variations and influenced biological, geological and even human activities in the environment (Gadd, 2010). For example, thermodynamically favourable oxidation of Fe(II) to Fe(III) in aerobic environments, which leads to the formation of insoluble iron oxides and hydroxides.

BIOCHEMICAL–MOLECULAR MECHANISM OF METAL TOXICITY IN MICROORGANISM

Metals directly and indirectly exhibit a range of hazardous effects by causing a series of serious antimicrobial and biochemical and molecular alterations in the metabolic reactions and exert toxicity in microbial cell. Many toxicity mechanisms have been discovered through the investigations in bacteria and yeast and some of the important mysteries are supposed to be remained uncovered in this regard. The growth inhibition and cellular death may or may not only the effect of one metal toxicity mechanism; it is more likely to be the result of a combined effect of different mechanisms. The mechanisms of toxicity might differ on the basis of metal chemistry. The toxicity concepts gained from biochemistry, molecular biology and toxicology so far are able to summarily draw the mechanistic framework for understanding how metals damage and kill microbial cells that can be simply represented in following categories (Lemire et al., 2013):

ROS (Reactive Oxygen Species) Stress Induced Toxicity

Reactive oxygen species (ROS) are chemically reactive molecules containing oxygen (for example, oxygen ions ($O_2^{\bullet-}$) and hydrogen peroxide (H_2O_2), hydroxyl radical ($OH^{\bullet}$) etc., which are usually produced as a normal byproduct of the normal metabolism of oxygen and have important roles in cell signaling and homeostasis; and an antioxidant (such as, thiols, ascorbic acid, polyphenols, etc.) is a molecule that inhibits the oxidation of other molecules. However, ROS levels can increase dramatically during environmental stress condition and may cause significant damages in cell such as, damage of DNA, oxidations of polyunsaturated fatty acids in lipids (lipid peroxidation), oxidations of amino acids in proteins and deactivation of specific enzymes by oxidation of co-factors through a range of biochemical routes (Brooker, 2011). These elevated ROS mediated adverse consequences are collectively known as "ROS stress". The cell uses antioxidant proportionately as increasing the level of ROS production which resulting in the "antioxidant depletion".

Several chemical and biological studies revealed that toxic doses of certain metal ions (such as Cr(VI), As(III) and Te(IV), but especially Fe(II) and Cu(II)) (Imlay, Chin, & Linn, 1988; Harrison et at., 2009; Perez et al, 2007; Nunoshiba et al., 1999; Touati, Jacques, Tardat, Bouchard, & Despied, 1995; Warnes, Caves, & Keevil, 2012) increase intracellular ROS (Figure 1). In *E. coli*, DNA damage and the inhibition of vital growth regulating enzyme activities are observed by exogenous introduction of hydrogen peroxide (H_2O_2) or agents that catalyze the production of superoxide ($O_2^{\bullet-}$), such as paraquat (Imlay, 2003). This similar damages are also pronounced in deregulation of Fe metabolism Damage (Nunoshiba et al., 1999; Touati et al., 1995) and Cu(II) or Te(IV) exposure in *E. coli.* In bacterial and *S. cerevisiae* mutants, lack of ROS-scavenging enzymes and other cellular antioxidants are frequently found in the stress condition of Cr, As, Te, Fe and Cu (Touati et al., 1995; Itoh et al. 1995; Geslin, Llanos, Prieur, & Jeanthon, 2001; Sumner et al. 2005*).* Additionally, a number of investigations proposed that ROS eliminating genes are upregulated in bacteria and yeast exposed to toxic doses of Cu(II), Cr(VI), As(III) or Te(IV) (Sumner et al., 2005; Perez et al., 2007). Therefore, ROS-mediated damages of cellular protein, membrane or DNA lead to microbial toxicity.

The exact mechanism for the development of ROS stress is initiated by entry of metal species and catalyzing the oxidation reaction following Fenton chemistry and other metal-catalyzed free radical chain reactions in the cell (Equation 1 - 9). Fenton reaction, a process of oxidation of organic molecules by

Figure 1. Production of reactive oxygen species (ROS) by the Fenton reaction (chemistry) and depletion of antioxidant

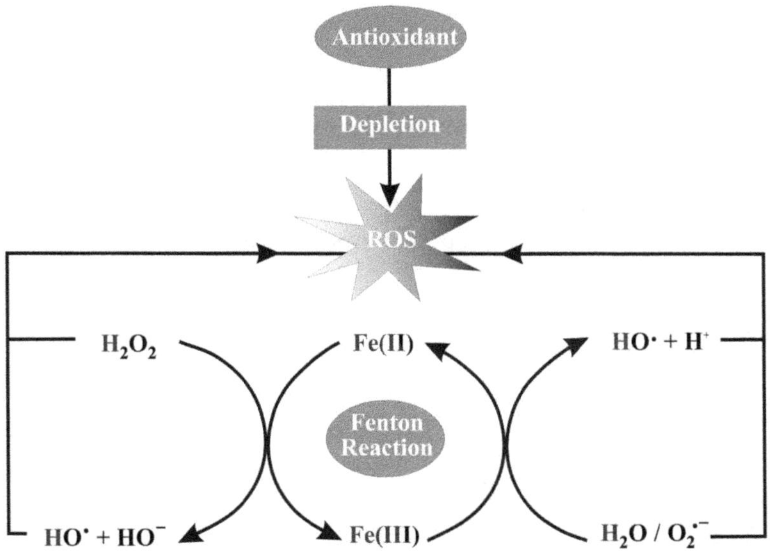

Fe(II) and hydrogen peroxide (H_2O_2), is very common in biological system first described by H. J. H. Fenton in 1894. In cellular system, partially reduced forms of molecular oxygen (O_2), such as H_2O_2 and superoxide (O_2•–) is inadvertently generated as by-products by aerobic respiration, this H_2O_2 oxidizes metal (Fe^{2+}) to produce hydroxide (OH^-) and highly reactive hydroxide (OH•) (Equation 3) and O_2•– reduces the oxidized metal (Fe^{3+}) (Equation 4). The net reaction is known as Haber–Weiss reaction (Equation 5). Fenton reaction generated free radicals are associated with the depletion of antioxidant (Figure 1) and production of other free radicals by reacting with various biomolecules (Equation 10 - 12). However, the following types of mechanism are responsible to produce increased rate of ROS by Fenton reaction that can cause metal poisoning in bacteria: (1) In addition to Fe, certain redox-active transition metals– such as Cu, Cr, Co, V and, to a lesser extent, Ni can catalyze Fenton chemistry *in vitro* (Valko, Morris, & Cronin, 2005; Stohs & Bagchi, 1995; Strlič, Kolar, Šelih, Kočar, & Pihlar, 2003). Similarly, some of these other metals might do the Fenton reaction. This reaction is influenced by some factors, such as pH, coordinating ligands and relative reduction potential. (2) The Fe coordinating donor ligands of cell might be disrupted by certain transition metals. The direct or indirect destruction of [4Fe–4S] clusters of protein ligand by metals could result in the release of additional Fenton-active Fe into the cytoplasm, which also resulting in increased ROS formation. (3) Thiol (is an organosulfur compound that contains a carbon-bonded sulfhydryl (–C–SH or R–SH) group)-mediated reduction of some metal species such as Fe(III), Cu(II), Cr(VI), As(III) and Te(IV) can produce ROS (Valko et al., 2005) (Figure 2). In the case of Cr(VI), thiol-mediated reduction produce Fenton active other Cr species, such as Cr(III), Cr(IV) and Cr(V) (Valko et al., 2005). In bacterial metal poisoning, the oxidation of cellular thiols is the important aspect of oxidative stress. Soft and borderline metal atoms (as well as metal oxyanions) are able to form the covalent bonds with S by electron-sharing affinities, which can lead to the formation of protein disulphides and the depletion of antioxidant reserves, particularly glutathione, within microbial cells (Harrison et at., 2009; Helbig, Grosse, & Nies, 2008; Liau, Read, Pugh, Furr, & Russell, 1997). For instance, depletion of total cellular thiols has been observed following the exposure of *E. coli* to toxic doses of Ag(I), Cd(II), Co(II), Zn(II), Cr(VI), As(III) and Te(IV) (Harrison et at., 2009).

Box 1.

$O_2 \rightarrow O_2^{\bullet-} + e$ (1)

$O_2^{\bullet-} + H+ \; = \; H_2O_2$ (2)

$Fe^{2+} + H_2O_2 \rightarrow Fe^{3+} + OH^- + OH^\bullet$ Fenton reaction (3)

$Fe^{3+} + O_2^{\bullet-} \rightarrow Fe^2+ + O_2$ (4)

$O_2^{\bullet-} + H_2O_2 \rightarrow O_2 + OH^- + OH^\bullet$ Haber–Weiss reaction (5)

$OH^\bullet + H_2O_2 \rightarrow HO_2^\bullet + H_2O$ (6)

$Fe^{3+} + HO_2^\bullet \rightarrow Fe^{2+} + H^+ + O_2$ (7)

$Fe^{2+} + HO_2^\bullet \rightarrow Fe^{3+} + HO_2^-$ (8)

$Fe^{2+} + OH^\bullet \rightarrow Fe^{3+} + OH^-$ (9)

OH• + R-SH → H2O + RS• Thiyl radical (10)

OH• + (R)3CH → H2O + (R)3C• Carbon-centered radical (11)

(R)3C• + O2 → (R)3COO• Peroxyl radical (12)

Figure 2. Depletion of thiol group by metals and production of ROS at increasing rate.

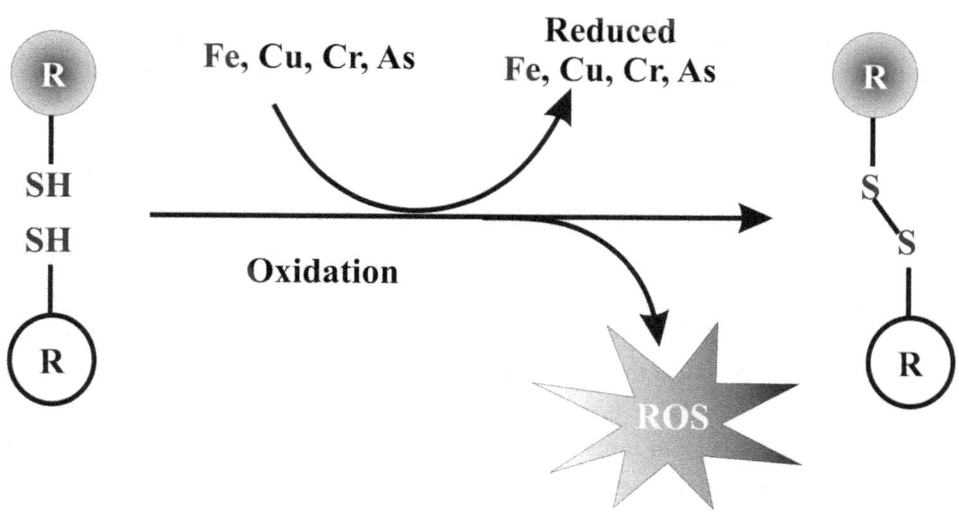

Alteration of Protein and Enzyme Functions Mediated Toxicity

Some metal species in lieu of essential inorganic ions can arrest the protein function and enzyme activity by biochemical alterations in the target molecule. The biochemical mechanisms in respect to metal mediated inhibition of protein and enzyme activities can be categorized as follows: (i) metal-catalyzed oxidation of amino acid, (ii) destruction of Fe–S cluster, (iii) ion mimicry, and (iv) metal substitutions at non-catalytic metal-binding sites.

i. Metal-Catalyzed Oxidation of Amino Acid

The amino acids of proteins are most sensitive to metal-catalyzed oxidation. The Fenton pathway is the basic mechanism for the oxidation of amino acids by metal ion-catalyzed reactions (Figure 3). Oxidation of amino acid in protein losses the catalytic activity and triggers an active process of protein degradation *in vivo*. In *E. coli,* it has reported that amino acid residue adjacent to metal-binding sites in any given protein is susceptible to metal-catalyzed oxidation (Stadtman & Levine, 2003, Stadtman, 1993). However, metals could catalyze site-specific damage to cellular proteins which in turn might be responsible for metal toxicity. Additionally, It is observed that carbonyl derivatives are products of the metal-catalyzed oxidation of several amino acid side chains (a reaction to which histidine, arginine, lysine and proline are particularly vulnerable), which is generally used as a marker of oxidative protein damage (Stadtman & Levine, 2003). For example, in *S. cerevisiae,* toxic doses of Cr(VI) cause a ROS-dependent increase in protein carbonyl levels within minutes. The most highly oxidized proteins are cytosolic enzymes involved either directly in glycolysis or in subsequent catabolic reactions, which could be a major factor influencing Cr(VI) toxicity (Sumner et al., 2005). The metal atoms such as Zn(II), Mn(II) or Co(II) are structural or functional constituents of mononuclear metalloenzymes which bind to Fe(II) *in vitro* and *in vivo* (Anjem & Imlay, 2012). In *E. coli,* these mononuclear Fe enzymes, such as peptide deformylase (PDF), threonine dehydrogenase (Tdh) and cytosine deaminase, are acutely sensitive to Fe-catalyzed Fenton chemistry (Anjem & Imlay, 2012) (Figure 3). Both PDF and Tdh use a cysteine residue to co-ordinate the Fe(II) atom, and this residue can become oxidized to sulphinic (RSO(OH)) or sulphonic ($RS(=O)_2$–OH) forms by the Fenton reaction (Anjem & Imlay, 2012). The Replacement of Fe(II) from these mononuclear metalloenzymes is found by Mn(II) in *E. coli* under conditions of H_2O_2 stress.

Figure 3. Metal-catalyzed oxidation of amino acid residue of protein. Oxidation of cysteine in Fe-binding site of amino acid leads to inactivation of protein function

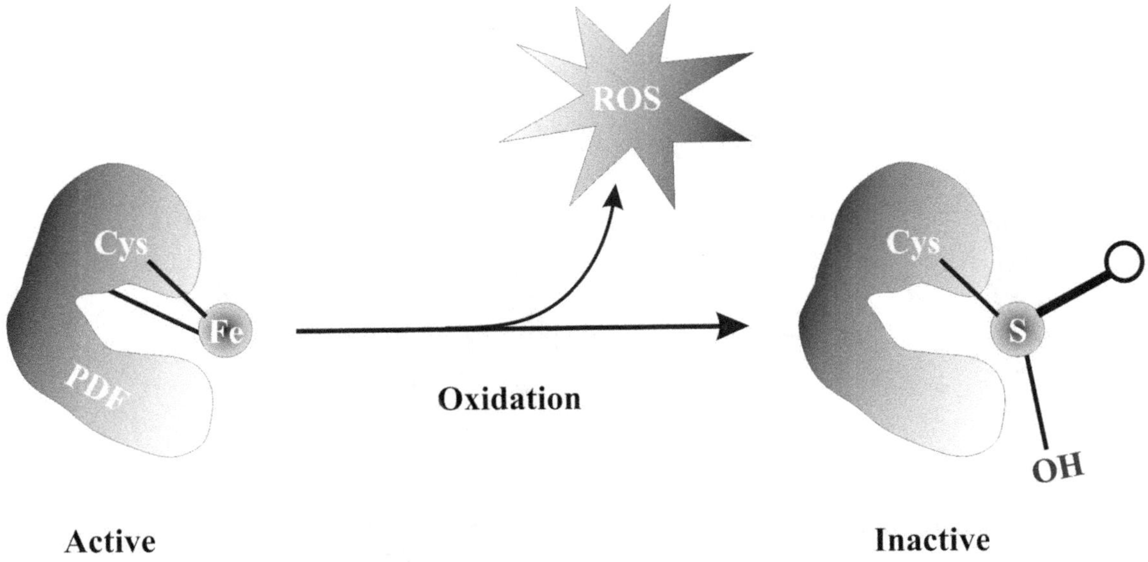

ii. Destruction of Iron-Sulfur (Fe–S) Cluster

Iron-sulfur (Fe-S) clusters, consisting of iron and elemental sulfur at various molar ratios, are often stable at multiple oxidation states and have physiologically relevant redox potentials (ranging from −500 to 150 mV) (Beinert & Kiley, 1999). The assembly of Fe and S is a carefully controlled process in cells to avoid toxicity from free iron and sulfide and it is required for critical biochemical pathways, including respiration (electron transfer), photosynthesis, and nitrogen fixation. Besides, Fe-S clusters are also involved in substrate binding and activation in dehydratases and radical-S-adenosylmethionine enzymes (Ruzicka & Beinert, 1978). The most common cluster forms, [2Fe-2S] and [4Fe-4S], are usually coordinated to proteins by Cys residues, although Asp, His, Ser, or backbone amides can coordinate clusters at single sites (Moulis, Davasse, Golinelli, Meyer, & Quinkal, 1996). The [4Fe–4S] clusters in Fe–S dehydratases enzymes such as dihydroxy-acid dehydratase and isopropylmalate isomerase, are required for branched-chain amino acid biosynthesis (Imlay, 2003). Therefore, destruction of Fe-S clusters sufficiently hampers various biochemical pathways.

The mechanism of Fe–S cluster destruction is metal dependent ROS toxicity. It is postulated from several studies that a family of bacterial Fe–S dehydratases is particularly vulnerable to site-specific inactivation by toxic metals (Macomber & Imlay, 2009; Xu & Imlay, 2012) (Figure 4). It has also revealed that Ag(I), Hg(II), Cd(II) and Zn(II) (but not Mn(II), Co(II), Ni(II) or Pb(II)) damage Fe–S-containing dehydratases *in vitro* independently of ROS (Xu & Imlay, 2012). The metal also inactivates the enzyme activity by damaging the Fe–S cluster and resulting in loss of cell growth. The activity of inactive enzyme of metal treated bacteria can be restore by proteins that repair Fe–S clusters, such as cysteine desulphurase (IscS) or the Fe–S cluster scaffold protein SufA (Xu & Imlay, 2012). Loss of activity of Fe–S-containing enzymes (such as fumarase A and aconitase B) has also been correlated with growth inhibition of *E. coli* by Te(IV).

Figure 4. Destruction of iron-sulfur (Fe–S) cluster and inactivation of protein function

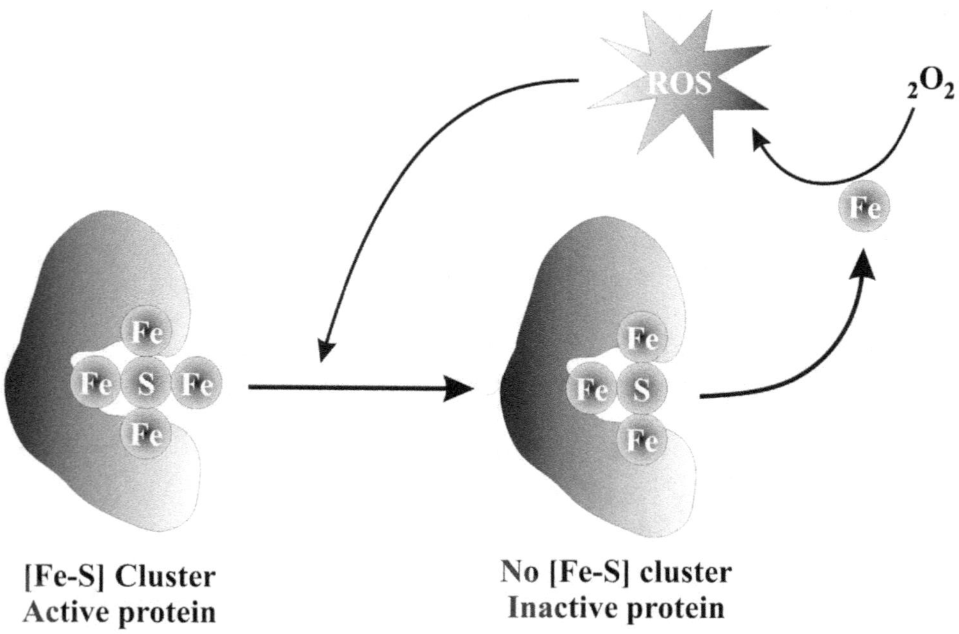

iii. Ion Mimicry and Metal Exchange at Catalytic Sites

The ionic mimicry is referred to the phenomena by which an unbound, cationic species of non-essential metal can behave or serve as a structural and/or functional homolog or mimic of another (usually an essential) element at the site of a carrier protein, ion channel, enzyme, structural protein, transcription factor and/or metal-binding protein. Such, the nonessential and toxic metal ions can get entry and cause lethal reactions especially inhibit the enzyme activity in different domains of cell resulting in the cellular metal poisoning or toxicity (Figure 5). A considerable amount of scientific data on molecular and ionic mimicry has been published recently. For example, many evidences have been obtained showing that the cationic species of certain toxic metals, such as Cd, can use calcium ion channels and certain membrane transporters, such as the divalent metal transporter to gain access into target cells of organisms. An another good example is the Zn(II), a Lewis acid catalyst, is replaced by by Pb(II) from the active site of δ-aminolevulinic acid dehydratase (ALAD) leads to strong inhibition of enzyme activity (Erskine et al., 1997) and to antimicrobial toxicity (Ogunseitan, Yang, & Ericson, 2000). Similarly, Ag(I) is inserted at the Cu(I)-binding active site of Cu and Zn superoxide dismutase (Cu–Zn SOD) in *S. cerevisiae* treated with silver nitrate, which abolishes the activity of this enzyme *in* vivo (Ciriolo et al., 1994).

iv. Metal Substitutions at Non-Catalytic Metal-Binding Sites

It is another kind of mechanism of non-essential metal mediated enzyme inactivation in addition to destruction of the active site of protein or enzyme. It includes the structural metal substitutions at non-catalytic metal-binding sites that can inhibit enzyme activity (Figure 6). For example, in *E. coli,* Ni(II) is substituted by Zn(II) at the non-catalytic Zn site of the glycolytic enzyme fructose-1,6-bisphosphate aldolase (FbaA) that leads to loss of FbaA activity (Macomber, Elsey, & Hausinger, 2011).

Figure 5. Ion mimicry and metal exchange at catalytic sites lead to inactivation of protein function

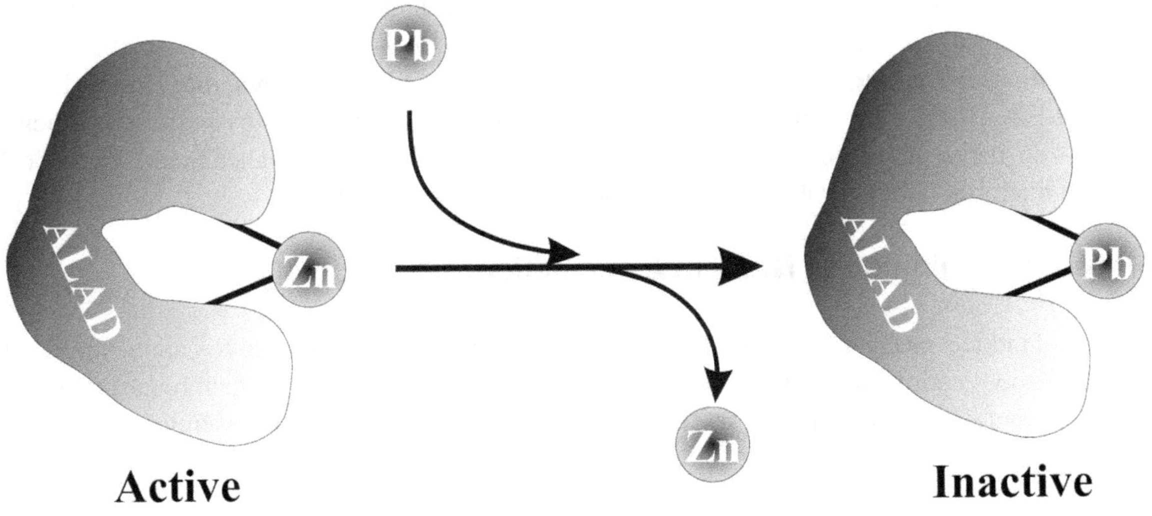

Figure 6. Metal substitutions at non-catalytic metal-binding sites and inactivation of protein function

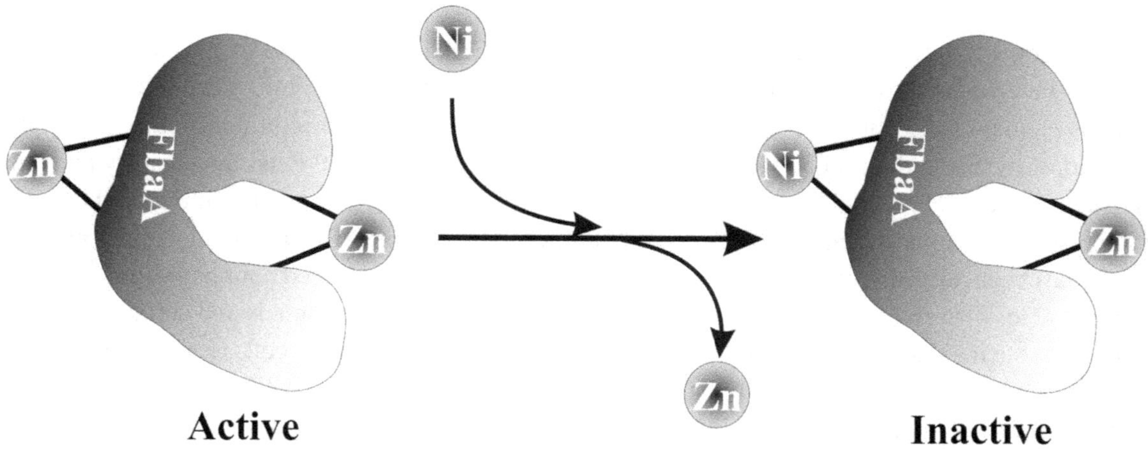

Alteration of Membrane Function Mediated Toxicity

Microbial cell membrane generally contains extracellular polymers with highly electronegative chemical groups. Therefore, membrane polymers act as efficient metal cations binding agents and are supposed to be responsible for metal accumulation at high concentration in membrane. As a result, the cellular membrane is one of the important susceptible sites for microbial metal toxicity. Various proteins and enzymes of membranes are the structural and functional constituents of the cell which is also damaged by the non-essential metals following the mechanisms, metal-catalyzed oxidation of amino acid, and destruction of Fe–S cluster, ion mimicry and metal substitutions at non-catalytic metal-binding sites as explained earlier (Figure 7). In addition, a member of investigations reported several impairment in the membrane functions such as, (i) some metals, particularly Ag, disrupt the activity of the bacterial electron transport chain (Gordon et al., 2010) and some metal oxyanions are reduced by the oxidoreductase DsbB, which draws electrons from the bacterial transport chain through the quinone pool, (ii) NADH:quinone oxidoreductase (NQR) is a component of the respiratory chain of some bacteria and generates a redox-driven transmembrane Na^+ potential by translocating Na^+, (iii) Ag(I) inhibits the activity of NQR (Figure 7), and site-specific mutations in this protein confer resistance to several metal cations (Fadeeva, Bertsova, Euro, & Bogachev, 2011) and is proposed to dissipate the chemiosmotic potential of the membrane by causing proton leakage through the membrane in *Vibrio cholerae* (Dibrov, Dzioba, Gosink, & Hase, 2002), (iv) certain toxic metal species starve microbial cells by indirectly siphoning electrons from the respiratory chain. Besides, lipid peroxidation has also been linked to Cu(II) and Cd(II) toxicity in microbes (Hong, Kang, Michels, & Gadura, 2012) (Figure 8).

Competitive Inhibition of Nutrient Assimilation

It is a kind of indirect mechanism of metal toxicity governed by starvation-induced cessation of microbial growth. Microbes assimilate essentials metal ions from environment across the membrane and used in different structural and function purposes. Due to ion mimic property, some non-essential metals are en-

Figure 7. Impairment of membrane function - various structural and functional proteins and enzymes of membranes are damaged by the non-essential metals following the mechanism of metal-catalyzed oxidation of amino acid

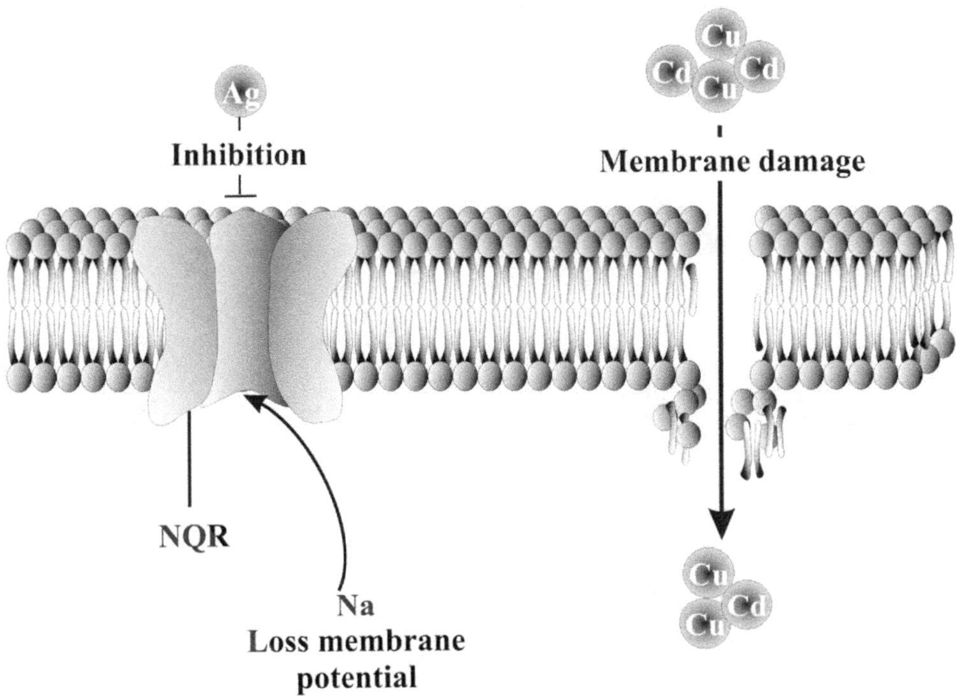

tered into the cell using the same proton motive force (PMF) or ATP pool dependent transporters system engaged for transportation of essential nutrients and participate in various biochemical reactions instead of essential metal ions. Consequently, the process of essential nutrients uptake is critically interfered, starvation is induced, growth is arrested and finally cell death is caused. Such, this energy dependent transporter-mediated uptake and assimilation of toxic metal species impair the membrane function and interfere the normal transport of essential substrates (Figure 8). This phenomenon can be termed as competitive inhibition of nutrient assimilation. For example, the sulphate uptake of *S. cerevisiae* is inhibited by the Cr(VI), which depletes intracellular S metabolite pools and induces enzymes of the S-sparing response (Pereira et al., 2008). Since, sulphate transporters (Sul1 and Sul2) of *S. cerevisiae* have similar affinities for transporting chromate and sulphate, the competitive inhibition of sulphate uptake is observed during the consistent chromate exposure of cell. (Pereira et al., 2008). It has been reported that supplementation of sulphate and other S sources mitigates toxicity in a concentration-dependent manner; therefore, chromate mediated S starvation is unable to mitigate such toxicity problem in cells. Similarly, in *P. aeruginosa,* Ga(III) inhibits the uptake of Fe(III) in a concentration-dependent manner leads to Fe starvation response in cell (Figure 8). In spite of competitive inhibition of transport, the toxicity of Ga in *P. aeruginosa* is governed by the deregulation in expression of gene involved in the selective uptake of Fe(III) and thereby interfere the Fe sensing.

Figure 8. Competitive inhibition of nutrient assimilation - some non-essential metals get entry into the cell by the same proton motive force (PMF) or ATP pool dependent transporters system engaged for transportation of essential nutrients and participate in various biochemical reactions instead of essential metal ions results in starvation of essential metal ions within cell

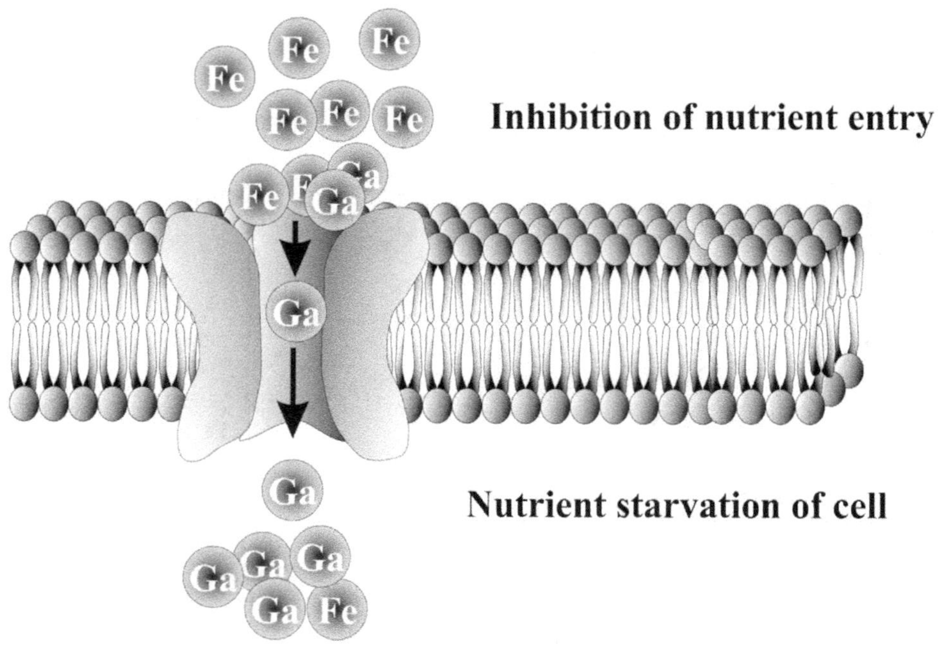

Metal Genotoxicity

Genotoxicity commonly refers to the property of chemical substances that damages the structure and function of genetic materials within a cell causing genetic mutations. Some metals act as a foreign mutagenic agent and can cause lethal damages in DNA of the microorganism (Figure 9). Both direct and indirect mutagenic effects of metals and their compounds have been evaluated by several investigations. Only a few metal compounds are supposed to produce direct DNA damage, but a number of metals are suspected to express mutagenic potential via indirect mechanism. Many studies also reported that Fe-mediated Fenton chemistry can catalyze (Linley, Denyer, McDonnell, Simons, & Maillard, 2012) the mechanism of lethal DNA damage in *E. coli* (Figure 9). For example, the amount of Fenton-active Fe in the cell is elevated by mutations that accelerate DNA damage and cause cell death (Touati et al., 1995). Although, most compounds of lead are not bacterial mutagens, lead chromate can act as mutagenic agent for *Salmonella typhimurium*. However, Ames test as well as other comparable analyses concerning genotoxicity assay proposed that toxic doses of Mn(II), Cr(vi), Co(II), Cd(II), Mo(IV), Sb(III) and As(III) (Nishioka, 1975; Green, Muriel, & Bridges, 1976; Asakura et al., 2009), as well as some organomercury complexes are mutagenic in bacteria. Additionally, some other metals (such as, Ni(II), Cu(II), Te(IV), Pb(II), Ag(I) and Al(III)) have not been associated with increased mutation rates in toxic dose (Nishioka, 1975; Green, Muriel, & Bridges, 1976; Asakura et al., 2009). In contrast, it has postulated that Cu protects *E. coli* from H_2O_2-mediated DNA damage, even though this metal is Fenton active and produces hydroxyl radicals *in vivo*.

Figure 9. Metal genotoxicity - toxic and non-essential metals damage the DNA by Fenton reaction

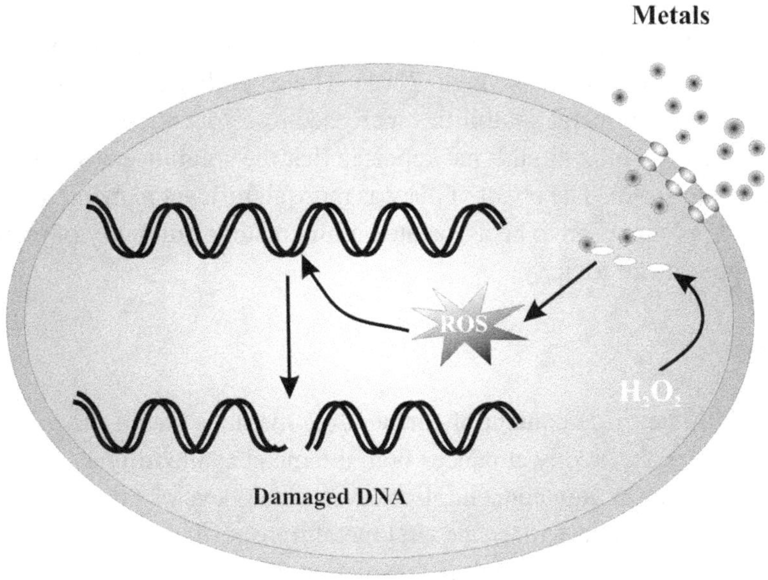

EFFECTS OF ENVIRONMENTAL FACTORS IN METAL TOXICITY

In addition to influence of various intracellular biochemical factors as described earlier, a number of environmental conditions also significantly play important roles in the metal toxicity of microbes. The concentration, bioavailability and chemical behaviour of the metal species and microbial communities are the important factors for metal toxicity to microbes in the environment. Metal bioavailability to microbes is greatly regulated by the physico-chemical nature of the environment (Gadd & Griffiths, 1978). A wide range of physico-chemical components of the environmental (water or soil) matrix, such as pH, reduction potential (E_h), free ions especially iron oxide content, water hardness, organic matter and clay content, colloids and particulates, various gases and metabolites, temperature, and various nutrient regulate the metal bioavailability in the environment surrounding the microbes. In spite of these, the exudates of plants root also play significant role in transforming the metals into bioavailable form. However, pH, $E_{h,}$ free metal ions, organic matters and clay and exudates of plants root are important as described below:

pH

pH is often found to be the important influencing factor due to its significant effects on solubility and speciation of metals in environment. The solubility, mobility and speciation of metals are increased with decreasing the pH value of medium which resulting in the significant acceleration of bioactive and bioavailable form of metal. The study suggested that Metal solubilities are higher when under slightly acidic conditions (pH = 5.0), and increased drastically when pH was kept at 3.3 (Chuan, Shu, & Liu, 1996). It has observed that each unit decrease in pH results in approximately 2-fold increase in the concentration of metals such as Zn, Ni, and Cd in the solution. Thus, pH is inversely correlated with toxic potentiality of metals. Additionally, pH is also supposed to be another critical factor for facilitating ion mimic process in metal transporting mechanism due to having the metal speciation property of pH.

Reduction Potential (E_h)

It is one of the important factors for metal solubility in the media/environment as well as availability to microorganism as that of pH (Kashem & Singh, 2001). The solubility of metal is varied with changing E_h. It is generally observed that metal solubilities increased as E_h decreased that is reducing condition, and is most favorable for soulubilization. It has reported that the solubility of Pb, Cd, and Zn greater in decreased E_h (Chuan et al., 1996). The effect of pH was more significant in metal solubilization than that of E_h. Thus, the lower E_h is supposed to be associated with greater availability of metals to microorganism which leads to higher metal toxicity.

Free Metal Ions

Free metal ions especially Fe oxide content of surrounding media is one of the prime factors for direct and indirect metal toxicities. It directly enhances both the metal availability as well as exposure to cell membrane of microorganisms at high concentration level which develops the incident of concentration gradient metal entry into cell along with other vital metal transporting mechanism. It is also generally assumed that the free ion is the chemical species, which is taken up by the microorganism and causes toxicity when present in excess (Checkai, Corey, & Helmke, 1987; Wolt, 1994). In addition, the concentration and bioavailability of free ions are further governed by the local pH of the medium. The free Fe ions of surrounding media supposed to be indirectly increased the Fenton active Fe in the cell which is well known cause of metal toxicity.

Organic Matters and Clay

Generally, organic matter and clay in medium (soil/water) can act as metal chelating agents, which inhibit metal to be bioavailable to the microorganism. The amount and types of organic matter present in the medium (soil) can also affect the metal solubility because of its complexation and chelation with metals. It is postulated that all part of the metal content in the medium are not bioavailable, because, a portion of total available metal is remained bound to organic molecules by means of chelation and complexation mechanisms, and the occurrence of various chemical forms which cannot directly be taken up by the microorganism. Some metals are more tightly bound by organics under anoxic or reducing conditions (Kashem & Singh, 2001). Such, these factors inhibit the metal uptake as well as reduce the toxicity problem of microbes.

Exudates of Plants Root

The roots of the plant excrete various extracellular biomolecules in the soil media, which is known as exudates of plants. The exudates affect metal availability by altering the rhizosphere chemistry in soil. As bacteria are present within colonies in soil, therefore, they may often expose to the solution of metals. Metals may become bound to bacteria or fungal cell walls or on extracellular polysaccharides (EPS) of bacteria (Gadd, 1990) and the ingestion of such bacteria by other organisms such as, protozoa and nematodes will result in vastly different exposure to metals in the predators than would result simply from exposure to the metals present in the soil solution. Microorganisms may also alter metal availability in their vicinity due to localized acidification of the environment, or production of compounds which complex metals i.e., known as bioleaching. For example, iron oxidizing bacteria, which reduce iron pyrites to $FeSO_4$, can cause extreme acidification causing increases in metal availability (Giller et al., 1998).

IMPACT AND FUTURE PROSPECTS

The metals contamination in microorganism leads to toxicity which is responsible for causing lethal damage of microbial cell, mutagenicity effects and/or cell death by various biochemical mechanisms discussed herein as literatures available. Innumerable and indefinite cellular biochemical reactions are also suggested that many other metabolic pathways of metal toxicity may still remain unknown to the researchers. Therefore, the biochemical and molecular mechanisms relating to metal toxicity extensively illustrated herein may open a new platform in order to further depth and critical metal toxicity study in microbes as well as other organisms for controlling and treating the metal toxicity and beneficial purposes of other organisms in future.

Metal toxicity to microorganism invites an immeasurable loss in microbial composition of environment. Since, it is responsible for damaging microorganism and normal microbial community structure and developing the microbial metal-resistant properties along with the acquisition of antibiotic-resistant properties which are severe alarming threat for environmental and human health. Nonetheless, the metal-resistant microbial strains have promising potentiality that can be applied as prospective agents for the benefit of environmental as well as human health perspectives (Bhakta et al., 2012a,b, 2014). Besides detrimental consequences, the novel metal-resistant microbes with specific beneficial characteristics could be employed for future biotechnological purposes (Hussein & Joo, 2013). Moreover, the vivid mechanistic pictures drawn herein in respect to microbial toxicity especially molecular mechanism of microbial metal toxicity would open new avenues in pharmaceutical and medicinal perspectives in future for human health benefits such as, controlling and treating the metal toxicity and in developing clean and sustainable environment.

REFERENCES

Andreini, C., Bertini, I., & Rosato, A. (2004). A hint to search for metalloproteins in gene banks. *Bioinformatics (Oxford, England)*, *20*(9), 1373–1380. doi:10.1093/bioinformatics/bth095 PMID:14962940

Anjem, A., & Imlay, J. A. (2012). Mononuclear iron enzymes are primary targets of hydrogen peroxide stress. *The Journal of Biological Chemistry*, *287*(19), 15544–15556. doi:10.1074/jbc.M111.330365 PMID:22411989

Asakura, K., Satoh, H., Chiba, M., Okamoto, M., Serizawa, K., Nakano, M., & Omae, K. (2009). Genotoxicity studies of heavy metals: Lead, bismuth, indium, silver and antimony. *Journal of Occupational Health*, *51*(6), 498–512. doi:10.1539/joh.L9080 PMID:19851040

Beinert, H., & Kiley, P. J. (1999). Fe–S proteins in sensing and regulatory functions. *Current Opinion in Chemical Biology*, *3*(2), 152–157. doi:10.1016/S1367-5931(99)80027-1 PMID:10226040

Bhakta, J. N. (2016). Microbial response against metal toxicity. In A. K. Rathoure & V. K. Dhatwalia (Eds.), Toxicity and Waste Management Using Bioremediation (pp. 75-96). IGI Global. doi:10.4018/978-1-4666-9734-8.ch004

Bhakta, J. N., Munekage, Y., Ohnishi, K., & Jana, B. B. (2012a). Isolation and identification of cadmium and lead resistant lactic acid bacteria for applying as metal removing probiotic. *International Journal of Environmental Science and Technology*, *9*(3), 433–440. doi:10.1007/s13762-012-0049-3

Bhakta, J. N., Munekage, Y., Ohnishi, K., Jana, B. B., & Balcazar, J. L. (2014). Isolation and characterization of cadmium and arsenic absorbing bacteria for bioremediation. *Water, Air, and Soil Pollution*, *225*(10), 2151. doi:10.1007/s11270-014-2151-2

Bhakta, J. N., Ohnishi, K., Munekage, Y., & Iwasaki, K. (2010). Isolation and probiotic characterization of arsenic-resistant lactic acid bacteria for uptaking arsenic. *International Journal of Chemical and Biological Engineering*, *3*(4), 167–174.

Bhakta, J. N., Ohnishi, K., Munekage, Y., Iwasaki, K., & Wei, M. (2012b). Characterization of lactic acid bacteria-based probiotics as heavy metals sorbents. *Journal of Applied Microbiology*, *112*(6), 1193–1206. doi:10.1111/j.1365-2672.2012.05284.x PMID:22404232

Brooker, R. J. (2011). *Genetics: Analysis and Principle*. New York, NY: McGraw-Hill Science.

Bruins, M. R., Kapil, S., & Oehme, F. W. (2000). Microbial resistance to metals in the environment. *Ecotoxicology and Environmental Safety*, *45*(3), 198–207. doi:10.1006/eesa.1999.1860 PMID:10702338

Checkai, R. T., Corey, R. B., & Helmke, P. A. (1987). Effects of ionic and complexed metal concentrations on plant uptake of cadmium and micronutrient metals from solution. *Plant and Soil*, *99*(2-3), 335–345. doi:10.1007/BF02370879

Chuan, M. C., Shu, G. Y., & Liu, J. C. (1996). Solubility of heavy metals in a contaminated soil: Effects of redox potential and pH. *Water, Air, and Soil Pollution*, *90*(3-4), 543–556. doi:10.1007/BF00282668

Ciriolo, M. R., Civitareale, P., Carri, M. T., De Martino, A., Galiazzo, F., & Rotilio, G. (1994). Purification and characterization of Ag, Zn-superoxide dismutase from *Saccharomyces cerevisiae* exposed to silver. *The Journal of Biological Chemistry*, *269*, 25783–25787. PMID:7929283

Collins, Y. E., & Stotzky, G. (1989). Factors affecting the toxicity of heavy metals to microbes. In T. J. Beveridge & R. J. Doyle (Eds.), *Metal ions and bacteria* (pp. 31–90). New York, NY: John Wiley & Sons, Inc.

Dibrov, P., Dzioba, J., Gosink, K. K., & Hase, C. C. (2002). Chemiosmotic mechanism of antimicrobial activity of Ag+ in *Vibrio cholerae. Antimicrob. Antimicrobial Agents and Chemotherapy*, *46*(8), 2668–2670. doi:10.1128/AAC.46.8.2668-2670.2002 PMID:12121953

Duffus, J. H. (2002). Heavy metals"—a meaningless term? *Pure and Applied Chemistry*, *74*(5), 793–807. doi:10.1351/pac200274050793

Erskine, P. T., Senior, N., Awan, S., Lambert, R., Lewis, G., Tickle, I. J., & Coope, J. B. (1997). X-ray structure of 5-aminolaevulinate dehydratase, a hybrid aldolase. *Nature Structural Biology*, *4*(12), 1025–1031. doi:10.1038/nsb1297-1025 PMID:9406553

Fadeeva, M. S., Bertsova, Y. V., Euro, L., & Bogachev, A. V. (2011). Cys377 residue in NqrF subunit confers Ag^+ sensitivity of Na^+-translocating NADH:quinone oxidoreductase from *Vibrio harveyi*. *Biochemistry*, *76*, 186–195. PMID:21568851

Finney, L. A., & OHalloran, T. V. (2003). Transition metal speciation in the cell: Insights from the chemistry of metal ion receptors. *Science*, *300*(5621), 931–936. doi:10.1126/science.1085049 PMID:12738850

Gadd, G. M. (1990). Heavy metal accumulation by bacteria and other microorganisms. *Experientia*, *46*(8), 834–840. doi:10.1007/BF01935534

Gadd, G. M. (2010). Metals, minerals and microbes: Geomicrobiology and bioremediation. *Microbiology*, *156*(3), 609–643. doi:10.1099/mic.0.037143-0 PMID:20019082

Gadd, G. M., & Griffiths, A. J. (1978). Microorganisms and heavy metal toxicity. *Microbial Ecology*, *4*(4), 303–317. doi:10.1007/BF02013274 PMID:24232222

Geslin, C., Llanos, J., Prieur, D., & Jeanthon, C. (2001). The manganese and iron superoxide dismutases protect *Escherichia coli* from heavy metal toxicity. *Research in Microbiology*, *152*(10), 901–905. doi:10.1016/S0923-2508(01)01273-6 PMID:11766965

Giller, K., Witter, E., & McGrath, S. P. (1998). Toxicity of heavy metals to microorganisms and microbial processes in agricultural soils: A review. *Soil Biology & Biochemistry*, *30*(10–11), 1389–1414. doi:10.1016/S0038-0717(97)00270-8

Gordon, O., Slenters, T. V., Brunetto, P. S., Villaruz, A. E., Sturdevant, D. E., Otto, M., & Fromm, K. M. et al. (2010). Silver coordination polymers for prevention of implant infection: Thiol interaction, impact on respiratory chain enzymes, and hydroxyl radical induction. *Antimicrobial Agents and Chemotherapy*, *54*(10), 4208–4218. doi:10.1128/AAC.01830-09 PMID:20660682

Green, M. H., Muriel, W. J., & Bridges, B. A. (1976). Use of a simplified fluctuation test to detect low levels of mutagens. *Mutation Research*, *38*(1), 33–42. doi:10.1016/0165-1161(76)90077-7 PMID:768758

Haas, K. L., & Franz, K. J. (2009). Application of metal coordination chemistry to explore and manipulate cell biology. *Chemical Reviews*, *109*(10), 4921–4960. doi:10.1021/cr900134a PMID:19715312

Harrison, J. J., Tremaroli, V., Stan, M. A., Chan, C. S., Vacchi-Suzzi, C., Heyne, B. J., & Turner, R. J. et al. (2009). Chromosomal antioxidant genes have metal ion-specifc roles as determinants of bacterial metal tolerance. *Environmental Microbiology*, *11*(10), 2491–2509. doi:10.1111/j.1462-2920.2009.01973.x PMID:19555372

Helbig, K., Grosse, C., & Nies, D. H. (2008). Cadmium toxicity in glutathione mutants of *Escherichia coli*. *Journal of Bacteriology*, *190*(15), 5439–5454. doi:10.1128/JB.00272-08 PMID:18539742

Hong, R., Kang, T. Y., Michels, C. A., & Gadura, N. (2012). Membrane lipid peroxidation in copper alloy-mediated contact killing of *Escherichia coli*. *Applied and Environmental Microbiology*, *78*(6), 1776–1784. doi:10.1128/AEM.07068-11 PMID:22247141

Hussein, K. A., & Joo, J. H. (2013). Heavy metal resistance of bacteria and its impact on the production of antioxidant enzymes. *African Journal of Microbiological Research*, *7*(20), 2288–2296. doi:10.5897/AJMR12.1764

Imlay, J. A. (2003). Pathways of oxidative damage. *Annual Review of Microbiology*, *57*(1), 395–418. doi:10.1146/annurev.micro.57.030502.090938 PMID:14527285

Imlay, J. A., Chin, S. M., & Linn, S. (1988). Toxic DNA damage by hydrogen peroxide through the Fenton reaction *in vivo* and *in vitro*. *Science*, *240*(4852), 640–642. doi:10.1126/science.2834821 PMID:2834821

Irving, H., & Williams, R. J. P. (1953). The stability of transition-metal complexes. *Journal of the Chemical Society*, *1953*, 3192–3210. doi:10.1039/jr9530003192

Itoh, M., Nakamura, M., Suzuki, T., Kawai, K., Horitsu, H., & Takamizawa, K. (1995). Mechanism of chromium(VI) toxicity in *Escherichia coli*: Is hydrogen peroxide essential in Cr(VI) toxicity? *Journal of Biochemistry*, *117*, 780–786. PMID:7592539

Jillian, E. G., Robert, S., & Rajakaruna, N. (2015). Transfer of heavy metals through terrestrial food webs: A review. *Environmental Monitoring and Assessment*, *187*(4). doi:10.1007/s10661-015-4436-3 PMID:25800370

Kashem, M. A., & Singh, B. R. (2001). Metal availability in contaminated soils: I. Effects of flooding and organic matter on changes in Eh, pH and solubility of Cd, Ni and Zn. *Nutrient Cycling in Agroecosystems*, *61*(3), 247–255. doi:10.1023/A:1013762204510

Kuperman, R. G., Siciliano, S. D., Römbke, J., & Oorts, K. (2014). Deriving Site-Specific Soil Clean-Up Values for Metals and Metalloids: Rationale for Including Protection of Soil Microbial Processes. *Integrated Environmental Assessment and Management*, *10*(3), 388–400. doi:10.1002/ieam.1513 PMID:24376192

Lemire, J. A., Harrison, J. J., & Turner, R. J. (2013). Antimicrobial activity of metals: Mechanisms, molecular targets and applications. *Nature Reviews. Microbiology*, *11*(6), 371–384. doi:10.1038/nrmicro3028 PMID:23669886

Lenart-Boroń, A., & Boroń, P. (2014). The Effect of Industrial Heavy Metal Pollution on Microbial Abundance and Diversity in Soils-A Review. In M. C. Hernandez-Soriano (Ed.), *Environmental Risk Assessment of Soil Contamination* (pp. 759–783). doi:10.5772/57406

Liau, S. Y., Read, D. C., Pugh, W. J., Furr, J. R., & Russell, A. D. (1997). Interaction of silver nitrate with readily identifiable groups: Relationship to the antibacterial action of silver ions. *Letters in Applied Microbiology*, *25*(4), 279–283. doi:10.1046/j.1472-765X.1997.00219.x PMID:9351278

Linley, E., Denyer, S. P., McDonnell, G., Simons, C., & Maillard, J. Y. (2012). Use of hydrogen peroxide as a biocide: New consideration of its mechanisms of biocidal action. *The Journal of Antimicrobial Chemotherapy*, *67*(7), 1589–1596. doi:10.1093/jac/dks129 PMID:22532463

Lippard, S. J., & Berg, J. M. (1994). *Principles of Bioinorganic Chemistry*. Mill Valley, CA: University Science Books.

Ma, Z., Jacobsen, F. E., & Giedroc, D. P. (2009). Coordination chemistry of bacterial metal transport and sensing. *Chemical Reviews*, *109*(10), 4644–4681. doi:10.1021/cr900077w PMID:19788177

Macomber, L., Elsey, S. P., & Hausinger, R. P. (2011). Fructose-1,6-bisphosphate aldolase (class II) is the primary site of nickel toxicity in *Escherichia coli*. *Molecular Microbiology*, *82*(5), 1291–1300. doi:10.1111/j.1365-2958.2011.07891.x PMID:22014167

Macomber, L., & Imlay, J. A. (2009). The iron-sulfur clusters of dehydratases are primary intracellular targets of copper toxicity. *Proceedings of the National Academy of Sciences of the United States of America, 106*(20), 8344–8349. doi:10.1073/pnas.0812808106 PMID:19416816

Moulis, J.-M., Davasse, V., Golinelli, M.-P., Meyer, J., & Quinkal, I. (1996). The coordination sphere of iron-sulfur clusters: Lessons from site-directed mutagenesis experiments. *Journal of Biological Inorganic Chemistry, 1*(1), 2–14. doi:10.1007/s007750050017

Nieboer, E., & Richardson, D. H. S. (1980). The replacement of the nondescript term 'heavy metals' by a biologically and chemically significant classification of metal ions. *Environmental Pollution, 1*(B), 3-26.

Nies, D. H. (1999). Microbial heavy-metal resistance. *Applied Microbiology and Biotechnology, 51*(6), 730–750. doi:10.1007/s002530051457 PMID:10422221

Nishioka, H. (1975). Mutagenic activities of metal compounds in bacteria. *Mutation Research, 31*(3), 185–189. doi:10.1016/0165-1161(75)90088-6 PMID:805366

Nunoshiba, T., Obata, F., Boss, A. C., Oikawa, S., Mori, T., Kawanishi, S., & Yamamoto, K. (1999). Role of iron and superoxide for generation of hydroxyl radical, oxidative DNA lesions, and mutagenesis in *Escherichia coli. The Journal of Biological Chemistry, 274*(49), 34832–34837. doi:10.1074/jbc.274.49.34832 PMID:10574955

Ogunseitan, O. A., Yang, S., & Ericson, J. (2000). Microbial d-aminolevulinate dehydratase as a biosensor of lead bioavailability in contaminated environments. *Soil Biology & Biochemistry, 32*(13), 1899–1906. doi:10.1016/S0038-0717(00)00164-4

Pearson, R. G. (1963). Hard and soft acids and bases. *Journal of the American Chemical Society, 85*(22), 3533–3539. doi:10.1021/ja00905a001

Pereira, Y., Lagniel, G., Godat, E., Baudouin-Cornu, P., Junot, C., & Labarre, J. (2008). Chromate causes sulfur starvation in yeast. *Toxicological Sciences, 106*(2), 400–412. doi:10.1093/toxsci/kfn193 PMID:18794233

Perez, J. M., Calderón, I. L., Arenas, F. A., Fuentes, D. E., Pradenas, G. A., Fuentes, E. L., & Vásquez, C. C. et al. (2007). Bacterial toxicity of potassium tellurite: Unveiling an ancient enigma. *PLoS ONE, 2*(2), e211. doi:10.1371/journal.pone.0000211 PMID:17299591

Ruzicka, F. J., & Beinert, H. (1978). The soluble "high potential" type iron-sulfur protein from mitochondria is aconitase. *The Journal of Biological Chemistry, 253*(8), 2514–2517. PMID:204652

Stadtman, E. R. (1993). Oxidation of free amino acids and amino acid residues in proteins by radiolysis and by metal-catalyzed reactions. *Annual Review of Biochemistry, 62*(1), 797–821. doi:10.1146/annurev.bi.62.070193.004053 PMID:8352601

Stadtman, E. R., & Levine, R. L. (2003). Free radical-mediated oxidation of free amino acids and amino acid residues in proteins. *Amino Acids, 25*(3-4), 207–218. doi:10.1007/s00726-003-0011-2 PMID:14661084

Stewart, E. J., Aslund, F., & Beckwith, J. (1998). Disulfide bond formation in the *Escherichia coli* cytoplasm: An *in vivo* role reversal for the thioredoxins. *The EMBO Journal, 17*(19), 5543–5550. doi:10.1093/emboj/17.19.5543 PMID:9755155

Stohs, S. J., & Bagchi, D. (1995). Oxidative mechanisms in the toxicity of metal ions. *Free Radical Biology & Medicine*, *18*(2), 321–336. doi:10.1016/0891-5849(94)00159-H PMID:7744317

Strlič, M., Kolar, J., Šelih, V.-S., Kočar, D., & Pihlar, B. (2003). A comparative study of several transition metals in Fenton-like reaction systems at circum-neutral pH. *Acta Chimica Slovenica*, *50*, 619–632.

Sumner, E. R., Shanmuganathan, A., Sideri, T. C., Willetts, S. A., Houghton, J. E., & Avery, S. V. (2005). Oxidative protein damage causes chromium toxicity in yeast. *Microbiology*, *151*(6), 1939–1948. doi:10.1099/mic.0.27945-0 PMID:15942001

Touati, D., Jacques, M., Tardat, B., Bouchard, L., & Despied, S. (1995). Lethal oxidative damage and mutagenesis are generated by iron in Δ*fur* mutants of *Escherichia coli*: Protective role of superoxide dismutase. *Journal of Bacteriology*, *177*(9), 2305–2314. doi:10.1128/jb.177.9.2305-2314.1995 PMID:7730258

Valko, M., Morris, H., & Cronin, M. T. D. (2005). Metals, toxicity and oxidative stress. *Current Medicinal Chemistry*, *12*(10), 1161–1208. doi:10.2174/0929867053764635 PMID:15892631

Waldron, K. J., & Robinson, N. J. (2009). How do bacterial cells ensure that metalloproteins get the correct metal? *Nature Reviews. Microbiology*, *7*(1), 25–35. doi:10.1038/nrmicro2057 PMID:19079350

Warnes, S. L., Caves, V., & Keevil, C. W. (2012). Mechanism of copper surface toxicity in *Escherichia coli* O157:H7 and *Salmonella* involves immediate membrane depolarization followed by slower rate of DNA destruction which differs from that observed for Gram-positive bacteria. *Environmental Microbiology*, *14*(7), 1730–1743. doi:10.1111/j.1462-2920.2011.02677.x PMID:22176893

Wolt, J. (1994). *Soil solution chemistry*. New York, NY: John Wiley.

Xu, F. F., & Imlay, J. A. (2012). Silver(I), mercury(II), cadmium(II), and zinc(II) target exposed enzymic iron-sulfur clusters when they toxify *Escherichia coli*. *Applied and Environmental Microbiology*, *78*(10), 3614–3621. doi:10.1128/AEM.07368-11 PMID:22344668

KEY TERMS AND DEFINITIONS

Antioxidant Depletion: It refers to the deficiency condition of antioxidant (such as, thiols, ascorbic acid, polyphenols, etc.) used in inhibiting the oxidation process of other molecules in cell.

Competitive Inhibition: In this chapter, it refers to the phenomenon by which the entry of essential metal ions into cell is interfered by other non-essential metal ions due to having ion mimic property that results in nutrient starvation of cell.

Donor Atom Selectivity: It is a specific principle of the atom of donor ligand that shows affinity to select specific metals and forms metal-ligand coordination complex.

Fenton Reaction: Fenton reaction, a process of oxidation of organic molecules by Fe(II) and hydrogen peroxide (H_2O_2), is very common in biological system first described by H. J. H. Fenton in 1894.

Free Radical: Free radical is highly reactive atomic or group of atom or molecular with unpaired (odd) number of electrons that have independent existence and capable to start a chain of reactions. Many transition metals are considered as free radical.

Hard–Soft Acid Base Theory: It is an empirically derived theory of chemistry that helps us to explain the stability of metal complexes and the mechanisms of their reactions in complex mixtures of inorganic and organic reactants.

Ion Mimicry: The ionic mimicry is referred to the phenomena by which an unbound, cationic species of non-essential metal can behave or serve as a structural and/or functional homolog or mimic of another (usually an essential) element at the site of a carrier protein, ion channel, enzyme, structural protein, transcription factor and/or metal-binding protein.

Iron-Sulfur (Fe–S) Cluster: Iron-sulfur (Fe-S) clusters, consisting of iron and elemental sulfur, are generally characteristically present in proteins and enzymes at various molar ratios which play important role to avoid toxicity in cell.

Metal Genotoxicity: It refers to the property of metals that damages the structure and function of genetic materials in a cell causing genetic mutations.

Metal Speciation: A metal can exist as different chemical forms/species [e.g. As(III) and As(V) species of arsenic (As)] in any single milieu which is known as metal speciation.

Reactive Oxygen Species (ROS) Stress: Reactive Oxygen Species (ROS) is referred to oxygen containing reactive molecules and free radicals derived from metabolism of oxygen which is responsible for causing significant damages in cell. These elevated ROS mediated adverse consequences in cell can be termed known as "ROS stress".

Reduction Potential: Reduction potential or redox potential or oxidation/reduction potential (i.e., ORP or pE or ε, or E_h) is a measure [by volts (V) or millivolts (mV)] of the tendency of a chemical species to acquire electrons and thereby be reduced.

Thiol: It is a carbon-bonded sulfhydryl (–C–SH or R–SH) group containing organosulfur compound that has capability to reduce some metal species.

ENDNOTES

1 Transition metals, other metals, and metalloids are collectively used as the term "metals" in this chapter.

Chapter 2
Implications of Molecular Docking Assay for Bioremediation

Zarrin Basharat
Fatima Jinnah Women University, Pakistan

Monazza Bibi
Fatima Jinnah Women University, Pakistan

Azra Yasmin
Fatima Jinnah Women University, Pakistan

ABSTRACT

Bioremediation utilizes microbes to control environmental pollution, primarily through diverse enzymatic processes. With the incorporation of computation in biological experimentation, bioremediation has also been influenced by computational techniques. Molecular docking assay is one such pedestal of computational assisted bioremediation, which has been elaborated in this chapter. It helps in inferring whether the active site accommodate the pollutant molecules or not, depending on the stearic hindrance of the residues and nature of the active site pocket. The spotting of consequential active site residues and binding characteristics of compounds under study can conceivably be employed for site-directed mutagenic testing. From a vantage point, no one had expected such a remarkable usefulness of molecular docking assay for environmental research. Positive shades of low cost and efficiency, combined with eco-friendliness have made it a valuable method for analyzing biodegradative properties of enzymes responsible for pollution remediation.

INTRODUCTION

The world is facing an ever increasing menace of pollution due to industrialization and other anthropogenic activities. It is, therefore, at the foremost agenda of environmental scientists to find better, fast and eco-friendly solutions to mitigate pollution. Bioremediation is one such endeavor that exploits microbes

DOI: 10.4018/978-1-5225-2325-3.ch002

to detoxify or degrade harmful contaminants in the environment. Bacterial enzymes are pollutant degradation workhorses that act to decontaminate environment. Molecular docking accelerates the procedure of choosing and improving enzymatic properties for a particular pollutant by quick computational scrutiny. The technique shows inordinate potential in its ability to predict enzyme adhesion and functionality in degrading or adsorbing the pollutant and presents a lucid picture of the relation and communication between pollutant and enzyme molecule. It helps in inferring whether the active site of certain enzymes accommodate the dye molecules or not, depending on the stearic hindrance of the residues and nature of the active site pocket. Protein-pollutant association is checked by assessing thousands of potential docking conformation poses. Picking poses that are reliable is an arduous task and one with lowest energy score is usually seen as the best match. Blind docking is attempted for proteins with unknown active site by entire protein surface scanning (Schnecke & Kuhn, 2000; Verdonk et al., 2005). Docking also provides information on mediator accommodation in the active site pocket to surmise about productive or non-favoured binding. The spotting of consequential active site residues and binding characteristics of compounds under study can conceivably be employed for site-directed mutagenic testing. These results then need to be tested in the lab experimentally and a critical evaluation of docking tools which gives better results might help define a set of rules and define a benchmark for such studies.

Molecular docking in remediation experienced a lag phase but now transitioned to alog phase in its-frequency of application and attention of environmental scientists. It is now a noticeable area of research methodology assisting bioremediation studies. Bioinformatics boasts of numerous successful techniques developed for protein-ligand docking but shape complementarity and simulation approaches are most popular. In shape complementarity approach a matching procedure portrays the protein and the ligand as complementary surfaces (Meng, Shoichet, & Kuntz, 1992). In the simulative approach, interaction energies of ligand and protein are calculated (Feig, Onufriev, Lee, Im, Case, & Brooks, 2004). Knowledge derived information from docking assays for bioremediation is critical for a reliable evaluation of the state of the enzymatic micro-environment besides the transformations taking place inside enzyme, as well as for crafting strategies for sozotechnical measures aimed at environment clean up. This chapter discusses foundation of eco-analytical information for bioremediation, gained from molecular docking assay and provides information on some studies conducted until now. Finally the challenges and barriers to the molecular docking based bioremediation studies have also been analyzed and future perspective of such studies is also provided. The predictions are valuable for environmental chemists perusing metabolic intermediates, supervisory bodies/controllers endeavoring probable toxic products, microbiologists seeking to perceive microbial biodegradation, besides others having wide array of fascinations.

OVERVIEW

Molecular docking is a technique which calculates the preferred binding positioning of one molecule relative to the other, at a minimum energy conformation during complex formation (Lengauer & Rarey, 1996; Brooijmans & Kuntz, 2003). Docking can be carried out among protein and a small molecule, protein-protein, Protein-DNA, protein-RNA, protein-lipid, protein-sugar, polymer-ligand etc. where larger moiety is treated as a substrate and smaller one as the ligand. In case of protein-ligand interaction, which is the focus of this chapter, ligand generally fits inside cavity of the protein, which is usually predicted by a search algorithm (Westhead, Clark, & Murray, 1997; Gohlke, Hendlich, & Klebe, 2000; Brooijmans & Kuntz, 2003). This protein cavity turns active as a result of contact with substrates. It is formed of evolutionarily conserved residues and can thus, be traced on the basis of comparison with similar proteins (Lichtarge, Bourne, & Cohen, 1996). Many computational methods exist for identifying ligand-binding residues or active site prediction such as LIGSITE, SURFNET, CAST and PASS (Huang, 2009). After participation in the catalytic reaction, active site or catalytic site residue can be reused (Alberts et al., 2013). Some enzymes may have catalytic sites capable of binding effector molecule other than active site. These are called allosteric sites and occur mostly in proteins with more than one sub-unit and allow a range of interactions in addition to specific active site (Dedecker, 2000).

MECHANISM AND METHODS OF DOCKING

Protein-pollutant interaction study via docking is basically protein-ligand association modeling relative to bioremediation. Pollutants are treated flexible and allowed to adapt to a wide range of surface and pocket conformations. The complementary interaction region is also smaller and portrays reduced discrimination. The focus is therefore on the flexibility of the pollutant and modeling of weak associations accurately. 3D structure of the enzyme is first analyzed to identify active sites for fitting moiety of interest (Schechter, 2005). Experimentally obtained protein structures are the best source of information for protein-ligand docking studies. Availability of structural data information and steady increase in amount of publicly available structures with time, not only permits the investigation of novel protein-ligand docking parameters but also improve the understanding of already resolved as well as new targets. Structures relating a protein bound to diverse substrates or ligands could assist the knowledge discovery regarding ligand-induced conformation variations. New structures can be modeled using the already available structures (Basharat, 2015a). Standard computational methods are utilized forsolving protein structures based on the templates available in Protein Data Bank. More than 100,000 structures are currently present in the database and the number of sequences being deposited in NCBI alone is far greater than the number of these structures which means the number of protein sequences exceeds far greater than the number of structures and there is a huge gap, wealth of hidden information and demand for structure modeling. Steady progress has been made in this regard to model the protein structures based on available templates using the fold prediction and geometry calculation method called 'homology modeling'. *In silico* methods assist even the experimentally determined structures for geometry refining, estimation of binding energies, modeling the effect of mutations, predict the location of binding surfaces for new molecules.

Gap between sequence-to-structure ratio increases manifold in case of membrane proteins. This also signifies that the total numbers of residues are many times larger than the corresponding unique chains from the Protein Data Bank (Werner, Morris, Dastmalchi, & Church, 2012). *De novo* or *ab initio* modeling is employed in such a case for computational modeling of completely novel protein structures. These modeling approaches rely on physicochemical and thermodynamic properties of amino acid residues of the protein sequence and may not have direct linkage to the parameters of experimentally solved structures. Employing the scoring of energy function approach, it resorts to matching protein conformation at global minimum to least energy through an iterative procedure. *De novo* methods use knowledge based scoring functions derived from experimental structures and their folds. Such methods are not template free *sensu stricto* and related to threading. *Ab initio* methods utilize force fields centered on first principles. This approach is template free and uses energy of the system for structure generation. Search space is large as all-atom model is used so this method works better in case of small sized proteins or peptides. The low-resolution structures obtained by computational approaches are further refined for docking studies through energy minimization and molecular dynamics (Werner et al., 2012). Protein structure assemblages are often implemented to obtain flexible models during protein-ligand docking. Structural conservation and catalytic site information is exploited for prediction of pollutant binding sites as catalytic residues and conformational changes in the active site region. Properly selected ensembles can improve docking performance considerably. Preprocessing of protein binding site ensembles may be carried out using dedicated webservers or databases like Astex Non-native Set (Verdonk, Mortenson, Hall, Hartshorn, & Murray, 2008), CoDNaS (Monzon, Juritz, Fornasari, & Parisi, 2013), Pocketome (An, Totrov, & Abagyan, 2005; Kufareva, Ilatovskiy, & Abagyan, 2012) and SIENA (Bietz & Rarey, 2016). The preparation of protein ensembles is quite intricate procedure due to unreliable annotation of many protein structures along with artifacts in structures. These impact the performance of ensemble employing applications. In case of missing or wrong information regarding binding site of a protein in the ensemble used for docking study, unintended bias may be introduced.

Ligand is usually obtained from a database source or drawn in a software or web interface (Exhaustive resource list available at http://www.click2drug.org/index.html) and saved. It is then processed and optimized before docking with the molecule. Missing residues and hydrogen atoms are added. Rotatable bonds are identified and partial atomic charges are calculated. As a ligand can move across six translational and rotational degrees of freedom, these movements are modeled. Initially both receptor and ligand molecules were treated as rigid bodies due to lack of computation power for solving the degrees of freedom. Maiden docking study of a protein and a small molecule (Kuntz, Blaney, Oatley, Langridge, & Ferrin, 1982) employed a negative receptor imageand six translational and rotational degrees of freedom. Spheres occupying the pockets and grooves on the receptor surface defined the prospective interaction sitesto clasp the ligand. Interactions were assessed by estimating the overlap as well as probablehydrogen bond formation between atoms in the protein-ligand complex. These types of rigid body calculations of protein-ligand studies have limitations as it does not account for induced fit ligand behavior (Brooijmans & Kuntz, 2013). With increment in computational power, algorithms and softwares were designed that could model interactions with one moiety as flexible and the other one as rigid (Leach, Shoichet, & Peishoff, 2006). Now, softwares exist for modeling the flexibility/degrees of freedom in both receptor molecule and ligand. The top results of active site accommodated ligands are judged on the basis of a statistical scoring function, which translates the unification energy into numerical value or the docking score (Leach et al., 2006). This is a critical step of docking, with correct prediction of the binding affinities/dissociation constant of a protein-ligand complex to separate true positives from false positives, a genuinely difficult task as its valuable outcome (Ambure & Roy, 2016; Yan & Wang, 2016). Complexes with least binding energies and maximum scores are then selected for further analysis. Calculation and scoring of interactions is a lot easier, less demanding and accurate in the gaseous phase as compared to that in the water phase due to small window of activity (Tirado-Rives & Jorgensen, 2006).

DOCKING ALGORITHMS AND SOFTWARES

For successful protein–ligand docking, speed and accuracy are the prime parameters. Development (Connolly, 1983) and usage of molecular surface calculations instead of explicit atom multi-configurations in docking have reduced the number of complexes and increased the speed of calculation (Jiang & Kim, 1991; Norel, Lin, Wolfson, & Nussinov, 1994). Ensuing this, grids to store physico-chemical properties of the receptor were integrated into docking procedures. Selected probe groups used were amino, methyl, and hydroxyl groups. The grids presented polar/nonpolar type of interactions, commonly preferred at each position in the binding site but presented difficulty in calculating energy. Later, tweaks in the algorithms led to hydrogen bonding, energy parameter and sophisticated scoring function calculation. This led to realistic scoring and the assessment of ligand flexibility in the active site during docking procedures. Ligand flexibility is usually calculated using systematic, stochastic, and deterministic search algorithms. Several algorithms use a sum of these approaches. Systematic search algorithms (Westhead et al., 1997) use grid values to calculate each degree of freedom with number of evaluations corresponding formal degrees of freedom. Cessation conditions are inserted to thwart the algorithm from sampling the space, identified to cause erroneous complex formation. Stochastic search algorithms check one degree of freedom at one time. Uncertainty of binding is rectified through independent runs. Examples include Monte Carlo methods and evolutionary algorithms. In deterministic searches, initial state leads to the next state, which is either equal or lower in energy to the previous state. These algorithms get trapped at local minima as they are unable to navigate across barriers. Different tactics are used to escalate their capacity to cross barriers or decrease the height of energy barriers. Examples include popular energy minimization and molecular dynamics simulation approaches. Other examples include Tabu search and Incremental construction etc. Some of these algorithms (Monte Carlo, Molecular dynamics and Genetic algorithms) are also applicable to receptor flexibility calculations (Brooijmans & Kuntz, 2013). For validation of the binding accuracy prediction of a protein with ligand, apart from the conventional root mean square deviation criteria, first principle, semi-empirical, empirical and knowledge-based methods exist and various softwares use these for calculating binding mode and affinities through estimation of parameters as van der Waals, electrostatic, hydrogen bond interactions etc (Kirchmair et al., 2012; Sliwoski, Kothiwale, Meiler, & Lowe, 2014). Classification and cataloguing of atom types, pairing this information with number of atom types and joint occurrence of specific class of atoms is essential part of knowledge based scoring functions. Free energy perturbation and thermodynamics incorporation are considered the high end procedures but computationally exhaustive. Molecular mechanics with Poisson Boltzmann surface area calculationapproach is also a recent popular method for free energy estimation of protein-ligand binding affinity (Weis, Katebzadeh, Söderhjelm, Nilsson, & Ryde, 2006; Liu, Liu, Zhu, Zhang, He, & Zhang, 2016). These approaches calculate the intermolecular, electrostatic, thermodynamic, stearic and solvent related forces or a combination of these forces. Algorithms/softwares for docking validation implement various amalgams of these forces for sieving out best results as neither of them is perfect in terms of precisionand broad applicability. Consensus scoring technique boosts the prospects of discovering accurately bound models by coalescing scores from several scoring functions (Huang, Grinter, & Zou, 2010).

More than fifty softwares are available for docking till to date, each varying in its approach or parameters but taking into account both these differing requirements. Speed of the fundamental steps has augmented due to availability of higher clock processing of modern computers along with the use of scoring grids. Latest developments in parallel computing domain have also enhanced the performance of docking algorithms and softwares considerably. Most popular of them are AutoDock (Huey, Morris, Olson, & Goodsell, 2007), Molegro Virtual Docker (Thomsen & Christensen, 2006), EADock (Grosdidier, Zoete, & Michielin, 2007), GOLD (Verdonk, Cole, Hartshorn, Murray, & Taylor, 2003; Joy, Nair, Hariharan, & Pillai, 2006), Flexx (Rarey, Kramer, Lengauer & Klebe, 1996; Kramer, Rarey, & Lengauer, 1999) etc. There are also dedicated web services available for docking like Molecular Docking Server (Bikadi & Hazai, 2009), SwissDock (Grosdidier, Zoete, & Michielin, 2011), PatchDock (Schneidman-Duhovny, Inbar, Nussinov, & Wolfson, 2005) etc for calculating geometry, site and energy of ligands interacting with proteins. The 3D pose of the bound pollutant-protein complex can be downloaded and visualized using different visualizing tools like PyMol, Rasmol etc which could help in inference of the best fit of ligand.

An increasing interest in human and animal enzyme interaction via molecular docking studies with pollutants is also of prime interest to environmental researchers working with occupational health hazard chemicals focused on bioaccumulation or toxicity property profiling of pollutants in human or animal bodies (Xie, Lü, & Chen, 2013; Zhou, Li, & Chen, 2013a; Liu, Chen, Jiang, & Song, 2014; Wang, Zhang, Cao, & Tang, 2014; Teng, Wang, Zou, Huang, & Du, 2015; Tripathi, Yasir, Gurjar, Bose, Dubey, & Shrivastava, 2015), Cytochrome P450 (Shimada et al., 2013; Zhou et al., 2013b; Shao, Chen, Li, & Zhao, 2014; Chushak, Chapleau, Frey, Mauzy, & Gearhart, 2015) and other xenobiotic degrading enzymes are of particular interest. Nanoparticle and polymer hazard has also been investigated through molecular docking assay for human and eco-toxicity bioindicator species (Lai et al, 2012; Hosseinzadeh, Moosavi-Movahedi, & Ghourchian, 2014; Liu et al., 2014; Yang, Hao, Li, Wei, Wang, & Liu, 2014; Ranjan, Dasgupta, Chinnappan, Ramalingam, & Kumar, 2015; Turabekova, Rasulev, Theodore, Jackman, Leszczynska, & Leszczynski, 2014). These are not being discussed in detail in this chapter as our focus is on bioremediation, and microbial enzyme docking investigation is best suited for such studies. Although, many of the enzymes in human and animal bodies also break down/detoxify the harmful pollutants but for environmental cleanup purpose, microbial enzymes are preferred owing to ease of availability and purification. Initiation and advancement of docking studies along the bioremediation line of research has been attempted by numerous researchers in the recent years (Table 1).

From the increasing number of studies including molecular docking for binding between pollutant and enzyme leading to degradation, it is evident that docking of protein-ligand is invaluable for studying the interactions of substrate and pollutants in bioremediation.

CASE STUDY: IBUPROFEN-COA LIGASE DOCKING

Sphyngomonas Ibu-2 has been reported to degrade ibuprofen [2-(4-Isobutylphenyl-Propionic acid)], an NSAID with an annual production of approximately 10 kilotonnes (Baumann & Baxendale, 2016), into catechols under aerobic conditions. CoA-ligase of *Sphyngomonas* sp. binds the said pharmaceutical and cleaves the acid side chain to yield isobutylcatechol metabolite (Murdoch & Hay, 2013). However, no report on the interaction of both these moieties exists. Here we describe a step by step method for protein-ligand study of this pollutant-enzyme interaction. The analysis shows that the CoA-ligase is capable of forming versatile conformations upon interaction with ibuprofen ligand. Modeling and abstraction played a noteworthy role in perceiving protein-ligand associations using computational *modus operandi*. Protein structural modeling and docking are helpful in inferring residues critical for interaction which might be mutated to obtain results of our interest. Detailed protocol and results are described as below:

Step 1: Modeling, Validation, and Optimization of Enzyme Structure

The structure of *Sphingomonas* sp. CoA ligase (Accession no: ABK92273) was modeled using I-TASSER as previously described (Basharat & Yasmin, 2015a; Basharat & Yasmin, 2016a, Basharat, Messaoudi, Ruba, & Yasmin, 2016a). Ten templates were used for threading based model generation of the structure. Fatty acyl CoA synthetase of *Mycobacterium tuberculosis* (PDB ID: 3R44) was the top template

Table 1. Comprehensive list of microbe-pollutant interaction studies carried out using molecular docking assay

Organism	Enzyme	Pollutant	Docking software	Reference
Pseudomonas sp.	Naphthalene dioxygenase	Indole, Naphthalene, Biphenyl	AutoDock program	Carredano, Karlsson, Kauppi, Choudhury, Parales, Parales, Lee, Gibson, Eklund, & Ramaswamy, 2000
Burkholderia sp.	Biphenyl dioxygenase	2,3′-Chlorobiphenyl, 3,4′-Chlorobiphenyl	Sybyl	Zielinski, Kahl, Hecht, & Hofer, 2003
Sphingomonas sp.	Naphthalene dioxygenase	Polyaromatic hydrocarbons (indole, naphthalene, biphenyl)	GRAMM	Librando & Forte, 2005
Rhizobium sp.	Haloalkanoic dioxygenase	Chloropropionic acid	AutoDock	Hamid, Hamid, Haziyamin, Wahab, & Huyop, 2006
Sinorhizobium meliloti, Pseudomonas diminuta	Amidohydrolase superfamily	Hydantoin, Paraoxon	DOCK	Hermann, Ghanem, Li, Raushel, Irwin, & Shoichet, 2006
Trametes versicolor, Bacillus subtilis	Laccase	Anthracene, Phenanthrene, Thiodicarb, Malathion, Captan, Atrazine, Remazol Red B, Chlorophenol	GOLD	Suresh, Kumar, Kumar, & Singh, 2008
Caldariomyces fumago	Chloroperoxidase	Limonene	GOLD	Aguila, Vazquez-Duhalt, Tinoco, Rivera, Pecchi, & Alderete, 2008
Pseudomonas sp.	Monooxygenase	Aromatic hydrocarbon	PyMol	Notomista, Cafaro, Bozza, & Di Donato, 2008
Coriolus versicolor, Pleurotus ostreatus, Pseudomonas sp.	Laccase, manganese peroxidase, lignin peroxidase	Polyaromatic hydrocarbon	FRED	Arun, Raja, Arthi, Ananthi, Kumar, & Eyini, 2008
Comamonas testosteroni	Biphenyl dioxygenase	Polychlorinated biphenyls	AutoDock 4.0	Baig & Manickam, 2009
Bacillus subtilis	Azoreductase	Azo dyes	PatchDock	Ramanathan, Shanthi, & Sethumadhavan, 2009
Sphingomonas paucimobills	Haloalkane dehalogenase	1,2,3-trichloropropane	Affinity	Zhao, Zheng, Zhang, Chu, & Sun, 2009
Rigidoporus lignosus	Laccases	2,5-xylidine	AutoDock version 3.5	Cambria, Di Marino, Falconi, Garavaglia, & Cambria, 2010
Comamonas testosteroni	Biphenyl-2,3-dioxygenase	Polychlorinated biphenyls	AutoDock	Baig & Manickam, 2010
Sphingomonas sp. CHY1	Ring-hydroxylating dioxygenase	Polyaromatic hydrocarbons	AutodockVina	Librando & Pappalardo, 2011
Burkholderia xenovorans, Pandoraea pnomenusa	Biphenyl dioxygenase	DDT	AutoDock4.2	L'Abbée, Tu, Barriault, & Sylvestre, 2011
Enterococcus faecalis	Azoreductase	Methyl red	AutodockVina	Feng, Kweon, Xu, Cerniglia, & Chen, 2012
Aeromonas hydrophila	Laccase	Aromatic/phenolic compounds	AutoDock	Singh, Mani, & Chaudhary, 2012
Burkholderia cepacia.	Catechol 2, 3-dioxygenase	Petroleum hydrocarbons	Autodock 4.0	Ajao et al., 2012

continued on following page

Table 1. Continued

Organism	Enzyme	Pollutant	Docking software	Reference
Trametes versicolor	Laccase	Phenol	Molegro Virtual Docker	Zhang, Zeng, Zeng, Liu, Liu, Chen, Liu, Li, & Xie, 2012
Pycnoporus cinnabarinus	Laccase	Acid blue 74, Reactive black 5, Reactive blue 19	GOLD	Prasad et al., 2012
Brevundimonas diminuta	N-acyl-L-homoserine Lactone	Paraoxon	AutoDock	Zhan, Zhou, Guan, & Han, 2013
Rigidoporus lignosus	Laccase	2,5-Xylidine	AutoDock	Cambria et al., 2010
Geobacteraceae	Mn-Fe superoxide dismutase	Bisphenol-A, aminotriazole, ethylene-diurea	Molegro Virtual Docker	Emtiazi & Hedayatkha, 2014
Pandoraea pnomenusa	Biphenyl dehydrogenase	Polychlorinated biphenyls	HADDOCK	Piccoli, Musiani, & Giorgetti, 2014
Bacillus safensis	Laccase	Procion blue, Procion green, Procion red, Reactive yellow 86	GOLD	Singh, Sharma, Jacob, & Gakhar, 2014
E. coli	Azoreductase	Azo dyes	Autodockvina	Thakuria, Jungai, & Adhikari, 2015a
Rhodococcus sp. ustb-1	Naphthalene dioxygenase	Pyrene	Sybyl7.3	Jin, Yao, Zhang, Yu, Chen, Liu, Peng, & Choi, 2015
Pseudomonas putida	Phenol hydroxylase	Phenol, 2-chloro phenol	AutodockVina	Ray & Banarjee, 2015
Pseudomonas putida	Azoreductase	Azo dyes (Azobenzene,p-Aminoazobenzene, Amaranth dye, Methyl Orange, Sudan IV, Amidoschwarz, Congo Red, C.I. Food Yellow 3,)	Swiss Dock	Thakuria, Singha, Maisnam, & Adhikari, 2015b
Bacillus sp. B29	Azoreductase	azo dyes (orange I, orange II, orange G and acid red 88)	Molegro Virtual Docker 6.0	Dehghanian, Kay, & Kahrizi, 2015
Trametes versicolor	Laccase	Medicarpin	AutoDock 4.2	Martínez-Sotres, Rutiaga-Quiñones, Herrera-Bucio, Gallo, & López-Albarrán, 2015
Ceriporiopsis subvermispora	Laccase	Herbicide diuron	AutoDock4.2	Vieira, Marschalk, Biavatti, Lorscheider, Peralta, & Seixas, 2015
Bacillus sp.	Azoreductase	Methyl orange, Congo red	PatchDock	Basharat, 2015b
Bacillus sp.	Azoreductase	Azo dyes (orange G, acid red 88, Sudan I, orange I, and methyl red).	AutoDock	Haghshenas, Kay, Dehghanian, & Tavakol, 2015
Yersinia enterocolitica	Laccase	Aspirin, Diclofenac	GOLD	Singh, Rawat, Waseem, Gupta, Lynn, Nitin, Ramchiary, & Sharma, 2016
Aeromonas hydrophila	Laccase	Azo dyes(Joyfix Red, Remazol Red, Reactive Red, Reactive Yellow	Glide	Kumar, Shantkriti, Muruganandham, Murugesh, Rane, & Govindwar, 2016
Lysinibacillus sphaericus	Azoreductase	Azo dyes, JoyfixRed, Remazol Red, Reactive Red, Reactive Yellow	Glide	Kumar, Shantkriti,, Muruganandham, Murugesh, Rane, & Govindwar, 2016
Sulfobacillus acidophilus	Barbiturase	Atrazine, Simazine, Hexazinone	PatchDock	Basharat & Yasmin, 2016b

with a structural similarity of more than 30%.Best generated model had aC-score of 0.44 (good quality model lies between -5 to 2 range). Estimated TM-score was 0.77±0.10 with a RMSD of 6.5±3.9Å. The structure was corrected for errors and missing residues. Fitting of intermolecular parameters to structural and thermodynamic data along with energy minimization was carried out until a gradient of 0.05 was achieved. The 3D protein structure (Figure 1) was then tested qualitatively using ERRAT (http://services.mbi.ucla.edu/ERRAT/) and Rampage (http://mordred.bioc.cam.ac.uk/~rapper/rampage.php).

The structure had an overall quality factor of 78.19. Good resolution structures produce a value of over 95 and around 91 for 2.5-3 Å resolution structures. Since this is a homology model and equivalent to low resolution structure so the quality was considered fine. Structure passed the VERIFY3D (services.mbi.ucla.edu/Verify_3D) check as 90.51% of the residues had an average 3D-1D score of more than or equivalent to 0.2. According to the standard, more than 80% residues should have 3D-1D score more than or equivalent to 0.2. Ramachandran plot showed 79.2% of the residues were in favored region, 15.6% in allowed region and 5.1% in outliers.

Step 2: Ligand Procurement and Preparation

Ligand i.e. ibuprofen molecule was obtained from PubChem in 2D format and converted into 3D .pdb format using OpenBabel software. The ligand was optimized for pH 7, energy minimized using MMFF94 forcefield and saved for further analysis.

Step 3: Molecular Docking and Analysis

Docking was carried out using Molecular Operating Environmentsoftware.This software uses Monte Carlo simulated annealing process for docking, allowing conformational space sampling for ligand. Screening of everypossible binding site in the selected region of the target macromolecule was carried out. Docking interaction energy (U_{total}) of a given conformation is estimated from a set of energy grids

Figure 1. 3D structure of the Sphyngomonas sp. ibuprofen-CoA ligase

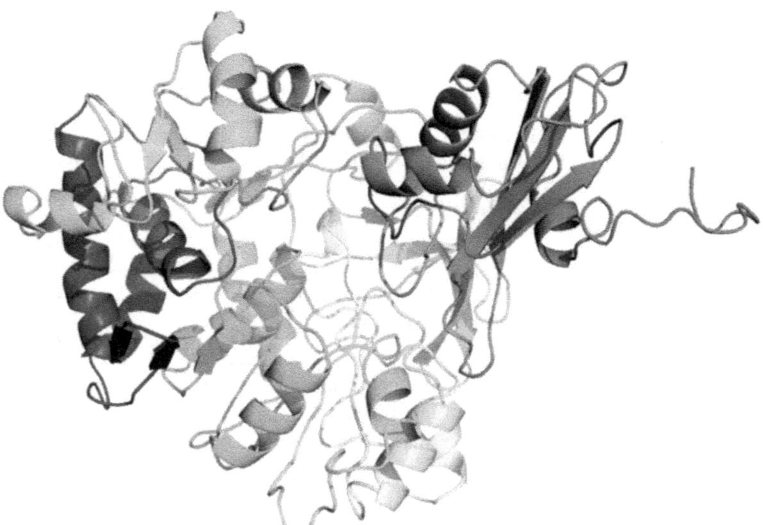

centered in the macromolecule binding site using the formula parameters $U_{total} = U_{ele} + U_{vdw} + U_{ligand} + U_{solv}$ where U_{ele} is the electrostatic, U_{vdw} is the van der Waal interaction between protein-ligand complex. U_{ligand} is the conformation energy whereas U_{solv} represents the solvation energy. Among the generated poses, pose with lowest U_{total} is considered optimal. Ligase protein coordinates remained rigid during the docking process, while the flexible, mobile ligand relocated along the grid to trace most favorable binding orientation and conformation based on the interaction energy. The top ranking ten poses were retained in .mdb database format. These were then categorized according to increasing S-value i.e. final score indicating binding free energy. The complex with least S value was studied to evaluate the interactions of ligand with the active site residues of the ligase. Bonding interactions were evaluated via ligX option of Molecular Operating Environment. A value of -7.6335 was obtained for the optimal protein-ligand complex. 3D and 2D plot of the docked complex are shown in Figure 2 and Figure 3 respectively.

PERSPECTIVE

With increasing computation power, improved hardware and better simulations, we are achieving better insights and accurate prediction nearing natural chemical accuracy of interactions. This type of computational experimentation can help in unearthing the role of these residues and compounds in substrate specificity and activity, which in turn can be used for designing improved mutants and recombinants for degradation of noxious compounds (Prasad, Vindal, Narayana, Ramakrishna, Kunal, & Srinivas, 2012). Computational assessment of mutated protein residues in the structure is an inexpensive approach to assess alteration in functionality (Basharat & Yasmin, 2016c; Basharat, Zeb, & Yasmin, 2016b). This type of analysis attempted previously for laccase enzyme of *Bacillus licheniformis* using site-directed and random mutagenesis (Koschorreck, Schmid, & Urlacher, 2009) and can be extended for other enzymes

Figure 2. 3D view of the protein-ligand complex. Docked ibuprofen is clearly visible in the pocket of the CoA-ligase of Sphyngomonas sp

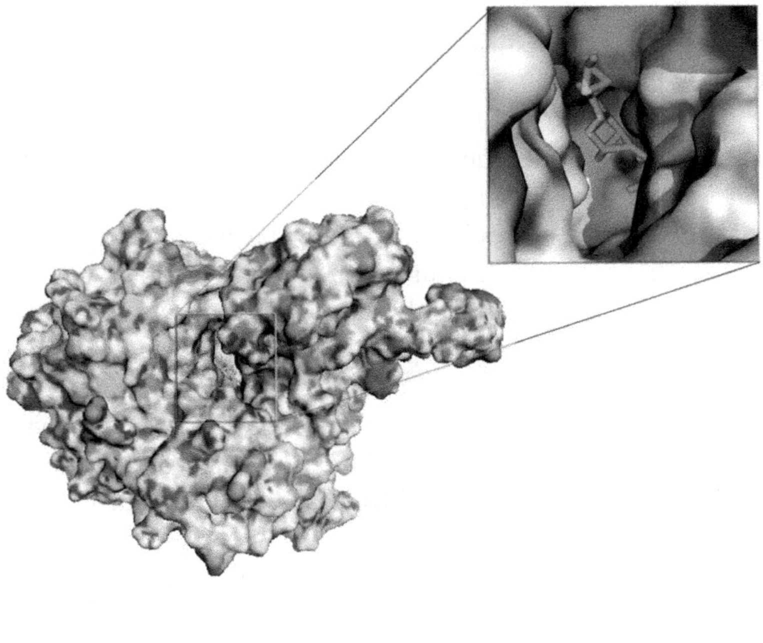

Figure 3. Molecular Operating Environment generated ligase CoA-ibuprofen interaction plot. It depicts 2D (flattened) spatial arrangement of ligand and CoA-ligase enzyme with respect to key interactions. The proximity contour shown by dashed lines. Trp210 donates electron to ibuprofen ligand OH group. The conformation of ligase-ibuprofen complex was energetically favored with the nucleophile and electrophile aptly positioned for catalysis

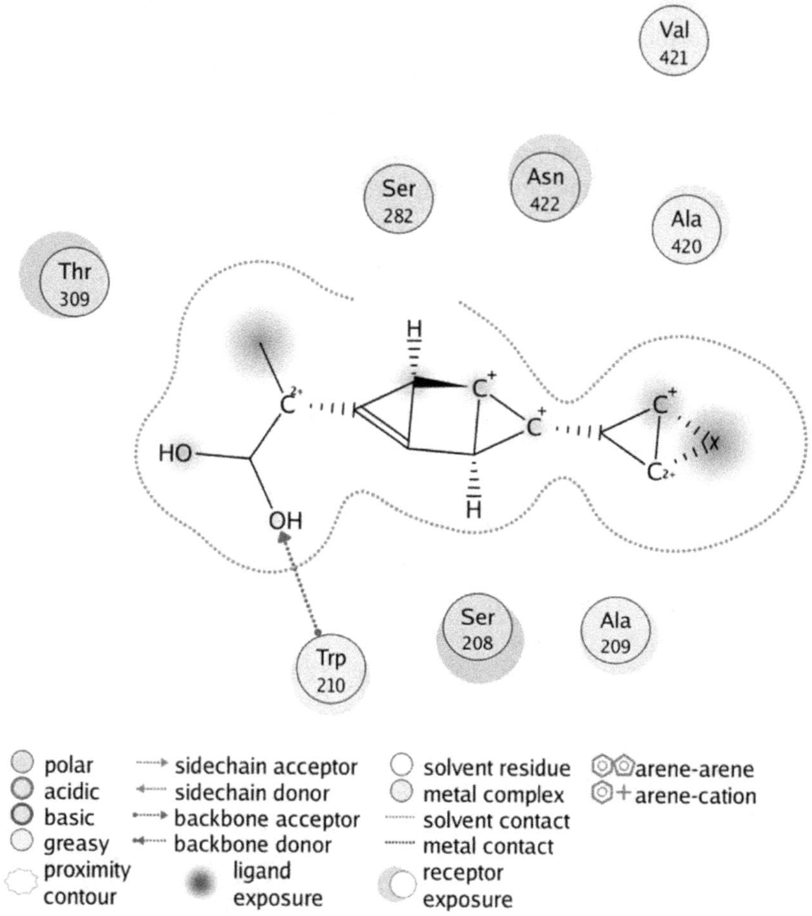

relevant to bioremediation. The existing algorithms and softwares are continually honed and perfected, as no one method functions fittingly for all protein systems.Paired occurrence of class of amino acids to more conjoined class of amino acids could help improve precision of scoring but is practically difficult to achieve due to scores of other undetermined parameters. Another challenge is the absence of a standard set of weighing coefficients for protein-ligand energy segment and forcefield scoring calculations. Presence of ions or co-factors in the pocket imposes parametrization difficulty in scoring as well. Evaluation of scoring functions is also a problem and each algorithm and software should be subject to CAPRI assessment for different datasets rather than simple root mean square deviation evaluation (for measuring quality of binding mode), before public release. Even though CAPRI (Janin, Henrick, Moult, Eyck, Sternberg, Vajda, Vakser, & Wodak, 2003; Méndez, Leplae, Lensink, & Wodak, 2005) exercises blind prediction feature that guarantees unbiased assessment but unrestricted usage of biological information like mutation dataand evolutionary analyses approach needs to be curtailed. Refined structural and binding datasets (including information of analogous pH and salt conditions) would be precious for parameter improvement. Systematic CASP-like ligand docking contests would do a good job of quality enhancement of algorithms.

Other challenges and barriers to the molecular docking include, but are not limited to issues like 'choosing the right approach', accuracy vs efficiency in assessing a particular algorithm, performance and prospective utility of a specific method over others in diverse situations etc. Throng of situations, factors, and data sets laid in front of the practitioner can confuse and the horde of pitfalls may render the docking approach less efficient. The shortcomings and technical traps of pollutant-enzyme docking methods could be avoided or reduced by knowledge based cautious approach which might make the practitioner aware and help make informed decisions which could limit probability of failure. Current scoring functions need to be trained rigorously on known and new datasets. In forcefield and knowledge based scoring methods, annexation of desolvation and entropic effects is indispensable for achieving precision.Scoring functions need to include fundamental compound measurements. Examples include transfer energies (for solvation and charges), dipoles/quadrupoles (for charge models), and pKa values for both small molecules and proteins (for solvation and binding energy). These can be calculated independently with softwares like DELPHI (Smith, Witham, Sarkar, Zhang, Li, Li, & Alexov, 2012; Sarkar, Witham, Zhang, Zhenirovskyy, Rocchia, & Alexov, 2013) but need to be integrated into the softwares and webservers.Such measurements are important as accurate scoring functions demands rigorous physical chemistry measurements for a wide variety of chemotypes.

Researchers have concentrated on pollutant binding to the protein, based upon structural knowledge of a known enzyme with degradation properties (Ajao, Kannan, Yakubu, Umoh. & Ameh, 2012; Basharat & Yasmin, 2016b). Conversely, designing a protein to bind a particular pollutant is also of interest to environmental scientists. It is intended for creating protein-based biosensors for pollutants. Computational design of pollutant-binding proteins is a perplexing inverse assessment of docking and scoring procedures, where experimental affinities are usually fed back into the algorithm development for refinement of complexes (Smith et al., 2016). Such protein design also entails a scoring function to rank and choose best conformations from the list of generated conformations. This would help in choosing relevant and active proteins to bioremediate/degrade pollutant.

Comparison between experimental and theoretical docking is not straightforward in biological entities but advent of combinatorial methods, growth in experimentally solved structures, theoretical advances, and improved computation power has definitely played a role in the success of docking studies and hopefully, will continue to do so in the near future. Computational efforts, particularly in combination with relevant chemical strategies couldfurther boost efficiency and quality of docking. Molecular docking has been in use for more than three decades now and will continue to evolve and benefit the research community, including environmental scientists.

CONCLUSION

We need innovative real-time analytical methods capable of monitoring variations of priority pollutants in both time and location, integrated simulative models for perceiving and improving the breakdown mechanism, ultimately leading to better formulation of bioremediation strategies. To discover the substrate overtness of 3D available protein structure or a homology modeled enzyme and for gaining insight into the catalytic site, important residues and mechanism of the pollutant remediation-comprehensive information of enzyme structure complexed with substrate/pollutant is obligatory. For this purpose, computational molecular docking analysis is a boon and has been previously attempted for several noxious pollutants with enzymes. Although, several research groups have attempted to add to the scanty information on biodegradation analysis using docking assays but there is a need for more studies reporting and reproducing the conformations of protein-bound pollutants. The acquired information could aid in designing enzyme mutants for better accommodation of dye pollutants, especially for targeting at-once for a whole class of pollutants.

REFERENCES

Aguila, S., Vazquez-Duhalt, R., Tinoco, R., Rivera, M., Pecchi, G., & Alderete, J. B. (2008). Stereoselective oxidation of R-(+)-limonene by chloroperoxidase from *Caldariomyces fumago*. *Green Chemistry*, *10*(6), 647–653. doi:10.1039/b719992a

Ajao, A. T., Kannan, M., Yakubu, S. E., Umoh, V. J., & Ameh, J. B. (2012). Homology modeling, simulation and molecular docking studies of catechol-2, 3-Dioxygenase from *Burkholderia cepacia:* Involved in degradation of Petroleum hydrocarbons. *Bioinformation*, *8*(18), 848–854. doi:10.6026/97320630008848 PMID:23144539

Alberts, B., Bray, D., Hopkin, K., Johnson, A., Lewis, J., Raff, M., & Walter, P. et al. (2013). *Essential Cell Biology*. Garland Science.

Ambure, P., & Roy, K. (2016). Scoring functions in docking experiments. In *Methods and Algorithms for Molecular Docking-Based Drug Design and Discovery* (pp. 54–98). IGI Global. doi:10.4018/978-1-5225-0115-2.ch003

An, J., Totrov, M., & Abagyan, R. (2005). Pocketome via comprehensive identification and classification of ligand binding envelopes. *Molecular & Cellular Proteomics*, *4*(6), 752–761. doi:10.1074/mcp.M400159-MCP200 PMID:15757999

Arun, A., Raja, P. P., Arthi, R., Ananthi, M., Kumar, K. S., & Eyini, M. (2008). Polycyclic aromatic hydrocarbons (PAHs) biodegradation by basidiomycetes fungi, *Pseudomonas* isolate, and their cocultures: Comparative *in vivo* and *in silico* approach. *Applied Biochemistry and Biotechnology*, *151*(2-3), 132–142. doi:10.1007/s12010-008-8160-0 PMID:18975143

Baig, M. S., & Manickam, N. (2010). Homology modeling and docking studies of *Comamonas testosteroni* B-356 biphenyl-2, 3-dioxygenase involved in degradation of polychlorinated biphenyls. *International Journal of Biological Macromolecules*, *46*(1), 47–53. doi:10.1016/j.ijbiomac.2009.10.014 PMID:19879892

Basharat, Z. (2015a). Computational sequence analysis and *in silico* modeling of a stripe rust resistance protein encoded by wheat TaHSC70 gene. *Journal of Computational Biology and Bioinformatics Research*, *7*(1), 1–7. doi:10.5897/JCBBR2013.082

Basharat, Z. (2015b). *In vitro and in silico analysis of enzymes from indigenous bacteria for remediation of azo dyes* (Unpublished M.Phil. dissertation). Fatima Jinnah Women University, Rawalpindi, Pakistan.

Basharat, Z., Messaoudi, A., Ruba, S., & Yasmin, A. (2016a). NQO1 rs1800566 polymorph is more prone to NOx induced lung injury: Endorsing deleterious functionality through informatics approach. *Gene*, *591*(1), 14–20. doi:10.1016/j.gene.2016.06.048 PMID:27349566

Basharat, Z., & Yasmin, A. (2015a). *In silico* assessment of phosphorylation and O-β-GlcNAcylation sites in human NPC1 protein critical for Ebola virus entry. *Infection, Genetics and Evolution*, *34*, 326–338. doi:10.1016/j.meegid.2015.06.001 PMID:26048414

Basharat, Z., & Yasmin, A. (2016a). *Understanding properties of the master effector of phage shock operon in Mycobacterium tuberculosis via bioinformatics approach*. DOI:.10.1101/050047

Basharat, Z., & Yasmin, A. (2016b). Molecular docking of *Sulfobacillus acidophilus* barbiturase with s-triazine compounds. *PeerJ Preprints, 4*, e2070v1. DOI:10.7287/peerj.preprints.2070v1

Basharat, Z., & Yasmin, A. (2016c). Bioinformatic analysis of human Gαq Q209 missense mutations associated with *uveal melanoma*. In *13th International Bhurban Conference on Applied Sciences and Technology (IBCAST)* (pp. 106-111). IEEE. DOI: doi:10.1109/IBCAST.2016.7429863

Basharat, Z., Zeb, S., & Yasmin, A. (2016b). *Computational study of some amoebicidal phytochemicals against heat shock protein of Naegleria fowleri*. Gene Reports; doi:10.1016/j.genrep.2016.09.003

Baumann, M., & Baxendale, I. R. (2016). Continuous photochemistry: The flow synthesis of ibuprofen via a photo-Favorskii rearrangement. *Reaction Chemistry & Engineering*, *1*(2), 147–150. doi:10.1039/C5RE00037H

Bietz, S., & Rarey, M. (2016). SIENA: Efficient compilation of selective protein binding site ensembles. *Journal of Chemical Information and Modeling*, *56*(1), 248–259. doi:10.1021/acs.jcim.5b00588 PMID:26759067

Bikadi, Z., & Hazai, E. (2009). Application of the PM6 semi-empirical method to modeling proteins enhances docking accuracy of AutoDock. *Journal of Cheminformatics*, *1*(1), 1. doi:10.1186/1758-2946-1-15 PMID:20150996

Brooijmans, N., & Kuntz, I. D. (2003). Molecular recognition and docking algorithms. *Annual Review of Biophysics and Biomolecular Structure*, *32*(1), 335–373. doi:10.1146/annurev.biophys.32.110601.142532 PMID:12574069

Cambria, M. T., Di Marino, D., Falconi, M., Garavaglia, S., & Cambria, A. (2010). Docking simulation and competitive experiments validate the interaction between the 2, 5-xylidine inhibitor and *Rigidoporus lignosus* laccase. *Journal of Biomolecular Structure & Dynamics*, *27*(4), 501–509. doi:10.1080/07391102.2010.10507334 PMID:19916571

Carredano, E., Karlsson, A., Kauppi, B., Choudhury, D., Parales, R. E., Parales, J. V., & Ramaswamy, S. et al. (2000). Substrate binding site of naphthalene 1, 2-dioxygenase: Functional implications of indole binding. *Journal of Molecular Biology*, *296*(2), 701–712. doi:10.1006/jmbi.1999.3462 PMID:10669618

Chushak, Y. G., Chapleau, R. R., Frey, J. S., Mauzy, C. A., & Gearhart, J. M. (2015). Identifying potential protein targets for toluene using a molecular similarity search, *in silico* docking and *in vitro* validation. *Toxicological Reviews*, *4*(2), 519–526.

Connolly, M. L. (1983). Solvent-accessible surfaces of proteins and nucleic acids. *Science*, *221*(4612), 709–713. doi:10.1126/science.6879170 PMID:6879170

DeDecker, B. S. (2000). Allosteric drugs: Thinking outside the active-site box. *Chemistry & Biology*, *7*(5), R103–R107. doi:10.1016/S1074-5521(00)00115-0 PMID:10801477

Dehghanian, F., Kay, M., & Kahrizi, D. (2015). A novel recombinant AzrC protein proposed by molecular docking and *in silico* analyses to improve azo dyes binding affinity. *Gene*, *569*(2), 233–238. doi:10.1016/j.gene.2015.05.063 PMID:26026905

Emtiazi, G., & Hedayatkhah, A. (2015). Comparative modelling of 3D-structure of *Geobacter* sp. M21 (a metal reducing bacteria) Mn-Fe superoxide dismutase and its binding properties with bisphenol-A, aminotriazole and ethylene-diurea. *Iranian Journal of Environmental Technology*, *1*(1), 19–29.

Feig, M., Onufriev, A., Lee, M. S., Im, W., Case, D. A., & Brooks, C. L. (2004). Performance comparison of generalized born and Poisson methods in the calculation of electrostatic solvation energies for protein structures. *Journal of Computational Chemistry*, *25*(2), 265–284. doi:10.1002/jcc.10378 PMID:14648625

Feng, J., Kweon, O., Xu, H., Cerniglia, C. E., & Chen, H. (2012). Probing the NADH-and Methyl Red-binding site of a FMN-dependent azoreductase (AzoA) from *Enterococcus faecalis*. *Archives of Biochemistry and Biophysics*, *520*(2), 99–107. doi:10.1016/j.abb.2012.02.010 PMID:22387379

Gohlke, H., Hendlich, M., & Klebe, G. (2000). Knowledge-based scoring function to predict protein-ligand interactions. *Journal of Molecular Biology*, *295*(2), 337–356. doi:10.1006/jmbi.1999.3371 PMID:10623530

Grosdidier, A., Zoete, V., & Michielin, O. (2007). EADock: Docking of small molecules into protein active sites with a multiobjective evolutionary optimization. *Proteins: Structure, Function, and Bioinformatics*, *67*(4), 1010–1025. doi:10.1002/prot.21367 PMID:17380512

Grosdidier, A., Zoete, V., & Michielin, O. (2011). SwissDock, a protein-small molecule docking web service based on EADock DSS. *Nucleic Acids Research*, *39*(suppl 2), W270–W277. doi:10.1093/nar/gkr366 PMID:21624888

Haghshenas, H., Kay, M., Dehghanian, F., & Tavakol, H. (2016). Molecular dynamics study of biodegradation of azo dyes via their interactions with AzrC azoreductase. *Journal of Biomolecular Structure & Dynamics*, *34*(3), 453–462. doi:10.1080/07391102.2015.1039585 PMID:26325128

Hamid, A. A. A., Hamid, T. A., Haziyamin, T., Wahab, R. A., & Huyop, F. (2015). Identification of functional residues essential for dehalogenation by the non-stereospecific α-haloalkanoic acid dehalogenase from Rhizobium sp. RC1. *Journal of Basic Microbiology*, *55*(3), 324–330. doi:10.1002/jobm.201300526 PMID:25727054

Hermann, J. C., Ghanem, E., Li, Y., Raushel, F. M., Irwin, J. J., & Shoichet, B. K. (2006). Predicting substrates by docking high-energy intermediates to enzyme structures. *Journal of the American Chemical Society*, *128*(49), 15882–15891. doi:10.1021/ja065860f PMID:17147401

Hosseinzadeh, R., Moosavi-Movahedi, A. A., & Ghourchian, H. (2014). Electrochemistry and molecular modeling of the hemoglobin–benzene interaction with a nanocrystalline mixed metal oxide. *RSC Advances*, *4*(90), 49128–49136. doi:10.1039/C4RA08018A

Huang, B. (2009). MetaPocket: A meta approach to improve protein ligand binding site prediction. *OMICS: A Journal of Integrative Biology*, *13*(4), 325–330. doi:10.1089/omi.2009.0045 PMID:19645590

Huang, S. Y., Grinter, S. Z., & Zou, X. (2010). Scoring functions and their evaluation methods for protein–ligand docking: Recent advances and future directions. *Physical Chemistry Chemical Physics*, *12*(40), 12899–12908. doi:10.1039/c0cp00151a PMID:20730182

Huey, R., Morris, G. M., Olson, A. J., & Goodsell, D. S. (2007). A semiempirical free energy force field with charge-based desolvation. *Journal of Computational Chemistry*, *28*(6), 1145–1152. doi:10.1002/jcc.20634 PMID:17274016

Janin, J., Henrick, K., Moult, J., Eyck, L. T., Sternberg, M. J., Vajda, S., & Wodak, S. J. et al. (2003). CAPRI: A critical assessment of predicted interactions. *Proteins: Structure, Function, and Bioinformatics*, *52*(1), 2–9. doi:10.1002/prot.10381 PMID:12784359

Jiang, F., & Kim, S. H. (1991). Soft docking: Matching of molecular surface cubes. *Journal of Molecular Biology*, *219*(1), 79–102. doi:10.1016/0022-2836(91)90859-5 PMID:2023263

Jin, J. N., Yao, J., Zhang, Q. Y., Yu, C., Chen, P., Liu, W. J., & Choi, M. M. et al. (2015). An integrated approach of bioassay and molecular docking to study the dihydroxylation mechanism of pyrene by naphthalene dioxygenase in *Rhodococcus* sp. ustb-1. *Chemosphere*, *128*, 307–313. doi:10.1016/j.chemosphere.2015.02.012 PMID:25747183

Joy, S., Nair, P. S., Hariharan, R., & Pillai, M. R. (2006). Detailed comparison of the protein-ligand docking efficiencies of GOLD, a commercial package and ArgusLab, a licensable freeware. *In Silico Biology*, *6*(6), 601–605. PMID:17518767

Kirchmair, J., Williamson, M. J., Tyzack, J. D., Tan, L., Bond, P. J., Bender, A., & Glen, R. C. (2012). Computational prediction of metabolism: Sites, products, SAR, P450 enzyme dynamics, and mechanisms. *Journal of Chemical Information and Modeling*, *52*(3), 617–648. doi:10.1021/ci200542m PMID:22339582

Koschorreck, K., Schmid, R. D., & Urlacher, V. B. (2009). Improving the functional expression of a *Bacillus licheniformis* laccase by random and site-directed mutagenesis. *BMC Biotechnology*, *9*(1), 1. doi:10.1186/1472-6750-9-12 PMID:19236694

Kramer, B., Rarey, M., & Lengauer, T. (1999). Evaluation of the FLEXX incremental construction algorithm for protein–ligand docking. *Proteins: Structure, Function, and Bioinformatics*, *37*(2), 228–241. doi:10.1002/(SICI)1097-0134(19991101)37:2<228::AID-PROT8>3.0.CO;2-8 PMID:10584068

Kufareva, I., Ilatovskiy, A. V., & Abagyan, R. (2012). Pocketome: An encyclopedia of small-molecule binding sites in 4D. *Nucleic Acids Research*, *40*(D1), D535–D540. doi:10.1093/nar/gkr825 PMID:22080553

Kumar, S. S., Shantkriti, S., Muruganandham, T., Murugesh, E., Rane, N., & Govindwar, S. P. (2016). Bioinformatics aided microbial approach for bioremediation of wastewater containing textile dyes. *Ecological Informatics*, *31*, 112–121. doi:10.1016/j.ecoinf.2015.12.001

Kuntz, I. D., Blaney, J. M., Oatley, S. J., Langridge, R., & Ferrin, T. E. (1982). A geometric approach to macromolecule-ligand interactions. *Journal of Molecular Biology, 161*(2), 269–288. doi:10.1016/0022-2836(82)90153-X PMID:7154081

LAbbée, J. B., Tu, Y., Barriault, D., & Sylvestre, M. (2011). Insight into the metabolism of 1, 1, 1-trichloro-2, 2-bis (4-chlorophenyl) ethane (DDT) by biphenyl dioxygenases. *Archives of Biochemistry and Biophysics, 516*(1), 35–44. doi:10.1016/j.abb.2011.09.016 PMID:22001737

Lai, L., Lin, C., Xu, Z. Q., Han, X. L., Tian, F. F., Mei, P., & Liu, Y. et al. (2012). Spectroscopic studies on the interactions between CdTe quantum dots coated with different ligands and human serum albumin. *Spectrochimica Acta. Part A: Molecular and Biomolecular Spectroscopy, 97*, 366–376. doi:10.1016/j.saa.2012.06.025 PMID:22797377

Leach, A. R., Shoichet, B. K., & Peishoff, C. E. (2006). Prediction of protein-ligand interactions. Docking and scoring: Successes and gaps. *Journal of Medicinal Chemistry, 49*(20), 5851–5855. doi:10.1021/jm060999m PMID:17004700

Lengauer, T., & Rarey, M. (1996). Computational methods for biomolecular docking. *Current Opinion in Structural Biology, 6*(3), 402–406. doi:10.1016/S0959-440X(96)80061-3 PMID:8804827

Librando, V., & Forte, S. (2005). Computer evaluation of protein segments removal effects from naphthalene 1, 2-dioxygenase enzyme on polycyclic aromatic hydrocarbons interaction. *Biochemical Engineering Journal, 27*(2), 161–166. doi:10.1016/j.bej.2005.08.011

Librando, V., & Pappalardo, M. (2011). Computational study on the interaction of a ring-hydroxylating dioxygenase from *Sphingomonas* CHY-1 with PAHs. *Journal of Molecular Graphics & Modelling, 29*(7), 915–919. doi:10.1016/j.jmgm.2011.03.001 PMID:21441049

Lichtarge, O., Bourne, H. R., & Cohen, F. E. (1996). An evolutionary trace method defines binding surfaces common to protein families. *Journal of Molecular Biology, 257*(2), 342–358. doi:10.1006/jmbi.1996.0167 PMID:8609628

Liu, X., Liu, J., Zhu, T., Zhang, L., He, X., & Zhang, J. Z. (2016). PBSA_E: A PBSA based free energy estimator for protein-ligand binding affinity. *Journal of Chemical Information and Modeling, 56*(5), 854–861. doi:10.1021/acs.jcim.6b00001 PMID:27088302

Liu, Y., Chen, M., Jiang, L., & Song, L. (2014). New insight into molecular interaction of heavy metal pollutant—cadmium (II) with human serum albumin. *Environmental Science and Pollution Research International, 21*(11), 6994–7005. doi:10.1007/s11356-014-2610-8 PMID:24522399

Martínez-Sotres, C., Rutiaga-Quiñones, J. G., Herrera-Bucio, R., Gallo, M., & López-Albarrán, P. (2015). Molecular docking insights into the inhibition of laccase activity by medicarpin. *Wood Science and Technology, 49*(4), 857–868. doi:10.1007/s00226-015-0734-8

Méndez, R., Leplae, R., Lensink, M. F., & Wodak, S. J. (2005). Assessment of CAPRI predictions in rounds 3–5 shows progress in docking procedures. *Proteins: Structure, Function, and Bioinformatics, 60*(2), 150–169. doi:10.1002/prot.20551 PMID:15981261

Meng, E. C., Shoichet, B. K., & Kuntz, I. D. (1992). Automated docking with grid-based energy evaluation. *Journal of Computational Chemistry*, *13*(4), 505–524. doi:10.1002/jcc.540130412

Monzon, A. M., Juritz, E., Fornasari, M. S., & Parisi, G. (2013). CoDNaS: A database of conformational diversity in the native state of proteins. *Bioinformatics (Oxford, England)*, *29*(19), 2512–2514. doi:10.1093/bioinformatics/btt405 PMID:23846747

Murdoch, R. W., & Hay, A. G. (2013). Genetic and chemical characterization of ibuprofen degradation by *Sphingomonas* Ibu-2. *Microbiology*, *159*(3), 621–632. doi:10.1099/mic.0.062273-0 PMID:23329679

Norel, R., Lin, S. L., Wolfson, H. J., & Nussinov, R. (1994). Shape complementarity at protein–protein interfaces. *Biopolymers*, *34*(7), 933–940. doi:10.1002/bip.360340711 PMID:8054472

Notomista, E., Cafaro, V., Bozza, G., & Di Donato, A. (2009). Molecular determinants of the regioselectivity of toluene/o-xylene monooxygenase from *Pseudomonas* sp. strain OX1. *Applied and Environmental Microbiology*, *75*(3), 823–836. doi:10.1128/AEM.01951-08 PMID:19074607

Piccoli, S., Musiani, F., & Giorgetti, A. (2014). Dynamic characterization and substrate binding of cis-2, 3-dihydrobiphenyl-2, 3-diol dehydrogenase—an enzyme used in bioremediation. *Journal of Molecular Modeling*, *20*(12), 1–6. doi:10.1007/s00894-014-2531-y PMID:25433599

Prasad, N. K., Vindal, V., Narayana, S. L., Ramakrishna, V., Kunal, S. P., & Srinivas, M. (2012). *In silico* analysis of *Pycnoporus cinnabarinus* laccase active site with toxic industrial dyes. *Journal of Molecular Modeling*, *18*(5), 2013–2019. doi:10.1007/s00894-011-1215-0 PMID:21877154

Ramanathan, K., Shanthi, V., & Sethumadhavan, R. (2009). *In silico* identification of catalytic residues in azobenzene reductase from *Bacillus subtilis* and its docking studies with azo dyes. *Interdisciplinary Sciences: Computational Life Sciences*, *1*(4), 290–297. PMID:20640807

Ranjan, S., Dasgupta, N., Chinnappan, S., Ramalingam, C., & Kumar, A. (2015). A novel approach to evaluate titanium dioxide nanoparticle–protein interaction through docking: An insight into mechanism of action. *Proceedings of the National Academy of Sciences, India Section B: Biological Sciences*, 1-7.

Rarey, M., Kramer, B., Lengauer, T., & Klebe, G. (1996). A fast flexible docking method using an incremental construction algorithm. *Journal of Molecular Biology*, *261*(3), 470–489. doi:10.1006/jmbi.1996.0477 PMID:8780787

Ray, S., & Banerjee, A. (2015). Molecular level biodegradation of phenol and its derivatives through dmp operon of *Pseudomonas putida:* A bio-molecular modeling and docking analysis. *Journal of Environmental Sciences (China)*, *36*, 144–151. doi:10.1016/j.jes.2015.03.035 PMID:26456616

Rivera-Hoyos, C. M., Morales-Álvarez, E. D., Poveda-Cuevas, S. A., Reyes-Guzmán, E. A., Poutou-Piñales, R. A., Reyes-Montaño, E. A., & Cardozo-Bernal, Á. M. et al. (2015). Computational analysis and low-scale constitutive expression of laccases synthetic genes GlLCC1 from *Ganoderma lucidum* and POXA 1B from *Pleurotus ostreatus* in *Pichia pastoris*. *PLoS ONE*, *10*(1), e0116524. doi:10.1371/journal.pone.0116524 PMID:25611746

Sarkar, S., Witham, S., Zhang, J., Zhenirovskyy, M., Rocchia, W., & Alexov, E. (2013). DelPhi Web Server: A comprehensive online suite for electrostatic calculations of biological macromolecules and their complexes. *Communications in Computational Physics*, *13*(01), 269–284. doi:10.4208/cicp.300611.201011s PMID:24683424

Schechter, I. (2005). Mapping of the active site of proteases in the 1960s and rational design of inhibitors/drugs in the 1990s. *Current Protein & Peptide Science*, *6*(6), 501–512. doi:10.2174/138920305774933286 PMID:16381600

Schnecke, V., & Kuhn, L. A. (2000). Virtual screening with solvation and ligand-induced complementarity. *Perspectives in Drug Discovery and Design*, *20*(1), 171–190. doi:10.1023/A:1008737207775

Schneidman-Duhovny, D., Inbar, Y., Nussinov, R., & Wolfson, H. J. (2005). PatchDock and SymmDock: Servers for rigid and symmetric docking. *Nucleic Acids Research*, *33*(Web Serversuppl 2), W363–W367. doi:10.1093/nar/gki481 PMID:15980490

Shao, J., Chen, J., Li, T., & Zhao, X. (2014). Spectroscopic and molecular docking studies of the *in vitro* interaction between puerarin and cytochrome P450. *Molecules (Basel, Switzerland)*, *19*(4), 4760–4769. doi:10.3390/molecules19044760 PMID:24743933

Shimada, T., Kim, D., Murayama, N., Tanaka, K., Takenaka, S., Nagy, L. D., & Guengerich, F. P. et al. (2013). Binding of diverse environmental chemicals with human cytochromes P450 2A13, 2A6, and 1B1 and enzyme inhibition. *Chemical Research in Toxicology*, *26*(4), 517–528. doi:10.1021/tx300492j PMID:23432429

Singh, D., Rawat, S., Waseem, M., Gupta, S., Lynn, A., Nitin, M., & Sharma, K. K. et al. (2016). Molecular modeling and simulation studies of recombinant laccase from *Yersinia enterocolitica* suggests significant role in the biotransformation of non-steroidal anti-inflammatory drugs. *Biochemical and Biophysical Research Communications*, *469*(2), 306–312. doi:10.1016/j.bbrc.2015.11.096 PMID:26631965

Singh, D., Sharma, K. K., Jacob, S., & Gakhar, S. K. (2014). Molecular docking of laccase protein from *Bacillus Safensis* DSKK5 Isolated from earthworm gut: A novel method to study dye decolorization potential. *Water, Air, and Soil Pollution*, *225*(11), 1–12. doi:10.1007/s11270-014-2175-7

Singh, V., Mani, I., & Chaudhary, D. K. (2012). Analysis of the multicopper oxidase gene regulatory network of *Aeromonas hydrophila*. *Systems and Synthetic Biology*, *6*(3-4), 51–59. doi:10.1007/s11693-012-9096-9 PMID:24294339

Sliwoski, G., Kothiwale, S., Meiler, J., & Lowe, E. W. (2014). Computational methods in drug discovery. *Pharmacological Reviews*, *66*(1), 334–395. doi:10.1124/pr.112.007336 PMID:24381236

Smith, N., Witham, S., Sarkar, S., Zhang, J., Li, L., Li, C., & Alexov, E. (2012). DelPhi web server v2: Incorporating atomic-style geometrical figures into the computational protocol. *Bioinformatics (Oxford, England)*, *28*(12), 1655–1657. doi:10.1093/bioinformatics/bts200 PMID:22531215

Smith, R. D., Damm-Ganamet, K. L., Dunbar, J. B. Jr, Ahmed, A., Chinnaswamy, K., Delproposto, J. E., & Doyle, L. et al. (2016). CSAR benchmark exercise 2013: Evaluation of results from a combined computational protein design, docking, and scoring/ranking challenge. *Journal of Chemical Information and Modeling*, *56*(6), 1022–1031. doi:10.1021/acs.jcim.5b00387 PMID:26419257

Suresh, P. S., Kumar, A., Kumar, R., & Singh, V. P. (2008). An *Insilco* approach to bioremediation: Laccase as a case study. *Journal of Molecular Graphics & Modelling*, *26*(5), 845–849. doi:10.1016/j.jmgm.2007.05.005 PMID:17606396

Teng, Y., Wang, X., Zou, L., Huang, M., & Du, X. (2015). Experimental and theoretical study on the binding of 2-mercaptothiazoline to bovine serum albumin. *Journal of Luminescence*, *161*, 14–19. doi:10.1016/j.jlumin.2014.12.046

Thakuria, B., Jungai, N., & Adhikari, S. (2015). Catalytic Site Prediction of azoreductase enzyme of *E. coli* with potentially important industrial dyes using molecular docking tools. *International Journal of Bioscience. Biochemistry and Bioinformatics*, *5*(2), 91.

Thakuria, B., Singha, C. J., Maisnam, P., & Adhikari, S. (2015). Functional and catalytic active sites prediction and docking analysis of azoreductase enzyme in *Pseudomonas putida* with a variety of commercially available azodyes. *African Journal of Biotechnology*, *14*(26), 2162–2169. doi:10.5897/AJB2015.14699

Thomsen, R., & Christensen, M. H. (2006). MolDock: A new technique for high-accuracy molecular docking. *Journal of Medicinal Chemistry*, *49*(11), 3315–3321. doi:10.1021/jm051197e PMID:16722650

Tirado-Rives, J., & Jorgensen, W. L. (2006). Contribution of conformer focusing to the uncertainty in predicting free energies for protein-ligand binding. *Journal of Medicinal Chemistry*, *49*(20), 5880–5884. doi:10.1021/jm060763i PMID:17004703

Tripathi, M. K., Yasir, M., Gurjar, V. S., Bose, P., Dubey, A., & Shrivastava, R. (2015). Insights from the molecular docking of hydrolytic products of methyl isocyanate (MIC) to inhibition of human immune proteins. *Interdisciplinary Sciences: Computational Life Sciences*, *7*(3), 287–294. PMID:26297312

Turabekova, M., Rasulev, B., Theodore, M., Jackman, J., Leszczynska, D., & Leszczynski, J. (2014). Immunotoxicity of nanoparticles: A computational study suggests that CNTs and C 60 fullerenes might be recognized as pathogens by Toll-like receptors. *Nanoscale*, *6*(7), 3488–3495. doi:10.1039/c3nr05772k PMID:24548972

Verdonk, M. L., Chessari, G., Cole, J. C., Hartshorn, M. J., Murray, C. W., Nissink, J. W. M., & Taylor, R. et al. (2005). Modeling water molecules in protein-ligand docking using GOLD. *Journal of Medicinal Chemistry*, *48*(20), 6504–6515. doi:10.1021/jm050543p PMID:16190776

Verdonk, M. L., Cole, J. C., Hartshorn, M. J., Murray, C. W., & Taylor, R. D. (2003). Improved protein–ligand docking using GOLD. *Proteins: Structure, Function, and Bioinformatics*, *52*(4), 609–623. doi:10.1002/prot.10465 PMID:12910460

Verdonk, M. L., Mortenson, P. N., Hall, R. J., Hartshorn, M. J., & Murray, C. W. (2008). Protein– ligand docking against non-native protein conformers. *Journal of Chemical Information and Modeling*, *48*(11), 2214–2225. doi:10.1021/ci8002254 PMID:18954138

Vieira, A. C., Marschalk, C., Biavatti, D. C., Lorscheider, C. A., Peralta, R. M., & Seixas, F. A. V. (2015). Modeling based structural insights into biodegradation of the herbicide diuron by laccase-1 from *Ceriporiopsis subvermispora*. *Bioinformation*, *11*(5), 224–228. doi:10.6026/97320630011224 PMID:26124565

Wang, Y. Q., Zhang, H. M., Cao, J., & Tang, B. P. (2014). Binding of a new bisphenol analogue, bisphenol S to bovine serum albumin and calf thymus DNA. *Journal of Photochemistry and Photobiology. B, Biology*, *138*, 182–190. doi:10.1016/j.jphotobiol.2014.06.002 PMID:24972352

Weis, A., Katebzadeh, K., Söderhjelm, P., Nilsson, I., & Ryde, U. (2006). Ligand affinities predicted with the MM/PBSA method: Dependence on the simulation method and the force field. *Journal of Medicinal Chemistry*, *49*(22), 6596–6606. doi:10.1021/jm0608210 PMID:17064078

Werner, T., Morris, M. B., Dastmalchi, S., & Church, W. B. (2012). Structural modelling and dynamics of proteins for insights into drug interactions. *Advanced Drug Delivery Reviews*, *64*(4), 323–343. doi:10.1016/j.addr.2011.11.011 PMID:22155026

Westhead, D. R., Clark, D. E., & Murray, C. W. (1997). A comparison of heuristic search algorithms for molecular docking. *Journal of Computer-Aided Molecular Design*, *11*(3), 209–228. doi:10.1023/A:1007934310264 PMID:9263849

Xie, X., Lü, W., & Chen, X. (2013). Binding of the endocrine disruptors 4-tert-octylphenol and 4-nonylphenol to human serum albumin. *Journal of Hazardous Materials*, *248*, 347–354. doi:10.1016/j.jhazmat.2013.01.036 PMID:23416871

Yan, Z., & Wang, J. (2016). Scoring Functions of Protein-Ligand Interactions. In *Methods and Algorithms for Molecular Docking-Based Drug Design and Discovery* (pp. 220–245). IGI Global. doi:10.4018/978-1-5225-0115-2.ch009

Yang, B., Hao, F., Li, J., Wei, K., Wang, W., & Liu, R. (2014). Characterization of the binding of chrysoidine, an illegal food additive to bovine serum albumin. *Food and Chemical Toxicology*, *65*, 227–232. doi:10.1016/j.fct.2013.12.047 PMID:24394486

Zhan, D., Zhou, Z., Guan, S., & Han, W. (2013). The Effect of conformational variability of phosphotriesterase upon N-acyl-L-homoserine lactone and paraoxon binding: Insights from molecular dynamics studies. *Molecules (Basel, Switzerland)*, *18*(12), 15501–15518. doi:10.3390/molecules181215501 PMID:24352010

Zhang, Y., Zeng, Z., Zeng, G., Liu, X., Liu, Z., Chen, M., & Xie, G. et al. (2012). Effect of Triton X-100 on the removal of aqueous phenol by laccase analyzed with a combined approach of experiments and molecular docking. *Colloids and Surfaces. B, Biointerfaces*, *97*, 7–12. doi:10.1016/j.colsurfb.2012.04.001 PMID:22580478

Zhao, Y. S., Zheng, Q. C., Zhang, H. X., Chu, H. Y., & Sun, C. C. (2009). Interaction between 1, 2, 3-trichloropropane and haloalkane dehalogenase LinB. *Chemical Research in Chinese Universities*, *25*(4), 556–559.

Zhou, X., Li, X., & Chen, X. (2013a). Binding mechanism of Orange G to human serum albumin: Saturation transfer difference-NMR, spectroscopic and computational techniques. *Dyes and Pigments*, *98*(2), 212–220. doi:10.1016/j.dyepig.2013.01.023

Zhou, X., Wang, Y., Hu, T., Or, P. M., Wong, J., Kwan, Y. W., & Yeung, J. H. et al. (2013b). Enzyme kinetic and molecular docking studies for the inhibitions of miltirone on major human cytochrome P450 isozymes. *Phytomedicine*, *20*(3), 367–374. doi:10.1016/j.phymed.2012.09.021 PMID:23102508

Zielinski, M., Kahl, S., Hecht, H. J., & Hofer, B. (2003). Pinpointing biphenyl dioxygenase residues that are crucial for substrate interaction. *Journal of Bacteriology*, *185*(23), 6976–6980. doi:10.1128/JB.185.23.6976-6980.2003 PMID:14617661

KEY TERMS AND DEFINITIONS

Ab initio: Method of estimation about biological features of a protein without any template structure or existing model.

Binding Mode: The positioning of ligand relative to the receptor and their coordination/conformation when bound/attached with each other.

De novo: Protein structure designed from the sequence alone.

Ligand: The molecule or entity which fits in/binds to the receptor. It might or might not be small in size compared to the receptor.

Pose: A specified position of ligand with respect to the receptor.

Ranking: Classification of poses based on scoring function.

Receptor: The molecule which acts as a receiving entity that might or might not be large in size as compared to the ligand.

Scoring: Evaluation of a pose through calculation of favorable interaction of parameters (such as hydrogen bonds, hydrophobic interactions).

Chapter 3
Insight of Proteomics and Genomics in Environmental Bioremediation

Raghvendra Pratap Singh
Shandong University, China

Zhi-Feng Li
Shandong University, China

Geetanjali Manchanda
D.A.V. University, India

Alok R. Rai
Seth Kasarimal Porwal College, India

ABSTRACT

Bioremediation of hazardous substances from environment is a major human and environmental health concern but can be managed by the microorganism due to their variety of properties that can effectively change the complexity. Microorganisms convey endogenous genetic, biochemical and physiological assets that make them superlative proxies for pollutant remediation in habitat. But, the crucial step is to degrade the complex ring structured pollutants. Interestingly, the integration of genomics and proteomics technologies that allow us to use or alter the genes and proteins of interest in a given microorganism towards a cell-free bioremediation approach. Resultantly, efforts have been finished by developing the genetically modified (Gm) microbes for the remediation of ecological contaminants. Gm microorganisms mediated bioremediation can affect the solubility, bioavailability and mobility of complex hazardous.

INTRODUCTION

The habitat is gifted with rich wealth of natural resources such as forests, wildlife, land, soil, air, water, wind, plants and animals. Rivalry, humans started civilization and started to use and overuse, and now the misuse of natural resources like exploitation of natural resources by the use of chemical fertilizers in agriculture, release of industrial waste, anthropogenic activities and burning of fossil fuels etc. Resultantly, a huge waste products generated in the natural niche. In current scenario, these waste products have varieties from raw sewage to nuclear waste. In the past disposal of these wastes intended to digging a hole and dumping then filling all wastage to it. However, these days, it has become insufficient due to heavy amount of wastage. Crucial stage has begun after the presence of toxic wastage among these

DOI: 10.4018/978-1-5225-2325-3.ch003

"dig and dump" sites. These toxic materials have begun to leak into water sources and into areas that sustain human life. This problem has led to modern-day bioremediation, always been a serious problem. Bioremediation is classically defined as the transformation or degradation of contaminants into non-hazardous or less hazardous chemicals (Hornung, 1997).

In general, millions of chemicals are commercially omnipresent in our daily activity (According to American database CHEMCATS). The minimization of the impact of potential pollutant is inevitability for environmental protection agencies, researchers, industries, and society as a whole. Deliberate releases of oil can also a major source of extensive contamination. Every year, 1,300,000 tonnes of oil enter the environment, for the utmost part through natural petroleum seeps. Gulf War (1991) was the critical example of oil as hazardous pollutant while it started menacing the desalination plants and coastal ecosystems of the Gulf after releasing of 0.82 megatonnes oil in Kuwait (Pearce, 1993). The facts that beach are not continuously covered with tar balls is due to the activity of microorganisms that can degrade the released petroleum. Unfortunately, some common environmental aromatic pollutants like BTEX (benzene, toluene, ethylbenzene and xylene), derived from petroleum derivatives such as gasoline. But, these hazardous compounds miraculously degraded commonly by the species of the genera *Alcanivorax, Cycloclasticus* and *Thalassolituus*, which can degrade various branched-chain and straight-chain saturated hydrocarbons and even polycyclic aromatic hydrocarbons; in fact, for some species hydrocarbons are the sole carbon source (Head et al., 2006). Although plants are the main producers of aromatic compounds, they lack degradation pathways that could recycle carbon from these substances.

Remarkably, Europe has been reflecting as examples to create the law for any type of uses of chemical, required the Registration, Authorization and restriction of Chemicals (REACH) regulations for biodegradability evaluation of high-production-volume organic chemicals (UE, 2006). Moreover, the abundance of hazardous going at high level risk due to their substance-specific assets, the outcome of each chemical in the environment diverges according to its mode of utilization and way of discharging (Federle & Vink et al., 1997).

Bioremediation of environmental niches such as soil, sediments and water are polluted with petrochemicals, metals or radionuclides can be accomplished through biologically encoded changes in the oxidation state (Wang et al., 2012) has concluded that changes in speciation like detoxification of mercury by methylation [$Hg(CH_3)_2$] can modify the solubility, transport properties and toxicity of radionuclides. The strategies of bioremediation of petrochemicals, metals as well as radionuclides hazardous depend on the active metabolizing capabilities of microorganisms. These pollutants can be solubilized by enzymatic reduction through oxidative reduction, change in pH and Eh (activity of electrons), biodegradation via organic complexes or biosorption by biomass (Law et al., 2010; Hegazy & Emam, 2011). Biotransformation of pollutants through microbial activity can be prominently influenced by electron donors and acceptors, nutrients, and environmental factors. Some possible mechanistic linkages of metalo-remediation via microorganism is represented in the Figure 1.

Therefore, we can say that degradation of aromatic substances is dominated by aerobic and anaerobic bacteria and aerobic fungi (Fuchs et. al., 2011). Regrettably, the aromatic compounds have less reactivity to chemicals and are basically attacked with the help of O_2 by oxygenases (Hayaishi, 2008) and derived the compounds like catechol (1,2-dihydroxybenzene) (in bacteria) or proto- catechuate (3,4-dihydroxy-benzoate) (in most fungi and some bacteria) as principal intermediates (Dagley et. al, 1960; 1964; 1971; Ornston &Stanier 1966; Stanier & Ornston, 1973; Harwood& Parales,1996). This peripheral degradation has been performed by ring-cleaving dioxygenases that catalyses the central-ring cleavage (Lipscomb,

Figure 1. Multiple ways for metalo-remediation via microorganisms

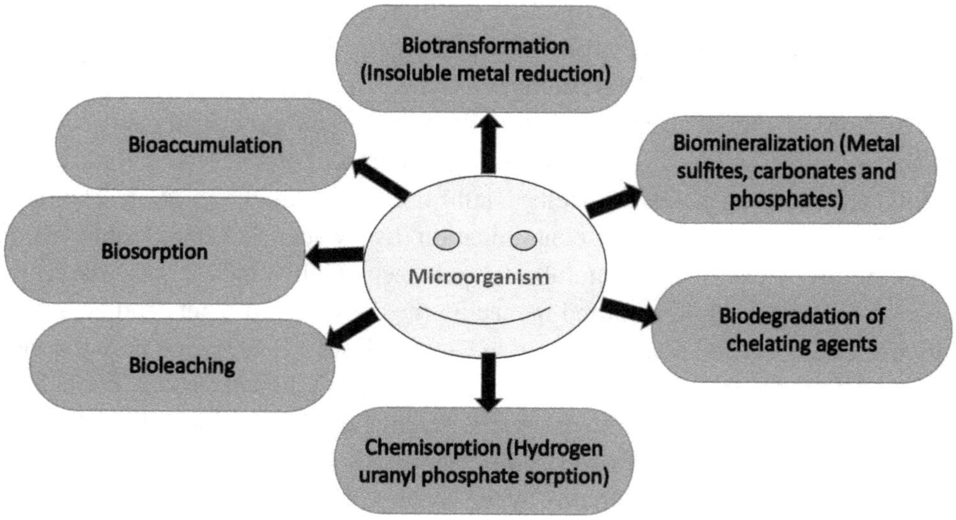

2008). The model pathway of this classical strategy has been completely understood by the well-known β-ketoadipate pathway (Stanier & Ornston, 1973; Harwood & Parales, 1996).

Undeniably, the numerousniche specific microbial communities are potentially played the mature role in the biodegradation of chemicals (Forney et al., 2001; Martiny, et al., 2006). These microbial communitiesarethe major factorsof the variation of environmental parameters linked with the conditions of *in-situ* life (Ranjard et al., 2013). Tackling with the unfeasibility to evaluatebiodegrade abilityin actual conditions (Mackay &Webster 2005), microorganisms are taken from a selected environment and used *in vitro* under laboratory conditions (Nyholm, 1991). Moreover, integrated testing strategy on biodegradation and Ready Biodegradation Tests (RBT) (OECD 1992) have performed to confirm their ability and susceptibility against various biohazardous.

The problem of remediation of hazardous material has been recurring for 60 years. However, the consideration of chemical persistence in the environment is not forthright and better screening tests are needed (Boethling et al., 2009). Bioremediation should address multiphasic, heterogenous environments, such as soils in which the contaminant is present in association with the soil particles, dissolved in soil liquids, and in the soil atmosphere. However, several factors distressing the bioremediation (Table 1) like Scientific factors (Energy sources, Bioavailability, Bioactivity and biochemistry) as well as Non-scientific factors (Regulation and laws). Because of these complexities, successful bioremediation is dependent on an interdisciplinary approach involving such disciplines as microbiology, engineering, ecology, geology, and chemistry.

Moreover, the fundamental resource of bioremediation (microorganisms) needs to be identified and characterized well at genomic as well as proteomic level. It is also important to know that how actual population of microorganisms that is potentially present in the environment and competent to degrade the released molecules under *in situ* conditions. Hence, the participation of genomics and proteomics (in various form of omics) needs to be employed to identify the dominant microbial communities and gene assets for bioremediation.

Table 1. Major factor affecting bioremediation

Microbial	• Growth until critical biomass is reached • Mutation and horizontal gene transfer • Enzyme induction • Enrichment of the capable microbial populations • Production of toxic metabolites
Environmental	• Depletion of preferential substrates • Lack of nutrients • Inhibitory environmental conditions
Substrate	• Too low concentration of contaminants • Chemical structure of contaminants • Toxicity of contaminants • Solubility of contaminants
Biological aerobic vs anaerobic process	• Oxidation/reduction potential • Availability of electron acceptors • Microbial population present in the site
Growth substrate vs co-metabolism	• Type of contaminants • Concentration • Alternate carbon source present • Microbial interaction (competition, succession, and predation)
Physico-chemical bioavailability of pollutants	• Equilibrium sorption • Irreversible sorption • Incorporation into humic matters
Mass transfer limitations	• Oxygen diffusion and solubility • diffusion of nutrients • Solubility/miscibility in/with water

MICROBIAL BIOREMEDIATION

Microbial bioremediation is defined as the process by which microorganisms are stimulated rapidly to degrade hazardous organic contaminants environmentally to safe levels in soils, subsurface materials, water, sludge, and residues. Microbes deal with poisonous chemicals by applying enzymes to convert one chemical into another form and taking energy or utilizable matter from this process. The chemical transformations generally involve in breaking of large molecules into several small molecules in simpler form and mediated by microorganisms is proving to be cost effective, eco-friendly and sustainable technology (Hivrale et al., 2016). In some cases, the by-products of bacterial bioremediation are not only harmless but may prove useful. For example, methane is derived from the bacterial degradation of sulphite liquor, a waste product in paper manufacturing.

Supported with sufficient nutrients and a terminal electron acceptor for metabolism, almost all the natural organics are biodegradable within a range of extent, but even simpler organic compounds fail to favour any microbial activity owing to unfavourable conditions like, extremes in temperature or pH, presence of toxicants or antimicrobial agents, lack of water, nutrient scarcity, absence of electron acceptor, etc. Microbial bioremediation is particularly useful for treatment of wastes and wastewater from municipal areas, food processing, agriculture sector, and more recently the hazardous wastes from various origins. Nonetheless, the capital theme of bioremediation or survival mechanisms depend on some change in metal speciation/hazardous complexity that leads to decrease or increased mobility. Collectively but not all, redox transformations, the production of metal-binding peptides and proteins organic and inorganic

Figure 2. Mechanisms involved in the detoxification and transformation of metals, including mechanisms that restrict entry into the cell and intracellular detoxification or organellar compartmentation
Altered from Gadd et al., 2010

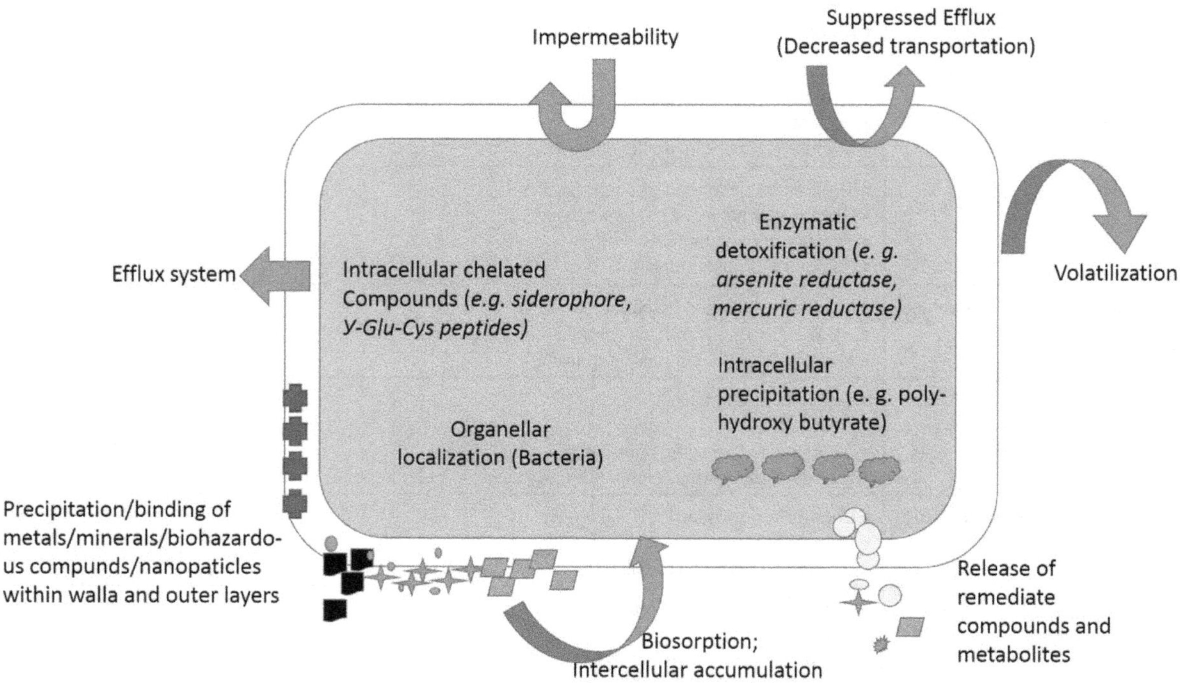

precipitation, active transport, efflux and intracellular compartmentalization are major mechanisms that involves in complex interactions involving rock, soil, water, air, and living organisms. Figure 2 reflects some possible mechanism of detoxification and transformation of metals/hazards during bioremediation.

Bioremediation of Xenobiotics

The term 'xenobiotics' is derived from the Greek word 'Xenon' that means 'a strange', and thus, denoted a material that is not naturally occurring in the biosphere. Major sources of xenobiotic compounds are agrochemical and petrochemical compounds. Aromatic compounds and pesticides are the most significant among them. In the bioremediation of xenobiotic compounds, the microbial dehalogenation is an important process as a large number of such compounds contain halogen group in their structure. Dehalogenation also facilitates further degradation of the compound due to cleavage of the carbon-halogen bond under the catalytic action of dehalogenases enzyme. It is important to mention that many oxygenases act as dehalogenases by actively catalyzing the dehalogenation step.

Lignases, a group of extracellular enzymes, include lignin peroxidase, arylmethoxy demethylase and phenol oxidase, well studied in white rot fungi *Phanerochaete chrysosporium.* Such enzymes are efficient it scavenging a range of hazardous polyaromatic hydrocarbons that are structurally diverse and normally considered resistant to microbial degradation. Detoxification mechanisms for organomercurials have also been reported well.

Desulphurization of Fossil Fuels

Inorganic sulphur removal from coal is achieved through the oxidation of sulphur by microbes and heterotrophic microorganisms. Direct oxidation of inorganic sulphur by *Thiobacillus* sp. is membrane bound reaction and requires direct contact of the substrate with the bacterium. Hence, attachment of the culture to coal particle is the absolute requirement. Studies have been undertaken for removal of organic sulphur from coal and oil using a variety of organisms in mixed and pure cultures of heterotrophic bacteria. However, sulphur removal has also been reported under anaerobic microbial action.

Precipitation or Biotransformation of Metals and Radionuclides

Microbial activity can biologically mediate the precipitation of metals from aqueous solutions. Certain bacterial extracellular products may interact with free or sorbed metal cations forming insoluble metal precipitates. The major mechanism for such precipitation is through the formation of hydrogen sulphide and the immobilization of metal cations as metal sulphides. Certain fungi that produce oxalic acid facilitate immobilization of metals as metal oxalate crystals. Microbes can also catalyze a range of metal transformations and are useful for waste treatment. The transformations include oxidation, reduction and alkylation reactions. Ferrous and manganese ions can be deposited by bacteria, fungi, algae or protozoa, in the oxidation reaction.

Bioremediation of the Industrial Effluents

Microorganisms like bacteria, protozoan, blue-green algae, are the main drivers of the biological treatment processes for industrial effluents and sewage streams. *Two* principal types of biological treatment are: (Wan et al., 2000) Percolating filter, also referred as trickling filter or biological filter for activated sludge processes. In the trickling filter process, the settled sewage or wastewater flows through interstices of the medium having a large surface area prone to establishment of microbial film. This gelatinous film contains bacteria, fungi, protozoa, and on the upper surface algae causing oxidation of biological oxygen demands in the sewage. The success of the activated sludge process is dependent on the ability of the microorganisms to form aggregates that can settle. Among the heterogeneous population, bacteria alone are responsible for removing the dissolved organic matter, whilst others play 'grazing' role for removing frees-swimming bacteria and help in sedimentation. Advanced biological reactors have been developed in recent decades for efficient removal of nitrogen and volatile components including phenols from waste streams. Several enzymes from bacteria, fungi, and plants have been reported to be involved in the biodegradation of toxic organic pollutants. (Karigar et al., 2011) has reviewed several enzymes for degradation of environmental pollutants. For example, monooxygenase plays vital role in the desulfurization, dehalogenation, denitrification, ammonification, hydroxylation, biotransformation, and biodegradation of various aromatic and aliphatic compounds. Figure 3 illustrate the active mechanism of monooxygenase and fungal cellulose for the degradation of pollutants from habitat.

Composting

The concept of composting, originally applicable for organic waste conversion into mulch and soil conditioner, is now being applied to the hazardous waste treatment. Composting requires that the material

Figure 3. Role of microbial enzymes in the bioremediation of pollutants

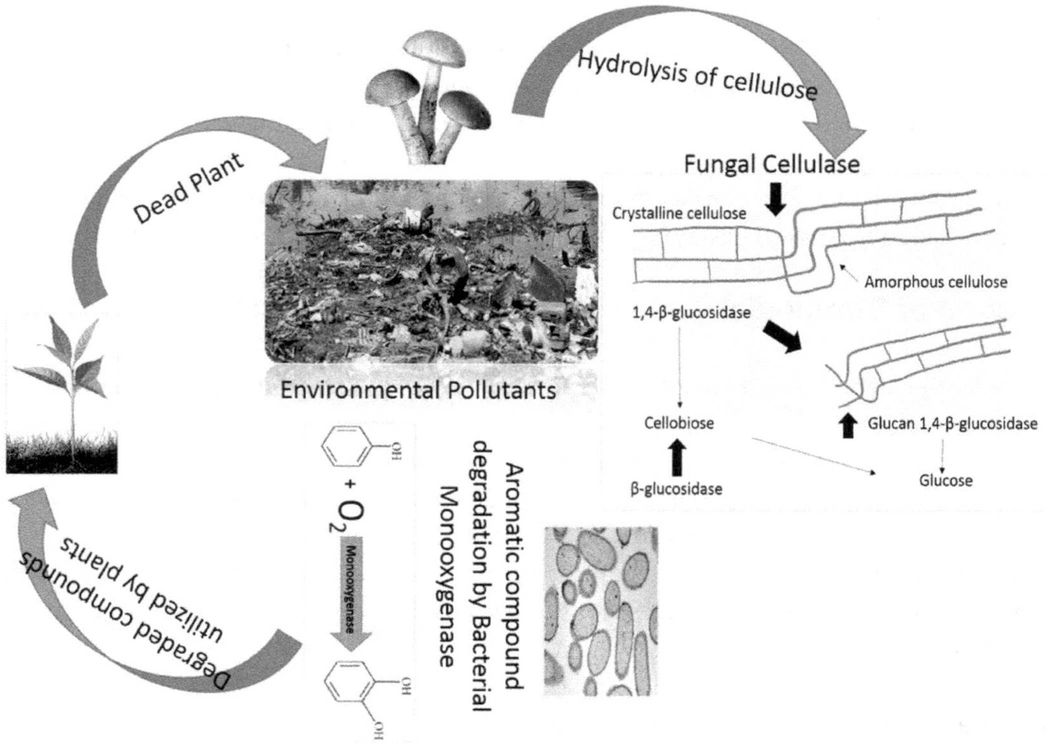

be biodegradable, with an adequate water content, sufficient quantity of porous substrate to retain heat and to allow gas exchange. For substrates that contain little nitrogen, nitrogen additions may be required. Composting leads to three prominent results, i.e., (a) waste is converted to less complex and relatively more stable material, (b) water content decreases significantly, and (c) reduction of mass of the residue. Materials that are amenable to the bioremediation composting process include sewage sludge, soils contaminated with diesel fuel or other petro-products, wastes from brewing, antibiotic fermentations, food processing, mineral oil, agriculture, etc. Bulking agents, for e.g., fibrous plant material, wood chips, bark, are added to increase the porosity so as to help aeration.

Bioaugmentation

Bioremediation operations may be made either on-site or off-site, *in-situ* or *ex-situ.* Bioaugmentation is the enhancement of decontamination in the media or waste-biodegradation by seeding of competent microflora and supplemented with desired level of nutrients. Faster decontamination has been achieved successfully by stimulation of existing microbial populations or augmentation with adapted strains. Thus, augmentation refers to establishing suitable conditions for bioremediation by means of adding nutrients for growth promotion, addition of terminal electron acceptor (oxygen or nitrate), moisture level adjustment or raising the temperature.

PROTEOMICS IN BIOREMEDIATION

Proteomics provides a global view of the protein complement of biological systems. It determines the protein expression profiles related to these research questions for communities as well as their expressed genetic properties. However, the cellular expression of proteins in an organism varies with environmental conditions. The changes in physiological response may occur due to the organism's adaptive responses to different external stimuli, such as the presence of toxic chemicals in the environment. The advent of proteomics has allowed an extensive examination of global changes in the composition, insight into bioremediation-related genes and their regulation, as well as identification of key proteins involved in the response of microorganisms in a given physiological state (Vasseur 1999, Wilkins 2001, Singh et al.2006). A number of reports have described sets of proteins that are up- or down-regulated in response to the presence of specific pollutants (Vasseur1999, Wilkins2001, Kim et al.2002). In-situ and ex-situ bioremediation of PAHs has been partially achieved using natural and genetically engineered microorganisms, as reviewed by (Samanta 2002& Johnsen et al. 2005). An 81-kDa protein similar to catalase–peroxidase that expressed in response to pyrene exposure was recovered using 2-DE from *Mycobacterium sp.* strain PYR-1 (Wang et al. 2000). Subsequently, six major proteins were significantly induced and overexpressed on 2-DE after exposer of *Mycobacterium sp.* strain PYR-1 against phenantherene, dibenzothiophene and pyrene (Khan et al. 2001). Several pyrene-specific polypeptides were identified by N-terminal and internal peptide sequencing as putative enzymes. Furthermore, the induction of two ring-hydroxylating dioxygenases, i.e. Pdo1 and Pdo2, in response to pyrene was proposed by (Krivobok etal., 2003) during pyrene catabolism by *Mycobacterium sp.* strain 6PY1. A composite profile for 20 PAH-induced proteins were presented when organism *M. vanbaabenii* PYR-1was grown in the presence of high-molecular-weight PAHs (Kim et al., 2004). The study of environmental sample to elucidate the role of a c-type cytochrome that was abundant during uranium bioremediation was done by (Yun et al., 2015) by using functional proteomics.

Microbial mediated bioremediation has a great potential to restore contaminated environment effectively but the lack of information about factors regulating the growth and metabolism of various microbial communities in polluted environment often limits its implementation. Newly seeded techniques such as transcriptomics, proteomics and interactomics offer remarkable promise as tools to address longstanding questions regarding the molecular mechanisms involved in the control of mineralization pathways. During mineralization, transcript structures and their expressions have been studied using high-throughput transcriptomic techniques with microarrays. Generally, however, transcripts have no ability to operate any physiological response; rather, they must be translated into proteins with significant functional impact. These proteins can be identified by proteomic techniques using powerful two-dimensional polyacrylamide gel electrophoresis (2-DE). Towards the establishment of functional proteomics, the current advances in mass spectrometry (MS) and protein microarrays play a central role in the proteomics approach. Exploring the differential expression of a wide variety of proteins and screening of the entire genome for proteins that interact with particular mineralization regulatory factors would help us to gain insights into bioremediation. Protein separation is the most important step in proteomics-based methodology. There are ample techniques to accomplish this and each one has its own advantages and disadvantages. The choice of strategy is based on experiment, availability of instruments and expertise.

Two-Dimension Electrophoresis (2-DE) for Protein Expression Profiling

2-DE is an indispensable part in protein expression profiling, where proteins are isolated by the correlative systems of IEF in the primary measurement and SDS-PAGE size fractionation in the second measurement. IPGs and bioinformatics have amazingly enhanced the reproducibility and comparability of 2-DE gels. Numerous studies have been demonstrated that 2-DE can recognize contrasts between proteins from typical and tainted environments. As per system biology, any physiological condition of a living being prompts a differential expression of proteins that affects a chain of enzymatic responses to catalyze the mineralization of contaminants in and around the cell. 2-DE can help to investigate the metabolic pathways of such living beings by recognizing the significant signature proteins and chemicals (Hobraud, 1999, Tomas-Gallardo, 2006)

The recognizable proof and portrayal of such proteins and chemicals that can be expelled and/or separated from cell are sought after for natural remediation. The cell lysate of Mycobacterium sp. was profiled on a 2-DE gel and an 81-kDa catalase-peroxidase like protein was reported to be significantly induced under pyrene exposure (Wang etal., 2000). Two ring-hydroxylating dioxygenases, Pdo1 and Pdo2, were observed to be induced during pyrene catabolism by another Mycobacteriumsp. (Krivobok et al.,2003). A proteome examination of*Acinetobacterradioresistens*S13 upon aromatic exposure revealed that an assortment of outer membrane proteinssuch as OmpA-like proteins, trimericporin, and glycosyltransferases, were up-regulated. Eighty exceptional proteins were recognized by 2-DE/MS from *P. putida* KT 2440 when exposed to various aromatic compounds (Kim et al., 2006). An enzymatic pathway has been proposed for phthalate digestion system in *Rhodococcus sp.* strain TFB (Tomas-Gallardo et al., 2006). A 3-D model homology of protein from psychrophilic bacterium *C. psychrerythraea* 34H uncovers critical changes in proteome that may advance chemicals' adequacy at low temperature (Methe et al., 2005). Other than 2-DE, MS-based proteomics methodologies were likewise used to profile an assortment of particular proteins (Mi et al., 2005). 2-DE based detection of these proteins and chemicals amid bioremediation makes them potential candidates for cell free remediation that can coordinate the utilizing biomolecules, separated or expelled from cell for bioremediation. Various highly sensitive fluorescent stains have been effectively utilized in differential in-gel electrophoresis (DIGE) to precisely measure protein expression in 2-DE (Marouga et al., 2005). In DIGE, protein tests covalently marked with red and/or green fluorescent colours are blended with a blue colour named inner standard protein (Alban et al., 2003). A 2-DE separation of these labelled protein samples helps avoid ambiguities in protein spot matching. The sensitivity of DIGE is tantamount to that of silver recoloring, and it is good with MS, which can be utilized to recognize the protein of interest after in-gel digestion. In environmental proteomics, DIGE tests have been performed to profile protein expression in the substrate dependent direction of aromatic degradation pathways for toluene and ethylbenzene in a denitrifying bacterium, strain EBN1 (Kuhner et al., 2005). More than 150 protein spots were found in expanded plenitude in the presence of toluene or ethyl-benzene-adapted cells. Utilizing DIGE, an assortment of compounds has been described for phathalate degradation pathway in *Rhodococcussp.* strain TFB (Tomas-Gallardo et al., 2006). These studies affirm DIGEs capacity to identify proteins of interest for bioremediation and exhibit the immense heterogeneity of bacterial cell lysate under stress conditions, fortifying the requirement for enhanced partition techniques.

Compartmental Proteomics in Bioremediation

A few very basic and hydrophobic membrane proteins are lost due to poor solubility in 2-DE-compatible lysis buffers or precipitation during first-dimension electrophoresis and it is the key limitation of 2-DE. Membrane proteins, which are particularly prone to precipitation, are of great interest for the biodegradation of aromatic pollutants such as PAHs, NACs, and organophosphates, where numerous alterations in the organism affect cell–surface proteins and receptors (Wang, 1995& Richins et. al., 1997). Membrane proteomics have recently been enhanced by the development of the multidimensional protein identification platform (MudPIT) where complex protein and mixtures of peptides can be separated using 2-D LC system (Paoletti, 2004 &Somiari et al., 2005) by omitting 2-DE to avoid the electro-separation of proteins. MudPIT directly interfaces with MS and can considerably reduce the analysis time-taken for 2-DE. In concise, MudPIT consists of strong cation exchange (SCX) material in-connection with an RP material inside fused-silica capillaries (Somiari et al., 2005). The chromatographic separation performed in cycles, every cycle involves araise in salt concentration to remove peptides of the SCX followed by a gradient of escalating hydrophobicity for dynamic elution of peptides from the RP (Somiari et al., 2005). The data-dependent acquisition of mass spectrometers isolates peptides as they elute and subject them to CID by recording the fragmented ions in a tandem mass spectrum. These spectra are compared to database peptide sequences by the SEQUEST algorithm (Somiari et al., 2005). The identified peptides from SEQUESTs are assembled and filtered into protein-level information using specific algorithm called DTA.As MS-compatible chromatographic systems only handle small peptides, the protein samples must be digested with protease (usually trypsin) prior to analysis. Fortunately, most proteins contain at least a few peptides that can be analyzed with MS, and MudPIT makes the detection of basic and hydrophobic proteins more likely. A better representation of low-abundance proteins and quantitative comparisons can be achieved by combining MudPIT with isotope-coded affinity tags (ICAT), mass tags that modify cysteine residues and permit the affinity purification of tagged pep-tides (Somiari et al., 2005). This diminishes the complexity of the peptide sample that needs to be analyzed; though, no proof-of-principle experiment has yet been reported to prove this simplicity for various cellular and compartmental proteomics in bio-remediation.

Mass Spectrometry (MS) in Proteomics

Progresses in MS have enhanced the examination of peptides for protein identification and made it possible to develop the field of ecological proteomics. MS deduces the composition of particles by deciding the specific peptide mass that outcomes from specific mixes of amino acids. The more precisely one can quantify peptide mass, the less vague the synthesis results will be. To think about these mixes of amino acids, proteins are routinely processed with proteases to create peptides and it is sufficiently little for MS analysis, allude as PMF. In this procedure, the controlled fracture of a peptide yields a progression of covering section particles that vary by the mass of a specific amino corrosive, which permits the full or fractional amino corrosive grouping of the peptide to be concluded. This marvel is likewise called MS/MS or coupled MS, as it normally utilizes one MS analyzer to choose particles for fracture and a second to quantify the section particles.

MALDI-TOF-MS is routinely used to recognize proteins of enthusiasm from 2-D gels, and in addition to distinguish and recognize vi-tricks, microbes, contagious spores, and low-mass mixes in natural examples (Lay, 2001, Ruelle 2004, Hu, 2005). The unpredictable mass spectra of natural examples can be utilized to make characteristic fingerprinting databases to identify numerous site-particular microorganisms. As far as bioremediation, MALDI-TOF-MS can identify particular bacterial mark proteins and bio-markers (essential and optional metabolites) from site-specific tests for the taxonomic distinguishing proof of potential microorganisms. A type of direct example examination on a microchip utilizing MALDI-TOF-MS – SELDI-TOF-MS – is another promising explanatory method for site-particular specimens. An assortment of differentially communicated signature proteins in blue mussels (Mytilusedulis) presented to PAHs and substantial metals were analyzed utilizing SELDI-TOF-MS (Knigge, 2004). In spite of the fact that SELDI investigation has been valuable for distinguishing potential biomarkers in clinical research, some have scrutinized its reproducibility and specificity (Jacobs & Menon, 2004; Diamandis, 2004). Another rising method, Fourier change ICR (FT-ICR) MS, may permit multistage MS tests (MSn) with an improved discovery point of confinement of, 30zmol for a 10 kDa protein (Laskin & Futrell 2005). Adding ESI and LC to MS has opened new expository windows for the discovery and ID of potential contaminants in nature (Bossi et al., 1999). An immediate examination of a peroxisomal protein design associated with marine toxin presentation was performed utilizing ESI MS/MS to distinguish different epoxide hydrolases, peroxisomal cell reinforcement catalysts, and SOX (Orbea et al.,2005). Henceforth, 2-DE-and MS-based distinguishing proof of proteins/chemicals from any site-particular bacterium would in the long run clear a path toward sans cell bioremediation.

Protein Aggregation and Protein–Protein Assembly

Huge scale investigations of protein accumulation, degradation, and protein-protein interaction (PPI) utilizing microarray stages have expanded as a part of prominence in the recent decade. The huge scale mapping of signalling pathways in microscopic organisms got to be possible when it was perceived that these pathways are sorted out as collaborating units of protein complexes that shape the physical machines for sign processing during chemotactic movement of an organism (Larsen et al., 2004). These particular protein buildings or peptides representing the proteins of interest can be set on a glass slide in an array-based approaches (Sauer et al., 2005). The significant hurdle here is producing the proteins themselves for printing libraries. Lamentably, this technique does not have some PTMs, and the yield of collapsed full-length proteins might be low (Predki, 2004). Proteins themselves can be printed by contact and noncontact strategies, then immobilized utilizing covalent, non-covalent, arranged, or random attachment procedures (Sauer et al., 2005). Labelling methodologies and recognition procedures for protein microarrays more complicated than the DNA clusters. For bioremediation purposes, proteo-clusters can distinguish authoritative of particular inhibitors or ligands of dioxygenase or mono-oxygenase enzymatic activities. Enzymatic movement can likewise be profiled to portray the neighbouring or corresponding enzymes based on qualitative and quantitative estimation amid in situ bioremediation. Nonetheless, no report has been publishing at this point to represent global PPI (interactomics) expression pattern of any organism for bioremediation.

GENOMICS IN BIOREMEDIATION

Genomics tools can be utilized to do an appraisal of the accessible natural capability of any eco-system applying particular biomarker or biosensors. Biosensors can screen an organic yield that can be changed over into a quantifiable signal, for instance, protein based biosensors can create the signal either by product formation, the disappearance of substrate, or co-catalyst transformation (Purohit,2003). Biomarkers are determined genotypes that can be utilized to track the survival and/or efficacy of particular microorganisms in bioremediation. For example, *luc* gene, encoding firefly luciferase and the *gfp* gene, encoding the green fluorescent protein (GFP). The *luc* quality has been exhibited to tag distinctive microorganisms utilized for bioremediation of gas or chlorophenols, where the microscopic organisms were checked on the premise of luciferase activity in cell separated from soil (Jansson et al., 2000). Bioremediation of contaminated sediments via genomics (in-situ) would be an efficiently stimulated owing to the coexistence of different electron acceptors and donors in the contaminated sediment by using metabolic versatile microorganisms. In case of nitrate toxicity, it has predicatively stated that the presence of nitrate would inhibit sulphate reduction because of higher reduction potential of nitrate than sulphate under anaerobic condition, increasing facts showed that the reduction of electron acceptors with lower reduction potential was not inhibited and even was enhanced by the presence of nitrate (Canfield, 2010, Gu, 2012 & Zhao et al., 2013). Many of the genera having broader electron-accepting capabilities have been enriched in systems containing multiple electron acceptors like Bacillus sp. and Ochrobactrum sp. from the radioactively contaminated sediment with high nitrate concentration (Thorpe et al., 2012) and Sulfuricurvum sp. and Desulfovibrio sp. from the system containing O2, ClO4-, sulphate and nitrate (Zhaoet al., 2013). For example, (Xu et al., 2014) has reported in his study that the amoA gene (239798446) from Desulfovibrio sp. and the van gene (198265864) from Ochrobactrum sp. were ranked in the top 10 stimulated genes after nitrate addition. (Hong et al., 2002) has reported several expressively stimulated C degradation genes from an important halogenated organic compound degradation organism Sphingomonas wittichii RW1, which could contribute a lot to the removal of PAHs and PBDEs from sediments

During bioremediation, diverse microorganisms render the key metabolic exercises, which set the procedures for removal of contaminations together. Investigation of these organisms can be completed utilizing sub-atomic instruments, for example, denaturing gradient gel electrophoresis (DGGE), temperature gradient gel electrophoresis (TGGE) and terminal restriction fragment length polymorphism (T-RFLP). DGGE investigations are applied for the division of double stranded DNA strands that are same length but diverge sequences. Blends of DNA fragment acquired by PCR reaction are electrophoresed in an acrylamide gel containing a gradient of expanding DNA denaturants. DNA fragments are wealthier in GC will be stable and stay double stranded until achieving higher denaturant fixations. Double stranded DNA fragments moves well in the acrylamide gel, while denatured DNA particles turn out to be efficaciously bigger and dawdling down or stop moving in the gel. Using this approach, DNA strands of contrasting sequence can be isolated in an acrylamide gel. DNA strand samplified by PCR can be isolated on the premise of their melting temperature, or separated by restriction digestions approaches. In T-RFLP, the amplification is performed with one or both the primers having their 5' end labelled with a fluorescent tag. Then the amplicon is subjected to restriction analysis, and isolated further using either capillary or polyacrylamide electrophoresis. The general pattern in mixture can be comprehended by utilizing general DNA profiling by DGGE/TGGE analysis (Muyzer, 1999) or T-RFLP (Pandey et al., 2007). By developing 16S rDNA library, the particular of the group and its fusion can be comprehended. These communities share the carbon source viz, the toxins and/or other organic molecules that are available in the eco-system. The community structure has been deciding by the transformation and utilization rates for various organic particles (Singh et al., 2006). Other than the organic particles, nitrogen and phosphorus play a vital role in survival of organisms in these challenging atmospheres. To survey the conceivable metabolic system or key microorganisms present in these situations, the metagenome extricated from that specimen is investigated. Among numerous methods that employed in metagenome examination, the most significant methodology is time dependent investigation for target loci. (Abbasian et al., 2016) has procured the microbial diversity from crude oil well and had mine the hydrocarbon degrading gene by metagenomics analysis.

Tracing Gene Loci in the Eco-Systems

The range of organic molecules present in nature and man-made is vast, as is the process of progression in organisms to use this extensive variety of particles. Naturally occurring microbial deeds are the beginning stage for all biotechnological applications. It is therefore essential to know about the metabolic capacities of the microbial population in the target niche and the understanding about degradative pathways, biochemically and hereditarily, is also important before planning for a prosperous bioremediation program. Bioremediation procedures depend on the catabolic limits of organisms to change harmful poisons to innocuous particles (Kumar et al., 2005). The vast majority of the characteristic soil microbes possess catabolic catalysts which are fit for using poisons as their development substrates. Microorganisms are the biggest repository of genetic and biochemical diversity. They have been evolving for the last 3.8 billion years and occupy almost every environment, regularly thriving out under extremes of supplements, pH, salinity, temperature and pressure (DeLong etal., 2001). The biochemical capability of common microflora can be surveyed by checking the accessible catabolic gene pool utilizing target catabolic genes or an objective genotype can be followed in ecological environments utilizing PCR. For instance, phenol degrading genotype was observed in environmental tests by designing primers from the dmpN and pheA operons (Kapley & Purohit, 2009). Likewise, the following of catabolic quality loci pheB, xylE, tod-isp, bedA and nahG in charge of the compounds catechol 1,2-dioxy-genase, catechol 2,3-dioxygenase, toluene dioxygenase-iron-sulfur protein segment, benzene dioxygenase and naphthalene dioxygenase in initiated biomass of gushing treatment plant treating wastewater created at refineries (Moharikar etal.,2005). The regular degradative capability of a domain can be resolved utilizing tools like PCR or DNA hybridization, and bioremediation procedures can be made for more viable degradation. If the target gene are in low numbers, advancement for degradative genotype can be a choice (Sutherland et. al., 2000, Siddique et al., 2003, Prakash, & Lal, 2006), or without target gene, bioaugmentation can be an option, as exhibited in treatment of waste-water containing hydrocarbons, where the bioaugmentation of the alkB genotype was appeared to enhance proficiency of treatment.

Observing the treatment of environmental contamination primarily included the evaluation of diminishing of the poison at the target area. Another situation is currently rising, where; a standard checking of various conceivable grids is getting to be crucial. In this procedure a few checking stations are chosen based on hydrogeology of that location. For instance, it is not just critical to analyse the contaminated soil and leachate around the site, be that as it may, taking into account the conceivable development of the pollution crest, investigation of the ground is also required. These investigations include the evaluation of the target chemical moieties and its possible metabolites. With the advancement of genetic tools in bioremediation, it is plausible to screen the key genes that help in transformation of a poisonous material to non-dangerous material. Rather than checking just expulsion of the contaminant, the genes required in degradation may likewise be observed to predict the proficiency of bioremediation. This investigation will let us know whether the bio-remediation program needs to follow the bioaugmentation or biostimulation path.

Isotope Based Examines

Recent years, micro-autoradiography has been used to envisage and count cells inside microbial communities applying labelled substrates. The microorganisms being analysed couldn't be recognized in all the cases. Fluorescence in situ hybridization (FISH) examines were produced to overcome these issues. Target microorganisms on various phylo-genetic levels could be distinguished by fluorescence microscopy employing fluorescently labelled oligonucleotide probes. This method has been accounted in the determination of the structure and elements of microbial population in ecological samples (Eickhorst & Tippkotter,2008). The technique permits a particular staining and count of soil microorganisms by applying fluorescent labelled oligonucleotide probes. Nevertheless, the recognition is mostly influenced by solid auto-fluorescence of the background predominantly in soil analysis. Various varieties of this tools are reported earlier (Eickhorst & Tippkotter, 2008, Martin et. al., 2008). Catabolic capability of the microbial population has additionally been reported using isotope tracer strategies. In the event that a substrate of interest is labelled with an isotope tracer, living beings that uptake the substrate can be marked in situ and followed employing unique identification devices (Radajewski et al., 2000).

Beyond the Culturable Microorganisms

Countless strains have been reported for in bioremediation by various scientists (Kutty et al., 2000, Qureshi, et al., 2001, 2007Khardenavis, et al., 2007), yet it has been understood that to change over lab-scale research into advances feasible in ecological corners we have to look beyond culturable microorganisms. Here is where the gnomic tools come in. The bioremediation field has a considerable measure to pick up from the advances in this rising range. Genomic tools are presently utilized for illustration of biodegradation pathways utilizing PCR, DNA hybridization, Microarray investigation, isotope distribution examination, sub-atomic network investigation, the assessment of mineralization procedure utilizing metabolic foot printing investigation, and the change of the biodegradation procedure through metabolic engineering (Villas, & Bruheim, 2007).

Now, the upgraded genomics insight study "metatranscriptomics" contrast the gene expression profile of microorganisms that has been deciphered given time that what microbial gene has activated and what are unregulated in the soil. For example, functional metatranscriptomic profiling of microbial activities in the rhizosphere has given an unprecedented level of details of expressed gene for bioremediation like hydrocarbon degradation genes, increased nitrogen inputs, increased plant exudation due to stressful conditions and exudation of salicylate and other PAH analogs (Yergeau et al., 2014).

CONCLUSION

The use of genetically modified microorganism to perceive and degrade environmental pollutants has the latent to modernize the field of bioremediation. This chapter reflects a wide range of biotechnological applications to remove or stabilize organic and inorganic contaminants by using the precise lifestyle of microbes and their biochemical potential through genomics and proteomics. Due to the actual demand for sustainable, cheap and tailor-made technologies for bioremediation, there is an ongoing trend towards passive remediation schemes, referred to as genetic engineered microbes, has employed. Prospectively, herbicide-responsive ribo-switch has been searched by (Sinha et al., 2010) that can be used to control motility, allowing bacteria to migrate in the presence of the herbicide atrazine. Likewise, the practice of designer RNA-based sensors to detect chemicals and trigger a regulated response holds much promise. Laboratory based selection techniques can also be used to screen large libraries for RNA sequences (known as aptamers) that can firmly and precisely bind to a range of ligands and can be assimilated into ribo-switches to regulate gene expression. However, the practice of genomics and proteomics in bioremediation has now started to come out from *in-vitro* to *in-vivo,* but it needs more refinement that allows the bacteria to survive in wide array of environmental condition and exhibit chemotaxis towards the hazardous articles. Nonetheless, these observations suggest that genomics and proteomics approaches to hold a great deal of promise.

REFERENCES

Abbasian, F., Palanisami, T., Megharaj, M., Naidu, R., Lockington, R., & Ramadass, K. (2016). Microbial diversity and hydrocarbon degrading gene capacity of a crude oil field soil as determined by metagenomics analysis. *Biotechnology Progress*, *32*(3), 638–648. doi:10.1002/btpr.2249 PMID:26914145

Ahuja, S. K., Ferreira, G. M., & Moreira, A. R. (2004). Utilization of enzymes for environmental applications. *Critical Reviews in Biotechnology*, *24*(2-3), 125–154. doi:10.1080/07388550490493726 PMID:15493529

Alban, A., David, S. O., Bjorkesten, L., Andersson, C., Sloge, E., Lewis, S., & Currie, I. (2003). A novel experimental design for comparative two-dimensional gel analysis: Two-dimensional difference gel electrophoresis incorporating a pooled internal standard. *Proteomics*, *3*(1), 36–44. doi:10.1002/pmic.200390006 PMID:12548632

Amann, R. I., Ludwig, W., & Schleifer, K. H. (1995). Phylogenetic identification and in situ detection of individual microbial cells without cultivation. *Microbiological Reviews*, *59*(1), 143–169. PMID:7535888

Boethling, R., Fenner, K., Howard, P., Klecka, G., Madsen, T., Snape, J. R., & Whelan, M. J. (2009). Environmental persis- tence of organic pollutants: Guidance for development and review of POP risk profiles. *Integrated Environmental Assessment and Management*, *5*(4), 39–556. doi:10.1897/IEAM_2008-090.1 PMID:19552498

Bossi, R., Seiden, P., Andersen, S. M., Jacobsen, C. S., & Streibig, J. C. (1999). Analysis of metsulfuron-methyl in soil by liquid chromatography/tandem mass spectrometry. Application to a field dissipation study. *Journal of Agricultural and Food Chemistry*, *47*(10), 4462–4468. doi:10.1021/jf981280t PMID:10552834

Cadet, C., Beteau, J. F., & Hernandez, S. C. (2004). Multicriteria control strategy for cost/quality compromise in wastewater treatment plants. *Control Engineering Practice*, *12*(3), 335–347. doi:10.1016/S0967-0661(03)00134-5

Canfield, D. E., Stewart, F. J., Thamdrup, B., De Brabandere, L., Dalsgaard, T., Delong, E. F., & Ulloa, O. et al. (2010). A cryptic sulfur cycle in oxygen-minimum–zone waters off the Chilean coast. *Science*, *330*(6009), 1375–1378. doi:10.1126/science.1196889 PMID:21071631

Chen, W. C., Chang, N. B., & Chen, J. C. (2003). Rough set-based hybrid fuzzy-neural controller design for industrial wastewater treatment. *Water Research*, *37*(1), 95–107. doi:10.1016/S0043-1354(02)00255-5 PMID:12465791

Cozzone, A. J. (1998). Post-translational modification of proteins by reversible phosphorylation in prokaryotes. *Biochimie*, *80*(1), 43–48. doi:10.1016/S0300-9084(98)80055-2 PMID:9587661

Dagley, S. (1971). Catabolism of aromatic compounds by micro-organisms. *Advances in Microbial Physiology*, *6*(0), 1–46. doi:10.1016/S0065-2911(08)60066-1 PMID:4950664

Dagley, S., Chapman, P. J., Gibson, D. T., & Wood, J. M. (1964). Degradation of benzene nucleus by bacteria. *Nature*, *202*(4934), 775–778. doi:10.1038/202775a0 PMID:14187616

Dagley, S., Evans, W. C., & Ribbons, D. W. (1960). New pathways in the oxidative metabolism of aromatic compounds by microorganisms. *Nature*, *188*(4750), 560–566. doi:10.1038/188560a0 PMID:13719300

DeLong, E. F., & Pace, N. R. (2001). Environmental diversity of bacteria and archaea. *Systematic Biology*, *50*(4), 470–478. doi:10.1080/106351501750435040 PMID:12116647

Diamandis, E. P. (2004). Mass spectrometry as a diagnostic and a cancer biomarker discovery tool opportunities and potential limitations. *Molecular & Cellular Proteomics*, *3*(4), 367–378. doi:10.1074/mcp.R400007-MCP200 PMID:14990683

Eickhorst, T., & Tippkötter, R. (2008). Improved detection of soil microorganisms using fluorescence in situ hybridization (FISH) and catalyzed reporter deposition (CARD-FISH). *Soil Biology & Biochemistry*, *40*(7), 1883–1891. doi:10.1016/j.soilbio.2008.03.024

Federle TW, Gasior SD, Nuck BA (1997) Extrapolating mineralization rates from the ready CO2 screening test to activated sludge, river water, and soil. *Environ. Toxicol. Chemist., 16*, 127–134. doi:10.1002/etc.5620160205

Forney, L. J., Liu, W.-T., Guckert, J. B., Kumagai, Y., Namkung, E., Nishihara, T., & Larson, R. J. (2001). Structure of microbial communities in activated sludge: Potential implications for assessing the biodegradability of chemicals. *Ecotoxicology and Environmental Safety*, *49*(1), 40–53. doi:10.1006/eesa.2001.2034 PMID:11386714

Fuchs, G., Boll, M., & Heider, J. (2011). Microbial degradation of aromatic compounds — from one strategy to four. *Nature Reviews. Microbiology*, *9*(11), 803–816. doi:10.1038/nrmicro2652 PMID:21963803

Gadd, G. M. (2010). Metals, minerals and microbes: Geomicrobiology and bioremediation. *Microbiology*, *156*(3), 609–643. doi:10.1099/mic.0.037143-0 PMID:20019082

Graham, J. H., Hodge, N. C., & Morton, J. B. (1995). Fatty Acid methyl ester profiles for characterization of glomalean fungi and their endomycorrhizae. *Applied and Environmental Microbiology*, *61*(1), 58–64. PMID:16534923

Gu, C., Laverman, A. M., & Pallud, C. (2012). Environmental controls on nitrogen and sulfur cycles in surficial aquatic sediments. *Frontiers in Microbiology*, *3*, 45. doi:10.3389/fmicb.2012.00045 PMID:22375137

Harwood, C. S., & Parales, R. E. (1996). The β-ketoadipate pathway and the biology of self-identity. *Annual Review of Microbiology*, *50*(1), 553–590. doi:10.1146/annurev.micro.50.1.553 PMID:8905091

Hayaishi, O. (2008). From oxygenase to sleep. *The Journal of Biological Chemistry*, *283*(28), 19165–19175. doi:10.1074/jbc.X800002200 PMID:18390546

Head, I. M., Jones, D. M., & Roling, W. F. M. (2006). Marine microorganisms make a meal of oil. *Nature Reviews. Microbiology*, *4*(3), 173–182. doi:10.1038/nrmicro1348 PMID:16489346

Hegazy, A. K., & Emam, M. H. (2011). Accumulation and soil-to-plant transfer of radionuclides in the Nile Delta coastal black sand habitats. *International Journal of Phytoremediation*, *13*(2), 140–155. doi:10.1080/15226511003753961 PMID:21598782

Hivrale, Pawar, Rane, & Govindwar. (2016). Application of Genomics and Proteomics in Bioremediation. *Toxicity and Waste Management Using Bioremediation*. IGI Global. doi:10.4018/978-1-4666-9734-8.ch005

Hobraud, M., Viande, E., & de Theix, A. (1999). Differential protein expression by Pseudomonas fragi submitted to various stresses. *Electrophoresis*, *20*(11), 2204–2213. doi:10.1002/(SICI)1522-2683(19990801)20:11<2204::AID-ELPS2204>3.0.CO;2-I PMID:10493125

Hong, H.-B., Chang, Y.-S., Nam, I.-H., Fortnagel, P., & Schmidt, S. (2002). Biotransformation of 2,7-dichloro- and 1,2,3, 4-tetrachlorodibenzo-p-dioxin by *Sphingomonas wittichii* RW1. *Applied and Environmental Microbiology*, *68*(5), 2584–2588. doi:10.1128/AEM.68.5.2584-2588.2002 PMID:11976140

Hornung, U. (1997). Soil Venting. Environmental Biotechnology, CRC Press, Jul 19, 2010 -. *Nature*.

Hsu, J. L., & Chen, S. H. (2005). Recent progress in quantitative proteomics using stable isotope labeling, multidimensional liquid chromatography and mass spectrometry. *Current Proteomics*, *2*(4), 287–302. doi:10.2174/157016405775201766

Hu, L., Xu, S., Pan, C., Yuan, C., Zou, H., & Jiang, G. (2005). Matrix-assisted laser desorption/ionization time-of-flight mass spectrometry with a matrix of carbon nanotubes for the analysis of low-mass compounds in environmental samples. *Environmental Science & Technology*, *39*(21), 8442–8447. doi:10.1021/es0508572 PMID:16294885

Jacobs, I. J., & Menon, U. (2004). Progress and challenges in screening for early detection of ovarian cancer. *Molecular & Cellular Proteomics*, *3*(4), 355–366. doi:10.1074/mcp.R400006-MCP200 PMID:14764655

Jansson, J. K., Björklöf, K., Elvang, A. M., & Jørgensen, K. S. (2000). Biomarkers for monitoring efficacy of bioremediation by microbial inoculants. *Environmental Pollution*, *107*(2), 217–223. doi:10.1016/S0269-7491(99)00140-2 PMID:15092998

Kapley, A., Prasad, S., & Purohit, H. J. (2007). Changes in microbial diversity in fed-batch reactor operation with wastewater containing nitroaromatic residues. *Bioresource Technology*, *98*(13), 2479–2484. doi:10.1016/j.biortech.2006.09.012 PMID:17070036

Kapley, A., & Purohit, H. J. (2009). Genomic tools in bioremediation. *Indian Journal of Microbiology*, *49*(2), 108–113. doi:10.1007/s12088-009-0012-2 PMID:23100758

Khan, A. A., Wang, R. F., Cao, W. W., Doerge, D. R., Wennerstrom, D., & Cerniglia, C. E. (2001). Molecular cloning, nucleotide sequence, and expression of genes encoding a polycyclic aromatic ring dioxygenase from *Mycobacterium sp.* strain PYR-1. *Applied and Environmental Microbiology*, *67*(8), 3577–3585. doi:10.1128/AEM.67.8.3577-3585.2001 PMID:11472934

Khardenavis, A. A., Kapley, A., & Purohit, H. J. (2007). Simultaneous nitrification and denitrification by diverse Diaphorobacter sp. *Applied Microbiology and Biotechnology*, *77*(2), 403–409. doi:10.1007/s00253-007-1176-5 PMID:17876578

Kim, S. I. I., Kim, S. J., Nam, M. H., Kim, S., Ha, K. S., Oh, K. H., & Park, Y. M. et al. (2002). Proteome analysis of aniline-induced proteins in *Acinetobacter lwoffi* K24. *Current Microbiology*, *44*(1), 61–66. doi:10.1007/s00284-001-0075-8 PMID:11727043

Kim, S. J., Jones, R. C., Cha, C. J., Kweon, O., Edmondson, R. D., & Cerniglia, C. E. (2004). Identification of proteins induced by polycyclic aromatic hydrocarbon in *Mycobacterium vanbaalenii* PYR-1 using two-dimensional polyacrylamide gel electrophoresis and de novo sequencing methods. *Proteomics*, *4*(12), 3899–3908. doi:10.1002/pmic.200400872 PMID:15540208

Kim, S. J., Jones, R. C., Cha, C. J., Kweon, O., Edmondson, R. D., & Cerniglia, C. E. (2004). Identification of proteins induced by polycyclic aromatic hydrocarbon in Mycobacterium vanbaalenii PYR-1 using two-dimensional polyacrylamide gel electrophoresis and de novo sequencing methods. *Proteomics*, *4*(12), 3899–3908. doi:10.1002/pmic.200400872 PMID:15540208

Kim, Y. H., Cho, K., Yun, S. H., Kim, J. Y., Kwon, K. H., Yoo, J. S., & Kim, S. I. (2006). Analysis of aromatic catabolic pathways in Pseudomonas putida KT 2440 using a combined proteomic approach: 2-DE/MS and cleavable isotope-coded affinity tag analysis. *Proteomics*, *6*(4), 1301–1318. doi:10.1002/pmic.200500329 PMID:16470664

Kirk, J. L., Beaudette, L. A., Hart, M., Moutoglis, P., Klironomos, J. N., Lee, H., & Trevors, J. T. (2004). Methods of studying soil microbial diversity. *Journal of Microbiological Methods*, *58*(2), 169–188. doi:10.1016/j.mimet.2004.04.006 PMID:15234515

Knigge, T., Monsinjon, T., & Andersen, O. K. (2004). Surface-enhanced laser desorption/ionization-time of flight-mass spectrometry approach to biomarker discovery in blue mussels (Mytilus edulis) exposed to polyaromatic hydrocarbons and heavy metals under field conditions. *Proteomics*, *4*(9), 2722–2727. doi:10.1002/pmic.200300828 PMID:15352246

Krivobok, S., Kuony, S., Meyer, C., Louwagie, M., Willison, J. C., & Jouanneau, Y. (2003). Identification of pyrene-induced proteins in *Mycobacterium sp.* strain 6PY1: Evidence for two ring-hydroxylating dioxygenases. *Journal of Bacteriology*, *185*(13), 3828–3841. doi:10.1128/JB.185.13.3828-3841.2003 PMID:12813077

Krivobok, S., Kuony, S., Meyer, C., Louwagie, M., Willison, J. C., & Jouanneau, Y. (2003). Identification of pyrene-induced proteins in Mycobacterium sp. strain 6PY1: Evidence for two ring-hydroxylating dioxygenases. *Journal of Bacteriology*, *185*(13), 3828–3841. doi:10.1128/JB.185.13.3828-3841.2003 PMID:12813077

Kühner, S., Wöhlbrand, L., Fritz, I., Wruck, W., Hultschig, C., Hufnagel, P., & Rabus, R. et al. (2005). Substrate-dependent regulation of anaerobic degradation pathways for toluene and ethylbenzene in a denitrifying bacterium, strain EbN1. *Journal of Bacteriology, 187*(4), 1493–1503. doi:10.1128/JB.187.4.1493-1503.2005 PMID:15687214

Kumar, M., Chaudhary, P., Dwivedi, M., Kumar, R., Paul, D., Jain, R. K., & Kumar, A. et al. (2005). Enhanced biodegradation of β-and δ-hexachlorocyclohexane in the presence of α-and γ-isomers in contaminated soils. *Environmental Science & Technology, 39*(11), 4005–4011. doi:10.1021/es048497q PMID:15984776

Kutty, R., Purohit, H. J., & Khanna, P. (2000). Isolation and characterization of a Pseudomonas sp. strain PH1 utilizing meta-aminophenol. *Canadian Journal of Microbiology, 46*(3), 211–217. doi:10.1139/w99-132 PMID:10749534

Larsen, M. H., Blackburn, N., Larsen, J. L., & Olsen, J. E. (2004). Influences of temperature, salinity and starvation on the motility and chemotactic response of Vibrio anguillarum. *Microbiology, 150*(5), 1283–1290. doi:10.1099/mic.0.26379-0 PMID:15133091

Laskin, J., & Futrell, J. H. (2005). Activation of large lons in FT-ICR mass spectrometry. *Mass Spectrometry Reviews, 24*(2), 135–167. doi:10.1002/mas.20012 PMID:15389858

Law, G. T., Geissler, A., Lloyd, J. R., Livens, F. R., Boothman, C., Begg, J. D., & Morris, K. et al. (2010). Geomicrobiological redox cycling of the transuranic element neptunium. *Environmental Science & Technology, 44*(23), 8924–8929. doi:10.1021/es101911v PMID:21047117

Lay, J. O. (2001). MALDI-TOF mass spectrometry of bacteria. *Mass Spectrometry Reviews, 20*(4), 172–194. doi:10.1002/mas.10003 PMID:11835305

Lipscomb, J. D. (2008). Mechanism of extradiol aromatic ring- cleaving dioxygenases. *Current Opinion in Structural Biology, 18*(6), 644–649. doi:10.1016/j.sbi.2008.11.001 PMID:19007887

Mackay, D., & Webster, E. (2005). Environmental persistence of chemicals. *Environmental Science and Pollution Research International, 13*(1), 43–49. doi:10.1065/espr2006.01.008 PMID:16417131

Mann, M., & Jensen, O. N. (2003). Proteomic analysis of post-translational modifications. *Nature Biotechnology, 21*(3), 255–261. doi:10.1038/nbt0303-255 PMID:12610572

Marouga, R., David, S., & Hawkins, E. (2005). The development of the DIGE system: 2D fluorescence difference gel analysis technology. *Analytical and Bioanalytical Chemistry, 382*(3), 669–678. doi:10.1007/s00216-005-3126-3 PMID:15900442

Martín, M., Gibello, A., Lobo, C., Nande, M., Garbi, C., Fajardo, C., & Martínez-Iñigo, M. J. (2008). Application of fluorescence in situ hybridization technique to detect simazine-degrading bacteria in soil samples. *Chemosphere, 71*(4), 703–710. doi:10.1016/j.chemosphere.2007.10.071 PMID:18082866

Martiny, J. B. H., Bohannan, B. J. M., Brown, J. H., Colwell, R. K., Fuhrman, J. A., Green, J. L., & Staley, J. T. et al. (2006). Microbial biogeography: Putting microorganisms on the map. *Nature Reviews. Microbiology, 4*(2), 102–112. doi:10.1038/nrmicro1341 PMID:16415926

Methé, B. A., Nelson, K. E., Deming, J. W., Momen, B., Melamud, E., Zhang, X., & Brinkac, L. M. et al. (2005). The psychrophilic lifestyle as revealed by the genome sequence of Colwellia psychrerythraea 34H through genomic and proteomic analyses. *Proceedings of the National Academy of Sciences of the United States of America, 102*(31), 10913–10918. doi:10.1073/pnas.0504766102 PMID:16043709

Mi, J., Orbea, A., Syme, N., Ahmed, M., Cajaraville, M. P., & Cristóbal, S. (2005). Peroxisomal proteomics, a new tool for risk assessment of peroxisome proliferating pollutants in the marine environment. *Proteomics, 5*(15), 3954–3965. doi:10.1002/pmic.200401243 PMID:16130170

Moharikar, A., Purohit, H. J., & Kumar, R. (2005). Microbial population dynamics at effluent treatment plants. *Journal of Environmental Monitoring, 7*(6), 552–558. doi:10.1039/b406576j PMID:15931414

Muyzer, G. (1999). DGGE/TGGE a method for identifying genes from natural ecosystems. *Current Opinion in Microbiology, 2*(3), 317–322. doi:10.1016/S1369-5274(99)80055-1 PMID:10383868

Nyholm, N. (1991). The European system of standardized legal tests for assessing the biodegradability of chemicals. *Environmental Toxicology and Chemistry, 10*(10), 1237–1246. doi:10.1002/etc.5620101002

OECD. (1992). *Test no. 301: ready biodegradability*. Paris: Organisation for Economic Co-operation and Development.

Ornston, L. N., & Stanier, R. Y. (1966). The conversion of catechol and protocatechuate to β-ketoadipate by Pseudomonas putida. *The Journal of Biological Chemistry, 241*, 3776–3786. PMID:5916391

Pandey, J., Sood, S. S., & Jain, R. K. (2007). Terminal restriction fragment length polymorphism (T-RFLP) analysis: Characterizing the unseen. *Indian Journal of Microbiology, 47*(1), 90–91. doi:10.1007/s12088-007-0017-7 PMID:23100647

Paoletti, A. C., Zybailov, B., & Washburn, M. P. (2004). Principles and applications of multidimensional protein identification technology. *Expert Review of Proteomics, 1*(3), 275–282. doi:10.1586/14789450.1.3.275 PMID:15966824

Pearce, F. (1993). What turns an oil spill into a disaster? *New Scientist, 1858*, 11–13.

Pessione, E., Giuffrida, M. G., Prunotto, L., Barello, C., Mazzoli, R., Fortunato, D., & Giunta, C. (2003). Membrane proteome of Acinetobacter radioresistens S13 during aromatic exposure. *Proteomics, 3*(6), 1070–1076. doi:10.1002/pmic.200300425 PMID:12833532

Prakash, O., & Lal, R. (2006). Description of Sphingobium fuliginis sp. nov., a phenanthrene-degrading bacterium from a fly ash dumping site, and reclassification of Sphingomonas cloacae as Sphingobium cloacae comb. nov. *International Journal of Systematic and Evolutionary Microbiology, 56*(9), 2147–2152. doi:10.1099/ijs.0.64080-0 PMID:16957112

Predki, P. F. (2004). Functional protein microarrays: Ripe for discovery. *Current Opinion in Chemical Biology, 8*(1), 8–13. doi:10.1016/j.cbpa.2003.12.005 PMID:15036150

Purohit, H. J. (2003). Biosensors as molecular tools for use in bioremediation. *Journal of Cleaner Production, 11*(3), 293–301. doi:10.1016/S0959-6526(02)00072-0

Purohit, H. J., Raje, D. V., Kapley, A., Padmanabhan, P., & Singh, R. N. (2003). Peer reviewed: Genomics tools in environmental impact assessment. *Environmental Science & Technology*, *37*(19), 356A–363A. doi:10.1021/es032594m PMID:14572077

Qureshi, A., Prabu, S. K., & Purohit, H. J. (2001). Isolation and Characterization of Pseudomonas Strain for Degradation of 4-nitrophenol. *Microbes and Environments*, *16*(1), 49–52. doi:10.1264/jsme2.2001.49

Qureshi, A., Verma, V., Kapley, A., & Purohit, H. J. (2007). Degradation of 4-nitroaniline by Stenotrophomonas strain HPC 135. *International Biodeterioration & Biodegradation*, *60*(4), 215–218. doi:10.1016/j.ibiod.2007.03.004

Radajewski, S., Ineson, P., Parekh, N. R., & Murrell, J. C. (2000). Stable-isotope probing as a tool in microbial ecology. *Nature*, *403*(6770), 646–649. doi:10.1038/35001054 PMID:10688198

Rani, A., Porwal, S., Sharma, R., Kapley, A., Purohit, H. J., & Kalia, V. C. (2008). Assessment of microbial diversity in effluent treatment plants by culture dependent and culture independent approaches. *Bioresource Technology*, *99*(15), 7098–7107. doi:10.1016/j.biortech.2008.01.003 PMID:18280146

Ranjard, L., Dequiedt, S., Chemidlin Prévost-Bouré, N., Thioulouse, J., Saby, N. P. A., Lelievre, M., & Lemanceau, P. et al. (2013). Turnover of soil bacterial diversity driven by wide-scale environ- mental heterogeneity. *Nature Communications*, *4*, 1434. doi:10.1038/ncomms2431 PMID:23385579

Richins, R. D., Kaneva, I., Mulchandani, A., & Chen, W. (1997). Biodegradation of organophosphorus pesticides by surface-expressed. *Nature Biotechnology*, 15. PMID:9335050

Rosselló-Mora, R., & Amann, R. (2001). The species concept for prokaryotes. *FEMS Microbiology Reviews*, *25*(1), 39–67. doi:10.1111/j.1574-6976.2001.tb00571.x PMID:11152940

Ruelle, V., Moualij, B. E., Zorzi, W., Ledent, P., & Pauw, E. D. (2004). Rapid identification of environmental bacterial strains by matrix-assisted laser desorption/ionization time-of-flight mass spectrometry. *Rapid Communications in Mass Spectrometry*, *18*(18), 2013–2019. doi:10.1002/rcm.1584 PMID:15378711

Santos, P. M., Benndorf, D., & Sá-Correia, I. (2004). Insights into Pseudomonas putida KT2440 response to phenol-induced stress by quantitative proteomics. *Proteomics*, *4*(9), 2640–2652. doi:10.1002/pmic.200300793 PMID:15352239

Sauer, S., Lange, B. M., Gobom, J., Nyarsik, L., Seitz, H., & Lehrach, H. (2005). Miniaturization in functional genomics and proteomics. *Nature Reviews. Genetics*, *6*(6), 465–476. doi:10.1038/nrg1618 PMID:15931170

Siddique, T., Okeke, B. C., Arshad, M., & Frankenberger, W. T. (2003). Enrichment and isolation of endosulfan-degrading microorganisms. *Journal of Environmental Quality*, *32*(1), 47–54. doi:10.2134/jeq2003.4700 PMID:12549541

Singh, O. M., Nagaraj, V., & Nagathihalli, S. (2006). Transcriptomics, proteomics and interactomics: Unique approaches to track the insights of bioremediation. *Briefings in Functional Genomics & Proteomics*, *4*(4), 355–362. doi:10.1093/bfgp/eli006 PMID:17202126

Singh, O. V. (2006). Proteomics and metabolomics: The molecular make-up of toxic aromatic pollutant bioremediation. *Proteomics*, *6*(20), 5481–5492. doi:10.1002/pmic.200600200 PMID:16972298

Singh, O. V., & Nagaraj, N. S. (2006). Transcriptomics, proteomics and interactomics: Unique approaches to track the insights of bioremediation. *Briefings in Functional Genomics & Proteomics*, *4*(4), 355–362. doi:10.1093/bfgp/eli006 PMID:17202126

Singh, R., Paul, D., & Jain, R. K. (2006). Biofilms: Implications in bioremediation. *Trends in Microbiology*, *14*(9), 389–397. doi:10.1016/j.tim.2006.07.001 PMID:16857359

Sinha, J., Reyes, S. J., & Gallivan, J. P. (2010). Reprogramming bacteria to seek and destroy an herbicide. *Nature Chemical Biology*, *6*(6), 464–470. doi:10.1038/nchembio.369 PMID:20453864

Somiari, R. I., Somiari, S., Russell, S., & Shriver, C. D. (2005). Proteomics of breast carcinoma. *Journal of Chromatography. B, Analytical Technologies in the Biomedical and Life Sciences*, *815*(1), 215–225. doi:10.1016/j.jchromb.2004.11.012 PMID:15652811

Stanier, R. Y., & Ornston, L. N. (1973). The β-ketoadipate pathway. *Advances in Microbial Physiology*, *9*, 89–151. doi:10.1016/S0065-2911(08)60377-X PMID:4599397

Sutherland, T. D., Horne, I., Lacey, M. J., Harcourt, R. L., Russell, R. J., & Oakeshott, J. G. (2000). Enrichment of an endosulfan-degrading mixed bacterial culture. *Applied and Environmental Microbiology*, *66*(7), 2822–2828. doi:10.1128/AEM.66.7.2822-2828.2000 PMID:10877774

Thorpe, C. L., Law, G. T. W., Boothman, C., Lloyd, J. R., Burke, I. T., & Morris, K. (2012). The synergistic effects of high nitrate concentrations on sediment bioreduction. *Geomicrobiology Journal*, *29*(5), 484–493. doi:10.1080/01490451.2011.581332

Tomás-Gallardo, L., Canosa, I., Santero, E., Camafeita, E., Calvo, E., López, J. A., & Floriano, B. (2006). Proteomic and transcriptional characterization of aromatic degradation pathways in Rhodoccocus sp. strain TFB. *Proteomics*, *6*(S1), S119–S132. doi:10.1002/pmic.200500422 PMID:16544280

UE. (2006). Regulation (EC) No 1907/2006 of the European Parliament and of the Council of 18 December 2006 concerning the Registration, Evaluation, Authorisation and Restriction of Chemicals (REACH), establishing a European Chemicals Agency, amending Directive 1999/45/EC and repealing Council Regulation (EEC) No 793/93 and Commission Regulation (EC) No 1488/94 as well as Council Directive 76/769/EEC and Commission Directives 91/155/EEC, 93/67/EEC, 93/105/EC and 2000/21/EC. *Official Journal of the European Union, L*, *136*, 1–280.

Unlu, M., Morgan, M. E., & Minden, J. S. (1997). Difference gel electrophoresis. A single gel method for detecting changes in protein extracts. *Electrophoresis*, *18*(11), 2071–2077. doi:10.1002/elps.1150181133 PMID:9420172

Vasseur, C., Labadie, J., & Hebraud, M. (1999). Differential protein expression by Pseudomonas fragi submitted to various stresses. *Electrophoresis*, *20*(11), 2204–2213. doi:10.1002/(SICI)1522-2683(19990801)20:11<2204::AID-ELPS2204>3.0.CO;2-I PMID:10493125

Villas-Boas, S. G., & Bruheim, P. (2007). The potential of metabolomics tools in bioremediation studies. *Omics: A Journal of Integrative Biology*, *11*(3), 305-313.

Vink, J. P. M., & Van Der Zee, S. E. A. T. M. (1997). Pesticide biotransformation in surface waters: Multivariate analyses of environmental factors at field sites. *Water Research*, *31*(11), 2858–2868. doi:10.1016/S0043-1354(97)00139-5

Wang, J., Feng, X., Anderson, C. W., Xing, Y., & Shang, L. (2012). Remediation of mercury contaminated sites – a review. *Journal of Hazardous Materials*, *221–222*, 1–18. PMID:22579459

Wang, R. F., Wennerstrom, D., Cao, W. W., Khan, A. A., & Cerniglia, C. E. (2000). Cloning, expression, and characterization of the katG gene, encoding catalase-peroxidase, from the polycyclic aromatic hydrocarbon-degrading bacterium *Mycobacterium sp.* strain PYR-1. *Applied and Environmental Microbiology*, *66*(10), 4300–4304. doi:10.1128/AEM.66.10.4300-4304.2000 PMID:11010873

Wang, R. F., Wennerstrom, D., Cao, W. W., Khan, A. A., & Cerniglia, C. E. (2000). Cloning, Expression, and Characterization of the katGGene, Encoding Catalase-Peroxidase, from the Polycyclic Aromatic Hydrocarbon-Degrading Bacterium *Mycobacterium sp.* Strain PYR-1. *Applied and Environmental Microbiology*, *66*(10), 4300–4304. doi:10.1128/AEM.66.10.4300-4304.2000 PMID:11010873

Wang, Y., Rawlings, M., Gibson, D. T., Labbé, D., Bergeron, H., Brousseau, R., & Lau, P. C. (1995). Identification of a membrane protein and a truncated LysR-type regulator associated with the toluene degradation pathway in Pseudomonas putida F1. *Molecular and General Genetics MGG*, *246*(5), 570–579. doi:10.1007/BF00298963 PMID:7535376

Wilkins, J. C., Homer, K. A., & Beighton, D. (2001). Altered protein expression of *Streptococcus oralis* cultured at low pH revealed by two-dimensional gel electrophoresis. *Applied and Environmental Microbiology*, *67*(8), 3396–3405. doi:10.1128/AEM.67.8.3396-3405.2001 PMID:11472910

Wilkins, J. C., Homer, K. A., & Beighton, D. (2001). Altered protein expression of Streptococcus oralis cultured at low pH revealed by two-dimensional gel electrophoresis. *Applied and Environmental Microbiology*, *67*(8), 3396–3405. doi:10.1128/AEM.67.8.3396-3405.2001 PMID:11472910

Xu, M., Zhang, Q., Xia, C., Zhong, Y., Sun, G., Guo Jun, Y., & Zhi, L. et al. (2014). Elevated nitrate enriches microbial functional genes for potential bioremediation of complexly contaminated sediments. *ISME*, *8*(9), 1932–1944. doi:10.1038/ismej.2014.42 PMID:24671084

Yergeau, E., Sanschagrin, S., & Maynard, C. S. (2014). Microbial expression profiles in the rhizosphere of willows depend on soil contamination. *ISME*, *8*(2), 344-3458.

Zelles, L., Bai, Q. Y., Beck, T., & Beese, F. (1992). Signature fatty acids in phospholipids and lipopolysaccharides as indicators of microbial biomass and community structure in agricultural soils. *Soil Biology & Biochemistry*, *24*(4), 317–323. doi:10.1016/0038-0717(92)90191-Y

Zhao, H. P., Ilhan, Z. E., Ontiveros-Valencia, A., Tang, Y., Rittmann, B. E., & Krajmalnik-Brown, R. (2013). Effects of multiple electron acceptors on microbial interactions in a hydrogen-based biofilm. *Environmental Science & Technology*, *47*(13), 7396–7403. doi:10.1021/es401310j PMID:23721373

KEY TERMS AND DEFINITIONS

Denaturing Gradient Gel Electrophoresis (DGGE): It is a kind of electrophoresis method for molecular fingerprinting that separates polymerase chain reaction (PCR)-generated DNA products on the basis of sequence differences that results in differential denaturing characteristics of the DNA.

Detoxification: Removal of toxic substances from environment.

Hazardous Material: Any item or agent (biological, chemical, radiological, and/or physical), which has the potential to cause harm to humans, animals, or the environment, either by itself or through interaction with other factors.

Isoelectric Focusing (IEF): It is also known as electrofocusing, is a technique for separating different molecules by differences in their isoelectric point (pI).

Metagenome: The study of collective genome of microorganisms from an environmental sample: to procured the information on the microbial diversity and ecology of a specific environment.

Metaproteome: Study of all protein samples recovered directly from environmental sources like soil, water etc.

Pollutant: This is a substance or condition that contaminates air, water, or soil and has adversely affects the usefulness of a resource.

Riboswitch: This is a critical feature of the hypothesized RNA world that would have been the ability to control chemical processes in response to environmental cues.

Chapter 4
Role of Microbes in Eco-Remediation of Perturbed Aquatic Ecosystem

Susmita Lahiri
University of Kalyani, India

Debarati Ghosh
University of Kalyani, India

Jatindra Nath Bhakta
University of Kalyani, India

ABSTRACT

Anthropogenic and natural hazards put aquatic ecosystem under severe stress due to organic and inorganic pollutants and calls for remediation. Eco-remediation, an ecological blessing for ecosystem stabilization refers to the restoration of the perturbing ecosystem to its original balanced state with a multi-purpose approaches for removing pollutants by vast array of biological components. Aquatic habitats and biodiversity are adversely affected by indiscriminate pollution along with rising levels of carbon dioxide. To overcome the aquatic stress of various pollutants, microorganisms significantly imparting crucial roles in environmentally sustainable bioremediation approaches by adapting themselves in order to survive in perturbed ecosystems resulting in the environmental remediation. This chapter attempts to deal with eco-remediation of ecosystems perturbed by various hazardous pollutants by focusing the degrading and transforming of contaminants into less toxic forms by microbes and carbon sequestration through heterotrophic pathway applying different ex situ or in situ detoxifying mechanisms. Considering the high eco-remedial efficiencies, microbial eco-remediation technologies could be applied as an emerging approach in restoring perturbed aquatic ecosystems in ecological engineering systems.

DOI: 10.4018/978-1-5225-2325-3.ch004

INTRODUCTION

Restoration of the perturbing ecosystem to its original balanced state or resilient state is a multi-purpose approach aimed to improve self-cleaning capacity of elements of ecosystem (Bhakta et al., 2016). Globally referred to as eco-remediation its main environmental benefits include stability of groundwater, enrichment of habitats, increased biotic diversity and, therefore, the growth in biomass and improved (Bardgett & van der Putten, 2014). The desperate need for ecosystem remediation originates as a result of introduction of diverse types of organic and inorganic pollutants both as a result of anthropogenic and natural hazards which put aquatic ecosystem under severe stress (Rombouts, Beaugrand, & Artigas, 2013; Boldt, Martone, & Samhouri, 2014). World aquatic ecosystems are increasingly being perturbed by various types of natural perturbations, including droughts, storms and floods (Hogan, 2010) affecting the organisms as well as the entire biosphere of aquatic ecosystems. A recently focused threat of aquatic ecosystem is nutrient pollution by phosphorus and nitrogen enrichment of aquatic bodies leading to algal bloom and eutrophication (Grizzetti, Bouraoui, & Aloe, 2012). Additionally, the greatest concern of the stress at the microbial level of the freshwater ecosystem where a bulk of biological production and nutrient cycling takes place, is due to the impact of human induced pollution and habitat alteration (Carpenter, Stanley, & Vander-Zanden, 2011; Aracic, Semenec, & Franks, 2014). Due to increasing organic and inorganic pollutants input the natural self-purification capacity of any aquatic environment is highly affected by dwindling decomposing abilities of microorganisms. Inorganic contaminants, in particular heavy metals, are also a prominent environmental concern because they are not biodegradable and can accumulate in living organisms (Bittsanszkya et al., 2005; Fu & Wang, 2011).

The ubiquitous nature of microorganisms and their ability to tolerate the variety of anthropogenic stresses (Crawford & Crawford, 2005) by degrading and transforming environmental contaminants into less toxic forms (Kot & Namieśńik, 2000) as well as carbon sequestration through heterotrophic pathway have made them primary candidate for exploitation in remediation and can be used as signature for eco-remediation. For their survival in heavy metal-polluted habitats, microorganisms have developed and adopted various different detoxifying mechanisms such as biosorption, bioaccumulation, biotransformation and biomineralization, which can be further exploited for bioremediation either *ex situ* or *in situ* (Lin & Lin 2005; Elekwachi, Andresen, & Hodgman, 2014). Pathogenic bacterial contamination of water can be effectively controlled by using bacteriophages (Aracic et al., 2014). It has been reported that microbes can be managed to enhance organic matter sequestration in soils, sediments and aquatic system to remove CO_2 from the atmosphere and reduce greenhouse warming (Lal, 2008; Savage, Afonso, Chen, & Silver, 2010; Gray & Engel, 2013).

Detoxification mechanism of bacterial and micro-algal populations surviving in metal contaminated environments and helping in metal bioremediation and nutrient reclamation from polluted aquatic system have been analysed by molecular tools (Jansson, Björklöf, Elvang, & Jørgensen, 2000; Gambhir, Kapoor, Nirola, Sohi, & Bansal, 2012). An extracellular polymeric substance (EPS) produced by a variety of bacteria has been demonstrated to be a potential bioemulsifier used in the degradation of hydrocarbons as well as biosorbing agents accumulating nutrients and heavy metals (Orsod, Joseph, & Huyop, 2012).

Traditional treatment technologies to produce high quality drinking water have to be modernized by developing materials and methods which are more efficient, cost-effective and environmentally sustainable such as eco-remediation (Schindler, 2012) or reuse of wastewater to restrict the worsening quality of drinking water (Gambhir et al., 2012).

Microorganisms already cultured for a variety of applications cover only a tiny fraction of those vast untapped and beneficially potential of environment remediation. Since eco-remediation or bioremediation process involves the enhancement of naturally occurring microbial activity it is pertinent to consider the deeper understanding of microbial communities and their response to the natural environment and pollutants, expanding the knowledge of the genetics of the microbes to increase capabilities to degrade pollutants, conducting field studies of new bioremediation techniques which are cost effective. Stemming from above points of view, the present chapter attempts to deal the eco-remediation of ecosystems perturbed by various hazardous pollutants by considering the different bioremediation approaches of microorganisms.

PERTURBATION OF AQUATIC ECOSYSTEM

Causes and Mechanism

Marine, groundwater, surface water resources are being vehemently threatened due to overuse and misuse and pollution by our very own human race as well as by natural factors. Human effects on aquatic ecosystems can result from pollution, changes to the landscape or hydrological systems, and larger-scale impacts such as global climate change. A six-tier classification of aquatic ecosystem based on functional hydrogeomorphic along with structural attributes (Ollis et al., 2015) has been done to understand the complexity of aquatic ecosystems and the linkages within them in order to predict effect of disturbances on them. These linkages mean that damage to one component of the ecosystem can lead to impacts on other ecosystem components. Bio-accumulation of pollutants in an aquatic ecosystem (Dixit et al., 2015) is an environmentally serious issue since they are being taken up by plankton and other aquatic organisms after being mineralized by microorganisms. Finally, several times biomagnified metals are taken up by man when he consumes fish from that contaminated water (El-Shahawi, Hamza, Bashammakhb, & Al-Saggaf, 2010). Climate change scenarios, on the other hand, predict an increase in physical stress (e.g. by storms) and organic matter. It is not yet known how these different impacts might combine to affect ecosystem processes (Heip et al., 2009).

Perturbations in different resources of our matrix of life are discussed as follows.

Marine Ecosystem

A range of human activities such as by-catch and run-off or pollution with plastics (Derraik, 2002) and other persistent toxins have reduced many species in abundance or driven to local extinction. Crude oil containing toxic polycyclic aromatic hydrocarbons (PAHs), spilled in aquatic ecosystem cascade across trophic levels, affects marine life (Perhar & Arhonditsis, 2014) by staying in the sediment and marine environment (Panetta, 2003).

The threat of eutrophication with algal blooms results in marine morbidity and mortality events by producing algal toxins due to their recent increase in frequency, intensity and geographic distribution (McKay & Mulvane, 2001). Sixteen percent coral decline due to bleaching and disease has emerged as large-scale disturbances as a result of fisheries, land-use changes increasing sedimentation, and nutrient pollution (Wilkinson, 2002).

The ocean absorbs approximately 30% of anthropogenic CO_2 (IPCC, 2013), mitigating global warming in a profound way. Around 50% of the photosynthetically produced particulate organic carbon is transformed through mechanisms including zooplankton grazing, excretion, the action of microbial ectohydrolases and viral lysis into dissolved organic carbon (Anderson & Tang, 2010).

Groundwater Ecosystem

Several parts of the world experiences groundwater contamination due to mixing of different minerals both naturally and due to human activities (Onodera et al., 2008). The use of nitrogen fertilizers in agriculture increases the nitrate concentration in groundwater leading to high nitrate levels in underground drinking water sources, thus causing life threatening "blue baby" syndrome (methemoglobinemia) in very young children, which is a serious problem in parts of rural Eastern Europe (National Research Council, 1992; Singh, Singh, & Sekhon, 1995; Sönmez, Kaplan, & Sönmez, 2007; Buitenkamp & Stintzing, 2008). On the other hand, pesticides directly applied on soil not only kill pests in the soil or on the ground but also can create leaching of pesticides to groundwater (Gambhir et al., 2012; Khan & Mohammad, 2014).

The cause of groundwater contamination is not only limited to this but is extended by on-site untreated or even treated sewage or wastewater wastewater disposal, from sanitation containing pathogenic microorganisms, especially various enteric viruses, bacteria, protozoa, helminthine eggs etc. Highly toxic heavy metals (e.g. mercury, cadmium, chromium, thallium and lead) are added to the ground water either naturally by slow leaching from soil to water or through various anthropogenic sources causing ill effects even at very low concentrations (Central Pollution Control Board [CPCB], 2008; Nimick, Caldwell, Skaar, & Selch, 2013). Various other man-made aberrant processes causing severe groundwater pollution are landfill leachate, hydraulic fracturing etc. (Figure 1) (Mclaughlin, Borch, & Blotevogel, 2016). These processes call for serious attention towards its remediation through intervention of scientists, engineers, local community and policy makers.

Figure 1. Different mechanisms of ground water deterioration by improper discharge of waste products
http://www.iloencyclopaedia.org/component/k2/90-55-environmental-pollution-control/water-pollution-control

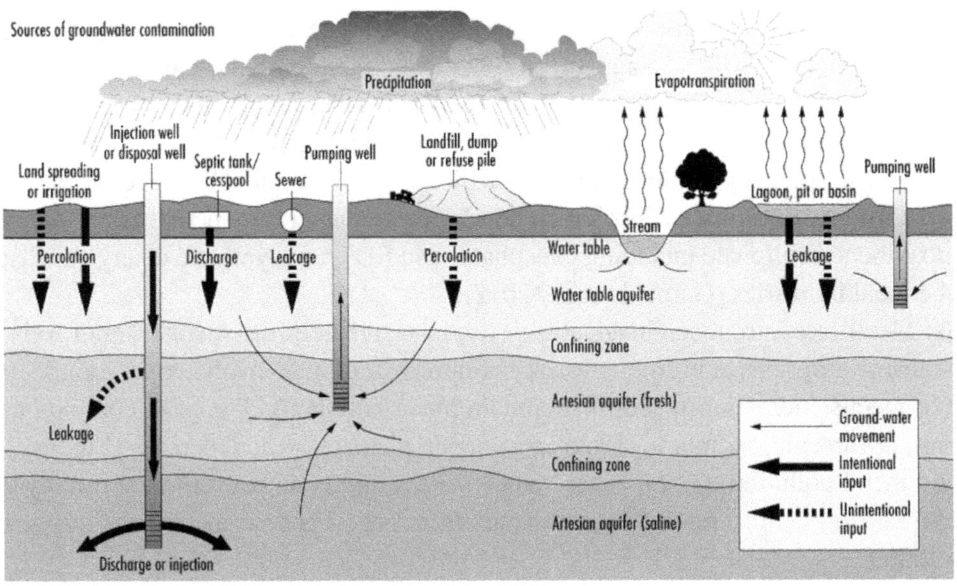

Arsenic and fluoride contamination of groundwater has recently been identified as a major environmental issue. In India West Bengal recorded highest arsenic toxicity (above the WHO's maximum permissible limit of 10 mcg/L) in at least 9 blocks out of 18 districts and affecting 26 million people (Chakraborti et al., 2009; Basu, Sen, & Jha, 2015). In India occurrence of fluoride in ground water has been detected in many places of Andhra Pradesh, Haryana, Jharkhand, Tamilnadu, Karnataka, Gujarat, Rajhasthan, Maharashtra, Madhya Pradesh, Chattisgarh, Punjab, Haryana, Uttar Pradesh, Kerala, Orissa, Jammu & Kashmir, Bihar and Delhi and in recent past West Bengal and Assam have been added in the list (Public Health Engineering Department [PHED], 2013).

Although fluoride is an essential element for animal species but very high concentration above 3.6 mg/l can cause disease like skeletal fluorosis (Bhattacharya & Chakrabarti, 2011; Datta, Chakrabortty, De-Dalal, & Lahiri, 2014). Bureau of Indian Standards (2003) has prescribed fluoride concentrations of 1 mg/l and 1.50 mg/l as desirable and permissible limits respectively in respect of drinking water. In India 62 million people including 6 million children are affected with fluoride related health diseases (Haldar & Ray, 2014).

Freshwater Ecosystem

Now-a-days, people dump their garbage into streams, lakes and rivers, thus turning these freshwater bodies into the final resting place of plastics (Iacurci, 2013), cans, bottles and other various household products (Harter, 2003). Industries also play a major role in polluting this important ecosystem by discharging their effluents in the form of toxic metals and radionuclides either in untreated or in partially treated forms. Agricultural runoffs or unplanned sewage disposal form important non-point source of freshwater pollution and contaminate the nearby water bodies to the level of irreversible condition. The primary pollutant types with their sources and polluting mechanism is described below:

- **Plastics:** Advanced life-style dependent booming consumer society generates microplastics (<5mm) which is emerging as a potentially damaging aquatic stressor in recent decades. The movement and potential deposition of microplastics depends largely on river flow, wave actionand releases of water from dams and hydropower plants. As such, microplastic concentrations are likely to be high in areas where water velocity is low, and the sediment it carries is suspended and deposited (Free et al., 2014). In aquatic environments, these particles can attract and concentrate hydrophobic contaminants, such as PAH, PCBs, DDT, DDE and PBDE (Matelich, 2015). These materials contain highly toxic chemicals such as organohalogen, styrene, acrylonitrile, formaldehyde, phthalates, diethanolamine, nonlphenolethoxylate, linear alkyl benzene sulfonates,1,4-dioxane, naphthotriazolystilbenes (linked with developmental and reproductive effects), benzoxazolyl, diaminostilbenedisulfonate including phosphates and EDTA of synthetic detergents discharged by petrochemical industries (Gambhir et al., 2012).
- **Metals:** Metals are introduced in aquatic systems as a result of the weathering of soils and rocks, from volcanic eruptions, and from a variety of human activities involving the mining, processing, or use of metals and/or substances that contain metal pollutants. The most common heavy metal pollutants are arsenic, cadmium, chromium, copper, nickel, lead and mercury (Figure 2). The most common metal pollution in freshwater comes from mining companies. They usually use an acid mine drainage system to release heavy metals from ores, because metals are very soluble in an acid solution.

Figure 2. Metal pollution scenario of various industries
Lokhande, Singare, & Pimple, 2011.

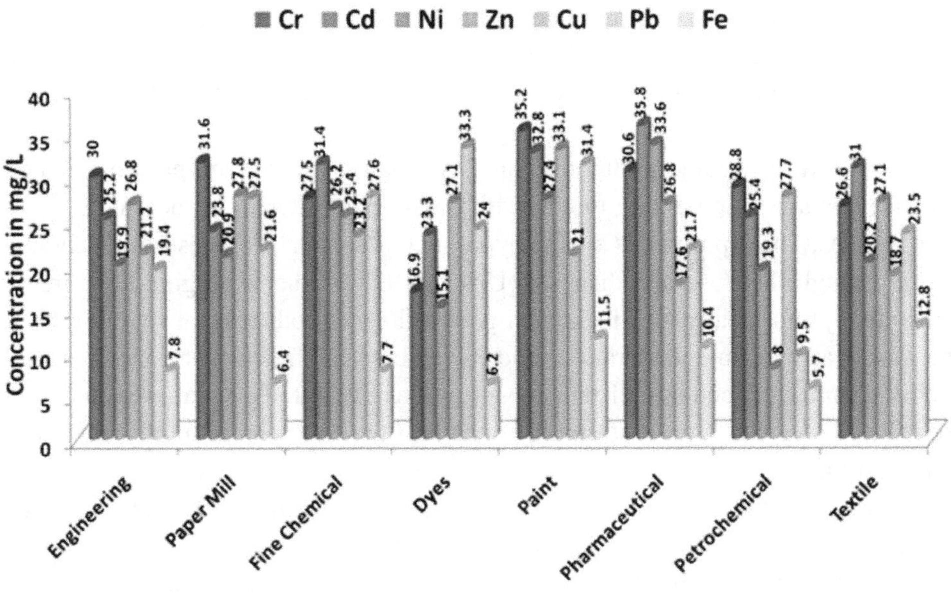

The most common metal pollution in freshwater comes from mining companies. They usually use an acid mine drainage system to release heavy metals from ores, because metals are very soluble in an acid solution. Iron pollution in water may be attributed to acid mine drainage, landfill leachate sewage or industries, the dissolution of rocks and minerals etc. The main sources of mercury (Hg) pollution in aquatic environment are industries like paints, oil refining, pulp and paper, fertilizer, chlor-alkali and rubber processing (Namasivayam & Periasamy, 1993), fluorescent light tubes and high intensity street lamps, batteries, pharmaceuticals, thermometers, cosmetics and pesticides (Krishnan & Anirudhan, 2002). Mercury sometimes transforms into highly neurotoxic compound methyl-mercury (ATSDR 2008; Nimick et al. 2013) and through the food chain biomagnifies upward in aquatic ecosystems (Tchounwon, Ayensu, & Ninashvili, 2003; Kim, Mason, Porter, & Soulen, 2004).

- **Nutrient:** Nutrient enrichment has become the planet's most widespread water quality problem (Pinckney, Paerl, & Harrington, 1999; United Nations World Water Assessment Programme [UNWWAP], 2009). It not only occurs with phosphorus and nitrogen from agricultural runoff, but also caused by industrial and human waste. In many regions, human activities have altered natural erosion rates and greatly altered the volume, rate, and timing of sediment entering streams and lakes, affecting physical and chemical processes and species' adaptations to pre-existing sediment regimes. Acidification of freshwater ecosystem is caused by nutrient enrichment which has impact on biodiversity (Millennium Ecosystem Assessment [MA], 2005). Water dilutes the pollutants and because of that superiority as a great solvent, water-soluble wastes easily pollute water. Various actions, often but not exclusively anthropogenic, can cause salts to build up in the water.
- **Pathogens:** One of the most a widespread and serious class of water quality contaminants, especially in areas where access to safe, clean water is limited, is pathogenic organisms: bacteria, protozoa, intestinal parasites and viruses (Pandey, 2015). These organisms pose one of the leading

global human health hazards. In general terms, the greatest microbial risks are associated with ingestion of water that is contaminated with human or animal feces. Wastewater discharges in fresh waters and costal seawaters are the major source of fecal microorganisms, including pathogens (World Health Organization [WHO], 2008). In recent years, 175 species of infectious agents from 96 different genera have been classified as emerging pathogens (WHO, 2003) threatening water quality.

- **Organic Pollutants:** Organic pollutants, such as dioxins, furans, and polychlorinated biphenyls (PCBs) have become an emerging threat, with possible long-term degradation of freshwater and other ecosystems. These pollutants are the byproduct of industrial processes and enter the environment both through their use and disposal (UNEP, 1998). Other emerging contaminants include endocrine disruptors, pharmaceuticals, and personal care products that may not be removed by existing wastewater treatment operations and end up entering freshwater systems. Invasive aquatic species has been introduced into distant ecosystems as a result of recreational, economic, or other purposes. Pesticides can migrate through water into the food chain, finally being consumed by humans or animals (De-Lorenzo, Scott, & Ross, 1999). Although diverse structurally, the properties of organic chemicals (USEPA, 2001) give them sufficient stability and impart a relatively high affinity for fatty tissues and resistance to breakdown in the body and thus allow them to accumulate in organisms and biomagnify in food chains (El-Shahawi et al., 2010). Alkyl phenols and acid pharmaceuticals present an exposure risk to aquatic life in receiving waters near municipal waste treatment plants (Metcalfe, Miao, Koenig, & Struger, 2003; Gan, Lau, & Ng, 2009). Pharmaceuticals ranging from painkillers and antidepressants to hormone-replacement therapies and chemotherapy agents (Carr & Neary, 2008) and personal care products originating from cosmetics, toiletries and detergents are also of increasing concern.

Impacts of Perturbation

Since water is the matrix of life, aquatic ecosystem pollution is leading to serious health problems and environmental hazard due to biomagnifications, bioconcentration phenomenon (Siwela, Nyathi, & Naik, 2009). Besides pollution, aquatic system is responding stressfully to current burning issue of temperature rise leading to global climate change (Huttermann, Orikiriza, & Agaba, 2009; Ma, Prasad, Rajkumar, & Freitas, 2011). Low dissolved oxygen levels below 5 ppm (Fayer & Trout, 2005) due to pollution in freshwater cause death of organisms. Below are the important hazardous perturbations of the aquatic ecosystem.

Heavy Metal Pollution

It has been reported that if the bioavailability of heavy metals shows a trend of Zn >Cd> Ni> Fe > Cu>Pb>Cr then it may pose a very high health risk to humans, due to the ability of these metals to enter the food chain (Lokeshwari & Chandrappa, 2007). Nickel pollution for example, decreases the nutrient removal capacities of indigenous protozoan species of *Aspidisca* sp., *Paramecium* sp., *Trachelophyllum* sp in wastewater (Kamika & Momba, 2015) and dermatitis (Ni-itch), cyanosis, extreme weakness, dry cough and shortness of breath, cancer of bone, lungs and nose in humans (Aquino et al., 2012). The key impacts of the important toxic heavy metals on human and aquatic life are given below:

Besides arsenicosis, other diseases like skin lesions, hyperkeratosis, hyper-pigmentation, cardiovascular diseases are caused by Arsenic in humans (WHO, 2006; Celik et al., 2008) and aggression behavior of fishes in aquatic ecosystem (Magellan et al., 2014). Low concentration of chromium may lead to respiratory problems, lung cancer, alteration of genetic material and also death in humans (Eastmond et al., 2008) and cause sub-lethal toxic effects in aquatic plants and animals. Children suffer from mental retardation and adults from impairment of pulmonary and kidney function along with abnormal distribution of chromosome and strong biomagnification in the aquatic food chain. Minamata disease in humans due to ingestion of methylmercury contaminated fish causing impairment of brain function. (Park & Zheng, 2012; Miller et al., 2013) was a significant inpact of heavy metal pollution. Cadmium in high concentration causes hepatotoxicity, nephrotoxicity and cancer in humans and fish skeletal disorder (Gonick, 2008; Sarkar et al., 2013) in aquatic bodies. Long-term exposure of lead in adults can result in decreased performance of the nervous and reproductive system (Levin et al., 2008; Martin & Griswold, 2009). Lead bioconcentrates in the skin, bones, kidneys, gill function and liver of fish rather than in muscle and does not biomagnify up the food chain. In algae, concentration> 500ppb causes inhibition of enzymes needed for photosynthesis (Taub, 2004).

Nutrient Load

Nutrient enrichment of water increases primary productivity rates to excessive levels due to overgrowth of algae, vascular plants and leading to depletion of dissolved oxygen in the water body. This can create stress on aquatic organisms due to change in water and habitat quality (Vilmi, Karjalainen, Kuoppala, Tolonen, & Heino, 2016) and kill them (Meorin, 1989). This typical condition is referred to as eutrophication of water body which leads to shift in species abundance and dominance of blue green algae. In long term, due to nutrient inputs some lakes and ponds have become so hyper-eutrophic that every single macro-organism has been eliminated. Eutrophication has caused significant changes also in coastal nutrient (C, N, P, Si) cycling, water quality, biodiversity, fisheries and overall ecosystem health (Jeppesen et al., 2015).

Pesticide Run-Off

Widespread use of the insecticides (Younis, Abo-El-Saad, Abdel-Razik, & Abo-Seda, 2002) pollutes water-bodies, contaminating fish, and ultimately poison bald eagles (and other animals) those ate the fish. In humans pesticides can damage the nervous, reproductive and endocrine system as well as liver (Richardson, Gordon, Muller, Pletschke, & Whitfield, 2010) and DNA causing a variety of cancers (Radosevich, Holt, & Ghersa, 2007). After entering water Glyphosate (Roundup), a most common herbicide of the world declined 70% of frog biodiversity and 86% of total mass of tadpoles (Hileman 2005; Relya 2005).

Natural Disaster

Due to an earthquake or a powerful volcanic eruption, or an underwater landslide, tsunami takes place which profoundly affect the aquatic ecosystem and forced it to reach a threshold beyond which different systems of processes predominate. Interestingly, freshwater lake converts to a near-brackish water ecosystem increasing average salinity abruptly from 2.82 to 10.96 ppt due to tsunami post-effect (Panigrahi, 2011). Physico-chemistry of groundwater changes after massive earthquake inducing a dramatic decrease in subterranean species and population turnover rates turning the aquifer zone into a zero life zone (Galassi et al., 2014)

Elevated Atmospheric CO_2 Level

Raising levels of carbon dioxide in the atmosphere due to climate change leads to persistent CO_2 storage which over time forms carbonic acid acidifying oceans (Haszeldine, 2009). Acids alter water quality drastically by increasing the leaching of various organic and inorganic contaminants. Harvell et al. (1999) reported rising temperature induced increased epizootics in pinnipeds. As a result aquatic ecosystems especially fish distributions are highly modified (Cheung, 2009) affecting the livelihoods of the fisheries dependent communities. Effects of ocean acidification due to high atmospheric CO_2 concentration on heterotrophic microbes have been predicted (Joint, Doney, & Karl, 2010). Healthy ocean ecosystems are important for the mitigation of climate change (Climate Change, Fisheries and Aquaculture [PACFA], 2009).

Oil-Spill

Response Of organisms to spilled oil depends on the quantity and chemical composition of the oil to which they are exposed of. Marine mammals and birds whose breathing takes place regularly at the air-water interface are most vulnerable to the oil-spill (Peterson et al., 2003), whereas pelagic fish species are minimally exposed (Paine et al., 1996). Toxicity due to oil spills may include accumulation in tissues, ingestion of oil, DNA damage, cardiac dysfunction, loss of buoyancy and insulation for birds, impacts to immune functioning, mass mortality of eggs and larvae, e.g., in fish, and inhalation of vapors (Aguilera, Méndez, Pásaro, & Laffon, 2010, Judson et al., 2010, Major & Wang, 2012).

Pathogens

Water borne pathogens cause human diseases like poliomyelitis (Polio virus), hepatitis (Hepatitis virus), diarrhea (*Escherichia coli*), cholera (*Vibrio cholerae*), shigellosis (*Shigella* spp.), typhoid (*Samonella typhi*) and paratyphoid (*Samonella paratyphi*), giardiasis (*Giardia* spp.), amaebiasis (*E. histalytica*) are mostly include in this category (Wolf et al., 2014; Wolf, Nick, & Cronin, 2015). The EPA reported that an average of 6 people per year die of a waterborne disease. Non-human pathogens can also pose a significant threat to aquatic ecosystems and biodiversity affecting wildlife fish and waterfowl (US Agriculture Centre, 2004).

ECO-REMEDIATION STRATEGIES OF MICROBES

Ecological restoration allows for the return of the perturbed aquatic ecosystem to the natural balance and function found in healthy ecosystems (National Research Council [NRC], 1992) ensuring the re-establishment of natural ecosystem, regulation of nutrient cycles and the hydrological cycle (Lovley & Lloyd, 2000). High cost and high energy utilizing contaminants removal processes include membrane-filtration, adsorption, sedimentation, chemical precipitation, ion-exchange and coagulation-flocculation (Fu & Wang, 2011).

Microbial eco-remediation is based on the co-metabolism action of one organism or a consortium of microorganisms with a multi-disciplinary, multi-trophic and an ecosystemic approach consisting of structural and functional indicators and endpoints developing innovative, sensitive, automated, and cost-effective procedures including computer-based techniques to reduce the problems of stress, healing, recovery and remediation (Cupples, Sanford, & Sims, 2005; Koukkou, 2011).

Effective bioremediation by microorganisms depends on its ability of interaction, both chemically and physically, with substances utilizing them for microbial metabolism leading to structural changes or complete degradation of the target molecule (Diez, 2010). The influencing factors of the process are:

1. Nature of specific contaminants,
2. Concentration of contaminant,
3. Oxygen supply,
4. Moisture,
5. Temperature,
6. pH,
7. Nutrient supply,
8. Bioaugmentation,
9. Co-metabolism.

The eco-remediation strategies may be classified into *ex situ* and in situ. While *ex situ* remediation involves extraction of the contaminants along with its disposal media to a secured environment for treatment, *in situ* bioremediation can be defined as the non invasive stimulation of native microflora to biodegrade target contaminants through the passive management of environmental parameters. The techniques include bioventing, biosparging, bioslurping, bioaugentation, biopiling, natural attenuation and phytoremediation along with physical, chemical and thermal processes. Many *in-situ* remediation processes such as ecological floating bed techniques and constructed wetlands have been developed for bioremediation of polluted surface water and have obtained satisfactory results (Cao, Zhang, Wang, & Pan, 2012). The inclusion of periphyton, as a stable micro-ecosystem, is a promising in situ strategy to restore decimated surface water ecosystems (Wu, Reynolds, Fourie, & Thomas, 2013). The first patent for a biological remediation agent was registered in 1974, being a strain of *Pseudomonas putida* (Prescott, Harley, & Klein, 2002) that was able to degrade petroleum.

Since numerous types of pollutants are to be encountered in a contaminated site, diverse types of microorganisms are likely to be required for effective mediation (Watanabe, Kodoma, Stutsubo, & Harayama, 2001) (Table 1).

Removal of Heavy and Other Metals

Bacterial, fungal and algal biomasses can form high metal sorbing substrates to remove heavy metals either in dead / inactive or in live forms. In aquatic system many bacterial species (e.g. *Bacillus, Lactobacillus, Escherichia, Micrococcus, Pseudomonas, Streptomyces,* etc.) can enhance their potential for heavy metal uptake with an addition of excess amount of citrate (Sahin, Keskin, & Keski, 2013; Bhakta, Ohnishi, Munekage, & Iwasaki, 2010; Bhakta, Munekage, Ohnishi, & Jana, 2012a; Bhakta, Ohnishi,

Table 1. Different groups of microorganisms and their activity in pollutant removal

Groups of Microorganism	Process	Services
Heterotrophs	• Organic matter break down, mineralization • Extracellular polymer production	• Decomposition, nutrient recycling, climate regulation, water purification • Carbon sequestration
Photoautotrophs	Photosynthesis	Primary production
Chemoautotrophs	Specific elemental transformation(e.g. Sulfate reduction, iron oxidation)	Nutrient recycling, climate regulation, water purification
Unicellular phytoplankton	Photosynthesis	Primary production
Archaea	Specific element transformations (e.g. Methanogenesis, nitrification)	Nutrient cycling, atmosphere and climate regulation
Protozoans	Consumption and mineralization of other microbes	Decomposition, nutrient cycling, soil formation
Fungi	Organic matter consumption and mineralization	Decomposition, nutrient cycling, soil formation
Fungi (mycorrhizal)	Nutrient recycling	Primary production (indirect)
Viruses	Lysis of hosts	Nutrient cycling

Munekage, Iwasaki, & Wei, 2012b; Bhakta, Munekage, Ohnishi, Jana, & Balcazar, 2014). Heavy-metal-tolerant bacteria containing the arsenic (III) S-adenosyl-methionine-methyltransferase (ArsM) are able to methylate toxic inorganic arsenic (III) to the less toxic arsenic (V) (Chen et al., 2014). Guo et al. (2010) found that effective removal of Cr (VI) and Ni (II) by powdered biomass of *Rnigricans* spp. at optimum pH 2. Electroactive microorganisms generally use heavy metals as terminal electron acceptors or reduce them through detoxification mechanisms. *Geobacter* spp. can reduce soluble uranium (VI) to insoluble uranium (IV), finally precipitating the ions out of the water (Gregory & Lovley, 2005). The list of microorganisms has been strategically used in bioremediation treatments for heavy metals (shown in Table 2).

Removal of Nutrients

Nitrosococcus, *Nitrospira*, *Nitrosomonas* and *Nitrobacter* are the most frequently isolated dominant groups responsible for nitrogen (Holben, Noto, Sumino, & Suwa, 1998) removal by molecular methods. The potentials of cyanobacteria in the removal of nutrients from wastewater rich in nitrogenous and phosphorus compounds, to oxidize oil components, complex organic compounds, and accumulate metal

Table 2. Microorganisms strategically used in bioremediation

Microbes	Species Names
Bacteria	*Arthrobacterspp.*, *Pseudomonas veronii*, *Burkholderia* spp., *Kocuriaflava*, *Bacillus cereus* and *Sporosarcinaginsengisoli*
Fungi	*Penicilliumcanescens*, *Aspergillusversicolor*, and *Aspergillusfumigatus*
Algae	*Cladophorafascicularis*, *Spirogyra* spp., *Cladophora* spp., *Spirogyra* spp. and *Spirullina* spp
Yeast	*Saccharomyces cerevisiae* and *Candida utili*

Source: Coelho, Rezende, de-Sousa, Melo, & Coelho, 2015.

ions of Zn, Co and Cu have been demonstrated (Mohapatra & Gupta, 2005). Phosphate solubilizing bacterial group Rhodocyclusin the beta subclass of Proteobacteria were abundantly used for efficient phosphate removal (Bond, Erhart, Wagner, Keller, & Blackall, 1999). Biogeochemical nutrient cycling bacteria have been effectively used as ecological signature to ascertain the nutrient reclamation efficiency of waste stabilization ponds utilized for aquaculture (Lahiri, Sarkar, & Jana, 2015). For marine eutrophication control one brown algae *Laminaria* and other red algae *Gracilaria* were selected (Jiang, Fang, Mao, &Wang, 2010).

Hydrocarbon Remediation

A particular group of microorganisms degrade oil and other petroleum products by molecular oxygen introduction into the hydrocarbon, transforming intermediates which are finally utilized for the general energy-yielding metabolic pathway of the cell (Coelho et al., 2015). Many bacteria including *Bacilli, Arthrobacter,Aeromonas, Pseudomonas, Moraxella, Beijerinckia,Flavobacteria, Chrbacteria, Nocardia, Corynebacteria, Modococci, Streptomyces, Aeromonas* and cyanobacteria and some yeast can degrade petroleum hydrocarbon products. Cyanobacteria play the role of facilitator by providing oxygen and nutrients to the associated hydrocarbon-degrading bacteria (Haritash & Kaushik, 2009). A bacterial strain *Pseudomonas putida* MHF 7109 isolated from cow dung microbial consortia has a high potential for biodegradation of a large group of petroleum compounds, such as benzene, toluene, and o-xylene (BTX) (Singh et al., 2014). A bacteria-algal consortium has been successfully developed and employed on a rotating biological contactor for efficient degradation of petroleum compounds (Abed & Köster, 2005).

Pesticide Biodegradation

Among the microbial communities, bacteria, fungi and actinomycetes are the main transformers and pesticide degraders. The species frequently isolated from pesticide contaminated sites are *Aspergillus niger, A. fumigatus, A. terreus, Pseudomonas* spp, *Rhizopus microspores, Absidia* spp, *Botrytis cinerea* and *Trichoder maviridae* has very high potential to remove herbicides like metroburon, linuron and pesticides like methyl parathion, endosulfan almost completely (Ortiz-Hernández, Sánchez-Salinas, Dantán-González, & Castrejón-Godínez, 2013). In the first metabolism stage for degradation involved three enzyme systems: hydrolases, esterases and the mixed function oxidases (MFO), and in the second phase glutathione S-transferases (GST) system activates (Li, Schuler, & Berenbaum, 2007).

CO_2 Sequestration

In order to mitigate ecosystem perturbation due to high levels of CO_2 emission, environmental scientists have tried to employ microorganisms such as microalgae cyanobacteria, green algae and some autotrophic bacteria who could potentially fix CO_2 more efficiently than higher plants, due to their faster growth (Bhakta, Lahiri, Pitman, & Jana, 2015; Rossi, Olguin, Diels, & De Philippis, 2015). A biological pump process involving carbon fixation into biomass by phytoplankton followed by their consumption by zooplankton and gravitational sedimentation into the deep ocean has further added to the removal of carbon dioxide from atmosphere (Ducklow, Steinberg, & Buesseler, 2001). Microbial carbon pump has evolved as a promising carbon sequestration strategy to combat global warming issue. Bacteria and protozoans can accelerate carbon sinking and storage by producing extracellular mucins that bind particles into larger aggregates (marine snow) that sink faster, or retard sinking and storage by decomposing the sinking materials (Azam & Malfatti, 2007). Potential for artificial iron additions (iron fertilization) to enhance CO_2 storage has emerged as a geo-engineering strategy for reducing anthropogenic CO_2 accumulation in the atmosphere (Boyd, Jickells, Law, & Blain, 2007). The studies demonstrate convincingly that iron addition stimulates photosynthesis, macronutrient utilization and CO_2drawdown (Buesseler & Boyd, 2003).

MECHANISM OF MICROBIAL ECO-REMEDIATION

Bioremediation of contaminated water can occur through innate processes performed by consortium of microbes (bacteria, yeast and fungi) or individually containing enzymes that allow them to utilize contaminants as a nutrient or energy source (Singh et al., 2014).

Biodegrading microbes may be indigenous to the site or exogenous species used for bioaugmentation. The genetic architecture of these microbial communities allows different mechanisms of biodegradation, biotransformation, biosorption and bioaccumulation of contaminants to occur in unison (Figure 3) (Pattanasupong et al., 2004; Nozari, Samaei, & Dehghani, 2014; Dixit et al., 2015; Pizarro-Tobías et al., 2015) involving a strong metal-microbe interaction pathways to strengthen the bioremediation capacity (Figure 4).

Figure 3. Different bioremediation mechanisms of microbe at different metabolic states

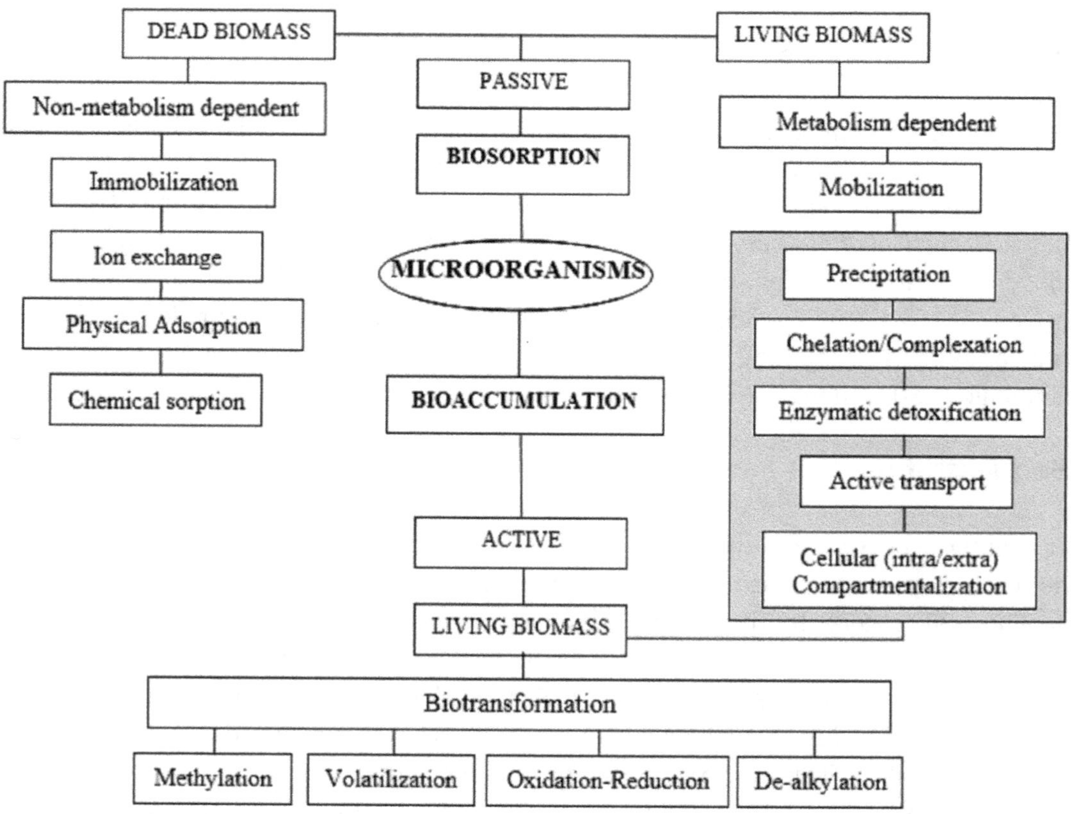

Figure 4. Diagram representing the various mechanisms contributing to microbial uptake and detoxification of toxic metals
https://www.omicsonline.org/role-of-rhizobacteria-in-reduction-of-arsenic-uptake-by-plants-a-review-2155-6199.1000126.php?aid=1950

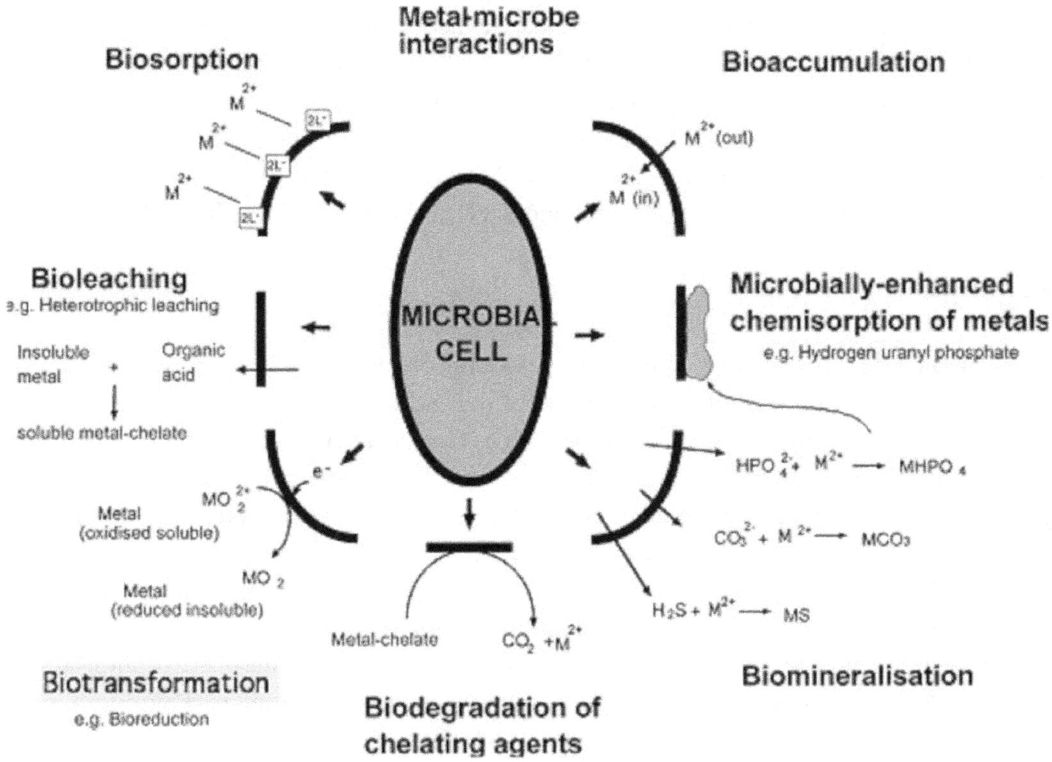

Biosorption

Biosorption means uptake by living or dead biomass via physico-chemical mechanisms such as adsorption or ion exchange process. The term sorption would refer to binding of a metal cation to a free site as opposed to one that was previously occupied by another cation through processes such as metal precipitation, complexation by siderophores and other metabolites, chelation by metal-binding proteins and peptides, transport and intracellular compartmentation. In addition, transformations of metal species can occur resulting in oxidation, reduction or methylation which is actively pH dependent.

Immobilization

Microorganisms can efficiently immobilize metal contaminants by reducing the mobility of contaminants through physically restricting contact between the contaminant or by chemically altering the contaminant (Evanko & Dzombak, 1997) precipitating as hydroxides carried out by chemical and bacterial reagents assist with the immobilization of metal contaminants. Metal contaminated sites are mixed or injected with chemical agents as a solidification approach to immobilize metals, whereas, in case of stabilization, the contaminated material is mixed with appropriate amounts of stabilizer material and water forming a solidified matrix with the waste. Volatile or semi volatile organics are treated with *in-situ* immobilization (Rani, Hemambika, Hemapriya, & Kannan, 2010).

Different bioreactors such as rotating biological contactors, fixed bed reactors, trickle filters, fluidized beds and air-lift bioreactors use packed immobilized living biomass which usually takes the form of biofilms on variable inert materials. For non-selective metal uptake *Bacillus* sp. from industrial waste was chemically treated with alkali, cross-linked, pelleted by extrusion or milling and dried to provide an indefinite shelf life and finally used to uptake Cd, Cr, Cu, Hg, Ni, Pb, U and Zn (singly or mixed) at a range of concentrations. Immobilized cells of *Rhodobacter sphaeroides, Alcaligenes eutrophus and Saccharomyces cerevisiae* have been found to uptake Cd and Pb and *Bacillus lentus, Aspergillus oryzae, Kluyceromyces marxiauns, Candida* spp. for Cr and Cu (Das, Vimala, & Karthika, 2008).

Ion Exchange

Ion-exchange is a process of exchange of metal ions with the counter bivalent metal ions of the microbial cell wall polyssacarides. For example, the alginates of marine algae occur as salts of K^+, Na^+, Ca^{2+} and Mg^{2+}. These ions can exchange with counter ions such as Co^{2+}, Cu^{2+}, Cd^{2+} and Zn^{2+}.The precise binding mechanism(s) may range from physical (i.e. electrostatic or London–van der Waals forces) to chemical binding (i.e. ionic, covalent and redox interaction). So the binding mechanism involves both physical and chemical adsorption processes whose capacity has close relationships with physical and chemical properties of cell surface, including specific surface area, cell wall volume, nitrogen-to-carbon (N:C) ratio, hydrophobicity, static force, hydrogen bonding and functional groups.

Cation binding ionizable functional groups involved in metal ion-exchange or adsorption are carboxyl, organic phosphate and organic sulfate whereas, anion exchange on biopolymers will take place on a variety of organic-nitrogen-based groupings. Hence, ion exchange mechanism is principally extracellular or occurring discretely at the cell wall.

The biosorption of copper by fungi *Ganoderm alucidium* and *Aspergillus niger*, high metal binding capacity of *Bacillus* genus are attributable to its significant negative charge density due to the attachment of teichoic and teichuronic acid to the peptidoglycan network of the cell wall. (Foucher, Battaglia-Brunet, Ignatiadis, & Morin, 2001).

Bioaccumulation

Microbial bioaccumulation is an energy-dependent contaminant transport system that includes metal pollutant influx across the bacterial membranes, ion pumps, ion channels, carrier mediated transport, endocytosis, complex permeation, and lipid permeation (Figure 4). Microbial extracellular polymeric substances (EPS) forming capsules or slime layers can bioaccumulate metals by a close association of their polysaccharide, glycoproteins and lipopolysaccharide with protein (Geesey & Jang, 1990) and deposit them in a chemically-altered form (Beveridge & Doyle, 1989).

Transmission electron microscopy (TEM) analysis of *Pseudomonas putida* 62BN demonstrated intracellular and periplasmic accumulation of cadmium (Rani & Goel, 2009). Heavy metal transport in many bacterial genera like *Citrobacter* spp. (lead and cadmium), *Thiobacillus ferrooxidans* (silver), *Bacillus cereus* (cadmium), *Bacillus subtilis* (chromium), *Pseudomonas aeruginosa* (uranium), *Micrococcus luteus* (strontium), *Rhizopus arrhizus* (mercury), *Aspergillus niger* (thorium), *Saccharomyces cerevisiae* (uranium) have been reported (Umrania, 2006; Ahemad & Malik, 2011).

Figure 5. Energy transfer in oxidation-reduction reaction by microbes
https://www.nap.edu/read/2131/chapter/4

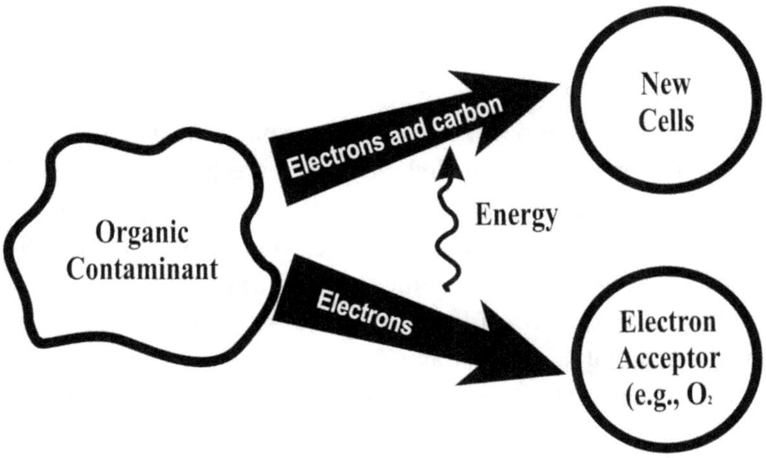

Biotransformation

Breaking of chemical bonds and transferring electrons away from the contaminant-electron donor and acceptance by microorganisms, thus gaining energy through these oxidation-reduction reactions. The energy gained is then "invested," along with some electrons and carbon from the contaminant, to produce more cells (Figure 5).

Abilities of these microbes can be enhanced through the addition of certain nutrients which promote proliferation of beneficial bacterial species or addition of terminal electron acceptors/donors enabling them to convert contaminants into non-hazardous or less hazardous forms through oxidation or reduction (Aulenta et al., 2009; Aracic et al., 2015). For microbial metalloid transformations reduction of selenate [Se (VI)] and selenite [Se (IV)] to elemental selenium is well known (Lovley & Lloyd, 2000).

Only prokaryotes can oxidize metals such as Fe^{2+}, Cu, Mn^{2+}, Co^{2+}, AsO^{2-}, SeO or SeO_3^{2-} can reduce Mn^{4+}, Fe^{3+}, Co^{3+}, AsO_4^{2-}, SeO_4^{2-} or SeO_3^{2} to obtain energy from the reactions. When Fe^{3+} or Mn^{4+} ion is reduced to Fe^{2+} or Mn^{2+}, this increases the solubility of the metals. Microbial species can efficiently reduce heavy metal ions to a lower oxidation state, and giving rise to metallic elements (load zero) which are less bioactive (Figure 4).

Microbial dealkylation of organometallic compounds such as organotins can result in the formation of ionic species which could possibly be removed using a biosorptive process (Ghani et al., 1993). Methylation involves methyl groups (CH_3) incorporation that are enzymatically transferred to a metal ions like Hg, As, Cd, Se, Sn, Te and Pb forming volatile methylated compounds resulting decrease in their toxicities. Bacterial species of *Aeromonas* spp., *Bacillus* spp. and *Pseudomonas* spp. produce volatile methyl derivatives of selenium such as dimethylselenide from SeO_4^{2-} and SeO_3^{2-}(Karlson & Frankenberger, 1993).

These innate abilities can be further enhanced through genetic modification of regulatory and/or metabolic genes (Dixit et al., 2015) following gene expression study through metagenomic and metatranscriptomic analyses of microbial communities in wastewater (Yergeau, Sanschagrin, Beaumier, &

Greer, 2012) as has been done for nitrification, denitrification, ammonification and nitrogen fixation processes (Yu & Zhang, 2012).

Mobilization

Metal mobilization by microorganisms occurs through autotrophic and heterotrophic leaching, chelation by microbial metabolites and siderophores, methylation, and redox transformations. Heterotrophic leaching is when microorganisms can acidify their environment by proton efflux thus leading to the acidification resulting in the release of free metal cations. Redox transformations can allow microorganisms to mobilize metals, metalloids, and organo-metallic compounds by reduction and oxidation processes.

Stock cultures of *Thiobacillus ferrooxidans* and *Thiobacillus thiooxidans* were added to fly-ash suspensions and it was found that significant amounts of copper were mobilized, in contrast to other elements, by metabolic products of *T. ferrooxidans* (Brandl, Bosshard, & Wegmann, 1999).

Precipitation

Living bacteria produce and/or excrete a substance that chemically reacts with metals present in solution (metal ions) to produce an insoluble metal compound by the process of bioprecipitation of metals. Soluble metal species and radionuclides are transformed to insoluble hydroxides, carbonates, phosphates and sulfides.

Microbial hydrogen sulfide gas production is widespread and confers resistance through the precipitation of insoluble metal sulfides (Figure 6). Enzymatically produced phosphate for example, *Citrobacter* spp. has been shown to precipitate and detoxify lead and copper (Crawford & Crawford, 2005).

Figure 6. Indirect mobilization of heavy metals and radionuclides by (a) metal reducing and (b) sulfate reducing bacteria
Tabak, HLens, van Hullebusch, & Dejonghe, 2005.

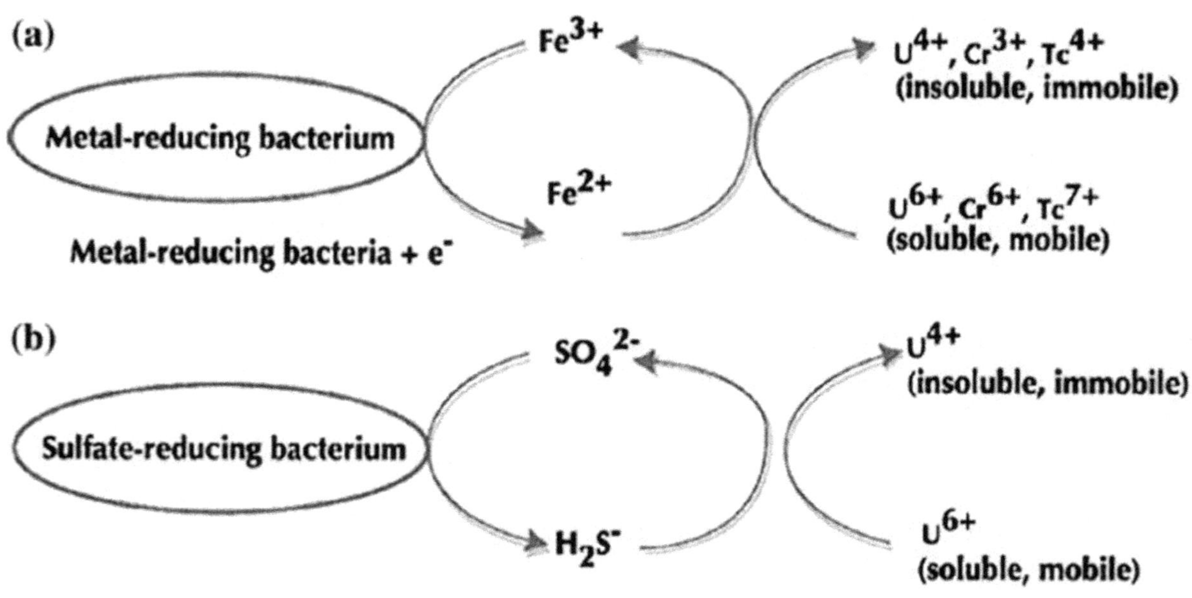

Complexation

Complex formation of metal ions with organic molecules involves ligand centers in the organic species i.e. the presence of an atom or atoms having lone pair electrons to donate. In order to elucidate biosorption mechanisms, metals have been classified into hard and soft according to Pearson's classification. Hard metals such as Na, K, Ca, Mg are often essential nutrients for microbial growth, they bind preferentially to oxygen-containing "hard" ligands such as OH^-, HPO_4^{2-}, CO_3^{2-}, R-COO^- and =C=O. Soft metals such as the precious metals Ag, Au, Pt, Pd are bound covalently to the cell wall by "soft" ligands containing nitrogen or sulfur. Borderline ions Mn^{2+}, Zn^{2+}, Cd^{2+}, Cu^{2+}, Pb^{2+} exhibit a significant degree of covalent bonding with nitrogen or sulfur containing "soft" ligands (Lin, Zhou, Wu, Zhou, & Wang, 2005; Park, Yun, Jo, & Park, 2005). Microorganisms' produced organic acids like citric, oxalic, fumaric, gluonic, lactic and malic acids, form complexation of toxic metals resulting metallo-organic molecules formation which solubilize metal compounds and lead to leach from their surfaces.

Enzymatic Detoxification

Microbial enzyme catalyzed detoxification of contaminants involve metabolic reactions of hydrolysis, oxidation, addition of an oxygen to a double bound, oxidation of an amino group (NH_2) to a nitro group, addition of a hydroxyl group to a benzene ring, dehalogenation, reduction of a nitro group (NO_2) to an amino group, replacement of a sulfur with an oxygen, metabolism of side chains, ring cleavage.

The detoxification of pesticides involve initial transformation of the properties of parent compound to a less toxic product through oxidation, reduction, or hydrolysis or through conjugation to a sugar or amino acid thus increasing the water solubility of the end product. All these fungi and bacteria mediated reactions are catalyzed by intracellular or extra cellular enzymes including hydrolytic enzymes, peroxidases, oxygenases, etc. (Van Eerd, Hoagland, Zablotowicz, & Hall, 2003). Hydrolases catalyze the hydrolysis of several major biochemical classes of pesticide (esters, peptide bonds, carbon-halide bonds, ureas, thioesters etc.) and generally operate in the absence of redox cofactors, making them ideal candidates for all of the current bioremediation strategies (Scott et al., 2008). The initial step in the aerobic catabolism of a PAH molecule occurs via oxidation of the PAH to a dihydrodiol by a multi-component enzyme system. The degradation pathway of carbofuran by hydrolase is represented in Figure 7:

The phosphotriesterases (PTE) encoded by a gene called *opd* (organophosphate-degrading) reduces the toxicity of organophosphate (OP) by decreasing the ability of OPs to inactivate *Ach*E. These enzymes specifically hydrolyze phosphoester bonds, such as P–O, P–F, P–NC, and P–S, and the hydrolysis mechanism involves a water molecule at the phosphorus center (Kapoor & Rajagopal, 2011).

Active Transport

Bioremediation through microbial active transport system involves a 'pump' system across the cell membrane resulting in the efflux of metal ions. Toxic metal ions that have entered the cell rapidly can be pumped out by these highly specific efflux systems deriving their energy from membrane potential or from ATP which may be specific for As, Cd and Cr (Crowford & Crowford, 2005). For cadmium, a plasmid gene, *cad*A, encodes a cadmium specific ATPase that transports cadmium out of cell found in *Staphylococcus aureus*, *Pseudomonas putida*, *E. coli* and *Bacillus subtilis*.

Figure 7. Microbial degradation of carbofuran by producing hydrolase enzyme
University of Minnesota. Biocatalysis/Biodegradation Database (http://www.umbbd.ethz.ch/cbf/cbf_image_map1.html).

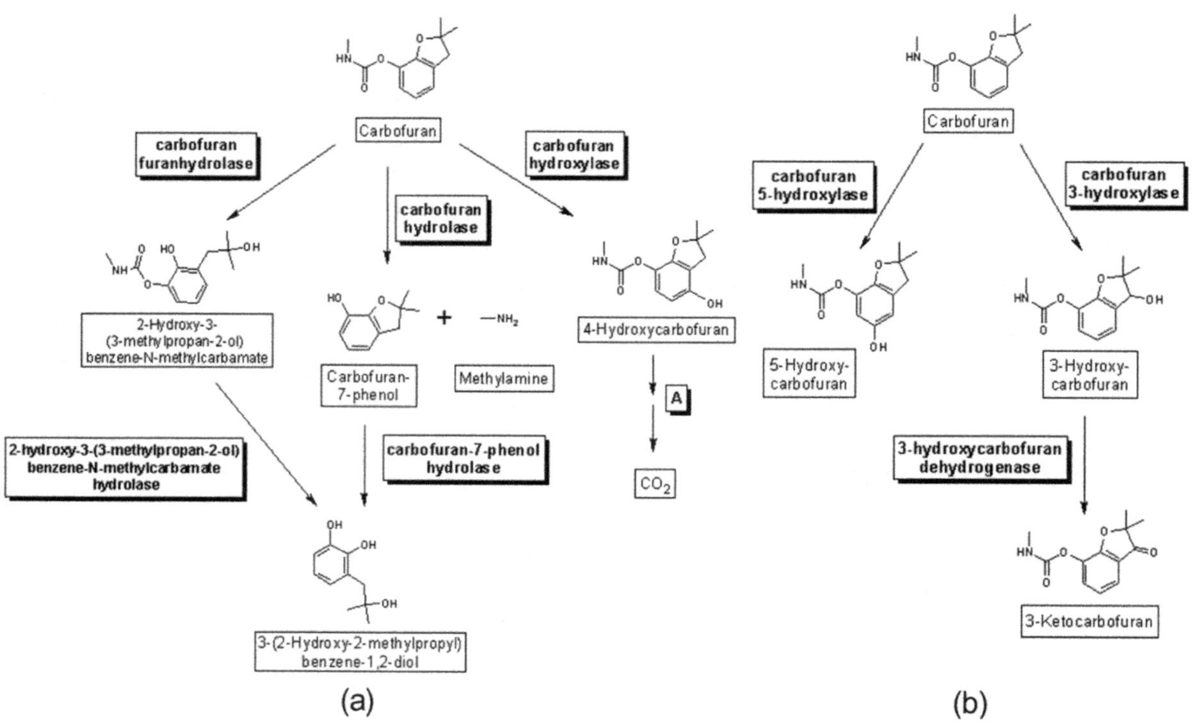

As a CO_2 sequestration measure, microbes transform the labile DOC to refractory DOC which is harder to digest and thus accumulate in a huge carbon reservoir in the 'microbial carbon pump' system (Jiao, 2013). Many marine heterotrophic prokaryotes produce extracellular polymeric substances (EPS) or transparent exopolymer particles (TEP) which cohere and interconnect cells, forming aggregates which then contribute to POC flux. Microbes colonize sinking aggregates, and can grow by means of exoenzymatic decomposition of the aggregated organic particles, which in turn could lead to a DOC plume following the sinking aggregate (Jiao et al., 2014).

Cellular Compartmentalization

There are few bacteria that developed a unique cytosolic sequestration mechanism to protect them from metal toxicity. In this process, after entering inside the bacterial cell, metal ions become compartmentalized in the cell's periplasm and the outer membrane and finally converted into more innocuous forms resulting high metal accumulation (Ahemad, 2012). Lin & Olson (1995) observed among 62% of the total copper resistant isolates 49% isolates had *cop* or *cop*-like gene systems for both compartmentalization and efflux systems. The protective mechanism against copper in a copper resistant plant pathogen *Pseudomonas syringae* was due to four types of proteins CopA and CopC (periplasm), CopB (outer membrane) and CopD (inner membrane) encoded by the *cop* operon in bacterial plasmid to compartmentalize copper away from bacterial cells (Spain, 2003).

Chelation

When more than one atom binds to a central metal ion at one time to form a ring structure, it is known as chelation. Since a chelating agent may bind to a metal ion in more than one place simultaneously, chelates are more stable than complexes involving unidentate ligands such as histidine and citrate that can bind metal ions making them inert. A variety of heavy metals can induce or accelerate melanin production in fungi and melanized cell forms, e.g. Chlamydospores can have high metal uptake capacities for metals and organometallic compounds, e.g. tributyltin chloride.

Siderophores are specific low molecular weight iron chelating ligands and are able to bind to other metals, such as magnesium, manganese, chromium, cobalt, gallium and radionuclide, such as plutonium (Figure 8) that are released under iron-limiting conditions by microorganisms, particularly bacteria and fungi, to enable accumulation of iron from the environment (Bellenger, Wichard, Kustka, & Kraepiel, 2008; Braud, Hoegy, Jezequel, Lebeau, & Schalk, 2009). The formation of Fe (III)–siderophore complexes are affected by pH because of the competition for the free siderophore ligands between free protons and Fe (Albrecht-Gary & Crumbliss, 1998).

Combining metagenomics with detailed chemical analysis will reveal important information that could be used to improve the current environmental applications and develop new applications for siderophores.

Figure 8. Siderophore mediated Pu accumulation by Microbacterium flavescens
John, 2001.

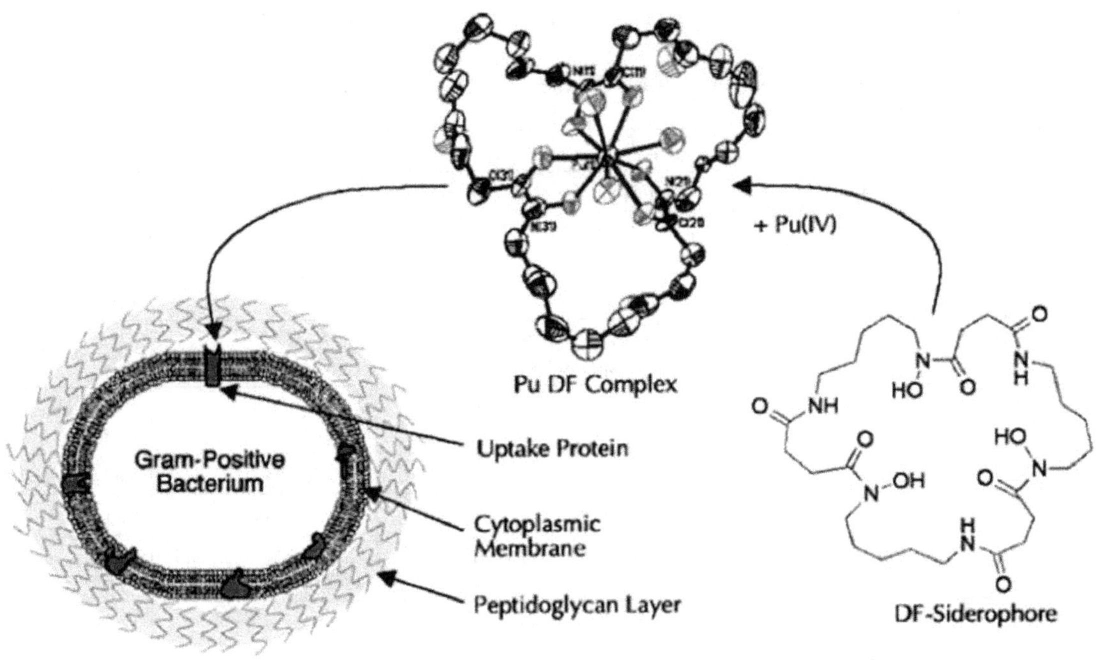

ECO-REMEDIATION EFFICIENCIES

The biodegradation of the herbicide Aroclor 1242 with a degradation percentage of 99.8, 89.4 and 98.4 respectively by three *Pseudomonas* species have been reported. Metal loadings are up to > 10% of the dry weight, giving a removal efficiency of > 99% and effluents with total metal concentrations around 1 and 50 ppb. The biosorption of immobilized cells of *Bacillus* spp. was 69.34% of Cu, *Pseudomonas* spp. was 90.41% of Cd and *Micrococcus* spp. was 84.27% of Pb, whereas the dead cells of *Bacillus* spp. was 44.73% of Cu, *Pseudomonas* spp. was 86.66% of Cd and *Micrococcus* spp. was 79.22% (Rani et al., 2010). Continuous cultures of Hg^{2+} resistant bacteria, which can reduce Hg^{2+} to Hg^0 with mercuric reductase, volatilized Hg^0 from contaminated sewage at a rate of 2.5 mg/h (98% removal). Table 3 shows the heavy metal removal efficiencies of genetically engineered bacteria collected from literature.

The removal efficiency of microbes is more functional if the strain is applied after isolation from previously contaminated site than application after experimentally testing for their pollutant tolerance capacity followed by isolation. This is because unscientific nutrient addition can aid heterotrophic population thus inadvertently triggering an antagonistic situation thereby limiting the degradation process. Therefore, some authors investigated a combination approach to further enhance the potential of a more robust technology.

Table 3. Pollutant removal efficiencies of genetically engineered bacteria (GMM)

Pollutants	Removal Efficiency (%)	Genetically Engineered Bacteria	Expressed Gene	Ref.
Arsenic	100	*E. coli* strain	ArsR protein	Kostal et al., 2004
Chromium	100	*Methylococcuscapsulatus*	CrR	Hasin et al., 2010
Mercury	96	*E. coli* JM109	Hg^{2+} transporter	Zhao et al., 2005
Nickel	80	*Pseudomonas fluorescens* 4F39	Phytochelatin synthase (PCS)	Lopez et al., 2002
Organophosphate	63 59	*Providencia* spp. BGW4, *Delftia* spp. BGW1	*opd*A	Ahire et al., 2011
Hydrocarbon	50	*Pseudomonas aeruginosa*(UG2)	*alk*B	Scheibenbogen et al., 1994
Naphthalene	40	*Pseudomonas fluorescens* HK44	*lux*CDABE	Sayler&Ripp, 2000
Toluene	70	*Burkholderiacepacia* L.S.2.4	pTOD plasmid	Barac et al., 2004
Nitrate nitrogen	97	*Pseudomonas stutzeri* ZF31	*nap*A, *nir*S, *nor*B, and *nos*Z.	Huang et al., 2015
Total nitrogen	99.6	*Geobacillusthermodenitrificans* 465T	*nos*gene cluster.	Yang et al., 2013

ADVANCED MICROBIAL TECHNOLOGIES AS ENVIRONMENTAL MANAGEMENT TOOL

Metagenomics and metatranscriptomics studies have been undertaken to indicate the presence of xenobiotic contaminants in the aquatic environment and link between the abundance of degradation genes and biodegradation rates (Yu & Zhang, 2012).The blueprint necessary for gene encoding for biodegradative enzymes is present in chromosomal and extra chromosomal DNA of such microbes (Fulekar, 2009) which leads to the generation of recombinant cells of Genetically Modified Microorganisms (GMM) through usual procedure of gene selection, DNA fragment insertion followed by cloning and finally screening for clones with desired properties.

In microorganism mediated CO_2 sequestration multiple factors such as optimization of culture conditions and concentration of CO_2 still have to be optimized for maximum rate of CO_2 fixation by these microorganisms. The use of metabolic engineering to develop a combined system of al wastewater treatment and biofuel production within a biorefinery is still in its infancy. The design of photobioreactors in the case of phototrophs is also an issue to be addressed and tested for cost-effective CO_2 sequestration.

Recent research has driven the growing interest for monitoring water quality utilizing microbial biosensors to detect phenols, non-ionic surfactants, pesticides, heavy metals, heavy nutrient load in domestic wastewater, natural waters and industrial effluents (Girotti, Ferri, Fumo, & Maiolini, 2008; Hnaien, Lagarde, Bausells, Errachid, & Jaffrezic-Renault, 2011). A measurable signal such as growth or enzymatic kinetics is produced by microbial whole-cell biosensors enabling detection and easy quantification of specific contaminants. Quantitative microbial risk assessment (QMRA) for pathogen contamination (Ramírez-Castillo, 2015), direct correlation of total phosphorus with abundance of benthic algae (Rott & Schneider, 2014), significant structural changes of cyanabacteria detecting absence of nitrogen or phosphorus (Whitton & Potts, 2007) and generation of electrical current or the luminescence in response to pollutants.

In microbial fuel cell systems, the measurement of electron flow from central metabolism can be used to detect changes of metabolic activities in an aquatic ecosystem over time (Aracic et al., 2014). Electro-active *Pseudomonas putida* cells immobilization on electrodes using carbon nanotubes increased response time to trichloroethylene about 2.8-fold (Hnaien et al., 2011).

CONCLUSION AND FUTURE PROSPECTS

In the light of above discussion it can be concluded that microbial eco-remediation has a great potential for sequestering heavy metal ions, nutrient load, organic pollutants, greenhouse gas from aquatic environment. Microbial technologies have of late been applied to a range of environmental problems, with considerable success. Presently, pollutants specific several novel microbial species have been discovered by researchers using advanced biotechnological approaches viz., Metagenomic culture-independent approaches to removal excess elements and toxic contaminants (Bai et al., 2014) in order to restore the perturbed environment resulting in marked remediation efficiencies.

Moreover, implication of microbial eco-remediation technology in practical field is still insufficient to restore the perturbed ecosystems due to lack of proper field oriented framework of effective and feasibly cost-effective methods. Besides, proper integration of potential microbial community as an ecological tool in designing the engineered ecosystem (Bhakta & Minekage, 2008) applying the ecological engineering approach is still in infancy stage. Significantly potential research in this concern that would play a pivotal role in recovering and restoring the perturbed ecosystems has potential future prospects. Microbial eco-remediation has potential implication in restoration of ecosystem contaminated by specific pollutants by designing specific engineering based ecological systems exploring and employing specific microbial components.

Present chapter explores the vast array of scope of microbial eco-remediation which may substitute chemical intensive conventional methods and directs towards eco-friendly greener technologies for environmental remediation.

REFERENCES

Abed, R. M. M., & Köster, J. (2005). The direct role of aerobic heterotrophic bacteria associated with cyanobacteria in the degradation of oil compounds. *International Biodeterioration & Biodegradation*, *55*(1), 29–37. doi:10.1016/j.ibiod.2004.07.001

Aguilera, F., Méndez, J., Pásaro, E., & Laffon, B. (2010). Review on the effects of exposure to spilled oils on human health. *Journal of Applied Toxicology*, *30*(4), 291–301. PMID:20499335

Ahemad, M. (2012). Implications of bacterial resistance against heavy metals in bioremediation: A review. *Institute of Integrative Omics and Applied Biotechnology Journal*, *3*, 39–46.

Ahemad, M., & Malik, A. (2011). *Bioaccumulation of heavy metals by Zinc resistant bacteria isolated from agricultural soils irrigated with waste water*. Bacteriology Journal; doi:10.3923/bj.2011

Ahire, K. C., Kapadnis, B. P., Kulkarni, G. J., Shouche, Y. S., & Deopurkar, R. L. (2012). Biodegradation of tributyl phosphate by novel bacteria isolated from enrichment cultures. *Biodegradation*, *23*(1), 165–176. doi:10.1007/s10532-011-9496-7 PMID:21755325

Albrecht-Gary, A. M., & Crumbliss, A. L. (1998). Coordination chemistry of siderophores: thermodynamics and kinetics of iron chelation and release. *Metal Ions in Biological Systems, 35*(2), 39–327.

Anderson, T. R., & Tang, K. W. (2010). Carbon cycling and POC turn-over in the mesopelagic zone of the ocean: Insights from a simple model. *Deep-sea Research. Part II, Topical Studies in Oceanography*, *57*(16), 1581–1592. doi:10.1016/j.dsr2.2010.02.024

Aquino, N. B., Sevigny, M. B., Sabangan, J., & Louie, M. C. (2012). The role of cadmium and nickel in estrogen receptor signaling and breast cancer: Metalloestrogens or not? *Journal of Environmental Science and Health. Part C, Environmental Carcinogenesis & Ecotoxicology Reviews*, *30*(3), 189–224. doi:10.1080/10590501.2012.705159 PMID:22970719

Aracic, S., Manna, S., Petrovski, S., Wiltshire, J. L., Mann, G., & Franks, A. E. (2015). Innovative biological approaches for monitoring and improving water quality. *Frontiers in Microbiology*, *6*, 826. doi:10.3389/fmicb.2015.00826 PMID:26322034

Aracic, S., Semenec, L., & Franks, A. E. (2014). Investigating microbial activities of electrode-associated microorganisms in real-time. *Frontiers in Microbiology*, *5*, 663. doi:10.3389/fmicb.2014.00663 PMID:25506343

Aulenta, F., Canosa, A., Reale, P., Rossetti, S., Panero, S., & Majone, M. (2009). Microbial reductive dechlorination of trichloroethene to ethene with electrodes serving as electron donors without the external addition of redox mediators. *Biotechnology and Bioengineering*, *103*(1), 85–91. doi:10.1002/bit.22234 PMID:19160378

Azam, F., & Malfatti, F. (2007). Microbial structuring of marine ecosystems. *Nature Reviews. Microbiology*, *5*(10), 782–791. doi:10.1038/nrmicro1747 PMID:17853906

Bai, Y., Qi, W., Liang, J., & Qu, J. (2014). Using high-throughput sequencing to assess the impacts of treated and untreated wastewater discharge on prokaryotic communities in an urban river. *Applied Microbiology and Biotechnology*, *98*(4), 1841–1851. doi:10.1007/s00253-013-5116-2 PMID:23912119

Barac, T., Taghavi, S., Borremans, B., Provoost, A., Oeyen, L., Colpaert, J. V., & van-der-Lelie, D. et al. (2004). Engineered endophytic bacteria improve phytoremediation of water-soluble, volatile, organic pollutants. *Nature Biotechnology*, *22*(5), 583–588. doi:10.1038/nbt960 PMID:15077119

Bardgett, R. D., & van der Putten, W. H. (2014). Belowground biodiversity and ecosystem functioning. *Nature*, *515*(7528), 505–511. doi:10.1038/nature13855 PMID:25428498

Basu, A., Sen, P., & Jha, A. (2015). Environmental arsenic toxicity in West Bengal, India: A brief policy review. *Indian Journal of Public Health*, *59*(4), 295–298. doi:10.4103/0019-557X.169659 PMID:26584169

Bellenger, J. P., Wichard, T., Kustka, A. B., & Kraepiel, A. M. L. (2008). Uptake of molybdenum and vanadium by a nitrogen-fixing soil bacterium using siderophores. *Nature Geoscience*, *1*(4), 243–246. doi:10.1038/ngeo161

Beveridge, T. J., & Doyle, R. J. (1989). *Metal Ions and Bacteria*. New York: J Wiley.

Bhakta, J. N., Jana, B. B., Lahiri, S., Panigrahi, A., Mandal, S. K., Rana, S., & Rokunuzzaman, M. et al. (2016). Ecological restoration: An emerging eco-technology for sustainable environmental conservation. *Int J Env Tech Sci*, *2*, 21–30.

Bhakta, J. N., Lahiri, S., Pitman, J. K., & Jana, B. B. (2015). Carbon dioxide sequestration in wastewater by a consortium of elevated carbon dioxide-tolerant microalgae. *Journal of CO_2 Utilization, 10,* 105–112.

Bhakta, J. N., & Munekage, Y. (2008). Role of ecosystem components in Cd removal process of aquatic ecosystem. *Ecological Engineering*, *32*(3), 274–280. doi:10.1016/j.ecoleng.2007.12.004

Bhakta, J. N., Munekage, Y., Ohnishi, K., & Jana, B. B. (2012a). Isolation and identification of cadmium and lead resistant lactic acid bacteria for applying as metal removing probiotic. *International Journal of Environmental Science and Technology*, *9*(3), 433–440. doi:10.1007/s13762-012-0049-3

Bhakta, J. N., Munekage, Y., Ohnishi, K., Jana, B. B., & Balcazar, J. L. (2014). Isolation and characterization of cadmium and arsenic absorbing bacteria for bioremediation. *Water, Air, and Soil Pollution*, *225*(10), 2151. doi:10.1007/s11270-014-2151-2

Bhakta, J. N., Ohnishi, K., Munekage, Y., & Iwasaki, K. (2010). Isolation and probiotic characterization of arsenic-resistant lactic acid bacteria for uptaking arsenic. *International Journal of Chemical and Biological Engineering*, *3*(4), 167–174.

Bhakta, J. N., Ohnishi, K., Munekage, Y., Iwasaki, K., & Wei, M. (2012b). Characterization of lactic acid bacteria-based probiotics as heavy metals sorbents. *Journal of Applied Microbiology, 112*(6), 1193–1206. doi:10.1111/j.1365-2672.2012.05284.x PMID:22404232

Bhattacharya, H. N., & Chakrabarti, S. (2011). Incidence of fluoride in the groundwater of Purulia District, West Bengal: A geo-environmental appraisal Scientific Report. *Current Science, 101*, 2.

Bittsanszkya, A., Kfmivesb, T., Gullnerb, G., Gyulaia, G., Kissa, J., Heszkya, L., & Rennenberg, H. et al. (2005). Ability of transgenic poplars with elevated glutathione content to tolerate zinc (2^+) stress. *Environment International, 31*(2), 251–254. doi:10.1016/j.envint.2004.10.001 PMID:15661291

Boldt, J. L., Martone, R., Samhouri, J., Perry, I., Itoh, S., Chung, I. K., & Yoshi, N. et al. (2014). Developing ecosystem indicators for responses to multiple stressors. *Oceanography (Washington, D.C.), 27*(4), 116–133. doi:10.5670/oceanog.2014.91

Bond, P. L., Erhart, R., Wagner, M., Keller, J., & Blackall, L. L. (1999). Identification of some of the major groups of bacteria in efficient and non-efficient biological phosphorus removal activated sludge systems. *Applied and Environmental Microbiology, 65*, 4077–4084. PMID:10473419

Boyd, P. W., Jickells, T., Law, C. S., Blain, S., Boyle, E. A., Buesseler, K. O., & Watson, A. J. et al. (2007). Mesoscale iron enrichment experiments 1993–2005: Synthesis and future directions. *Science, 315*(5812), 612–617. doi:10.1126/science.1131669 PMID:17272712

Brandl, H., Bosshard, R., & Wegmann, M. (1999). Computer-munching microbes: Metal leaching from electronic scrap by bacteria and fungi. *Process Metallurgy, 9B*, 569–576. doi:10.1016/S1572-4409(99)80146-1

Braud, A., Hoegy, F., Jezequel, K., Lebeau, T., & Schalk, I. J. (2009). New insights into the metal specificity of the *Pseudomonas aeruginosa* pyoverdine–iron uptake pathway. *Environmental Microbiology, 11*(5), 1079–1091. doi:10.1111/j.1462-2920.2008.01838.x PMID:19207567

Buesseler, K. O., & Boyd, P. W. (2003). Will ocean fertilization work? *Science, 300*(5616), 67–68. doi:10.1126/science.1082959 PMID:12677048

Buitenkamp, M., & Stintzing, R. A. (2008). *Europe's sanitation problem - 20 million Europeans need access to safe and affordable sanitation. Women in Europe for a Common Future*. Netherlands: WECF.

Cao, W., Zhang, H., Wang, Y., & Pan, J. (2012). Bioremediation of polluted surface water by using biofilms on filamentous bamboo. *Ecological Engineering, 42*, 146–149. doi:10.1016/j.ecoleng.2012.02.018

Carpenter, S. R., Stanley, E. H., & Vander-Zanden, M. J. (2011). State of the worlds freshwater ecosystems: Physical, chemical and biological changes. *Annual Review of Environment and Resources, 36*(1), 75–99. doi:10.1146/annurev-environ-021810-094524

Carr, G. M., & Neary, J. P. (2008). *Water Quality for Ecosystem and Human Health* (2nd ed.). United Nations Environment Programme Global Environment Monitoring System.

Celik, I., Gallicchio, L., Boyd, K., Lam, T. K., Matanoski, G., Tao, X., & Alberg, A. J. et al. (2008). Arsenic in drinking water and lung cancer: A systematic review. *Environmental Research, 108*(1), 48–55. doi:10.1016/j.envres.2008.04.001 PMID:18511031

Central Pollution Control Board (CPCB). (2008). Status of water quality in India 2007. New Delhi, India: Author.

Chakraborti, D., Das, B., Rahman, M. M., Chowdhury, U. K., Biswas, B., Goswami, A. B., & Das, D. et al. (2009). Status of groundwater arsenic contamination in the state of West Bengal, India: A 20-year study report. *Molecular Nutrition & Food Research*, *53*(5), 542–551. doi:10.1002/mnfr.200700517 PMID:19382148

Climate Change, Fisheries and Aquaculture (PaCFA). (2009). *Strategic framework for fisheries, aquaculture and climate change*. Author.

Chen, J., Sun, G. X., Wang, X. X., Lorenzo, V. C. D., Rosen, B. P., & Zhu, Y. G. (2014). Volatilization of arsenic from polluted soil by *Pseudomonas putida* engineered for expression of the arsM arsenic (III) S-adenosine methyltransferase gene. *Environmental Science & Technology*, *48*(17), 10337–10344. doi:10.1021/es502230b PMID:25122054

Cheung, W. W. L., Lam, V. W. Y., Sarmiento, J. L., Kearney, K., Watson, R., Zeller, D., & Pauly, D. (2009). *Re-distribution of fish catch by climate change: A summary of a new scientific analysis*. Global Change Biology, Pew Ocean Science Series.

Coelho, L. M., Rezende, H. C., de-Sousa, P. A. R., Melo, D. F. O., & Coelho, N. M. M. (2015). Bioremediation of Polluted Waters Using Microorganisms. In N. Shiomi (Ed.), *Advances in Bioremediation of Wastewater and Polluted Soil*. doi:10.5772/60770

Crawford, R. L., & Crawford, D. L. (Eds.). (2005). *Bioremediation: Principles and applications*. Cambridge University Press.

Cupples, A. M., Sanford, R. A., & Sims, G. K. (2005). Dehalogenation of Bromoxynil (3,5-Dibromo-4-Hydroxybenzonitrile) and Ioxynil (3,5-Diiodino-4-Hydroxybenzonitrile) by *Desulfitobacterium chlororespirans*. *Applied and Environmental Microbiology*, *71*(7), 3741–3746. doi:10.1128/AEM.71.7.3741-3746.2005 PMID:16000784

Das, N., Vimala, R., & Karthika, P. (2008). Biosorption of heavy metals - An overview. *Indian Journal of Biotechnology*, *7*, 159–169.

Datta, A. S., Chakrabortty, A., De-Dalal, S. S., & Lahiri, S. C. (2014). Fluoride contamination of ground water in West Bengal, India. *Research report. Fluoride*, *47*(3), 241–248.

De-Lorenzo, M. E., Scott, G. I., & Ross, P. E. (1999). Effects of the agricultural pesticides atrazine, deethylatrazine, endosulfan, and chlorpyrifos on an estuarine microbial food web. *Environmental Toxicology and Chemistry*, *18*(12), 2824–2835. doi:10.1002/etc.5620181224

Derraik, J. G. B. (2002). The pollution of the marine environment by plastic debris: A review. *Marine Pollution Bulletin*, *44*(9), 842–852. doi:10.1016/S0025-326X(02)00220-5 PMID:12405208

Diez, M. C. (2010). Biological Aspects involved in the degradation of organic pollutants. *Journal of Soil Science and Plant Nutrition*, *10*(3), 244–267. doi:10.4067/S0718-95162010000100004

Dixit, R., Malaviya, D., Pandiyan, K., Singh, U. B., Sahu, A., & Shukla, R. (2015). Bioremediation of heavy metals from soil and aquatic environment: An over view of principles and criteria of fundamental processes. *Sustainability*, *7*(2), 2189–2212. doi:10.3390/su7022189

Ducklow, H. W., Steinberg, D. K., & Buesseler, K. O. (2001). Upper ocean carbon export and the biological pump. *Oceanography (Washington, D.C.)*, *14*(4), 50–58. doi:10.5670/oceanog.2001.06

Eastmond, D. A., MacGregor, J. T., & Slesinski, R. S. (2008). Trivalent Chromium: Assessing the Genotoxic Risk of an Essential Trace Element and Widely Used Human and Animal Nutritional Supplement. *Critical Reviews in Toxicology*, *38*(3), 173–190. doi:10.1080/10408440701845401 PMID:18324515

El-Shahawi, M. S., Hamza, A., Bashammakhb, A. S., & Al-Saggaf, W. T. (2010). An overview on the accumulation, distribution, transformations, toxicity and analytical methods for the monitoring of persistent organic pollutants. *Talanta*, *80*(5), 1587–1597. doi:10.1016/j.talanta.2009.09.055 PMID:20152382

Elekwachi, C. O., Andresen, J., & Hodgman, T. C. (2014). Global use of bioremediation technologies for decontamination of ecosystems. *Journal of Bioremediation and Biodegradation*, *5*, 1–9. doi:10.4172/2155-6199.1000225

Evanko, C. R., & Dzombak, D. A. (1997). Remediation of metals-contaminated soil and groundwater. *Environmental Sciences*, *412*, 1–45.

Fayer, R., & Trout, J. M. (2005). Zoonotic protists in the marine environment. In S. Belkin & R. R. Colwell (Eds.), *Oceans and Health: Pathogens in the Marine Environment* (pp. 143–163). Springer. doi:10.1007/0-387-23709-7_7

Foucher, S., Battaglia-Brunet, F., Ignatiadis, I., & Morin, D. (2001). Treatment by sulfate-reducing bacteria of Chessy acid-mine drainage and metals recovery. *Chemical Engineering Science*, *56*(4), 1639–1645. doi:10.1016/S0009-2509(00)00392-4

Free, C. M., Jensen, O. P., Mason, S. A., Eriksen, M., Williamson, N. J., & Boldgiv, B. (2014). High-levels of microplastic pollution in a large, remote, mountain lake. *Marine Pollution Bulletin*, *85*(1), 156–163. doi:10.1016/j.marpolbul.2014.06.001 PMID:24973278

Fu, F., & Wang, Q. (2011). Removal of heavy metal ions from wastewaters: A review. *Journal of Environmental Management*, *92*(3), 407–418. doi:10.1016/j.jenvman.2010.11.011 PMID:21138785

Fulekar, M. H. (2009). Bioremediation of fenvalerate by *Pseudomonas aeruginosa* in a scale up bioreactor. *Romanian Biotechnological Letters*, *14*(6), 4900–4905.

Galassi, D. M. P., Lombardo, P., Fiasca, B., Di-Cioccio, A., Di-Lorenzo, T., Petitta, M., & Di-Carlo, P. (2014). Earthquakes trigger the loss of groundwater biodiversity. *Scientific Reports*, *4*, 6273. doi:10.1038/srep06273 PMID:25182013

Gambhir, R. S., Kapoor, V., Nirola, A., Sohi, R., & Bansal, V. (2012). Water Pollution: Impact of Pollutants and New Promising Techniques in Purification Process. *Journal of Human Ecology (Delhi, India)*, *37*(2), 103–109.

Gan, S., Lau, E. V., & Ng, H. K. (2009). Remediation of soils contaminated with polycyclic aromatic hydrocarbons (PAHs). *Journal of Hazardous Materials*, *172*(2-3), 532–549. doi:10.1016/j.jhazmat.2009.07.118 PMID:19700241

Geesey, G., & Jang, L. (1990). Extracellular polymers for metal binding. In H. L. Ehrlich & C. L. Brierley (Eds.), *Microbial Mineral Recovery* (pp. 223–247). New York: Mc-GrawHill.

Ghani, B., Takai, M., Hishani, N. Z., Kishimoto, N., Ismail, A. K., Tano, T., & Sugio, T. (1993). Isolation and Characterisation of a Mo^{6+}-reducing bacterium. *Applied and Environmental Microbiology*, *59*, 1176–1180. PMID:16348915

Girotti, S., Ferri, E. N., Fumo, M. G., & Maiolini, E. (2008). Monitoring of environmental pollutants by bioluminescent bacteria. *Analytica Chimica Acta*, *608*(1), 2–29. doi:10.1016/j.aca.2007.12.008 PMID:18206990

Gonick, H. C. (2008). Nephrotoxicity of cadmium & lead. *The Indian Journal of Medical Research*, *128*, 335–352. PMID:19106433

Gray, C. J., & Engel, A. S. (2013). Microbial diversity and impact on carbonate geochemistry across a changing geochemical gradient in a karst aquifer. *International Society for Microbial Ecology*, *7*, 325–337. PMID:23151637

Gregory, K. B., & Lovley, D. R. (2005). Remediation and recovery of uranium from contaminated subsurface environments with electrodes. *Environmental Science & Technology*, *39*(22), 8943–8947. doi:10.1021/es050457e PMID:16323798

Grizzetti, B., Bouraoui, F., & Aloe, A. (2012). Changes of nitrogen and phosphorus loads to European seas. *Global Change Biology*, *18*(2), 769–782. doi:10.1111/j.1365-2486.2011.02576.x

Guo, H., Luoa, S., Chena, L., Xiaoa, X., Xi, Q., Wei, W., & Yejuan, H. et al. (2010). Bioremediation of heavy metals by growing hyper-accumulator endophytic bacterium *Bacillus* spp. L14. *Bioresource Technology*, *101*(22), 8599–8605. doi:10.1016/j.biortech.2010.06.085 PMID:20637605

Haldar, A., & Ray, S. K. (2014). Arsenic and Fluoride Problems of Groundwater in West Bengal and Available Technologies for Remediation. *International Journal of Innovative Research in Science*, *3*(6).

Haritash, A. K., & Kaushik, C. P. (2009). Biodegradation aspects of polycyclic aromatic hydrocarbons (PAHs): A review. *Journal of Hazardous Materials*, *169*(1-3), 1–15. doi:10.1016/j.jhazmat.2009.03.137 PMID:19442441

Harter, T. (2003). *Groundwater Quality and Groundwater Pollution*. University of California, Division of Agriculture and Natural Resources, Publication 8084.

Harvell, C. D., Kim, K., Burkholder, J. M., Colwell, R. R., Epstein, P. R., Grimes, D. J., & Vasta, G. R. et al. (1999). Emerging marine diseases: Climate links and anthropogenic factors. *Science*, *285*(5433), 1505–1510. doi:10.1126/science.285.5433.1505 PMID:10498537

Hasin, A. A., Gurman, S. J., Murphy, L. M., Perry, A., Smith, T. J., & Gardiner, P. E. (2010). Remediation of chromium (VI) by a methane-oxidizing bacterium. *Environmental Science & Technology*, *44*(1), 400–405. doi:10.1021/es901723c PMID:20039753

Haszeldine, R. S. (2009). Carbon capture and storage: How green can black be? *Science*, *325*(5948), 1647–1652. doi:10.1126/science.1172246 PMID:19779187

Heip, C., Hummel, H., van-Avesaath, P., Appeltans, W., Arvanitidis, C., Aspden, R., & Nash, R. et al. (2009). *Marine Biodiversity and Ecosystem Functioning*. Dublin, Ireland: Printbase.

Hileman, B. (2005). Common Herbicide Kills Tadpoles. *Chemical and Engineering News*, *83*(15), 11.

Hnaien, M., Lagarde, F., Bausells, J., Errachid, A., & Jaffrezic-Renault, N. (2011). A new bacterial biosensor for trichloroethylene detection based on a three dimensional carbon nanotubes bio-architecture. *Analytical and Bioanalytical Chemistry*, *400*(4), 1083–1092. doi:10.1007/s00216-010-4336-x PMID:21052645

Hogan, C. M. (2010). Abiotic factor in Encyclopedia of Earth. Washington, DC: National Council for Science and Environment.

Holben, W. E., Noto, H., Sumino, T., & Suwa, Y. (1998). Molecular analysis of bacterial communities in a three-compartment granular activated sludge system indicates community-level control by incompatible nitrification processes. *Applied and Environmental Microbiology*, *64*, 2528–2532. PMID:9647825

Huang, T., Guo, L., Zhang, H., Su, J., Wen, G., & Zhang, K. (2015). Nitrogen-removal efficiency of a novel aerobic denitrifying bacterium, *Pseudomonas stutzeri* strain ZF31, isolated from a drinking-water reservoir. *Bioresource Technology*, *196*, 209–216. doi:10.1016/j.biortech.2015.07.059 PMID:26241840

Huttermann, A., Orikiriza, L. J. B., & Agaba, H. (2009). Application of superabsorbent polymers for improving the ecological chemistry of degraded or polluted lands. *Clean Soil Air Water*, *37*(7), 517–526. doi:10.1002/clen.200900048

Iacurci, J. (2013). *Plastic Particles Harm Freshwater Organisms Too. Nature World News*. New Orleans, LA: Americal Chemical Society.

Intergovernmental Panel on Climate Change (IPCC). (2013). The Physical Science Basis. In T. F. Stocker, D. Qin, G. Plattner, M. Tignor, S. K. Allen, J. Boschung, & P. M. Midgley et al. (Eds.), *Working Group I Contribution to the Fifth Assessment Report of the Intergovernmental Panel on Climate Change, 1535*. Cambridge, UK: Cambridge University Press.

Jansson, J. K., Björklöf, K., Elvang, A. M., & Jørgensen, K. S. (2000). Biomarkers for monitoring efficacy of bioremediation by microbial inoculants. *Environmental Pollution*, *107*(2), 217–223. doi:10.1016/S0269-7491(99)00140-2 PMID:15092998

Jeppesen, E., Brucet, S., Naselli-Flores, L., Papastergiadou, E., Stefanidis, K., Nõges, T., & Beklioğlu, M. et al. (2015). Ecological impacts of global warming and water abstraction on lakes and reservoirs due to changes in water level and related changes in salinity. *Hydrobiologia*, *750*(1), 201–227. doi:10.1007/s10750-014-2169-x

Jiang, Z. J., Fang, J. G., Mao, Y. Z., & Wang, W. (2010). Eutrophication assessment and bioremediation strategy in a marine fish cage culture area in Nansha Bay, China. *Journal of Applied Phycology*, *22*(4), 421–426. doi:10.1007/s10811-009-9474-1

Jiao, N. (2013). Microbial Carbon Sequestration in the Ocean: The Big Role of Small Bugs in Mitigating Global Climate Change. Bulletin of the Chinese Academy of Sciences. *Special Reports*, *27*, 3.

Jiao, N., Robinson, C., Azam, F., Thomas, H., Baltar, F., Dang, H., & Zhang, R. et al. (2014). Mechanisms of microbial carbon sequestration in the ocean – future research directions. *Biogeosciences*, *11*(19), 5285–5306. doi:10.5194/bg-11-5285-2014

John, S. G., Ruggiero, C. E., Hersman, L. E., Tung, C.-S., & Neu, M. P. (2001). Siderophore mediated plutonium accumulation by *Microbacterium flavescens* (JG-9). *Environmental Science & Technology*, *35*(14), 2942–2948. doi:10.1021/es010590g PMID:11478246

Joint, I., Doney, S. C., & Karl, D. M. (2010). Will ocean acidification affect marine microbes? *International Society for Microbial Ecology*, *5*, 1–7. PMID:20535222

Judson, R. S., Martin, M. T., Reif, D. M., Houck, K. A., Knudsen, T. B., Rotroff, D. M., & Dix, D. J. et al. (2010). Analysis of eight oil spill dispersants using rapid, in vitro tests for endocrine and other biological activity. *Environmental Science & Technology*, *44*(15), 5979–5985. doi:10.1021/es102150z PMID:20602530

Kamika, I., & Momba, M. N. B. (2015). Effect of nickel on nutrient removal by selected indigenous protozoan species in wastewater systems. *Saudi Journal of Biological Sciences*, *22*(2), 147–156. doi:10.1016/j.sjbs.2014.09.010 PMID:25737645

Kapoor, M., & Rajagopal, R. (2011). Enzymatic bioremediation of organophosphorus insecticides by recombinant organophosphorous hydrolase. *International Biodeterioration & Biodegradation*, *65*(6), 896–901. doi:10.1016/j.ibiod.2010.12.017

Karlson, U., & Frankenberger, W. T. (1993). Biological alkylation of selenium and tellurium. In H. Sigel & A. Sigel (Eds.), Metal ions in Biological Systems: Biological Properties of metal Alkyl derivatives, (pp. 185-227). New York: Academic Press.

Khan, M. N., & Mohammad, F. (2014). Eutrophication: Challenges and Solutions. In A. A. Ansari & S. S. Gill (Eds.), *Eutrophication: Causes, Consequences and Control*. Springer Science & Business, Media Dordrecht. doi:10.1007/978-94-007-7814-6_1

Kim, E. H., Mason, R. P., Porter, E. T., & Soulen, H. L. (2004). The effect of resuspension on the fate of total mercury and methyl mercury in a shallow estuarine ecosystem: A mesocosm study. *Marine Chemistry*, *86*(3-4), 121–137. doi:10.1016/j.marchem.2003.12.004

Kostal, J. R. Y., Wu, C. H., Mulchandani, A., & Chen, W. (2004). Enhanced arsenic accumulation in engineered bacterial cells expressing ArsR. *Applied and Environmental Microbiology*, *70*(8), 4582–4587. doi:10.1128/AEM.70.8.4582-4587.2004 PMID:15294789

Kot, A., & Namiesńik, J. (2000). The role of speciation in analytical chemistry. *Trends in Analytical Chemistry*, *19*(2-3), 69–79. doi:10.1016/S0165-9936(99)00195-8

Koukkou, A. I. (Ed.). (2011). *Microbial Bioremediation of Non-metals: Current Research*. Caister Academic Press.

Krishnan, A. K., & Anirudhan, T. S. (2002). Removal of mercury (II) from aqueous solutions and chlor-alkali industry effluent by steam activated and sulphurised activated carbons prepared from bagasse pith: Kinetics and equilibrium studies. *Journal of Hazardous Materials*, *92*(2), 161–183. doi:10.1016/S0304-3894(02)00014-6 PMID:11992701

Lahiri (Ganguly), S., Sarkar (Paria), D., & Jana, B. B. (2015). Evaluating the Ecological Resilient Driven Performance of a Tropical Waste Stabilization Pond System using Ecological Signature of Biological Integrity. *Journal of Ecological Engineering*, *16*(3), 97–107.

Lal, R. (2008). Carbon sequestration. *Philosophical Transactions of the Royal Society of London. Series B, Biological Sciences*, *363*(1492), 815–830. doi:10.1098/rstb.2007.2185 PMID:17761468

Levin, R., Brown, M. J., Kashtock, M. E., Jacobs, D. E., Whelan, E. A., Rodman, J., & Sinks, T. et al. (2008). Lead Exposures in U.S. Children, 2008: Implications for Prevention. *Environmental Health Perspectives*, 116. PMID:18941567

Li, X., Schuler, M. A., & Berenbaum, M. R. (2007). Molecular mechanisms of metabolic resistance to synthetic and natural xenobiotics. *Annual Review of Entomology*, *52*(1), 231–253. doi:10.1146/annurev.ento.51.110104.151104 PMID:16925478

Lin, C., & Olson, B. H. (1995). Occurrence of cop-like resistance genes among bacteria isolated from a water distribution system. *Canadian Journal of Microbiology*, *41*(7), 642–646. doi:10.1139/m95-087

Lin, C. C., & Lin, H. L. (2005). Remediation of soil contaminated with the heavy metal (Cd^{2+}). *Journal of Hazardous Materials*, *122*(1-2), 7–15. doi:10.1016/j.jhazmat.2005.02.017 PMID:15943924

Lin, Z., Zhou, C., Wu, J., Zhou, J., & Wang, L. (2005). A further insight into the mechanism of Ag^{+} biosorption by *Lactobacillus* spp. strain A09. *Spectrochimica Acta. Part A: Molecular and Biomolecular Spectroscopy*, *61*(6), 1195–1200. doi:10.1016/j.saa.2004.06.041 PMID:15741121

Lokeshwari, H., & Chandrappa, G. T. (2007). Effects of heavy metal contamination from anthropogenic sources on Dasarahalli tank, India. *Lakes and Reservoirs: Research and Management*, *12*(3), 121–128. doi:10.1111/j.1440-1770.2007.00337.x

Lokhande, R. S., Singare, P. U., & Pimple, D. S. (2011). Toxicity Study of Heavy Metals Pollutants in Waste Water Effluent Samples Collected from Taloja Industrial Estate of Mumbai, India. *Resources and Environment*, *1*(1), 13–19. doi:10.5923/j.re.20110101.02

Lopez, A., Lazaro, N., Morales, S., & Margues, A. M. (2002). Nickel biosorption by free and immobilized cells of *Pseudomonas fluorescens*4F39: A comparative study. *Water, Air, and Soil Pollution*, *135*(1/4), 157–172. doi:10.1023/A:1014706827124

Lovley, D. R., & Lloyd, J. R. (2000). Microbes with a metal for bioremediation. *Nature Biotechnology*, *18*(6), 600–601. doi:10.1038/76433 PMID:10835594

Ma, Y., Prasad, M. N., Rajkumar, M., & Freitas, H. (2011). Plant growth promoting rhizobacteria and endophytes accelerate phytoremediation of metalliferous soils. *Biotechnology Advances*, *29*(2), 248–258. doi:10.1016/j.biotechadv.2010.12.001 PMID:21147211

Magellan, K., Barral-Fraga, L., Rovira, M., Srean, P., Urrea, G., García-Berthou, E., & Guasch, H. (2014). Behavioural and physical effects of arsenic exposure in fish are aggravated by aquatic algae. *Aquatic Toxicology (Amsterdam, Netherlands), 156*, 116–124. doi:10.1016/j.aquatox.2014.08.006 PMID:25190483

Major, D. N., & Wang, H. (2012). How public health impact is addressed: A retrospective view on three different oil spills. *Toxicological and Environmental Chemistry, 94*(3), 442–467. doi:10.1080/02772248.2012.654633

Martin, S., & Griswold, W. (2009). Human health effects of heavy metals. *Environmental Science and Technology Briefs for Citizens, 15*, 1–6.

Matelich, G. (2014). *Freshwater: The new frontier in micro-plastics research.* AMB Express.

McKay, B., & Mulvaney, K. (2001). A Review of Marine Major Ecological Disturbances (MMEDs). *Endangered Species Update, 18*(1), 14–24.

McLaughlin, M. C., Borch, T., & Blotevogel, J. (2016). Spills of hydraulic fracturing chemicals on agricultural topsoil: Biodegradation, sorption, and co-contaminant interactions. *Environmental Science & Technology, 50*(11), 6071–6078. doi:10.1021/acs.est.6b00240 PMID:27171137

Meorin, E. C. (1989). Urban run-off treatment in a fresh/brakish water marsh in Fremont, California. In D. A. Hammer (Ed.), Constructed Wetlands for Wastewater Treatment, (pp. 667-685). Industrial and Agricultural Lewis Publishers.

Metcalfe, C. D., Miao, X. S., Koenig, B. G., & Struger, J. (2003). Distribution of acidic and neutral drugs in surface waters near sewage treatment plants in the lower Great Lakes, Canada. *Environmental Toxicology and Chemistry, 22*(12), 2881–2889. doi:10.1897/02-627 PMID:14713027

Millennium Ecosystem Assessment (MA). (2005). *Ecosystems and Human Well-being: Synthesis.* Washington, DC: Island Press.

Miller, S., Pallan, S., Gangji, A. S., Lukic, D., & Clase, C. M. (2013). Mercury-associated nephrotic syndrome: A case report and systematic review of the literature. *American Journal of Kidney Diseases, 62*(1), 135–138. doi:10.1053/j.ajkd.2013.02.372 PMID:23602193

Mohapatra, H., & Gupta, R. (2005). Concurrent sorption of Zn(II), Cu(II) and Co(II) by *Oscillatoria angustissima* as a function of pH in binary and ternary metal solutions. *Bioresource Technology, 96*(12), 1387–1398. doi:10.1016/j.biortech.2004.11.004 PMID:15792587

Namasivayam, C., & Periasamy, K. (1993). Bicarbonate treated peanut hull carbon for Hg (II) removal from aqueous solution. *Water Resources, 27*, 1663–1668.

National Research Council (NRC). (1992). *Restoration of aquatic ecosystems: Science, technology and public policy.* Washington, DC: National Academy Press.

Nimick, D. A., Caldwell, R. R., Skaar, D. R., & Selch, T. M. (2013). Fate of geothermal mercury from Yellowstone National Park in the Madison and Missouri Rivers, USA. *The Science of the Total Environment, 443*, 40–54. doi:10.1016/j.scitotenv.2012.10.080 PMID:23178889

Nozari, M., Samaei, M. R., & Dehghani, M. (2014). Investigation of the Effect of Co-Metabolism on Removal of Dodecane by Microbial Consortium from Soil in a Slurry Sequencing Bioreactor. *Journal of Bioremediation and Biodegradation*, *5*, 253.

Ollis, D. J., Ewart-Smith, J. L., Day, J. A., Job, N. M., Macfarlane, D. M., Snaddon, C. D., & Mbona, N. et al. (2015). The development of a classification system for inland aquatic ecosystems in South Africa. *Water Science and Technology*, *41*, 5.

Onodera, S. M., Saito, M., Sawano, T., Hosono, M., Taniguchi, J., Shimada, Y., & Buapeng, R. D. et al. (2008). Effects of intensive urbanization on the intrusion of shallow groundwater into deep groundwater: Examples from Bangkok and Jakarta. *The Science of the Total Environment*, *404*(2-3), 401–410. doi:10.1016/j.scitotenv.2008.08.003 PMID:18804843

Orsod, M., Joseph, M., & Huyop, F. (2012). Characterization of exopolysaccharides produced by *Bacillus cereus* and *Brachybacterium* spp. isolated from Asian Sea Bass (*Lates calcarifer*). *Malaysian Journal of Microbiology*, *8*(3), 170–174.

Ortiz-Hernández, M. L., Sánchez-Salinas, E., Dantán-González, E., & Castrejón-Godínez, M. L. (2013). Pesticide Biodegradation: Mechanisms, Genetics and Strategies to Enhance the Process. In R. Chamy & F. Rosenkranz (Eds.), Biodegradation - Life of Science. Academic Press.

Paine, R. T., Ruesink, J. L., Sun, A., Soulanille, E. L., Wonham, M. J., Harley, C. D. G., & Secord, D. L. et al. (1996). Trouble on oiled waters: Lessons from the Exxon Valdez oil spill. *Annual Review of Ecology and Systematics*, *27*(1), 197–235. doi:10.1146/annurev.ecolsys.27.1.197

Pandey, P. K., Kass, P. H., Soupir, M. L., Biswas, S., & Singh, V. P. (2015). Contamination of water resources by pathogenic bacteria. *Pathogens*, *4*(2), 307–334. PMID:26011827

Panetta, L. E. (2003). *America's living oceans: charting a course for sea change*. Pew Oceans Commission.

Panigrahi, S., Hussain, J., Ramesh, T., & Mohanty, A. K. (2011). Natural disaster compels biota to adapt. *Current Science*.

Park, D., Yun, Y. S., Jo, J. H., & Park, J. M. (2005). Mechanism of hexavalent chromium removal by dead fungal biomass of *Aspergillus niger*. *Water Research*, *39*(4), 533–540. doi:10.1016/j.watres.2004.11.002 PMID:15707625

Park, J. D., & Zheng, W. (2012). Human exposure and health effects of inorganic and elemental mercury. *Journal of Preventive Medicine and Public Health = Yebang Uihakhoe Chi*, *45*(6), 344–352. doi:10.3961/jpmph.2012.45.6.344 PMID:23230464

Pattanasupong, A., Nagase, H., Inoue, M., Hirata, K., Tani, K., Nasu, M., & Miyamoto, K. (2004). Ability of a microbial consortium to remove pesticide, carbendazim and 2,4-dichlorophenoxyacetic acid. *World Journal of Microbiology & Biotechnology*, *20*(5), 517–522. doi:10.1023/B:WIBI.0000040405.56291.52

Perhar, G., & Arhonditsis, G. B. (2014). Aquatic ecosystem dynamics following petroleum hydrocarbon perturbations: A review of the current state of knowledge. *Journal of Great Lakes Research*, *40*(3), 56–72. doi:10.1016/j.jglr.2014.05.013

Peterson, C. H., Rice, S. D., Short, J. W., Esler, D., Bodkin, J. L., Ballachey, B. E., & Irons, D. B. (2003). Long-term ecosystem response to the Exxon Valdez oil spill. *Science*, *302*(5653), 2082–2086. doi:10.1126/science.1084282 PMID:14684812

Pinckney, J. L., Paerl, H. W., & Harrington, M. B. (1999). Responses of the phytoplankton community growth rate to nutrient pulses in variable estuarine environments. *Journal of Phycology*, *35*(6), 1455–1463. doi:10.1046/j.1529-8817.1999.3561455.x

Pizarro-Tobías, P., Niqui, J. L., Roca, A., Solano, J., Fernández, M., Bastida, F., & Ramos, J. L. et al. (2015). Field trial on removal of petroleum-hydrocarbon pollutants using a microbial consortium for bioremediation and rhizoremediation. *Environmental Microbiology Reports*, *7*(1), 85–94. doi:10.1111/1758-2229.12174 PMID:25870876

Prescott, L. M., Harley, J. P., & Klein, D. A. (2002). *Microbiology* (5th ed.). New York: McGrawHill.

Public Health Engineering Department (PHED). (2013). *A Note on Fluoride Contamination of Ground Water in West Bengal*. Author.

Radosevich, S. R., Holt, J. S., & Ghersa, C. M. (2007). Ecology of Weeds and Invasive Plants: Relationship to Agriculture and Natural Resource Management. Wiley-Interscience. doi:10.1002/9780470168943

Ramírez-Castillo, F. Y., Loera-Muro, A., Jacques, M., Garneau, P., Avelar-González, F. J., Harel, J., & Guerrero-Barrera, A. L. (2015). Waterborne Pathogens: Detection Methods and Challenges. *Pathogens*, *4*(2), 307–334. doi:10.3390/pathogens4020307 PMID:26011827

Rani, A., & Goel, R. (2009). Strategies for crop improvement in contaminated soils using metal-tolerant bioinoculants. In M. S. Khan, A. Zaidi & J. Musarrat, (Eds.), Microbial strategies for crop improvement, (pp. 105–132). Springer. doi:10.1007/978-3-642-01979-1_5

Rani, M. J., Hemambika, B., Hemapriya, J., & Kannan, V. R. (2010). Comparative assessment of heavy metal removal by immobilized and dead bacterial cells: A biosorption approach. *African Journal of Environmental Science and Technology, 4*(2), 77-83.

Relya, R. A. (2005). The lethal impact of Roundup® on aquatic and terrestrial amphibians. *Ecological Applications*, *15*, 618.

Richardson, N., Gordon, A. K., Muller, W. J., Pletschke, B. I., & Whitfield, A. K. (2010). The use of liver histopathology, lipid peroxidation and acetylcholine esterase assays as biomarkers of contaminant induced stress in the Cape stumpnose (*Rhabdosargus holubi*) (Teleostei: Sparidae), from selected South African estuaries. *Water S.A.*, *36*(4), 407–415. doi:10.4314/wsa.v36i4.58409

Rombouts, I., Beaugrand, G., Artigas, L. F., Dauvin, J.-C., Gevaert, F., Goberville, E., & Kirby, R. R. et al. (2013). Evaluating marine ecosystem health: Case studies of indicators using direct observations and modelling methods. *Ecological Indicators*, *24*, 353–365. doi:10.1016/j.ecolind.2012.07.001

Rossi, F., Olguin, E., Diels, L., & De Philippis, R. (2015). Microbial fixation of CO_2 in water bodies and in dry lands to combat climate change, soil loss and desertification. *New Biotechnology*, *32*(1), 109–120. doi:10.1016/j.nbt.2013.12.002 PMID:24355428

Sahin, I., Keskin, S. Y., & Keski, C. S. (2013). Biosorption of cadmium, manganese, nickel, lead, and zinc ions by *Aspergillus tamari. Desalination and Water Treatment*, 1–6.

Sarkar, A., Ravindran, G., & Krishnamurthy, V. (2013). A brief review on the effect of cadmium toxicity: From cellular to organ level. *International Journal of Bio-Technology and Research*, *3*(1), 17–36.

Savage, D. F., Afonso, B., Chen, A. H., & Silver, P. A. (2010). Spatially ordered dynamics of the bacterial carbon fixation machinery. *Science*, *327*(5970), 1258–1261. doi:10.1126/science.1186090 PMID:20203050

Sayler, G. S., & Ripp, S. (2000). Field applications of genetically engineered microorganisms for bioremediation processes. *Current Opinion in Biotechnology*, *11*(3), 286–289. doi:10.1016/S0958-1669(00)00097-5 PMID:10851144

Scheibenbogen, K., Zybner, R. G., Lee, H., & Trevors, J. J. (1994). Enhanced removal of selected hydrocarbons from soil by *Pseudomonas aeruginosa* UG 2 biosurfactants and some chemical surfactants. *Journal of Chemical Technology and Biotechnology (Oxford, Oxfordshire)*, *59*(1), 55–59. doi:10.1002/jctb.280590109

Scott, C., Pandey, G., Hartley, C. J., Jackson, C. J., Cheesman, M. J., Taylor, M. C., & Oakeshott, J. G. et al. (2008). The enzymatic basis for pesticide bioremediation. *Indian Journal of Microbiology*, *48*(1), 65–79. doi:10.1007/s12088-008-0007-4 PMID:23100701

Singh, B., Singh, Y., & Sekhon, G. S. (1995). Fertilizer-N use efficiency and nitrate pollution of groundwater in developing countries, Integrated Nitrogen Management in Relation to Leaching and Groundwater Quality. *Journal of Contaminant Hydrology*, *20*(3–4), 167–184. doi:10.1016/0169-7722(95)00067-4

Singh, R., Singh, P., Sharma, R., Selvalakshmi, S., Jayakumar, S., & Ramachandran, V. S. (2014). Microorganism as a tool of bioremediation technology for cleaning environment: A review. *Proceedings of* International Academy *of* Ecology *and* Environmental. *The Sciences*, *4*, 1–6.

Siwela, A. H., Nyathi, C. B., & Naik, Y. S. (2009). Metal accumulation and antioxidant enzyme activity in *C. gariepinus*, Catfish, and *O. mossambicus*, tilapia, collected from lower Mguza and Wright Dams, Zimbabwe. *Bulletin of Environmental Contamination and Toxicology*, *83*(5), 648–651. doi:10.1007/s00128-009-9861-y PMID:19768358

Sönmez, İ., Kaplan, M., & Sönmez, S. (2007). An investigation of seasonal changes in nitrate contents of soils and irrigation waters in greenhouses located in antalya-demre region. *Asian Journal of Chemistry*, *19*(7), 5639–5646.

Spain, A., & Alm, E. (2003). Implications of microbial heavy metal tolerance in the environment. *Review of Undergraduate Research*, *2*, 1–6.

Tabak, H. H., Lens, P., van Hullebusch, E. D., & Dejonghe, W. (2005). Developments in bioremediation of soils and sediments polluted with metals and radionuclides – 1. Microbial processes and mechanisms affecting bioremediation of metal contamination and influencing metal toxicity and transport. *Reviews in Environmental Science and Biotechnology*, *4*(3), 115–156. doi:10.1007/s11157-005-2169-4

Taub, F. B. (2004). *Fish 430 lectures: Biological Impacts of Pollutants on Aquatic Organisms*. Seattle, WA: University of Washington College of Ocean and Fishery Sciences.

Tchounwon, P. B., Ayensu, W. K., & Ninashvili, N. (2003). Environmental exposure to mercury and its toxico-pathologic implications for public health. *Environmental Toxicology*, *18*(3), 149–175. doi:10.1002/tox.10116 PMID:12740802

Umrania, V. V. (2006). Bioremediation of toxic heavy metals using acidothermophilic autotrophes. *Bioresource Technology*, *97*(10), 1237–1242. doi:10.1016/j.biortech.2005.04.048 PMID:16324838

United Nations Environmental Programme (UNEP). (1998). Environmental effects of ozone depletion: 1998 Assessment. Nairobi, Kenya: UNEP.

United Nations World Water Assessment Programme (UNWWAP). (2009). *The World Water Development Report 3: Water in a Changing World. United Nations World Water Assessment Programme*. Paris, France: UNESCO.

US Agriculture Centre. (2004). *A Farmer's Guide to Agriculture and Water Quality Issues*. Author.

Van Eerd, L. L., Hoagland, R. E., Zablotowicz, R. M., & Hall, J. C. (2003). Pesticide metabolism in plants and microorganisms. *Weed Science*, *51*(4), 472–495. doi:10.1614/0043-1745(2003)051[0472:PMIPAM]2.0.CO;2

Vilmi, A., Karjalainen, S. M., Kuoppala, M., Tolonen, K. T., & Heino, J. (2016). Taxonomic distinctness along nutrient gradients: More diverse, less diverse or not different from random? *Ecological Indicators*, *61*(2), 1033–1041. doi:10.1016/j.ecolind.2015.10.061

Watanabe, K., Kodoma, Y., Stutsubo, K., & Harayama, S. (2001). Molecular characterization of bacterial populations in petroleum contaminated ground water discharge from undergoing crude oil storage cavities. *Applied and Environmental Microbiology*, *66*(11), 4803–4809. doi:10.1128/AEM.66.11.4803-4809.2000 PMID:11055927

Whitton, B. A., & Potts, M. (2007). *The ecology of cyanobacteria: their diversity in time and space*. Kluwer Academic Publishers.

Wilkinson, C. (2002). Status of Coral Reefs of the World: 2002. Global Coral Reef Monitoring Network and Australian Institute of Marine Science.

Wolf, J., Prüss-Ustün, A., Cumming, O., Bartram, J., Bonjour, S., Cairncross, S., & Higgins, J. P. T. et al. (2014). Systematic review: Assessing the impact of drinking water and sanitation on diarrhoeal disease in low- and middle-income settings: systematic review and meta-regression. *Tropical Medicine & International Health*, *19*(8), 928–942. doi:10.1111/tmi.12331 PMID:24811732

Wolf, L., Nick, A., & Cronin, A. (2015). *How to keep your groundwater drinkable: Safer siting of sanitation systems – Working Group 11 Publication*. Sustainable Sanitation Alliance.

World Health Organization (WHO). (2003). *Guidelines for Safe Recreational Water Environments, 1*. Geneva, Switzerland: Coastal and Freshwaters.

World Health Organization (WHO). (2006). Guidelines for drinking-water quality, 1, Recommendations. Geneva, Switzerland: WHO.

World Health Organization (WHO). (2008). Guidelines for Drinking-water Quality, 1, Incorporating 1st and 2nd Addenda, Recommendations (3rd ed.). Geneva, Switzerland: WHO.

Wu, M. Z., Reynolds, D. A., Fourie, A., & Thomas, D. G. (2013). Optimal field approaches for electrokinetic *in situ* oxidation remediation. *Ground Water Monitoring and Remediation*, *33*(1), 62–74. doi:10.1111/j.1745-6592.2012.01410.x

Yang, S. H., Cho, J. K., Lee, S. Y., Abanto, O. D., Kim, S. K., Ghosh, C., & Hwang, S. G. et al. (2013). Isolation and Characterization of Novel Denitrifying Bacterium *Geobacillus* spp. SG-01 Strain from Wood Chips Composted with Swine Manure. *Asian-Australasian Journal of Animal Sciences*, *26*(11), 1651–1658. doi:10.5713/ajas.2013.13184 PMID:25049754

Yergeau, E., Sanschagrin, S., Beaumier, D., & Greer, C. W. (2012). Metagenomic analysis of the bioremediation of diesel-contaminated Canadian high arctic soils. *PLoS ONE*, *7*(1), e30058. doi:10.1371/journal.pone.0030058 PMID:22253877

Younis, H. M., Abo-El-Saad, M. M., Abdel-Razik, R. K., & Abo-Seda, S. A. (2002). Resolving the DDT target protein in insects as a subunit of the ATP synthase. *Biotechnology and Applied Biochemistry*, *35*, 9–17. PMID:11834125

Yu, K., & Zhang, T. (2012). Metagenomic and metatranscriptomic analysis of microbial community structure and gene expression of activated sludge. *PLoS ONE*, *7*(5), e38183. doi:10.1371/journal.pone.0038183 PMID:22666477

Zhao, X. W., Zhou, M. H., Li, Q. B., Lu, Y. H., He, N., Sun, D. H., & Deng, X. (2005). Simultaneous mercury bioaccumulation and cell propagation by genetically engineered *Escherichia coli*. *Process Biochemistry*, *40*(5), 1611–1616. doi:10.1016/j.procbio.2004.06.014

KEY TERMS AND DEFINITIONS

Bioaccumulation: A process by which certain toxic substances (such as pesticides, heavy metals and polychlorinated biphenyl) accumulated and keep on accumulating in various tissues of living organisms, when the rate of intake of a substance is greater than the rate of excretion or metabolic transformation of that substance, posing a threat to health, life and to the environment.

Carbon Sequestration: Carbon sequestration is the process involved carbon capture and removal and long-term storage of atmospheric carbon dioxide or other forms of carbon in carbon sinks (such as oceans, forests or soils) through physical or biological processes, such as photosynthesis to either mitigate or defer global warming and avoid dangerous climate change.

Detoxification: Detoxification may be defined as the removal or sequester of poisonous or toxic substances from a system by the neutralization them using various physical, chemical or biological process.

Eco-Restoration: Ecological restoration is an intentional activity that initiates or accelerates the recovery of an ecosystem with respect to its health, integrity and sustainability.

Eco-Technology: An eco-technology is a class of technology that includes a living ecosystem consists of an interacting set of "wild" or natural species that are allowed to self-organize within specified boundary conditions in order to function as a part of the solution to a problem. An eco-technology is therefore a hybrid system with a non-living part that is designed by humans and a living part (the ecosystem) that self-designs.

Ecosystem Perturbation: Ecosystem perturbation is a temporary quick change in environmental conditions that causes a pronounced great change in an ecosystem, to alter the physical structure or arrangement of biotic and abiotic elements.

Ecosystem Resilience: The capacity of ecosystem to absorb disturbance, reorganize and return to an equilibrium or steady-state while undergoing some change or perturbation so that still retain essentially the same function, structure, identity, and feedbacks.

Microbial Diversity: Microbial diversity encompasses the spectrum of variability among all types of microorganisms including all bacteria, archaea, fungi, algae, protozoa, rotifers and viruses the natural world and as altered by human intervention.

Chapter 5
Microbiological Carbon Sequestration:
A Novel Solution for Atmospheric Carbon – Carbon Sequestration through Biological Approach

Mohammad Oves
King Abdulaziz University, Saudi Arabia

I.M.I. Ismail
King Abdulaziz University, Saudi Arabia

Fohad M. Hussain
King Saud University, Saudi Arabia

Nadeen M. Felemban
King Abdulaziz University, Saudi Arabia

Huda A. Qari
King Abdulaziz University, Saudi Arabia

ABSTRACT

Global modernization demands a huge energy for developing new technologies to meet the increasing human needs. The initial source of energy was fossil fuels that might release some harmful gases. Burning of fossil fuels was practiced since the 18th century, which consequently resulted in 40% increase in atmospheric CO_2 according researcher. Presently, the atmospheric level of C is increasing day by day. However, this level would reach ~1000 ppm at the end of this century. The consistent discharge of CO_2 into the atmosphere is a major threat to global warming; hence, it is important to minimize the atmospheric CO_2 by using innovative carbon sequestration technologies. Therefore, soil and water can act as a sink for C storage with the assistance of microorganisms. Here, an attempt is made to discuss these processes in details, interactions between plant and microbes, and the requirement of C sources for C sequestration in the context of climate change. We also conferred about the development of microbial inoculants for C sequestrations and their application for sustainability of agro-ecosystems.

DOI: 10.4018/978-1-5225-2325-3.ch005

INTRODUCTION

Carbon sequestration is a process of capturing and assimilation of C from CO_2 or other C associated gases as well as their transformation into organic and inorganic materials. In another word, C sequestration can be defined as;

1. "The process of removing C from the atmosphere to be deposited in a reservoir;"
2. "Capturing and storage of C where CO_2 is removed from fuel gases (such as on power stations) before it is stored in underground reservoirs" (UNFCCC, 2007), or
3. Long-term storage of CO_2 or other forms of C in reservoirs to either mitigate or diminish the global warming and avoid the risk of climate change.

Furthermore, C sequestration has been proposed as a strategy to enhance terrestrial status (soil, sediments, and vegetation), geological, and marine accumulation of the atmospheric greenhouse gases. These gases are mostly generated by anthropogenic activities particularly by burning fossil fuels either in natural or deliberate processes. Various biological, chemical or physical processes contribute to building up the global 'C-cycle', in which CO_2 is naturally captured and secured and its equilibrium is maintained in the atmosphere to sustain life. Similarly, some anthropogenic sequestration techniques (intentional geological or ocean storage) exploit these natural processes to capture CO_2 from its source of emission, and then stored in the ocean or underground channels such as old oil reservoirs, aquifers, and coal seams (Akinola, 2014).

Global industrial revolution is responsible for generating a huge amount of C emitted into the atmosphere. Also, rapid growth of anthropogenic activities is increasing the atmospheric level of CO_2 rapidly. Moreover, studies have shown that C emission is strongly combined with world economy and country development (Canadell et al., 2007). When comparing the growth rate of CO_2 emission from fossil fuel consumption during the period 1990 to 2006 it was found thatthe rate was increased from 1.3 to 3.3% $y^{-1,}$ (Canadell et al., 2007). In another study, there was a significant increase of C emission (about 29%) between 2000 and 2008 from fossil fuel consumption by emerging economies while C emission from land-use changes remained constant without any change (Canadell et al., 2007; Le Quéré et al., 2009). According to Mona Loa Observatory (USA), the average of CO_2 emission rate was observed from 2005 to 2014 and it was about 2.1 ppm y^{-1}. While the emission rate was almost two times the observed rate in1960, the overall average of annual increasing of concentration in 2014 was 3.98.55 ppm (http://co2now.org/, 2015). European Commission and Netherland Environmental Assessment Agency (EU-NEAA) collected the data of CO_2 emission from worldwide locations. The agency has reported that China, America, European Union, India, and Russia are the major countries that emit a significant amount of CO_2; however, Australia, America, Saudi Arabia, Canada and South Korea released the highest amount of CO_2 per capita (http://edgar.jrc.ec.europa.eu/news, 2014).

Since then, the global emissions of C that resulted from burning fossil fuels, land-use change, and depleting soil organic C into atmospheric C was approximately 270±30 Pg, 136±55 Pg, and 78±12 Pg (Pg = petagram = 10^{15} g = 1 billion ton), respectively (Lal, 2004). Another estimation has shown a significant increase of CO_2 emission from anthropogenic activities that increased from 6.3 Gt C to 8.7 Gt C during 1994 and 2009 respectively (Boden & Marland, 2014), which represents about 39% increase over that period (Sommer and Bossio 2014). Moreover, land-use changes such as burning forest, deforestation, soil degradation, and soil organic C (SOC) depletion, released annually additional 2

Gt of C as CO_2 into the atmosphere (Sommer and Bossio 2014). In 2030, the expected emission of CO_2 into the atmosphere at this rate without using mitigation technologies will be grown to 600-1500 ppm (IPCC, 2013). In a more recent report, IPCC stressed the need to reduce the global C concentration in the atmosphere by 41–72% until 2050 and by 78–118% until 2100 compared to its level in 2010 (IPCC, 2013). To achieve this goal, there is an urgent need to develop innovative technologies that can sequester C from the atmosphere. In this regard, Leung et al., (2014) have reviewed different technologies of CO_2 capture and storage (CCS) process, which have been used throughout the world to mitigate the increased emission of CO_2. These technologies have been categorised into 5 main groups, as: i) improvement in the energy conservation and efficiency of current CO_2 emitting systems, ii) enhancing the usage of low C fuels (natural gas, hydrogen or nuclear power), iii) deployment of renewable energy sources, iv) application of geoengineering approaches including CO_2 removal, solar radiation management, afforestation and reforestation, and v) CO_2 capture and storage system. However, it has been difficult to apply most of these technologies because they are expensive and can not be applied on a commercial scale (Leung et al., 2014). Moreover, Vaughan et al., (2009) concluded that any geoengineering method developed to date could not provide a simple or acceptable solution to resolve climate change problem. Recently, a report (IPCC, 2013) has demonstrated that most of the CO_2 arise from natural sources due to rising temperatures from the "Little Ice Age" and increased CO_2 level from fossil fuel consumption; thereby both aspects lead to increase the temperature of the atmosphere. However, the report has also concluded that the temperature is mostly rising by natural means, which is a very interesting finding to improve our understanding of the CO_2 effect on global temperature (Gosselin 2013). In this regard, we have also plotted a graph based on observational data of CO_2 level, and increasing the level of global temperature from the year 2000 to 2015 (Figure 1).

Figure 1. Time scale with increasing CO_2 level and temperature yearly (according Mauna Loa observatory)

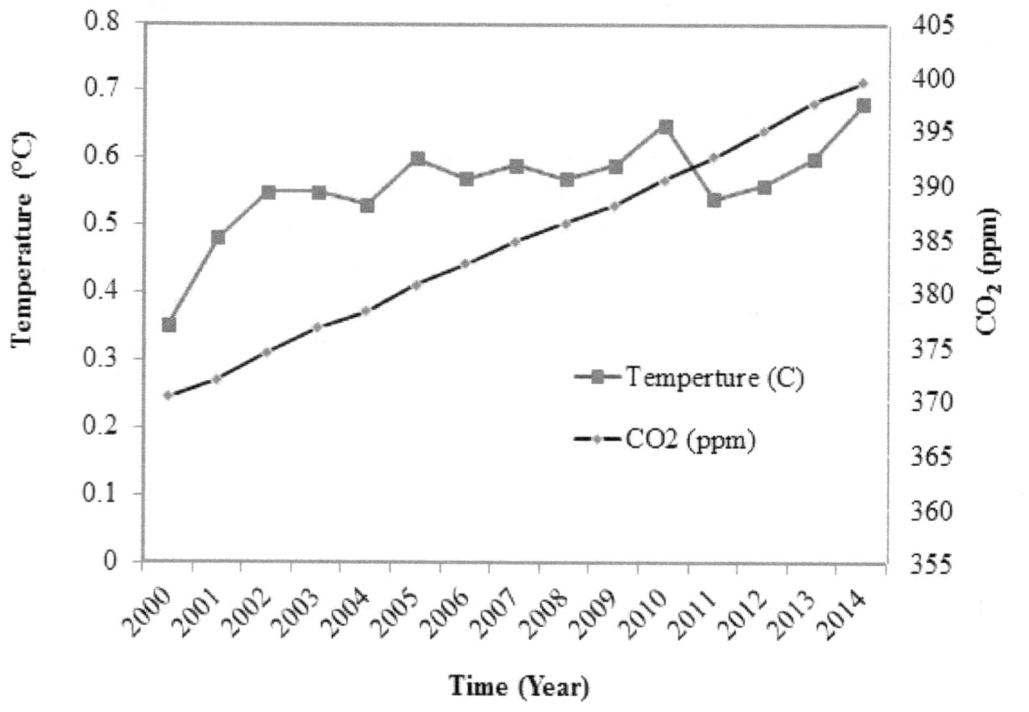

Soil-born microbes and plant-associated microorganisms can carry out C stabilization processes. Also, some species of autotrophic microbes such as bacteria (obligate and facultative autotrophs, and Synechococcus and Prochlorococcus) and algae, both can fix the atmospheric CO_2 and store it in plant tissues as well as in soil (Tabita, 1999). For that purpose, they have used specific pathways known as Calvin-Benson-Bassham cycle, where CO_2 fixation is catalysed by ribulose 1,5-bisphosphate carboxylase/oxygenase (RuBisCO) secreted by this biota to fix the atmospheric CO_2 (Chen et al., 2014). Hence, this biota annually transform 10^2 Pg of inorganic atmospheric CO_2 into organic cellular constituents through RuBisCO enzyme (Siegenthaler and Sarmiento 1993). However, Yuan et al., (2012) have found that high RuBisCO activity in autotrophic bacteria and algae was related to high organic C synthesis rate in soil. Thus, a large number of microbial diversity with specific genetic potential has the ability to overcome the atmospheric CO_2 load and assimilate or fix CO_2 in soils. In this book chapter, we highlighted the role of microbial diversity in fixing the atmospheric CO_2 levels. Moreover, we will also discuss the development of bio-inoculant for C sequestration and applications in order to overcome the global atmospheric CO_2 levels.

Background

In general, the level of CO_2 in Earth ecosystems is maintained by making a balance between photosynthesis and respiration processes (Singh et al., 2009). Continued increase of anthropogenic activities may disturb the balance between these processes, which leads to increase in the global CO_2 level rapidly. Moreover, one of the major fears of scientist is the uncertainty of understanding C–climate system that arises due to the imbalance of soil C: N ratio (Fernández-Martínez et al. 2014). However, some studies pointed out that due to a shift of nitrogen-induced plant species, the response of terrestrial ecosystems to elevated levels of CO_2 would be restricted (Langley and Megonigal 2010). In this regard, there are some major concerns about peat lands and permafrost soils that need to be considered, since the authors have observed some differences in the climatic conditions that could result in organic material accumulations in the soil of both ecosystems. As a result, the accumulated organic matter might lead to release a huge amount of C into the atmosphere under unfavourable changes in future climate (Davidson and Janssens, 2006). Therefore, further investigations need to be conducted in this area particularly on predicting the impacts and feedbacks between climate change and the global C cycle (King, 2011).

BIOLOGICALLY CARBON SEQUESTRATION IN OCEAN

Recent reports suggested that ocean is one of the biggest reservoirs for C sequestration process (SCBD, 2009). One-third of total C emission is housing in Ocean and resulted in a significant reduction of atmospheric carbon. Moreover, scientists have also been working on the mechanisms of marine CO_2 sequestration, which includes:

1. C pump is a process of CO_2 absorption into the ocean, and
2. Solubility pump is a phenomenon of CO_2 dissolved in ocean surface layer and transported to different layers.

While some studies have already been done on the accumulation of atmospheric CO_2 on the ocean surface, however, these studies are less effective for the present scenario of high anthropogenic activities. The dissolution of total anthropogenic-induced CO_2 in the ocean using solubility pump was approximately 80% in the last century whereas the solubility pump rate has decreased to < 50% recently (BCAS, 2013). A new concept of microbial C pump (MCP) is arising with great efforts of Prof. Jiao Nianzhi (Xiamen University). MCP is a marine-derived microbial diversity that used for maintenance of C cycle in the ocean from thousands of years ago. Such a pump is driven by bacteria, archaea, phytoplankton and viruses, which involves the transformation of dissolved organic C in the water from labile to refractory states. The dissolved labile organic C is biologically available while refractory dissolved organic C is difficult to metabolise and thus acts as a reservoir for Cin the ocean.

STABILIZATION OF CARBON BY SOIL MICROBIAL DIVERSITY

The exact function of soil microbial diversity in soil is not clearly identified because of their diverse functions in natural conditions (Gans et al., 2005). For that reason, C cycle model has been known as Black box model (Tiedje 2009). Many soil scientists have found that only 1-10% of soil microbial communities are able to grow in laboratory condition whereas 90–99% of them are non-cultured by modern technologies (Hill et al., 2000). Due to limited knowledge of soil microbial diversity, we described the soil processes based on the identified soil microorganisms. Three specific biological domains contribute to balance soil biogeochemical cycles, which include:

1. Bacteria,
2. Fungi, and
3. Actinomycetes.

These domains are predominant according to soil factors and climatic conditions. However, bacteria and fungi are the most dominant domains, which play a valuable role in C stabilization. Contradictory reports have discussed about C utilization efficiency by bacteria and fungi. Initial reports have shown that both (bacteria and fungi) have similar C utilization efficiency (Payne, 1970). However, later reports suggested that fungi have shown higher efficiency of C utilization than bacteria (Adu and Oades 1978; Thiet et al., 2006). Even when both fungi and bacteria metabolize equal amounts of C, the experimental results identified that the bacterial domain contributed less to C pool (Adu and Oades 1978; Claus et al., 1991). Bacteria and fungi are very different metabolically and structurally as they produce various extracellular enzymes, which significantly influence the soil stability and function. Fungi can secret a number of enzymes with massive amounts such as laccases, phenoloxidases, and peroxidases. These enzymes demonstrated the capability to enhance condensation reaction in the presence of lignitic materials in soil. Similarly, bacteria can also secrete enzymes such as cellulose and lipase, which react easily with non-lignitic materials. Soil humification is a process in which soil-born fungi start growing on lignite materials, and subsequently decompose those materials into monomers to be easily mixed with soil particles. Initially, fungal growth and biomass formation can lead to C stabilization in soil. Fungal cell wall constitutes of complex materials like chitin and melanin whereas bacterial cell wall is mainly derived from less complex compound like peptidoglycan and phospholipids. While the compositions of the bacterial cell membrane can be easily degraded and become available for next generation (Frey et al., 2001), the fungal cell wall is recalcitrant, and its composition is unavailable for next generation (Guggenberger et al., 1999). The fungal domain is able to store 26 times higher C than bacterial domain when both were grown separately on litter of leaves (Suberkropp & Weyers, 1996).

A microbiological active zone on soil surface is also known as a detritusphere, which is influenced by various factors including C source and litter (Gupta et al., 1998). According to Gaillard et al., (2003) detritusphere is a biogeochemical interface in the vertical dimension with several millimetres range with a heterogeneous mixture of litter-derived C containing substrates for microbiological reactions. These indigenous microorganisms can utilize the available substrates, which enhance microbial biomass and stimulate releasing enzymes. Therefore, changing the organic C of the surface soil from the bulk soil modifies the soil constituents. Berg and Matzner (1997) has reported that microbial degradation of litter is mainly divided into three stages:

1. Initially, in wet condition the degradation of litter without involvement of extracellular microbial enzymes,
2. Enzymatic degradation of litter due to lack of water soluble, and
3. Finally start recalculation of carbonaceous materials, which slow down decomposition rate.

A number of studies have verified all of the above steps and observed the depolymerisation of enzymes succession (Carreiro et al., 2000; Sinsabaugh & Shah, 2010) during microbial litter decomposition.

NUTRIENT INPUT IMPACTS ON SOIL CARBON

In a terrestrial ecosystem, global anthropogenic activities are continuously increasing in parallel to the rising of nutrient input, which may directly affect the global nutrient cycle (Vitousek 2004). When a huge amount of nutrients are added into soil ecosystems, then an overdose of nutrients may result in disturbance of the ecosystem functioning. These disturbances may lead to delay in the decomposition of organic matter (Vitousek 2004). The terrestrial ecosystems are major sinks for the overloaded concentration of atmospheric C. A small alteration in the pool of organic C resulted in a small change in the flux rates, which in turn may have a strong influence on atmospheric CO_2 concentrations. Some nutrients are required for decomposition of soil organic matter, which can change the actual rate of decomposition. Hence, it is very difficult to express a general model about long-lasting effects on the regular input of nutrient in soil for soil organic matter decomposition.

In terrestrial ecosystem, the stability between photosynthesis and respiration processes is responsible for maintaining C cycle (Prentice, 2001). Carbonfixation or assimilation is a process of conversion of atmospheric CO_2 (inorganic C) to organic compounds through photosynthesising plants, autotrophic organisms, and chemo/photo autotrophic microbes (Lu, 2005). Moreover, the organisms fix atmospheric CO_2 to organic C, which is returned to the atmosphere by heterotrophic microorganisms through respiration or decomposition (Liang and Balser, 2011). The resulted organic compounds (fixed carbon) were used by heterotrophic microorganisms in order to release energy as well as for their growth. However, anoxic conditions in paddies, peatlands or waterlogged soils may lead to reduced CO_2 by methanogenesis in hydrogenotrophic archaea (Lu, 2005). However, the amount of organic and inorganic C, which is required by soil biota to be utilized as energy sources is very important. Some other key factors are essential for soil microbial growth, which includes N, P, pH, temperature, soil structure, texture, moisture (water) content and mineral contents. These factors influence the microbial functions and thereby control the rate of C sequestration and respiration (Davidson and Janssens, 2006). For primary production, complicated interactions between biological processes and environmental conditions exist, which control the distribution and stability of organic matter in soil ecosystems (Schmidt, 2011).

In general, soil microbial communities can actively maintain soil fertility due to the decomposition of organic materials, in which essential nutrients for soil structure, maintenance, and plant growth promotion are released. However, organic materials are rich sources of C, can be decomposed by microorganisms and subsequently supplied to the primary producers of the ecosystem like plants. Soil microbial diversity depends on the presence of soil organic matter and mineralisation of organic compounds; hence, their availability is essential for plant growth and nutrition. Moreover, it has been suggested that the function of ecosystem is regulated by the abundance, composition, and diversity of soil microorganisms. The process of ecosystem is maintained by C cycle, which mainly involves decomposition and C sequestration (Figure 2).

Figure 2. The CO_2 cycle in ecosystem maintenance by decomposition and C sequestration

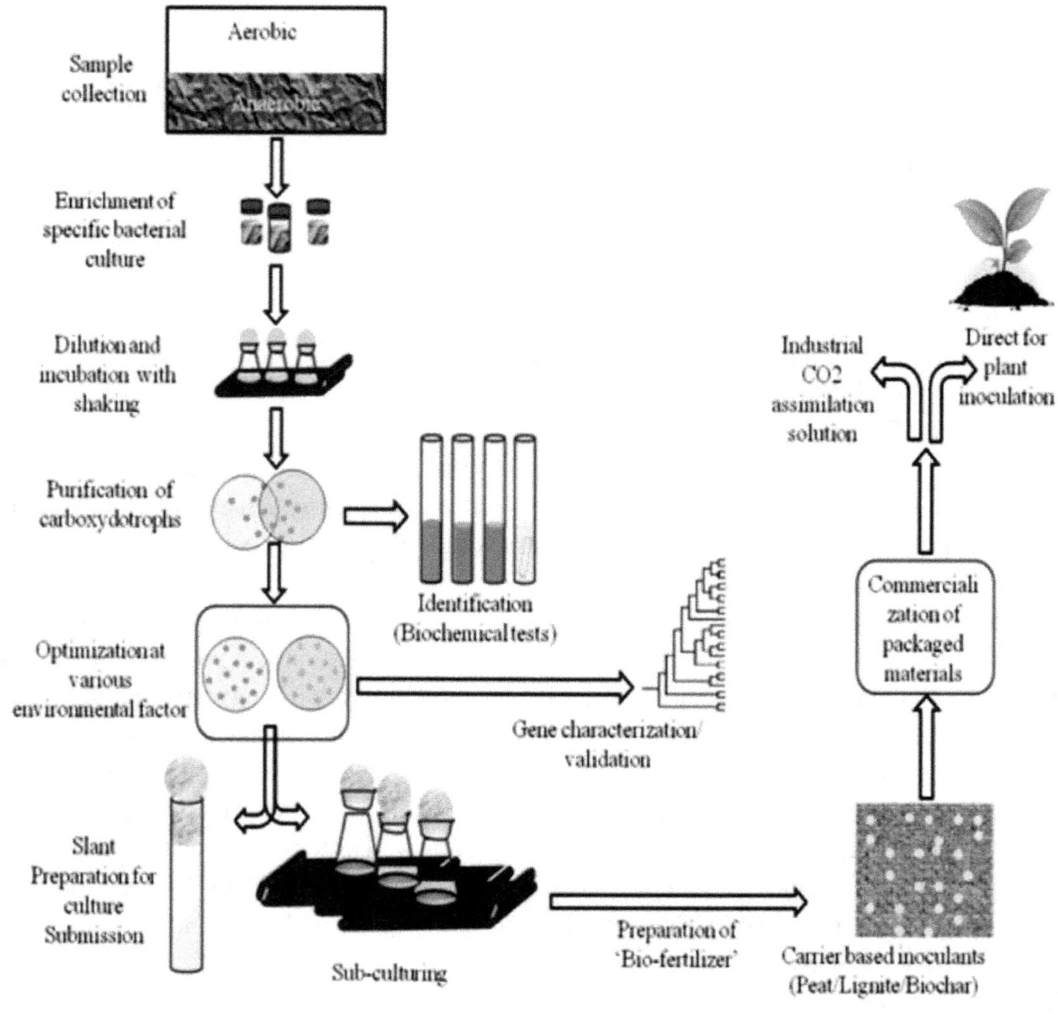

MICROBIAL INOCULANTS FOR CARBON SEQUESTRATIONS

In modern agro-ecosystem, free-living and symbiotic microorganisms are necessary for C sequestration. Some free-living organisms can directly fix the atmospheric C, and significant amount of symbiotic organisms are found in plant root zone that are able to absorb root exudates and solubilize soil organic matter. An elevated atmospheric CO_2 level is responsible for high rhizodeposition, and the subsequent increase of C/N ratio may prohibit the rate of decomposition (Norby and Cotrufo 1998). Number of researchers have explored the possibility of C sequestration by different land management practices. Likewise, Bailey et al., (2002) compared ex-situ incubations of soil samples from five different ecosystems:

1. Desert,
2. Restored tall grass prairie,
3. Douglas fir forest,
4. Globlolly pine forest, and
5. Agricultural land.

Therefore, the observed results demonstrated that tall grass prairie soil samples had the highest fungal-to-bacterial activities and maximum total soil C compared to other samples. Based on this study, the researcher suggested that increasing ratio of fungi to bacteria in soil was directly correlated with increasing C storage in soil. Verburg et al., (2006) have designed two-year experiment to compare total ecosystem C exchange between the atmosphere and cheat grass (*Bromus tectorum*) varieties in the Desert Research Institute's (Nevada, USA). The observed results demonstrated that the initial C uptake was increased by fertilization and subsequently was loosed by soil respiration. Interestingly, a contradictory report from grassland ecosystem in Eschikon, Switzerland, had shown that when N nutrient was supplied to the plants, then fixed C allocation was enhanced in shoots and reduced in root due to declined rhizodeposition (Bazot et al., 2009). Increased temperatures affect the availability, microbial activities as well as net primary production. The variation in C sequestration potential of plant and microbes was mainly dependent on the species and C: N ratio along with the soil moisture content. C sequestration through microbial inoculant is under nascent phase, which requires global efforts with the collaborative interdisciplinary research programme. Researchers are hardly working to explore C sequestration process by microorganisms in agro-ecosystem to overcome the enrichment of atmospheric. C. Soil microbial communities have a direct or indirect impact on global C cycle where a balance is established between formation and degradation of microbial products. Indirectly, soil microbes release metabolites, aggregate soil particles, and physically protect soil organic matter. Microbial biomass also influences the soil properties and dynamics though change texture, clay mineralogy, pore-size distribution, and aggregate dynamics (Six et al., 2006). Furthermore, soil microorganisms utilize C as a substrate for cell metabolism, growth, and secondary metabolites production. The chemical nature of the resulted by-products helps in stabilizing soil organic C and enhancing C sequestration in agro-ecosystem, which leads to microbial community shifts from bacteria to fungal dominance (Six et al., 2006).

Development of microbial inoculants for C sequestration in agro-ecosystem positively influences primary productivity, and it is a very complicated system to understand the interaction between plant–soil-microbe. Because of continues increase of global food demands, thus effective potential biological approaches need to be developed as alternative solutions of chemical fertilizers for sustainable agriculture (Ashrafuzzaman et al., 2009; Piromyou et al., 2011). From the last two decades, there was a significant increase of yield of pulses and cereals crop due to improvement in soil fertility by inoculation of plant growth promoting rhizobacteria in agro-ecosystem (Piromyou et al., 2011; Richardson 2008). Similarly, application of C sequestering organisms with plant growth promoting property will be an effective strategy to compensate both food demand and C level in atmosphere.

MICROBIAL ENZYME FOR CARBON SEQUESTRATION

In a report, a greater N availability will enhance the C sequestration in soil. As a result of high N concentration, the rate of degradation of decomposable litter is accelerated and the degradation of recalcitrant litter is declined by stimulation and repression of extracellular enzymes of soil microbial diversity (Waldrop et al., 2014). Natural system of C capture is mainly dependent on the enzymatic catalytic reaction in living being, and this mechanism can be mimicked and utilized for the development of low-cost C sequestration technology (Di Fiore et al., 2015). Recently, University of North Dakota Energy & Environmental Research Centre collaborates with some companies to research and develop an enzyme-based noble solution for C capture. Government of Canada's Industrial Research Assistance Program has developed an enzyme 1T1 with considering its stability, efficiency, cost and catalytic activity for Ccapture (http://co2now.org/, 2015)., in a study carbonic anhydrase enzyme, was immobilized for transformation of CO_2. The results revealed that genetically modified *E. coli* was able to produce carbonic anhydrase enzyme, which was able to transform CO_2 into bicarbonate. This technology has provided a noble strategy of C capture that could be converted to beneficial product like backing soda, chalk, and limestone. Furthermore, carbonic anhydrase enzyme immobilized based bioreactor was developed for sustainable and eco-friendly technology (Lu et al., 2005). Both CO_2 and dissolved inorganic C in the form of bicarbonate are essential components for microbial growth (Kusian et al., 2002). Essential CO_2 is transported in a microbial cell by hydration, which depends on the presence of dissolved inorganic C or bicarbonate. This reaction has occurred very fast in cell, and it was catalysed by carbonic anhydrase, and catalytic rate of this enzyme was approximately 10^5 reactions per second (Tu et al., 2012). Carbonic anhydrase enzyme exists in most of living organisms, and it was reported previously from plants, animals, and organisms (Karlsson et al., 1998). Based on gene sequence analysis, this enzyme was present in microorganism before Archaea and Bacteria were divided into two separate groups. Carbonic anhydrase is metalloenzyme with zinc ion, and it has been categorised by protein sequence into five classes: α, β, γ, δ and ε. Enzyme class α and β are the most common in bacteria and archaea (Smith et al., 1999).

Carbonic anhydrase forward reaction is proceeded with constant rate 6.2×10^{-2} s at 25°C (Druckenmiller and Maroto-Valer 2005), and formation of carbonic acid reaction rate is limited by CO_2 hydration. Two other enzymes also play a potential role in C sequestration process, includes (i) isocitrate dehydrogenase and (ii) aconitase. In which ketoglutarate is converted to isocitrate by isocitrate dehydrogenase, and then isocitrate is converted into citrate by aconitase, and both reactions are individually reversible (Beh et al., 1993; Xia et al., 2014). In addition, researchers are also working on the potential enzyme cascade for CARBON sequestration in one step reaction (Xia et al., 2015).

Reaction catalysed by carbonic anhydrase is as follows.

$$CO_2 + H_2O \xrightarrow{Carbonicanhydrase} HCO_3^- + H^+$$

SELECTION AND PREPARATION OF POTENTIAL BACTERIAL INOCULANTS

Soil is a complex of minerals and many organic materials in a layered structure that contain a diverse range of microorganisms. It is a big challenge to screen and isolate potential microbes responsible for C sequestration. Soil samples should be collected from different layers into sterile sampling bottles to be carefully transported to the laboratory for purification and identification of potential C sequestering microbes. Post initial enrichment step, the soil microbes (present in the enriched soil samples) are inoculated into appropriate flask media, which then incubated overnight on a shaker. This incubation will help to enhance the growth of all microbes present in the soil sample. Hence, it is important to purify properly and identify the potential microbes (i.e., carboxydotrophs) from the mixed culture pool. This task can be achieved by plating the cultured growth media (broth) onto agar plates, followed by biochemical analysis. After incubation, colonies of carboxydotrophs are identified based on their morphological characteristics (size, shape, consistency, and transparency) and Gram staining results. Subsequently, carboxydotrophs colonies are then sub-cultured onto selective media (agar plates) to enhance the identification of potential carboxydotrophs. Furthermore, the growth response of potential carboxydotrophs under various environmental conditions (growth temperature, pH, salinity, etc.) should also be investigated. In order to complete the identification process, the inoculants of potential carboxydotroph should be characterized at the gene level by using molecular techniques such as gene sequencing and phylogenetic mapping. Upon completion of the identification and validation process, the isolated carboxydotroph should be sub-cultured onto slant media so that it can be stored for further application. When carboxydotrophs are used along with carrier materials (such as peat, lignite or biochar), they can act as a 'bio-fertilizer'. However, it is necessary to optimize the carrier material for carboxydotrophs, and field experiments should be conducted in this respect. Upon successful optimization, in-house produced bio-fertilizer that is responsible for C sequestration can be packed for commercial and other industrial purposes (Figure 3).

C/N EFFECTS ON BACTERIACYANOBACTERIA CO-INOCULATION

Recently, microbiologists are using different microbial strains to obtain the maximum benefits of coculturing. Co-inoculation is a process of mixing two diverse biological inoculants for various biotechnological beneficial purposes like phosphate and minerals solubilization and plant growth promoting (Khan et al., 2009; Zaidi et al., 2010). Recently, application of co-inoculation techniques for C sequestration in agro-ecosystem was developed. Mainly in paddy fields, co-inoculation method was applied for different applications, such as rice crop production, enhancement, and C-N sequestration. Three bacterial strains *Providencia sp.*, *Brevundimonas sp.* and *Ochrobacterium sp.* were co-inoculated, and three cyanobacterial strains namely, *Anabaena sp.*, *Calothrix sp.* and *Anabaena sp.* were used for maximum nitrogen fixation and crop production . Organic C was also significantly increased in the form of microbial biomass (Prasanna et al., 2012). A variety of beneficial bacteria attracted by root exudates colonized on root hairs and aerial root part (Mehnaz et al., 2001; Yanni and Dazzo, 2010), which act as a plant growth promoter and provide protection from the harsh condition. Similarly, beneficial microbes derived from cereal roots also play a role in nitrogen fixing. These microbes fix nitrogen and enhance phytohormones production, which in turn lead to efficient uptake of water and nutrient to promote plant growth in soil ecosystem (Ashrafuzzaman et al., 2009; Keyeo et al., 2011; Prasanna et al., 2011). Previous studies also supported the plant growth promoting activities that enhanced by diverse bacterial spe-

Figure 3. Potential microbes isolation and characterization for bio-fertilizer preparation especially for C sequestration and plant growth promotion to control global warming

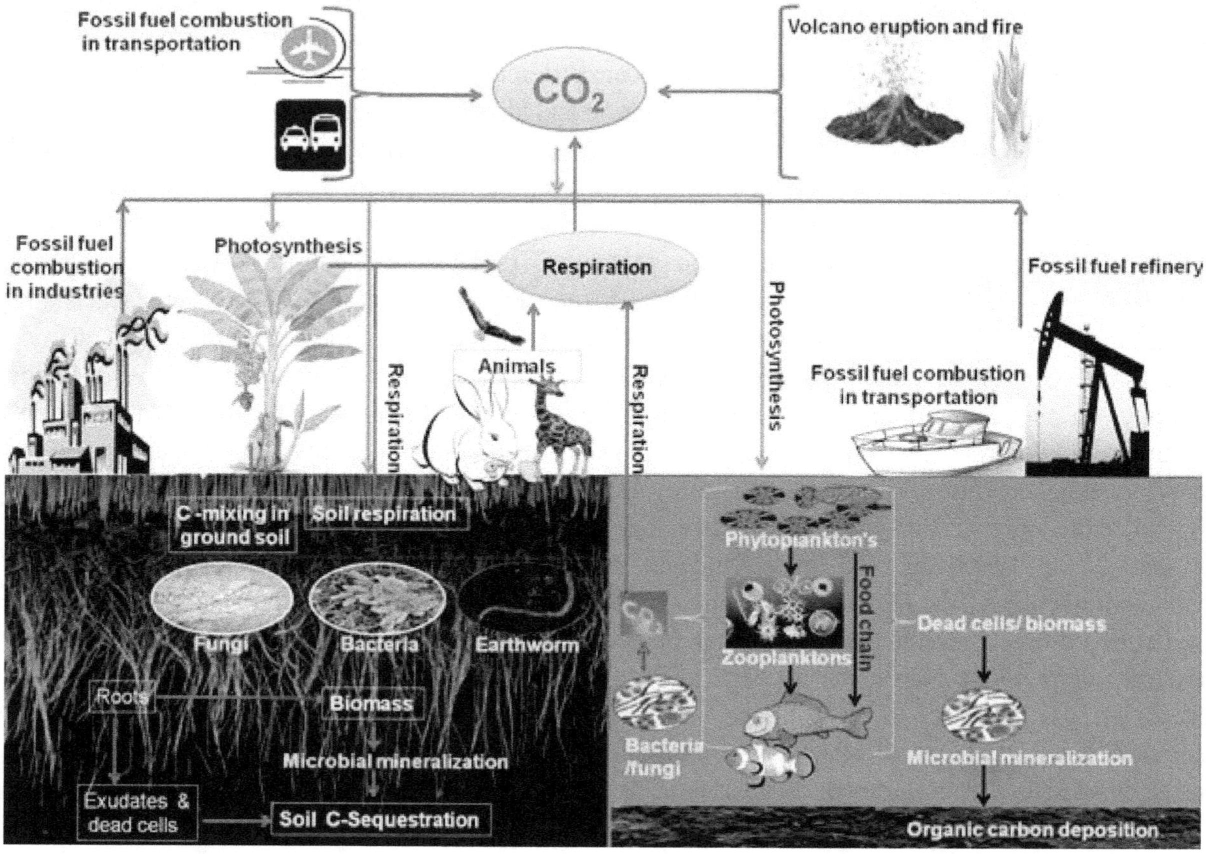

cies like *Azotobacter, Acetobacter, Azospirillum, Burkholderia, Bacillus* and *Pseudomonas* (Oves et al., 2013; Herman et al., 2008). Cyanobacterial species are also releasing plant growth promoting substances for crop development, and it is well known for nitrogen supplements and tolerance in agro-ecosystem (Prasanna et al., 2008). Moreover, cyanobacteria may also play an important role in C sequestration with crop improvement when both plant growth promoting rhizobacteria and cyanobacteria are applied as co-inoculants. The experimental findings suggested that cyanobacteria played a significant role in crop production, C-N fixation, and C: N ration balance in agro-ecosystem. Balanced C: N ratio in soil had a significant impact on microbial diversity promotion, which acts as a major sink of organic C (Averill and Finzi, 2011). The development of new techniques to control the balance of C:N ratio in soil for C sequestration are required to improve crop and C levels.

MYCORRHIZAL FUNGI FOR C SEQUESTRATION

Mycorrhizal fungi symbiotically colonize the roots of terrestrial plants and can be associated with 80% of different plant species (Smith & Read, 1997). These fungi are indirectly involved in C accumulation, symbiotically received plant nutrients through the roots that are covered by mycorrhizal hyphae in soil, and also obtained photosynthate-C from the plant. Additionally, mycorrhizal fungi also provide organic phosphorus to plant, restore greater amount of C in soil and assist in the subsequent growth of plants. Zhu and Miller (2003) estimated the annual input of mycorrhizal fungi with approximately 54–900 kg C/ha, which indicates the existence of a significant impact on climate change. Some mycorrhizal fungi are also the sole producers of glycoprotein, known as "glomalin," which can interact with soil minerals and persist for a long time in soil with organic C (Steinberg & Rillig, 2003). Substantial amount of soil organic C is stabilized by mycorrhizal fungi through formation of soil aggregates (Miller and Jastrow 1990). Mycorrhizal hyphae with roots can create a "sticky-string bag" that enmeshes and entangles soil particles to stabilize macroaggregates (Miller & Jastrow, 2000). In addition, the deposition of glomalin may also contribute to stabilize aggregates (Rillig, 2004). Mycorrhizal fungi play an active role in soil C sequestration through nutrient transfer from soil to the host plant, but they are unable to remove any free-living microorganisms (Rillig and Allen, 1999). Analytical studies have been conducted to investigate the use of amino acids in the bacterial peptidoglycan cell wall for tracing the assimilation of organic compounds. The experimental findings have shown that the role of fungi in C sequestration cannot be completed without bacterial decomposition of organic compounds due to soil organic C dynamics and storage (Pelz et al., 1998). In soil dynamics, decomposer organisms such as bacteria, fungi, and soil invertebrates play essential roles in controlling ecosystem processes such as decomposition of plant litter. Beare et al. (1995) have reported that soil microbes either can directly or indirectly affect the ecosystem processes. The direct microbial effects could be through consumption of debris and subsequently releasing inorganic compounds to the soil. On the other hand, the indirect effects could be through inducing some modifications on structure, biomass, and activities of soil microbial communities (Beare et al., 1995; Mikola et al 2002). In general, an elevated level of ambient CO_2 increases the growth response of plants (Poorter., 1993), root: shoot ratio of crops (Rogers et al., 1996), and subsequently stimulate the translocation of a greater amount of C to mycorrhizal structures (O'Neill. et al., 1987). Moreover, several studies have been conducted to investigate the impact of elevated CO_2 on mycorrhizal growth particularly on inducing changes in both percentage and total root length colonized per each plant (O'Neill. 1994Mycorrhizal fungi might act as rich sources of C under extreme environmental changes such as elevated CO_2 levels and Nitrogen deposition (Treseder and Allen 2000). The researchers observed the effect of both elevated CO_2 and N deposition on growth or biomass of mycorrhizal hyphae. The experimental findings have shown no significant change in hyphal biomass or growth response in both mycorrhizal species ECM and AM fungi when CO_2 level was raised (Table 1).

CARBON SEQUESTRATION BY MICROALGAE

Recently, a number of researchers investigated C sequestration by microalgae to overcome the global C load with the support of cheap and fast growing biological resources. In comparison to other microorganisms, microalgae are more efficient in CO_2 fixation, and its efficiency ranged from 28 to 53% in the presence of flue gases with CO_2 (Goswami et al., 2012). Microalgae can grow faster than other terrestrial plants with significant C sequestration efficiency (Fulke et al., 2010). Maximum C sequestration rate was reported at appropriate nutrient availability, pH, temperature, humidity, and light intensity (Cheng et al., 2013; Jacob-Lopes et al., 2008). C sequestration ability of microalgae species *Chlorella* and *Scenedesmus* have been widely reported (Toledo-Cervantes et al., 2013). The CO_2 sequestration

Table 1. Microalgae survival at different level of CO_2 for biomass production

Microalgae	Grow at CO2 %	Reference
Chlorella sp.	5-54	Yadav et al., 2015
Chlorella sp.	10-30	Li et al., 2015
Scenedesmus dimorphus	20	Giannetto et al., 2015
Scenedesmus obtusiusculus	5	Cervantes et al., 2013
Cyanidium caldarium	100	Seckbach et al., 1970
Scenedesmus sp.	80	Hanagata et al., 1992
Eudorina sp.	20	
Synechococcus elongatus	60	Miyairi, 1995
Chlorococcum littorale	60	Kodama et al., 1994
Euglena gracilis	45	Nakano et al., 1996
Chlorella sp.	40	Sakai et al., 1995
Chlorococcum littorale	40	Iwasaki et al., 1998
Haematococcus pluvialis	34	Huntley & Redalje, 2007
Scenedesmus obliquus	20	Li et al., 2011
Chlorella kessleri	18	deMorais & Costa, 2007b
Scenedesmus obliquus	18	deMorais & Costa, 2007a
Chlorella vulgaris	15	Yun et al., 1997
Nannochloropsis sp	15	Jiang et al., 2011
Spirulina platensis 1	15	Kumar et al., 2010
Chlorella sp. UK001	15	Murakami & Ikenouchi, 1997
Chlorella vulgaris	10	Sydney et al., 2010
Chlorella pyrenoidosa	10	Tang et al., 2011
Synechocystis aquatilis	10	Janssen et al., 2000
Botryococcus brauni	10	Sydney et al., 2010
Anabaena sp	10	Chiang et al., 2011
Chlorella sorokiniana.	16	Kumar et al., 2014

of Microalgae spp., *Chlorella pyrenoidosa* and *Scenedesmus abundans,* was significantly affected by continuous decrease in flow rate of CO_2 when grown in a tubular bath photo-bioreactor with continuous flow of CO_2 (Kargupta et al., 2015). Several researchers reported the C fixation in terms of efficient removal of CO_2 with biomass productivity (Table 1). Chiu et al. (2008) reported the CO_2 sequestration in algal biomass at different hydrodynamic conditions. In large scale, C sequestration requires proper design of structure with light, aeration facility, and flow rate of CO_2, which should be less than 10% achieved for maximum biomass productivity (Kargupta et al., 2015). The flow rate is one of the most important parameters because it may indirectly move the bubble volume, size, and residence time of gas bubbles inside the liquid phase of the reactor, and subsequently may affect the gas–liquid mass transfer. When CO_2 enriched air is bubbled up through the liquid phase, some of the CO_2 would be absorbed into the liquid phase while the remaining CO_2 would leave with the outgoing gas stream. Moreover, mass flow rate of gas is expected to influence the degree of turbulence in media, distribution of cells in the reactor, nutrient availability to cells as well as CO_2 absorption rate (Scoparo et al., 2010). Recent experimental studies have evaluated the CO_2 sequestration potential of microalgae, *C. pyrenoidosa*, which was grown in a batch photobioreactors. When bubbling of CO_2 last for more than 20 h, it means no CO_2 sequestration has occurred due to depletion of essential nutrients in the growing media and thus resulted in algal growth termination. Maximum biomass was observed at the highest growth rate after 15h incubation period when microalgal strains grown in a tubular batch photobioreactors (Kargupta, 2012). Microalgae can fix $CO_2$from three different sources, including (i) CO_2 from the atmosphere, (ii) CO_2 from industries, and (iii) in the form of soluble carbonates (e.g., $NaHCO_3$ and Na_2CO_3). Microalgae can be cultivated in closed systems or open ponds, but it requires aeration of CO_2 for microalgae cell growth and development. Industrial exhaust gases contain up to 15% CO_2, when it flows in an open or closed system, it provides a CO_2-rich source for microalgal cultivation and potentially more efficient route for CO_2 bio-fixation (Wang et al., 2008). From this study it was concluded that optimum flow rate of CO_2 and microalgae inoculants would be beneficial for C sequestration and for preparation of fair amount of biomass for human use.

ENERGY GENERATION THROUGH CARBON SEQUESTRATION

A certain class of carbon-dependent microorganisms known as carboxydotrophs, utilize the atmospheric CO and CO_2 as an energy/C source for their growth. Carboxydotrophs microorganisms are grown and kept for long time at higher CO concentration (~100ppm) (Weber and King 2007). Most of the carboxydotrophs grew aerobically and oxidized to CO and CO_2 by using Carbon monoxide dehydrogenase enzyme. The majority of carboxydotrophs can grow in diverse climatic conditions where CO is available at various concentrations from low to high CO concentration. For instance, carboxydotrophs can grow under thermophilic conditions, such as near volcano eruption where CO level is too high. Interestingly, carboxydotrophs are also able to grow in compost, marine environment, plant nodules and mammalian blood where CO level is too low. These microorganisms have a specific enzyme "C monoxide dehydrogenase" for oxidation of CO and CO_2. The enzyme Carbon monoxide dehydrogenase is able to catalyse the reversible reaction in the presence of water molecules.

$$CO + H_2O \xleftrightarrow{Co-dehydrogenase} CO_2 + 2H^+ + 2e^-$$

Interestingly, Carbon monoxide dehydrogenase catalysed reaction plays an important role in biofuel production. In the chemical reaction stated above, a proton is produced and moved toward cathode, where it combines with oxygen to generate electrons for energy production (Anand and Satyanarayana, 2012). In the table below, we have grouped thee CO utilizing microbes responsible for catalytic conversion of CO (Table 2).

Table 2. Bacterial diversity types and containing enzymes for carbon sequestration

Organisms	Type	Enzyme	Reference
Serratia sp.	Chemolithotrophic	CA & RuBisCO	Bharti et al., 2014; Srivastava et al., 2015
Halothiobacillus neapolitanus	Chemolithotrophic	RubisCO	Menon et al., 2008
Streptococcus pneumoniae	Respiratory track	carbonic anhydrase	Burghout et al., 2010
Bacillus schlegelli	Aerobic	carbonic anhydrase Mo-CODH	Muley et al., 2014
Bacillus subtilis	tropical soil	carbonic anhydrase	Oviya et al., 2012
Bacillus mucilaginosus	Aerobic	carbonic anhydrase	Zhang et al., 2011
Bacillus megaterium	highly alkaline soil	carbonic anhydrase	Achal & Pan, 2011
Neisseria gonorrhoeae	Aerobic	carbonic anhydrase	Chirica et al., 1997
Citrobacter freundii		Carbonic Anhydrase	Ramanan et al., 2009
Microcoleus cthonoplastes	Khilganta soda lake	carbonic anhydrase	Kupriyanova et al., 2007
Rhabdoderma lineare	hypersaline lake	carbonic anhydrase	Dudoladova et al., 2004
Streptomyces thermoautotrophicus	Aerobic/ Aerobic	Mo-CODH	Bell et al., 1987
Pseudomonas thermocarbxydovorans	Sewage waste water	Ni-CODH	Meyer & Schlegel, 1978
Pseudomonas carboxydovorans Streptomyces thermoautotrophicus	chemolithoautotrophic	CODH	Hugendieek & Meyer, 1992
Oligotropha carboxydovorans	Sewage waste water	CODH	Roberts et al., 2004
Mycobacterium sp.		CODH	Oh et al., 2010
Mycobacterium tuberculosis	Mammals samples	CODH+NODH	Song et al., 2010
Mycobacterium sp.	Mammals samples	CODH+NODH	Song et al., 2010
Stappia sp.	Marine sample	CODH	Weber & King 2007
Bradyrhizobium Japonicum	Root nodules	CODH	Lorite et al., 2000
Pseudonocardia carboxy divorans	---	Mo-CODH	Park et al., 2008
Carboxydothermus hydrogenoformans	Hot swamp mud	CoA-CODH	Svetlichnyi et al., 1991
Thermosinus carboxydivvorans; Thermoterrabacterium ferrireducens	Thermal deep vent	CODH	Henstra et al., 2004
Carboxydocella thermoautotrophica	Thermal deep vent	CODH	Sokolova et al., 2002
Methanopyrus kundleri	Black smok	CODH	Sin et al., 1999
Methanothermobacter thermoautotrophicus	Natural environment	Methyl coenzyme M reductase	Luo et al., 2002
zospirillum lipoferum, Rhodopseudomonas palustris, Bradyrhizobium japonicum,	Soil sample	RubisCO	Yuan et al., 2012

FUTURE RESEARCH DIRECTIONS AND CONCLUSION

The interpretations of obtained data from laboratory and field experiments quantify the C fractions and temperature of earth. Most of the studies represent C dynamics in ecosystem, but they are unable to mitigate the effect and overall increasing of C-level in earth atmosphere. A remarkable increase in anthropogenic activities is responsible for C emission, which may lead to climate change and possesses a significant threat to natural system. All individual people, governmental bodies, and decision makers must consider the consequences of increasing C-level and thus direct their efforts to control the C-level by finding alternative methods and applying the advanced technologies for controlling C emissions. Implementation of innovative microbial inoculants-based technology may enhance the C uptake through increasing biomass or productivity and reducing the C losses by sequestering C in the form of primary productivity. Undoubtedly, number of microbes has a crucial role in global C sequestration through their metabolic activities, enzymatic conversation, and transformation of atmospheric C dioxide into cell metabolites for cell viability. The screening and development of microbial inoculants for atmospheric C solution is an excellent approach to overcome the atmospheric C load. A form of bioreactor with potential C sequestrating bacteria and fungi has been established with C emission site, which helps to contain the emitted C from industries. Further, microbial inoculants with plant growth-promoting activities are also highly beneficial to improve the environmental C-level and the performance of crops. Based on the observations, the application of microbial technology to improve C level in atmosphere for sustainable ecosystem will be fruitful.

ACKNOWLEDGMENT

This chapter writing was supported by the Ministry of Higher Education, Kingdom of Saudi Arabia, Centre of Excellence in Environmental Studies, King Abdulaziz University, Jeddah, Kingdom of Saudi Arabia.

REFERENCES

Achal, V., & Pan, X. (2011). Characterization of urease and carbonic anhydrase producing bacteria and their role in calcite precipitation. *Current Microbiology*, *62*(3), 894–902. doi:10.1007/s00284-010-9801-4 PMID:21046391

Adu, J. K., & Oades, J. M. (1978). Utilization of organic materials in soil aggregates by bacteria and fungi. *Soil Biology & Biochemistry*, *10*(2), 117–122. doi:10.1016/0038-0717(78)90081-0

Akinola, O. O. (2014). Review of the role of plant in carbondioxide sequestration globally using chlorophy II or leaf index. *J. of Environ & Earth Sci*, *4*, 22–30.

Anand, A., & Satyanarayana, T. (2012). Applicability of carboxydotrophic bacterial Cmonoxide dehydrogenase (CODH) in Csequestration and bioenergy generation. *Journal of Scientific and Industrial Research*, *71*, 381–384.

Ashrafuzzaman, M., Hossen, F. A., Razi, I. M., Hoque, M. A., Islam, M. Z., Shahidullah, S. M., & Meon, S. (2009). Efficiency of plant growthpromoting rhizobacteria (PGPR) for the enhancement of rice growth. *African Journal of Biotechnology*, *8*, 1247–1252.

Averill, C., & Finzi, A. (2011). Increasing plant use of organic nitrogen with elevation is reflected in nitrogen uptake rates and ecosystem δ15N. *Ecology*, *92*(4), 883–891. doi:10.1890/10-0746.1 PMID:21661551

Bailey, V. L., Peacock, A. D., Smith, J. L., & Bolton, H. Jr. (2002). Relationships between soil microbial biomass determined by chloroform fumigation–extraction, substrate-induced respiration and phospholipid fatty acid analysis. *Soil Biology & Biochemistry*, *34*(9), 1385–1389. doi:10.1016/S0038-0717(02)00070-6

Baldrian, P. (2003). Interactions of heavy metals with white-rot fungi. *Enzyme and Microbial Technology*, *32*(1), 78–91. doi:10.1016/S0141-0229(02)00245-4

Bazot, S., Blum, H., & Robin, C. (2008). Nitrogen rhizodeposition assessed by a 15NH3 shoot pulse-labelling of Lolium perenne L. grown on soil exposed to 9 years of CO2 enrichment. *Environmental and Experimental Botany*, *63*(1-3), 410–415. doi:10.1016/j.envexpbot.2007.10.020

BCAS. (2013). Special Reports[Chinese Academy of Sciences.]. *Bulletin of the Chinese Academy of Sciences*, *27*, 3.

Beare, M. H., Coleman, D. C., Crossley, D. A. Jr, Hendrix, P. F., & Odum, E. P. (1995). A hierarchical approach to evaluating the significance of soil biodiversity to biogeochemical cycling. *Plant and Soil*, *170*(1), 5–22. doi:10.1007/BF02183051

Beh, M., Strauss, G., Huber, R., Stetter, K. O., & Fuchs, G. (1993). Enzymes of the reductive citric acid cycle in the autotrophic eubacterium Aquifex pyrophilus and in the archaebacterium Thermoproteus neutrophilus. *Archives of Microbiology*, *160*(4), 306–311. doi:10.1007/BF00292082

Beilby, J. P. (1983). Effects of inhibitors on early protein, RNA, and lipid synthesis in germinating vesicular-arbuscular mycorrhizal fungal spores of Glomus caledonium. *Canadian Journal of Botany*, *29*, 596–601. PMID:6883219

Berg, B., & Matzner, E. (1997). Effect of N deposition on decomposition of plant litter and soil organic matter in forest systems. *Environ Rev*, *5*(1), 1–25. doi:10.1139/a96-017

Bharti, R. K., Srivastava, S., & Thakur, I. S. (2014). Proteomic Analysis of CConcentrating Chemolithotrophic Bacteria Serratia sp. for Sequestration of CDioxide. *PLoS ONE*, *9*(3), e91300. doi:10.1371/journal.pone.0091300 PMID:24619032

Boden, A. B., & Marland, T. G. (2014). *Climate change 2014 – Impacts, adaptation and vulnerability: Global and sectoral aspect*. New York: Cambridge university press.

Burghout, P., Cron, L. E., Gradstedt, H., Quintero, B., Simonetti, E., Bijlsma, J. J., & Hermans, P. W. et al. (2010). Carbonic anhydrase is essential for Streptococcus pneumoniae growth in environmental ambient air. *Journal of Bacteriology*, *192*(15), 4054–4062. doi:10.1128/JB.00151-10 PMID:20525828

Canadell, J. G., Le Quéré, C., Raupach, M. R., Field, C. B., Buitenhuis, E. T., Ciais, P., & Marland, G. et al. (2007). Contributions to accelerating atmospheric CO2 growth from economic activity, Cintensity, and efficiency of natural sinks. *Proceedings of the National Academy of Sciences of the United States of America*, *104*(47), 18866–18870. doi:10.1073/pnas.0702737104 PMID:17962418

Carreiro, M. M., Sinsabaugh, R. L., Repert, D. A., & Parkhurst, D. F. (2000). Microbial enzyme shifts explain litter decay responses to simulated nitrogen deposition. *Ecology*, *81*(9), 2359–02365. doi:10.1890/0012-9658(2000)081[2359:MESELD]2.0.CO;2

Chen, X. J., Wu, X. H., Jian, Y., Yuan, H. Z., Zhou, P., Ge, T. D., & Wu, J. S. et al. (2015). Cdioxide assimilation potential, functional gene amount and RubisCO activity of autotrophic microorganisms in agricultural soils. *Huan Jing Ke Xue*, *35*, 1144–1150. PMID:24881409

Cheng, J., Huang, Y., Feng, J., Sun, J., Zhou, J., & Cen, K. (2013). Improving CO2 fixation efficiency by optimizing Chlorella PY-ZUI culture conditions in sequential bioreactors. *Bioresource Technology*, *144*, 321–327. doi:10.1016/j.biortech.2013.06.122 PMID:23891832

Chiu, S. Y., Kao, C. Y., Chen, C. H., Kuan, T. C., Ong, S. C., & Lin, C. S. (2008). Reduction of CO2 by a high density culture of Chlorella sp. in a semicontinuous photobioreactor. *Bioresource Technology*, *99*(9), 3389–3396. doi:10.1016/j.biortech.2007.08.013 PMID:17904359

Claus, H., Gleixner, G., & Filip, Z. (1999). Formation of humic-like substances in mixed and pure cultures of aquatic microorganisms. *Acta Hydrochimica et Hydrobiologica*, *27*(4), 200–207. doi:10.1002/(SICI)1521-401X(199907)27:4<200::AID-AHEH200>3.0.CO;2-Q

Clemmensen, K., Bahr, A., Ovaskainen, O., Dahlberg, A., Ekblad, A., Wallander, H., & Lindahl, B. et al. (2013). Roots and Associated Fungi Drive Long-Term CSequestration in Boreal Forest. *Science*, *339*(6127), 1615–1618. doi:10.1126/science.1231923 PMID:23539604

Current Data for Atmospheric CO2. 2015. (http://co2now.org/,)

Davidson, E. A., & Janssens, I. A. (2006). Temperature sensitivity of soil Cdecomposition and feedbacks to climate change. *Nature*, *440*(7081), 165–173. doi:10.1038/nature04514 PMID:16525463

Di-Fiore, A., Alterio, V., Monti, S. M., DeSimone, G., & DAmbrosio, K. (2015). Thermostable Carbonic Anhydrases in Biotechnological Applications. *International Journal of Molecular Sciences*, *16*(7), 15456–15480. doi:10.3390/ijms160715456 PMID:26184158

Douds, J. D. D., Pfeffer, P. E., & Shachar-Hill, Y. (2000). Cpartitioning, cost, and metabolism of arbuscular mycorrhizas. In Arbuscular mycorrhizas: physiology and function. 107-129.

Drigo, B., Pijl, A. S., Duyts, H., Kielak, A. M., Gamper, H. A., Houtekamer, M. J., & vanVeen, J. A. et al. (2010). Shifting Cflow from roots into associated microbial communities in response to elevated atmospheric CO2. *Proceedings of the National Academy of Sciences of the United States of America*, *107*(24), 10938–10942. doi:10.1073/pnas.0912421107 PMID:20534474

Druckenmiller, T., & Maroto-Valer, M. M. (2005). Csequestration using brine of adjusted pH to form mineral carbonates. *Matthew L. Fuel Processing Technology*, *86*(14-15), 1599–1614. doi:10.1016/j.fuproc.2005.01.007

EDGAR. (2014). *Eemissions-report-93171*. Retrieved from http://edgar.jrc.ec.europa.eu/news_docs/jrc-2014-trends-in-global-co2-

Fernández-Martínez, M., Vicca, S., Janssens, I. A., Sardans, J., Luyssaert, S., Campioli, M., & Peñuelas, J. et al. (2014). Nutrient availability as the key regulator of global forest Cbalance. *Nature Climate Change*, *4*(6), 471–476. doi:10.1038/nclimate2177

Frey, S. D., Gupta, V. V. S. R., Elliott, E. T., & Paustian, K. (2001). Protozoan grazing affects estimates of Cutilization efficiency of the soil microbial community. *Soil Biology & Biochemistry*, *33*(12-13), 1759–1768. doi:10.1016/S0038-0717(01)00101-8

Fulke, B. A., Mudliar, S. N., Yadav, R., Shekh, A., Srinivasan, N., Ramanan, R., & Chakrabarti, T. et al. (2010). Bio-mitigation of CO2, calcite formation and simultaneous biodiesel precursors production using Chlorella species. *Bioresource Technology*, *101*(21), 8473–8476. doi:10.1016/j.biortech.2010.06.012 PMID:20580227

Gaillard, V., Chenu, C., & Recous, S. (2003). Cmineralisation in soil adjacent to plant residues of contrasting biochemical quality. *Soil Biology & Biochemistry*, *35*(1), 93–99. doi:10.1016/S0038-0717(02)00241-9

Gans, J., Wolinsky, M., & Dunbar, J. (2005). Computational improvements reveal great bacterial diversity and high metal toxicity in soil. *Science*, *309*(5739), 1387–1390. doi:10.1126/science.1112665 PMID:16123304

Gosselin, P. (2013). *Most of the rise in CO2 Likely Comes From Natural Sources*. Retrieved from http://notrickszone.com/2013/03/02/most-of-the-rise-in-co2-likely-comes-from-natural-sources/# sthash.p7H9e59p.dpbs

Goswami, R. C. D., Kalita, N., & Kalita, M. C. (2012). A study on growth and CO2 mitigation by microalgae Selenastrum sp.: Its growth behaviour under different nutrient environments and lipid production. *Ann Biol Res*, *3*, 499–510.

Guggenberger, G., Frey, S. D., Six, J., Paustian, K., & Elliot, E. T. (1999). Bacterial and fungal cell-wall residues in conventional and no-tillage agroecosystems. *Soil Science Society of America Journal*, *63*(5), 1188–1198. doi:10.2136/sssaj1999.6351188x

Gupta, V. S. S. R., & Ponge, J. F. (1998). Regulation of microbial activity and organic matter dynamics by macro-organisms in the detritusphere.*Proc World Cong Soil Sci,*20–26.

Herman, M. A. B., Nault, B. A., & Smart, C. D. (2008). Effects of plant growth promoting rhizobacteria on bell pepper production and green peach aphid infestations in New York. *Crop Protection (Guildford, Surrey)*, *27*(6), 996–1002. doi:10.1016/j.cropro.2007.12.004

Hill, G. T., Mitkowski, N. A., Aldrich-Wolfe, L., Emele, L. R., Jurkonie, D. D., Ficke, A., & Nelsona, E. B. et al. (2000). Methods for assessing the composition and diversity of soil microbial communities. *Applied Soil Ecology*, *15*(1), 25–36. doi:10.1016/S0929-1393(00)00069-X

Ho, I., & Trappe, J. M. (1973). Translocation of 14C from Festuca plants to their endomycorrhizal fungi. *Nature*, *224*, 30–31.

IPCC. (2013). Fourth assessment report: climate change (AR5). Stockholm, Sweden: IPCC.

Jacob-Lopes, E., Lacarda, L. M. C. F., & Franco, T. T. (2008). Biomass production and Cdioxide fixation by Aphanothecemicroscopica Nageli in a bubble column photobioreactor. *Biochemical Engineering Journal*, *40*(1), 27–34. doi:10.1016/j.bej.2007.11.013

Jennings, D. H. (1995). *The Physiology of Fungal Nutrition*. Cambridge, UK: Cambridge University Press. doi:10.1017/CBO9780511525421

Kargupta, W. (2012). *CSequestration by Microalgae* (M.Tech. thesis). IIT Bombay, Mumbai, India.

Kargupta, W., Ganesh, A., & Mukherji, S. (2015). Estimation of Cdioxide sequestration potential of microalgae grown in a batch photobioreactor. *Bioresource Technology*, *180*, 370–375. doi:10.1016/j.biortech.2015.01.017 PMID:25616748

Karlsson, J., Clarke, A. K., Chen, Z. Y., Hugghins, S. Y., Park, Y. I., Husic, H. D., & Samuelsson, G. et al. (1998). A novel alpha-type carbonic anhydrase associated with the thylakoid membrane in Chlamydomonas reinhardtii is required for growth at ambient CO2. *The EMBO Journal*, *17*(5), 1208–1216. doi:10.1093/emboj/17.5.1208 PMID:9482718

Keyeo, F., Aishah, N., & Amir, H. G. (2011). The effect of nitrogen fixing activity and phytohormone production of diazotroph in promoting growth of rice seedlings. *Biotechnol*, *10*(3), 267–273. doi:10.3923/biotech.2011.267.273

Khan, M. S., Zaidi, A., Wani, P. A., & Oves, M. (2009). Role of plant growth promoting rhizobacteria in the remediation of metal contaminated soils. *Environmental Chemistry Letters*, *7*(1), 1–19. doi:10.1007/s10311-008-0155-0

King, G. M. (2011). Enhancing soil Cstorage for Cremediation: Potential contributions and constraints by microbes. *Trends in Microbiology*, *19*(2), 75–84. doi:10.1016/j.tim.2010.11.006 PMID:21167717

Klironomos, J. N., Rillig, M. C., & Allen, M. F. (1996). Below-ground microbial and microfaunal responses to Artemisia tridentata grown under elevated atmospheric CO2. *Functional Ecology*, *10*(4), 527–534. doi:10.2307/2389946

Klironomos, J. N., Rillig, M. C., Allen, M. F., Zak, D. R., Kubiske, M., & Pregitzer, K. S. (1997). Soil fungal–arthropod responses to Populus tremuloides grown under enriched atmospheric CO# under field conditions. *Global Change Biology*, *3*(6), 473–478. doi:10.1046/j.1365-2486.1997.00085.x

Klironomos, J. N., Ursic, M., Rillig, M. C., & Allen, M. F. (1998). Interspecific differences in the response of arbuscular mycorrhizal fungi to Artemisia tridentata grown under elevated atmospheric CO2. *The New Phytologist*, *138*(4), 599–605. doi:10.1046/j.1469-8137.1998.00141.x

Kusian, B., Sultemeyer, D., & Bowien, B. (2002). Carbonic anhydrase is essential for growth of Ralstonia eutropha at ambient CO2concentrations. *Journal of Bacteriology*, *184*(18), 5018–5026. doi:10.1128/JB.184.18.5018-5026.2002 PMID:12193617

Lal, R. (2014). Abating climate change and feeding the world through soil Csequestration. In Soil as world heritage. Springer.

Lal, R. (2014). Soil Csequestration to mitigate climate change. *Geoderma*, *123*(1-2), 1–22. doi:10.1016/j.geoderma.2004.01.032

Lal, R., Kimble, J. M., Follett, R. F., & Cole, C. V. (1998). *The potential of US cropland to sequester Cand mitigate the greenhouse effect*. Boca Raton, FL: CRC Press, LLC.

Langley, J. A., & Megonigal, J. P. (2010). Ecosystem response to elevated CO2 levels limited by nitrogen-induced plant species shift. *Nature*, *466*(7302), 96–99. doi:10.1038/nature09176 PMID:20596018

LeQuéré, C., Raupach, M. R., Canadell, J. G., Marland, G., Bopp, L., Ciais, P., & Foster, P. et al. (2009). Trends in the sources and sinks of Cdioxide. *Nature Geoscience*, *2*(12), 831–836. doi:10.1038/ngeo689

Leung, D. Y., Caramanna, G., & Maroto-Valer, M. M. (2014). An overview of current status of Cdioxide capture and storage technologies. *Renewable & Sustainable Energy Reviews*, *39*, 426–443. doi:10.1016/j.rser.2014.07.093

Liang, C., & Balser, T. C. (2011). Microbial production of recalcitrant organic matter in global soils: Implications for productivity and climate policy. *Nature Reviews. Microbiology*, *9*(1), 75. doi:10.1038/nrmicro2386-c1 PMID:21113179

Lu, Y., & Conrad, R. (2005). In situ stable isotope probing of methanogenic archaea in the rice rhizosphere. *Science*, *309*(5737), 1088–1090. doi:10.1126/science.1113435 PMID:16099988

Lu, Y., Lueders, T., Friedrich, M. W., & Conrad, R. (2005). Detecting active methanogenic populations on rice roots using stable isotope probing. *Environmental Microbiology*, *7*(3), 326–336. doi:10.1111/j.1462-2920.2005.00697.x PMID:15683393

Lussenhop, J., Treonis, A., Curtis, P. S., Teeri, J. A., & Vogel, C. S. (1998). Response of soil biota to elevated atmospheric CO! in poplar model systems. *Oecologia*, *113*(2), 247–251. doi:10.1007/s004420050375

Mauna Loa. (2015). *Hawaii observatory, Earth System Research Laboratory, Global monitoring division*. Retrieved from http://www.esrl.noaa.gov/gmd/obop/mlo/

Mehnaz, S., Mirza, M. S., Haurat, J., Bally, R., Normand, P., Bano, A., & Malik, K. A. (2001). Isolation and 16S rRNA sequence analysis of the beneficial bacteria from the rhizosphere of rice. *Canadian Journal of Microbiology*, *47*(2), 110–117. doi:10.1139/w00-132 PMID:11261489

Menon, B., Dou, Z., Heinhorst, S., & Cannon, G. C. (2008). Halothiobacillus neapolitanus Carboxysomes Sequester Heterologous and Chimeric RubisCO Species. *PLoS ONE*, *3*(10), e3570. doi:10.1371/journal.pone.0003570 PMID:18974784

Mikola, J., Bardgett, R. D., & Hedlund, K. (2002). Biodiversity, ecosystem functioning and soil decomposer food webs. In M. Loreau, S. Naeem, & P. Inchausti (Eds.), *Biodiversity andecosystem functioning–synthesis and perspectives* (pp. 169–180). Oxford, UK: Oxford University Press.

Millard, P., & Singh, B. K. (2010). Does grassland vegetation drive soil microbial diversity? *Nutrient Cycling in Agroecosystems*, *88*(2), 147–158. doi:10.1007/s10705-009-9314-3

Miller, R. M., & Jastrow, J. D. (1990). Hierarchy of root and mycorrhizal fungal interactions with soil aggregation. *Soil Biology & Biochemistry*, *22*(5), 579–584. doi:10.1016/0038-0717(90)90001-G

Miller, R. M., & Jastrow, J. D. (2000). Mycorrhizal fungi influence soil structure. In Y. Kapulnik & D. D. Douds Jr., (Eds.), *Arbuscular mycorrhizas: physiology and function* (pp. 3–18). Dordrecht, The Netherlands: Kluwer. doi:10.1007/978-94-017-0776-3_1

Muley, P., Dhumal, M., & Vora, D. (2014). Sequestration of atmospheric Cdioxide by microbial carbonic anhydrase. *IOSR J Environ Sci, Toxicol &. Food Technology*, *8*, 45–48.

Norby, R. J., & Cotrufo, M. F. (1998). Global change: A question of litter quality. *Nature*, *396*(6706), 17–18. doi:10.1038/23812

ONeill, E. G. (1994). Responses of soil biota to elevated atmospheric Cdioxide. *Plant and Soil*, *165*(1), 55–65. doi:10.1007/BF00009962

ONeill, E. G., Luxmoore, R. J., & Norby, R. J. (1987). Increases in mycorrhizal colonization and seedling growth in Pinus echinata and Quercus alba in an enriched CO! atmosphere. *Canadian Journal of Forest Research*, *17*(8), 878–883. doi:10.1139/x87-139

Oves, M., Khan, M.S., & Zaidi, A. (2013). Biosorption of heavy metals by Bacillus thuringiensis strain OSM29 originating from industrial effluent contaminated north Indian soil. *Saudi Journal of Biological Sciences, 20,* 121-129.

Oviya, M., Giri, S.S., Sukumaran, V., Natarajan, P., (2012). Immobilization of carbonic anhydrase enzyme purified from Bacillus subtilis VSG-4 and its application as CO2 sequesterer. *Prep Biochem & Iotechnol, 42,* 462–475.

Payne, W. J. (1970). Energy yields and growth of heterotrophs. *Annual Review of Microbiology*, *24*(1), 17–52. doi:10.1146/annurev.mi.24.100170.000313 PMID:4927132

Pelz, O., Cifuentes, L. A., Hammer, B. T., Kelley, C. A., & Coffin, R. B. (1998). Tracing the assimilation of organic compounds using δ13C analysis of unique amino acids in the bacterial peptidoglycan cell wall. *FEMS Microbiology Ecology*, *25*, 229–240.

Piromyou, P., Buranabanyat, B., Tantasawat, P., Tittabutr, P., Boonkerd, N., & Teaumroong, N. (2011). Effect of plant growth promoting rhizobacteria (PGPR) inoculation on microbial community structure in rhizosphere of forage corn cultivated in Thailand. *European Journal of Soil Biology*, *11*(47), 44–54. doi:10.1016/j.ejsobi.2010.11.004

Poorter, H. (1993). Interspecific variation in the growth response of plants to an elevated ambient CO2 concentration. *Vegetatio*, *140*(1), 77–97. doi:10.1007/BF00048146

Prasanna, R., Jaiswal, P., Singh, Y. V., & Singh, P. K. (2008). Influence of biofertilizers and organic amendments on nitrogenase activity and phototrophic biomass of soil under wheat. *Acta Agron Hung*, *56*(2), 149–159. doi:10.1556/AAgr.56.2008.2.4

Prasanna, R., Joshi, M., Rana, A., & Nain, L. (2010). Modulation of IAA production in cyanobacteria by tryptophan and light. *Polish Journal of Microbiology / Polskie Towarzystwo Mikrobiologow = the Polish Society of Microbiologists*, *59*, 99–105. PMID:20734754

Prasanna, R., Joshi, M., Rana, A., Shivay, Y. S., & Nain, L. (2012). Influence of co-inoculation of bacteria-cyanobacteria on crop yield and C–N sequestration in soil under rice crop. *World Journal of Microbiology & Biotechnology*, *28*(3), 1223–1235. doi:10.1007/s11274-011-0926-9 PMID:22805842

Prentice, I. C., Farquhar, G. D., Fasham, M. J. R., Goulden, M. I., Heinmann, M., Jaramillo, V. J., … Wallace, D.W. (2001). The Ccycle and atmospheric Cdioxide, in Climate Change. The Scientific Basis. Contribution of Working Group I to the Third Assessment Report of the Intergovernmental Panel on Climate Change. Cambridge University Press.

Read, D. J. (1998b). The role of mycorrhiza in determining the response of Plantago lanceolata to CO! enrichment. *The New Phytologist*, *139*(2), 367–373. doi:10.1046/j.1469-8137.1998.00200.x

Reicosky, D. (2003). Conservation Agriculture: Global Environmental Benefits of Soil CManagement. In L. García-Torres, J. Benites, A. Martínez-Vilela, & A. Holgado-Cabrera (Eds.), *Conservation agriculture* (pp. 291–300). Springer. doi:10.1007/978-94-017-1143-2_35

Richardson, C. J. (2008). *The Everglades Experiments: Lessons for Ecosystem Restoration*. New York: Springer-Verlag. doi:10.1007/978-0-387-68923-4

Rillig, M. C. (2004). Arbuscular mycorrhizae, glomalin, and soil aggregation. *Canadian Journal of Soil Science*, *84*(4), 355–363. doi:10.4141/S04-003

Rillig, M. C., & Allen, M. F. (1998). Arbuscular mycorrhizae of Gutierrezia sarothrae and elevated Cdioxide: Evidence for shifts in C allocation to and within the mycobiont. *Soil Biology & Biochemistry*, *30*(14), 2001–2008. doi:10.1016/S0038-0717(98)00073-X

Rillig, M. C., & Allen, M. F. (1999). What is the role of arbuscular mycorrhizal fungi in plant-to-ecosystem responses to elevated atmospheric CO2? *Mycorrhiza*, *9*(1), 1–8. doi:10.1007/s005720050257

Rillig, M. C., Field, C. B., & Allen, M. F. (1999a). Soil biota responses to long-term atmospheric CO! enrichment in two California annual grasslands. *Oecologia*, *119*(4), 572–577. doi:10.1007/s004420050821

Rogers, H. H., Prior, S. A., Runion, G. B., & Mitchell, R. J. (1996). Root to shoot ratio of crops as influenced by CO2. *Plant and Soil*, *187*(2), 229–248. doi:10.1007/BF00017090

Schleppi, P., Wallin, I. B., Hagedorn, F., & Korner, C. (2012). Increased nitrate availability in the soil of a mixed mature temperate forest subjected to elevated CO2 concentration (canopy FACE). *Global Change Biology*, *18*(2), 757–768. doi:10.1111/j.1365-2486.2011.02559.x

Schmidt, M. W. I., Torn, M. S., Abiven, S., Dittmar, T., Guggenberger, G., Janssens, I. A., & Trumbore, S. E. et al. (2011). Persistence of soil organic matter as an ecosystem property. *Nature*, *478*(7367), 49–56. doi:10.1038/nature10386 PMID:21979045

Scoparo, C. H. G., Queiroz, M. I., & Franco, T. T. (2010). Biotransformations of Cdioxide in photobioreactors. *Energy Conversion and Management*, *51*(5), 894–900. doi:10.1016/j.enconman.2009.11.027

Secretariat of the Convention on Biological Diversity (SCBD). (2009). Scienti- c Synthesis of the Impacts of Ocean Fertilization on Marine Biodiversity. Technical Series No. 45. Author.

Siegenthaler, U., & Sarmiento, J. (1993). Atmospheric Cdioxide and the ocean. *Nature*, *365*(6442), 119–125. doi:10.1038/365119a0

Singh, B. K., Dawson, L. A., Macdonald, C. A., & Buckland, S. M. (2009a). Impact of biotic and abiotic interaction on soil microbial communities and functions: A field study. *Applied Soil Ecology*, *41*(3), 239–248. doi:10.1016/j.apsoil.2008.10.003

Sinsabaugh, R. L., & Shah, J. J. (2010). Integrating resource utilization and temperature in metabolic scaling of riverine bacterial production. *Ecology*, *91*(5), 1455–1465. doi:10.1890/08-2192.1 PMID:20503877

Six, J., Frey, S. D., Thiet, R. K., & Batten, K. M. (2006). Bacterial and Fungal Contributions to CSequestration in Agroecosystems. *Soil Science Society of America Journal*, *70*(2), 555–569. doi:10.2136/sssaj2004.0347

Smith, J. L., & Paul, E. A. (1990). The significance of soil microbial biomass estimations. In J. Bollag & G. Stotsky (Eds.), *Soil biochemistry* (Vol. 8, pp. 357–393). Mercel Dekker, New York.

Smith, K. S., Jakubzick, C., Whittam, T. S., & Ferry, J. G. (1999). Carbonic anhydrase is an ancient enzyme widespread in prokaryotes. *Proceedings of the National Academy of Sciences of the United States of America*, *96*(26), 15184–15189. doi:10.1073/pnas.96.26.15184 PMID:10611359

Smith, S. E., & Read, D. J. (1997). *Mycorrhizal symbiosis*. San Diego, CA: Academic Press.

Sommer, R., & Bossio, D. (2014). Dynamics and climate change mitigation potential of soil organic Csequestration. *Journal of Environmental Management*, *144*, 83–87. doi:10.1016/j.jenvman.2014.05.017 PMID:24929498

Staddon, P. L., Fitter, A. H., & Graves, J. D. (1999). Effect of elevated atmospheric CO! on mycorrhizal colonization, external mycorrhizal hyphal production and phosphorus inflow in Plantago lanceolata and Trifolium repens in association with the arbuscular mycorrhizal fungus Glomus mosseae. *Global Change Biology*, *5*(3), 347–358. doi:10.1046/j.1365-2486.1999.00230.x

Steinberg, P. D., & Rillig, M. C. (2003). Differential decomposition of arbuscular mycorrhizal fungal hyphae and glomalin. *Soil Biology & Biochemistry*, *35*(1), 191–194. doi:10.1016/S0038-0717(02)00249-3

Suberkropp, K., & Weyers, H. (1996). Application of fungal and bacterial production methodologies to decomposing leaves in streams. *Applied and Environmental Microbiology*, *62*, 1610–1615. PMID:16535312

Tabita, F. R. (1999). Microbial ribulose-1,5-bisphosphate carboxylase/oxygenase: A different perspective. *Photosynthesis Research*, *60*(1), 1–28. doi:10.1023/A:1006211417981

Thieta, R. K., Freya, S. D., & Six, J. (2006). Do growth yield efficiencies differ between soil microbial communities differing in fungal:bacterial ratios? Reality check and methodological issues. *Soil Biology & Biochemistry*, *38*(4), 837–844. doi:10.1016/j.soilbio.2005.07.010

Tiedje, J. M., Asuming-Brempong, S., Nusslein, K., Marsh, T. L., & Flynn, S. J. (1999). Opening the black box of soil microbial diversity. *Applied Soil Ecology*, *13*(2), 109–122. doi:10.1016/S0929-1393(99)00026-8

Toledo-Cervantes, A., Morales, M., Novelo, E., & Revah, S. (2013). Cdioxide fixation and lipid storage by Scenedesmus obtusiusculus. *Bioresource Technology*, *130*, 652–658. doi:10.1016/j.biortech.2012.12.081 PMID:23334023

Treseder, K. K., & Allen, M. F. (2000). Mycorrhizal fungi have a potential role in soil Cstorage under elevated CO2 and nitrogen deposition. *The New Phytologist*, *147*(1), 189–200. doi:10.1046/j.1469-8137.2000.00690.x

Tu, C., Foster, L., Alvarado, A., McKenna, R., Silverman, D. N., & Frost, S. C. (2012). Role of Zinc in Catalytic Activity of Carbonic Anhydrase IX. *Archives of Biochemistry and Biophysics*, *521*(1-2), 90–94. doi:10.1016/j.abb.2012.03.017 PMID:22465027

UNFCCC. (2007). *Report of the conference of parties on its thirteenth session*. Bali, Indonesia: Author.

Vaughan, N. E., Lenton, T. M., & Shepherd, J. G. (2009). Climate change mitigation: Trade-offs between delay and strength of action required. *Climatic Change*, *96*(1-2), 29–43. doi:10.1007/s10584-009-9573-7

Verburg, P. H., Kok, K., Pontius, R. G., & Veldkamp, A. J. (2004). Modeling Land-Use and Land-Cover Change. Part of the series Global Change - The IGBP Series.

Vitousek, P. M. (2004). *Nutrient cycling and limitation: Hawai'i as a model system*. Princeton University Press.

Waldrop, M. P., Zak, D. R., Sinsabaugh, R. L., Gallo, M., & Lauber, C. (2004). Nitrogen deposition modifies soil Cstorage through changes in microbial enzymatic activity. *Ecological Applications*, *14*(4), 1172–1177. doi:10.1890/03-5120

Wang, B., Li, Y., Wu, N., & Lan, C. Q. (2008). CO2 bio-mitigation using microalgae. *Applied Microbiology and Biotechnology*, *79*(5), 707–718. doi:10.1007/s00253-008-1518-y PMID:18483734

Weber, C. F., & King, G. M. (2007). Physiological ecological, and phylogenetic characterization of stappia, a Marine CO-Oxidizing Bacterial Genus. *Applied and Environmental Microbiology*, *73*(4), 1266–1276. doi:10.1128/AEM.01724-06 PMID:17142374

Xia, S., Frigo-Vaz, B., Zhao, X., Kim, J., & Wang, P. (2014). Biocatalytic Ccapture via reversible reaction cycle catalyzed by isocitrate dehydrogenase. *Biochemical and Biophysical Research Communications*, *452*(1), 147–150. doi:10.1016/j.bbrc.2014.08.058 PMID:25152403

Xia, S., Zhao, X., Frigo-Vaz, B., Zheng, W., Kim, J., & Wang, P. (2015). Cascade enzymatic reactions for efficient Csequestration. *Bioresource Technology*, *182*, 368–372. doi:10.1016/j.biortech.2015.01.093 PMID:25708541

Yanni, Y. G., & Dazzo, F. B. (2010). Enhancement of rice production using endophytic strains of Rhizobium leguminosarum bv. trifolii in extensive field inoculation trials within the Egypt Nile delta. *Plant and Soil*, *336*(1-2), 129–142. doi:10.1007/s11104-010-0454-7

Yuan, H., Ge, T., Chen, C., ODonnell, A. G., & Wu, J. (2012). Significant role for microbial autotrophy in the sequestration of soil carbon. *Applied and Environmental Microbiology*, *78*(7), 2328–2336. doi:10.1128/AEM.06881-11 PMID:22286999

Zaidi, A., Ahemad, M., Oves, M., Ahmad, E., & Khan, M. S. (2010). Role of phosphate solubilizing bacteria in legume improvement.Microbes for Legume Improvement, 273-292. doi:10.1007/978-3-211-99753-6_11

Zhang, Z., Lian, B., Hou, W., Chen, M., Li, X., & Li, Y. (2011). Bacillus mucilaginosus can capture atmospheric CO2 by carbonic anhydrase. *African Journal of Microbiological Research*, *5*, 106–112.

Zhu, Y. G., & Miller, R. M. (2003). Ccycling by arbuscular mycorrhizal fungi in soil–plant systems. *Trends in Plant Science*, *8*(9), 407–409. doi:10.1016/S1360-1385(03)00184-5 PMID:13678905

KEY TERMS AND DEFINITIONS

Carbon Sink: It is natural and artificial system and material that suck and adsorbe/absorbe carbon dioxide from the medium or atmosphere.

C-Emission: Coversion of carbon to carbon dioxide is known as carbon emission. For example 91% of total CO_2 emission cused by burning of fossil fuel in 2014.

Climate: The metrological condition, including temperature, precipitation, and wind that characteristically significantly prevail in a particular region.

CO_2: A heavy odorless colorless gas formed during the respiration and decomposition of organic materials, and absorbed from the air by the plants in photosynthesis reaction.

C-Sequestration: Carbon sequestration describes long-term storage of carbon dioxide or other forms of carbon to either mitigate or defer global warming and avoid dangerous climate change.

Ecological Risk Assessment: An ecological risk assessment is the process for evaluating how likely it is that the environment may be impacted as a result of exposure to one or more environmental stressors such as chemicals, land change, disease, invasive species and climate change.

Energy Generation: Dedicated to various fuel sources to meet our new generation' *energy* demands.

Global Warming: A gradual increase in the overall temperature of the earth's atmosphere generally attributed to the greenhouse effect caused by increased levels of carbon dioxide, cfcs, and other pollutants.

Microbial Inoculants: Initially, it is commonly known as soil inoculants to promote plant health but now days many beneficial microbes known for industrial and natural process.

Soil: The portion of earth's surface consisting of disintegrated rock and humus.

Chapter 6
Technological Approach of Bioremediation Using Microbial Tools:
Bacteria, Fungi, and Algae

Mostafa M. El-Sheekh
Tanta University, Egypt

Yehia A-G. Mahmoud
Tanta University, Egypt & Albaha University, Saudi Arabia

ABSTRACT

Bioremediation is applied to eliminate various contaminants, such as organic, inorganic or other pollutants from the environment. Environment worldwide is under great stress due to industrialization and human interfering on the limited natural resources. The release of chemicals pollution needs several techniques to treat some of these chemicals, but due to their cost, new technologies should be developing in order to create cost-effective and eco-friendly bioremediation technologies for environmental conversions. Bioremediation is an increasingly popular using microbial and algae strains for degrading waste contaminants. It is using of microorganisms and its enzymes to protect the environment from severe pollution. Bioremediation may be employed in order to eliminate specific contaminants, such as chlorinated pesticides or other pollutants from the environment. Microorganisms degrade the different pollutants in a natural environment but some modifications can be done to enhance its degradation efficiency at a faster rate in a limited time frame by using the genetically engineered microorganisms and microalgae. In this chapter, the role of the bacteria, fungi and algae in bioremediation of different environmental pollutants was highlighted.

INTRODUCTION

Bioremediation is a process in which organisms or microbes are used to remove or neutralize harmful pollutants from contaminated sites including water and soil. It requires a technique to select organisms capable of uptake or release some enzymes which can degrade these pollutants. Using biological ma-

DOI: 10.4018/978-1-5225-2325-3.ch006

terials, it is more effective than the traditional strategies because bioremediation strategies can be used directly at the site of contaminant without the need to transfer contaminant materials (Singh & Gupta, 2016). In the bioremediation process, bacteria alone, consortia, or combination of bacteria and fungi can be used for effective biodgradation (Fuentes, Barra, Caporaso, & Seeger, 2016; Ma, Ding, Peterson, & Daugulis, 2016). Bacteria, Fungi and algae are known for bioremediation of toxic hazardous materials including heavy metals. In order to increase its ability to degrade bioremediation, efforts have been done for exploiting diverse bacterial detoxification genes. Genetic engineering is a tool for manipulation of bacterial genome which can adapt and enhance the bioremediation potential of toxic metal detoxification that is not usually performed by normal bacteria (Das, Dash, & Chakraborty, 2016). Deshmukh, Khardenavis, & Purohit (2016) discussed the role of fungi in bioremediation due to their robust morphology and diverse metabolic capacity. Fungi possess multiple modes employed for detoxification of different toxic pollutants through the powerful enzymatic machinery. Algae also play an important role in the bioremediation of wastewater and decontaminate pollutants such as organic and inorganic compounds. Algae can be used in the free form or immobilized form for bioremediation of toxic heavy metals. Some treatment technologies involve the use of microalgae for effluent treatment, either as single species (Voltolina, Gómez-Villa, & Correa, 2005) or as mixed cultures/consortia (Tarlan, Dilek, & Yetis, 20021; Bhakta, Lahiri, Pitman, & Jana, 2015) to treat and remove nitrogen, phosphorus and chemical oxygen demand, from different types of effluents. Microalgae are also able to remove heavy metals from contaminated wastewater (Dwivedi, 2012; Prabha, Soni, Gupta, & Sonal, 2016).

WHY USE BIOLOGICAL SOURCES IN BIOREMEDIATION?

Microorganisms in general have effective ability to rapidly multiply and increase in huge numbers when be inoculated to decontaminate polluted area, when compared with chemical means. Antizar-Ladislao, Spanova, Beck, & Russell, (2008) indicated that microorganisms possess efficient enzymatic pathways able to eliminate or modify different pollutants. Moreover, different microorganisms could be added to polluted area (called inoculation) to enhance biodegradation rate. The inoculums may be a mixture of nonindigenous microbes from various polluted sites (specially selected and cultivated for its various pollutant degrading capabilities) or it may be a mixture of microbes selected from the site to depolluted or mass-cultured in the laboratory. Addition of nutrients along with inoculated process shows enhanced results for bioremediation (Boopathy, 2000).

This chapter deals with the significance of bioremediation, as it plays crucial role in the restoration of degraded land which is an important conservation effort for sustainable development and environmental management. The current chapter will summarize the available information on various attributes of microbial utilization for bioremediation. The gathered information is valuable to design reactors for contaminants elimination at regional and national levels.

MICROORGANISMS AND BIOREMEDIATION

The recent classification of living organisms based upon DNA and RNA has classified it; into three domains: eubacteria, Archaea, and Eukarya. Bacteria are prokaryotic microorganisms lacking true nucleus and other eukaryotic organelles such as mitochondria...etc. There are two types of bacteria according to Gram reaction; one is positive and the other is negative. However, fungi have considered as kingdom in "Eukarya" domain, hence it has true nucleus, forming hyphae and cell wall. Whereas, algae came in this recent classification system as a member of Kingdom Protista of "Eukarya" domain based on RNA and DNA. The term "microbial consortiums" is used to describe the group of microorganisms which have the power or the ability to degrade or modify toxic contaminants.

Some literatures indicated that the following microbes are powerful in this respect. Concerning Prokaryotic bacterial species that performed efficiency in pollutants degradation, there are *Pseudomonas, Actinobacter, Acaligenes, Arthrobacter, Bacillins, Berijerinckia, Flavobacterium, Methylosinus, Mycococcus, Nitrosomonas, Nocardia, Serratia* and *Xanthofacter, Exiguobacterium, Geobacter* and *Mycobacterium* as listed by many researchers.

On the other hand, *Penicillium, Phanerochaete, Trametes Polyporus, Rhizoctonia, Aspergillus niger, Ganoderma lucidium, Pleurotus ostrateous* and another laccase synthesizing enzyme as eukaryotic fungal species that have been manipulated and studied with different environmental pollutants.

Moreover, the following algal species such as *Chlorella vulgaris, Rhizoclonium hieroglyphicum* and mixed algae culture (*Microspora* sp., *Navicula* sp., *Lyngbya* sp., *Cladophora* sp., *Spirogyra* sp. and *Rhizoclonium* sp.) have proved efficiency in waste waters treatment. Furthermore, various species of algae such as *Anabaena inaequalis, Chlorella* sp., *Stigeoclonium tenue, Synechococcus* sp. and *Westiellopsis prolifica* developed resistant to heavy metals and thus have been used for the removal of heavy metals (Rahman et al., 2011; Bajguz, 2011).

Exposing of microorganisms to different contaminants switch certain enzyme genes on in order to synthesis different proteins of certain enzymes that able to attack theses area contaminated compounds. These contaminants might be lead to microorganism's enzyme induction or repression. These enzymes occurring in bacteria, fungi and/or plants have the ability to degrade or modify these toxic contaminants. Microorganisms used in polluted area treatments might be area inhibiting biota or isolated elsewhere. Algae and fungi play important role in returning the environment to its original state altered by various contaminants. The physiology, biochemistry and genetics of the selected microorganisms should be well studied in order to enhance the microbial process to achieve high percentage bioremediation. For e.g. the genes encoding for enzyme which involve in the breakdown of the organics pollutants to simple non-toxic compounds must be identified which will provide good understanding of the microbial capability to degrade the desired pollutant in a short time (Singh, 2008).

The main two mechanisms of bioremediation by microorganisms are biosorption and bioaccumulation. Biosorption is a passive adsorption mechanism which is reversible and fast. Biosorption needs inexpensive cost because the biomass can be obtained from industrial waste, and it can be regenerated and reused in many cycles. In biosorption, either the living or dead biomass can be used in bioremediation process, however, bioaccumulation involves only living biomass which cannot be reused and therefore it needs cost expensive (Coelho et al., 2015). Other technology used the metabolic potential of microorganisms to decompose contaminants into simple no toxic compounds as compared with other physicochemical methodologies (Perpetuo, Souza, & Nascimento, 2011). Microorganisms can also survive in contaminated habitats because they are metabolically able to exploit contaminants as potential energy sources (Ilyina, Castillo Sanchez, Villarreal Sanchez, Ramirez Esquivel, & Candelas Ramirez, 2003). In all techniques mentioned above either employing one type of organism or a consortium of microorganisms, high toxic chemicals are converted into less toxic chemicals by biological means (Nicolaou, Gaida, & Papoutsakis, 2010). Biosorption of pollutants especially toxic metals can be affected by several environmental conditions such as pH, ionic strength, biomass concentration, temperature, particle size, and presence of

Figure 1. Microorganisms and mechanisms employed in the bioremediation involved in the case of dead and living biomass
Coelho et al., 2015.

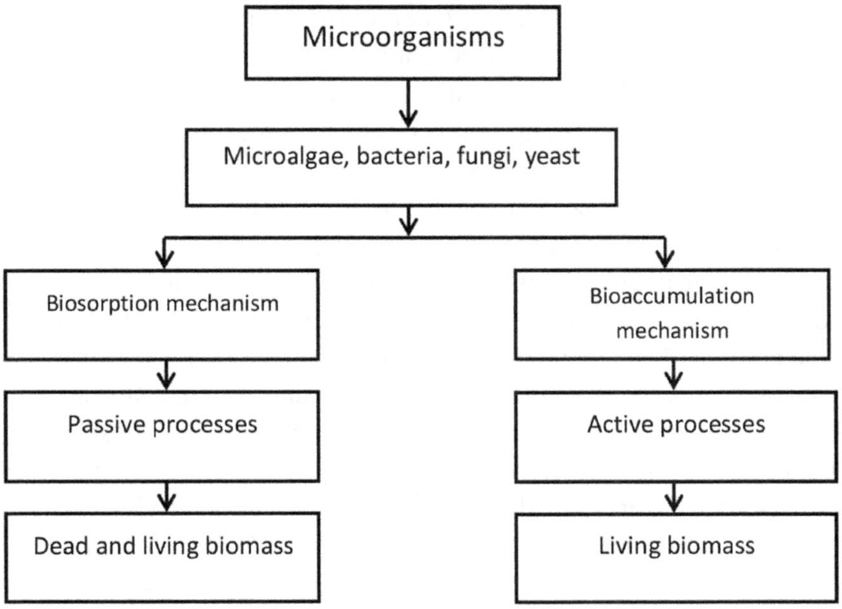

other ions in the solution (Velosky, 2004). The major groups of microorganisms commonly used for the bioremediation of metals, which include bacteria, microalgae, fungi, and yeast as well as the mechanisms for bioremediation are summarized in Figure 1.

BACTERIAL BIOREMEDIATION

It has been indicated that heavy metal cations showed adsorption to the cells walls of Gram-positive bacteria (Doyle, Matthews, & Streips, 1980). Many bacteria, such as *Azotobacter, Actinomycetes* and *Pseudomonas* could synthesize different substances to capture Fe^{2+} which they require for their metabolic activity and biosorption (Pattus & Abdallah 2000). Jayashree, Nithya, Rajesh, & Krishnaraju, (2012) used *Pseudomonas* as fuel eating bacteria which can degrade the hydrocarbons. *Pseudomonas syringae* also showed the formation of bond which plays an important role in the accumulation of cadmium, calcium, magnesium, zinc, copper and mercury. Bhakta, Munekage, Ohnishi, Jana, & Balcazar, (2014) isolated and characterized 40 cadmium (Cd)-and arsenic (As)-resistant bacterial isolates from coastal sediments. Out the isolated strains, four isolates showed higher growth rate in Cd- or As-supplemented culture media and greater magnitude in absorbing Cd compared to As from water. They concluded that the isolated strong metal(loid)-specific resistant bacterial isolates would be potential agents of bioremediation for the treatment of specific hazardous metal(loid)-contaminated aquatic environments. *Geobacter metallireducens* is a Fe (III) reducing organisms that can oxidize a variety of aromatic pollutants (benzene and naphthalene) and eliminate or decrease uranium (a radioactive waste) from drainage water in mining operations and from contaminated groundwater (Lovely, 2002). *Exiguobacterium aurantiacum* has showed capability for phenol degradation and PAHs in laboratory applying a batch culture when provided with pure compounds as a sole source for life (Jeswani & Mukhergi, 2012). This activated sludge consortia and batch culture form a good biofilm on the rotating discs in a Rotating Biological Contractor (RBC) resulting in removal of COD and TOC (Mukherji & Jeswani, 2011). Microorganisms have the ability to uptake heavy metals from contaminated water (Babel & Kurniawan, 2003). The mechanisms of metal binding with microorganism could be through sorption into the cell surface (both live and died cells), intracellular accumulation within living microbial and/or precipitation within living cells (Kujan et al., 2005). Microbial cells have the power to uptake metals by the production of metal-binding proteins in living cells. Bishnoi & Garima, (2005) have recommended the use of dead biomass over living where it has no toxicity limitations, doesn't need growth media and nutrients requirements, and high ability to bind metals. In solutions the process of metal removal takes place by the formation of complex between the pollutants (metal) and the active groups on cell surface after interaction.

However, many researchers mentioned that any microorganisms isolated from soil are capable of degrading the crude oil. It has been found that applying mix cultures versus the use of pure cultures led to synergistic interactions in bioremediation (http://en.wikipedia.org/wiki/ Biostimulation, retrieved on 28 Mar. 2011.). Bacteria for hydrocarbon decomposition are commercially available in freeze dried form, which can be used for bioremediation after propagation to a minimum of 2×10^{8} CFU/ml (http://www.eugris.info/newsdownloads/ Green Remediation/ pdf /D05 AnneLouise Gimsing_Paper.pdf.). Many lteratures have listed the use of these bacteria for degradation of petroleum products: *Pseudomonas, Aeromonas, Moraxella, Beijerinckia, Flavobacteria, chrobacteria, Nocardia, Corynebacteria, Atinetobacter, Mycobactena, Modococci, Streptomyces, Bacilli, Arthrobacter, Aeromonas,* Cyanobacteria etc.

The Role of Genetic Engineering in Bacterial Bioremediation

Nowadays applications of the genetic engineering approach to produce microorganisms capable for degradation of pollutants and wastes have taken great attention. The genetic engineered bacteria proved to more effective in hazardous wastes and hydrocarbon bioremediation (Das & Chandran, 2011). The presence of additional genes in bacteria provides them with extraordinary metabolic diversity. The development of the genetic engineered bacteria approaches includes the optimization of enzymes and some metabolic pathways to be effective in bioremediation (Pieper, & Reineke, 2000). The genetic engineered microorganisms including fungi, algae and bacteria were used for removal of heavy metals from hazardous wastes. Chen, & Wilson, (1997) examined the efficiency of the genetic engineered cells *Escherichia coli strain* for their selectivity for Hg^{2+} accumulation. They results showed that the genetically engineered cells of *Escherichia coli* accumulated Hg^{2+}effectively in suspension at low concentrations (0-20 μM) over a broad range of pH (3 to 11).

FUNGAL BIOREMEDIATION

Fungal bioremediation is the bioremediation or biodegradation of contaminants and hazardous pollutants in the environment using fungi. This achieved through stimulating enzyme activity, which break down the toxic materials into simple and non- toxic compounds. Some fungi are also capable of accumulating, or absorbing and concentrating heavy metals in the mycelia. The mycoremediation of contaminants from the environment either in soil or aquatic habitats includes yeast, micro fungi and higher fungi. The first step to achieve efficient mycoremediation is the selection of the right fungal species to degrade a specific pollutant. The extracellular enzymes secreted by mycelia can break down many organic compounds to simple no-toxic compounds (Aglave, Jagtap, Kalal, & Lokhande, 2009). It uses fungal mycelium to decontaminate or filter the toxic waste from contaminated area. The fungal mycelia secrete various extracellular enzymes and acids that break down the lignin and cellulose. The key to mycoremediation is getting the more suitable fungal species to target a specific pollutant. Fungi (*Ligninolytic fungi*) such as *Phanaerochaete chrysosporium* (white rot fungus) and *Polyporus* sp. are promising candidates for bioremediation, as it performed high ability to degrade an extremely diverse range of persistent or toxic environmental pollutants such as petroleum hydrocarbons, explosives, polychlorinated biphenyls (PCBs), polycyclic aromatic hydrocarbons (PAHs), and organochlorine pesticides (Wu, & Yu, 2007; Favero, Costa, & Massimino, 1991; Xueheng, Hardin, & Huey-Min, 2006; Bai & Abraham 2001; Ayu, Tony, Tadashi, Yasuhiro, & Kazuhiro, 2011). These fungal organisms also have shown ability to degrade a wide variety of organopollutants because of lignin-degrading or wood rotting enzymes. The extracellular enzymes secreted by white-rot fungi during lignin decay have been found as promising agents for oxidizing pollutants. *Cyathus bulleri*, a bird's nest ligninolytic fungus that colonizes with dead herbaceous stems, wood chips, dung, sticks and other woody debris are found ecologically suitable for lignin degradation. It produces a single laccase the internal peptides. Laccases are copper containing enzymes which catalyze the oxidation of a broad range of aromatic amines and phenolic compounds by using molecular oxygen as the electron acceptor (Garg, Baranwal, Mishra, Chaudhuri, & Bisaria, 2008). Many researchers have showed that the extracellular enzymes secreted by these fungi oxidizes woody plant material i.e. lignin which shows structural similarity with PAHs (Levin, Viale, & Forchiassin, 2003). Contaminants degradation is called mineralization, and the end product is carbon dioxide (Figure 2, Malik 2006). The mechanisms of degradation are not yet conclusively mapped but it is likely that intracellular and extracellular enzymes are responsible for oxidizing different pollutants under different conditions.

Figure 2. Contaminants mineralization by microbial consortium
Malik, 2006.

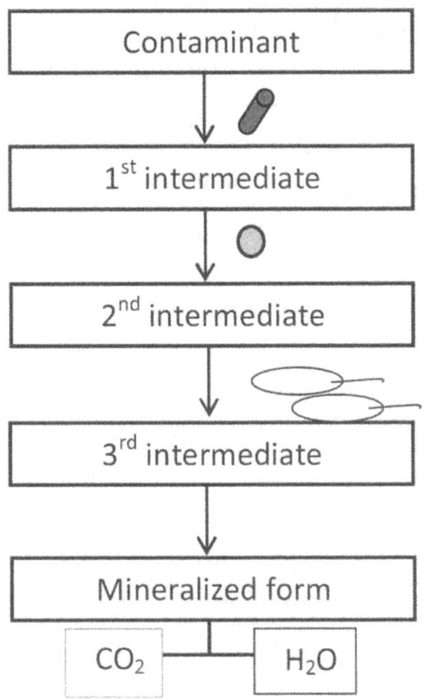

The biosorption of uranium, cobalt, cadmium, zinc and copper by dead biomass of microorganisms goes through electrostatic interactions between the metal ions in solutions and cell walls (Kuyucak & Volesky, 1989, Muraleedharan & Venkobachar, 1990; Mahmoud, 2004). The fungi *Aspergillus niger* and *Ganoderma lucidum* are used for biosorption of chromium through ion exchange mechanism (Muraleedharan & Venkobachar, 1990; Ahluwalia & Goyal, 2006). *Pleurotus ostreatus* (tasty oyster mushroom) has large range of application in bioremediation. Using efficiency was proved when a plot of soil contaminated with diesel oil was inoculated with mycelia of oyster mushrooms. After four weeks, when the soil was analyzed more than 95% of PAH (polycyclic aromatic hydrocarbons) was reduced to non-toxic components in the mycelial-inoculated plots (Favero et al., 1991). Particularly, wood-degrading fungi are effective degrading aromatic pollutants (toxic components of petroleum), as well as chlorinated compounds (Arica, Bayramoglu, Yılmaz, Genc, & Bektas, 2004, Abd El-Zaher, Mahmoud, & Magda, 2011). Many researchers listed that the natural microbial community participates with the fungi to degrade contaminants, eventually into carbon dioxide and water. Recently an extensive research on using fungi for bioremediation has been started in different countries worldwide.

YEAST BIOREMEDIATION

Yeasts could accumulate metals and metalloids, where some of them are essential for structural and catalytic functions, whereas others are of no metabolic value (Chatterjee, 2011). Yeasts are also known as a good agent for dye decolourization from food industry effluents mainly by three mechanisms such as bioaccumulation, biosorption, and biodegradation. The removing mechanisms of dyes color from industrial effluents, with cells of yeasts are either through absorbtion or adsorbtion at the cell surface (Donmez, 2002). The yeast species such as *Candida*, *Clavispora*, *Debaryomyces*, *Leucosporidium*, *Trichosporon*, *Pichia*, *Rhodosporidium*, *Rhodotorula*, *Sporidiobolus*, *Sporobolomyces*, *Stephanoascus*, and *Yarrowia* are applied in bioremediation process and shows biodegrading properties (Csutak, Stoica, Ghindea, Ana-Maria, & Vassu, 2010). Microbial biomass derived from yeasts, particularly *Candida utilis* has the ability to accumulate metal ions and radionuclides from the environment. A study by Kujan et al., (2006) also showed that *C. utilis* biomass can conveniently be used for cadmium biosorption from aqueous solutions. The yeast *Rhodotorula mucilaginosa* is efficient in bioadsorption of lead and is also known to accumulate free and complexed silver ions by metabolism dependent and independent processes (Gomes, Rosa, Pimentel, & Mendonça-Hagler, 2002). A comparative study was made by Ksheminska et al., (2003) on the sensitivity of yeast *Pichia guilliermondii* to Cr (III) and Cr (VI) as well as on the uptake potential of Cr. The results indicated accumulation of Cr (III) and Cr (IV) by *Pichia* sp. and also showed increase in Cr tolerance by the addition of riboflavin. Besides used for metal removal, microbes are also finding their application in removing various hydrocarbons and hazardous chemicals. Burning of fossil fuels produced a chemical that have been successfully bioremediated at contaminated sites using bioremediation procedures. Gasoline and the other components such as benzene, tolune, ethylbenzene and xylene (BTEX) in particular, have been taken great attention of substantial biodegradation and bioremediation research. Many researchers have indicated that extent of oil and total petroleum hydrocarbon (TPH) biodegradation is depending upon oil type and its molecular composition (Sarma, Bhattacharya, Krishnan, & Lal, 2004; Mandal et al., 2011; Mandal et al., 2012; Mohanty & Mukherji, 2013). Yeast has been reported to be a good candidate for removal of alkanes present in petroleum, hydrocarbons, organic contaminant substances and heavy metals from industrial effluents. Trama et al., (2014) isolated 23 yeast isolate capable of utilizing n-hexadecane and one of them was able to tolerate high concentration of lead (Pb) with a high biosorption index as compared with control reference yeast strains.

ALGAL BIOREMEDIATION

The process of phycoremediation or algal bioremediation is the use of algae to remove pollutants from the environment or to convert them into harmless form. Phycoremediation in a much broader sense is the use of micro or macroalgae for degrading or biotransformation of different pollutants, including nutrients and xenobiotics from wastewater and CO_2 from air. Algae are highly adaptive and can grow autotrophically, heterotrophically or mixotrophically in any environment. In natural habitats, algae play a major role in controlling metal concentration of lakes and oceans. It possess the ability to degrade or accumulate toxic heavy metals and organic pollutants such as phenolics, hydrocarbons, pesticides and biphenyls from the environment, resulting in higher concentrations within themselves as compared to surrounding water (Shamsuddoha, Bulbul, & Huq, 2006).

Algal Bioremediation of Heavy Metal

Wang & Fenglian (2011) reviewed the different technologies for removal of heavy metal from wastewater. These techniques include chemical precipitation, ion-exchange, adsorption, membrane filtration, coagulation–flocculation, flotation and electrochemical methods. In spite of all methods and techniques used, still the biological treatment by algae is more effective and low-cost compared to the other methods.

Pollutant degrading mixotrophic algae are excellent agents for remediation and carbon sequestration (Subashchandrabose, Ramakrishnan, Megharaj, Venkateswarlu, & Naidu, 2013). An alga fixes CO_2 and produce O_2 by the process of photosynthesis and increases BOD level in contaminated water along with the efficient removal of excess of nutrients (Fathi, Azooz, & Al-Fredan, 2013). Uptake of metals by living microalgae occurs in two steps. The first step is independent of cell metabolism and involves physical adsorption to the cell surface) afterwards these ions are transported slowly into the cytoplasm known as chemisorption (Vymazal, 1984). The second step is dependent on cell metabolism and involves absorption or intracellular uptake of heavy metals. Many researchers demonstrated that various metals such as Pb, Cu, Cd, Co, Hg, Zn, Mg, Ni and Ti are sequestered in polyphosphate bodies of algae and perform two functions i.e. storage and detoxification of metals (Dwivedi, 2012) . Due to its role in sequestration of heavy metals by algal cell wall, these are considered as an ideal source of multifunctional polymers (Seufferheld & Curzi, 2010). Recently El-Sheekh, Farghl, Galal, & Bayoumi, (2016) studied the ability of two species of green unicellular microalgae namely *Chlorella vulgaris* (freshwater alga) and *Chlorella salina* (marine alga) to bioremdiate different types of polluted water and they found that algae were able to reduce pH, total dissolved solids (TDS), biological oxygen demand (BOD), chemical oxygen demand (COD), nitrate, ammonia, phosphate, sulphate, calcium, magnesium, sodium, potassium, heavy metals (Zn, Cu, Mn, Ni, Co, Fe and Cr) and the number of total Coli-form bacteria from different mixtures of water samples. They also found that the removal efficiency of heavy metals ranged between 13.6-100%. Various species of algae such as *Anabaena inaequalis, Chlorella* sp., *Stigeoclonium tenue, Synechococcus* sp. and *Westiellopsis prolifica* are resistant to heavy metals and thus used for the removal of heavy metals (Rahman et al., 2011; Bajguz, 2011; Kannan, Vijayasanthi, & Chinnasamy, 2012). El-Sheekh, El-Shouny, Osman, & El-Gammal, (2005) cultivated the cyanobacteria *Nostoc muscorum* and

Anabaena subcylindrica in sewage and industrial wastewater effluents and they found that the metal sorption efficiency depended on the type of biosorbent, the physiological state of the cells, availability of heavy metal, concentration of heavy metal and chemical composition of wastewater. Olgui, (2003) stated that the use of filamentous microalgae with a high autoflocculation capacity as well as the immobilized cells can develoe an efficient nutrient removal algal systems. He also stated that use of algae in this process require strains with special characteristics such as tolerance to extreme environmental growth conditions such as temperature, chemical composition, a quick sedimentation behavior, or a capacity for growing mixotrophically.

Many algae are used for biosorption of heavy metals either live or dead cells. The dead algal cells can sorb more metal than live cells (Mehta & Gaur, 2005). The aforementioned authors also emphasized the requirements for commercial application of algal technology for metal removal from different sources of wastewaters. They stated that the algae strains should be of high metal sorption efficiency, besides the economic feasibility and low-cost methods of cell immobilization.

Bioremediation of Wastewater by Algae

Ahmed, Khan, & Yasser, (2013) investigated the ability of various algae such as *Chlorella vulgaris, Rhizoclonium hieroglyphicum* and mixed algae culture (*Microspora* sp., *Navicula* sp., *Lyngbya* sp., *Cladophora* sp., *Spirogyra* sp. and *Rhizoclonium* sp.) which belonging to different algae divisions for their efficiency of sewage water treatment. The results showed that *Chlorella vulgaris* could reduce the different parameters of polluted water such as chemical oxygen demand (98.3%), biochemical oxygen demand (98.7%), total nitrogen (93.1%), total phosphorus (98.0%), nitrate (98.3%), phosphate (98.6%), chloride (94.2%), total coliforms (99.0%), faecal coliforms (99.0%) and total dissolved solids (98.2%). When the mixed of algae was used, the total suspended solids was reduced by 92.0%.

Algae are also known for effective removal of nitrogen from soil or water through the process of absorption and store it as biomass. As the time passes, the biomass decomposes and releases the nitrogen back into the soil (ammonia or urea) or atmosphere (N_2O), where it may be recycled or lost (Woodward, Fellows, Conway, & Hunter, 2009).

Algae plays important role in pH correction, sludge removal and TDs reduction, whereas in conventional treatment, separate methods or stages are required. Amongst algae, blue green algae or cyanobacteria are susceptible to various physical and chemical alterations of temperature, light, salinity, and nutrient composition. Recently, there has been increasing worldwide interest in using cyanobacteria as an economic and low-maintenance remediation technology for contaminated and polluted sites. Cyanobacteria have been used efficiently for remediating dairy waste water by converting the dissolved nutrients into biomass. However, the beneficial application of cyanobacteria in bioremediation of contaminated waters, either water environments or industrial effluents, still needs further research.

The wastewater treatment by algae is not restricted to only removal of pollutants and enhancement of water quality but also used to cultivate algae for biofuel production. Abou-shanab et al., (2014) studied the possible cultivation of microalgae *Micractinium reisseri* and *Scenedesmus obliquus* which were isolated from municipal wastewater mixed with agricultural drainage for production of biomass, lipids and fatty acids for feedstock of biodiesel. The results supported the growth of the green microalga *Micractinium reisseri* on secondary and tertiary effluents of municipal wastewater for high biomass production. The results also showed that the algae removed nitrogen and phosphate efficiently and this supports the use of this alga as good candidate for oil and biofuel production using the wastewater. Recntly, El-Sheekh et al., (2016) used the freshwater alga, *Chlorella vulgaris* and the marine alga *Chlorella salina* to reduce the toxic pollutants from different mixtures of water samples such as sewage, sea and well water.

Algal Bioremediation of Oil Spills

The expanding in the petrochemicals industry caused serious environmental problems and contamination to all habitats. The hydrocarbon compounds of these contaminants have been classified as carcinogenic and neurotoxic organic pollutants (Das & Chandran, 2011). Some algae can grow on crude oil and have the ability to degrade it to simple compounds. Walker, Colwell, Vaituzis, & Meyer, (1975) found that the achlorophyllous alga *Prototheca zopfii* was able to degrade 10% of motor oil and 40% of crude oil and the produced byproducts were not harmful. Further attempts of Cerniglia, Gibson, &Van Baalen, (1980) who used brown algae (*Petalonia*), green algae (*Dunaliella* sp., *Chlorella* sp. *Chlamydomonas* sp., *and Ulva* sp.*)*, cyanobacteria (*Oscillatoria* sp., *Anabaena* sp., *Coccochloris* sp., *Nostoc* sp., *Cylindretheca* sp., *Aphanocapsa* sp. *Cylindretheca* sp., *Aphanocapsa* sp.) and diatom (*Amphora*) to oxidize naphthalene.

Some other algal species such as *Aphanocapsa* sp., *Oscillatoria salina, Plectonema terebrans* and *Synechococcus* sp. develop mats in aquatic environments and have been successfully used in bioremediation of oil spills in different parts of the world (Rahman et al., 2011; Brahmbhatt, Patel, & Jasrai, 2012; Nakajima & Tsuruta, 2004). El-Sheekh & Hamouda (2014) cultivated the two cyanobacteria *Nostoc punctiforme* and *Spirulina platensis* isolated from polluted waters on media containing different concentrations of crude oil. The two cyanobacteria could degrade the crude oil under heterotrophic conditions. The analyses of crude oil residue by GC–MS showed that Decane, Pentacosane, Hexacosane, Octacosane, Nonacosane totally removed from the medium.

Organic Pollutants Bioremediation

The algae also play an important role in the bioremediation and biodegradation of the organic pollutants other than crude oil. The organic pollutants such as dyes and phenolic compounds can be degraded to simple compounds such as urea and CO_2 through enzymatic pathway. It has been documented that algae have the ability to grow on industrial effluents, therefore, micro-algae are a good option for the bioremediation of colored waste water; in addition, microalgae do not need added carbon sources.

Bioremediation of Dyes by Algae

Azo dyes are the largest class in number and amount produced of synthetic dyes used in industry because of their ease of synthesis and versatility. Azo dyes are organic compounds which are characterized by the presence of one or more azo bonds (-N=N-) in combination with one or more aromatic systems (Bafana, Kirishnamurthi, Devi, & Chakrabarti, 2008). Their color is attributed to the azo-bond, the associate auxochromes and a system of conjugated double bonds (aromatic hydrocarbons). Many studies indicated that most of the azo dyes are toxic and carcinogenic (Moutaouakkil, et al., 2003). Azo dye compounds have been found to be linked to bladder cancer in humans and to hepato-carcinoma cells in mice (Puvaneswari, Muthukrishnan, & Gunasekaran, 2006). Thus, a number of azo dyes have been classified as carcinogenic (Mc Alinden, 2008). El-Sheekh Gharieb, & Abou-El-Souod, (2009) studied the ability of green and blue algae (cyanobacteria) isolated from different polluted sites in an industrial region, to degrade or decolorize high ratios of azo dyes toxicants. The results revealed the ability of

Chlorella vulgaris, Lyngbya lagerlerimi, Nostoc lincki, Oscillatoria rubescens, Elkatothrix viridis and *Volvox aureus* to decolorize and remove methyl red, orange II, G-Red (FN-3G), basic cationic, and basic fuchsin by ratios ranged between 4-95%, and this ratio depends on the alga type, growth conditions and the dye molecular weight. The study also revealed that the algae degrade or decolorize the dye toxicants through induction of certain enzymes implicated in degradation of these compounds (Figure 3). The release of such dyes in channels water after the the dyeing process may exert a toxic effect on the whole environment (Marungruenga & Pavasant, 2006). Recently, Abou-El-Souod & El-Sheekh, (2016) studied the biodegradation of basic fuchsin and methyl red by the blue green algae (cyanobacteria) *Hydrocoleum oligotrichum* and *Oscillatoria limnetica* and they found that the biodegradation depends on the type of the alga used and also the type of the dye to be degraded (Figure 4). The investigation of the biodegradable byproducts of the dyes measured by the modern techniques proved that the two algae are a good model for the bioremediation of hazardous materials and its removal from water.

Figure 3. Activity of azo dye reductase before and after treatment of Chlorella vulgaris with G-Red (A), and Nostoc lincki with methyl red (B), (El-Sheekh ,Gharieb, & Abou-El-Souod, 2009)

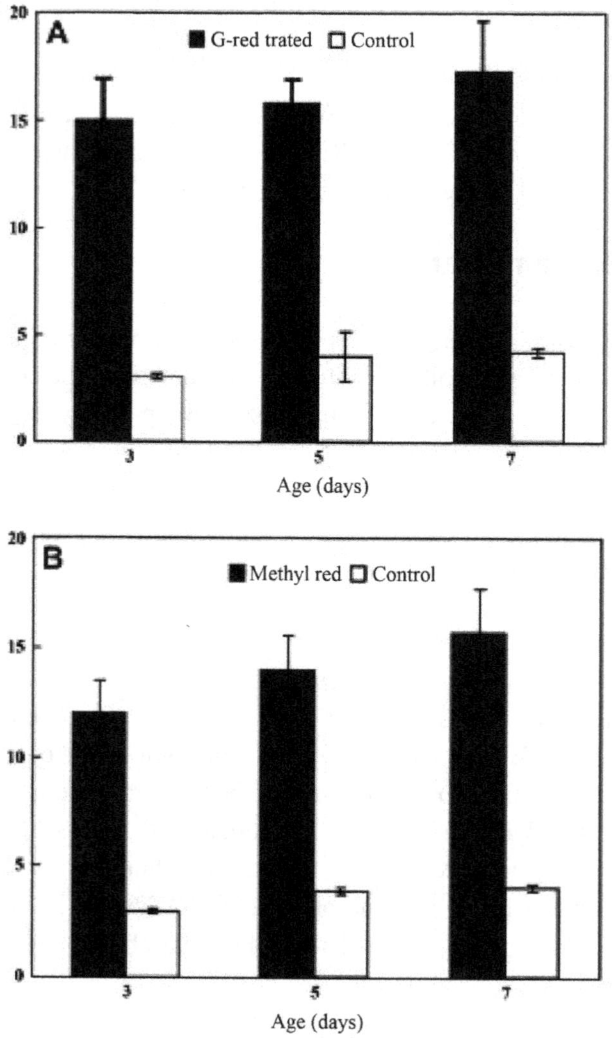

Figure 4. Biodegradation of basic fuchsin and methyl red by the blue green algae (Cyanobacteria) Hydrocoleum oligotrichum and Oscillatoria limnetica
Abou-El-Souod & El-Sheekh, (2016).

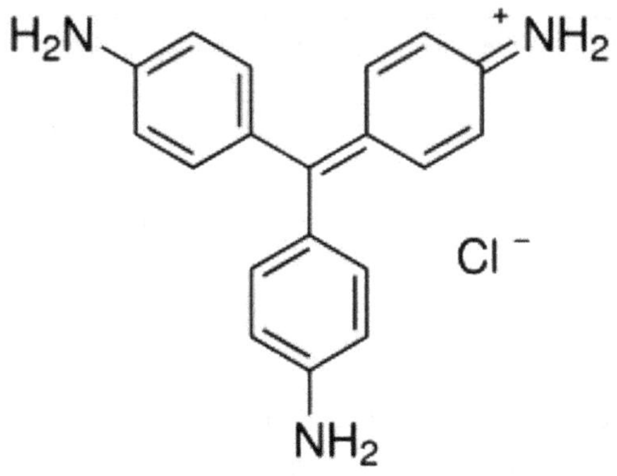

Basic fuchsin m/z 323.83

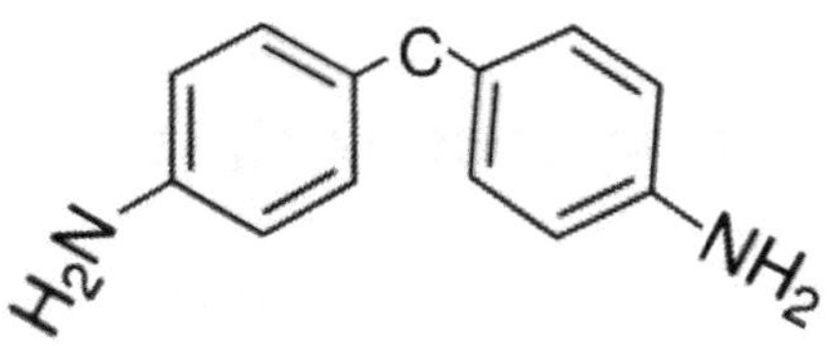

Basic fuchsin after degradation m/z 118

Bioremediation of Polycyclic Aromatic Compounds by Algae and Cyanobacteria

The biodegradation of polycyclic aromatic hydrocarbons (PAHs) by bacteria, fungi and algae is well documented and this process require the isolation of microorganisms capable of degrading low molecular weight polycyclic aromatic hydrocarbon compounds that contain three or less fused benzene rings. However, there is some bacterial strains which capable to biodegrade high molecular weight polycyclic aromatic hydrocarbons which contain four or more fused benzene rings (Juhasz & Naidu, 2000). The isolation of some algae and cyanobacteria from industrial polluted waters is an effective way to degrade industrial pollutants including polycyclic aromatic compounds. Cerniglia, (1992) stated that eukaryotic algae and cyanobacteria are capable for transforming naphthalene to simple compounds which are water soluble. The green alga *Scendesmus obliques* was effective in biodegradation of naphthalene sulphonic acids by utilizing it as sulfur source for biomass with releasing the carbon ring into the medium. *Scendesmus obliques* also could dehalogenate chlorobenzoates and the chloride being accumulated by the algae cells (Luther & Shabaan, 1990). El-Sheekh, Gharieb, & Abou-El-Souod, (2012) studied the

Table 1. Degradation of different heterocyclic aromatic compounds by different algae after 7 days incubation

Compound	Degradation %					
	Lyngbya lagerlerimi	*Nostoc linckia*	*Oscillatoria rubescens*	*Chlorella vulgaris*	*Elkatothrix virdis*	*Volvox aureus*
2- Methylthie.3 phenyl quinazlin (4- 3H) one	77.11	54.21	80.45	40.51	71.23	83.39
2- phenyl 3,1 benzexazin. 4- one	33.4	46.47	71.44	73.03	79.74	53.24

Modified from El-Sheekh et al. 2012.

biodegradation of phenoloic and polycyclic aromatic compounds by some eukaryotic algae and cyanobacteria. The results (Table 1) revealed that the green alga *Volvox aureus* and the cyanobacteria *Nostoc Linckia* and *Oscillatoria rubescens* coild oxidize and decolorize phenol to catechol. The results also showed the oxidation of β-naphthol by *Volvox aureus*, *Lyngbya lagerlerimi* and *Nostoc linckia*, and the oxidation of catechol by *Chlorella vulgaris* and *Volvox aureus*. The results revealed that the degradation of polycyclic aromatic hydrocarbons depend on the alga used, its physiological state and the molecular weight of the compound used.

Bioremediation of Phenolic Compounds by Algae

Phenolic compounds are considered of highly toxic pollutants which released from the industrial activity into surface water and affect human health and biota. Microorganisms including fungi, bacteria and algae could detoxify these phenolic compounds. In this connection Pinto, Pollio, Previtera, & Temussi, (2002) studied the ability of the green microalgae *Scenedesmus quadricauda* and *Ankistrodesmus falcatus* to degrade phenols at concentration 400 mg ml^{-1} from olive oil mill wastewaters. The results showed that the microalgae are good candidates to degrade phenolic compounds and they attained more than 70% degradation efficiency for both algae used. Klekner & Kosaric, (1992) also tested the algae *Chlorella* sp., *Scenedesmus obliquus* and *Spirulina maxima* for the degradation of 2,4-dimethylphenol at a concentration 1000 mg l^{-1}. It was found that *Chlorella* converted phenols to an isomer of dimethylbenzenediol that was accumulated in the medium. They also stated that the complete degradation of the toxic phenolic compounds depends on the biomass accumulation and the adaptation period of the alga. Semple & Cain, (1996) found that the Chrysopyte alga *Ochromonas danica* could grow well on phenol as a sole carbon source and removed the phenol carbon completely from the growth medium. *Ochromonas danica* converted phenols to catechol, which was further metabolized by the meta-cleavage pathway and then it converted to pyruvate and CO_2. Recently, Alfawwaz, Jacob, & Al-Wahishe, (2016) isolated and identified two unicellular green algae *Chlamydomonas* sp. and *Desmodesmus* sp. The two algae species removed phenol from the medium effectively and the highest removal percentage was found (75%) by *Desmodesmus* sp. after 25 days at 25 mg/L. The phenol bioremoval capacity was dependent on the physicochemical factors such as initial phenol concentration, contact time, and the synergistic effect.

CONCLUSION AND FUTURE PROSPECTS

In this chapter, we have drawn an overview of the potential of microorganisms for bioremediation of different pollutants in aquatic ecosystems. Bioremediation is an effective process over the chemical methods since bioremediation is eco-friendly and low-cost as compared with other methods. The release of organic and inorganic pollutants in the environment cause many problems to the soil, aquatic systems and consequently affects the human health. Therefore, it is necessary to find out specific strains of microorganism which are capable of biodegradation or detoxification of such pollutants from the environment. The isolated microorganisms (bacteria, fungi and algae) from the polluted sites proved to be more effective in bioremediation and this depend on the adaptation of the organism to grow for long time on the pollutants. The application of microorganisms in bioremediation still encounter with some difficulties since some organic pollutants are difficult to be degraded by all bacteria or fungi and algae. The genetic engineering can overcome this problem and improve the bioremediation of many organic, inorganic pollutants and increase microorganism tolerance to these pollutants. The environmental condition of are necessary to be understood for each strain in order to increase the absorption, accumulation and or biodegradation of the pollutants. It is also necessary for understanding the mechanism of biodegradation of microorganisms for the different pollutants (oil spills, organic hydrocarbons, dyes, phenols or polycyclic aromatic compounds) which cause serious problems to the environment and are recorded as carcinogenic compounds for safe and clean environment. Therefore, based on the present review, it may be concluded that microbial (Bacteria, Fungi and algae) degradation can be considered as a key component in the cleanup strategy for different pollutants bioremediation.

REFERENCES

Abd El-Zaher, E. H. F., Mahmoud, Y. A.-G., & Magda, M. A. (2011). Effect of different concentrations of phenol on growth of some fungi isolated from contaminated soil. *African Journal of Biotechnology*, *8*, 1384–1392.

Abou-El-Souod, G., & El-Sheekh, M. M. (2016). Biodegradation of basic fuchsin and methyl red by the blue green algae *Hydrocoleum oligotrichum* and *Oscillatoria limnetica. Environmental Engineering and Management Journal*, *15*(2), 279–286.

Abou-shanab, R. A. I., El-Dalatony, M., El-Sheekh, M. M., Ji, M.-K., Salam, E., Kabra, A., & Jeon, B.-H. (2014). Cultivation of a new microalga, *Micractinium reisseri*, in municipal wastewater for nutrient removal, biomass, lipid, and fatty acid production. *Biotechnology Bioprocess Engineering*, *19*(3), 510–518. doi:10.1007/s12257-013-0485-z

Aglave, B. A., Jagtap, K. U., Kalal, P. R., & Lokhande, M. O. (2009). Helping the ecosystem through mycoremediation. *Asian Sciences*, *4*, 100–101.

Ahluwalia, S. S., & Goyal, D. (2006). Microbial and plant derived biomass for removal of heavy metals from wastewater. *Bioresource Technology*, *98*(12), 2243–2257. doi:10.1016/j.biortech.2005.12.006 PMID:16427277

Ahmed, F., Khan, A. U., & Yasser, A. (2013). Comparative phycoremediation of sewage water by various species of algae. *Proceeding of Pakistan Academy of Science*, *50*(2), 131–139.

Alfawwaz, A. T., Jacob, A. J., & Al-Wahishe, T. E. (2016). Bioremoval capacity of phenol by green micro-algal and fungal species isolated from dry environment. *International Journal of Scientific & Technology Research*, *5*(2), 155–160.

Antizar-Ladislao, B., Spanova, K., Beck, A. J., & Russell, N. J. (2008). Microbial community structure changes during bioremediation of PAHs in an aged coal-tar contaminated soil by invessel composting. *International Biodeterioration & Biodegradation*, *61*(4), 357–364. doi:10.1016/j.ibiod.2007.10.002

Arıca, M. Y., Bayramoglu, G., Yılmaz, M., Genc, O., & Bektas, S. (2004). Biosorption of Hg, Cd and Zn by Ca-alginate and immobilized wood rotting fungus *Funalia trogii*. *Journal of Hazardous Materials*, *109*(1-3), 191–199. doi:10.1016/j.jhazmat.2004.03.017 PMID:15177759

Ayu, K. R., Tony, H., Tadashi, T., Yasuhiro, T., & Kazuhiro, M. (2011). Bioremediation of crude oil by white rot fungi *Polyporus* sp. S133. *Journal of Microbiology and Biotechnology*, *21*(9), 995–1000. doi:10.4014/jmb.1105.05047 PMID:21952378

Babel, S., & Kurniawan, T. A. (2003). Low-cost adsorbents for heavy metals uptake from contaminated water: A review. *Journal of Hazardous Materials*, *97*(1-3), 219–243. doi:10.1016/S0304-3894(02)00263-7 PMID:12573840

Bafana, A., Kirishnamurthi, K., & Devi, S. & Chakrabarti T. (2008). Biological decolorization of C.I. Direct Black38 by *E. gallinarum*. *Journal of Hazardous Materials*, *157*(1), 183–193. doi:10.1016/j.jhazmat.2007.12.085 PMID:18280646

Bai, S. R., & Abraham, T. E. (2001). Biosorption of Cr(VI) from aqueous solution by *Rhizopus nigricans*. *Bioresource Technology*, *79*(1), 73–81. doi:10.1016/S0960-8524(00)00107-3 PMID:11396911

Bajguz, A. (2011). Suppression of *Chlorella vulgaris* growth by cadmium, lead, and copper stress and its restoration by endogenous brassinolide. *Archives of Environmental Contamination and Toxicology*, *60*(3), 406–416. doi:10.1007/s00244-010-9551-0 PMID:20523975

Bhakta, J. N., Lahiri, S., Pitman, J. K., & Jana, B. B. (2015). Carbon dioxide sequestration in wastewater by a consortium of elevated carbon dioxide-tolerant microalgae. *Journal of CO_2 Utilization*, *10*, 105–112.

Bhakta, J. N., Munekage, Y., Ohnishi, K., Jana, B. B., & Balcazar, J. L. (2014). Isolation and characterization of cadmium and arsenic absorbing bacteria for bioremediation. *Water, Air, and Soil Pollution*, *225*(10), 2151. doi:10.1007/s11270-014-2151-2

Biostimulation. (n.d.). In *Wikipedia, the free encyclopedia*. Retrieved from http://en.wikipedia.org/wiki/

Bishnoi, N. R., & Garima. (2005). Fungus: An alternative for bioremediation of heavy metal containing wastewater: a review. *Journal of Scientific and Industrial Research*, *64*, 93–100.

Boopathy, R. (2000). Factors limiting bioremediation technologies. *Bioresource Technology*, *74*(1), 63–67. doi:10.1016/S0960-8524(99)00144-3

Brahmbhatt, N. H., Patel, R. V., & Jasrai, R. T. (2012). Bioremediation potential of *Spirogyra* sp. and *Oscillatoria* sp. for cadmium. *Asian Journal of Biochemical and Pharmaceutical Research*, *2*, 102–107.

Cerniglia, C. E. (1992). Biodegradatation of polycyclic aromatic hydrocarbons. *Biodegradation*, *3*(2-3), 351–368. doi:10.1007/BF00129093

Cerniglia, C. E., Gibson, D. T., & Van Baalen, C. (1980). Oxidation of naphthalene by cyanobacteria and microalgae. *Journal of General Microbiology*, *116*(2), 495–500. PMID:7003059

Chatterjee, S., Gupta, D., Roy, P., Chatterjee, N. C., Saha, P., & Dutta, S. (2011). Study of a lead tolerant yeast strain BUSCY1 (MTCC9315). *African Journal of Microbiological Research*, *5*, 5362–5372.

Chen, S., & Wilson, D. W. (2008). Genetic engineering of bacteria and their potential for Hg^{2+} bioremediation. *Bioremediation*, *8*(2), 97–103. PMID:9342882

Coelho, M. L., Rezende, C. H., Coelho, M. L., de Sousa, A. R. P., Melo, F. O. D., & Coelho, M. M. N. (2015). Bioremediation of polluted waters using microorganisms. In N. Shiomi (Ed.), *Advances in Bioremediation of Wastewater and Polluted Soil* (pp. 1–22). InTech Publisher. doi:10.5772/60770

Csutak, O., Stoica, I., Ghindea, R., Ana-Maria, T., & Vassu, T. (2010). Insights on yeast bioremediation processes. *Romanian Biotechnological Letters*, *15*(2), 5066–5071.

Das, N., & Chandran, P. (2011). Microbial Degradation of Petroleum Hydrocarbon Contaminants: An Overview. *Biotechnology Research International*. doi:10.4061/2011/941810

Das, S., Dash, H. R., & Chakraborty, J. (2016). Genetic basis and importance of metal resistant genes in bacteria for bioremediation of contaminated environments with toxic metal pollutants. *Applied Microbiology and Biotechnology*, *100*(7), 2967–2984. doi:10.1007/s00253-016-7364-4 PMID:26860944

Deshmukh, R., Khardenavis, A. A., & Purohit, H. J. (2016). Diverse Metabolic Capacities of Fungi for Bioremediation. *Indian Journal of Microbiology*, *56*(3), 247–264. doi:10.1007/s12088-016-0584-6 PMID:27407289

Donmez, G. (2002). Bioaccumulation of the reactive textile dyes by *Candida tropicalis* growing in molasses medium. *Enzyme and Microbial Technology*, *20*(3), 363–366. doi:10.1016/S0141-0229(01)00511-7

Doyle, R. J., Matthews, T. H., & Streips, U. N. (1980). Chemical basis for selectivity of metal ions by the Bacillus subtilis cell wall. *Journal of Bacteriology*, *143*, 471–480. PMID:6772632

Dwivedi, S. (2012). Bioremediation of heavy metal by algae: Current and future perspective. *Journal of Advanced Laboratory Research in Biology*, *3*(3), 195–199.

El-Sheekh, M. M., El-Shouny, W. A., Osman, M. E., & El-Gammal, E. W. (2005). Growth and heavy metals removal efficiency of *Nostoc muscorum* and *Anabaena subcylindrica* in sewage and industrial wastewater effluents. *Environmental Toxicology and Pharmacology*, *19*(2), 357–365. doi:10.1016/j.etap.2004.09.005 PMID:21783496

El-Sheekh, M. M., Farghl, A. A., Galal, H. R., & Bayoumi, H. S. (2016). Bioremediation of Different Types of Polluted Water Using Microalgae. *Rendconti Lincei-Scienze Fisiche e Naturali*, *27*(2), 401–410. doi:10.1007/s12210-015-0495-1

El-Sheekh, M. M., Gharieb, M. M., & Abou-El-Souod, G. W. (2009). Biodegradation of dyes by some green algae and cyanobacteria. *International Biodeterioration & Biodegradation, 63*(6), 699–704. doi:10.1016/j.ibiod.2009.04.010

El-Sheekh, M. M., Gharieb, M. M., & Abou-El-Souod, G. W. (2012). Biodegradation of Phenolic and Polycyclic Aromatic Compounds by Some Algae and Cyanobacteria. *Bioremediation & Biodegradation, 3*, 1–9.

El-Sheekh, M. M., & Hamouda, R. (2014). Biodegradation of crude oil by some cyanobacteria underheterotrophic conditions. *Desalination and Water Treatment, 52*(7-9), 1448–1454. doi:10.1080/19443994.2013.794008

Fathi, A. A., Azooz, M. M., & Al-Fredan, M. A. (2013). Phycoremediation and the potential of sustainable algal biofuel production using wastewater. *American Journal of Applied Sciences, 10*(2), 189–194. doi:10.3844/ajassp.2013.189.194

Favero, N., Costa, P., & Massimino, M. L. (1991). *In vitro* uptake of cadmium by basidiomycete *Pleurotus ostreatus. Biotechnology Letters, 10*(10), 701–704. doi:10.1007/BF01088172

Fuentes, S., Barra, B., Caporaso, J. G., & Seeger, M. (2016). From rare to dominant: A fine-tuned soil bacterial bloom during petroleum hydrocarbon bioremediation. *Applied and Environmental Microbiology, 82*(3), 888–896. doi:10.1128/AEM.02625-15 PMID:26590285

Garg, S. N., Baranwal, R. M., Mishra, S. C., Chaudhuri, T. K., & Bisaria, V. S. (2008). Laccase of *Cyathusbulleri*: Structural, catalytic characterization and expression in *Escherichia coli. Biochimica et Biophysica Acta, 1784*(2), 259–268. doi:10.1016/j.bbapap.2007.11.006 PMID:18083129

Gomes, N. C. M., Rosa, C. A., Pimentel, P. F., & Mendonça-Hagler, L. C. S. (2002). Uptake of free and complexed silver ions by different strains of *Rhodotorula mucilaginosa. Brazilian Journal of Microbiology, 33*(1), 62–66. doi:10.1590/S1517-83822002000100013

Ilyina, A., Castillo Sanchez, M. I., Villarreal Sanchez, J. A., Ramirez Esquivel, G., & Candelas Ramirez, J. (2003). Isolation of soil bacteria for bioremediation of hydrocarbon contamination. *Bulletin of Moscow University, 44*(1), 88–91.

Jayashree, R., Nithya, S. E., Rajesh, P. P., & Krishnaraju, M. (2012). Biodegradation capability of bacterial species isolated from oil contaminated soil. *Journal of Academia Industrial Research, 1*(3), 140–143.

Jeswani, H., & Mukherji, S. (2012). Degradation of phenolics, nitrogen-heterocyclics and polynuclear aromatic hydrocarbons in a rotating biological contactor. *Bioresource Technology, 111*, 12–20. doi:10.1016/j.biortech.2012.01.157 PMID:22382298

Juhasz, A. L., & Naidu, R. (2000). Bioremediation of high molecular weight polycyclic aromatic hydrocarbons: A review of the microbial degradation of benzo[a]pyrene. *International Biodeterioration & Biodegradation, 45*(1-2), 57–88. doi:10.1016/S0964-8305(00)00052-4

Kannan, V., Vijayasanthi, M., & Chinnasamy, M. (2012). Bioremediation of chromium in tannery effluent by filamentous Cyanobacteria *Anabaena flos-aquae* West. *International Journal of Environmental Sciences, 2*(4), 2360–2366.

Klekner, V., & Kosaric, N. (1992). Degradation of phenols by algae. *Environmental Technology, 13*(5), 493–502. doi:10.1080/09593339209385176

Ksheminska, H., Jaglarz, A., Fedorovych, D., Babyak, L., Yanovych, D., Kaszycki, P., & Koloczek, H. (2003). Bioremediation of chromium by the yeast *Pichia guilliermondii*: Toxicity and accumulation of Cr (III) and Cr (VI) and the influence of riboflavin on Cr tolerance. *Microbiological Research, 158*(1), 59–67. doi:10.1078/0944-5013-00178 PMID:12608581

Kujan, P., Prell, A., Šafa, H., Sobotka, M., Ezanka, T., & Holler, P. (2005). Removal of copper ions from dilute solutions by Streptomyces noursei mycelium. Comparison with yeast biomass. *Folia Microbiologica, 50*(4), 309–314. doi:10.1007/BF02931411 PMID:16408849

Kujan, P., Prell, A., Safar, H., Sobotka, M., Rezanka, T., & Holler, P. P. (2006). Use of the industrial yeast *Candida utilis* for cadmium sorption. *Folia Microbiologica, 51*(4), 257–260. doi:10.1007/BF02931807 PMID:17007420

Kuyucak, N., & Volesky, B. (1989). Accumulation of cobalt by marine alga. *Biotechnology and Bioengineering, 33*(7), 809–814. doi:10.1002/bit.260330703 PMID:18587987

Levin, L., Viale, A., & Forchiassin, A. (2003). Degradation of organic pollutants by the white rotbasidiomycete *Trametes trogii. International Biodeterioration & Biodegradation, 52*(1), 1–5. doi:10.1016/S0964-8305(02)00091-4

Lovely, D. R. (2002). Dissimulatory metal reduction: From early life to bioremediation. *ASM News, 68*(5), 231–237.

Luther, M., & Shaaban, M. M. (1990). Algal degradation of aromatic compounds. *Forum Mikrobiology, 1*, 42.

Ma, X., Ding, N., Peterson, E. C., & Daugulis, A. J. (2016). Heavy metals species affect fungal-bacterial synergism during the bioremediation of fluoranthene. *Applied Microbiology and Biotechnology, 100*(17), 7741–7750. doi:10.1007/s00253-016-7595-4 PMID:27178182

Mahmoud, Y. A.-G. (2004). Uptake of radionuclides by some fungi. *Mycobiology, 32*(3), 110–114. doi:10.4489/MYCO.2004.32.3.110

Malik, A. (2006). Bioremediation. In *Environmental Microbiology*. National Science Digital Library.

Mandal, A. K., Sarma, P. M., Singh, B., Jeyaseelan, C. P., Channashettar, V. A., Lal, B., & Datta, J. (2011). Bioremediation: A sustainable eco-friendly biotechnological solution for environmentalpollution in oil industries. *Journal of Sustainable Develeopment and Environmental Protection, 1*(3), 5–23.

Mandal, A. K., Sarma, P. M., Singh, B., Jeyaseelan, C. P., Channashettar, V. A., Lal, B., & Datta, J. (2012). Bioremediation: An environment friendly sustainable biotechnological solution for remediation of petroleum hydrocarbon contaminated waste. *Journal of Science and Technology, 2*, 1–12.

Marungruenga, K., & Pavasant, P. (2006). Removal of basic dye (Astrazon blue FGRL) using macroalgae *Caulerpa lentillifera. Journal of Environmental Management, 78*(3), 268–274. doi:10.1016/j.jenvman.2005.04.022 PMID:16112338

Mc Alinden, J. (2008). *The use of liquid dye penetrants containing the azo compound CI Solvent Red 164 in the detection of flaws or cracks in metal components occupational hygiene, Bootle (VPN 523 4545).2008*. Retrieved from http://www.britishmetalforming.com/uploads/file/pdf/Azo%20dyes%20-%20Nov%2008.pdf

Mehta, S. K., & Gaur, J. P. (2005). Use of algae for removing heavy metal ions from wastewater: Progress and prospects. *Critical Reviews in Biotechnology*, *25*(3), 113–152. doi:10.1080/07388550500248571 PMID:16294830

Mohanty, S., & Mukherji, S. (2013). Surfactant aided biodegradation of NAPLs by Burkholderiam-ultivorans: Comparison between Triton X-100 and rhamnolipid JBR-515. *Colloids and Surfaces. B, Biointerfaces*, *102*, 644–652. doi:10.1016/j.colsurfb.2012.08.064 PMID:23104033

Moutaouakkil, A., Zeroual, Y., Dzayri, F. Y., & Talbi, M., Lee, K., & Blaghen, M. (2003). Bacterial decolorization of the azo dye methyl red by *Enterobacter agglomerans. Annals of Microbiology*, *53*, 161–169.

Mukherji, S., & Jeswani, H. (2011). Treatment of biomass gasifier wastewater in a rotating biological contactor. *Conference proceedings,IWA Biofilm Conference, Processes in Biofilms*, 215-216.

Muraleedharan, T. R., & Venkobachar, C. (1990). Mechanism of biosorption of copper (II) by *Ganoderma lucidum. Biotechnol Bioenginering*, *5*(3), 320–325. doi:10.1002/bit.260350314

Nakajima, A., & Tsuruta, T. (2004). Competitive biosorption of thorium and uranium by Micrococcus luteus. *Journal of Radioanalytical and Nuclear Chemistry*, *260*(1), 13–18. doi:10.1023/B:JRNC.0000027055.16768.1e

Nicolaou, S. A., Gaida, S. M., & Papoutsakis, E. T. (2010). A comparative view of metabolite and substrate stress and tolerance in microbial bioprocessing: From biofuels and chemicals, to biocatalysis and bioremediation. *Metabolic Engineering*, *12*(4), 307–331. doi:10.1016/j.ymben.2010.03.004 PMID:20346409

Olgui, E. J. (2003). Phycoremediation: Key issues for cost-effective nutrient removal processes. *Biotechnology Advances*, *22*(1-2), 81–91. doi:10.1016/S0734-9750(03)00130-7 PMID:14623045

Pattus, F., & Abdallah, M. (2000). Siderophores and iron-transport in microorganisms. *Review Journal of Chinese Chemical Society*, *47*(1), 1–20. doi:10.1002/jccs.200000001

Perpetuo, E. A., Souza, C. B., & Nascimento, C. A. O. (2011). Engineering bacteria for bioremediation. In A. Carpi (Ed.), *Progress in Molecular and Environmental Bioengineering. From Analysis and Modeling to Technology Applications* (pp. 605–632). Rijeka: InTech.

Pieper, D. H., & Reineke, W. (2000). Engineering bacteria for bioremediation. *Current Opinion in Biotechnology*, *11*(3), 262–270. doi:10.1016/S0958-1669(00)00094-X PMID:10851148

Pinto, G., Pollio, A., Previtera, L., & Temussi, F. (2002). Degradation of phenols by microalgae. *Biotechnology Letters*, *24*(24), 2047–2051. doi:10.1023/A:1021367304315

Prabha, Y., Soni, S. K., Gupta, S., & Sonal, . (2016). Potential of Algae in Bioremediation of Wastewater: Current Research. *International Journal of Current Microbiology and Applied Sciences*, *5*(2), 693–700. doi:10.20546/ijcmas.2016.502.076

Puvaneswari, N., Muthukrishnan, J., & Gunasekaran, P. (2006). Toxicity assessment and microbial degradation of azo dyes. *Indian Journal of Experimental Biology*, *44*, 618–626. PMID:16924831

Rahman, M. A., Soumya, K. K., Tripathi, A., Sundaram, S., Singh, S., & Gupta, A. (2011). Evaluation and sensitivity of cyanobacteria, *Nostoc muscorum* and *Synechococcus* PCC 7942 for heavy metals stress - a step toward biosensor. *Toxicological and Environmental Chemistry*, *93*(10), 1982–1990. doi :10.1080/02772248.2011.606110

Sarma, P. M., Bhattacharya, D., Krishnan, S., & Lal, B. (2004). Assessment of intra-species diversity among strains of *Acinetobacter baumannii* isolated from sites contaminated with petroleum hydrocarbons. *Canadian Journal of Microbiology*, *50*(6), 405–414. doi:10.1139/w04-018 PMID:15284886

Semple, K. T., & Cain, R. B. (1996). Biodegradation of Phenols by the Alga *Ochromonas danica*. *Applied and Environmental Microbiology*, *62*(4), 1265–1273. PMID:8919787

Seufferheld, M. J., & Curzi, M. J. (2010). Recent discoveries on the roles of polyphosphates in plants. *Plant Molecular Biology Reporter*, *28*(4), 549–559. doi:10.1007/s11105-010-0187-z

Shamsuddoha, A. S. M., Bulbul, A., & Huq, S. M. I. (2006). Accumulation of arsenic in green algae and its subsequent transfer to the soil-plant system. *Bangladesh Journal of Medical Microbiology*, *22*(2), 148–151.

Singh, D. (2008). Biodegradation and bioremediation of pesticide in soil: Concept, method and recent developments. *Indian Journal of Microbiology*, *48*(1), 35–40. doi:10.1007/s12088-008-0004-7 PMID:23100698

Singh, S., & Gupta, V. K. (2016). Biodegradation and bioremediation of pollutants: Perspectives strategies and applications. *International Journal of Pharmacology and Biological Sciences*, *10*(1), 53–65.

Subashchandrabose, S. R., Ramakrishnan, B., Megharaj, M., Venkateswarlu, K., & Naidu, R. R. (2013). Mixotrophic cyanobacteria and microalgae as distinctive biological agents for organic pollutant degradation. *Environment International*, *51*, 59–72. doi:10.1016/j.envint.2012.10.007 PMID:23201778

Trama, B., Fernandes, J., Labuto, G., Oliveira, J., Viana-Niero, C., Pascon, R., & Vallim, M. (2014). The Evaluation of Bioremediation Potential of a Yeast Collection Isolated from Composting. *Advances in Microbiology*, *4*(12), 796–807. doi:10.4236/aim.2014.412088

Volesky, B. (2004). *Sorption and biosorption*. Quebec: BV-Sorbex, Inc.

Voltolina, D., Gómez-Villa, H., & Correa, G. (2005). Nitrogen removal and recycling by *Scenedesmus obliquus* in semi continuous cultures using artificial wastewater and simulated light and temperature cycle. *Bioresource Technology*, *96*(3), 359–362. doi:10.1016/j.biortech.2004.04.004 PMID:15474938

Vymazal, J. (1984). Short term uptake of heavy metals by periphyton algae. *Hydrobiologia*, *119*(3), 171–179. doi:10.1007/BF00015208

Walker, J. D., Colwell, R. R., Vaituzis, Z., & Meyer, S. A. (1975). Petroleum degrading achlorophyllous alga *Prototheca zopfii*. *Nature*, *254*(5499), 423–424. doi:10.1038/254423a0 PMID:1118032

Wang, Q., & Fenglian, F. (2011). Removal of heavy metal ions from wastewaters: A review. *Journal of Environmental Management*, *92*(3), 407–418. doi:10.1016/j.jenvman.2010.11.011 PMID:21138785

Woodward, K. B., Fellows, C. S., Conway, C. L., & Hunter, H. M. (2009). Nitrate removal, denitrification and nitrous oxide production in the riparian zone of an ephemeral stream. *Soil Biology & Biochemistry*, *41*(4), 671–680. doi:10.1016/j.soilbio.2009.01.002

Wu, J., & Yu, H. Q. (2007). Biosorption of 2,4-dichlorophenol by immobilized white-rot fungus *Phanerochaete chrysosporiyum* from aqueous solutions. *Bioresource Technology*, *98*(2), 253–259. doi:10.1016/j.biortech.2006.01.018 PMID:16517157

Xueheng, Z., Hardin, I. R., & Huey-Min, H. (2006). Biodegradation of a model azo disperse dye bythe white rot fungus *Pleurotus ostreatus. International Biodeterioration & Biodegradation*, *57*(1), 1–6. doi:10.1016/j.ibiod.2005.10.008

KEY TERMS AND DEFINITIONS

Algal Bioremediation: It is the use of algae to remove pollutants from the environment or to convert them into harmless form.

Azo dye: It is the largest class of synthetic dyes.

Bacterial Bioremediation: It includes the bioremediation or biodegradation of contaminants and hazardous pollutants in the environment using bacteria.

Dye Decolourization: It is a process of colour removal from the waste or effluents by means of physic-chemical and biological mechanisms (bioaccumulation, biosorption, and biodegradation).

Fungal Bioremediation: It is the bioremediation or biodegradation of contaminants and hazardous pollutants in the environment using fungi.

Genetically Engineered Microorganisms: These are the microorganisms which genetic makeup has been changed through biotechnology process.

Microbial Consortium: Two or more microbial groups living symbiotically.

PAH: It is a group of polycyclic aromatic hydrocarbons.

Yeast Bioremediation: It refers to the bioremediation or biodegradation of contaminants and hazardous pollutants in the environment using yeast.

Chapter 7
Hydrocarbon Biodegradation Using Agro-Industrial Wastes as Co-Substrates

Abdullah Mohammed El Mahdi
Arabian Gulf Oil Co. (AGOCO), Libya

Hamidi Abdul Aziz
Universiti Sains Malaysia, Malaysia

ABSTRACT

The diversity of agro industrial wastes makes them an attractive group of organic wastes for potential use in a wide variety of industrial and biotechnological applications. The new stimulating development in this current area of research approaches in combination with the technologies of large-scale production and biotechnology engineering, agro industrial wastes will be economically successful materials of the future. Increased public awareness of issues related to hydrocarbon pollution strongly influences the development of technologies that speed up cleaning hazardous contaminants. The cost of biodegradation technology and the low bioavailability including mass transfer limitations of hydrocarbons, especially those recalcitrant components, from various mediums into the aqueous phase for effective enzyme-based microbial biodegradation still constitute major challenges. Sustainable replacement of traditional microbiological media with agroindustrial wastes as substrates for biosurfactant production holds great potential; thereby decrease numerous management problems of handling industrial waste. These organic nitrogen-rich nutrients (biostimulation) are an effective means to enhance the bioremediation process and widely available as wastes in the environment; hence, they can serve as "natural waste-to-environmental clean-up." However, current chapter have focused on the combined use of biosurfactants and enzymes produced from renewable resources such as agro-industrial waste, through assisted biostimulation and bioaugmentation, for hydrocarbon biodegradation.

DOI: 10.4018/978-1-5225-2325-3.ch007

INTRODUCTION

Petroleum hydrocarbons are important energy resources and also a major pollutant of the environment. This in turn can cause irremediable damage to economy causing millions of losses in revenues instead to raise substantial revenues. Moreover, oil spills during exploration, transportation, and refining, have caused serious environmental problems (Zhang et al., 2011; Silva, Rufino, Luna, Santos, & Sarubbo, 2014). Hydrocarbon Oil spills is of serious concern particularly in small inland seas having very high sea traffic. In general, the most susceptible part of the world to oil spill was the Mediterranean coastal regions. This semi-enclosed Mediterranean Sea is the most active transport route for offshore oil fields connecting many continents. Expectedly, it is also the most polluted zones, with estimated annual oil dumped to be about 883,000 tons (Daffonchio et al., 2013). Although, this region constitutes only about 0.82% of the world ocean surface, it accounts for a whopping 20% of the global oil pollution (Daffonchio et al., 2013).

Environmental regulatory agencies constantly encourage companies to come up with clean and green technology. These regulations have made the oil producing companies, to develop environmental friendly strategies to detect pollution. Among all the techniques, natural cleaning method of hydrocarbons is one of the most environmental friendly routes for cleaning of oil spills. One of the most important components of natural attenuation is degraded, the change in the form of compounds carried out by living microorganisms. Under the right conditions, microorganisms can cause or assist chemical reactions that change the form of the contaminants so that little or no health risk remains (Dave & Ghaly, 2011). Hence, remedy of petroleum hydrocarbon polluted sites can be carried out by bioremediation, thereby enhance the natural process of biodegradation using biosurfactant producing and oil degrading bacterial cultures (Macaulay, 2015). Studies have shown that bioremediation is one of the most effective and best technologies available, to solve pollution of the environment due to petroleum components (Thapa, Kc, & Ghimire, 2012). Although biodegradation was revealed to be successful in naturally remediating oil contamination, much remains to be learned about the environmental controls of hydrocarbon degradation in the environment. There is a general trend to replace synthetic microbiological media with agro industrial wastes which holds great potential. Various low-cost substrates such as solid waste date (SWD), and corn steep liquor (CSL) not only act as nutrients for the microbial growth, but also act as an important source for the isolation of novel biosurfactant producing microorganisms and a higher production yield is also achieved (Gudina et al., 2015; El Mahdi, Aziz, El-Gendy, Amr, & Nassar, 2015a).

In environmental bioremediation applications, microorganisms can be supported on solid agrowaste to provide the required macro- and micro-nutrients required for biofilm formation. This usually enhances the metabolic activities of the microorganisms for solubilization and biodegradation of hydrocarbon contaminants. In the recent industrialization, the exploration of natural resources has served as a source of experimentation for scientific and advanced biotechnologies. This is giving rise to the utilizing of by products with high aggregate value in the world market, such as biosurfactants. Biosurfactants are easily degraded by microorganisms in water and soil, making these biomolecules compounds adequate for bioremediation and waste treatment (D.Santos, Rufino, Luna, J. M., A.Santos, & Sarubbo, 2016). The future success of biosurfactant knowledge of bioremediation invention is promising, but will require the specific targeting of the biosurfactant systems to reduce production costs and increase product yield (Sobrinho, Luna, Rufino, Porto, & Sarubbo, 2013). In addition extending a strong positive impact on the main global environmental problems, biosurfactant production has considerable importance to the linking of sustainable industrial processes, such as the use of renewable resources and "green" products. Biodegradability and low toxicity have led to the increase of scientific studies on a wide range of industrial applications for biosurfactants.

Like increase in the field of bioremediation as well as the petroleum, food processing, health, chemical, agricultural, and cosmetic industries. Hence, the greater emphasis is being laid on procurement of various cheap agro-industrial substrates including vegetable oils, distillery and dairy wastes, soya molasses, animal fat, waste and starchy waste as raw materials. These wastes can be used as substrates for large-scale production of biosurfactants with advanced technology, which is the matter of future research development (Saharan, Sahu, & Sharma, 2011; R. Panesar, P. Panesar, & Bera, 2011). This chapter revealed that utilization of agroindustrial wastes for microbial growth and effective surface active materials production is feasible and has promising application with a view to enhancing the bioavailability and bioremediation of recalcitrant hydrocarbons. Moreover, this will reduce many management problems of processing agro industrial wastes. Although it would be presuming to believe that waste would be able to wholly meet the requirements for use as substrates, it is still achievable if a merged approach to recycle and reuse waste is developed, where in the waste coming from one industrial or sector converts the raw material for a different.

AGRO WASTE TO ENHANCE HYDROCARBON BIODEGRADATION

All substances in nature ultimately submit to decay. Much of this phenomenon is a natural significance of the laws of thermodynamics. Many molecules degrade by the action of oxygen, halogens and radicals naturally found in the environment. While a large proportion of materials degrade because their components are subject to the action of enzymes (Fitzpatrick, Champagne, Cunningham, & Whitney, 2010; Zelzer, Todd, Hirst, McDonald, & Ulijn, 2013). Most organic chemicals and many inorganic ones are subject to enzymatic attack through the activities of living organisms. Most of environmental pollutants are included within these chemicals, and the actions of enzymes on them are usually taken under the term biodegradation. The productive use of biodegradative processes to remove or clean up pollutants that have found their way into the surrounding environment and threaten public health, usually as pollutants of soil, water, or sediments is bioremediation (Thapa, Kc, & Ghimire, 2012). The principle of the degradation process is related to the genetic potential of the particular microorganism to introduce molecular oxygen into hydrocarbon and to create the intermediates that later enter the general energy- yielding metabolic pathway of the organism cell (Thapa, Kc, & Ghimire, 2012; Röling & van Bodegom, 2014).

Typically, individual organisms degrade only a limited range of hydrocarbons. However, there is increasing research on the isolation of individual organisms that can not only degrade the major components of crude oils but also demonstrate versatility for other more recalcitrant hydrocarbons (Obayori, Salam, & Ogunwumi, 2014). Bioremediation is a waste management technique that involves the use of organisms to remove or neutralize pollutants from a contaminated site. According to the EPA, bioremediation is a “treatment that uses naturally occurring organisms to break down hazardous substances into less toxic or nontoxic substances” (Maheshwari, U. Singh, P. Singh, N. Singh, & Rani., 2014).The beauty of bioremediation is the end product of breaking down the hydrocarbons into carbon dioxide and water (Fernández-Luqueño et al., 2011; Macaulay and Rees, 2014). The main challenge in bioremediation of hydrocarbon compounds is their extremely low solubility in water (O. Chikere, Okpokwasili, & B. Chikere, 2011; Tanase et al., 2012). Increasing substrate solubilization through the use of a biosurfactant is a way to enhance the bioavailability and metabolism of contaminants (Rita Silva et al., 2014; Macaulay and Rees, 2014). As the insolubility of hydrocarbons decreases the efficiency and rate of degradation, this limitation can be overcome by either the addition of a surface active compound (surfactant) to the growth medium, thereby making hydrocarbons more water soluble and available for cells to degrade, or by the addition of degrading microorganisms producing biosurfactants. Biosurfactants also stimulate the catabolism of hydrocarbons through a co-metabolism process, as these organic compounds readily degradable by microorganisms (Rita Silva et al., 2014).

The bioavailability of a hydrocarbon in general is controled by physical-chemical processes such as sorption and desorption, diffusion and dissolution. Microorganisms improve bioavailability of potential biodegradable nutrients by production of biosurfactants (Reis, Pacheco, Pereira, & Freire, 2013; Rita Silva et al., 2014). Biosurfactant are directly involved in the process of hydrocarbon removal from the environment through the bioavailability and subsequent of the biodegradation of hydrocarbons by direct cell contact. For this reason, is considered that biosurfactant production is closely related to the ability of microorganism to grow and degrade the hydrocarbons and fuels (Montero-Rodríguez et al., 2014; Rita Silva et al., 2014; Pacwa-Płociniczak, Płaza, Poliwoda, & Piotrowska-Seget, 2014; Patowary, Saikia, Kalita, & Deka, 2014).

The biodegradation of oil-derived hydrocarbons by biosurfactants occurs through two mechanisms. The first involves an increase in the bioavailability of the hydrophobic substrate to microorganisms, with a consequent reduction in surface tension of the medium around the bacterium as well as a reduction in interfacial tension between the cell wall and hydrocarbon molecules. The other mechanism involves the interaction between the biosurfactant and cell surface, leading to changes in the membrane, facilitating hydrocarbon adherence (increase in hydrophobicity) and reducing the lipopolysaccharide index of the cell wall without damaging the membrane (Aparna, Srinikethan, & Hedge, 2011; F. Santos, Rufino, Luna, A. Santos, & Sarubbo, 2016). Thus, biosurfactants block the formation of hydrogen bridges and allow hydrophobic-hydrophilic interactions, which cause molecular rearrangements and reduce the surface tension of the liquid by increasing its surface area as well as promoting bioavailability and consequent biodegradability (Pacwa-Płociniczak, Płaza, Piotrowska-Seget, & Cameotra, 2011; Aparna et al., 2011).

Such biodegradation is known to be time consuming and new technologies have been developed; for example bio-augmentation, biosurfactants etc. While, the effectiveness of enhancing hydrocarbon degradation through addition of microbial inocula of non-indigenous populations has been ambiguous, the use of a combination of new approaches as shown much success. For instance, the addition of biosurfactant helps to stimulate the indigenous microbial population to degrade hydrocarbons at rates higher than those which could be achieved through addition of nutrients alone (Das & Chandran, 2010).

All these approaches can make the bioremediation process an economically and environmentally viable mitigation technology. The progress of sustainable technologies has driven the search for natural, biodegradable compounds to remediate sites contaminated by hydrocarbons. This has led to the discovery of surfactants of a natural origin. Most of these surfactants are synthesized by living organisms, such as, saponins produced by plants, glycolipids produced by microorganisms, and bile salts produced by animals. Compounds with surfactant properties produced by microorganisms are denominated biosurfactants. (Aparna et al., 2011). Although biosurfactants exhibit such important advantages, they have not yet been employed extensively in industry because of relatively high production costs. One potential approach for reducing costs is the utilization of alternative substrates such as agro industrial waste (Makkar, Cameotra, & Banat, 2011; Noparat et al., 2014). The use of cheaper raw materials (obtained from agro wastes) as biostimulating agents resulted in the enhanced bioremediation of petroleum hydrocarbon. Hence, the agro wastes greatly and significantly enhance the degradation of petroleum hydrocarbon in environmental contaminated with hydrocarbons. However, bioremediation of hydrocarbon using agro-wastes in oil hydrocarbons contaminated environment offers a better and more environmentally friendly technique that if properly and thoroughly explored in optimized process will lead to a safer environment for both public and surrounded sphere.

The capability of biosurfactants and biosurfactant-producing bacterial strains to enhance organic contaminants' availability and biodegradation rates was reported by many researchers (Rita Silva et al., 2014; Pacwa-Płociniczak et al., 2014; Patowary et al., 2014). Obayori et al. (2009) investigated the biodegradative properties of biosurfactant produced by *Pseudomonas sp.* LP1 strain on crude oil and diesel. The results obtained confirmed the ability of strain LP1 to metabolize the hydrocarbon components of crude and diesel oil. They reported 92.34% degradation of crude oil and 95.29% removal of diesel oil. Al-Bahry et al. (2013) has reported that the biosurfactant produced using date molasses is suitable for use in the petroleum industry, such as oil, bioremediation clean up and enhanced oil recovery.

Silva et al. (2014) reported that the bacterium *Pseudomonas cepacia* CCT6659 cultivated with 2% soybean waste frying oil and 2% corn steep liquor as substrates produced a biosurfactant with potential application in the bioremediation of hydrocarbons. Significant oil biodegradation activity (83%) occurred in the first 10 days. Thavasi et al. (2011) and Khalifeh, Roozbehani, Moradi, Imani Moqadam, & Mirdrikvand (2013) reported that high emulsification activity at low concentration observed with biosurfactant produced by *P. aeruginosa* and maximum biodegradation was found (85%), (89%) in their studies that could be used for environmental and crude oil tanker cleaning applications.

The biosurfactants seem to enhance biodegradation by influencing the bioavailability of the contaminant. Due to their biodegradability and low toxicity, they are very promising for use in remediation technologies (Soberón-Chávez &Maier, 2011; Pacwa-Płociniczak et al., 2011). Perfumo, Smyth, Marchant, & Banat (2010) have reported that one of the most important features to be considered for the useful application of biosurfactants is the ability to form and stabilize emulsions, mostly for environmental applications such as bioremediation and enhanced oil recovery.

IN SITU CO-SUBSTRATE TECHNIQUE OF SOLID WASTE DATE

Tons of date palm fruit wastes are discarded daily by the date processing industries leading to environmental problems. Wastes such as date pits represent an average of 10% of the date fruits. Thus, there is a vital need to find appropriate applications for this waste (Chandrasekaran & Bahkali, 2013). Date trees produce huge quantities of agricultural waste and according to one estimate; each date tree produces about 20 kg of dry leaves annually. Other wastes such as date pits represent an average of 10% of the date fruits and date press cakes are being generated by the date palm agro industry and date processing industries (Chandrasekaran & Bahkali, 2013; El May, Dorge, Jeguirim, Trouvé, & Said, 2012).

In fact, all organic waste materials arising from agro waste processing have potential for use as a natural wastes and inexpensive raw materials for biosurfactant production. Date molasses was used as a novel substrate and sole source of carbon for biosurfactant production (Al-Bahry et al., 2012). However, their low production cost enhances their use in bioremediation processes thereby increasing the amount of bioavailability during metabolic degradation process. As such, great importance is being given on the procurement of various cheap agro-industrial substrates, because the goal of any bioremediation protocol is the destruction of the pollutant by using cheap and available nutrient sources. Promising outcomes from such study accessible solid waste date (SWD) as an effective bioremediation tool for the cleanup of hydrocarbons contaminated environments (El Mahdi et al., 2014). Inexpensive and environment friendly options to enhance petroleum hydrocarbon degradation are necessary. El Mahdi et al. (2015 b) has concluded that corn steep liquor (CSL) and solid waste dates (SWD), as natural, readily available low cost agro-industrial wastes were enhanced the bacterial growth and biodegradation efficiency. Molasses is a low-cost by-product generated during the crystallization of sugar from liquid extracts of dates, sugarcane, or sugar beet. It contains a high concentration of carbohydrates (usually about 50%), as well as other valuable compounds such as vitamins. Corn steep liquor (CSL) is a liquid by-product generated by the corn wet milling industry. It is rich in vitamins, minerals, amino acids, and proteins, being an important source of nitrogen for many biotechnological processes (Henkel et al., 2012; Maddipati, Atiyeh, Bellmer, & Huhnke, 2011; R. Panesar, P. Panesar, & Bera, 2011).

Gudiña et al. (2015) has reported that the high nutritional content of both substrates, together with their low cost and availability, make them valuable products to be used as culture medium or nutrient enhancements for microbial growth in various industrial fermentation processes. Gudiña et al. (2015) in their work reported that a low-cost culture medium was settled using two agro-industrial by-products (corn steep liquor and molasses) for biosurfactant production by a *P. aeruginosa* strain. This biomolecules was characterized as a blend of eight different rhamnolipid congeners, and revealed a well performance in oil assays when compared with the commercially chemical surfactants, making it a favorable candidate for application in the oil industry or bioremediation processes.

UTILIZATION OF AGRO-INDUSTRIAL TO INTEGRATED BIOTECHNOLOGY

The abundant agro-industrial waste, such as corn steep liquor (CSL) and solid waste date (SWD) are the potential candidates as they are available throughout the year. The integrated approach of emerging efficient microbial strains capable of growing on these alternative substrates nutritionally rich low value industrial by-products and designing of low cost nutritionally balanced optimized medium could significantly enhance the hydrocarbon biodegradation efficiency.

In recent study, twenty different bacterial strains were isolated from oily polluted water samples with different degradation ability. Only one showed potentially ability to utilize crude oil as a sole carbon source. Molecular identification of isolates was performed by amplification and sequencing the 16SrDNA gene sequencing and matching it to the database of known 16SrDNA sequence. These isolated bacteria belonged to the genera *Pseudomonas* sp. (designated as NAF1) and *Kocuria* sp. (designated as SAR3) among which the strain *Psudomonase* sp. (NAF1) had the highest potential for degradation of light crude oil as the sole carbon source. It is related to the well-known *Pseudomonas aerugonisa* bacterium strain PAO1 (obtained throughout the successive enrichment cycles in basal salt medium) (El Mahdi et al., 2014, 2015a).The agroindustrial wastes such as CSL and SWD were used to improve biodegradation efficiencies for crude oil hydrocarbon.

The growth bacterial strains NAF1, SAR3, and NAF4 increased steadily upon supplementation of the two co-substrates as shown in (Figure 1). Total bacterial count increased with the supplementation of 0.2% (w/v) CSL and SWD (Figure 1). SWD was more efficient in supporting bacterial growth than CSL. Similarly, in addition to the extensive bacterial growth observed (from 10^{11} CFU/mL to 10^{14} CFU/mL and 10^{12} CFU/mL to 10^{16} CFU/mL), the results strongly support that the addition of CSL and SWD significantly enhanced bacterial growth and resulted in increased degradation rates. (Silva et al., 2014; El-Gendy et al., 2014).

Figure 1. Biodegradation comparisons of abiotic, natural attenuation, and different single strains and consortium with CSL and SWD 0.2-0.5% (w/v) in concentration of 5% crude oil within 28 days incubation time

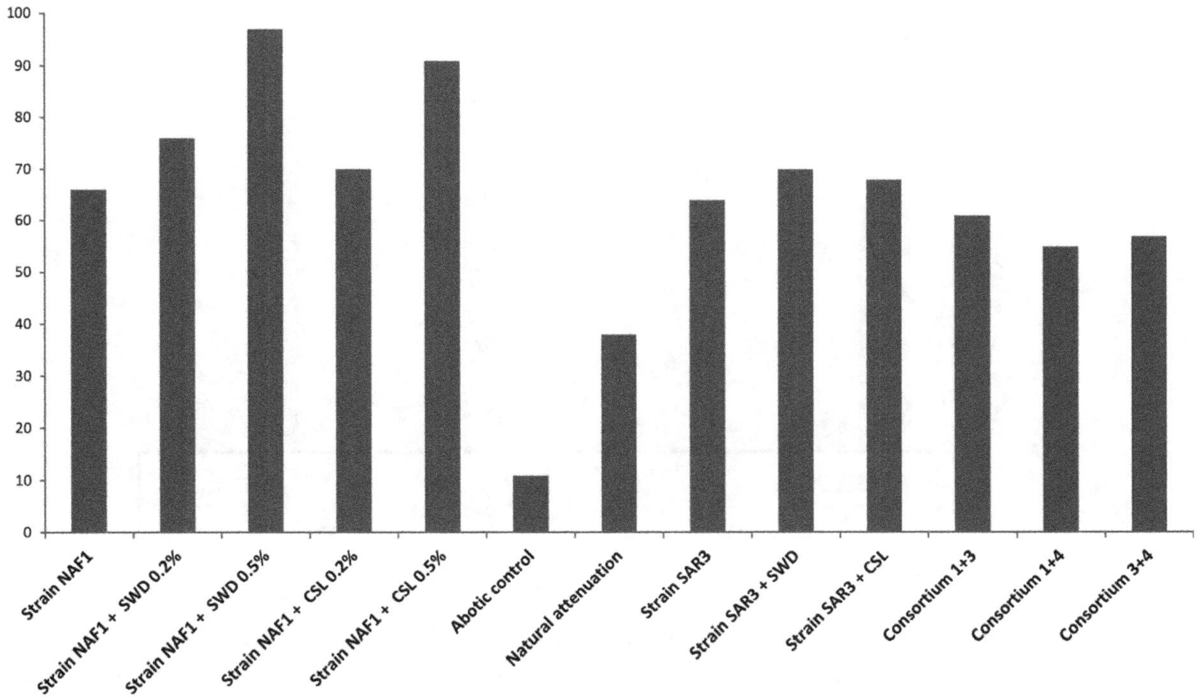

According to El-Gendy et al. (2014), and Silva et al. (2014), CSL can serve as a good source of nutrients and act as an emulsifier, thereby increasing the amount of bioavailability during metabolic degradation process.

Furthermore, waste dates have been found to be a carbon source for microbial production of a variety of biochemical (Al-Bahry et al., 2013). Results in the current study suggests that the addition of SWD has improved the biodegradation of crude oils by *Pseudomonas* sp. NAF1. It could be concluded that in the production of these biomolecules, *Pseudomonas* sp. NAF1 can convert crude oil into cell mass. The potential biodegradation of crude oil was assessed based on the development of visible a fermentative process with a strain of *Pseudomonas aeruginosa* which produced rhamnolipids when cultured in a basal salt medium using vegetable oil or glycerol as the sole carbon source (El Mahdi et al., 2015 a). Degradation of Libyan light crude oils (CO35) (API gravity, 35) and (CO39) (API gravity, 39) by bacterial strain *Pseudomonas* sp. NAF1 was studied. Within 28 days culture flasks containing CO35 and CO39 exhibited 91.74, 98.01% and 97.15, 99.10% (Figure 2) enhanced with corn steep liquor and solid waste dates were degraded respectively. Subsequently the final growth rate of 2.3 x 10^{11}, 3.2 x 10^{11} and 2.9 x 10^{11}, 1.1 x 10^{12} TCFU/mL-^{1}, (CFU: colony forming unit) were measured for CO35 and CO39 respectively. Thus concluded the presence of visual emulsification was found to be growth associated in strain NAF1, where a parallel relationship exists between growth substrate utilization and emulsion activity.

Design Expert software (version 6.0.7) was used for the statistical design of biodegradation experiments using agroindustrial wastes and data analysis based on CCD and response surface RSM. This mathematical and statistical technique was applied to optimize and assess the relationship among three

Figure 2. Comparison for biodegradation removal of CO35 and CO39 and natural attenuation at concentration of 0.63% (w/v) with SWD / CSL conc. 0.5% (w/v)

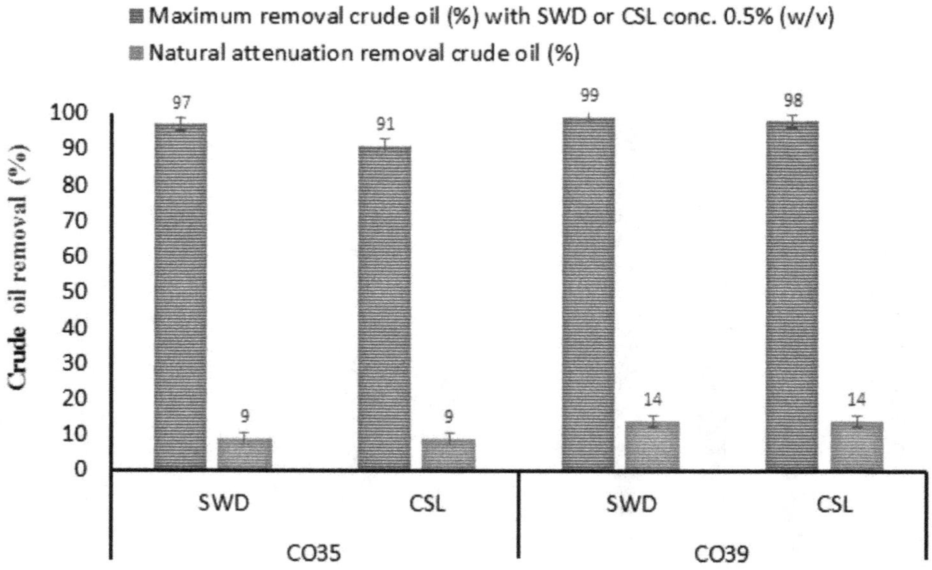

significant independent variables: (1) Crude oil concentration (CO35 and CO39), (2) Corn steep liquor concentration, and (3) Incubation period.This study has been applied to confirm the optimum concentration of corn steep liquor (CSL) and solid waste date (SWD); as a commercial and readily available co-substrate and nutrient to obtain maximum biodegrading activity of indigenous microbial population. Study summary results of biodegradation enhanced with corn steep liquor and solid waste date presented in Tables 1 and 2.

The strain plays an important role during their ability in biodegradation of crude oil components lead to oxidation of carbon source in the crude oil components. Three indicators led to the ability of this strain in biodegradation process: change in color of culture media from brownish to colorless, disappearance of crude oil from the medium, and development of a mass of bacteria growth at the bottom of the culture medium (Al-Nasrawi, 2012). These results, in addition to the extensive bacterial growth observed, indicated that the addition of CSL and SWD resulted in improved degradation rates. This indication is probably because of the organic nitrogen compounds present in waste dates, which could

Table 1. Summary results for CO35 and CO39 biodegradation with corn steep liquor

Factors			CO35 Removal %		CO39 Removal %	
Crude Oil Conc. % (w/v)	CSL Conc. % (w/v)	Time (Day)	Observed	Predicted	Observed	Predicted
0.63	0.50	28.00	91. 74	93.40	98.01	93.74
0.25	0.20	07.00	55.31	59.27	65.46	67.71
1.0	0.20	15.00	81.00	76.00	-	-
1.0	0.20	10.00	-	-	88.00	84.00

Table 2. Summary results for CO35 and CO39 biodegradation with Solid waste dates

Factors			CO35 Removal %		CO39 Removal %	
Crude Oil Conc.(w/v)	SWD Conc.(w/v)	Time (Day)	Observed	Predicted	Observed	Predicted
0.63	0.50	28.00	97.15	95.76	99.10	92.96
0.25	0.20	07.00	75.80	76.61	88.65	88.30
1.00	0.21	11.00	79.00	82.00	-	-
1.00	0.20	14.00	-	-	94.00	95.00

lead to significantly increased growth and reduced process time (El-Gendy et al., 2014; Silva et al., 2014; El Mahdi et al., 2015 a). *Pseudomonas* can grow on numerous low-cost raw substrates and synthesize sufficient amounts of biosurfactant (Makkar et al., 2011; Thavasi, Nambaru, Jayalakshmi, Balasubramanian, & Banat, 2011; Khalifeh et al., 2013). Considering their complex enzymatic system, the groups of *Pseudomonas* present a wide kind of metabolic and physiological attributes. They are also the most predominant group of microorganisms that degrade xenobiotic compounds (Pacwa-Płociniczak, Płaza, Poliwoda, & Piotrowska-Seget, 2014; Patowary, Saikia, Kalita, & Deka, 2014).

Numerous bacteria from *Pseudomonas*, including *P. aeruginosa* strains, inhabit oil-contaminated water and can degrade more than 100 different organic compounds (Das and Chandran 2011; Saikia, S. Deka, M. Deka, & Banat, 2012; Puškárová et al. 2013).Significant oil biodegradation activity (83%) occurred in the first 10 days. It is evident from the results that the biosurfactant alone and its producer species are both capable of promoting biodegradation to a large extent. This result is in accordance with the findings of Silva et al. (2014). The results obtained are also comparable to those reported by Thavasi et al. (2011), and Khalifeh et al. (2013), who investigated the effect of biosurfactants and nutrient supplementation from fertilizers on the biodegradation of crude oil by marine isolates of *Bacillus megaterium, Corynebacterium kutscheri* and *Pseudomonas aeruginosa*. They observed that *Pseudomonas aeruginosa* and the biosurfactants produced by it resulted in the maximum crude oil degradation of about 89%.

EXPLOIT AND MANAGE MICROBIAL DIVERSITY

Microbial biodiversity contributes to ecosystem resilience and sustainability, while understanding of this correlation is still developing (Chatzipavlidis, Kefalogianni, Venieraki, & Holzapfel, 2013). Useful microorganisms are of great importance in agro-industrial processes. There is also a need for gathering information on the diversity of microorganisms in order to develop approaches for expanding and enhancing their utilization in agro-industrial processes, where applicable. According to BCC, Business Communication Co` (2011) the overall global market for microbes and microbial products was value more than 144 billion US$ in 2010, 156 billion US$ in 2011, and was projected to amount to 259 billion US$ in 2016 (Figure 3).

Over the years, scientists have manipulated environmental keys in order to understand the factors that encourage microbial metabolism and hydrocarbon degradation, a knowledge which could help return the environment to a pre-pollution state as early as possible (Macaulay & Rees, 2014). As microbial communities play a significant role in biogeochemical cycles, it is important to analyze the community

Figure 3. Trends in the global market for microbes and its products
B CC, 2011.

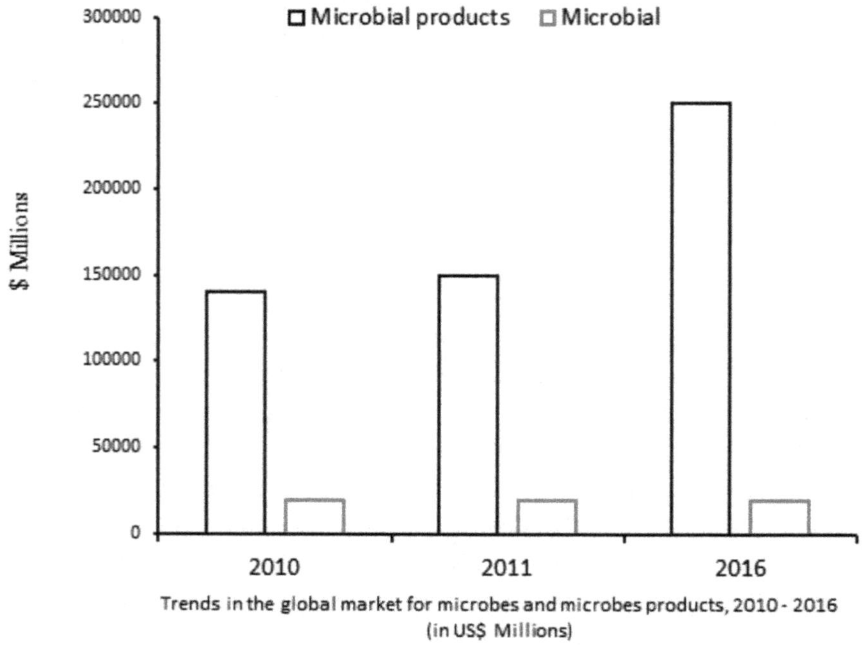

structure and its changes during bioremediation processes. They are playing a vital role in bioremediation and restoration biology of environment and aquatic systems. Furthermore, Microorganisms should be well-preserved for their important sources of knowledge about the strategies and limits of life and their serious significance to the sustainability of lifecycle on our sphere. Their diversity is a great source for biotechnological exploration of novel organisms, products, and processes (Gupta, Barrow, & Puri, 2012). Since the majority of bioremediation processes rely on the activities of complex microbial populations, researchers have much to learn about the interactive and interdependent roles played by individual species in these communities.

Hydrocarbon pollutant environments can be studied and exploited by coupling advanced microbial cultivation through the high input cultivation systems and the advanced 'omics'-based research technologies. Isolation of novel hydrocarbon degrading microbes and a sequences of state of the art 'meta-omics' technologies are the baseline tools for improving our knowledge on biodegradation capacities mediated by microbes under different environmental settings and for designing novel site-tailored bioremediation approaches (Daffonchio et al,. 2013). Recently, network researchers exploits microbial resource mining by the isolation of novel microorganisms as well as by novel advanced 'metaomics' technologies for solving pollution of major high priority pollutant classes, among them petroleum hydrocarbons (Daffonchio et al., 2012). Microbial consortia and their ecology, their components or products are used for designing novel pollutant- and site tailored bioremediation approaches (Daffonchio et al., 2012; Daffonchio et al., 2013).

Such consortia and their single species microbial components can be directly used as inoculants or as source of biological products, like catabolic enzymes (Daffonchio et al., 2012) or biosurfactant that can be exploited in pollutant- and site-tailored bioremediation approaches (Paniagua-Michel & Rosales, 2015). In this context, it is important to understand the mechanism of bacterial agents and their effectiveness in oil-impacted ecosystems. Thus, it is necessary to study the bacterial ecology and its associated biodegradation process in order to develop newer and effective remedial measures (McGenity, 2014). Advances in next generation sequencing technologies and the use of stable isotope tracers have greatly improved our ability to question the phylogenetic and functional diversity of hydrocarbon-degrading microorganisms in the field.

The development and application of omics approaches have led to the characterization of novel biochemical pathway of biogeochemical significance (Kostka, Teske, Joye, & Head, 2014).The producer organism genetics is an important factor affecting the yield of all biotechnological products, since the capability to produce a metabolite is given by the genes of the organism. The potential use of these hyper producers in addition to novel, cost-effective biosurfactant covers the real challenges and offers great opportunities for making industrial production of biosurfactants a success story (Chatzipavlidis et al., 2013). Xu, Nakajima, Liu, & Shiina (2011) shown that the growing significance in microbially produced surfactants is based on their varied diversity in structure and function and on their low cost, which is facilitated by their production from low-cost agro-based substrates and waste materials.

METABOLIC PROCESS MAP

The microorganisms used for bioremediation produce enzymes in the presence of carbon. These enzymes help in reduction of the hydrocarbon molecules. Various enzymes are used to facilitate this process as they have different metabolic pathways in reducing the petroleum based hydrocarbon. Correct selection of microorganism based on the enzyme it produces is very crucial as it allows to reduce the hydrocarbons or act as barrier (Iordanis et al., 2013; Kostka et al., 2014). The biodegradation of various petroleum compounds happens at the same instant but at different rates. This depends on the microbial activity time required to break down the hydrocarbons (Scott et al., 2010). Das and Chandran (2011) showed that under aerobic conditions the degradation achieved is faster for most of the pollutant. Aerobic biodegradation enhances the oxidation process forming oxygenases and peroxidases. Oxygenases uses O_2, incorporating it into the substrate. As the organisms need oxygen for metabolism at the initial attack stage and at the end of the respiratory chain as shown in Figure 4 (a).

In order to reduce the aliphatic and cycloaliphatic hydrocarbons the presence of molecular oxygen is required. The enzymatic reaction takes place in 2 ways as shown above in Figure 4 (b) and also dependent on the substrate to achieve the enzymatic reaction (Das & Chandran, 2011; Hassanshahian & Cappello, 2013). Controlling the bioremediation process is a complex process involving many factors which needs to be considered for optimization. The main factor being the energy required for cell growth which depends on the metabolism rate of the microorganism. The cell growth is dependent on these factors:

- **Primary Organic Substrates:** If the cell continuously metabolizes and draws energy for further multiplication then the contaminant is referred as primary substrate.
- **Secondary Organic Substrates:** In some cases, even though the enzymes metabolize and help the cells in draw energy but without cell growth this type of contaminant is known as secondary substrate.
- **Co-Metabolisms:** In this process, the cell won't use any energy from metabolism of hydrocarbons but it obtains energy from other transformable compounds. Thus the process is known as co-metabolic or co-oxidation where in it oxidizes other compounds to support the microbial growth (Pandey & Fulekar, 2012).

Figure 4. (a) Main principle of Aerobic degradation of hydrocarbons and (b) Initial attacks on xenobiotic by oxygenases. Monooxygenases incorporate one atom of oxygen of O2 into the substrate; the second atom is reduced to H2O. Dioxygenases incorporate both atoms into the substrate
Das and Chandran, 2011.

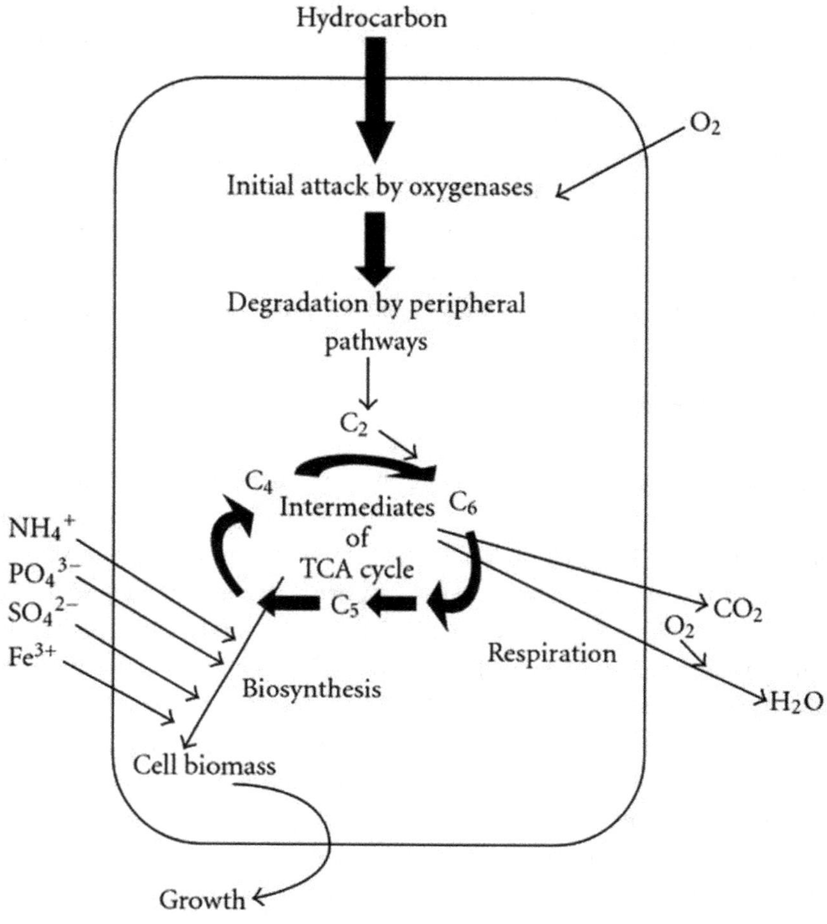

The alkanes are the most abundant compounds and more simples to oxide. Aliphatic hydrocarbons are degraded with greater speed, but the key step involves oxidation of the molecule to increase their solubility. The first step of alkane degradation (Figure 5) is the oxidation of the methyl group to the alcohol by the alkane hydroxylase system, which involve monooxygenase enzyme (Ji et al., 2013). The complete oxidation reaction can be represented as $C_nH_{2n} + 3/2\ N\ O_2 \rightarrow N\ CO_2 + N\ H_2O$. The pathway for degradation of n-alkanes has been extensively studied in *Pseudomonas putida* and is codified by alkgenes. These genes are expressed in n-alkanes presence. Alkanes are greatly distributed in the environment, and microorganisms are likely to utilize these highly reduced compounds as carbon and energy sources. The alkanes degraders including strains of *Pseudomonas, Acinetobacter, Alcaligenes,* and *Rhodoccocus* (Gunasekera, Striebich, Mueller, Strobel, & Ruiz, 2013).

Figure 5. Metabolic pathway of hydrocarbon degradation by Pseudomonas aeruginosa. (I) n-alkanes, (II) naphthalene and (III) toluene. The last step in the n-alkanes degradation involves beta-oxidation and the aromatics and polyaromatics ring degradation finish in TCA cycle.
Source: Adapted from (Silva et al., 2010).

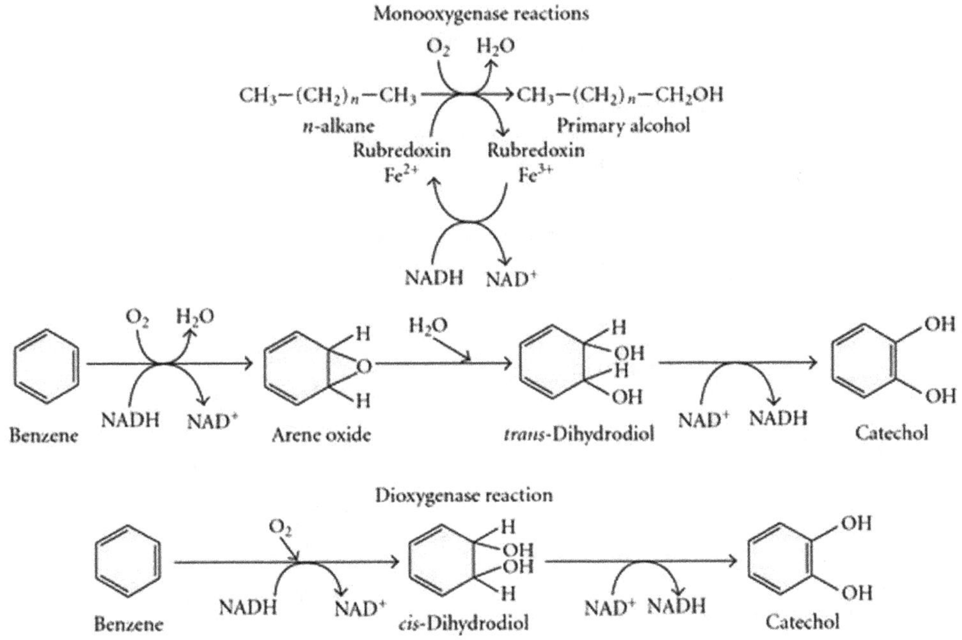

NATURAL CHOICE: BIOSURFACTANTS

Increasing public awareness of environmental pollution influences the search and development of technologies that help in cleanup of organic and inorganic contaminants such as hydrocarbons. An alternative and eco-friendly method of remediation technology of environments contaminated with these pollutants is the use of biosurfactants and biosurfactant-producing microorganisms.

The diversity of biosurfactants makes them an attractive group of compounds for potential use in a wide variety of industrial and biotechnological applications (Pacwa-Płociniczak et al., 2011). Biosurfactants are amphiphilic compounds with both hydrophilic and hydrophobic moieties that act between fluids of different polarities such as oil/water or air/water interfaces. The polar moiety is often a hydrocarbon chains, whereas the polar moiety may be ionic or neutral, whereas only some are cationic, such as those that contain amine groups. (Mao, Jiang, Xiao, & Yu, 2015; Johal, 2011), as showed in Figure 6.

Surfactants increase the solubility of hydrophilic molecules such as hydrocarbons, thereby reducing both surface and interfacial tensions at the oil/water (Aparna et al., 2011). The critical micelle concentration (CMC) is the concentration of surfactant at which organized molecular assemblies, known as micelles, are formed Figure 7 (a) and corresponds to the point at which the tensioactive agent achieves the lowest stable surface tension Figure 7 (b) (Campos et al., 2013).

The naturally occurring surface-active compounds derived from microorganisms are called biosurfactants and they hold several advantages over chemical surfactants, such as low toxicity, inherent good biodegradability, and ecological acceptability. Several biosurfactant are produced by a diversity

Figure 6. Surfactant molecule with a polar (hydrophobic) and polar (hydrophilic) moieties
Rufino et al., 2016.

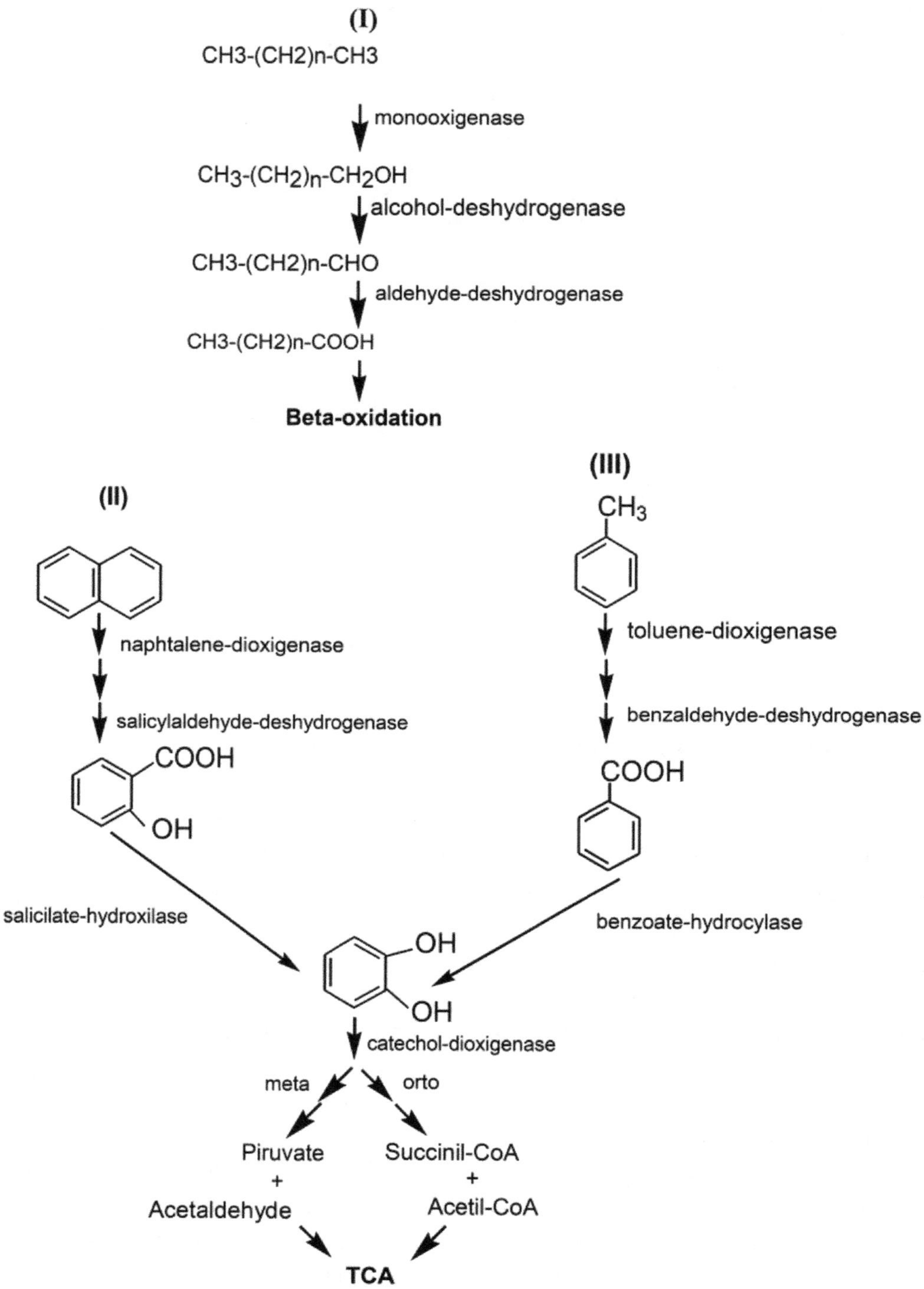

Figure 7. (a) Tensioactive agent and micelle formation and (b) Regions in which micelle formation occurs (CMC)
Campos et al., 2013.

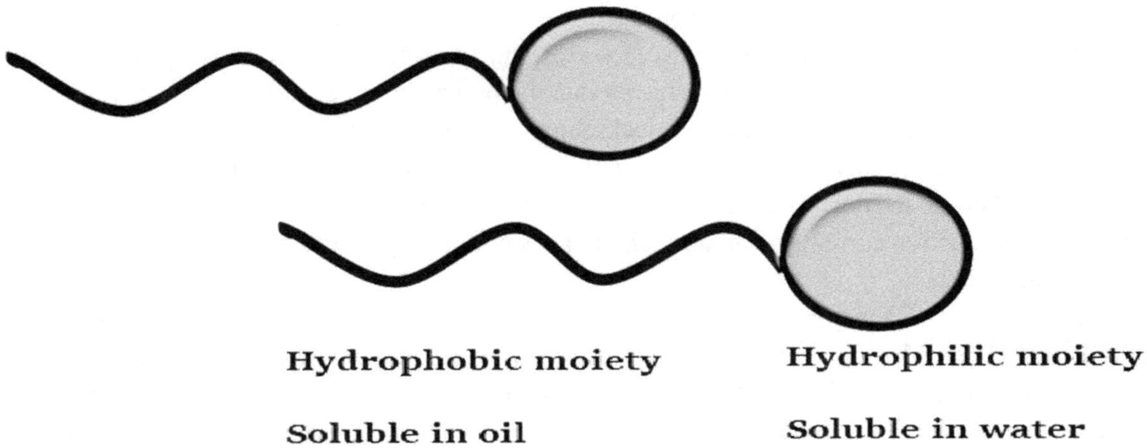

of microorganisms in order to survive in an oil-rich environment, and this adaptation process selected for surfactants with highly adaptable physiochemical properties. Biosurfactants are, consequently, very suitable for applications in the oil recovery and bioremediation, thereby revealed in the market, where the large majority of biosurfactants produced are in petroleum-based applications (Reis, Pacheco, Pereira, & Freire, 2013; Rufino et al., 2016). The applications are, in general, in oil recovery, oil spill management, and as oil dispersants and demulsifiers (Pacwa-Płociniczak et al., 2014). In the recent years, an increase in environmental awareness has led to much more interest in the use of renewable-based, biodegradable, and more eco- friendly surfactants. The advance of sustainable technologies has driven the search for natural, biodegradable compounds to remediate sites contaminated by hydrocarbons. This has led to the discovery of surfactants of a natural origin. Most of these surfactants are synthesized by living organisms, such as, saponins produced by plants, glycolipids produced by microorganisms, and bile salts produced by animals.

Dominated compounds with surfactant properties produced by microorganisms are denominated biosurfactants (Aparna et al., 2011; Souza, Vessoni-Penna, & de Souza, 2014; Rocha e Silva, Rufino, Luna, Santos, & Sarubbo, 2014). Biomolecules are mainly produced by aerobic microorganisms in aqueous media with a carbon source feedstock, such as carbohydrates, hydrocarbons, fats, and oils. It is assumed that biomolecules are secreted into the culture medium to support in the growth of the microorganism by simplifying the transport of insoluble substrates thru cell membranes (Campos et al., 2013). The most common role of biosurfactants is to enhance the dispersal of contaminants in the aqueous phase and increase the bioavailability of the hydrophobic substrate to microorganisms, with following removal of such pollutants through biodegradation (Aparna et al., 2011). Biosurfactants play an important role in remediation processes due to their efficacy as dispersion and remediation agents as well as their environmentally friendly characteristics, such as low toxicity and high biodegradability (Rocha e Silva et al., 2014). The global market for these ''green'' alternatives to synthetic surfactants was 344 kilo tons in 2013, and it is expected to reach 462 kilo tons and 2.308 billion USD by 2020 (Grand View Research, 2014). Currently, the major market for biosurfactants is the petroleum industry, in which these compounds can be used in the cleanup of oils spills, the removal of oil residue from storage tanks, microbial-enhanced oil recovery, and the bioremediation of soil and water (Sobrinho, Luna, Rufino, Porto, & Sarubbo, 2013).

In most recent years, biosurfactants refers exclusively to surfactants from microbial origin gained more and more attention, since sustainable production processes became more considerable, which show the potential to replace synthetic surfactants. However, the application of biotechnologically produced surfactants is limited to certain areas, since the production process is not economically competitive with synthetic surfactants derived from petrochemical sources. This can be explained mostly due to the use of high-priced substrates, relatively low product-yields, and expensive downstream-processing (Henkel, et al., 2012). In addition, the downstream cost, low productivity and intense foaming formation during the biosurfactant production currently is an obstacle for an economically viable production of biosurfactant (Henkel, et al., 2012).

Therefore, recent researches have been focusing on increasing the production yield, reducing raw material cost, and developing strategies to minimize foaming formation (Amodu, Ntwampe, & Ojumu, 2013; Sobrinho et al., 2013; Kang, B. Kim, Shin, & K. Kim, 2010). Moreover, biosurfactants can be obtained with the use of agro-industrial waste as substrate, which helps reduce overall production costs (Rocha e Silva et al., 2014). There are several advantages of biosurfactants when compared to chemical surfactants, mainly biodegradability, low toxicity, biocompatibility, and ability to be synthesized from renewable sources. In spite of having many commercially attractive properties and clear advantages compared with their synthetic equivalents, biosurfactants have not yet been employed extensively in industry because of their low yields and relatively high production and recovery costs. However, the use of mutants and recombinant hyper producing microorganisms along with the use of cheaper raw materials and optimal growth and production conditions and more efficient recovery processes, the production of biosurfactant can be made economically feasible (Kapadia & Yagnik, 2013). Therefore, future research aiming for high-level production of biosurfactants must be focused towards the development of appropriate combinations of hyper producing microbial strains, optimized cheaper production media, and optimized process conditions, which will lead to economical commercial level biosurfactant production (Dubey et al., 2012; Joshi & Desai, 2010). The use of cheaper, renewable substrates from various industries such as agricultural (sugars, molasses, plant oils, oil wastes, starchy substances, lactic whey), distillery wastes, animal fat, oil industries have been reported and reviewed thoroughly by several researchers (Makkar, Cameotra, & Banat, 2011). From an environmental viewpoint, these biomolecules are more acceptable for the remediation process both at sea and land (Hazra, Kundu, & Chaudhari, 2012; Cameotra & Makkar, 2010).

Biosurfactants help in containing the problem by rapidly emulsifying the oil and increasing the microbial growth (Thavasi, Jayalakshmi, & Banat, 2011). One group among these promising biosurfactants is rhamnolipids, surface-active glycolipids mainly known from *Pseudomonas aeruginosa*. Rhamnolipids may potentially be used for bioremediation of contaminated soil and water by improving degradation of hydrocarbons by enhancing their bioavailability to degrading microorganisms (Saharan, Sahu, & Sharma, 2011; Reis et al., 2013). Rhamnolipids are usually synthesized by the Gram-negative bacterium *Pseudomonas aeruginosa* and related species as complex mixtures and more than 60 different congeners have been identified to date (Abbasi et al., 2012). Figure 8, shows the involvement of biosurfactant (rhamnolipids) produced by *Pseudomonas* sp. and the mechanism of formation of micelles in the uptake of hydrocarbons (Das & Chandra, 2011).

Rhamnolipids are produced by *P. aeruginosa* strains as mixtures of different congeners, being the most common and produces two major types of rhamnolipid in liquid cultures: the monorhamnolipid,

Figure 8. Involvement of biosurfactant (rhamnolipid) produced by Pseudomonas sp. in the uptake of hydrocarbons
Das and Chandra, 2011.

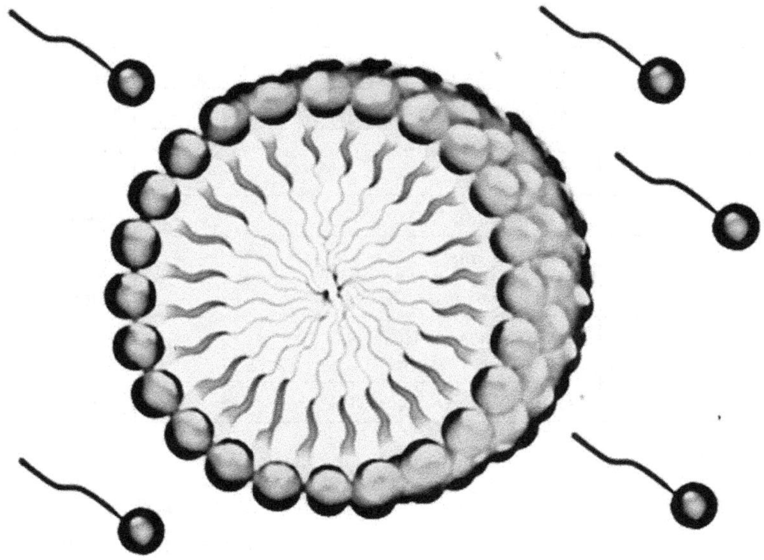

rhamnosyl-β-hydroxydecanoyl-β-hydroxydecanoate (Rha-C10-C10) and the dirhamnolipid, rhamnosyl-rhamnosyl-β-hydroxydecanoyl-β-hydroxydecanoate (Rha-Rha-C10-C10) (Zhang, Pemberton, & Maier, 2014; Abbasi et al., 2012; Aparna, Srinikethan, & Smitha, 2012). The surface-active properties of rhamnolipids are affected by the size of their hydrophilic head (one or two rhamnoses), the length of the alkyl chains and the presence of double bonds. These properties affect the stability of rhamnolipids in the aqueous phase or their capability to solubilize hydrophobic organic compounds (Abbasi et al., 2012). According to Muller and Hausmann (2011), the main objectives of such strategies with *P. aeruginosa* would be: (a) metabolic engineering and further strain improvement to enhance the metabolic fluxes towards the product, (b) removal/adaptation of autologous genes to avoid byproduct formation and (c) introduction of heterologous genes for the use of alternative substrates and enhanced metabolism. The biosurfactant compounds can be natural or chemically synthesized surfactants such as parts of cellular structures or membranes (Aparna et al., 2011, 2012; Montero-Rodríguez et al., 2014) and play a major role in hydrocarbon biodegradation of contaminated sites (Nikolopoulou, Eickenbusch, Pasadakis, Venieri, & Kalogerakis, 2013 a, b; Silva et al., 2014). Biosurfactants are of many types and are classified based on their chemical nature such as glycolipids, lipopolysaccharides, oil-gosaccharides and lipopeptides and are known to produce diverse bacterial population (Franzetti et al., 2010; Makkar et al.,2011; Pacwa-Płociniczak et al., 2011; Kapadia and Yagnik, 2013; Geys, Soetaert, & Van Bogaert, 2014).

Biosurfactants increases the efficacy as dispersion thus increases the bioremediation activity. It has shown to be biodegradable with low toxicity thus helps in achieving sustainable and green approach (Sivapathasekaran, Mukherjee, Ray, Gupta, & Sen, 2010; Satpute, Banat, Dhakephalkar, Banpurkar, & Chopade, 2010). These biosurfactants are used to reduce amphiphilic compounds (Aparna et al., 2011). *Pseudomonas* bacteria is known to produce high quantities of glycolipids and is classified as rhamnolipids reducing the surface tension of water in between 25 and 30 mN/m along with CMC (critical

Table 3. Waste and byproducts used for biosurfactant production and respective producing microorganisms

Waste/By-Product	Biosurfactant-Producing Microorganism	Reference
Paneer whey	*Pseudomonas aeruginosa* strain, SR17	Patowary et al., 2016
Orange peel	*Bacillus licheniformis (KC710973)*	Kumar et al., 2016
Solid waste dates and corn steep liquor	*Pseudomonas sp.* NAF1	El Mahdi et al., 2015
CSL and Molasses	*Pseudomonas aeruginosa and Bacillus subtillis*	Rodrigues et al., 2015
Canola waste frying oil and corn steep liquor	*Pseudomonas cepacia* CCT6659	da Silva et al., 2013
Clarified cashew apple juice	*Bacillus subtilis* LAMI005	Oliveira et al., 2013
Vinasse and waste frying oil	*Bacillus pumilus*	Oliveira et al., 2013
Animal fat and corn steep liquor	*Candida lipolytica* UCP0988	Santos et al., 2013
Soybean oil refinery residue and corn steep liquor	*Candida sphaerica* UCP0995	Luna et al., 2011
Molasses	*Pseudomonas aeruginosa* (P.A.)	Santos et al., 2010

Rita Silva et al., 2014.

micelle concentration) in the range of 10 and 230 mg/L, depending on environmental factors, pH, and salinity conditions. Other mechanism involves the interaction of biosurfactant and cell surface. Table 3, lists wastes and byproducts studied by researchers to help reduce the cost of biosurfactant production.

These promote hydrocarbon adherence and reducing the lipopolysaccharide index of the cell wall without damaging the membrane (Aparna et al., 2011). Reduction of surface tension increases the surface area and promotes bioavailability as the molecular rearrangements takes place. Natural surfactants have advantages compared to chemically synthesized ones, effectiveness at extremes of temperature, pH, salinity, and widespread applicability, and their unique nature which provides new properties those conventional surfactants may nonappearance (Santos et al., 2013).

Unlike chemical surfactants, which are mostly derived from petroleum sources, these molecules can be produced by microbial fermentation processes using cheaper agro-based substrates and waste materials (Jeevitha, Keserkar, & Hemapriya, 2015). They have lower toxicity, are biodegradable and environmental friendly. The only drawback being the production cost which is high for biosurfactants compared to the chemical surfactants. Agro industrial waste has provided a solution to this problem (Nikolopoulou and Kalogerakis, 2011; Makkar et al., 2011; Macaulay and Rees, 2014). Recent studies have shown that the following waste can be used as substrates which permit the microbial cell growth. Molasses (Makkar et al., 2011; Saimmai, Sobhon, & Maneerat, 2011; Aparna et al., 2012; Al-Bahry et al., 2013), corn step liquor (El-Gendy et al., 2013, 2014; Silva et al., 2014) peanut oil cake (Thavasi, 2011a) and potato process effluents (Makkar et al., 2011). The above substrates can be used along with *Pseudomonas aerugonisa* and *Bacillus megaterium* (Thavasi et al., 2011 b; Pacwa-Płociniczak et al., 2014; Patowary et al., 2014). Due to the availability of right amount of carbon and nitrogen combination utilization of agro industrial waste has been on the rise (Amodu et al., 2014; Sobrinho et al., 2013) and (Sousa, Melo, Rodrigues, Sant'ana, & Gonçalves, 2012). Research on few of the agro waste are carried out are cassava wastewater (Saravanan and Subramaniyan, 2014), ground-nut oil refinery residue and corn steep liquor (Silva et al., 2014), molasses (Saimmai et al., 2011; Parna et al., 2012; Al-Bahry et al., 2013), potato peels (Vidhyalakshmi, 2012) and solid waste date and corn steep liquor (El Mahdi et al., 2015 a, b).

The applications of biosurfactants are huge; it can be used in bioremediation of dispersed oil spill, enhanced oil recovery and transfer, food industry, cosmetic and health care industry, in chemical plants (Saharan et al., 2011; Reis et al., 2013). According to Chatzipavlidis et al. (2013) Microbial products comprise the largest segment of the market, with a value of 151 billion US$ in 2011 and predictable value of 252 billion US$ in 2016. On the other hand, the potential of rhamnolipids production from renewable resources is so far not fully exploited. According to Henkel et al (2012), use of waste substrates in the production processes is likely to increase its influence on the field of microorganism-based production, since they are usually cheaper, maximize the utilization efficiency regarding the overall production process, and makes the process more environmentally friendly with wide potential in industrial applications.

AGRO-INDUSTRIAL WASTES MEET THE REQUIREMENT AS SUBSTRATES

Over the past years, agro-industries have been one of the major contributors in the world economy, especially for developing nations which often rely on existence farming and income generation from recognized and/or familiar agro-industries. At the same time, this industry is also one of the main contributors to the overall industrial pollution problem (Teh, Wu, & Juan, 2014). Even though several agricultural residues can be disposed of safely (due to biodegradable nature) in the environment, the massive quantities in which they are generated as a result of diverse agricultural and industrial practices, needs the requirement to look for some avenues where these could be utilized for some beneficial use. Since these are rich in organic nature, they consider one of the most available energy-rich resources on the planet (Makkar &Cameotra, 2011).

Furthermore it is often quite expensive to dispose of these wastes this is not to mention the economic loss of not exploiting them properly in the first place to obtain products of economic importance. Continually, the industrial processing of agricultural raw materials and animal products are generating huge amounts of organic wastes every year. About 1.6 billion tons of agrowaste are reportedly generated globally per year (FAO, 2014). Organic wastes derived from agricultural activities include corn, solid dates, straw, seed/stones, pulp, or stubble from fruits, beans and cereals (e.g. rice, wheat, barley, etc.), bagasses from sugarcane and others. Sugars, fibers, proteins, vitamins, and minerals represent essential constituents of the agro wastes. The presence of sugars, proteins, vitamins, minerals, and water make the agro-industrial wastes suitable for the growth of microorganisms, almost fungi and bacteria strains, adapted to these wastes. The agro-industrial wastes may also contain phenolic compounds or other potentially toxic compounds, which may cause deterioration of the environment when the waste is disposed. Furthermost organic waste materials used either as animal feed or burned, although digestibility and nutritional values were found to be unsatisfactory (Chatzipavlidis et al., 2013). The choice of inexpensive raw materials is important to the overall economics of the biosurfactants process because they account for 50% of the final product cost and also decrease the expenses with waste management.

Thus, millions tons of hazardous and non-hazardous wastes are generated each year throughout the world (Azar & Safa, 2016). The best way to reduce substrate cost for Bioengineering at present is to use wastes with the right balance of carbohydrates and lipids to support optimum bacterial growth and biosurfactants production. Agro-industrial wastes with a high content of carbohydrates or lipids meet the requirements for use as substrates for bio-surfactant production. According to Banat et al. (2014), the use of agro-industrial waste is one of the steps towards the implantation of sustainable biosurfactant production on an industrial scale, for which the optimization of the different components involved is required. The studies contributed by researchers suggest that the biosurfactant produced from agro industrial wastes has an important role for hydrocarbon biodegradation in a contaminated environment. Banat et al. (2014) has shown such studies have proved the importance of agro industrial wastes in bioremediation processes.

A. Rodrigues, Gudiña, Teixeira, & L. Rodrigues, (2015) experienced corn steep liquor and molasses for biosurfactant production from *Pseudomonas* and *Bacillus subtillis* strains and showed that the highest biosurfactant production (3.2 g/L) by the *P. aeruginosa* strain was obtained using a culture medium containing corn steep liquor (CSL) (10% (v/v)) and molasses (10% (w/v)), whereas the best biosurfactant production by the *B. subtillis* isolate (1.3 g/L) was obtained using a culture medium consisting of 10% (v/v) of CSL.

These wastes are existing at low cost from the respective processing industries and are as primary as low-cost substrates for industrial level biosurfactant production. Several other starchy waste substrates, such as rice water (effluent from rice processing industry and domestic cooking), corn steep liquor, and wastewater from the processing of cereals, pulses and molasses, have remarkable potential to support microbial growth and biosurfactant production. Extensive research is needed to establish the suitability of these carbohydrate-rich substrates in industrial-level biosurfactant production process. According to Henkel et al. (2012), it is evident that this agro-industrial waste could completely serve as potential feedstocks for the production of rhamnolipids. In addition, the fact that the application of sugar containing substrates (e.g. Sucrose) for biotechnological process results in a direct competition with food, pure sugar (e.g. Glucose) is also quite expensive. The better management of these wastes through the concept: reduce, reuse, and recycle holds great potential use (White, Dranke, & Hindle, 2012). Agricultural sector activities as well as this industrial processing leads to agro-industrial residues, which should be considered now also as valuable nutrient substrates, rather than as a waste. Furthermore, revealed that indeed, these nutrient substrates, which are the "noble worker" ingredients in biotechnology field.

AGROWASTE TO ENVIRONMENTAL CLEANUP: PROMOTING GREEN TECHNOLOGY

Sometimes these agro-industrial residues have seen as waste and pose disposal problems for the associated industries. This can be solved through its utilization, turning a valueless waste into a valuable substrate for biomolecules production (Arancon, Lin, Chan, Kwan, & Luque, 2013). Yearly increasing production of agro-industrial waste is a major concern of environmental pollution. The mixed characteristics of the massive quantity of wastes generated annually lead to complexity in recycling and utilization. The problem of wastage is currently on an increase, involving all sectors of waste management from collection to disposal; the identifying of sustainable solutions extends to all providers to the food supply chains, agricultural and industrial sectors, as well as stores and last consumers (Girotto, Alibardi, & Cossu, 2015). The shift from 'dirty' to 'clean' technologies could generate economic growth and justifies strong public support for clean technology development. Environmental related issues are becoming more prevalent among the public and many have now realized the need and importance of its protection, conservation, and care.

Furthermore, industries and companies have already started to act on the green measures with the aim to incorporate green practices in their manufacturing processes and to create sustainable products. Agro-industrial residues are the most abundant renewable resources on earth, as they constitute a significant proportion (amounting to over 30%) of the worldwide agricultural output (Chatzipavlidis et al., 2013). Accumulation of this biomass in large quantities every year results not only in the deterioration of the environment, but also in the loss of potentially valuable material which can be processed to yield a number of value added products such as food, fuel, feed and a variety of byproducts. Strategies for remediation of polluted sites are necessary to accelerate the degradation and natural attenuation of hydrocarbons.

The utilization of Agro-industrial wastes for generation of value added products will promote bioeconomy developments, in addition to help reducing environmental pollution (Wadhwa, Bakshi, & Makkar, 2015). Numerous organic materials and products are derived from microbes, often when cultured on agro-industrial residues with great commercial significance. Microbial biosurfactants are organic materials have widely applications in environmental protection, which includes oil spills control, biodegradation, and detoxification of oil-contaminated industrial effluents and soils. They can be obtained with the use of agro-industrial wastes as substrates, which helps reduce overall production costs (Montero- Rodríguez et al., 2015). According to Syldatk and Hausmann, (2010) the limited use of microbial surfactants in industry due to the use of expensive substrates, limited product concentrations, low yields and formation of product blends instead of pure compounds. The main strategy to achieve this are through (i) valuation of the substrate and product yield with focus on right organism, nutritional balance and the use of cheap or waste substrates to lower the initial raw material costs involved in the process; (ii) development of efficient bioprocesses, including optimization of the culture conditions and cost-effective separation processes to maximize recovery; and (iii) development and use of overproducing mutant or hybrid strains for enhanced yields (Becker & Wittmann, 2015; Makkar et al., 2011).

The use of the alternative substrates such as agro based industrial wastes is one of the attractive strategies for economical biosurfactants production. The reduction of production costs of biosurfactants requires enhancement of biosynthesis efficiency and the selection of inexpensive medium components since they constitute 50% of the total production costs. Unique promising approach to reduce these costs is the utilization of alternative substrates, such as waste or by-products from the agro-industry (Makkar et al., 2011; Banat, Satpute, Cameotra, Patil, & Nyayanit, 2014). Hence, It is necessary to evaluate the culturing conditions that optimize their production, assess the economic use of new substrates, such as those arising from industrial waste, and to evaluate techniques of isolation and purification to make production more economically feasible (Bustamante, Durán, & Diez, 2012).The knowledge of novel microbes and any information on their genomics, or about their communities, will pose an enormous potential to provide industry with unique products and processes. This would be useful for reducing the organic waste pollution and increasing interest towards efficient utilization of agro industrial wastes for biosurfactant production. Given the scope of research directed towards biodegradation of hydrocarbons and the understanding of the fundamental mechanisms gained, further improvements in the cost of biosurfactant production, mass transfer, reactor design, and metabolic rate can be expected in the near future. The chemical industry will be challenged with the switch to utilize this huge biomass sooner than anticipated.

CONCLUSION AND FUTURE PROSPECTS

Mother nature provides great immense possibilities for isolation of novel biomolecules producing microbial communities and products that can be utilized in the various application fields such as petroleum industry among them bioremediation. The use of economically cheaper substrates is paving the way for cost effective Biosurfactants production process in an industries. Large scale production of these surface active compounds is promising; however, product with pure quality needs further simplified approaches. Use in bioremediation technology has received better treatment as hydrocarbon-polluted sites can be treated effectively with Biosurfactant products or the producing organisms growing on agroindustrial wastes. Although number of developments have taken place, it is important to note that Biosurfactants production should be followed by minimum economic input using cheap low cost waste materials while maintaining quality and quantity wherever possible.

In future, research on Biosurfactants must be targeted on the economics of the fermentation processes, primarily carried out through the practice of alternative low-cost effective production media and recovery processes. In future, the establishment of database for agricultural and industrial residues substrates will help to essay the range different compositions and quality of substrates and their influence on the biosurfactants' types and purity. This would involve the selection of suitable strains with the desired properties, use of inexpensive alternative substrates. Therefore, more efforts are required to assess biosurfactant production in situ and their effect on the indigenous microorganisms, and to evaluate the efficiency and effectiveness of a bioremediation process in situ. In addition, most studies involving rhamnolipids focus mainly on evaluating the degradation efficiency of oil hydrocarbons.

REFERENCES

Abbasi, H., Hamedi, M. M., Lotfabad, T. B., Zahiri, H. S., Sharafi, H., Masoomi, F., & Noghabi, K. A. (2012). Biosurfactant-producing bacterium, *Pseudomonas aeruginosa* MA01 isolated from spoiled apples: Physicochemical and structural characteristics of isolated biosurfactant. *Journal of Bioscience and Bioengineering*, *113*(2), 211–219. doi:10.1016/j.jbiosc.2011.10.002 PMID:22036074

Al-Bahry, S. N., Al-Wahaibi, Y. M., Elshafie, A. E., Al-Bemani, A. S., Joshi, S. J., Al-Makhmari, H. S., & Al-Sulaimani, H. S. (2013). Biosurfactant production by *Bacillus subtilis* B20 using date molasses and its possible application in enhanced oil recovery. *International Biodeterioration & Biodegradation*, *81*, 141–146. doi:10.1016/j.ibiod.2012.01.006

Al-Nasrawi, H. (2012). Biodegradation of crude oil by fungi isolated from Gulf of Mexico. *Journal of Bioremediation & Biodegradation*.

Amodu, O. S., Ntwampe, S. K., & Ojumu, T. V. (2014). Emulsification of Hydrocarbons by Biosurfactant: Exclusive Use of Agrowaste. *BioResources*, *9*(2), 3508–3525. doi:10.15376/biores.9.2.3508-3525

Aparna, A., Srinikethan, G., & Hedge, S. (2011). Effect of addition of biosurfactant produced by *Pseudomonas* ssp. on biodegradation of crude oil. *International Proceedings of Chemical Biological & Environmental Engineering*, *6*, 71–75.

Aparna, A., Srinikethan, G., & Smitha, H. (2012). Production and characterization of biosurfactant produced by a novel Pseudomonas sp. 2B. *Colloids and Surfaces. B, Biointerfaces*, *95*, 23–29. doi:10.1016/j.colsurfb.2012.01.043 PMID:22445235

Arancon, R. A. D., Lin, C. S. K., Chan, K. M., Kwan, T. H., & Luque, R. (2013). Advances on waste valorization: New horizons for a more sustainable society. *Energy Science & Engineering*, *1*(2), 53–71. doi:10.1002/ese3.9

Azar, S. K., & Azar, S. S. (2016). Waste Related Pollutions and Their Potential Effect on Cancer Incidences in Lebanon. *Journal of Environmental Protection*, *7*(06), 778–783. doi:10.4236/jep.2016.76070

Banat, I. M., Satpute, S. K., Cameotra, S. S., Patil, R., & Nyayanit, N. V. (2014). Cost effective technologies and renewable substrates for biosurfactants' production. *Frontiers in Microbiology, 5*, 697.

BCC Research. (2015). *Trends in the global market for microbes and microbial products, 2010-2016 (in US$ Millions)*. Retrieved August 13, 2016, from http://www.bccresearch.com/report/microbial-productsmarkets-bio086a.html

Becker, J., & Wittmann, C. (2015). Advanced Biotechnology: Metabolically Engineered Cells for the Bio-Based Production of Chemicals and Fuels, Materials, and Health-Care Products. *Angewandte Chemie International Edition*, *54*(11), 3328–3350. doi:10.1002/anie.201409033 PMID:25684732

Bezerra de Souza Sobrinho, H., de Luna, J. M., Rufino, R. D., Figueiredo Porto, A. L., & Sarubbo, L. A. (2013). Assessment of toxicity of a biosurfactant from *Candida sphaerica* UCP 0995 cultivated with industrial residues in a bioreactor. *Electronic Journal of Biotechnology*, *16*(4), 4–4.

Bustamante, M., Durán, N., & Diez, M. C. (2012). Biosurfactants are useful tools for the bioremediation of contaminated soil: a review. *Journal of Soil Science and Plant Nutrition, 12*(4), 667-687.

Cameotra, S. S., & Makkar, R. S. (2010). Biosurfactant-enhanced bioremediation of hydrophobic pollutants. *Pure and Applied Chemistry*, *82*(1), 97–116. doi:10.1351/PAC-CON-09-02-10

Campos, J. M., Montenegro Stamford, T. L., Sarubbo, L. A., de Luna, J. M., Rufino, R. D., & Banat, I. M. (2013). Microbial biosurfactants as additives for food industries. *Biotechnology Progress*, *29*(5), 1097–1108. doi:10.1002/btpr.1796 PMID:23956227

Chandrasekaran, M., & Bahkali, A. H. (2013). Valorization of date palm (Phoenix dactylifera) fruit processing by-products and wastes using bioprocess technology –[Review]. *Saudi Journal of Biological Sciences*, *20*(2), 105–120. doi:10.1016/j.sjbs.2012.12.004 PMID:23961227

Chatzipavlidis, I., Kefalogianni, I., Venieraki, A., & Holzapfel, W. (2013). Status and Trends of the Conservation and Sustainable Use of Microorganisms in Agroindustrial Processes. Commission on Genetic Resources for Food and Agriculture, Food and Agriculture Organization of the United Nations.

Chikere, C. B., Okpokwasili, G. C., & Chikere, B. O. (2011). Monitoring of microbial hydrocarbon remediation in the soil. *3 Biotech, 1*(3), 117-138.

Daffonchio, D., Ferrer, M., Mapelli, F., Cherif, A., Lafraya, A., Malkawi, H. I., & Kalogerakis, N. et al. (2013). Bioremediation of Southern Mediterranean oil polluted sites comes of age. *New Biotechnology*, *30*(6), 743–748. doi:10.1016/j.nbt.2013.05.006 PMID:23727339

Daffonchio, D., Mapelli, F., Cherif, A., Malkawi, H. I., Yakimov, M. M., Abdel-Fattah, Y. R., & Boon, N. et al. (2012). ULIXES, unravelling and exploiting Mediterranean Sea microbial diversity and ecology for xenobiotics and pollutants cleanup. *Reviews in Environmental Science and Biotechnology*, *11*(3), 207–211. doi:10.1007/s11157-012-9283-x

Das, N., & Chandran, P. (2010). Microbial degradation of petroleum hydrocarbon contaminants: An overview. *Biotechnology Research International*. PMID:21350672

Dave, D., & Ghaly, A. E. (2011). *Remediation technologies for marine oil spills: A critical review and comparative analysis*. Academic Press.

de Cássia, F. S., da Silva, R., Rufino, R. D., Luna, J. M., Farias, C. B., Filho, H. J., & Sarubbo, L. A. et al. (2013). Enhancement of biosurfactant production from *Pseudomonas cepacia* CCT6659 through Optimisation of Nutritional Parameters Using Response Surface Methodology. *Tenside Surfactants Detergents*, *50*(2), 137–142. doi:10.3139/113.110241

de Oliveira, D. W. F., França, Í. W. L., Félix, A. K. N., Martins, J. J. L., Giro, M. E. A., Melo, V. M. M., & Gonçalves, L. R. B. (2013). Kinetic study of biosurfactant production by *Bacillus subtilis* LAMI005 grown in clarified cashew apple juice. *Colloids and Surfaces. B, Biointerfaces*, *101*, 34–43. doi:10.1016/j.colsurfb.2012.06.011 PMID:22796769

Dubey, K.V., Charde, P.N., Meshram, S.U., Yadav, S.K., Singh, S., & Asha, A. J. (2012). Potential of New Microbial Isolates for Biosurfactant Production using Combinations of Distillery Waste with other Industrial Wastes. *J Petrol Environ Biotechnol S1:002*. doi:10.4172/2157-7463.S1-002

El-Gendy, N. S., Ali, H. R., El-Nady, M. M., Deriase, S. F., Moustafa, Y. M., & Roushdy, M. I. (2014). Effect of different bioremediation techniques on petroleum biomarkers and asphaltene fraction in oil-polluted sea water. *Desalination and Water Treatment*, *52*(40-42), 7484–7494. doi:10.1080/19443994.2013.831784

El Mahdi, A. M., Abdul Aziz, H., Amr, A., Salem, S., El-Gendy, N. S., & Nassar, H. (2015a). Performance of some New Bacterial Isolates on Biodegradation of Libyan Light Crude Oil Using Agro-Industrial Wastes as Co-Substrates. In Applied Mechanics and Materials (Vol. 802, pp. 496-500). Trans Tech Publications.

El Mahdi, A. M., Aziz, H. A., & Amr, A. (2015b). Performance of Isolated Kocuria sp. SAR1 in Light Crude Oil Biodegradation. *Journal of Bioremediation & Biodegradation*, *6*(4), 1. doi:10.4172/2155-6199.1000303

El Mahdi, A. M., Aziz, H. A., El-Gendy, N. S., Amr, S. S. A., & Nassar, H. N. (2014). Optimization of Libyan Crude Oil Biodegradation by Using Solid Waste Date as a Natural Low-Cost Material. *J Bioremed Biodeg*, *5*(7), 252.

El May, Y., Dorge, S., Jeguirim, M., Trouvé, G., & Said, R. (2012). Measurement of gaseous and particulate pollutants during combustion of date palm wastes for energy recovery. *Aerosol and Air Quality Research*, *12*(5), 814–825.

FAO. (2014). *Food wastage: Key facts and figures*. Retrieved August 13, 2016 from “http://www.fao.org/news/story/en/item/196402/icode/

Fernández-Luqueño, F., Valenzuela-Encinas, C., Marsch, R., Martínez-Suárez, C., Vázquez-Núñez, E., & Dendooven, L. (2011). Microbial communities to mitigate contamination of PAHs in soil—possibilities and challenges: *A review. Environmental Science and Pollution Research International*, *18*(1), 12–30. doi:10.1007/s11356-010-0371-6 PMID:20623198

FitzPatrick, M., Champagne, P., Cunningham, M. F., & Whitney, R. A. (2010). A biorefinery processing perspective: Treatment of lignocellulosic materials for the production of value-added products. *Bioresource Technology*, *101*(23), 8915–8922. doi:10.1016/j.biortech.2010.06.125 PMID:20667714

Franzetti, A., Tamburini, E., & Banat, I. M. (2010). Applications of biological surface active compounds in remediation technologies. In *Biosurfactants* (pp. 121–134). Springer New York. doi:10.1007/978-1-4419-5979-9_9

Geys, R., Soetaert, W., & Van Bogaert, I. (2014). Biotechnological opportunities in biosurfactant production. *Current Opinion in Biotechnology*, *30*, 66–72. doi:10.1016/j.copbio.2014.06.002 PMID:24995572

Girotto, F., Alibardi, L., & Cossu, R. (2015). Food waste generation and industrial uses: *A review. Waste Management (New York, N.Y.)*, *45*, 32–41. doi:10.1016/j.wasman.2015.06.008 PMID:26130171

Grand View Research. (2014). *Biosurfactants market by product (rhamnolipids, sophorolipids, MES, APG, sorbitan esters, sucrose esters) expected to reach USD 2308.8 Million by 2020.* Retrieved August 13, 2016 from http://www.grandviewresearch.com/industryanalysis/biosurfactants-industry

Gudiña, E. J., Rodrigues, A. I., Alves, E., Domingues, M. R., Teixeira, J. A., & Rodrigues, L. R. (2015). Bioconversion of agro-industrial by-products in rhamnolipids toward applications in enhanced oil recovery and bioremediation. *Bioresource Technology*, *177*, 87–93. doi:10.1016/j.biortech.2014.11.069 PMID:25479398

Gunasekera, T. S., Striebich, R. C., Mueller, S. S., Strobel, E. M., & Ruiz, O. N. (2013). Transcriptional profiling suggests that multiple metabolic adaptations are required for effective proliferation of Pseudomonas aeruginosa in jet fuel. *Environmental Science & Technology*, *47*(23), 13449–13458. doi:10.1021/es403163k PMID:24164330

Gupta, A., Barrow, C. J., & Puri, M. (2012). Omega-3 biotechnology: Thraustochytrids as a novel source of omega-3 oils. *Biotechnology Advances*, *30*(6), 1733–1745. doi:10.1016/j.biotechadv.2012.02.014 PMID:22406165

Hassanshahian, M., & Cappello, S. (2013). Crude oil biodegradation in the marine environments. *Chamy R. Biodegradation-Engineering and Technology,* 101-135.

Hazra, C., Kundu, D., & Chaudhari, A. (2012). Biosurfactant-assisted bioaugmentation in bioremediation. In *Microorganisms in Environmental Management* (pp. 631–664). Springer Netherlands. doi:10.1007/978-94-007-2229-3_28

Henkel, M., Müller, M. M., Kügler, J. H., Lovaglio, R. B., Contiero, J., Syldatk, C., & Hausmann, R. (2012). Rhamnolipids as biosurfactants from renewable resources: Concepts for next-generation rhamnolipid production. *Process Biochemistry*, *47*(8), 1207–1219. doi:10.1016/j.procbio.2012.04.018

Jeevitha, P., Keserkar, J. V., & Hemapriya, J. (2015). Economically Feasible-Bacterial Mediated Biosurfactant Production using Agro Wastes as Carbon Source. *International Journal of Scientific Engineering & Technology Research*, *04*(9), 5797–5803.

Ji, Y., Mao, G., Wang, Y., & Bartlam, M. (2013). Structural insights into diversity and n-alkane biodegradation mechanisms of alkane hydroxylases. *Frontiers in Microbiology, 4*, 58.

Johal, M. S. (2011). *Understanding nanomaterials.* CRC Press.

Joshi, S. J., & Desai, A. J. (2010). Biosurfactant's role in bioremediation of NAPL and fermentative production. In Biosurfactants (pp. 222-235). Springer New York. doi:10.1007/978-1-4419-5979-9_17

Kang, S.-W., Kim, Y.-B., Shin, J.-D., & Kim, E.-K. (2010). Enhanced biodegradation of hydrocarbons in soil by microbial biosurfactant, sophorolipid. *Applied Biochemistry and Biotechnology*, *160*(3), 780–790. doi:10.1007/s12010-009-8580-5 PMID:19253005

Kapadia, S. G., & Yagnik, B. N. (2013). Current trend and potential for microbial biosurfactants. Asian J. Exp. *Biol. Sci*, *4*(1), 1–8.

Khalifeh, A., Roozbehani, B., Moradi, A. M., Imani Moqadam, S., & Mirdrikvand, M. (2013). Isolation of Crude Oil from Polluted Waters Using Biosurfactants Pseudomonas Bacteria: Assessment of Bacteria Concentration Effects. Engineering, *Technology & Applied Science Research, 3*(2).

Kostka, J. E., Teske, A. P., Joye, S. B., & Head, I. M. (2014). The metabolic pathways and environmental controls of hydrocarbon biodegradation in marine ecosystems. *Frontiers in Microbiology, 5*, 471. doi: 10.3389/fmicb.2014.00471

Kumar, A. P., Janardhan, A., Viswanath, B., Monika, K., Jung, J. Y., & Narasimha, G. (2016). Evaluation of orange peel for biosurfactant production by *Bacillus licheniformis* and their ability to degrade naphthalene and crude oil. *3 Biotech, 6*(1), 1-10.

Luna, J. M., Rufino, R. D., Albuquerque, C. D., Sarubbo, L. A., & Campos-Takaki, G. M. (2011). Economic optimized medium for tensio-active agent production by *Candida sphaerica* UCP0995 and application in the removal of hydrophobic contaminant from sand. *International Journal of Molecular Sciences*, *12*(4), 2463–2476. doi:10.3390/ijms12042463 PMID:21731452

Macaulay, B. M. (2015). Understanding the behaviour of oil-degrading micro-organisms to enhance the microbial remediation of spilled petroleum. *Applied Ecology and Environmental Research*, *13*(1), 247–262.

Macaulay, B. M., & Rees, D. (2014). Bioremediation of Oil Spills: A Review of Challenges for Research Advancement. *Annals of Environmental Science*, *8*(1), 2.

Maddipati, P., Atiyeh, H. K., Bellmer, D. D., & Huhnke, R. L. (2011). Ethanol production from syngas by *Clostridium* strain P11 using corn steep liquor as a nutrient replacement to yeast extract. *Bioresource Technology*, *102*(11), 6494–6501. doi:10.1016/j.biortech.2011.03.047 PMID:21474306

Maheshwari, R., Singh, U., Singh, P., Singh, N., Lal, J., & Rani, B. (2014). To decontaminate wastewater employing bioremediation technologies. *J Adv Sci Res*, *5*(2), 7–15.

Makkar, R. S., Cameotra, S. S., & Banat, I. M. (2011). Advances in utilization of renewable substrates for biosurfactant production. *AMB Express*, *1*(5), 1–19. PMID:21906330

Mao, X., Jiang, R., Xiao, W., & Yu, J. (2015). Use of surfactants for the remediation of contaminated soils: A review. *Journal of Hazardous Materials*, *285*, 419–435. doi:10.1016/j.jhazmat.2014.12.009 PMID:25528485

McGenity, T. J. (2014). Hydrocarbon biodegradation in intertidal wetland sediments. *Current Opinion in Biotechnology*, *27*, 46–54. doi:10.1016/j.copbio.2013.10.010 PMID:24863896

Montero-Rodríguez, D., Andrade, R. F. S., Ribeiro, D. L. R., Lima, R. A., Araujo, H. W. C., & Campos-Takaki, G. M. (2014). Ability of Serratia marcescens UCP/WFCC 1549 for biosurfactant production using industrial wastes and fuels biodegradation In Industrial, medical and environmental applications of microorganisms: Current status and trends, (pp. 211-216). Academic Press.

Müller, M. M., & Hausmann, R. (2011). Regulatory and metabolic network of rhamnolipid biosynthesis: Traditional and advanced engineering towards biotechnological production. *Applied Microbiology and Biotechnology*, *91*(2), 251–264. doi:10.1007/s00253-011-3368-2 PMID:21667084

Nikolopoulou, M., Eickenbusch, P., Pasadakis, N., Venieri, D., & Kalogerakis, N. (2013 a). Microcosm evaluation of autochthonous bioaugmentation to combat marine oil spills. *New Biotechnology*, *30*(6), 734–742. doi:10.1016/j.nbt.2013.06.005 PMID:23835403

Nikolopoulou, M., & Kalogerakis, N. (2011). Petroleum spill control with biological means. *Comprehensive Biotechnology*, *6*, 263–274. doi:10.1016/B978-0-08-088504-9.00392-5

Nikolopoulou, M., Pasadakis, N., & Kalogerakis, N. (2013 b). Evaluation of autochthonous bioaugmentation and biostimulation during microcosm-simulated oil spills. *Marine Pollution Bulletin*, *72*(1), 165–173. doi:10.1016/j.marpolbul.2013.04.007 PMID:23660443

Noparat, P., Maneerat, S., & Saimmai, A. (2014). Application of biosurfactant from *Sphingobacterium spiritivorum* AS43 in the biodegradation of used lubricating oil. *Applied Biochemistry and Biotechnology*, *172*(8), 3949–3963. doi:10.1007/s12010-014-0829-y PMID:24590892

Obayori, O. S., Ilori, M. O., Adebusoye, S. A., Oyetibo, G. O., Omotayo, A. E., & Amund, O. O. (2009). Degradation of hydrocarbons and biosurfactant production by *Pseudomonas* sp. strain LP1. *World Journal of Microbiology & Biotechnology*, *25*(9), 1615–1623. doi:10.1007/s11274-009-0053-z

Obayori, O. S., Salam, L. B., & Ogunwumi, O. S. (2014). Biodegradation of Fresh and Used Engine Oils by *Pseudomonas aeruginosa* LP5. *J Bioremed Biodeg*, *5*(1), 213. doi:10.4172/2155-6199.1000213

Oliveira, J. G. D., & Garcia-Cruz, C. H. (2013). Properties of a biosurfactant produced by Bacillus pumilus using vinasse and waste frying oil as alternative carbon sources. *Brazilian Archives of Biology and Technology*, *56*(1), 155–160. doi:10.1590/S1516-89132013000100020

Pacwa-Płociniczak, M., Płaza, G. A., Piotrowska-Seget, Z., & Cameotra, S. S. (2011). Environmental applications of biosurfactants: Recent advances. *International Journal of Molecular Sciences*, *12*(1), 633–654. doi:10.3390/ijms12010633 PMID:21340005

Pacwa-Płociniczak, M., Płaza, G. A., Poliwoda, A., & Piotrowska-Seget, Z. (2014). Characterization of hydrocarbon-degrading and biosurfactant-producing *Pseudomonas* sp. P-1 strain as a potential tool for bioremediation of petroleum-contaminated soil. *Environmental Science and Pollution Research International*, *21*(15), 9385–9395. doi:10.1007/s11356-014-2872-1 PMID:24743958

Pandey, B., & Fulekar, M. H. (2012). Bioremediation technology: A new horizon for environmental clean-up. *Biologie Medicale*, *4*(1), 51–59.

Panesar, R., Panesar, P. S., & Bera, M. B. (2011). Development of low cost medium for the production of biosurfactants. *Asian J. Biotechnol*, *3*(4), 388–396. doi:10.3923/ajbkr.2011.388.396

Paniagua-Michel, J., & Rosales, A. (2015). Marine Bioremediation - A Sustainable Biotechnology of Petroleum Hydrocarbons Biodegradation in Coastal and Marine Environments. *J Bioremed Biodeg*, *6*, 273. doi:10.4172/2155-6199.1000273

Patowary, K., Saikia, R. R., Kalita, M. C., & Deka, S. (2014). Degradation of polyaromatic hydrocarbons employing biosurfactant-producing Bacillus pumilusKS2. *Annals of Microbiology*, 1–10.

Patowary, R., Patowary, K., Kalita, M. C., & Deka, S. (2016). Utilization of Paneer Whey Waste for Cost-Effective Production of Rhamnolipid Biosurfactant. *Applied Biochemistry and Biotechnology*, 1–17. PMID:27142272

Perfumo, A., Smyth, T. J. P., Marchant, R., & Banat, I. M. (2010). Production and roles of biosurfactants and bioemulsifiers in accessing hydrophobic substrates. In *Handbook of hydrocarbon and lipid microbiology* (pp. 1501–1512). Springer Berlin Heidelberg. doi:10.1007/978-3-540-77587-4_103

Puškárová, A., Bučková, M., Chovanová, K., Harichová, J., Karelová, E., Godočíková, J., & Pangallo, D. et al. (2013). Diversity and PAH growth abilities of bacterial strains isolated from a contaminated soil in Slovakia. *Biologia*, *68*(4), 587–591. doi:10.2478/s11756-013-0193-3

Reis, R. S., Pacheco, G. J., Pereira, A. G., & Freire, D. M. G. (2013). Biosurfactants: production and applications. *Biodegradation: Life of science,* 978-953.

Rodrigues, A. I., Gudiña, E. J., Teixeira, J. A., & Rodrigues, L. R. (2015). Development of low-cost culture media for effective biosurfactant production. *SINAFERM 2015 XX Simpósio Nacional de Bioprocessos*, 1-5.

Röling, W. F., & van Bodegom, P. M. (2014). Toward quantitative understanding on microbial community structure and functioning: A modeling-centered approach using degradation of marine oil spills as example. *Front. Microbiol*, *5*, 125. doi:10.3389/fmicb.2014.00125 PMID:24723922

Saharan, B. S., Sahu, R. K., & Sharma, D. (2011). A review on biosurfactants: fermentation, current developments and perspectives. *Genetic Engineering and Biotechnology Journal, 2011*(1), 1-14.

Saikia, R. R., Deka, S., Deka, M., & Banat, I. M. (2012). Isolation of biosurfactant-producing *Pseudomonas aeruginosa* RS29 from oil-contaminated soil and evaluation of different nitrogen sources in biosurfactant production. *Annals of Microbiology*, *62*(2), 753–763. doi:10.1007/s13213-011-0315-5

Saimmai, A., Sobhon, V., & Maneerat, S. (2011). Molasses as a whole medium for biosurfactants production by Bacillus strains and their application. *Applied Biochemistry and Biotechnology*, *165*(1), 315–335. doi:10.1007/s12010-011-9253-8 PMID:21509601

Santos, A. C., Bezerra, M. S., dos Santos Pereira, H., Dos Santos, E. S., & De Macedo, G. R. (2010). Production and recovery of rhamnolipids using sugar cane molasses as carbon source. *Journal of Chemistry and Chemical Engineering*, *4*(11), 27–33.

Santos, D. K., Rufino, R. D., Luna, J. M., Santos, V. A., Salgueiro, A. A., & Sarubbo, L. A. (2013). Synthesis and evaluation of biosurfactant produced by *Candida lipolytica* using animal fat and corn steep liquor. *Journal of Petroleum Science Engineering*, *105*, 43–50. doi:10.1016/j.petrol.2013.03.028

Santos, D. K. F., Rufino, R. D., Luna, J. M., Santos, V. A., & Sarubbo, L. A. (2016). Biosurfactants: Multifunctional Biomolecules of the 21st Century. *International Journal of Molecular Sciences*, *17*(3), 401. doi:10.3390/ijms17030401 PMID:26999123

Saravanan, V., & Subramaniyan, V. (2014). Production of biosurfactant by *Pseudomonas aeruginosa* PB3A using agro-industrial wastes as a carbon source. *Malays. J. Microbiol.*, *10*, 57–62.

Satpute, S. K., Banat, I. M., Dhakephalkar, P. K., Banpurkar, A. G., & Chopade, B. A. (2010). Biosurfactants, bioemulsifiers and exopolysaccharides from marine microorganisms. *Biotechnology Advances*, *28*(4), 436–450. doi:10.1016/j.biotechadv.2010.02.006 PMID:20172021

Scott, C., Lewis, S. E., Milla, R., Taylor, M. C., Rodgers, A. J., Dumsday, G., & Russell, R. J. et al. (2010). A free-enzyme catalyst for the bioremediation of environmental atrazine contamination. *Journal of Environmental Management*, *91*(10), 2075–2078. doi:10.1016/j.jenvman.2010.05.007 PMID:20570036

Silva, E. J., Silva, N. M. P. R., Rufino, R. D., Luna, J. M., Silva, R. O., & Sarubbo, L. A. (2014). Characterization of a biosurfactant produced by *Pseudomonas cepacia* CCT6659 in the presence of industrial wastes and its application in the biodegradation of hydrophobic compounds in soil. *Colloids and Surfaces. B, Biointerfaces*, *117*, 36–41. doi:10.1016/j.colsurfb.2014.02.012 PMID:24613853

e. Silva, N. M. P. R., Rufino, R. D., Luna, J. M., Santos, V. A., & Sarubbo, L. A. (2014). Screening of Pseudomonas species for biosurfactant production using low-cost substrates. *Biocatalysis and Agricultural Biotechnology*, *3*(2), 132–139. doi:10.1016/j.bcab.2013.09.005

Silva, R. D. C. F., Almeida, D. G., Rufino, R. D., Luna, J. M., Santos, V. A., & Sarubbo, L. A. (2014). Applications of biosurfactants in the petroleum industry and the remediation of oil spills. *International Journal of Molecular Sciences*, *15*(7), 12523–12542. doi:10.3390/ijms150712523 PMID:25029542

Sivapathasekaran, C., Mukherjee, S., Ray, A., Gupta, A., & Sen, R. (2010). Artificial neural network modeling and genetic algorithm based medium optimization for the improved production of marine biosurfactant. *Bioresource Technology*, *101*(8), 2884–2887. doi:10.1016/j.biortech.2009.09.093 PMID:19914826

Soberón-Chávez, G., & Maier, R. M. (2011). Biosurfactants: a general overview. In Biosurfactants (pp. 1-11). Springer Berlin Heidelberg. doi:10.1007/978-3-642-14490-5_1

Sobrinho, H. B., Luna, J. M., Rufino, R. D., Porto, A. L. F., & Sarubbo, L. A. (2013). Biosurfactants: Classification, properties and environmental applications. *Recent Developments in Biotechnology*, *11*, 1–29.

Sousa, M., Melo, V. M. M., Rodrigues, S., Santana, H. B., & Gonçalves, L. R. B. (2012). Screening of biosurfactant-producing Bacillus strains using glycerol from the biodiesel synthesis as main carbon source. *Bioprocess and Biosystems Engineering*, *35*(6), 897–906. doi:10.1007/s00449-011-0674-0 PMID:22218992

Souza, E. C., Vessoni-Penna, T. C., & de Souza Oliveira, R. P. (2014). Biosurfactant-enhanced hydrocarbon bioremediation: An overview. *International Biodeterioration & Biodegradation*, *89*, 88–94. doi:10.1016/j.ibiod.2014.01.007

Syldatk, C., & Hausmann, R. (2010). Microbial biosurfactants. *European Journal of Lipid Science and Technology*, *112*(6), 615–616. doi:10.1002/ejlt.201000294

Tanase, A. M., Vassu, T., Csutak, O., Pelinescu, D., Robertina, I., & Stoica, I. (2012). Phylogenetic analysis of oil polluted soil microbial strains. *Rom. Biotechnol. Lett*, *17*, 7093–7103.

Teh, C. Y., Wu, T. Y., & Juan, J. C. (2014). Optimization of agro-industrial wastewater treatment using unmodified rice starch as a natural coagulant. *Industrial Crops and Products*, *56*, 17–26. doi:10.1016/j.indcrop.2014.02.018

Thapa, B., Kc, A. K., & Ghimire, A. (2012). A review on bioremediation of petroleum hydrocarbon contaminants in soil. *Kathmandu University Journal of Science, Engineering and Technology*, *8*(1), 164-170.

Thavasi, R., Jayalakshmi, S., & Banat, I. M. (2011). Application of biosurfactant produced from peanut oil cake by *Lactobacillus delbrueckii* in biodegradation of crude oil. *Bioresource Technology*, *102*(3), 3366–3372. doi:10.1016/j.biortech.2010.11.071 PMID:21144745

Thavasi, R., Nambaru, V. S., Jayalakshmi, S., Balasubramanian, T., & Banat, I. M. (2011). Biosurfactant production by *Pseudomonas aeruginosa* from renewable resources. *Indian Journal of Microbiology*, *51*(1), 30–36. doi:10.1007/s12088-011-0076-7 PMID:22282625

Vidhyalakshmi, R., Vallinachiyar, C., & Radhika, R. (2012). Production of Xanthan from Agro-Industrial Waste. *Journal of Advanced Scientific Research*, *3*(2).

Vijayakumar, S., & Saravanan, V. (2015). Biosurfactants-Types, Sources and Applications. Research. *Journal of Microbiology (Seoul, Korea)*, *10*(5), 181.

Wadhwa, M., Bakshi, M. P., & Makkar, H. P. (2015). Waste to worth: Fruit wastes and by-products as animal feed. *CAB Reviews*, *10*(031), 1–26.

White, P., Dranke, M., & Hindle, P. (2012). *Integrated solid waste management: A lifecycle inventory*. Springer Science & Business Media.

Xu, Q., Nakajima, M., Liu, Z., & Shiina, T. (2011). Biosurfactants for microbubble preparation and application. *International Journal of Molecular Sciences*, *12*(1), 462–475. doi:10.3390/ijms12010462 PMID:21339998

Zelzer, M., Todd, S. J., Hirst, A. R., McDonald, T. O., & Ulijn, R. V. (2013). Enzyme responsive materials: Design strategies and future developments. *Biomaterials Science*, *1*(1), 11–39. doi:10.1039/C2BM00041E

Zhang, L., Pemberton, J. E., & Maier, R. M. (2014). Effect of fatty acid substrate chain length on *Pseudomonas aeruginosa* ATCC 9027 monorhamnolipid yield and congener distribution. *Process Biochemistry*, *49*(6), 989–995. doi:10.1016/j.procbio.2014.03.003

Zhang, Z., Hou, Z., Yang, C., Ma, C., Tao, F., & Xu, P. (2011). Degradation of n-alkanes and polycyclic aromatic hydrocarbons in petroleum by a newly isolated *Pseudomonas aeruginosa* DQ8. *Bioresource Technology*, *102*(5), 4111–4116. doi:10.1016/j.biortech.2010.12.064 PMID:21227683

KEY TERMS AND DEFINITIONS

Aliphatic Hydrocarbon: A compound built from carbon and hydrogen joined in a linear chain. Petroleum products are composed primarily of aliphatic hydrocarbons.

Bacterium: A single-celled organism of microscopic size (bacteria).

Bioaugmentation: The addition of nonnative microorganisms to a site.

Biodegradation: Biologically mediated conversion of one compound to another.

Biomass: Total mass of microorganisms present in a given amount of water or soil.

Bioremediation: Use of microorganisms to control and destroy contaminants.

Cometabolism: A reaction in which microbes transform a contaminant even though the contaminant cannot serve as an energy source for the organisms. To degrade the contaminant, the microbes require the presence of other compounds (primary substrates) that can support their growth.

Enzyme: A protein created by living organisms to use in transforming a specific compound. The protein serves as a catalyst in the compound's biochemical transformation.

Hydrocarbon: A chemical composed of carbon and hydrogen in any of a wide variety of configurations. Petroleum products, as well as many synthetic industrial chemicals, contain many different hydrocarbons.

Hydrophobic Compound: A "water-fearing" compound, such as oil, that has low solubility in water and tends to form a separate phase.

In Situ: Latin term meaning "in place"—in the natural or original position.

Metabolism: The chemical reactions in living cells that convert food sources to energy and new cell mass.

Sorption: Collection of a substance on the surface of a solid by physical or chemical attraction.

Substrate: A compound that microorganisms can use in the chemical reactions catalyzed by their enzymes.

Surfactant: Soap or a similar substance that has a hydrophobic and a hydrophilic end. Surfactants can bond to oil and other immiscible compounds to aid their transport in water.

Chapter 8
Biodegradation of Xenobiotic Compounds:
An Overview

Sunil Kumar Narwal
Himachal Pradesh University, India

Reena Gupta
Himachal Pradesh University, India

ABSTRACT

The continuous accumulation of recalcitrant xenobiotic compounds into the ecosystem released from various sources caused a serious global concern. Xenobiotics compounds are carcinogenic, mutagenic, causing teratogenic effect and persist over a long period of time in the environment. Therefore there is an urgent need for the detoxification of these compounds. Biodegradation is a technique that employs natural biological processes to completely degrade toxic contaminants from the environment. The microorganisms possess a wide range of catabolic biodegradation pathways and, thus, use these toxic xenobiotics as the sole source of carbon and energy. Bacteria and fungi are source of xenobiotic degradation. For the development of successful and improved bioremediation processes, understanding of the biochemical and molecular aspects of xenobiotics biodegradation is required. The chapter aims to provide an overview of xenobiotic compounds, factors affecting biodegradation, the metabolic pathways and genetic adaptation in microorganisms for degradation of recalcitrant xenobiotic compounds.

INTRODUCTION

Environmental pollution caused by the release of a wide range of compounds as a consequence of industrial progress has now assumed serious proportions. It has dramatically changed every aspect of human life and lifestyles (Jain et al., 2005). The industries develop and manufacture diverse group of compounds that improve human's life, but most of them are highly toxic and have adverse effects on biomes. Therefore, the removal of such chemicals from industrial effluents is of great concern (Das & Kumar, 2015). For quite a long time, humans assumed that atmospheric, terrestrial, and aquatic systems were sufficient to

DOI: 10.4018/978-1-5225-2325-3.ch008

absorb and break down wastes from population centres, industry, and farming. In any case, this is not valid. Today there are two major issues: the dumping of vast amounts of wastes that are being continuously produced and evacuation of the harmful toxic compounds that have been accumulating at dump sites in the soil and in water systems over the last few decades. Thus in this way natural conditions have enormous impact on the living beings. Both natural and manmade activities result in accumulation of wide ranges of harmful xenobiotic compounds in the earth, and along these lines cause a worldwide concern. In the past, wastes were disposed by digging hole and filling it with waste material (Versha, Ch, & Chenna, 2011). This mode of waste disposal was difficult to manage owing to lack of new place every time to dump. New advancements for waste disposal that use high-temperature incineration and chemical decomposition (e.g., base-catalyzed dechlorination, UV oxidation) have developed. Although they can be very effective at reducing wide range of contaminants but at the same time have several disadvantages. These strategies are complex, uneconomical, and also generate other toxic byproducts. The associated inadequacies in these techniques have centered efforts towards harnessing advanced bioremediation process as a suitable alternative. Biodegradation is a viable bioremediation technology for organic pollutants by using microorganisms to remove or detoxify toxic or unwanted chemicals in an environment. A goal of bioremediation is to transform organic pollutants into harmless metabolites or mineralize the pollutants into carbon dioxide and water. It has been considered as a cost-effective, safe, and promising method for the removal or detoxification of DDT residues in the environment (Pan et al., 2016). The present chapter emphasizes an outline of the current knowledge in the area of biodegradation of wide range of pollutants with emphasis on the biodegradation processes for better exploitation of microbial abilities in bioremediation challenges.

XENOBIOTIC COMPOUNDS

The word, xenobiotic is a combination of two distinct roots "xeno" and "biotic." The word xeno is derived from Greek which means strange, unnatural, or different while biotic is a word that implies life. Xenobiotic, therefore, refers to an organic compound that imitates natural biochemicals that are crucial for life, but are strange and unnatural. The xenobiotic compounds are highly thermostable thus persists in the environment. Xenobiotics are often toxic to life. Also, they may not be recognized by biochemical processes in plants and microorganisms and are thus resistant to degradation in the environment. Primarily, xenobiotics are those compounds that are alien to a living individual and have a propensity to accumulate in the environment. Xenobiotics include many compounds that are used in both industrial and agricultural activities. The major xenobiotics include pesticides, fuels, solvents, alkanes, polycyclic hydrocarbons (PAHs), antibiotics, synthetic azo dyes, pollutants (dioxins and polychlorinated biphenyls), polyaromatic, chlorinated and nitro- aromatic compounds. The main issue with xenobiotic compounds is the toxicity threat they pose to public health. It is quite shocking that some xenobiotic compounds (phenols, biphenyl compounds, phthalates, etc.) act as endocrine disruptors (Borgeest, Greenfeld, Tomic, & Flaws, 2002).

There is growing public concern over the wide range of xenobiotic compounds being introduced inadvertently or deliberately into soil. Such contamination can be long term and have a significant impact on both decomposition processes and thus nutrient cycling. For example, the widespread incorporation of herbicides into soil every year constitutes a major concern since they can potentially pose a threat to our health as well as to the quality of soil, surface water, and groundwater resources.

Use of pesticides has benefited the modern society by improving the quantity and quality of the world's food production. Gradually, pesticide usage has become an integral part of modern agriculture system. Many of the artificially made complex compounds *i.e.* xenobiotics persist in environment and do not undergo biological transformation. Thus microorganisms play an important role in degradation of xenobiotics, and maintenance of steady state concentrations of chemicals in the environment. Ever since it was discovered that microbes have the ability to transform or degrade xenobiotics, scientists have been exploring the microbial diversity, particularly of contaminated areas in search for organisms that can degrade a wide range of pollutants. In addition, they play a crucial role in biogeochemical cycles and in sustainable development of the biosphere. Microorganisms have evolved towards ecological fitness rather than biotechnological efficiency; thus, it would take a long time for bacteria capable of cleaning up anthropogenic pollution to evolve by natural selection. Hence, studying the physiology, biochemistry and genetics of the catabolic pathways become crucial to recreate and accelerate the natural processes in the test tube as well as to accomplish their rational manipulation to design more efficient biocatalysts for different biotechnological applications.Biological decontamination methods are preferable to conventional approaches because in general, microorganisms degrade numerous environmental pollutants without producing toxic intermediates. A huge number of bacterial and fungal genera possess the capability to degrade organic pollutants.They transform toxic contaminants into non-hazardous or less hazardous substances.It is one of the most cost-effective methods amongst remedial approaches. Biodegradation of xenobiotic compounds primarily depends upon many of the enzymes used in the pathways for degrading unusual substrates catalyze novel reactions. The initial intracellular attack of organic pollutants is an oxidation process, the activation and incorporation of oxygen is the enzymatic key reaction catalyzed by oxygenases and peroxidases. After those peripheral degradation pathways convert organic pollutants step by step into intermediates of the central intermediary metabolism, e.g., the tricarboxylic acid cycle followed by the biosynthesis of cell biomass from the central precursor metabolites, e.g., acetyl- CoA, succinate and pyruvate. Although many bacteria are able to metabolize organic pollutants, a single bacterium does not possess the enzymatic capability to degrade all or even most of the organic compounds in a polluted soil. Mixed microbial communities have the most powerful biodegradative potential because the genetic information of more than one organism is necessary to degrade the complex mixtures of organic compounds present in contaminated areas.Microorganisms commonly exist in the soil but may not be enough to remove contaminants. It may be useful sometimes to inoculate the polluted sites with effective hydrocarbon-degrading microbial consortia to support the existing microbes. The complete degradation of a pesticide molecule to its inorganic components that can be eventually used in an oxidative cycle removes its potential toxicity from the environment. However, there are two objectives in relation to biodegradation of xenobiotics: *(i)* how biodegradation activity arises, evolves and transferred among the members of the soil microflora, and *(ii)* to device bioremediation methods for removing or detoxifying high concentration of dangerous pesticide residues. The characters of pesticide degradation of microorganisms are located on plasmids and transposons, and are grouped in clusters on a chromosome. Understanding of the characters provides clues to the evolution of degradative pathways and makes the task of gene manipulation easier to construct the genetically engineered microbes capable of degrading the pollutants.

Figure 1. Types of xenobiotic compounds

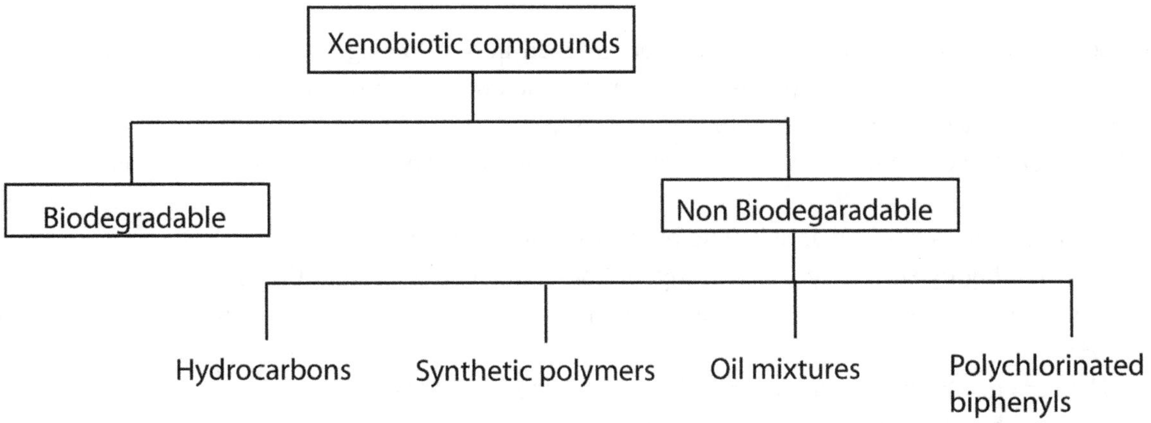

Types of Xenobiotic Compounds

There are two types of xenobiotic compounds. They may be biodegradable or non-degradable (recalcitrant) (Figure 1). Biodegradable xenobiotic compounds are those that get degraded by the action of microbes or other reactions while recalcitrant compounds are resistant to degradation by any reactions.

Table 1. Duration of persistence of insecticides and herbicides in soil

Biocides	Time Taken for 75-100% Disappearance
A. Chlorinated Insecticides	
DDT (l,l,l-trichloro-2,2-bis-(p-chlorophenyl) ethane)	4 years
Aldrin	3 years
Chlordane	5 years
Heptachlor	2 years
Lindane (hexachloro-cyclohexane)	3 years
B. Organophosphate Insecticides	
Diazinon	12 years
Malathion	1 week
Parathion	1 week
C. Herbicides	
2,4-D (2,4-dichlorophenoxyacetic acid)	4 weeks
2,4,5-T	30 weeks
Atrazine	40 weeks
Simazine	48 weeks
Propazine	1.5 years

Madigan, Martinko, & Parker, 1997.

Types of Recalcitrant Xenobiotic Compounds

The recalcitrant xenobiotic compounds can be grouped into following six types: hydrocarbons, polychlorinated biphenyls, synthetic polymer, oil mixture and others. The xenobiotic compounds take much time to get degraded into harmless compounds (Table 1).

The xenobiotic compounds may be recalcitrant due to one or more of the following reasons:

1. They are not recognized as substrate by the existing degradative enzymes.
2. They are highly stable, i.e., chemically and biologically inert due to the presence of substitution groups like halogens, nitro-, sulphonate, amino-, methoxy- and carbamyl groups (substitution of –H).
3. They are insoluble in water, or are adsorbed to external matrices like soil.
4. They are highly toxic or give rise to toxic products due to microbial activity.
5. Their large molecular size prevents entry into microbial cells.
6. Inability of the compounds to induce the synthesis of degrading enzymes, and
7. Lack of the permease needed for their transport into the microbial cells.

Factors Affecting Rate of Biodegradation

The rate of biodegradation in soil depends on four variables.

1. Availability of pesticide or metabolite to the microorganisms.
2. Physiological status of the microorganisms.
3. Survival and/or proliferation of recalcitrant compounds degrading microorganisms at contaminated site.
4. Sustainable population of these microorganisms.

The fates of xenobiotics in soil include (i) complete mineralization or (ii) stabilization of the parent compound, or some metabolite of the compound, in soil. If the xenobiotics are toxic and the rate of degradation is very slow, adverse effects on human and on ecological health are possible. Therefore, it is desirable to maintain xenobiotic concentrations in soil at as low a level as possible.

BIODEGRADATION PATHWAY OF XENOBIOTICS COMPOUND

Depending on the nature of xenobiotic compounds, biodegradation pathways can be categorized into two parts: Aerobic biodegradation and anaerobic biodegradation as shown in Figure 2.

- **Aerobic Biodegradation:** In aerobic biodegradation, oxygen is used as an electron acceptor. Some of the xenobiotics like petroleum hydrocarbons, chlorinated aliphatics, benzene, toluene, phenol, naphthalene, fluorine, pyrene, chloroanilines, pentachlorophenol and dichlorobenzenes are rapidly and potentially degraded by the aerobic degradation process. The microbial consortia degrade the toxic compounds to nontoxic compounds by producing specific enzymes.

Figure 2. Strategies for biodegradation of xenobiotic compounds

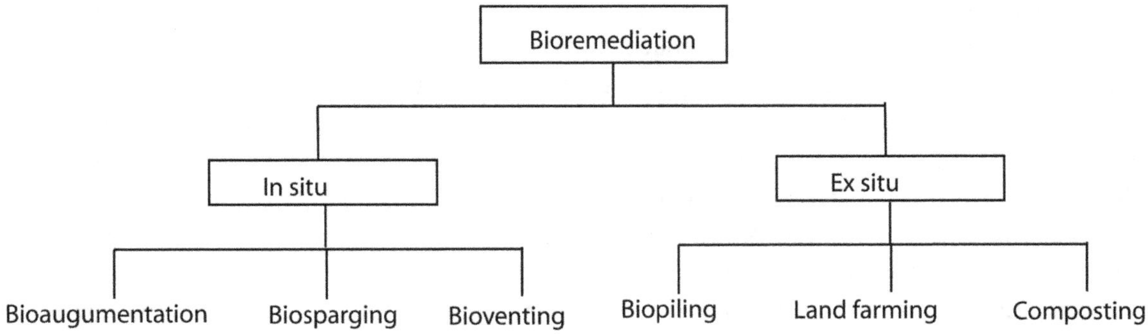

Xenobiotic compound + O_2----------------CO_2 + H_2O + biomass + residue (Shimao, 2001).

Examples of aerobic bacteria recognized for their degradative abilities are *Pseudomonas, Alcaligenes, Sphingomonas, Rhodococcus*, and *Mycobacterium*. These microbes have often been reported to degrade pesticides and hydrocarbons, both alkanes and compounds. Many of these bacteria use the contaminant as the sole source of carbon and energy.

- **Anaerobic Biodegradation:** In anaerobic biodegradation, microbes degrade xenobiotic compounds to gases like carbon dioxide, methane, and nitrogen compounds using various terminal electron acceptors such as NO_3^-, SO_4^- or CO_2.

Xenobiotic compound----------------------- CO_2 + CH4 + H_2O + biomass + residue (Jayasekara, Harding, Bowater, & Lonergan, 2005).

The polychlorinated biphenyls (PCBs), chlorinated dioxins and some pesticides like DDT are metabolized by anaerobic bacteria such as *Acidovorax, Bordetella, Pseudomonas, Sphingomonas, Variovorax, Veillonella alkalescens, Desulfovibrio* spp., *Desulfuromonas michiganensis, and Desulfitobacterium halogenans, D. oleovorans, G. metallireducens, D. Acetonicum.* Anaerobic bacteria are not as frequently used as aerobic bacteria. There is an increasing interest in anaerobic bacteria used for bioremediation of polychlorinated biphenyls (PCBs) in river sediments, dechlorination of the solvent trichloroethylene (TCE), and chloroform. Environmental conditions have tremendous effect on the living organisms especially microorganisms. They adapt to the climatic conditions very fast. Their versatility makes them useful for various biotechnological applications. They are widely used for biodegradation or biosorption for removal of pollutants from the environmental matrices. In this process it is observed that certain microorganisms metabolize toxic pollutants and use it as a source of energy, carbon, nitrogen, or other nutrient, or as final electron acceptor in a respiratory process.

Finally, the chemicals are transformed into harmless compounds such as carbon dioxide and water. Organic pollutants are generally degraded aerobically or anaerobically as shown. It has also been observed that biosorption has a great potential for the removal of xenobiotic compounds from industrial effluents. The surface properties of bacteria, yeasts, fungi and algae enable them to adsorb different kinds of pollutants from aqueous solutions allowing the recovery and/or environmentally acceptable disposal of the pollutants. Biodegradation is the process by which organic substances are broken down by the enzymes produced by living organisms. Later it undergoes mineralization thereby converting organic matter to minerals. However, the great versatility of microbes offers a simpler, economical and more environmental friendly strategy to reduce environmental pollution and to help in biodegradation of xenobiotic compounds. In addition, they play a crucial role in biogeochemical cycles and in sustainable development of the biosphere. Biodegradation of complex organic compounds generally require the concerted efforts of several different enzymes (Karigar, & Rao, 2011). The recalcitrance of many synthetic chemicals to biodegradation is mainly due to a lack of enzymes that can carry out critical steps in a catabolic pathway. These xenobiotic chemicals are relatively water soluble and bioavailable, and theoretically could be converted by short metabolic routes to intermediates that support cellular growth under aerobic conditions. Yet, no organisms have been found that oxidatively degrade and use as a carbon source important environmental chemicals such as chloroform, trichloroethylene, 1,1,1-trichloroethane, 1,2-dichloropropane and 1,2,3-trichloropropane. The bacterial genera found to undergo bio-fix, a wide range of xenobiotic chemicals include both aerobic e.g. *Pseudomonas, Escherichia, Sphingobium, Pandoraea, Rhodococcus, Gordonia, Bacillus, Moraxella, Micrococcus* and anaerobic types e.g. *Pelatomaculum, Desulfotomaculum, Syntrophobacter, Syntrophus, Desulphovibrio, Methanospirillum, Methanosaeta.*

BIOREMEDIATION STRATEGIES

Feasibility of bioremediation depends on the location of contaminants. Approaches for implementation of bioremediation depend on whether the impacted soil to be treated is intact in the environment or it is to be excavated for treatment in an offsite facility (Adams, Fufeyin, Okoro, & Ehinomen, 2015). If on site, the term *In-situ* remediation suffices and if offsite, it is described as *Ex-situ* (Figure 3).

Figure 3. Pathways for degradation of different xenobiotic compounds

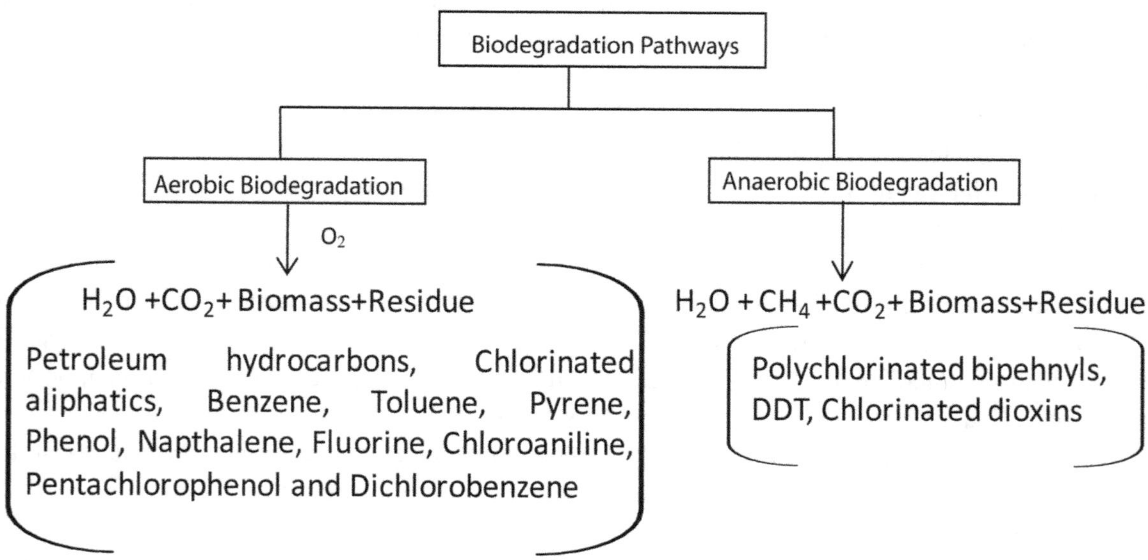

In-Situ Bioremediation

In situ bioremediation is the application of biological treatment to the cleanup of hazardous chemicals present in the subsurface. This is because equipment is not needed to unearth the contaminated soil leading it to be less expensive and cleaner since it does not send dust and contaminants into the surrounding area. Some disadvantages to this method of bioremediation are that it may take longer to decontaminate, it is less manageable and it is mostly effective in loose soil.

Bioaugumentation

Bioaugmentation is a process where selected, standardized bacteria (microbes) are added to an area that has been contaminated with an unwanted substance. These bacteria finally degrade the contaminants. Bioaugmentation should be applied in soils (1) with low or non-detectable number of contaminant-degrading microbes, (2) containing compounds requiring multi-process remediation, including processes detrimental or toxic to microbes and (3) for small-scale sites on which cost of non-biological methods exceed cost for bioaugmentation. One of the most difficult issues is survival of strains introduced to soil. It has been observed that the number of exogenous microorganisms has decreased shortly after soil inoculation (Niu et al., 2009). Many studies have shown that both abiotic and biotic factors influence the effectiveness of bioaugmentation. The most important abiotic factors such as temperature, moisture, pH and organic matter content are discussed; however, aeration, nutrient content and soil type also determine the efficiency of bioaugmentation. Biotic factors, including competition between indigenous and exogenous microorganisms for limited carbon sources as well as antagonistic interactions and predation by protozoa and bacteriophages, also play essential roles in the final results of bioaugmentation. The selection of proper culture should take into consideration the following features of microorganisms: fast growth, easy cultured, to withstand high concentrations of contaminants and the ability to survive in a wide range of environmental conditions. There are several approaches that allow selection of microorganisms useful for bioaugmentation. Bacteria for this purpose may be isolated from given contaminated soils and after culturing under laboratory conditions pre-adapted pure bacterial strains return to the same soil. This approach is called reinoculation of soil with indigenous microorganisms.

Biosparging

To degrade large amounts of a petroleum product, a substantial bacterial population is required which, in turn, requires oxygen for both metabolic processes and an increase in the overall bacterial population. Biosparging involves the injection of air under pressure below the water table to increase groundwater oxygen concentrations and enhance the rate of biological degradation of contaminants by naturally occurring bacteria. In biosparging, air (or oxygen) and nutrients (if needed) are injected into the saturated zone to increase the biological activity of the indigenous microorganisms. It is injected by using compressors with pressure valves (up to 10 bar) or blowers (up to 0.8 to 1 bar), whereby the injection pressure (mostly varied between 0.5 and 2 bar) is determined by the depth beneath the ground-water table and the permeability of the soil. Biosparging is effective in reducing petroleum products at underground storage tank (UST) sites. Biosparging is most often used at sites with mid-weight petroleum products (*e.g.*, diesel fuel, jet fuel); lighter petroleum products (*e.g.*, gasoline) tend to volatilize readily and to be removed more rapidly using air sparging. The effectiveness of biosparging depends on two primary factors- permeability of the soil and the biodegradability of the constituents. Fine-grained soils (e.g., clays and silts) have lower permeability than coarse-grained soils (e.g., sands and gravels). When volatile compounds are present; biosparging is often combined with other remedial technologies such as soil vapour extraction (SVE) or bioventing. When biosparging is combined with SVE, the vapour extraction system creates a negative pressure in the unsaturated zone through a series of extraction wells that control the migration of the vapour plume.

Bioventing

Bioventing is an *in-situ* remediation technology that uses indigenous microorganisms to biodegrade organic constituents adsorbed to soils in the unsaturated zone. In bioventing, the activity of the indigenous bacteria is enhanced by inducing air (or oxygen) flow into the unsaturated zone (using extraction or injection wells) and, if necessary, by adding nutrients. It is effective in remediating petroleum products including gasoline, jet fuels, kerosene, and diesel fuel. In many cases, the promotion of biological activity through the addition of nutrients and moisture, and optimization of the bioventing flow rates, may achieve greater contaminant reductions than venting alone. The volatilization of low molecular weight, high vapor pressure compounds can also be maximized using a greater volume of air or short travel distances between injection and extraction wells. Bioventing is not appropriate for sites with groundwater tables located less than 3 feet below the land surface. While bioventing is relatively inexpensive, this method can take a few years to clean up a site, depending on contaminant concentrations and site-specific removal rates.

Ex-Situ Bioremediation

Ex-situ bioremediation is a biological process in which excavated soil is placed in a lined above-ground treatment area and aerated following processing to enhance the degradation of organic contaminants by the indigenous microbial population. Although this technique for decontamination requires the removal of soil prior to the treatment, it can be faster, easier to control and usually more able to treat a wider range of toxins and soils. The main advantage of ex situ treatment is that it generally requires shorter time periods, and there is more certainty about the uniformity of treatment because of the ability to monitor and continuously mix the groundwater.

Biopiling

Biopiling is a full-scale technology in which excavated soils are piled and typically constructed in a treatment area that consists of a leachate collection and aeration system. It is commonly applied to reduce concentrations of petroleum constituents in excavated soils through the use of biodegradation. This technology involves heaping contaminated soils into piles (or "cells") and stimulating aerobic microbial activity within the soils through the aeration and/or addition of minerals, nutrients, and moisture. Typically, this is a composting process which utilizes forced air via injection or vacuum extraction, moisture control, nutrient addition and environmental monitoring. The contaminants are reduced to carbon dioxide and water (Wu & Craper, 2009).

Land Farming

Land farming is a bioremediation treatment process that is performed in the upper soil zone or in biotreatment cells. Contaminated soils, sediments, or sludges are incorporated into the soil surface and periodically turned over or tilled to aerate the mixture. Contaminants are degraded, transformed, and immobilized by microbiological processes and by oxidation. This technique has been successfully used for years in the management and disposal of oily sludge and other petroleum refinery wastes. Soil conditions are controlled to optimize the rate of contaminant degradation.

Composting

Organic solid waste management is a major challenge in high population density areas like apartments, educational institutions, hospitals, etc., as their disposal with other wastes could not only lead to issues like vector menace, odour generation but also would lead to loss of resources that could be reused such as nutrients which will benefit soil and significantly improves soil properties and land productivity. (Sangamithirai, Jayapriya, Hema, & Manoj, 2015) . Composting involves the accelerated degradation of organic matter by microorganisms under controlled conditions (aeration, temperature and humidity), in which the organic material undergoes a characteristic thermophilic stage that allows sanitization of the waste by the elimination of pathogenic microorganisms. In addition to soil amendment, composting food scraps reduces greenhouse gas emissions such as carbon dioxide (CO_2), methane (CH_4) and nitrous oxide (N_2O). The Intergovernmental Panel on Climate Change (IPCC) has identified organic waste as a key contributor to climate change, releasing 7% of total gas emissions in developing countries (Wolka & Melaku, 2015). Compost bioremediation has proved effective in degrading or altering many contaminants such as chlorinated and non-chlorinate hydrocarbons, wood preserving chemicals, solvents, heavy metals, pesticides, petroleum products and explosives. Fungi are especially good at digesting complex organic compounds that otherwise is not degraded by other organisms. Under ideal conditions, composting proceeds through three major phases:

1. In initial, mesophilic phase, in which the decomposition is carried out under moderate temperatures by mesophilic microorganisms.
2. As the temperature rises, a second, thermophilic phase starts, in which the decomposition is carried out by various thermophilic bacteria under high temperatures.
3. As the supply of high-energy compounds dwindles, the temperature starts to decrease, and the mesophiles once again predominate in the maturation phase.

BIODEGRADATION OF PESTICIDES

A pesticide can be defined as any substance or mixture of substances intended for preventing, destroying, repelling, or mitigating any pest (insects, mites, nematodes, weeds, rats, etc.). Pesticides like insecticides, herbicides, fungicides, and various other substances are used to control pests. Although wide-scale application of pesticides and herbicides is an essential part of augmenting crop yields; excessive use of these chemicals leads to the microbial imbalance, environmental pollution and health hazards. Globally, an estimated 1 to 2.5 million tons of active pesticide ingredients are used each year, mainly in agriculture.

Forty percent are herbicides, followed by insecticides and fungicides. Pesticides are applied over large areas in agriculture and urban settings. Pesticide use therefore represents an important source of diffuse chemical environmental inputs.The extensive use of pesticides causes serious environmental concerns, as only 5% or less from the applied pesticides reach the target organisms which resulted in contamination of soil and water bodies (Javaid, Ashiq, & Tahir, 2016). In present age, more than 500 different formulations of pesticides are used mainly in agricultural tricks. These formulations are in general artificially synthesized substances which are non-biodegradable and enhance environmental toxicity. Pesticides reaching the soil in significant quantities have direct effect on soil microbiological aspects, which in turn influence plant growth. Some of the most important effects caused by pesticides are: a) accumulation of the chemicals in soil to highly toxic levels, b) may be assimilated by the plants and get accumulated in edible plant products, c) accumulation in the edible portions of the root crops, d) to be get eroded with soil particles and may enter into the water streams, and finally leading to the soil, water and air pollutions. The increased absorption of such toxic chemicals by plants to the level at which the consumption of plant products may prove deleterious / hazardous to human beings as well as livestock's. There are chronic threats to human life, caused by long term, low dose exposure to pesticides. It can cause hormonal disruption, diminished intelligence, and reproductive abnormalities. Degradation involves both biotic and abiotic transformation processes. Biotic transformation is mediated by microorganisms, while abiotic transformation involves processes such as chemical and photochemical reactions. Microorganisms plays major role in the degradation of pesticides. Complete biodegradation of a pesticide involves the oxidation of parent compound to form carbon dioxide and water. This process provides both carbon and energy for the growth and reproduction of microbes. Many soil microorganisms have the ability to act upon pesticides and convert them into simpler non-toxic compounds (Table 2). This process of degradation of pesticides and conversion into non-toxic compounds by microorganisms is known as "biodegradation". Biodegradation methodology is widely used for the treatment of xenobiotics such as pesticides in soil. It is employed in many countries due to its low cost and being ecofriendly. Under anaerobic conditions, the enzymatic reactions common to many pesticides include dechlorination, hydrolysis, nitro reduction, and dealkylation. A bacterium may be partially responsible for these metabolic activities, and in some cases the bacteria may have a metabolic shift from one pathway to another. Degradation of pesticide by either external or internal enzyme will stop at any step if an appropriate enzyme is not present. Absence of an appropriate enzyme is one of the common reasons for persistence of any pesticide. If an appropriate microorganism is absent in soil or if biodegrading microbial population has been reduced due to toxicity of pesticide in that case a specific microorganism can be added or introduced in soil to enhance the activity of the existing population. Several bacteria competent of degrading carbofuran, carbarly, baygon or aldicarb have been isolated and characterized. Endosulfan (pesticide) can be removed from environment by applying strains of microbes (*Aspergillus*). The specific genes and enzymes are very critical for the cleavage of specific functional groups of the pesticide. Pesticide-degrading bacteria and *Rhizobium meliloti* coating on *Medicago sativa* seeds are effective for repairing soil, polluted by organic phosphorus pesticide. This approach is very efficient, possessing several advantages, that is, rapid soil repairing rate, simple operation, and high treating capability for removal of organic phosphorus pesticide.

Table 2. List of organisms known to degrade pesticides

Organisms	Target Compound
Enterobacter aerogenes	bifenthrin, cypermethrin
Acinetobacter johnsonii (MA-19)	organophosphate pesticides
Rhodococcus	para-nitrophenol
Bacillus thuringiensis	malathion
Acinetobacter	Esbiothrin
Stenotrophomonas maltophilia	DDT and endosulfan
Pseudomonas	Endosulfan, Atrazine
Sphingomonas	DDT
Rhodobacter sphaeroides	OPs, chlorinated pesticides
Vibrio and *Shewanella*	methyl parathion
Ochrobactrum	triazophos
Acidomonas sp	Allethrin
Escherichia coli	organochlorine
Paracoccus sp.	Carbofuran

BIODEGRADATION OF AZO DYES

Azo compounds are compounds bearing the functional group R–N=N–R′, in which R and R′ can be either aryl or alkyl. As a consequence of π-delocalization, aryl azo compounds have vivid colors, especially reds, oranges, and yellows (Table 3). Therefore, they are used as dyes, and are commonly known as azo dyes. The textile mills daily discharge millions of liters of untreated effluents in the form of wastewater into public drains that eventually empty into rivers (Chacko & Subramaniam, 2011). Azo dyes contribute 84%, of which sulphonated azo dyes predominate. About 10-15% (128 tons/day globally) of the dyes are lost at various finishing steps of the printed cloths. Most of them are recalcitrant in nature, especially azo dyes. Azo dyes play an important role as colouring agents in the textile, food, and pharmaceutical industry. Medical importance of these compounds is well known for their antibiotic, antifungal and anti-HIV properties. On the other hand they bring a certain danger for health and environment because of the toxicity, mutagenicity and carcinogenicity of azo dyes as given in Table 3, thus their breakdown products; their removal from industrial wastewaters has been an urgent challenge. The main source of azo dyes in human organism is food. The stability and their xenobiotic nature of reactive azo dyes makes them recalcitrant hence they are not totally degraded by conventional wastewater treatment processes that involve light, chemicals or activated sludge. This also alters the pH, increases the biochemical oxygen demand (BOD) and chemical oxygen demand (COD), and gives the rivers intense colourations and public is greatly concerned about water quality. The presence of unnatural colours is aesthetically unpleasant and tends to be associated with contamination. It has been demonstrated that several microorganisms like lignolytic fungi or bacteria are able to enhance the bioremediation of azo dyes, under certain environmental conditions, to transform azo dyes to non-colored products or even to completely mineralize them. Biotreatment offers a cheaper and environmentally friendlier alternative for colour removal in textile effluents. A number of microorganisms have been found to be able to decolourize textile dyes including bacteria, fungi, and yeasts. Furthermore, azo-dye degrading microbial communities are sensitive to high concentrations of salts that are used in the dye process. This can limit growth and activity of the degrader bacteria such that the process treatment times become impractical. With the discovery and isolation of very efficient, salt-tolerant azo-dye degrading bacteria, bioaugmentation of biotreatment systems with specific microbial strains has now become an effective strategy to improve wastewater treatment systems. They have developed enzyme systems for the decolourization and mineralization of azo dyes under certain environmental conditions. The chemical structure of dyes in general is comprised of a conjugated system of double bonds and aromatic rings. The azo aromatic one is the most widespread dye class in the industry. They may have one or more azo (N=N) groups. The main drawback of this class of dyes is that they are not easily degraded by aerobic bacteria and with the action of anaerobic or microaerobic reductive bacteria. Degradation of dyes by bacteria is mostly based on reduction of the azo bond, which results in formation of usually colorless amines. The electron withdrawing nature of the azo linkages obstructs the susceptibility of azo dye molecules to oxidative reaction. Therefore, azo dyes generally resist aerobic bacterial biodegradation. Only bacteria with specialized azo dye reducing enzymes (azoreductase) were found to degrade azo dyes under fully aerobic conditions. This anaerobic reduction implies decolourizations of the dyes to potentially harmful aromatic amines. Aromatic amines are generally not further degraded under anaerobic conditions. Anaerobic treatment must therefore be considered merely as the first stage of the complete degradation of azo dyes. The second stage involves conversion of the produced aromatic amines. For several aromatic amines, this can be achieved by biodegradation under aerobic conditions. Azo reductases and laccases seem to be the most promising enzymes. Azo reductases catalyze the reaction only in presence of reducing equivalents like FADH and NADH.

Laccases are multicopper phenol oxidases that decolourize azo dyes through a highly nonspecific free radical mechanism forming phenolic compounds, thereby avoiding the formation of toxic aromatic amines (Han, Choi, & Song, 2004). Laccase oxidizes the phenolic group of the phenolic azo dye, with the participation of one electron generating a phenoxy radical which is sequentially followed by oxidation to a carbonium ion. A nucleophilic attack by water on the phenolic ring carbon bearing the azo linkage to produce 3-diazenyl-benzenesulfonic acid (III) and 1, 2- naphthoquinone then takes place (Camarero, Ibarra, Martínez, & Martínez, 2005). The radicals which are formed by the one-electron oxidation of dye by laccase will react with 1, 2-naphthoquinone rather than be oxidized. The formed radicals can undergo oxidation to yield compound VIII, reduction to yield compound X, or further polymerization and again oxidation to form compound IX. According to the reported mechanism, cross-coupling between the reactive species results in the formation of C-C and C-O bonds between phenolic molecules and of C-N and N-N bonds between aromatic amines. By phenolic cross-coupling, an electron is removed from the hydroxyl group, generating an alkoxy radical. Alkoxy free radical forms dimers in the *ortho* and *para* positions with the hydroxyl groups. Phenolic radicals can be further oxidized to yield oligomeric products. Under certain conditions, the C-C-formed dimers can take part in coupling reactions to form extended quinines. The capacity of fungi to reduce azo dyes is related to the formation of exo enzymes such as peroxidases and phenol oxidases. Peroxidases are hem proteins that catalyze reactions in the presence of hydrogen peroxide. The first ligninolytic peroxidases were isolated from *P. chrysosporium* and called lignin peroxidase (LiP) and manganese peroxidase (MnP). LiP catalyzes the oxidation of non-phenolic aromatic compounds like veratryl alcohol. MnP preferentially oxidizes Mn^{2+} to Mn^{3+}, and the Mn^{3+} is responsible for the oxidation of many phenolic compounds.

Table 3. Types of dyes and their effects

S. No.	Name of Dye	Effects	Structure
1	Reactive brilliant red	Inhibit human serum albumin	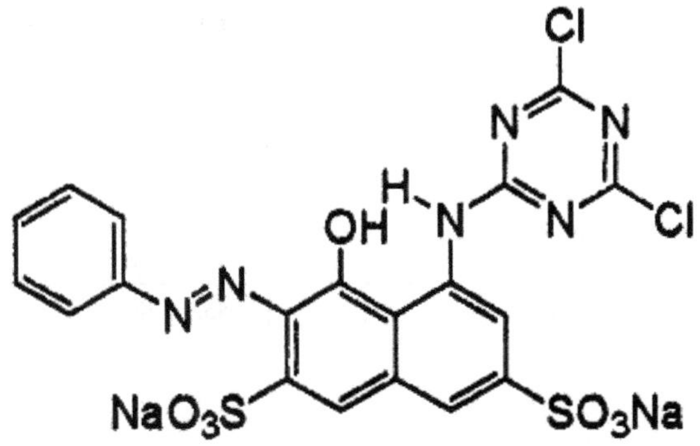
2	Acid violet 7	Induce chromosome aberration, lipid peroxidation	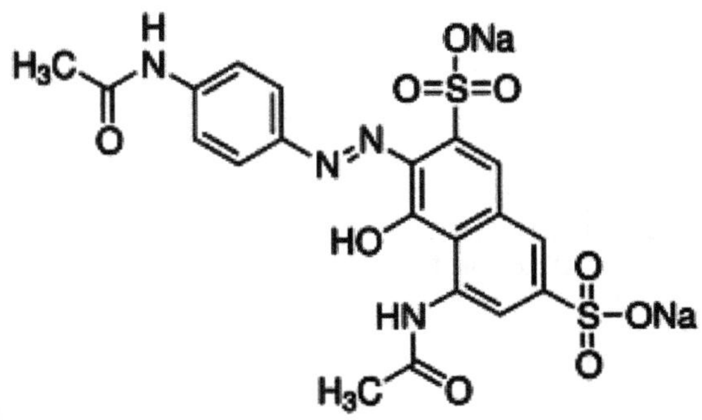
3	Disperse red-1 & Disperse orange - 1	Increase the frequency of micronuclei in human lymphocytes	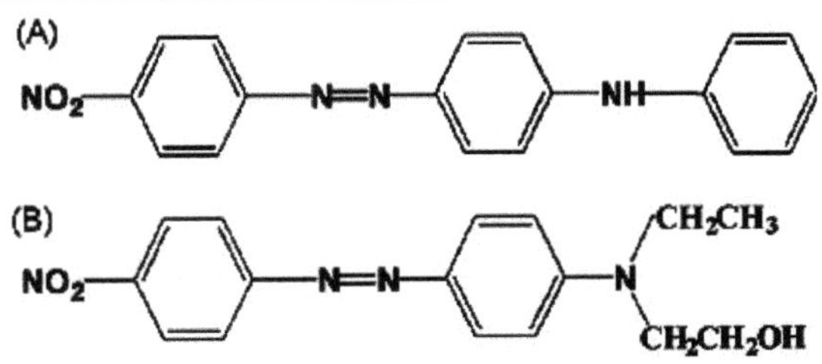
4	Reactive black 5	Decrease urease activity, arginine ammonification rate in terrestrial ecosystem	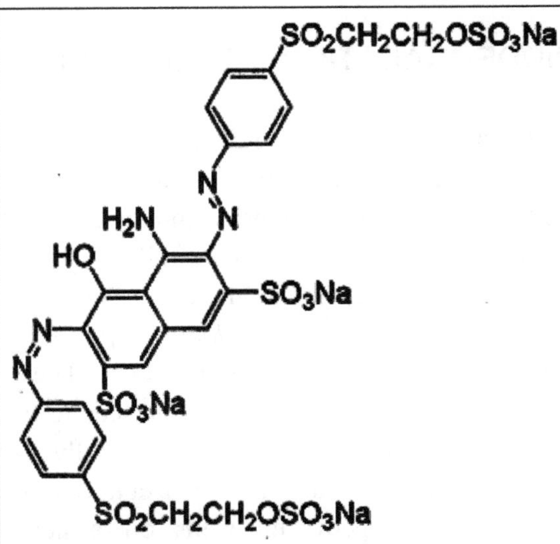

continued on following page

Table 3. Continued

S. No.	Name of Dye	Effects	Structure
5	Disperse blue 291	Mutagenic, cytotoxic, genotypic effect, DNA fragmentation in human hepatoma cells.	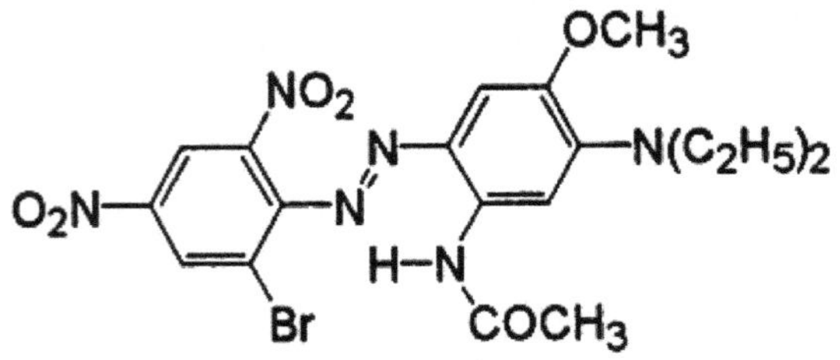
6	Direct black 38	Urinary bladder cancer in human	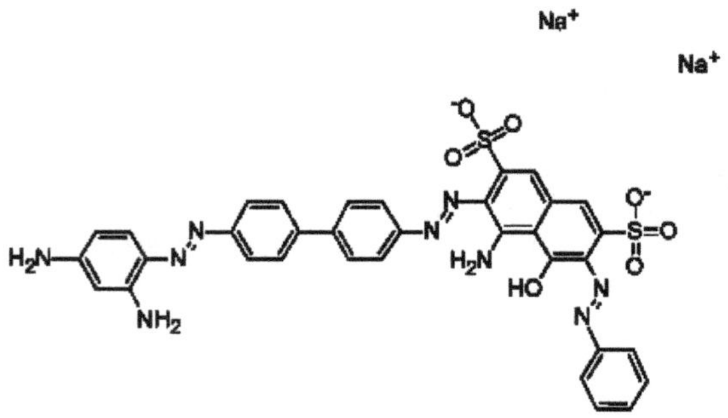
7	Direct blue 15	Mutagenic	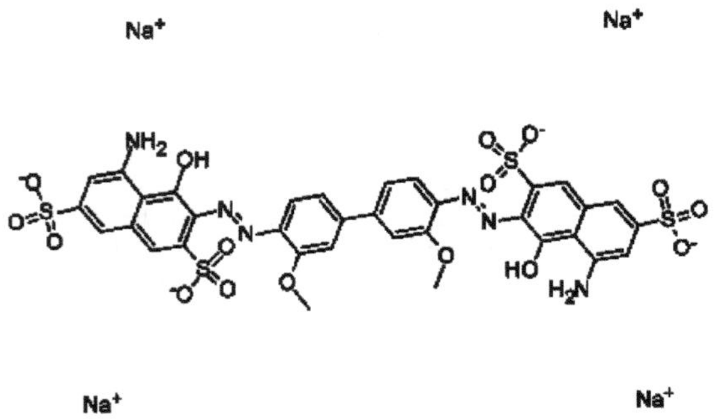

BIODEGRADATION OF ALIPHATIC HYDROCARBONS

Hydrocarbons in crude petroleum can be classified as alkanes, cycloalkanes, aromatics, polycyclic aromatics, asphaltines, and resins. Hydrocarbons such as butanes, cycloalkanes, diolefins, xylenes, heptanes, methylnaphthalene, pentane, camphor and octanes are the components of petroleum and petroleum products. Leakage and spilling of oil have become more common in marine transport. This affects the life of many coastal organisms. The bacteria, yeast, and filamentous fungi are known to degrade hydrocarbons in the oil spill. *Pseudomonas putida*, *P. oleo varans*, *P. fluorescent*, *Corynebacteria*, etc. are the examples of oil-degrading microbes. These microbes convert hydrocarbons into alcohols that are in turn converted into aldehydes and then to fatty acids. Among the petroleum HCs, n-alkanes are the most amenable to biodegradation. Normally alkanes in the range of C_5 to C_{10} are inhibitory to majority of the HC degraders at higher concentration as they disrupt lipid membrane when present as solvent. The effluents of petroleum wells, oil refinery operations, industrial wastewaters, fuel consumption, wood-processing activities, detergents, pesticides, paints and other chemicals release huge quantities of hydrocarbons into soil and air. n-alkanes are oxidized by an electron carrier dependent monooxygenase system, the alkane hydroxylases, to their corresponding alkan-1-ol. Based on the substrate, these enzymes are called toluene mono/dioxygenase, methane monooxygenase, ammonia monooxygenase, propanemonooxygenase and alkane oxygenase. In eukaryotes, including fungi, endoplasmic nucleotide-dependent cytochrome P450 reductases (CPR) transfer two electrons from NADPH to cytochrome P450 which uses these electrons for a variety of molecular biotransformation reactions such as the detoxification or biosynthesis of secondary compounds In addition to these two prototype reactions for the oxidation of alkanes, there are some other types of alkane metabolism which can be seen in special microorganisms. Through an alternative terminal pathway, some yeast such as *Mortierella* sp. utilize alcohol oxidases instead of alcohol dehydrogenases, producing either or both ω-hydroxic acids and α-ω-dioic acids.

Alkene Metabolism

Unsaturated hydrocarbons (alkenes) are more sensitive than alkanes and aromatic hydrocarbons to degradation due to the higher reactivity of their double bonds (Grossi et al., 2011). Several microorganisms can insert an oxygen atom into an alkene substrate via a monooxygenase activity, producing an epoxide molecule. Epoxidation is a process in which O_2-/NADH-dependent monooxygenases, namely alkene monooxygenases (AMOs), add an oxygen atom to the double bonds of the unsaturated aliphatic hydrocarbon (Krishnakumar, 2007). Iron core that interacts with glutamate and histidine residues. This core binds the O_2 molecule necessary for the hydroxylation and epoxidation of a broad range of unsaturated alkenes as well as a variety of chlorinated alkenes, such as tri-chloro-ethene, vinyl chloride and 1,3-dichloropropene. Since the components of hydrocarbons are highly recalcitrant to degradation (due to their low water solubility), they remain for a long time. Direct or indirect exposure of living organisms to hydrocarbon compounds can have varying ecological and/or economic impacts, as described below:

Direct exposure to oil spill: A variety of ill effects may develop when the oil spill occurs, mainly to those who live and work in such areas. They may come in contact with gaseous oil compounds or oil compounds adsorbed on particulate matter (dispersed in air) through breathing. Another exposure pathway may relate to activities carried out, in contaminated ground (soil) or directly through skin adsorption on touching the spilled contaminated material.

Indirect exposure: This happens through consumption of contaminated food or water especially relevant in the case of consumption of fish that was from an oil spill polluted environment. As some oil components can accumulate in living organisms, if a fish lives in a polluted environment, it will keep accumulating in its body some oil components (without excretion) which may reach concentrations several orders of magnitude higher than those of the surrounding waters. Through consumption of such polluted fish meat, humans may become seriously exposed to concentration of oil components higher than in the surrounding environment or as compared to ingestion of the polluted water or bathing in the polluted water.

Economic impact: Oil spills have hampered many industries such as fishing, tourism, oil industries, etc.

Aliphatic hydrocarbon may be saturated or unsaturated. Saturated hydrocarbons can be degraded more easily than unsaturated ones and branched chain N-alkanes. Hydrocarbons of carbon chain length of 10-24 are most deadly biodegradable. Similarly aliphatics are easier to degrade show decreased biodegradation. Alkanes in range of C_{20} to C_{40}, also referred to as waxes, are less biodegradable as they being hydrophobic solids have low solubility in water. During degradation the alkanes are converted to alcohol by the action of oxygenase enzymes that attack the terminal methyl group. The alcohol is further oxidized to aldehyde and then to fatty acids. Further utilization of fatty acid occurs by β-oxidation of aliphatic chain. Higher the methyl branching lower is the extent of β-oxidation.Cyanide (HCN): is toxic to biological system and even microorganism capable of degrading cyanide cannot withstand a high concentration of HCN. Some of the cyanides e.g. HCN and CH_3CN are volatile. Therefore disposal of strictly controlled. It is degraded by fungal hydratase into $HCONH_2$ and by pseudomonas fluorescence into CO_2 and NH_2.

Factors Affecting HC Biodegradation

Several environmental factors influence biodegradation of petroleum HC such as temperature, pH, nutrient and oxygen availability, salinity, pressure, and light. Bioavailability of contaminant is also an important aspect of biodegradation. Presence of HC degrading populations of microbes at sufficiently high levels is a prerequisite for an effective bioremediation. Occurrence and abundance of microorganisms in environment depend on availability and diversity of carbon sources. Surface soil has high organic matter but subsurface and deep layers have lower organic content. Because of a low amount of organic matter organisms in these regions are often dormant. Various physicochemical factors affecting HC degradation are described below (Maier, 2000).

- **Oxygen Availability:** HC degradation takes place both in presence and absence of oxygen. However aerobic conditions are more favorable as oxygenases are the primary enzymes needed for degradation to occur. Oxygenases function in presence of oxygen, so degradation rates are higher in aerobic conditions as compared to those under anaerobic conditions.
- **Nutrient Availability:** The process of biodegradation can be enhanced by addition of essential nutrients such as nitrogen and phosphorous. In case of petroleum oil spills where nitrogen shortages can be acute; carbon, nitrogen and phosphorous are added in the ratio of approximately 100: 10: 1 (C: N: P) (Kothari, Panchal, & Srivastava, 2013).

DEGRADATION OF AROMATIC COMPOUNDS

Aromatic compounds can be defined as organic molecules containing one or more aromatic rings, specifically benzene rings. Different aromatic compounds co-exist as complex mixtures in petroleum refinery and distillation sites (Seo, Keum, & Li, 2009). Aromatic compounds are among the most prevalent and persistent pollutants in the environment. The commercial, industrial, natural activities generate huge amount of aromatic hydrocarbons causes a great concern due to potential hazard to flora and fauna. Benzene, toluene, ethyl benzene, styrene and the xylenes are among the 50 largest-volume industrial chemicals produced with production figures of millions of tons per year. There are three major categories: polycyclic aromatic hydrocarbons (PAHs), heterocyclics, and substituted aromatics (Cheung, & Kinkle, 2001). As compared to most other cyclic compounds, aromatic compounds are more stable because of sharing of delocalized electrons by pi bonds. The crucial step in their degradation is overcoming the resonance energy that stabilizes the ring structure. Hence, such aromatic compounds are believed to be hazardous to the biosphere. There are two major steps involved in biodegradation of an aromatic molecule (1) activation of the ring, (2) ring cleavage Activation is achieved by the incorporation of molecular oxygen into the aromatic ring leading to dihydroxylation of aromatic nucleus, and the enzymes responsible for this are oxygenases. Monooxygenases, characteristic of fungi and other eukaryotes, catalyze the incorporation of a single atom of oxygen to form an epoxide which can then undergo hydration to yield transdihydrodiols. Dioxygenases catalyze simultaneous incorporation of two atoms of oxygen to form a dihydrodiol. These dioxygenase reactions have been shown to occur for benzene, halogenated benzenes, toluene, para-chlorotoluene, xylenes, biphenyls, napthelene, anthracene, etc. These dihydrodiols are further oxidized to catechols which are precursors to ring cleavage. Catechol can be oxidized either via ortho-cleavage pathway which involves cleavage of the bond between carbon atoms of the two hydroxyl groups to yield muconic acid, or via the meta- cleavage pathway which involves cleavage of the bond between a carbon atom with a hydroxyl group and the adjacent carbon atom to yield 2-hydroxymuconic semi aldehyde. These compounds are further degraded to form organic acids which are then utilized by microorganisms for their cell synthesis and energy generation. However, microorganisms possess a wide range of catabolic biodegradation pathways and, thus, use these hazardous xenobiotics as the sole source of carbon and energy. The breakdown of aromatic compounds by ring cleavage is an essential biochemical step in Nature's 'carbon' cycle and is performed by several kinds of microorganism. Bacteria are the most versatile in this respect, but several yeasts and fungi are able to degrade a more limited range of benzenoid structures. In the case of aerobic biodegradation, the original substrate undergoes mostly oxygenation reactions and is converted to dihydroaromatic derivatives such as catechol, protocatechuate, gentisate, homoprotocatechuate, homogentisate, hydroquinone and hydroxyquinone. The key enzymes catalyzing the reactions are monooxygenases and/or hydroxylating dioxygenases. These intermediates may act as substrates for enzymes that cleave the aromatic ring in reactions involving molecular oxygen. The aromatic ring may be opened by cleavage of the bond between two neighbouring carbon atoms carrying hydroxyls. In this case, the process is termed *ortho* – cleavage, and the enzymes involved are intradiol dioxygenases (Lipscomb, 2008). On the other hand, if other bonds taking a meta-position are cleaved in order to open the aromatic ring, the term used is meta-cleavage with the enzymes extradiol dioxygenases involved in the process. Central metabolic pathways involve series of reactions that culminate in the formation of intermediates of the Krebs cycle.These microbes produce, mostly as a result of induction, a whole sequence of enzymes which convert aromatic substrates into an *ortho* or *para* dihydroxyphenol derivative, followed by cleavage of the ring to aliphatic acids; these ring fission products are then funnelled into the Krebs cycle through a variety of pathways, depending on the organism and cultural conditions (Sahoo, Ramesh, & Pakshirajan, 2012).

DEGRADATION OF HALOGENATED COMPOUNDS

Organic halogen compounds are a large class of natural and synthetic chemicals that contain one or more halogens (fluorine, chlorine, bromine, or iodine). Greater the number of halogens in the molecule, the less biodegradable the compound will be in aerobic systems and the more degradable it will be in anaerobic systems. The biodegradation rate is also dependent on the type of halogen in the compound. In general the order of halogen compounds is Br>Cl>F. Halogenated organic compounds constitute one of the largest and diverse groups of environmental chemicals. Although some of these chemicals are generated by naturally occurring biotic and abiotic processes in the oceans and atmosphere, the widespread use of halogen-based chemistry in industrial-scale chemical processing over the past 100 years has introduced many additional man-made halocarbons into the environment. These compounds are manufactured and used as pesticides, plasticizers, paint and printing-ink components, adhesives, flame retardants, hydraulic and heat transfer fluids, refrigerants, solvents, additives for cutting oils, and textile auxiliaries. The hazardous chemicals enter the environment through production, commercial application, and waste. As a result of bioaccumulation in the food chain and groundwater contamination, they pose public health problems because many of them are toxic, mutagenic, or carcinogenic. Generally, the potential for bioremediation requires that a halo-organic compound can be biodegraded, partly or completely destroyed by metabolism.

It appeared that specific bacterial enzymes exist, dehalogenases, which can degrade halogenated compound, discovered and firstly named by Jensen (1957). Dehalogenases normally catalyse the key initial step in the catabolism of halogenated compounds. The factors affecting microbial dehalogenation reactions are: stereospecificity, the number of halogens per molecule and position of halogen on the molecule. It is easier to dehalogenate compounds substituted in the C2 position than on other position. Also increasing chain length of the substituted compound decreases dehalogenase reactivity. Different organisms such as *Pseudomonas*, *Rhizobium*, *Alcaligenes*, *Bacillus* and *Mycobacterium* have been shown to produce dehalogenases.These enzymes make use of a variety of distinctly different catalytic mechanisms to cleave carbon-halogen bonds. Dehalogenases catalyse the hydrolysis of halogen-substituted alkanoic acids yielding either hydroxyalkanoic acids from mono-halogenated acids or oxo-alkanoic acids from di-halogenated compounds products which may be readily metabolized.The recalcitrant nature of these compounds is due to low electron density at the aromatic ring, and the enzyme oxygenase that usually initiates the degradation of other aromatic compounds, cannot attack these halogenated compounds. This reaction does not proceed through an electrophilic attack but via a nucleophilic attack. Despite the metabolic challenges posed by halo-organic compounds, microorganisms have demonstrated a remarkable capacity to biodegrade such compounds. Many synthetic compounds have a xenobiotic molecular structure that did not occur on earth in biologically significant concentrations before the development of industrial organic synthesis some 70-100 years ago. Many chlorinated and brominated chemicals that were originally expected to be recalcitrant to biodegradation now appear to be able to serve as a carbon source for some specialized microorganisms. Such organisms have developed a catabolic pathway that harvests energy and carbon from the halocarbon compound when degrading it. Other organohalogens can only be degraded under aerobic conditions by cometabolism or by anaerobic, reductive reactions. The former reactions are particularly well carried out by methanotrophic bacteria that produce a soluble type of methane monooxygenase. This enzyme has a very broad substrate range and can cooxidize various recalcitrant organohalogens, including trichloroethylene and dichloroethylenes. A drawback is that these cometabolic conversions often are accompanied by toxic effects, which are cause by the fact that various reactive transformation products can be formed. In view of these effects, the isolation of organisms that can degrade and grow on organohalogens still is an important bottleneck towards the degradation of an efficient treatment process.

BIODEGRADATION OF POLYMERS

A polymer is a large molecule, or macromolecule, composed of many repeated subunits. Polymers are of two types:

Natural polymeric materials such as shellac, amber, wool, silk and natural rubber have been used for centuries. A variety of other natural polymers exist, such as cellulose, which is the main constituent of wood and paper. The list of synthetic polymers includes synthetic rubber, phenol formaldehyde resin (or

Bakelite), neoprene, nylon, polyvinyl chloride (PVC or vinyl), polystyrene, polyethylene, polypropylene, polyacrylonitrile, PVB, silicone, and many more. Polyethylene (abbreviated PE) or polyethene is the most common plastic. The name Plastic is derived from the Greek word plastikos, which means able to be shaped. Plastic materials consist of long chain molecules (polymers). They are usually synthetic, most commonly derived from petrochemicals, but many are partially natural. The plastics of various forms such as nylon, polycarbonate, polyethylene-terephthalate, polyethylene, polypropylene, polystyrene, polytetraflouroethylene, polyurethane, polyvinyl chloride are being continuously used in our day-to-day life. Most conventional plastics such as polyethylene, polypropylene, polystyrene, poly (vinyl chloride) and poly (ethylene terephthalate), are non-biodegradable, and their increasing accumulation in the environment has been a threat to the planet. As per the reports the most commonly used non-degradable solid waste is polythene which is a linear hydrocarbon polymers consisting of long chains of the ethylene monomers (C_2H_4). The general formula of polyethylene is C_nH_{2n}, where 'n' is the number of carbon atoms. Polythene is made from the cheap petrochemical stocks extracted from oil or gas through efficient catalytic polymerization of ethylene monomers. Extensive use of non-biodegradable thermoplastics and the rate at which they accumulate in the environment, makes the humankind to realize the necessity to find its environmental impact. As the polymer usage is unavoidable, ways have to be found to Enhance the biodegradability of the polymers by blending them with biodegradable natural polymers such as starch or cellulose etc. Mixing with prooxidants so that they are easily degraded and Isolate and improve microorganisms that can efficiently degrade these polymers (Shimao, 2000). When plastics are used as substrates for microorganisms, evaluation of their biodegradability should not only be based on their chemical structure, but also on their physical properties (melting point, glass transition temperature, crystallinity, storage modulus etc.). Polymers do not consist simply of only one chemical homogenous component, but contain different polymers (blends) or low molecular weight additives (plasticizers) which can serve as good nutrients for the ambient microorganisms developed on polymer surfaces. Biodegradability is primarily dependent on hydrolyzable and oxidizable chemical structures, balance of hydrophobicity, and molecular weights. In order to attempt the third option the mechanism of biodegradation should be understood. Most of the polymers are resistant to degradation, research over the past couple of decades has focused on developing biodegradable polymers, which are degraded and catabolized ultimately to carbon dioxide and water by bacteria, fungi under natural environment. The biodegradable plastics can be produced by addition of biodegradable groups into the main chain during the preparation of industrial polymers prepared by free radical copolymerization.

Mechanism of Polymers Biodegradation

Biodegradation of polymers involves following steps:

1. Surface attachment of microorganism.
2. Use of polymer as a carbon source by microorganisms for their growth and development.
3. Degradation of polymers to intermediate compounds.
4. Final degradation.

Microorganisms are not able to transport the polymers directly through their outer cell membranes into the cells where most of the biochemical processes take place due to the lack of water-solubility and the length of the polymer molecules. Microorganisms attaches to the hydrophilic surface of the polymer. Since Polypropylene and Polyethylene have only CH_2 groups, the surfaces are hydrophobic. Initial physical or chemical degradation leads to the insertion of hydrophilic groups on the polymer surface making it more hydrophilic (insertion of hydrophilic groups also decreases the surface energy). Once the organism gets attached to the surface, they start growing by using the polymer as the carbon source. The microbes excrete extracellular enzymes which depolymerize the polymers outside the cells. Extracellular and intracellular depolymerases enzymes are actively involved in biological degradation of polymers. In the primary degradation, the main chain cleaves, leading to the formation of low-molecular weight fragments (oligomers), dimers or monomers. The microbes excrete extracellular enzymes which depolymerize the polymers outside the cells. Extracellular and intracellular depolymerases enzymes are actively involved in biological degradation of polymers. During degradation, exoenzymes from microorganisms break down complex polymers yielding short chains or smaller molecules, e.g., oligomers, dimers, and monomers, that are smaller enough (water soluble) to pass the semi-permeable outer bacterial membranes and then to be utilized as carbon and energy sources Anaerobic microorganisms can also degrade these polymers under anoxic conditions. The primary products then are CO_2, H_2O, CH_4 and biomass under methanogenic condition or H_2S, CO_2 and H_2O under sulfidogenic condition. The environmental conditions decide the group of microorganisms and the degradative pathway involved in the ultimate degradation of recalcitrant (Arutchelvi et al., 2008). The microorganisms present in polyethylene dumped areas can be interacted with polythene and undergo changes in mechanical properties of tensile strength, optical changes of cracking, erosion and decolorization. It is clear that the polymers can be degraded to some extent by microbes (Singh, Singh, & Bhatt, 2016).

BIODEGRADATION OF DETERGENTS

Detergents were developed in response to the shortage of the animal and vegetable fats used to make soap during World War I and World War II. Detergents are primarily surfactants, which could be produced easily from petrochemicals. Surfactants (surface-active agents) are a diverse group of chemicals consisting of a polar, water-soluble head group and a non-polar hydrocarbon tail group, which is not as soluble in water. Surfactants lower the surface tension of water, essentially making it ‘wetter’ so that it is less likely to stick to itself and more likely to interact with oil and grease. Massive quantities of surfactants are being used in households and industry every day, and most end up dispersed in different environmental compartments (soil, water, sediment). The elevated levels of surfactants in the environment can greatly affect the ecosystem; their toxicity to organisms from mammals to bacteria is well known. Excessive use of any type of surfactants and their disposal in the environment, especially in aquatic bodies, could seriously affect the ecosystem. Surfactants show a marked biological activity. Anionic surfactants can bind to bioactive macromolecules such as peptides, enzymes, and DNA. Binding to proteins and peptides may change the folding of the polypeptide chain and the surface charge of a molecule. This may modify biological function. The primary target site of cationic surfactants is the cytoplasmic (inner) membrane of bacteria. QACs bind to inner membranes and disorganize them via their long alkyl chain. Nonionic surfactants exert antimicrobial activity by binding to various proteins and phospholipid membranes. Such binding increases the permeability of membranes and vesicles, causing leakage of low molecular mass compounds. This can result in cell death or damage through loss of ions or amino acids (Ivanković, 2009). Degradation of surfactants through microbial activity is the primary transformation occurring in the environment. Biodegradation is an important process to treat surfactants in raw sewages in sewage treatment plants, and it also enhances the removal of these surfactants in the environment, thus reducing their impact on biota. During biodegradation, microorganisms can either utilize surfactants as substrates for energy and nutrients or cometabolize the surfactants by microbial metabolic reactions. There are many chemical and environmental factors that affect biodegradation of a surfactant in the environment. The most important influencing factors are the physiochemical characteristics of the compounds (solubility, concentration, structure, etc.), the physiochemical conditions of the environmental media (dissolved oxygen, temperature, pH, light, nutrient concentration, etc.) and the microorganisms present in the aquatic environment. Linear alkyl benzene sulfonates (LAS) is one of the major xenobiotic anionic surfactants. In 2011, linear alkylbenzene sulphonates (LAS) holds 97.9% of the market. About 82% - 87% of LAS produced is used for the production of household detergents and cleaning products. LAS are also used for the manufacture of heavy duty laundry liquids, laundry powders, and light duty dish washing liquids and industrial cleaners (Asok, Fathima, & Jisha, 2015).

GENETIC AND METABOLIC ENGINEERING IN BIODEGRADATION

Many pollutants are still persistent in the environment reflects the inadequacy of the current microbial catabolic capacity to deal with such pollutants. This is where molecular biologists step in to engineer bacteria for more efficient remediation of contaminated areas, and to degrade hitherto recalcitrant organic pollutants. Advances in genetic engineering have enabled a far more rational approach for the development of microbes for the degradation of xenobiotics, than the classical approach via mutagenesis and screening. It offers opportunity to create artificial combination of genes that do not exist together in nature. Genetically engineered microorganisms have higher degradative capacity and have been demonstrated successfully for the degradation of various pollutants under defined conditions. Certain plasmids play an important role in adaptation of natural microbial populations to oil and other hydrocarbons. Some of the microbial catabolic pathways responsible for the degradation, including the *alk* (C_5 to C_{12}*n*-alkanes), *nah* (naphthalene) and *xyl* (toluene) pathways have been extensively characterized and are generally located on large catabolic plasmids. These plasmids can then be introduced into a host organisms resulting in a recombinant or genetically engineered microorganisms (GEM) with new degradative capabilities. The first step in GMM construction is selection of suitable gene/s. Next, the DNA fragment to be cloned is inserted into a vector and introduced into host cells. The modified bacteria are called recombinant cells. The following step is production of multiple gene copies and selection of cells containing recombinant DNA. The final step includes screening for clones with desired DNA inserts and biological properties. This has led to the development of a new field of metabolic engineering that involves the improvement of cellular activities by manipulation of enzymatic, transport and regulatory functions of the cell using recombinant DNA technology (Wasilkowski, Swędzioł, & Mrozik, 2012). Genes responsible for degradation of environmental pollutants, for example, toluene, chlorobenzene acids, and other halogenated pesticides and toxic wastes have been identified. For every compound, one separate plasmid is required. It is not like that one plasmid can degrade all the toxic compounds of different groups. The plasmids are grouped into four categories: 1) OCT plasmid which degrades octane, hexane and decane; 2) XYL plasmid which degrades xylene and toluenes, 3) CAM plasmid that decompose camphor and 4) NAH plasmid which degrades naphthalene (Joutey, Bahafid, Sayel, & Ghachtouli, 2013). Bacteria, especially from genus *Pseudomonas*, are the major object of genetic manipulations. There are ubiquitous inhabitants of many environment and are known as efficient degraders of many toxic substances. While genetic engineering has produced numerous strains able to degrade otherwise intractable pollutants in a petridish or in a bioreactor, the practical translation of this research into actual in situ bioremediation practices has been quite scanty. One major issue in this respect is the growing realization that the strains and bacterial species that most frequently appear in traditional enrichment procedures are not the ones performing the bulk of biodegradation in natural niches and may not even be any good as bioremediation mediators.

LIMITATIONS OF BIODEGRADATION

There are several limitations of biodegradation technology. One major constraint is the nature of the organisms. The degradation of xenobiotic compounds by microorganisms is not their feature. Rather, it is a strategy for survival. Most of organisms do their job under environmental conditions that suit their needs. Therefore, environmental modification is needed to encourage the organisms to degrade or take up the toxic compounds at an acceptable rate (Singh, 2008). The organism must be given low levels of the pollutant over a period of time. This lead the organism to switch on the metabolic pathways expected to digest the pollutant. In case of bacteria and fungi, it is usually necessary to add fertilizer or oxygen to the polluting site. This can be disruptive to other organisms when done in situ. In situations where simple compounds and metals are being taken up it is likely that these pollutants are at toxic levels for the organisms. Overall, the organisms do not always live as well on the pollutant diet as on other nutrients found more commonly in their environment. This causes problem when doing in situ remediation. Other limitations of immediate concern are cost/benefit ratios i.e. cost versus overall environmental impact. The petroleum industries are embroiled currently in a battle with the EPA about the added costs of maintaining new Clean Air Act standards. Bioremediation is generally very costly, labor intensive, and can take several months for the remediation to achieve acceptable levels. Air bioremediation in particular is very inefficient, considering the volume of polluted air generated by industry. Another problem is that both *ex situ* and *in situ* technologies can cause environmental disruption beyond the damage done by the pollution. The long-term effects of introducing naturally occurring non-native bioremediation organisms into an area are not fully understood. The impact of genetically altered bioremediation organisms is even less understood. Bioremediation has been proven to work effectively under laboratory conditions. Short-term studies show that it also works under several field conditions. Like many technologies with good scientific foundations its merits are marred by over-optimistic speculations and fraudulent claims. Bioremediation's popularity is further enhanced because it is perceived as being more "green" than other remediation technologies. Companies and individuals are investing in biotechnology futures in spite of the high risks. As a result bioremediation companies have a viable future regardless of its long-term effectiveness.

CONCLUSION

The toxic xenobiotics are progressively more concentrated in each step of a food chain, a process called bio-magnification, thus affecting the ecological balance. The emerging science and technology of bioremediation offers an alternative method to detoxify contaminants. There are different methods such as physic-chemical and biological which have been employed in the degradation of xenobiotics. Since physic-chemical methods are very expensive therefore biological methods employing microbes are widely used. Microorganisms have always been the simplest and most efficient tools used for removal of harmful toxic chemicals from environment. They have the capability of degrading most of the natural and manmade compounds using enzymatic activity. The transformation of a xenobiotic compounds by microbial community is determined by its entry into the cell and the degree of structural analogy between the synthetic compound and the natural compound. The application of bioremediation technology in the restoration of ecosystem and soil management is used less, hence, we need extensive research programs to increase the capabilities of bioremediation to deep, extensive, subsurface contamination due to chlorinated hydrocarbons and complex mixed wastes, including soils and groundwater. The ongoing microbial genomics studies will deliver more robust technologies for the bioremediation of metal – contaminated waters and land. Exciting developments in the use of microorganisms for the recycling of metal waste, with the formation of novel biominerals with unique properties are also predicted in the near future. The search for new fungal taxa and taxon groups with biotechnological potential and their implementation in naturally occurring organismic alliances will continue to play a significant role for the elucidation of cryptic or novel natural products and their application in various fields such as bioremediation. As more insights are gained in understanding the chemical nature of recalcitrance, enhanced transformation and, hopefully, complete mineralization of xenobiotic effluents could be achieved by modular multi-step processes, coupling reduction and oxidation reactions by abiotic or biotic means. This principle seems to be bearing fruit for an increasing range of recalcitrant molecules. More work is required to design a combination of various techniques for remediation purposes. Further, because complex microbial communities rather than only bacteria are important for effective bioremediation of contaminated environment, more research is required to elucidate the interactive and symbiotic roles played by individual species in these communities.

REFERENCES

Adams, G. O., Fufeyin, P. T., Okoro, S. E., & Ehinomen, I. (2015). Bioremediation, biostimulation and bioaugmention: A review. *International Journal of Environmental Bioremediation & Biodegradation*, *3*(1), 28–39.

Arutchelvi, J., Sudhakar, M., Arkatkar, A., Doble, M., Bhaduri, S., & Uppara, P. V. (2008). Biodegradation of polyethylene and polypropylene. *Indian Journal of Biotechnology*, *7*(1), 9.

Asok, A. K., Fathima, P. A., & Jisha, M. S. (2015). Biodegradation of Linear Alkylbenzene Sulfonate (LAS) by Immobilized Pseudomonas sp. *Advances in Chemical Engineering and Science*, *5*(04), 465–475. doi:10.4236/aces.2015.54048

Borgeest, C., Greenfeld, C., Tomic, D., & Flaws, J. A. (2002). The effects of endocrine disrupting chemicals on the ovary. *Frontiers in Bioscience: A Journal and Virtual Library, 7*, d1941-8.

Camarero, S., Ibarra, D., Martínez, M. J., & Martínez, Á. T. (2005). Lignin-derived compounds as efficient laccase mediators for decolorization of different types of recalcitrant dyes. *Applied and Environmental Microbiology*, *71*(4), 1775–1784. doi:10.1128/AEM.71.4.1775-1784.2005 PMID:15812000

Chacko, J. T., & Subramaniam, K. (2011). Enzymatic degradation of azo dyes-A review. *International Journal of Environmental Sciences*, *1*(6), 1250.

Cheung, P. Y., & Kinkle, B. K. (2001). Mycobacterium diversity and pyrene mineralization in petroleum-contaminated soils. *Applied and Environmental Microbiology*, *67*(5), 2222–2229. doi:10.1128/AEM.67.5.2222-2229.2001 PMID:11319104

Das, N. R., & Kumar, N. (2015). Assessment of phenol biodegradation capacity of indigenous bacteria isolated from sewage treatment plant. *African Journal of Microbiological Research*, *9*(16), 1150–1161. doi:10.5897/AJMR2013.5909

Grossi, V., Cravo-Laureau, C., Rontani, J. F., Cros, M., & Hirschler-Réa, A. (2011). Anaerobic oxidation of n-alkenes by sulphate-reducing bacteria from the genus Desulfatiferula: N-ketones as potential metabolites. *Research in Microbiology*, *162*(9), 915–922. doi:10.1016/j.resmic.2011.07.004 PMID:21810468

Han, M. J., Choi, H. T., & Song, H. G. (2004). Degradation of phenanthrene by Trametes versicolor and its laccase. *JOURNAL OF MICROBIOLOGY-SEOUL*, *42*(2), 94–98. PMID:15357301

Ivanković, T., & Hrenović, J. (2010). Surfactants in the environment. *Arhiv za Higijenu Rada i Toksikologiju*, *61*(1), 95–109. doi:10.2478/10004-1254-61-2010-1943 PMID:20338873

Jain, R. K., Kapur, M., Labana, S., Lal, B., Sarma, P. M., Bhattacharya, D., & Thakur, I. S. (2005). Microbial diversity: Application of micro-organisms for the biodegradation of xenobiotics. *Current Science*, *89*(1), 101–112.

Javaid, M. K., Ashiq, M., & Tahir, M. (2016). Potential of Biological Agents in Decontamination of Agricultural Soil. *Scientifica, 2016.*

Jayasekara, R., Harding, I., Bowater, I., & Lonergan, G. (2005). Biodegradability of a selected range of polymers and polymer blends and standard methods for assessment of biodegradation. *Journal of Polymers and the Environment*, *13*(3), 231–251. doi:10.1007/s10924-005-4758-2

Jensen, H. L. (1957). Decomposition of chloro-substituted aliphatic acids by soil bacteria. *Canadian Journal of Microbiology*, *3*(2), 151–164. doi:10.1139/m57-019

Joutey, N. T., Bahafid, W., Sayel, H., & El Ghachtouli, N. (2013). *Biodegradation: involved microorganisms and genetically engineered microorganisms. In Biodegradation-life of science* (pp. 289–320). Rijeka: InTech.

Karigar, C. S., & Rao, S. S. (2011). Role of microbial enzymes in the bioremediation of pollutants: A review. *Enzyme Research*, *2011*, 1–11. doi:10.4061/2011/805187 PMID:21912739

Kothari, V., Panchal, M., & Srivastava, N. (2013). *Microbial Degradation of Hydrocarbons*. Ahmedabad, Gujarat, India: Institute of Science: Nirma University.

Krishnakumar, A. M. (2007). *Structural studies of enzymes involved in propylene and acetone metabolism in Xanthobacter autotrophicus*. Academic Press.

Lipscomb, J. D. (2008). Mechanism of extradiol aromatic ring-cleaving dioxygenases. *Current Opinion in Structural Biology*, *18*(6), 644–649. doi:10.1016/j.sbi.2008.11.001 PMID:19007887

Madigan, M. T., Martinko, J. M., & Parker, J. (1997). Brock biology of microorganisms (Vol. 514). Upper Saddle River, NJ: Prentice Hall.

Maier, R. M. (2000). Microorganisms and organic pollutants. In R. M. Maier, I. L. Pepper, & C. P. Gerba (Eds.), *Environmental Microbiology* (pp. 363–402). San Diego, CA: Academic Press.

Niu, G. L., Zhang, J. J., Zhao, S., Liu, H., Boon, N., & Zhou, N. Y. (2009). Bioaugmentation of a 4-chloronitrobenzene contaminated soil with Pseudomonas putida ZWL73. *Environmental Pollution*, *57*(3), 763–771. doi:10.1016/j.envpol.2008.11.024 PMID:19108939

Pan, X., Lin, D., Zheng, Y., Zhang, Q., Yin, Y., Cai, L., & Yu, Y. (2016). Biodegradation of DDT by Stenotrophomonas sp. DDT-1: Characterization and genome functional analysis. *Scientific Reports*, 6. PMID:26888254

Sahoo, N. K., Ramesh, A., & Pakshirajan, K. (2012). *Bacterial degradation of aromatic xenobiotic compounds: an overview on metabolic pathways and molecular approaches. In Microorganisms in Environmental Management* (pp. 201–220). Springer Netherlands.

Sangamithirai, K. M., Jayapriya, J., Hema, J., & Manoj, R. (2015). Evaluation of in-vessel co-composting of yard waste and development of kinetic models for co-composting. *International Journal of Recycling of Organic Waste in Agriculture*, *4*(3), 157–165. doi:10.1007/s40093-015-0095-1

Seo, J. S., Keum, Y. S., & Li, Q. X. (2009). Bacterial degradation of aromatic compounds. *International Journal of Environmental Research and Public Health*, *6*(1), 278–309. doi:10.3390/ijerph6010278 PMID:19440284

Shimao, M. (2001). Biodegradation of plastics. *Current Opinion in Biotechnology*, *12*(3), 242–247. doi:10.1016/S0958-1669(00)00206-8 PMID:11404101

Singh, D. K. (2008). Biodegradation and bioremediation of pesticide in soil: Concept, method and recent developments. *Indian Journal of Microbiology*, *48*(1), 35–40. doi:10.1007/s12088-008-0004-7 PMID:23100698

Singh, G., Singh, A. K., & Bhatt, K. (n.d.). Biodegradation of polythenes by bacteria isolated from soil. *International Journal of Research and Development in Pharmacy and Life Sciences, 5*(2), 2056-2062.

Varsha, Y. M., & Chenna, S. (2012). An emphasis on xenobiotic degradation in environmental cleanup. *Journal of Bioremediation & Biodegradation.*

Wasilkowski, D., Swędzioł, Ż., & Mrozik, A. (2012). The applicability of genetically modified microorganisms in bioremediation of contaminated environments. *CHEMIK nauka-technika-rynek, 1*(66), 817-826.

Wolka, K., & Melaku, B. (2015). Exploring selected plant nutrient in compost prepared from food waste and cattle manure and its effect on soil properties and maize yield at Wondo Genet, Ethiopia. *Environmental Systems Research*, *4*(1), 1–7. doi:10.1186/s40068-015-0035-0

Wu, T., & Crapper, M. (2009). Simulation of biopile processes using a hydraulics approach. *Journal of Hazardous Materials*, *171*(1), 1103–1111. doi:10.1016/j.jhazmat.2009.06.131 PMID:19631454

KEY TERMS AND DEFINITIONS

Biomagnification: It is a process in which chemical substances become more concentrated at each higher trophic level. Such chemical substances tend to be difficult to be removed from animal and plant tissues.

Bioremediation: Bioremediation is a waste management technique that involves the use of organisms to remove or neutralize pollutants from a contaminated site.

Biosorption: It is a physiochemical process that occurs naturally in certain biomass which allows it to passively concentrate and bind contaminants onto its cellular structure.

Co-Metabolism: The simultaneous degradation of two compounds, in which the degradation of the second compound (the secondary substrate) depends on the presence of the first compound (the primary substrate).

Mesophile: Mesophile is an organism that grows best in moderate temperature, neither too hot nor too cold, typically between 20 and 45 °C.

Mineralization: The decomposition or oxidation of the chemical compounds in organic matter into plant-accessible forms.

Surfactants: The compounds that lower the surface tension (or interfacial tension) between two liquids or between a liquid and a solid. Surfactants may act as detergents, wetting agents, emulsifiers, foaming agents, and dispersants.

Thermophile: Thermophile is an organism a type of extremophile that thrives at relatively high temperatures, between 41 and 122 °C.

Xenobiotic: Xenobiotics are those compounds that are alien to a living individual and have a propensity to accumulate in the environment.

Chapter 9

Role of Rhizoremediation in Decontaminating Some Hazardous Pollutants

Hossein Farraji
University Saiens Malaysia, Malaysia

Mohammad Ali Zahed
Kharazmi University, Iran

Nastaein Qamaruz Zaman
Universiti Sains Malaysia, Malaysia

Hamed Faraji
Islamic Azad University – Shiraz, Iran

ABSTRACT

Rhizoremediation is a natural biological base, multifactorial operation treatment method which highly depends to environmental factors especially in terrestrial environment as operation of remediation process. Interaction and relationship between microbial community and plant roots is specific attraction and advantages of this advanced treatment method. Decontamination efficiency and time could be highly manageable by operation and rhizoaugmentation. Rhizoremediation is advanced compact combination of phytoextraction, phytovolatilization, phytotransformation and phytostabilization. High efficiency of pollutants removal, adaptive and wide range of operation and augmentation factors cause of emerging application of this for biodegradable organic compounds and heavy metals. Selection of suitable plant and joining them with comfortable microbial couples is a function which can plane simultaneously removal system. This chapter will present an overview on PAH, TPH, PCB and heavy metal removal and mechanism of decontamination, plant selection and augmentation process.

INTRODUCTION

Sustainable development requires the serious national and international development and wide promotion of environmental care and management as well as a constant search for eco-friendly techniques to treat terrestrial and aquatic habitats contaminated by rapid increasing anthropogenic activities. Traditional methods for waste management are switching a problem to another side or aspect of environmental

DOI: 10.4018/978-1-5225-2325-3.ch009

footprint. Microorganisms are the baseline of bioremediation of pollutants and plants can be considered as a leader in phytoremediation process. Interdisciplinary cooperation between microorganisms, plants, augmentation factors and operational factor are essential to earning higher efficiency in a shorter time. More than one century ago a German scientist Prof Hiltner in 1904 coined the term “Rhizosphere” for distinctive plant root’s environment.

Good success in rhizoremediation has been attempted to suitable organism selected for this process. Survival of selected bacteria in contaminated soil and concentration on a specific generation of degrading bacteria to converting them to commercial application. Plant root colonization can be inoculated by selecting degrading bacteria. Host plants with effective, supplying rhizosphere in one hand and concentration of contamination in other hand, are the main aspects of rhizoremediation. Genetically modified microorganisms can be applied after application of wild types of microorganism. Some of the well-known organic compounds such as polycyclic aromatic hydrocarbons (PAH) (Maila et al., 2005; Techer et al., 2012), total petroleum hydrocarbon (TPH), (Wojtera-Kwiczor et al., 2014), polychlorinated biphenyls (PCBs) (Mackova et al., 2009; Slater et al., 2011), trichloroethylene (TCE), (Schöftner et al., 2015; Yee et al., 1998) have been degraded and removal of high toxic heavy metal (Masood & Malik, 2013) in extremely high concentration, have been associated by rhizoremediation. Environmental impacts of hydrocarbons could be managed by this treatment method. Improving fertilization of soils with nitrogen fixing is traditional application of this green technology.

Finding a couple partners contain plant and bacteria in order to form a combined degradation society is the fundamental step to effective pollutant removal from contaminated soils. Interaction between plant and microbial community will present decontamination process as well as environmental condition for growth and development. Because of a high correlation between organic pollutants degradation and volume of carbon dioxide evolved, CO_2 as a mineralization production through the rhizoremediation can be used for collecting efficiency of organic compounds degradation furthermore, the positive correlation between pollutants degradation and CO_2 evolving recognized as dehydrogenase activity index (Namkoong et al., 2002).

CONCEPTUAL OVERVIEW OF RHIZOREMEDIATION

Rhizoremediation is a specific phytoremediation which rhizosphere of plants drive main action on remediation of contaminations by preparing suitable condition for especial microorganisms. Knowledge triangle of bioremediation could be arranged by efficiency of removal as: natural attenuation < phytoremediation < rhizoremediation (Prasad, M. N. V., & Prasad, R., 2012). Figure 1 illustrate rhizoremediation as microbial augmented phytoremediation and phytoremediation. Soil is the media that rhizoremediation carried out there and the main contaminants which degraded through the rhizoremediation are organic compound. High hydrophobicity of organic compounds is the major limitation that prohibits their entrance to plants. The main mode of the remediation body in rhizoremediation are not plants, meanwhile, bacteria are the practical, active organs which do the remediation process. Plants are solar-powered pump for microorganisms to arranging contaminant degradation. These cooperation cause to production of suitable substrates for microorganisms and performing pollutants. In fact, microbial community in symbiotic life with root of plant species could be considered as effective combination of living organisms for enhancing the process of contaminated media. Amino acid, sugar and organic acid supplied by plants for microbial community and numerous promoting reaction will present by microbes as consequence of living activities. The most important and well-recognized reactions related to microbial rhizosphere are:

- Metal (loid) mobilization (Bolan et al., 2014).
- Nitrogen fixing for plant application (Nevita et al., 2013; Aslam et al., 2016).
- Nutrient of plant by Phosphate solubility enhancement (Gadd, G. M, 2004).
- Plant growth promotion by hormone production (Weyens et al., 2009; Nevita et al., 2013).
- Controlling pathogenic bacteria by antibiotic making (Segura et al., 2009).
- Enhancing mineral solubilization and uptaking (Ganesan, V., 2008).
- Reduction of specific metallic elements (As, Se, Hg, Cr) (Aslam et al., 2016; Pepi et al., 2011).

Organic compounds such as polycyclic aromatic hydrocarbon, total petroleum hydrocarbon, polychlorinated biphenyls, which are known as biodegradable pollutants, recognized as the most capable compound for rhizoremediation process. Rhizotransformation, rhizodegradation, phytooxidation, phytoreduction, phytotransformation and phytoassimilation are the most effective treatment techniques which have been reported through the rhizoremediation (Prasad, M. N. V., & Prasad, R., 2012). Meanwhile, heavy metals as non-degradable pollutants have been removed from terrestrial media or their toxicity decreased by augmented rhizoremediation such as Cr (Aslam et al., 2016), Cd (Wu et al., 2006), Ni, Pb and Zn (Burd et al., 2000). It seems that not only, rhizoremediation is not a single specific treatment method for a spectacular pollutant but also, it could be considered as managing operating on a series of action and reactions between plant, microorganism, pollutant and soil condition for achieving highest decontamination.

Natural recovery of soil after derangement is very slow and it causes to developing, combining microbial communities and with host phytoremediator plant roots in order to accelerate soil property's recovery (Kuiper et al., 2004; Segura & Ramos, 2013; Wood, 2008) furthermore additional treatment such as nutrient food elements, enzyme activity and pH adjusting could be used as an additional aug-

Figure 1. Microbial augmented phytoremediation comparing with phytoremediation without microbes
Rosenblueth et al., 2006; Mohsenzadeh et al., 2010; Bolan et al., 2014; Ullah et al., 2015; Agnello et al., 2016.

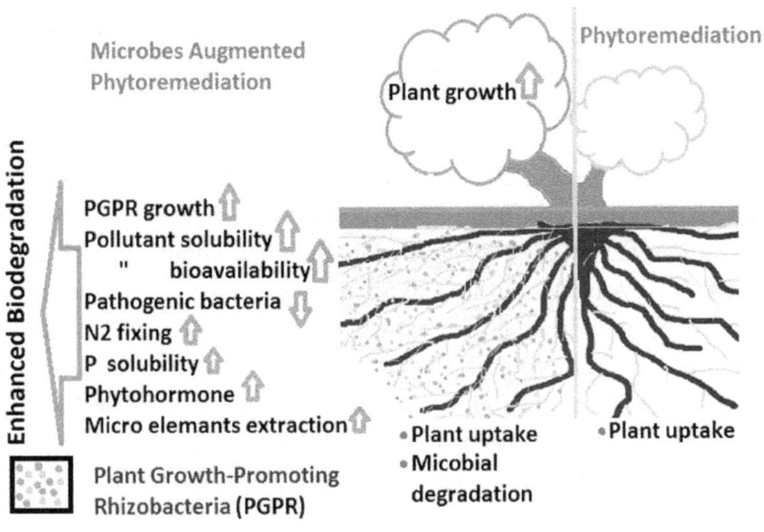

mentation for rhizoremediation. In a current field scale study of a burned land, microorganism inoculation of soil could improve the resilience and accelerating of Burnet ecosystem. The ability of inoculated bacteria community for survival in burnt ecosystem and promoting growth and development in plants were detected by (Pizarro et al., 2015). It shows the ability and power of bacterial communities, even as a single treatment in a natural ecosystem after fire. Since bacterial communities have a critical effect on remediation process and emerging trend of phytoremediation is going to be faster than the past decade, consequently, future studies may concern, on rhizoremediation augmentation as practical green method for enhancing efficiency of biodegradation of toxic or hazardous pollutants.

SOME RELEVANT TERMS

Most of the remediation studies only refer to a specific part of their methods, applied material or augmented process, meanwhile, other studies may carry out in the same case without concerning on the quarters of effective responses. Presenting a standard method of classification or categorizing pathway has not been fully addressed yet. For achieving a worldwide accepted operating and comparing system in remediation studies, some of the baseline and frameworks should be described.

Generally, the main effective technique could be considered as the title of studies. If rhizoremediation plays the main role in pollutant removal (maximum volume removal) and planting augmentation cause increasing in rhizoremediation, the combination should be named as "rhizoremediation augmented by phytoremediation". If a plant species are controlling pollutant removal and this system can be more effective (but no more than plant efficiency) with additional microorganisms consequently this combination may name" phytoremediation augmented by rhizosphere microbial community". Table 1 is a summary of techniques which related to remediation.

RHIZOREMEDIATION OF SOME HAZARDOUS POLLUTANTS

Polycyclic Aromatic Hydrocarbon (PAH)

Polycyclic aromatic hydrocarbon classified as an organic pollutant in the soil and recognized as one of the main origin of soil-persistent with specific physicochemical properties which review by (Mah et al., 2013). Based on the report of (Eisler, 1987) there was no regulation limit for PAH till 1987 and first reported standard for PAH published by the US. Environmental Protection Agency in 2000 for aquatic media in the 0.0001 mg /L3 (Kim et al., 2013) and it clearly illustrate the hazardous rate of these toxic, carcinogenic and mutagenic compounds for human with such extremely low concentration standard limit. The area of spreading could reach to kilometer scale and their decontamination is a real practical project (Villanneau et al., 2013), these polycyclic compounds are eligible to present a heterogeneous distribution causes to spread hotspot especially in small scales (De Fouquet, 2011; Hybholt et al., 2011). Even after 14 years passed from pollutants removing, there can be a lasting impact on microbial population and organic matter in soil (Ji et al., 2013). Tar balls which are lumps of solidified in PAHs play main role in soil texture as main physiochemical properties of growing media (Amellal et al., 2001), soil mineralization (Choudhary & Mantri, 2000) and content or availability of organic matter in soil (Cornelissen et al., 2005). Degradation of PAH highly depended to microbial enzymes, which, have a critical role

Table 1. Applied techniques through the remediation process

Technique	Description	Reference
Natural attenuation	Reduction of pollutants concentration and mass by relying on the assimilative capacity of groundwater	(Wiedemeier, 1999)
Biomineralization	The process in which living organisms form minerals that cause metal removal and detoxification	(Gadd & Pan, 2016)
Cyanoremediation	Tinny plants for metabolize organic compounds and metal, nitrate and phosphates absorption	(Gothalwal & Chillara, 2012)
Mycoremediation	Bioaccumulation of heavy metals with fungi in water and soil	(Aalapti Singh & Gauba, 2014)
Phytostabilization	Decreasing mobility of toxic compounds by plants roots rhizosphere	(Yang et al., 2014)
Biostimulation	Organic matter amendment in the rhizosphere for increasing microbial activities then hydrocarbon degradation	(Silva-Castro et al., 2013)
Bioaugmentation	The addition of biocatalysts material(s) to promote degradation of target pollutant(s)	(Lyon & Vogel, 2013)
Phytoextraction	Accumulation of pollutants in harvestable parts of biomass	(Bhargava, 2012)
Phytostabilization	The use of metal tolerant plants for reduction of solubility, mobility, and toxic metal inactivating in rhizosphere area	(Rajkumar & Freitas, 2013)
Phytofiltration	Sequestration of pollutions by aquatic plants species from contaminated water	(Ghosh & Singh, 2005)
Dendroremediation	Application of trees for cleaning terrestrial environment	(Erakhrumen, 2015)
Phytovolatilization	Converting pollutants to volatile form and their subsequent liberate them to the atmosphere	(Marques et al., 2009)
Phytodegradation	Organic xenobiotic compounds through this method degraded by plant tissue enzymes	(Khan, 2005)
Rhizofiltration	accumulation of contaminants in Aquatic plants tissues	(M. Yang, Jawitz, & Lee, 2015)
Rhizodegradation	Degradation of organic compound by microbial communities in the rhizosphere	(Tangahu et al., 2011)
Phytodegradation	Degradation organic pollutants by plant roots usually in aquatic media	(Al-Baldawi, Abdullah, et al., 2015)
Phytotronsformation	Highly toxic contaminants through this technique transformed to low toxic compounds	(McCutcheon & Schnoor, 2003)

in catalytic actions. Plant selection for oil, hydrocarbon biodegradation should be done by considering the high activity of basic oxidative enzymes (Khatisashvili et al., 2015). Some of the enzymes in PAH degradation are, dioxygenase (Pieper et al., 2004), monooxygenase (McLean et al., 2005), nitroreductase (Liu et al., 2007), peroxidase (Aken et al., 2000) and dehalogenase (Bastida et al., 2015; Mena-Benitez et al., 2008). Figure 2 shows a schematic process and mechanism of PAH biodegradation in rhizosphere based on the literature. Through these processes which start by releasing amino acid, sugars and organic acid from plant roots, application as carbon source by microorganism community, microorganism growth and development, increasing bioavailability of PAH by microbe and final selective biodegradation of PAH contaminants.

Total Petroleum Hydrocarbon (TPH)

Total petroleum hydrocarbon could be considered as one of major environmental persistent organic pollution, which caused by human activities. These compounds are toxic to many organisms in terrestrial, aquatic and air. Contamination of soil and aquatic media with TPH caused by all process of extraction raw petroleum, transportation of raw and produced petroleum, refining used related products of petroleum and final consumption of this fossil fuel. Since of a wide range of transportation and extremely application in social living area and industries all worldwide, these pollutants have deep and huge environmental footprint. The remediation process for these organic, persistent contaminants is a time taker and expensive

Figure 2. Mechanism of PAH biodegradation in rhizosphere
Collected from (Bezza & Chirwa, 2016; Blyth et al., 2015; Cébron et al., 2011; Chaudhry et al., 2005; Gerhardt et al., 2009; E. Joner et al., 2002; E. J. Joner & Leyval, 2003; Kuiper et al., 2004; Kumar et al., 2006; Storey et al., 2014; Sun et al., 2013; Vančura & Hovadik, 1965)

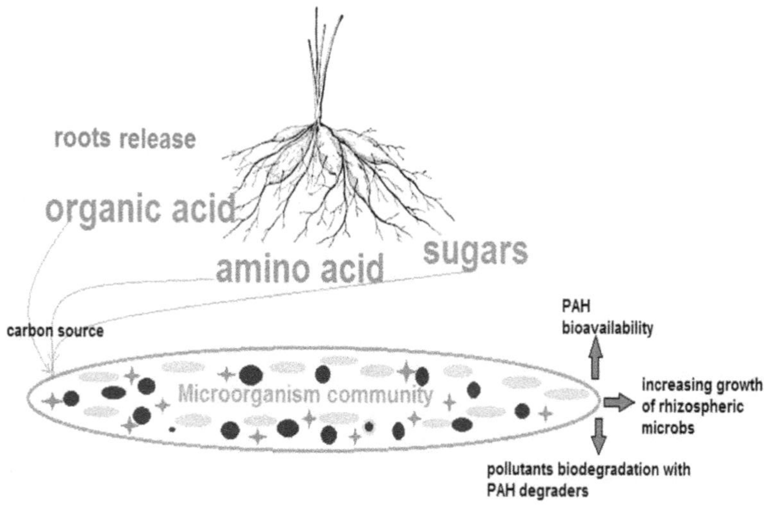

process (Huang et al., 2005). Rhizoremediation is well fitted for TPH-contaminated soils furthermore, this eco-friendly technology will present perfect results for low concentration of TPH (Zhuang et al., 2007). Table 2 shows numbers of recent studies on organic pollutant removal by rhizoremediation process. The major removal mechanism for TPH is rhizosphere positive effect on enhancing biodegradation of contaminants around the rhizosphere (Gerhardt et al., 2009; Irvine et al., 1993). Bioaugmentation assisted

Figure 3. Comparison of the efficiency of microorganisms and plants in heavy metals removal
Improved from (Agnello et al., 2015).

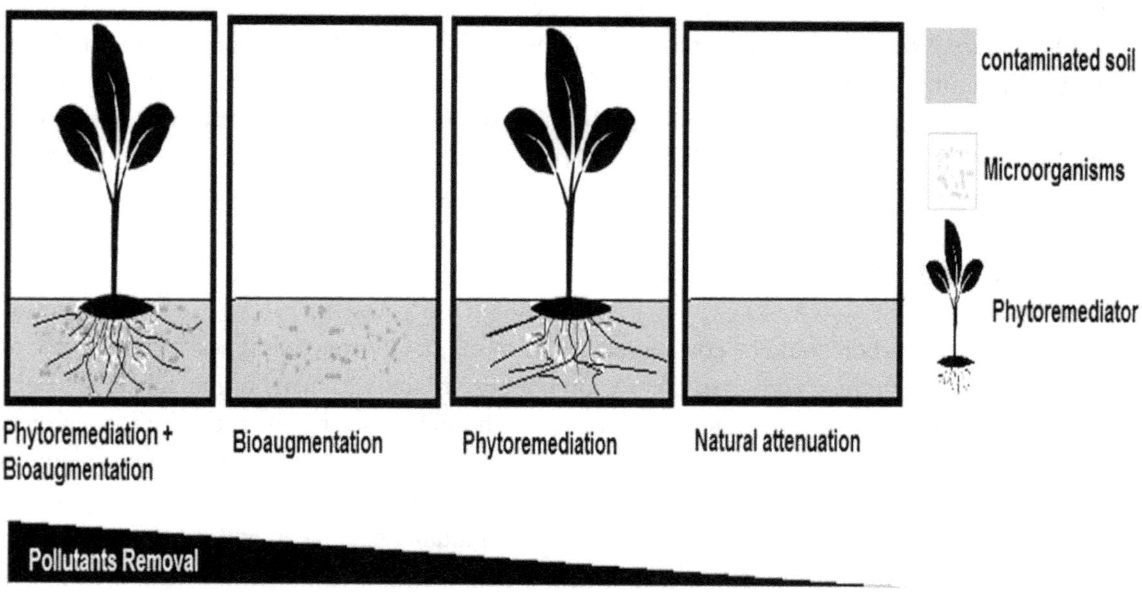

phytoremediation (plant+microorganisms) can cause highest degradation followed by bioaugmentation (only microorganism), phytoremediation (only plant) and control (no plant + no microorganism). A fungal rhizoaugmentation study on petroleum pollution (Mohsenzadeh et al., 2010) which carried out as first report on *P. aviculare* ability in petroleum–polluted soil treatment. Research of Agnello et al. (2016) confirmed aforementioned order. Figure 3 shows a summary of improved pollutant removal study by combination phytoremediation and bioaugmentation in comparison with single treatment techniques.

Polychlorinated Biphenyls (PCBs)

Production of PCBs is going to reach about 100 years historical background from 1929. These compounds widely used as coolants and dielectric fluids at home and industrial electrical apparatus (Robertson & Hansen, 2015). Polychlorinated biphenyls (PCBs) are oily liquids synthesized by 1 to 10 chlorine atoms on carbon-carbon bond aromatic rings. This synthetic chemical compounds are serious environmental pollutants which have a hazardous effect on both human and natural ecosystem and their production banned in many countries (Ross, 2004). In the commercial market, there are 20-60 configuration of these versatile compositions which theoretically could be 209 congeners (Ohtsubo et al., 2004). Figure 4 shows a general chemical structure of PCBs.

Table 2. Numbers of recent studies on organic pollutants removal by rhizoremediation process

Plant	Augmentation	Pollutant	Comments	Reference
Zea mays L.	Microbial community exposed for roots	PAH	Xenobiotic degradation by root exudates beside soil microorganism growth	(Yoshitomi & Shann, 2001)
Legume mixture Grass mixture Wooden trees	Microbial enrichment NPK fertilizer Compost extract	Diesel fuel	Legume plant removed all pollutants Soil amendment have no significant effect	(Palmroth et al., 2002)
Herbal mixture Tree (poplar)	Microbial combination Carbon source	Diesel fuel	Trees are more sensitive than herbal mixture Some trees have high tolerance to pollutants	(Tesar et al., 2002)
Trifolium repens L. *Lolium penne* L.	Microbial inoculants	PAH	Plant growth and vigor improved	(Johnson et al., 2004)
Festuca arundinacea (Tree)	Land-farming Degrading bacteria PGPR	TPH	Plant species and PGPR are key factor in successful phytoremediation	(Huang et al., 2005)
Festuca arundinacea	Land-farming Degrading bacteria PGPR	PAH	Rapid and high removal carried out for PAH	(Huang et al., 2004a)
Medicago sativa L.	Rhizobacteria	Diesel fuel	Plant roots presence in contaminated soil have greater effect on bacterial activity than absence of diesel contamination	(Kim et al., 2006)
Echinochloa crusgalli (grass)	Rhizosphere treatment	Diesel fuel	Rhizospher positively influenced bacterial community density	(J. Kim et al., 2007)
Zea mays L.	*Gordonia* sp. S2RP-17	TPH	Diesel-degrading and growth promoting done by rhizobacteria	(Hong et al., 2011)

Figure 4. General chemical structure of chlorinated biphenyls
Passatore et al., 2014.

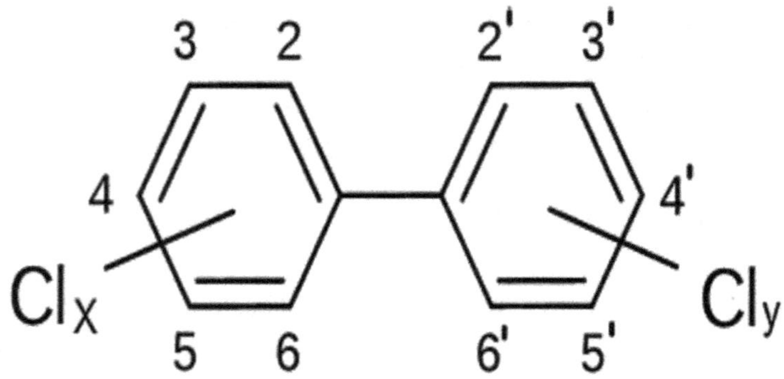

PCBs are specific contaminants with following characteristics:

- Low solubility in water,
- High tendency to adhere to soil,
- Recognized as toxic compounds,
- Biologically recalcitrant,
- High accumulation ability in human adipose,
- Widely dispersed in a natural environment,
- Stable hydrophobic, thermal and chemical,
- Co-metabolically transformed to chlorobenzoic acid bacteria.

The most common, widely practically applied treatment method for PCBs is incineration in high temperature, which depends on high expensive energy and materials furthermore, this method decreases soil fertility and contain toxic compound formation (dioxin) as well as low efficiency of treatment (Ross, 2004). Bioremediation is one of the most effective alternations for degradation of PCBs contaminated soils. For bioremediation of contaminated environment with PCB compounds, several treatment approaches have been developed based on the microbial augmentation. These approaches are designed to reach higher solubility of PCBs and eligible them for cell entrance and enzyme engineering expression. The process of dechlorination highly polluted soils arranged as follows anaerobic steps (Vasilyeva & Strijakova, 2007; Zanaroli et al., 2006):

1. Presence of chloroorganic compounds as pollutant in soil,
2. Available anaerobic bacteria switch to dechlorinating,
3. The lag period of time (one to several months),
4. Obligate anaerobic performance (sulfate reducer, methanogens),
5. Presence of: optimum level of hydrogen, maintains the anaerobic condition and carbon source,
6. Dehalogenase replace chlorine atoms one by one with hydrogen in the PCB.

Cause of the high toxicity of PCBs and aforementioned characteristic, bioremediation treatment techniques as an emerging issue, concentrate on enhancing removal efficiency and decreasing the time of the degradation process. Based on the literature analysis, efficiency of PCBs bioremediation strongly depends on contamination degree and characters. Aerobic biodegradation in moderate contaminated soils can reach to 40-75% removal efficiency (Vasilyeva & Strijakova, 2007). Meanwhile, for fresh and high concentration of PCB pollution, anaerobic condition will be the suitable treatment method. Figure 5 shows anaerobic and aerobic dechlorinating through the other remediation process such as rhizoremediation, phytoremediation, and evapotranspiration. The process that plants can increase PCB degradation are rhizodegradation and translocation of organic contaminants (Leigh et al., 2006; Singer et al., 2003; Xu et al., 2010). Table 3 presents the numbers of PCB researches characteristics in literatures. Development stage of the plant species and characteristics of any species, cause changes in root exudate composition. Consequently, these changes which provide specific exudate, will have high promoting or inhibiting effects on microbial communities around the plant rhizosphere (Berg et al., 2002; Smalla et al., 2001). Thus, proper plant selection and isolation of microbial community are critical stage for both targets of their own specific effect on decontamination and their interaction effects on each other.

Table 3. Numbers of PCB researches characteristics in literatures

Bacteria/Plants	Effective Augmentation	Duration/Finding	Reference
Arthrobacter sp. BIB *Ralstonia eutrophus* H850	34 times application of surfactant	18 weeks 55-59% removal	(Singer et al., 2000)
Medicago sativa	*Sinorhizobum meliloti*, strain A-025	Positive effect of augmentation in early stage of transformation	(Mehmannavaz et al., 2002)
Brassica nigra	PCB degrading bacteria	*B. nigra* reached to 61% removal alone	(Singer et al., 2003)
Medicago sativa L. *Lathyrus sylvestris* L. *Lespedeza cuneate* *Wagner pea*	Plant-bacteria interaction	38-82% removal	(Chekol et al., 2004)
Nicotiana tabacum *Solanum nigrum* *Armoracia rusticana* *Medicago sativa* *Salix*	Plant-bacteria interaction	PCBs Removal % depends on plants choice of suitable Plant species is crucial for effective cleaning PCBs. 10 years	(Mackova et al., 2009)
Nicotiana tabacum *Solanum nigrum*	Plant-bacteria interaction	Different strategy of bacterial isolation reported	(Kurzawova et al., 2012)
Glomus caledonium L *Lolium perenne*	Earthworms Plant-bacteria interaction	180 days 58.4-79.5% removal	(Lu, Lu et al., 2014)
Medicago sativa *Brassica juncea* *Pseudomonas* sp. *Stenotrophomonas* sp.	Plant secondary metabolites Plant-bacteria interaction	Positive effect of Plant secondary metabolites Confirmed	(Pino et al., 2016)
Mesorhizobium sp.ZY1 *Austragalus sinicus*	Synergistic augmentation of bacteria and plant stimulate degrade	Bacteria 20.5% Plant 23% Plant + bacteria 53.1% removal	(Teng et al., 2016)
Panicum virgatum *Populus deltoids* x *nigra* DN34	Marginally aerobic dechlorination	32 week Planted coupled with bacteria had greater removal	(Meggo et al., 2013)
Panicum virgatum *Dehalogenimonas* sp.	Geobacter identified in rhizosphere	Organohalide-respiring bacteria have specific effect even with 1% of total bacteria	(Liang et al., 2015)

Figure 5. Anaerobic and aerobic decholorination through the other remediation process
Horwat et al., 2015.

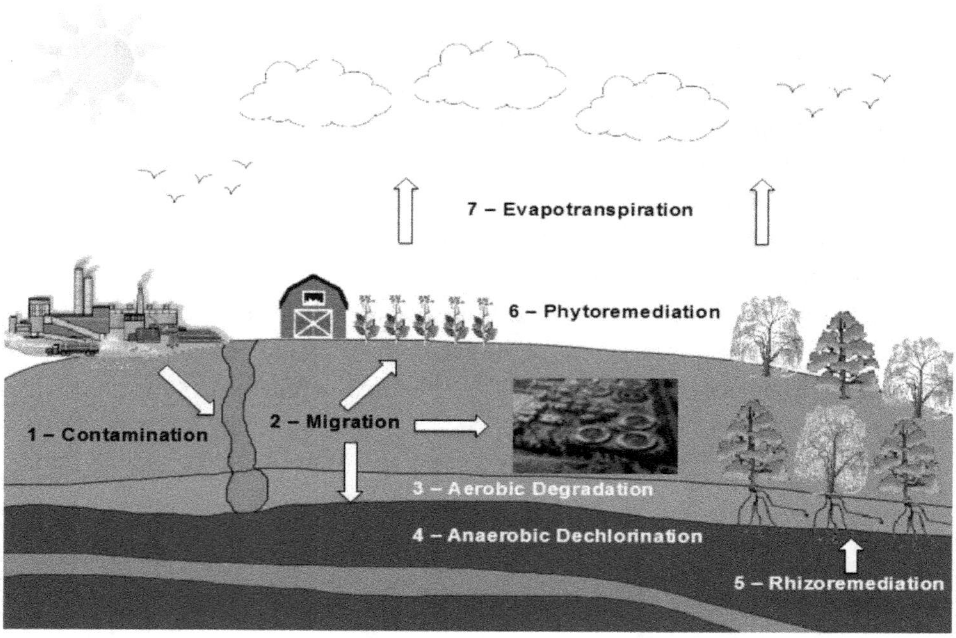

Heavy Metals (Trace Elements)

Heavy metals (both metals and metalloids) or recently known as track elements are specific no degradable pollutants with density higher than 5-6 gr/cm3RR. These elements are biologically essential (Co, Cr, Mn, Cu, Zn) or non-essential elements (Cd, Hg, Pb). All track elements in exorbitant concentration are toxic for human, animals and plants (Park et al., 2011). Mining, smelting, pesticides, fertilizers batteries, medical waste, and fuel combustion (coal/petrol) are the main sources of track elements. Metallic elements cause pollutant for soil, water and air. Application of microorganisms and plants in order to remove hazardous contaminants from air, soil and water living, environmental system, known as biological treatment (Singh et al., 2009). Environmentally friendly, natural process, low side effect environmental footprint, low cost of operation and maintenance and positive acceptance by the public are attraction and merits of biological methods (Walton & Anderson, 1990) and long operation period, low efficiency, highly dependent to environmental condition and sensitivity to toxic compounds especially in high concentration are drawbacks of biological treatment (Ford et al., 2007). This is the main reason that biostimulation and bioaugmentation are the central character in successful lab scale studies and bystander in field application (Lin et al., 2016). Interdisciplinary cooperation between microorganisms (as rhizosphere operator leading), plants, augmentation factors (as a condition optimizer for optimal remediation and degradation) and operational factor (as process management) are essential to earning higher efficiency in a shorter time. The solubility, mobility, and bioaccumulation of metallic elements depend on proliferation of microbial community in the soil, and plant-related factors as one side and trace element properties as the other side of a rhizoremediation coin (B. Robinson et al., 2005). As far as literatures indicate, hyperaccumulator plant species (HPS) are the ideal selection for track element removal even in very high concentration. Therefore, these plants could be considered as central leader in rhizoremediation. Nevertheless, their slow growing rate, single element's hyperaccumulating ability and low biomass production are major disadvantages in the decontamination process. On the other hand, adaptation with rhizosphere microorganism community, root release substrate rate, final depth of roots growing, are other major requirements for joining HPS in rhizoremediation process. Proper integration of HPS and arbuscular mycorrhiza fungi have been reported as a key solution to harnesses of metallic elements by (Miransari, 2011), right bacteria/consortia and inoculation to seed/roots of proper HPS (Kamaludeen & Ramasamy, 2008) as a wide perspective for rhizoremediation in future studies. By considering the advantages of phytoremediation (Farraji et al., 2016), recent researchers are narrowing down on effective parameters through the rhizoremediation such as enzymes (Shukla et al., 2013), root exudation (Rohrbacher & St-Arnaud, 2016), plant growth promoting bacteria (Ullah et al., 2015) and organic amendment (Park et al., 2011). Phytoremediation by edible vegetable such as spinach (Farraji et al., 2014) just for heavy metal removal, improved to mixed culturing HPS with edible cabbage (Niu et al., 2015) or Welsh onion (Wang et al., 2015) to safe and economic vegetables production this improvement earned for supplying of the rhizosphere. Thus, rhizoremediation is converting to a management issue in environmental science. Extensively reviewed literature shows multi-objective studies are going to indicate future arrow for decontamination technology by rhizoremediation. Expecting high removal efficiency in dream range or amazing short time treatment is future vision of this emerging area of environmental science. Nevertheless, aquatic phytoremediation recently could reach to 80% (Wang et al., 2015) and 99% (Jian-guang et al., 2014) removal of arsenic with a hyperaccumulator fern. Table 4 illustrates most recent researches on heavy metal removal by rhizoremediation. Rhizofiltration, hyperaccumulation, dendroremediation, phytostabilization, phytodegradation, mycoremediation, cyanoremediation, biostimulation, biomineralization is the methods that could be used as amended in shallow contamination pollutants sites through the microbial communities assisted phytoremediation (Mani & Kumar, 2014). Planning for amendment through these green eco-friendly treatment methods depends on several circumstances and practical factors. There are two major factors which the other process and responses highly depended to presence and suitable operation of them. Plants and microorganisms are aforementioned factors as fundamental of these treatment technologies. In all rhizoremediation research, the first objective should be pollutants removal in sustainable and rapid method, then concentrating on enhancing efficiency by organic amendment, adjusting pH, and inoculation of plant growth promoting bacteria.

Through the process of rhizoremediation, some of specific promoting mechanism could be distinguished as mentioned in literatures. Microbial resistance to cadmium as a metallic element well-known as efflux mechanism which is energetically defended. Effect of two genes (czr and cadA) as a resistance element to cadmium and zinc reported for *Pseudomonas aeruginosa* (van der Lelie et al., 1999) and for cadmium (Lee et al., 2001). Cadmium phosphate, cadmium sulfide and cadmium carbonate are nontoxic compounds of cadmium. Microbial transformation is another mechanism which toxic form of heavy metal (cadmium) converts to a nontoxic form (Aiking, Govers, & Van't Riet, 1985). The historical background of these mechanisms belongs to heavy metal bioremediation, which is a highly cited issue with remediation technique. A mechanism that involved through enhancing resistance to toxic metallic elements are Microbial detoxification to mercury salts (Gavis & Ferguson, 1972; J. B. Robinson & Tuovinen, 1984; Summers & Silver, 1978), volatilization to Hg after chemical transformation (Ichirö KŌMURA et al., 1971; Ichirö KŌMURA & Izaki, 1971; Meissner & Falkinham, 1984; Olson, Porter, Rubinstein, & Silver, 1982; J. B. Robinson & Tuovinen, 1984), volatilization to methyl mercury after chemical transformation (Hamdy & Noyes, 1975; Pan-Hou & Imura, 1982; J. B. Robinson & Tuovinen, 1984; Vonk & Sijpesteijn, 1973) and membrane permeability decreasing (Pan-Hou, 1981). Finally, in

Table 4. Most recent researches on heavy metal removal by rhizoremediation 1) Dichlorodiphenyltrichloroethane (DDT) 2) 2, 4, 5-trichlorophenol (TCP)

Hyperaccumulator	Microorganism/Augmentation	Pollutants	Description	Reference
Sedum alfredii	*Bacillus subtilis* *Paracoccus* sp. *Flavobacterium* *Pseudomonas* sp.	Cadmium Carbendazim	Carbendazim degrading bacteria strains	(Xiao et al., 2013)
Sedum alfredii	*Brkholderia cepacia*	Cd / Zn Phenanthrene	Metal tolerance Translocation	(Li & Wong, 2012)
Sedum alfredii	*Pseudomonas* sp. DDT-1	Cadmium DDT	DDT-degrading microbes	(Zhu et al., 2012)
Pteris vittata *Pteris multifida*	Not mentioned	Arsenic	As (III) oxidation As (III) and As (V) Absorb	(Chien et al., 2015)
Pteris vittata	*Psedumonas* sp.	PAH Arsenic	Enhance uptake and Translocation for arsenic Degrading PAH	(Feng et al., 2014)
Pteris vittata	Not mentioned	Arsenic	Non-aeration Low-nutrition Cause 80% rem	(Wang et al., 2015)
Brassica napus	*Pantoea* sp. FC1.	Cr (VI) Phenol	Cr (VI) to Cr (III) PGPB effect solubilizing	(Ontañon et al., 2014)
Solanum nigrum L	Glycine+ glutamic acid+ cysteine	Cadmium	Edible vegetables production in polluted soil	(Niu et al., 2015)
Sedum alfredii co-planting with rygrass or castor	Not mentioned	PAHs Cd, Zn, Pb	Edible production in polluted soil	(Wang et al., 2015)
Clitocybe maxima	Not mentioned	Cu, Cd + TCP	Metal extraction TCP dissipation	(H. Liu et al., 2015)
Astragalus bisulcatus	*Stenotrophomonas* sp. (SelTE02)	Selenite	Reduce soluble Se (IV) to insoluble & unavailable Se (0)	(Di Gregorio et al., 2005)

the last decade of the past century (Prasad, 1995), following mechanisms mentioned for heavy metal tolerance in polluted media which planted for decontamination:

- Binding trace elements to cells of plants,
- Decreasing transportation of metallic elements through the cell wall,
- Forming chelation,
- Compartmentalization,
- Efflux inactive position.

Rhizoaugmentation

There is no single method which could be considered as the solution for rhizoremediation of organic and inorganic pollutants (Gomes et al., 2013). Even combining several adsorbent and plant species in recirculated aquatic media could not reach high efficiency in decontamination (Mojiri et al., 2016). Presenting successful remediation of contaminated site highly depend on suitable bacterial communities and plant species selection, proper design and on time bacterial inoculation, different implementations of plant species, and concentration on indigenous combination effective responses should be the pathway of rhizoaugmentation. Through the amendment, which will prepare effective rhizoremediation for pollutant removal, organic amendment has a fundamental role. Organic amendment recognized as a suitable method for enhancing rhizoremdiation efficiency as well as energy saving and easily practicable (Chen et al., 2016), even compost addition can be considered as perfect bioaugmentation as a partner for degrading microorganism community through the organic pollutant treatments that well reviewed by (Kästner & Miltner, 2016). The mode of action in organic amendment may consider enzyme activity with application of sewage sludge (Balík et al., 2007), municipal solid waste (Crecchio et al., 2004), humic acid and (Canellas et al., 2002). Organic augmentation may carry out for enhancing microbial activity (Baudoin et al., 2003). Table 5 shows organic amendment for rhizosphere augmentation in relation to heavy metal bioremediation.

Different treatment method in the rhizosphere media depends on soil texture and physiochemical characteristics such as redox and pH, contaminants concentration and bioavailability and key substrate absence (Kuiper et al., 2004). Any amendments for rhizosphere will cause enhancing efficiency of bioremediation. Organic amendments as a well-known augmentation for soil properties could be a significant and suitable through the bioremediation process such as immobilization, volatilization, reduction and rhizosphere modification. Figure 6 shows organic amendments in rhizosphere.

Table 5. Rhizoaugmentation in order to heavy metals bioremediation by bioaugmentation

Bacteria	Pollutant	Plant	Augmentation	Reference
Pseudomonas aeruginosa	Cd	Not mentioned	Plant growth promoting	(Ganesan, 2008)
Rhodobacter fascians	Petroleum	*Medicago sativa Lolium perenne*	Biosurfactants producer	(Akhundova & Atakishiyeva, 2015)
Fungal strains (Fusarium species)	Petroleum	*Polygonum aviculare*	Fusarium species is the most effective factor in biodegradation	(Mohsenzadeh et al., 2010)
Pseudomonas sp. *Rhodococcus* sp.	Oil hydrocarbon	12 species (grasses, shrubs and trees)	Biosurfactants	(Khatisashvili et al., 2015)
Microbacterium sp. KL5 *Candida tropicalis* C10	PAH Cd, Zn	*Lolium perenne* with *Seduce alfredii*	Enzyme activity Removal efficiency	(F. Chen et al., 2016)
Bacilus sp.SBER3	PAH	*Populus deltoides*	Inhibiting growth of inhibitor fungi and Growth promotion	(Bisht et al., 2014)
Bacteria Fungi	PAH	Alfalfa *Lolium perenne*	Root fungi and bacteria growth	(Bourceret et al., 2015)
Pseudomonas sp. *Cupriavidus* sp. *Rhodococcus* sp.	Nitrophenol	*Spirodela polyrhiza*	Growth promoter	(Kristanti et al., 2014)
Endophytic bacteria L14 (EBL14)	Cu, Cd, Pb	*Solanum nigrum* L.	Uptake and Precipitation	(Guo et al., 2010)
Gordonia sp.S2RP-17	Diesel	*Zea mays*	Growth promoter	(Hong et al., 2011)

Figure 6. Organic amendment through the rhizoremeiation and its consequences on plants, microorganisms and pollutants
Park et al., 2011.

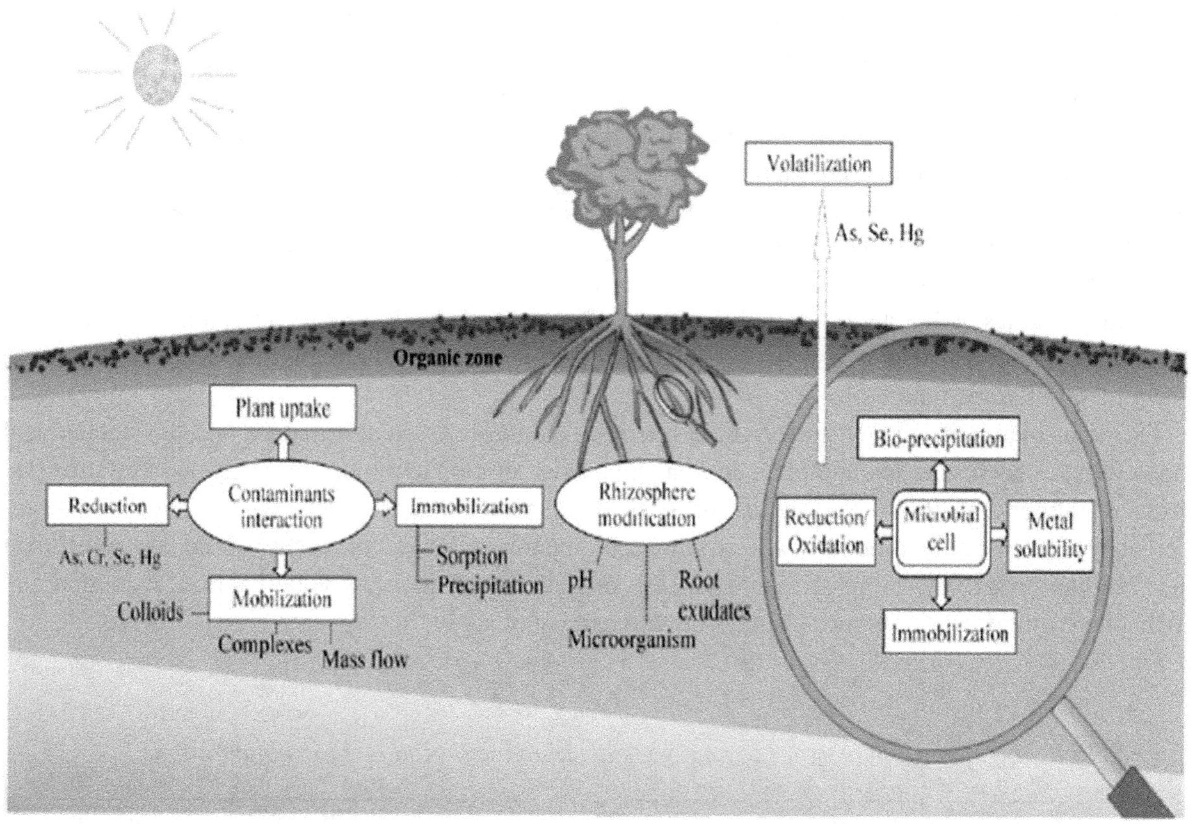

PLANT SELECTION FOR RHIZOREMEDIATION

Plants are the main base of phytoremediation and rhizoremediation as an expert method which supplied by rhizosphere microbial communities highly depended to plants as host of biodegradation bacteria. The results of many researches on plant application for phytoremediation indicate that the fundamental components of a plant species response to metallic elements including up taking accumulating process and toxic effect decreasing could be changed depending on the condition of experimental materials and method of application in aquatic media or contaminated soils. Furthermore, many plant species are quite sensitive to most of contaminants, including TPH (Huang et al., 2004), rhizoremediation properties are varied among the plants (Mackova et al., 2006) and hyperaccumulator plant species cannot be applied for media contain several pollutants (Wu et al., 2006). Thus plant selection for rhizoremediation by itself is a case of multifactorial decision and it will be a hot case of managing when pollutant specification, climate condition, economical factors should be included. In lab scale study's objective concentrate on finding specific parts of knowledge and after full assessment by several researchers, we can make practical plan. Enhancing removal efficiency, decreasing treatment required time, accumulating on harvestable part of plants and finding suitable augmentation for aforementioned objectives. The last subject is a specific subtitle in rhizoremediation which is emerging topic for current and future studies entitle to rhizoaugmentation or engineering plant-microbe. Following tips are mentioned in literature as specific characteristics of phytoremediators in rhizoremediation process.

- Plant selection should be carried out from natural polluted media (Farraji.H, 2014).
- Contain transgenic ability for reaching higher quality of phytoremediation (Hsieh et al., 2009).
- Capable for engineering plant-microbe symbiosis (Wu et al., 2006).
- High tolerance to climate variation and adaptive to field and greenhouse studies (Danh et al., 2009; Sylvestre et al., 2009).
- Mine species (*Fragaria vesca, Epilobium parviflorum, Calamagrostis arundinacea* and *Stachys sylvatica*) are qualify nominates (Zabłudowska et al., 2009).
- High accumulate rate minimum 1-3% metals in their dry weight (Cunningham & Berti, 1993; Cunningham & Ow, 1996).
- Combining artificial adsorbent in growing media of wetland macrophytes (Mojiri et al., 2016).
- High biomass production ability (Kvesitadze et al., 2006).
- Advantages and disadvantages of decontamination by macrophytes (Farraji et al., 2016).

Application suitable plant species and plant type have a main role in the success of rhizoremediation. Easy application, propagation, culturing, sampling and short life period in herbs caused to heavily present in literatures. There are about 45 plant families that categorized as hyperaccumulator (Ghosh & Singh, 2005). Most of these hyperaccumulator species belongs to herbal plant species (Brassicaceae, Asteraceae, Lamiaceae and Fabaceae) or most of plant species in family are herbal (Euphorbiaceae and Scrophulariaceae). Consequently, most of the researches have been carried out by herbal plant species. Meanwhile, higher biomass production and deeper root in trees make them as eligible nominated for phytoremediation as a perennial phytoremediator which every year increases metabolites around the roots and present higher stability improving terrestrial media. See for review e.g. (Cook & Hesterberg, 2013; Donzelli et al., 2013). These review indicate that trees are great in nitrogen participant and for water competitors, grasses are perfect. There are specific tips for plant application as following:

- **Terrestrial Plants:** In most of the polluted area.
- **Halophyte Plants:** Seaside phytodesalination (*Suaeda salsa*).
- **Crop Plants:** In agricultural area (*Glycine max, Zea mays, Oryza sativa, Vigna radiate*).
- **Ornamental Plants:** In cities, living area (*Mirabilis jalapa, Impatiens balsamina, Canna indica*).

To sum it up, considering local climate, available water sources, concentration of contaminants, deep of polluted soil and operational limits should be assessed in decontamination plan. As a matter of fact, rhizoremediation process strictly should be started from polluted sites with considering all local adaptations and available resources. Plant application in recent researches of hydrocarbon polluted soils collected in Table 6.

Table 6. Terrestrial plants for hydrocarbon (petroleum) phytoremediation

Scientific Name	Herb	Tree	Country	Reference
Brachiaria brizantha Centrosema	X		Venezuela	(Merkl et al., 2004)
Pennisetum glaucum *Sorghum sudanense* *Brachiaria ramose* *Aeschynomene Americana*	X X X X		USA	(Kirkpatrick et al., 2006)
Lolium arundinaceum *Lolium multiflorum* *Cynodon dactylon*	X X X		USA	(White Jr et al., 2006)
Zea mays *Glycine max*	X X		USA	(Issoufi et al., 2006)
Tectona grandis		X	Nigeria	(Mary Agbogidi et al., 2007)
Polygonum aviculare	X		Iran	(Mohsenzadeh et al., 2010)
Lolium multiflorum	X		USA	(Alarcón & Zuberer, 2008)
Astaragalus membranaceus *Aeschynomene indica* *Panicum bisulcatum*	X X X		Republic of Korea	(S.-H. Lee et al., 2008)
Sebastiania commersoniana		X	Brazil	(Ramos et al., 2009)
Mirabilis jalapa	X		China	(Peng et al., 2009)
Impatiens balsamina	X		China	(Cai et al., 2010)
Canna indica	X		Thailand	(Boonsaner, 2011)
Chromolaena odorata	X		South Africa	(Atagana, 2011)
Cyperus rotundus	X		India	(Basumatary et al., 2012)
Cyperus brevifolius	X		India	(Basumatary et al., 2012)
Biden pilosa	X		Taiwan	(Kuo et al., 2014)
Azadirachta indica *Gmelina araborea* *Michelia champaca* *Tectona grandis*		X X X X	India	(Yenn et al., 2014)
Lolium perenne *Leptochloa fusca* *Brachiaria mutica* *Lecucaena leucocephala* *Acacia ampliceps*	X X X	X X	Pakistan	(Fatima et al., 2015)
Suaeda salsa	X		China	(Gagaoo et al., 2014)
Leucanthemum vulgare	X		Iran	(Noori, 2014)
Bassia scoparia	X		Egypt	(Moubasher et al., 2015)
Vigna radiate *Oryza sativa*	X X		India	(Roy et al., 2014)
Leucanthemum vulgare	X		Iran	(Noori, 2015)
Cynodon dactylon *Frstuca arundinacea* *Lolium multiflorum*	X X X		USA	(Savin et al., 2015)

ROLE OF PLANT GROWTH PROMOTING RHIZOBACTER (PGPR) IN RHIZOREMEDIATION

The unique organism which can live almost everywhere are microorganisms (Maier et al., 2009). For fungi, bacteria protozoa and algae rhizosphere is the specific ecosystem with a specific association of plant roots (Whipps, 2001). Forming biofilm of microbial communities on the plant root surface starts relationship between plants and microorganisms. Rhizobacteria which play main role in rhizoremediation can be classified to deleterious rhizobacteria (DRB) that may inhibit plant growth and plant growth-promoting rhizobacteria (PGPR) (Nehl et al., 1997). Rhizosphere bacteria can provide nutrients for plant roots and may directly operate bacterial degradation contain transformation, volatilization and rhizodegradation. Plant growth promotion microorganisms can effect on some specific factors such as soil chemistry, solubility of metals and contamination level. Klebsiella, Serratia, Flavobacterium, Bacillus, Erwinia, Enterobacter, Beijerinckia, Burkholderia and Gluconacetobacter are some of well-known genera of PGPR bacteria. Plant growth-promoting rhizobacteria (PGPR) which classified as types of rhizobacteria play a critical role in rhizoremediation process (Ullah et al., 2015). A clear process explained by (Johnson et al., 2005; Koo & Cho, 2009; Zhuang et al., 2007) for PGPR role through the rhizoremediation. These process summarized in Figure 7.

CONCLUSION AND FUTURE PROSPECT

Phytoremediation has several types of application which depends on pollutants specification and characteristics of phytoremediator plants. Consideration of non-biodegradable (inorganic) and degradable

Figure 7. Mechanism of PGPR effect on plant roots rhizoremediation process

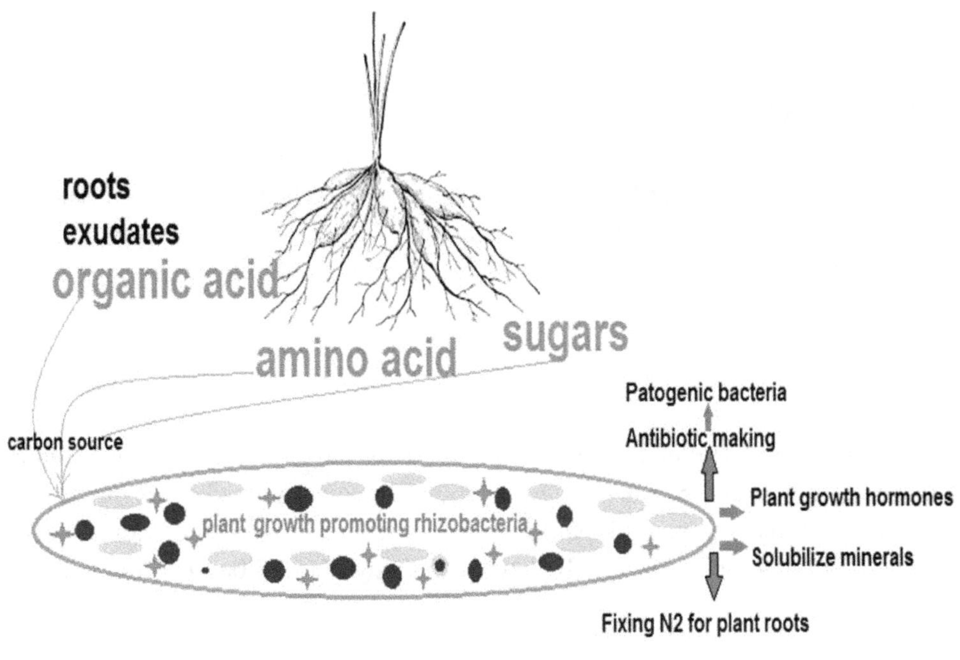

(organic) pollutants versus sensitive or HPS in the presence or absence of promoter and inhibitor factors in remediation media, target of studies could be planned. On the other hand, rhizosphere is the fundamental interaction media for plant roots, microorganisms and physicochemical characteristics of soil in the presence of pollutants and biosurfactants. Several live organisms with a wide range of responses to effective elements and numerous interaction with other live partners gather together to presenting rhizoremediation as an environmentally friendly pollutant removal technique. Providing optimum operation for effective factors is very important. It seems rhizoremediation as a single treatment method will not concern of future research. With bearing in mind all or most effective factors and responses through the rhizoremediation, it will considered as "rhizomanagement" which is optimization of all involved (promoter and inhibitor) factors for achieving the highest efficiency in pollutant removal as well as concentrating on shortest required time in the cost effective pathway. Following prospects could be considered as future research areas:

- Enhancing efficiency of rhizoremediation for decreasing required treatment time period.
- Applicability of rhizoremediation for multi-contaminants (organic and inorganic).
- Concentrating on specific microorganism for macrophytes in aquatic media.
- Improving biomass production of hyperaccumulation plant species by finding particular rhizobacteria partner, plant growth promoter and nutrient management.
- Classification of plant species, rhizobacteria and growing media based on the most common pollutants.
- More concentration on ornamental plant species as suitable macrophytes for rhizoremediation in urban constructed wetlands.
- Application of multicultural system (perennial and herbal phytoremediator) in order to sustainable treatment and low-cost maintenance decontamination method.
- Improving selective macrophytes for safe food production in polluted rhizosphere with combining plant growth condition and soil decontamination as Agro-rhizoremediation.

ACKNOWLEDGMENT

This work is funded by Universiti Sains Malaysia under Iconic grant scheme (Grant no. 1001/CKT/870023) for research associated with the Solid Waste Management Cluster. Institute of Standard and Industrial Researches of Iran (ISIRI) highly appreciate for providing Ph.D. research opportunity.

REFERENCES

Agnello, A., Bagard, M., van Hullebusch, E., Esposito, G., & Huguenot, D. (2016). Comparative bioremediation of heavy metals and petroleum hydrocarbons co-contaminated soil by natural attenuation, phytoremediation, bioaugmentation and bioaugmentation-assisted phytoremediation. *The Science of the Total Environment*, *563-564*, 693–703. doi:10.1016/j.scitotenv.2015.10.061 PMID:26524994

Aiking, H., Govers, H., & Van't Riet, J. (1985). Detoxification of mercury, cadmium, and lead in Klebsiella aerogenes NCTC 418 growing in continuous culture. *Applied and Environmental Microbiology, 50*(5), 1262–1267. PMID:3911898

Aken, B. V., Stahl, J. D., Naveau, H., Agathos, S. N., & Aust, S. D. (2000). Transformation of 2, 4, 6-trinitrotoluene (TNT) reduction products by lignin peroxidase (H8) from the white-rot basidiomycete Phanerochaete chrysosporium. *Bioremediation Journal, 4*(2), 135–145. doi:10.1080/10889860091114194

Akhundova, E., & Atakishiyeva, Y. (2015). *Interaction Between Plants and Biosurfactant Producing Microorganisms in Petroleum Contaminated Absheron Soils. In Phytoremediation for Green Energy* (pp. 115–122). Springer.

Al-Baldawi, A., Abdullah, S. R. S., Anuar, N., Suja, F., & Mushrifah, I. (2015). Phytodegradation of total petroleum hydrocarbon (TPH) in diesel-contaminated water using Scirpus grossus. *Ecological Engineering, 74*, 463–473. doi:10.1016/j.ecoleng.2014.11.007

Alarcón, A., & Zuberer, D. A. (2008). Arbuscular mycorrhiza and petroleum-degrading microorganisms enhance phytoremediation of petroleum-contaminated soil. *International Journal of Phytoremediation, 10*(4), 251–263. doi:10.1080/15226510802096002 PMID:19260211

Amellal, N., Portal, J.-M., Vogel, T., & Berthelin, J. (2001). Distribution and location of polycyclic aromatic hydrocarbons (PAHs) and PAH-degrading bacteria within polluted soil aggregates. *Biodegradation, 12*(1), 49–57. doi:10.1023/A:1011909107858 PMID:11693295

Arnold, C. W., Parfitt, D. G., & Kaltreider, M. (2007). Phytovolatilization of oxygenated gasoline-impacted groundwater at an underground storage tank site via conifers. *International Journal of Phytoremediation, 9*(1), 53–69. doi:10.1080/15226510601139409 PMID:18246715

Atagana, H. I. (2011). Bioremediation of co-contamination of crude oil and heavy metals in soil by phytoremediation using Chromolaena odorata (L) King & HE Robinson. *Water, Air, and Soil Pollution, 215*(1-4), 261–271. doi:10.1007/s11270-010-0476-z

Balík, J., Pavlíková, D., Vanek, V., Kulhánek, M., & Kotková, B. (2007). The influence of long-term sewage sludge application on the activity of phosphatases in the rhizosphere of plants. *Plant, Soil and Environment, 53*(9), 375.

Bastida, F., Jehmlich, N., Lima, K., Morris, B., Richnow, H., Hernández, T., & García, C. et al. (2016). The ecological and physiological responses of the microbial community from a semiarid soil to hydrocarbon contamination and its bioremediation using compost amendment. *Journal of Proteomics, 135*, 162–169. doi:10.1016/j.jprot.2015.07.023 PMID:26225916

Basumatary, B., Bordoloi, S., & Sarma, H. P. (2012). Crude oil-contaminated soil phytoremediation by using Cyperus brevifolius (Rottb.) Hassk. *Water, Air, and Soil Pollution, 223*(6), 3373–3383. doi:10.1007/s11270-012-1116-6

Basumatary, B., Saikia, R., & Bordoloi, S. (2012). Phytoremediation of crude oil contaminated soil using nut grass, Cyperus rotundus. *Journal of Environmental Biology, 33*(5), 891. PMID:23734455

Baudoin, E., Benizri, E., & Guckert, A. (2003). Impact of artificial root exudates on the bacterial community structure in bulk soil and maize rhizosphere. *Soil Biology & Biochemistry, 35*(9), 1183–1192. doi:10.1016/S0038-0717(03)00179-2

Bedard, D. L., & Quensen III, J. F. (1995). Microbial reductive dechlorination of polychlorinated biphenyls. *Microbial Transformation and Degradation of Toxic Organic Chemicals*, 127-216.

Berg, G., Roskot, N., Steidle, A., Eberl, L., Zock, A., & Smalla, K. (2002). Plant-dependent genotypic and phenotypic diversity of antagonistic rhizobacteria isolated from different Verticillium host plants. *Applied and Environmental Microbiology, 68*(7), 3328–3338. doi:10.1128/AEM.68.7.3328-3338.2002 PMID:12089011

Bezza, F. A., & Chirwa, E. M. N. (2016). Biosurfactant-enhanced bioremediation of aged polycyclic aromatic hydrocarbons (PAHs) in creosote contaminated soil. *Chemosphere, 144*, 635–644. doi:10.1016/j.chemosphere.2015.08.027 PMID:26408261

Bhargava, C., Carmona, F. F., Bhargava, M., & Srivastava, S. (2012). Approaches for enhanced phytoextraction of heavy metals. *Journal of Environmental Management, 105*, 103–120. doi:10.1016/j.jenvman.2012.04.002 PMID:22542973

Bisht, S., Pandey, P., Kaur, G., Aggarwal, H., Sood, A., Sharma, S., & Bisht, N. et al. (2014). Utilization of endophytic strain Bacillus sp. SBER3 for biodegradation of polyaromatic hydrocarbons (PAH) in soil model system. *European Journal of Soil Biology, 60*, 67–76. doi:10.1016/j.ejsobi.2013.10.009

Blyth, W., Shahsavari, E., Morrison, P. D., & Ball, A. S. (2015). Biosurfactant from red ash trees enhances the bioremediation of PAH contaminated soil at a former gasworks site. *Journal of Environmental Management, 162*, 30–36. doi:10.1016/j.jenvman.2015.07.041 PMID:26217887

Boonsaner, M., Borrirukwisitsak, S., & Boonsaner, A. (2011). Phytoremediation of BTEX contaminated soil by Canna× generalis. *Ecotoxicology and Environmental Safety, 74*(6), 1700–1707. doi:10.1016/j.ecoenv.2011.04.011 PMID:21497398

Bourceret, A., Leyval, C., De Fouquet, C., & Cébron, A. (2015). Mapping the Centimeter-Scale Spatial Variability of PAHs and Microbial Populations in the Rhizosphere of Two Plants. *PLoS ONE, 10*(11), 1–22. doi:10.1371/journal.pone.0142851 PMID:26599438

Burd, G. I., Dixon, D. G., & Glick, B. R. (2000). Plant growth-promoting bacteria that decrease heavy metal toxicity in plants. *Canadian Journal of Microbiology, 46*(3), 237–245. doi:10.1139/w99-143 PMID:10749537

Cai, Z., Zhou, Q., Peng, S., & Li, K. (2010). Promoted biodegradation and microbiological effects of petroleum hydrocarbons by Impatiens balsamina L. with strong endurance. *Journal of Hazardous Materials, 183*(1), 731–737. doi:10.1016/j.jhazmat.2010.07.087 PMID:20724074

Canellas, L. P., Olivares, F. L., Okorokova-Façanha, A. L., & Façanha, A. R. (2002). Humic acids isolated from earthworm compost enhance root elongation, lateral root emergence, and plasma membrane H+-ATPase activity in maize roots. *Plant Physiology, 130*(4), 1951–1957. doi:10.1104/pp.007088 PMID:12481077

Cébron, A., Louvel, B., Faure, P., France-Lanord, C., Chen, Y., Murrell, J. C., & Leyval, C. (2011). Root exudates modify bacterial diversity of phenanthrene degraders in PAH-polluted soil but not phenanthrene degradation rates. *Environmental Microbiology*, *13*(3), 722–736. doi:10.1111/j.1462-2920.2010.02376.x PMID:21087382

Chaudhry, Q., Blom-Zandstra, M., Gupta, S. K., & Joner, E. (2005). Utilising the synergy between plants and rhizosphere microorganisms to enhance breakdown of organic pollutants in the environment (15 pp). *Environmental Science and Pollution Research International*, *12*(1), 34–48. doi:10.1065/espr2004.08.213 PMID:15768739

Chekol, T., Vough, L. R., & Chaney, R. L. (2004). Phytoremediation of polychlorinated biphenyl-contaminated soils: The rhizosphere effect. *Environment International*, *30*(6), 799–804. doi:10.1016/j.envint.2004.01.008 PMID:15120198

Chen, F., Tan, M., Ma, J., Zhang, S., Li, G., & Qu, J. (2016). Efficient remediation of PAH-metal co-contaminated soil using microbial-plant combination: A greenhouse study. *Journal of Hazardous Materials*, *302*, 250–261. doi:10.1016/j.jhazmat.2015.09.068 PMID:26476312

Chen, W.-Y., Wu, J.-H., Lin, S.-C., & Chang, J.-E. (2016). Bioremediation of polychlorinated-p-dioxins/dibenzofurans contaminated soil using simulated compost-amended landfill reactors under hypoxic conditions. *Journal of Hazardous Materials*, *312*, 159–168. doi:10.1016/j.jhazmat.2016.03.060 PMID:27037469

Chien, M. F., Makita, R., Sugawara, K., & Inoue, C. (2015). Study on As Uptake and Rhizobacteria of Two as Hyperaccumulators Forward to As Phytoremediation. Paper presented at the *Advanced. Materials Research*, *1130*, 568–571.

Choudhary, V. R., & Mantri, K. (2000). Adsorption of aromatic hydrocarbons on highly siliceous MCM-41. *Langmuir*, *16*(17), 7031–7037. doi:10.1021/la991714u

Cook, R. L., & Hesterberg, D. (2013). Comparison of trees and grasses for rhizoremediation of petroleum hydrocarbons. *International Journal of Phytoremediation*, *15*(9), 844–860. doi:10.1080/15226514.2012.760518 PMID:23819280

Cornelissen, G., Gustafsson, Ö., Bucheli, T. D., Jonker, M. T., Koelmans, A. A., & van Noort, P. C. (2005). Extensive sorption of organic compounds to black carbon, coal, and kerogen in sediments and soils: Mechanisms and consequences for distribution, bioaccumulation, and biodegradation. *Environmental Science & Technology*, *39*(18), 6881–6895. doi:10.1021/es050191b PMID:16201609

Crecchio, C., Curci, M., Pizzigallo, M. D., Ricciuti, P., & Ruggiero, P. (2004). Effects of municipal solid waste compost amendments on soil enzyme activities and bacterial genetic diversity. *Soil Biology & Biochemistry*, *36*(10), 1595–1605. doi:10.1016/j.soilbio.2004.07.016

Cunningham, S. D., & Berti, W. R. (1993). Remediation of contaminated soils with green plants: An overview. *In Vitro Cellular & Developmental Biology. Plant*, *29*(4), 207–212. doi:10.1007/BF02632036

Cunningham, S. D., & Ow, D. W. (1996). Promises and prospects of phytoremediation. *Plant Physiology*, *110*(3), 715–719. doi:10.1104/pp.110.3.715 PMID:12226213

Danh, L. T., Truong, P., Mammucari, R., Tran, T., & Foster, N. (2009). Vetiver grass, Vetiveria zizanioides: A choice plant for phytoremediation of heavy metals and organic wastes. *International Journal of Phytoremediation*, *11*(8), 664–691. doi:10.1080/15226510902787302 PMID:19810597

De Fouquet, C. (2011). From exploratory data analysis to geostatistical estimation: Examples from the analysis of soil pollutants. *European Journal of Soil Science*, *62*(3), 454–466. doi:10.1111/j.1365-2389.2011.01374.x

Di Gregorio, S., Lampis, S., & Vallini, G. (2005). Selenite precipitation by a rhizospheric strain of Stenotrophomonas sp. isolated from the root system of Astragalus bisulcatus: A biotechnological perspective. *Environment International*, *31*(2), 233–241. doi:10.1016/j.envint.2004.09.021 PMID:15661289

Donzelli, D., De Michele, C., & Scholes, R. (2013). Competition between trees and grasses for both soil water and mineral nitrogen in dry savannas. *Journal of Theoretical Biology*, *332*, 181–190. doi:10.1016/j.jtbi.2013.04.003 PMID:23639405

Eisler, R. (1987). Polycyclic aromatic hydrocarbon hazards to fish, wildlife, and invertebrates: a synoptic review. *US Fish and Wildlife Service Biological Report, 85*(1.11), 1-55.

Erakhrumen, A. (2015). Assessment of In-Situ Natural Dendroremediation Capability of Rhizophora racemosa in a Heavy Metal Polluted Mangrove Forest, Rivers State, Nigeria. *Journal of Applied Sciences and Environmental Management*, *19*(1), 21–27. doi:10.4314/jasem.v19i1.3

Farraji, H. (2014). Wastewater Treatment by Phytoremediation Methods. In M. A. H. A. Aziz (Ed.), *Wastewater Engineering: Types, Characteristics and Treatment Technologies* (pp. 205–218). Malaysia: IJSRPUB.

Farraji, H., Aziz, H. A., Tajuddin, R. M., & Mojiri, A. (2014). Optimization of Phytoremediation of Lead-Contaminated Soil by Spinach (Spinacia oleracea L). *International Journal of Scientific Research in Knowledge*, *2*(10), 480–486. doi:10.12983/ijsrk-2014-p0480-0486

Farraji, H., Zaman, N. Q., Tajuddin, R. M., & Faraji, H. (2016). Advantages and disadvantages of phytoremediation: A concise review. *Int J Env Tech Sci*, *2*, 69–75.

Fatima, K., Afzal, M., Imran, A., & Khan, Q. M. (2015). Bacterial rhizosphere and endosphere populations associated with grasses and trees to be used for phytoremediation of crude oil contaminated soil. *Bulletin of Environmental Contamination and Toxicology*, *94*(3), 314–320. doi:10.1007/s00128-015-1489-5 PMID:25661008

Feng, T., Lin, H., Guo, Q., & Feng, Y. (2014). Influence of an arsenate-reducing and polycyclic aromatic hydrocarbons-degrading Pseudomonas isolate on growth and arsenic accumulation in Pteris vittata L. and removal of phenanthrene. *International Biodeterioration & Biodegradation*, *94*, 12–18. doi:10.1016/j.ibiod.2014.06.005

Ford, C. I., Walter, M., Northcott, G. L., Di, H. J., Cameron, K. C., & Trower, T. (2007). Fungal Inoculum Properties: Extracellular Enzyme Expression and Pentachlorophenol Removal by New Zealand Species in Contaminated Field Soils. *Journal of Environmental Quality*, *36*(6), 1749–1759. doi:10.2134/jeq2007.0007 PMID:17965377

Gadd, G. M. (2004). Microbial influence on metal mobility and application for bioremediation. *Geoderma*, *122*(2), 109–119. doi:10.1016/j.geoderma.2004.01.002

Gadd, G. M., & Pan, X. (2016). Biomineralization, Bioremediation and Biorecovery of Toxic Metals and Radionuclides. *Geomicrobiology Journal*, *33*(3-4), 175–178. doi:10.1080/01490451.2015.1087603

Ganesan, V. (2008). Rhizoremediation of cadmium soil using a cadmium-resistant plant growth-promoting rhizopseudomonad. *Current Microbiology*, *56*(4), 403–407. doi:10.1007/s00284-008-9099-7 PMID:18239844

Gao, Y., Guo, S., Wang, J., Li, D., Wang, H., & Zeng, D.-H. (2014). Effects of different remediation treatments on crude oil contaminated saline soil. *Chemosphere*, *117*, 486–493. doi:10.1016/j.chemosphere.2014.08.070 PMID:25240723

Gavis, J., & Ferguson, J. F. (1972). The cycling of mercury through the environment. *Water Research*, *6*(9), 989–1008. doi:10.1016/0043-1354(72)90053-X

Gerhardt, K. E., Huang, X.-D., Glick, B. R., & Greenberg, B. M. (2009). Phytoremediation and rhizoremediation of organic soil contaminants: Potential and challenges. *Plant Science*, *176*(1), 20–30. doi:10.1016/j.plantsci.2008.09.014

Ghosh, M., & Singh, S. (2005). A review on phytoremediation of heavy metals and utilization of it's by products. *Asian J Energy Environ*, *6*(4), 214–231.

Gomes, H. I., Dias-Ferreira, C., & Ribeiro, A. B. (2013). Overview of in situ and ex situ remediation technologies for PCB-contaminated soils and sediments and obstacles for full-scale application. *The Science of the Total Environment*, *445*, 237–260. doi:10.1016/j.scitotenv.2012.11.098 PMID:23334318

Gothalwal, R., & Chillara, S. (2012). *Cyanoremediation: A Green Clean Technology. In Microorganisms in Environmental Management* (pp. 767–786). Springer. doi:10.1007/978-94-007-2229-3_34

Guo, H., Luo, S., Chen, L., Xiao, X., Xi, Q., Wei, W., & Chen, J. et al. (2010). Bioremediation of heavy metals by growing hyperaccumulaor endophytic bacterium Bacillus sp. L14. *Bioresource Technology*, *101*(22), 8599–8605. doi:10.1016/j.biortech.2010.06.085 PMID:20637605

Hamdy, M., & Noyes, O. (1975). Formation of methyl mercury by bacteria. *Applied Microbiology*, *30*(3), 424–432. PMID:1180551

Hong, S. H., Ryu, H., Kim, J., & Cho, K.-S. (2011). Rhizoremediation of diesel-contaminated soil using the plant growth-promoting rhizobacterium Gordonia sp. S2RP-17. *Biodegradation*, *22*(3), 593–601. doi:10.1007/s10532-010-9432-2 PMID:21082332

Horwat, M., Tice, M., & Kjellerup, B. V. (2015). Biofilms at work: Bio-, phyto-and rhizoremediation approaches for soils contaminated with polychlorinated biphenyls. *AIMS Bioengineering*, *2*(4), 224–234. doi:10.3934/bioeng.2015.4.324

Hsieh, J.-L., Chen, C.-Y., Chiu, M.-H., Chein, M., Chang, J.-S., Endo, G., & Huang, C.-C. (2009). Expressing a bacterial mercuric ion binding protein in plant for phytoremediation of heavy metals. *Journal of Hazardous Materials*, *161*(2), 920–925. doi:10.1016/j.jhazmat.2008.04.079 PMID:18538925

Huang, X.-D., El-Alawi, Y., Gurska, J., Glick, B. R., & Greenberg, B. M. (2005). A multi-process phytoremediation system for decontamination of persistent total petroleum hydrocarbons (TPHs) from soils. *Microchemical Journal, 81*(1), 139–147. doi:10.1016/j.microc.2005.01.009

Huang, X.-D., El-Alawi, Y., Penrose, D. M., Glick, B. R., & Greenberg, B. M. (2004). A multi-process phytoremediation system for removal of polycyclic aromatic hydrocarbons from contaminated soils. *Environmental Pollution, 130*(3), 465–476. doi:10.1016/j.envpol.2003.09.031 PMID:15182977

Hybholt, T. K., Aamand, J., & Johnsen, A. R. (2011). Quantification of centimeter-scale spatial variation in PAH, glucose and benzoic acid mineralization and soil organic matter in road-side soil. *Environmental Pollution, 159*(5), 1085–1091. doi:10.1016/j.envpol.2011.02.028 PMID:21396755

Irvine, R., Earley, J., Kehrberger, G. J., & Tod Delaney, B. (1993). Bioremediation of soils contaminated with bis-(2-ethylhexyl) phthalate (BEHP) in a soil slurry-sequencing batch reactor. *Environment and Progress, 12*(1), 39–44. doi:10.1002/ep.670120108

Issoufi, I., Rhykerd, R., & Smiciklas, K. (2006). Seedling growth of agronomic crops in crude oil contaminated soil. *Journal Agronomy & Crop Science, 192*(4), 310–317. doi:10.1111/j.1439-037X.2006.00212.x

Ji, X., Ripp, S. A., Layton, A. C., Sayler, G. S., & DeBruyn, J. M. (2013). Assessing long term effects of bioremediation: Soil bacterial communities 14 years after polycyclic aromatic hydrocarbon contamination and introduction of a genetically engineered microorganism. *Journal of Bioremediation & Biodegradation, 4*(8), 1–8.

Jian-guang, Y., Jun-yuan, L., Jian-ying, Y., Xu-liang, Z., & Zi-xiang, D. (2014). Hydrothermal processing of arsenic containing bioremediation biomass: Pteris vittata. *Journal of Environmental Chemical Engineering, 2*(3), 1358–1364. doi:10.1016/j.jece.2014.04.011

Johnson, D., Anderson, D., & McGrath, S. (2005). Soil microbial response during the phytoremediation of a PAH contaminated soil. *Soil Biology & Biochemistry, 37*(12), 2334–2336. doi:10.1016/j.soilbio.2005.04.001

Johnson, D., Maguire, K., Anderson, D., & McGrath, S. (2004). Enhanced dissipation of chrysene in planted soil: The impact of a rhizobial inoculum. *Soil Biology & Biochemistry, 36*(1), 33–38. doi:10.1016/j.soilbio.2003.07.004

Joner, E., Corgie, S., Amellal, N., & Leyval, C. (2002). Nutritional constraints to degradation of polycyclic aromatic hydrocarbons in a simulated rhizosphere. *Soil Biology & Biochemistry, 34*(6), 859–864. doi:10.1016/S0038-0717(02)00018-4

Joner, E. J., & Leyval, C. (2003). Rhizosphere gradients of polycyclic aromatic hydrocarbon (PAH) dissipation in two industrial soils and the impact of arbuscular mycorrhiza. *Environmental Science & Technology, 37*(11), 2371–2375. doi:10.1021/es020196y PMID:12831019

Kamaludeen, S., & Ramasamy, K. (2008). Rhizoremediation of metals: Harnessing microbial communities. *Indian Journal of Microbiology, 48*(1), 80–88. doi:10.1007/s12088-008-0008-3 PMID:23100702

Kästner, M., & Miltner, A. (2016). Application of compost for effective bioremediation of organic contaminants and pollutants in soil. *Applied Microbiology and Biotechnology*, *100*(8), 3433–3449. doi:10.1007/s00253-016-7378-y PMID:26921182

Khan, A. G. (2005). Role of soil microbes in the rhizospheres of plants growing on trace metal contaminated soils in phytoremediation. *Journal of Trace Elements in Medicine and Biology*, *18*(4), 355–364. doi:10.1016/j.jtemb.2005.02.006 PMID:16028497

Khatisashvili, G., Matchavariani, L., & Gakhokidze, R. (2015). Improving Phytoremediation of Soil Polluted with Oil Hydrocarbons in Georgia. In K. R. Hakim & M. Sabir et al. (Eds.), *Soil Remediation and Plants* (pp. 547–570). doi:10.1016/B978-0-12-799937-1.00019-X

Kim, J., Kang, S.-H., Min, K.-A., Cho, K.-S., & Lee, I.-S. (2006). Rhizosphere microbial activity during phytoremediation of diesel-contaminated soil. *Journal of Environmental Science and Health Part A*, *41*(11), 2503–2516. doi:10.1080/10934520600927658 PMID:17000542

Kim, J., Min, K., Cho, K., & Lee, I. (2007). Enhanced bioremediation and modified bacterial community structure by barnyard grass in diesel-contaminated soil. *Environmental engineering research-seoul*, *12*(2), 37–45. doi:10.4491/eer.2007.12.2.037

Kim, K.-H., Jahan, S. A., Kabir, E., & Brown, R. J. (2013). A review of airborne polycyclic aromatic hydrocarbons (PAHs) and their human health effects. *Environment International*, *60*, 71–80. doi:10.1016/j.envint.2013.07.019 PMID:24013021

Kirkpatrick, W. D., White, P. Jr, Wolf, D., Thoma, G., & Reynolds, C. (2006). Selecting plants and nitrogen rates to vegetate crude-oil–contaminated soil. *International Journal of Phytoremediation*, *8*(4), 285–297. doi:10.1080/15226510600992840 PMID:17305303

Kōmura, I., Funaba, T., & Izaki, K. (1971). Mechanism of Mercuric Chloride Resistance in Microorganisms: II. NADPH-Dependent Reduction of Mercuric Chloride and Vaporization of Mercury from Mercuric Chloride by a Multiple Drug Resistant Strain of Escherichia coli. *Journal of Biochemistry*, *70*(6), 895–901. PMID:4401341

Kōmura, I., & Izaki, K. (1971). Mechanism of Mercuric Chloride Resistance in Microorganisms: I. Vaporization of a Mercury Compound from Mercuric Chloride by Multiple Drug Resistant Strains of Escherichia coli. *Journal of Biochemistry*, *70*(6), 885–893. PMID:4947308

Koo, S.-Y., & Cho, K. S. (2009). Isolation and characterization of a plant growth-promoting rhizobacterium, Serratia sp. SY5. *Journal of Microbiology and Biotechnology*, *19*(11), 1431–1438. PMID:19996698

Kristanti, R. A., Toyama, T., Hadibarata, T., Tanaka, Y., & Mori, K. (2014). Sustainable Removal of Nitrophenols by Rhizoremediation Using Four Strains of Bacteria and Giant Duckweed (Spirodela polyrhiza). *Water, Air, and Soil Pollution*, *225*(4), 1–10. doi:10.1007/s11270-014-1928-7

Kuiper, I., Lagendijk, E. L., Bloemberg, G. V., & Lugtenberg, B. J. (2004). Rhizoremediation: A beneficial plant-microbe interaction. *Molecular Plant-Microbe Interactions*, *17*(1), 6–15. doi:10.1094/MPMI.2004.17.1.6 PMID:14714863

Kumar, R., Pandey, S., & Pandey, A. (2006). Plant roots and carbon sequestration. *Current Science*, *91*(7), 885–890.

Kuo, H.-C., Juang, D.-F., Yang, L., Kuo, W.-C., & Wu, Y.-M. (2014). Phytoremediation of soil contaminated by heavy oil with plants colonized by mycorrhizal fungi. *International Journal of Environmental Science and Technology*, *11*(6), 1661–1668. doi:10.1007/s13762-013-0353-6

Kurzawova, V., Stursa, P., Uhlik, O., Norkova, K., Strohalm, M., Lipov, J., & Mackova, M. et al. (2012). Plant–microorganism interactions in bioremediation of polychlorinated biphenyl-contaminated soil. *New Biotechnology*, *30*(1), 15–22. doi:10.1016/j.nbt.2012.06.004 PMID:22728721

Kvesitadze, G., Khatisashvili, G., Sadunishvili, T., & Ramsden, J. J. (2006). *Biochemical mechanisms of detoxification in higher plants: basis of phytoremediation*. Springer Science & Business Media.

Lee, S.-H., Lee, W.-S., Lee, C.-H., & Kim, J.-G. (2008). Degradation of phenanthrene and pyrene in rhizosphere of grasses and legumes. *Journal of Hazardous Materials*, *153*(1), 892–898. doi:10.1016/j.jhazmat.2007.09.041 PMID:17959304

Lee, S.-W., Glickmann, E., & Cooksey, D. A. (2001). Chromosomal locus for cadmium resistance in Pseudomonas putida consisting of a cadmium-transporting ATPase and a MerR family response regulator. *Applied and Environmental Microbiology*, *67*(4), 1437–1444. doi:10.1128/AEM.67.4.1437-1444.2001 PMID:11282588

Leigh, M. B., Prouzová, P., Macková, M., Macek, T., Nagle, D. P., & Fletcher, J. S. (2006). Polychlorinated biphenyl (PCB)-degrading bacteria associated with trees in a PCB-contaminated site. *Applied and Environmental Microbiology*, *72*(4), 2331–2342. doi:10.1128/AEM.72.4.2331-2342.2006 PMID:16597927

Li, W., & Wong, M. (2012). Interaction of Cd/Zn hyperaccumulating plant (Sedum alfredii) and rhizosphere bacteria on metal uptake and removal of phenanthrene. *Journal of Hazardous Materials*, *209*, 421–433. doi:10.1016/j.jhazmat.2012.01.055 PMID:22309655

Liang, Y., Meggo, R., Hu, D., Schnoor, J. L., & Mattes, T. E. (2015). Microbial community analysis of switchgrass planted and unplanted soil microcosms displaying PCB dechlorination. *Applied Microbiology and Biotechnology*, *99*(15), 6515–6526. doi:10.1007/s00253-015-6545-x PMID:25820643

Lin, Z., Zhen, Z., Wu, Z., Yang, J., Zhong, L., Hu, H., & Zhang, D. et al. (2016). The impact on the soil microbial community and enzyme activity of two earthworm species during the bioremediation of pentachlorophenol-contaminated soils. *Journal of Hazardous Materials*, *301*, 35–45. doi:10.1016/j.jhazmat.2015.08.034 PMID:26342149

Liu, H., Guo, S., Jiao, K., Hou, J., Xie, H., & Xu, H. (2015). Bioremediation of soils co-contaminated with heavy metals and 2, 4, 5-trichlorophenol by fruiting body of Clitocybe maxima. *Journal of Hazardous Materials*, *294*, 121–127. doi:10.1016/j.jhazmat.2015.04.004 PMID:25863026

Liu, L., Jiang, C. Y., Liu, X. Y., Wu, J. F., Han, J. G., & Liu, S. J. (2007). Plant–microbe association for rhizoremediation of chloronitroaromatic pollutants with Comamonas sp. strain CNB-1. *Environmental Microbiology*, *9*(2), 465–473. doi:10.1111/j.1462-2920.2006.01163.x PMID:17222144

Lu, Y.-F., Lu, M., Peng, F., Wan, Y., & Liao, M.-H. (2014). Remediation of polychlorinated biphenyl-contaminated soil by using a combination of ryegrass, arbuscular mycorrhizal fungi and earthworms. *Chemosphere*, *106*, 44–50. doi:10.1016/j.chemosphere.2013.12.089 PMID:24457052

Lyon, D. Y., & Vogel, T. M. (2013). *Bioaugmentation for Groundwater Remediation: An Overview. In Bioaugmentation for Groundwater Remediation* (pp. 1–37). Springer. doi:10.1007/978-1-4614-4115-1_1

Mackova, M., Dowling, D., & Macek, T. (2006). *Phytoremediation and rhizoremediation* (Vol. 9). Springer Science & Business Media. doi:10.1007/978-1-4020-4999-4

Mackova, M., Prouzova, P., Stursa, P., Ryslava, E., Uhlik, O., Beranova, K., & Macek, T. et al. (2009). Phyto/rhizoremediation studies using long-term PCB-contaminated soil. *Environmental Science and Pollution Research International*, *16*(7), 817–829. doi:10.1007/s11356-009-0240-3 PMID:19823887

Mah, H., Boyd, D., Gadaginamath, G., Wood, A., Buening, M., Chang, R., . . . Conney, A. (2013). *Polycyclic aromatic hydrocarbons.* Paper presented at the Toxicology: Proceedings of the 7th International Congress of Pharmacology, Paris, France.

Maila, M. P., Randima, P., & Cloete, T. E. (2005). Multispecies and monoculture rhizoremediation of polycyclic aromatic hydrocarbons (PAHs) from the soil. *International Journal of Phytoremediation*, *7*(2), 87–98. doi:10.1080/16226510590950397 PMID:16128441

Mani, D., & Kumar, C. (2014). Biotechnological advances in bioremediation of heavy metals contaminated ecosystems: An overview with special reference to phytoremediation. *International Journal of Environmental Science and Technology*, *11*(3), 843–872. doi:10.1007/s13762-013-0299-8

Marques, A. P., Rangel, A. O., & Castro, P. M. (2009). Remediation of heavy metal contaminated soils: Phytoremediation as a potentially promising clean-up technology. *Critical Reviews in Environmental Science and Technology*, *39*(8), 622–654. doi:10.1080/10643380701798272

Mary Agbogidi, O., Dickens Dolor, E., & Mercy Okechukwu, E. (2007). Evaluation of Tectona grandis (Linn.) and Gmelina arborea (Roxb.) for phytoremediation in crude oil contaminated soils.[ACS]. *Agriculturae Conspectus Scientificus*, *72*(2), 149–152.

Masood, F., & Malik, A. (2013). *Current Aspects of Metal Resistant Bacteria in Bioremediation: From Genes to Ecosystem. In Management of Microbial Resources in the Environment* (pp. 289–311). Springer.

McCutcheon, S., & Schnoor, J. (2003). Overview of phytotransformation and control of wastes. In Phytoremediation: Transformation and control of contaminants (pp. 3-58). doi:10.1002/047127304X.ch1

McLean, K., Sabri, M., Marshall, K., Lawson, R., Lewis, D., Clift, D., & McVey, J. et al. (2005). Biodiversity of cytochrome P450 redox systems. *Biochemical Society Transactions*, *33*(4), 796–801. doi:10.1042/BST0330796 PMID:16042601

Meggo, R. E., Schnoor, J. L., & Hu, D. (2013). Dechlorination of PCBs in the rhizosphere of switchgrass and poplar. *Environmental Pollution*, *178*, 312–321. doi:10.1016/j.envpol.2013.02.035 PMID:23603468

Mehmannavaz, R., Prasher, S. O., & Ahmad, D. (2002). Rhizospheric effects of alfalfa on biotransformation of polychlorinated biphenyls in a contaminated soil augmented with Sinorhizobium meliloti. *Process Biochemistry*, *37*(9), 955–963. doi:10.1016/S0032-9592(01)00305-3

Meissner, P., & Falkinham, J. (1984). Plasmid-encoded mercuric reductase in Mycobacterium scrofulaceum. *Journal of Bacteriology*, *157*(2), 669–672. PMID:6693354

Mena-Benitez, G. L., Gandia-Herrero, F., Graham, S., Larson, T. R., McQueen-Mason, S. J., French, C. E., & Bruce, N. C. et al. (2008). Engineering a catabolic pathway in plants for the degradation of 1, 2-dichloroethane. *Plant Physiology*, *147*(3), 1192–1198. doi:10.1104/pp.108.119008 PMID:18467461

Merkl, N., Schultze-Kraft, R., & Infante, C. (2004). Phytoremediation in the tropics—the effect of crude oil on the growth of tropical plants. *Bioremediation Journal*, *8*(3-4), 177–184. doi:10.1080/10889860490887527

Miransari, M. (2011). Hyperaccumulators, arbuscular mycorrhizal fungi and stress of heavy metals. *Biotechnology Advances*, *29*(6), 645–653. doi:10.1016/j.biotechadv.2011.04.006 PMID:21557996

Mohsenzadeh, F., Nasseri, S., Mesdaghinia, A., Nabizadeh, R., Zafari, D., Khodakaramian, G., & Chehregani, A. (2010). Phytoremediation of petroleum-polluted soils: Application of Polygonum aviculare and its root-associated (penetrated) fungal strains for bioremediation of petroleum-polluted soils. *Ecotoxicology and Environmental Safety*, *73*(4), 613–619. doi:10.1016/j.ecoenv.2009.08.020 PMID:19932506

Mojiri, A., Ziyang, L., Tajuddin, R. M., Farraji, H., & Alifar, N. (2016). Co-treatment of landfill leachate and municipal wastewater using the ZELIAC/zeolite constructed wetland system. *Journal of Environmental Management*, *166*, 124–130. doi:10.1016/j.jenvman.2015.10.020 PMID:26496842

Moubasher, H., Hegazy, A., Mohamed, N., Moustafa, Y., Kabiel, H., & Hamad, A. (2015). Phytoremediation of soils polluted with crude petroleum oil using Bassia scoparia and its associated rhizosphere microorganisms. *International Biodeterioration & Biodegradation*, *98*, 113–120. doi:10.1016/j.ibiod.2014.11.019

Namkoong, W., Hwang, E.-Y., Park, J.-S., & Choi, J.-Y. (2002). Bioremediation of diesel-contaminated soil with composting. *Environmental Pollution*, *119*(1), 23–31. doi:10.1016/S0269-7491(01)00328-1 PMID:12125726

Nehl, D., Allen, S., & Brown, J. (1997). Deleterious rhizosphere bacteria: An integrating perspective. *Applied Soil Ecology*, *5*(1), 1–20. doi:10.1016/S0929-1393(96)00124-2

Niu, M., Wei, S., Bai, J., Wang, S., & Ji, D. (2015). Remediation and safe production of Cd contaminated soil via multiple cropping hyperaccumulator Solanum nigrum L. and low accumulation Chinese cabbage. *International Journal of Phytoremediation*, *17*(7), 657–661. doi:10.1080/15226514.2014.955168 PMID:25976879

Noori, A., et al. (2015). Leucanthemum vulgare Lam. crude oil phytoremediation. *International Journal of Phytoremediation*.

Noori, A., Zare Maivan, H., & Alaie, E. (2014). Leucanthemum vulgare Lam. germination, growth and mycorrhizal symbiosis under crude oil contamination. *International Journal of Phytoremediation*, *16*(9), 962–970. doi:10.1080/15226514.2013.810577 PMID:24933896

Ohtsubo, Y., Kudo, T., Tsuda, M., & Nagata, Y. (2004). Strategies for bioremediation of polychlorinated biphenyls. *Applied Microbiology and Biotechnology*, *65*(3), 250–258. doi:10.1007/s00253-004-1654-y PMID:15248039

Olson, G. J., Porter, F., Rubinstein, J., & Silver, S. (1982). Mercuric reductase enzyme from a mercury-volatilizing strain of Thiobacillus ferrooxidans. *Journal of Bacteriology*, *151*(3), 1230–1236. PMID:6286594

Ontañon, O. M., González, P. S., Ambrosio, L. F., Paisio, C. E., & Agostini, E. (2014). Rhizoremediation of phenol and chromium by the synergistic combination of a native bacterial strain and Brassica napus hairy roots. *International Biodeterioration & Biodegradation*, *88*, 192–198. doi:10.1016/j.ibiod.2013.10.017

Palmroth, M. R., Pichtel, J., & Puhakka, J. A. (2002). Phytoremediation of subarctic soil contaminated with diesel fuel. *Bioresource Technology*, *84*(3), 221–228. doi:10.1016/S0960-8524(02)00055-X PMID:12118697

Pan-Hou, H. S., & Imura, N. (1982). Involvement of mercury methylation in microbial mercury detoxication. *Archives of Microbiology*, *131*(2), 176–177. doi:10.1007/BF01054003 PMID:6462125

Pan-Hou, H. S., Nishimoto, M., & Imura, N. (1981). Possible role of membrane proteins in mercury resistance of Enterobacter aerogenes. *Archives of Microbiology*, *130*(2), 93–95. doi:10.1007/BF00411057 PMID:6459062

Park, J. H., Lamb, D., Paneerselvam, P., Choppala, G., Bolan, N., & Chung, J.-W. (2011). Role of organic amendments on enhanced bioremediation of heavy metal (loid) contaminated soils. *Journal of Hazardous Materials*, *185*(2), 549–574. doi:10.1016/j.jhazmat.2010.09.082 PMID:20974519

Passatore, L., Rossetti, S., Juwarkar, A. A., & Massacci, A. (2014). Phytoremediation and bioremediation of polychlorinated biphenyls (PCBs): State of knowledge and research perspectives. *Journal of Hazardous Materials*, *278*, 189–202. doi:10.1016/j.jhazmat.2014.05.051 PMID:24976127

Peng, S., Zhou, Q., Cai, Z., & Zhang, Z. (2009). Phytoremediation of petroleum contaminated soils by Mirabilis Jalapa L. in a greenhouse plot experiment. *Journal of Hazardous Materials*, *168*(2), 1490–1496. doi:10.1016/j.jhazmat.2009.03.036 PMID:19346069

Pieper, D. H., dos Santos, V. A. M., & Golyshin, P. N. (2004). Genomic and mechanistic insights into the biodegradation of organic pollutants. *Current Opinion in Biotechnology*, *15*(3), 215–224. doi:10.1016/j.copbio.2004.03.008 PMID:15193329

Pino, N. J., Muñera, L. M., & Peñuela, G. A. (2016). Root exudates and plant secondary metabolites of different plants enhance polychlorinated biphenyl degradation by rhizobacteria. *Bioremediation Journal*, *20*(2), 108–116. doi:10.1080/10889868.2015.1124065

Pizarro-Tobías, P., Fernández, M., Niqui, J. L., Solano, J., Duque, E., Ramos, J. L., & Roca, A. (2015). Restoration of a Mediterranean forest after a fire: Bioremediation and rhizoremediation field-scale trial. *Microbial Biotechnology*, *8*(1), 77–92. doi:10.1111/1751-7915.12138 PMID:25079309

Prasad, M. (1995). Cadmium toxicity and tolerance in vascular plants. *Environmental and Experimental Botany*, *35*(4), 525–545. doi:10.1016/0098-8472(95)00024-0

Rajkumar, M., Ma, Y., & Freitas, H. (2013). Improvement of Ni phytostabilization by inoculation of Ni resistant Bacillus megaterium SR28C. *Journal of Environmental Management*, *128*, 973–980. doi:10.1016/j.jenvman.2013.07.001 PMID:23895909

Ramos, D. T., Maranho, L. T., Godoi, A. F. L., da Silva Carvalho Filho, M. A., Lacerda, L. G., & de Vasconcelos, E. C. (2009). Petroleum hydrocarbons rhizodegradation by Sebastiania commersoniana (BAILL.) LB SM. & Downs. *Water Air and Soil Pollution Focus*, *9*(3-4), 293–302. doi:10.1007/s11267-009-9208-z

Robertson, L. W., & Hansen, L. G. (2015). *PCBs: recent advances in environmental toxicology and health effects*. University Press of Kentucky.

Robinson, B., Bolan, N., Mahimairaja, S., & Clothier, B. (2005). Solubility, mobility, and bioaccumulation of trace elements: Abiotic processes in the rhizosphere. In *Trace elements in the environment: Biogeochemistry, Biotechnology and Bioremediation* (pp. 97–110). Boca Raton, FL: CRC Press. doi:10.1201/9781420032048.sec2

Robinson, J. B., & Tuovinen, O. H. (1984). Mechanisms of microbial resistance and detoxification of mercury and organomercury compounds: Physiological, biochemical, and genetic analyses. *Microbiological Reviews*, *48*(2), 95. PMID:6377034

Rohrbacher, F., & St-Arnaud, M. (2016). Root Exudation: The Ecological Driver of Hydrocarbon Rhizoremediation. *Agronomy*, *6*(1), 19. doi:10.3390/agronomy6010019

Ross, G. (2004). The public health implications of polychlorinated biphenyls (PCBs) in the environment. *Ecotoxicology and Environmental Safety*, *59*(3), 275–291. doi:10.1016/j.ecoenv.2004.06.003 PMID:15388267

Roy, A. S., Baruah, R., Borah, M., Singh, A. K., Boruah, H. P. D., Saikia, N., & Bora, T. C. et al. (2014). Bioremediation potential of native hydrocarbon degrading bacterial strains in crude oil contaminated soil under microcosm study. *International Biodeterioration & Biodegradation*, *94*, 79–89. doi:10.1016/j.ibiod.2014.03.024

Savin, M. C., Wolf, D. C., Davis, K. J., Gbur, E. E., & Thoma, G. J. (2015). Nematodes as Bioindicators of Ecosystem Recovery During Phytoremediation of Crude Oil Contaminated Soil. *International Journal of Phytoremediation*, *17*(2), 182–190. doi:10.1080/15226514.2013.876964 PMID:25361231

Schöftner, P., Watzinger, A., Holzknecht, P., Wimmer, B., & Reichenauer, T. G. (2015). Transpiration and metabolisation of TCE by willow plants–a pot experiment. *International Journal of Phytoremediation*.

Segura, A., & Ramos, J. L. (2013). Plant–bacteria interactions in the removal of pollutants. *Current Opinion in Biotechnology*, *24*(3), 467–473. doi:10.1016/j.copbio.2012.09.011 PMID:23098915

Shukla, K. P., Sharma, S., Singh, N. K., Singh, V., Bisht, S., & Kumar, V. (2013). Rhizoremediation: A Promising Rhizosphere Technology. In Y. B. Patil & P. Rao (Eds.), *Applied Bioremediation-active and passive approaches, 331*. doi:10.5772/56905

Silva-Castro, G. A., Rodelas, B., Perucha, C., Laguna, J., González-López, J., & Calvo, C. (2013). Bioremediation of diesel-polluted soil using biostimulation as post-treatment after oxidation with Fenton-like reagents: Assays in a pilot plant. *The Science of the Total Environment*, *445*, 347–355. doi:10.1016/j.scitotenv.2012.12.081 PMID:23354375

Singer, A., Gilbert, E., Luepromchai, E., & Crowley, D. (2000). Bioremediation of polychlorinated biphenyl-contaminated soil using carvone and surfactant-grown bacteria. *Applied Microbiology and Biotechnology*, *54*(6), 838–843. doi:10.1007/s002530000472 PMID:11152078

Singer, A. C., Smith, D., Jury, W. A., Hathuc, K., & Crowley, D. E. (2003). Impact of the plant rhizosphere and augmentation on remediation of polychlorinated biphenyl contaminated soil. *Environmental Toxicology and Chemistry*, *22*(9), 1998–2004. doi:10.1897/02-471 PMID:12959523

Singh, A., & Gauba, P. (2014). Mycoremediation: A Treatment for Heavy Metal Pollution of Soil. *Journal of Civil Engineering and Environment Technology*, *1*(4), 59–61.

Singh, A., Kuhad, R. C., & Ward, O. P. (2009). Biological remediation of soil: an overview of global market and available technologies. In A. Singh & C. Ramesh et al. (Eds.), *Advances in applied bioremediation* (pp. 1–19). Springer. doi:10.1007/978-3-540-89621-0_1

Slater, H., Gouin, T., & Leigh, M. B. (2011). Assessing the potential for rhizoremediation of PCB contaminated soils in northern regions using native tree species. *Chemosphere*, *84*(2), 199–206. doi:10.1016/j.chemosphere.2011.04.058 PMID:21596420

Smalla, K., Wieland, G., Buchner, A., Zock, A., Parzy, J., Kaiser, S., & Berg, G. et al. (2001). Bulk and rhizosphere soil bacterial communities studied by denaturing gradient gel electrophoresis: Plant-dependent enrichment and seasonal shifts revealed. *Applied and Environmental Microbiology*, *67*(10), 4742–4751. doi:10.1128/AEM.67.10.4742-4751.2001 PMID:11571180

Storey, S., Ashaari, M., McCabe, G., Harty, M., Dempsey, R., Doyle, O., & Doyle, E. et al. (2014). Microbial community structure during fluoranthene degradation in the presence of plants. *Journal of Applied Microbiology*, *117*(1), 74–84. doi:10.1111/jam.12518 PMID:24712542

Summers, A. O., & Silver, S. (1978). Microbial transformations of metals. *Annual Review of Microbiology*, *32*(1), 637–672. doi:10.1146/annurev.mi.32.100178.003225 PMID:360977

Sun, B., Ling, W., & Wang, Y. (2013). Can root exudate components influence the availability of pyrene in soil? *Journal of Soils and Sediments*, *13*(7), 1161–1169. doi:10.1007/s11368-013-0712-4

Sylvestre, M., Macek, T., & Mackova, M. (2009). Transgenic plants to improve rhizoremediation of polychlorinated biphenyls (PCBs). *Current Opinion in Biotechnology*, *20*(2), 242–247. doi:10.1016/j.copbio.2009.01.006 PMID:19250817

Tangahu, B. V., Sheikh Abdullah, S. R., Basri, H., Idris, M., Anuar, N., & Mukhlisin, M. (2011). A review on heavy metals (As, Pb, and Hg) uptake by plants through phytoremediation. *International Journal of Chemical Engineering*, *2011*, 1–31. doi:10.1155/2011/939161

Techer, D., Martinez-Chois, C., Laval-Gilly, P., Henry, S., Bennasroune, A., DInnocenzo, M., & Falla, J. (2012). Assessment of Miscanthus× giganteus for rhizoremediation of long term PAH contaminated soils. *Applied Soil Ecology*, *62*, 42–49. doi:10.1016/j.apsoil.2012.07.009

Teng, Y., Li, X., Chen, T., Zhang, M., Wang, X., Li, Z., & Luo, Y. (2016). Isolation of the PCB-degrading bacteria Mesorhizobium sp. ZY1 and its combined remediation with Astragalus sinicus L. for contaminated soil. *International Journal of Phytoremediation, 18*(2), 141–149. doi:10.1080/15226514.2015.1073667 PMID:26292091

Tesar, M., Reichenauer, T. G., & Sessitsch, A. (2002). Bacterial rhizosphere populations of black poplar and herbal plants to be used for phytoremediation of diesel fuel. *Soil Biology & Biochemistry, 34*(12), 1883–1892. doi:10.1016/S0038-0717(02)00202-X

Ullah, A., Heng, S., Munis, M. F. H., Fahad, S., & Yang, X. (2015). Phytoremediation of heavy metals assisted by plant growth promoting (PGP) bacteria: A review. *Environmental and Experimental Botany, 117*, 28–40. doi:10.1016/j.envexpbot.2015.05.001

van der Lelie, D., Springael, D., Römling, U., Ahmed, N., & Mergeay, M. (1999). Identification of a gene cluster, czr, involved in cadmium and zinc resistance in Pseudomonas aeruginosa. *Gene, 238*(2), 417–425. doi:10.1016/S0378-1119(99)00349-2 PMID:10570969

Vančura, V., & Hovadik, A. (1965). Root exudates of plants: II. Composition of root exudates of some vegetables. *Plant and Soil, 22*(1), 21–32. doi:10.1007/BF01377686

Vasilyeva, G., & Strijakova, E. (2007). Bioremediation of soils and sediments contaminated by polychlorinated biphenyls. *Microbiology, 76*(6), 639–653. doi:10.1134/S002626170706001X PMID:18069324

Villanneau, E. J., Saby, N. P., Orton, T. G., Jolivet, C. C., Boulonne, L., Caria, G., & Arrouays, D. et al. (2013). First evidence of large-scale PAH trends in French soils. *Environmental Chemistry Letters, 11*(1), 99–104. doi:10.1007/s10311-013-0401-y

Vonk, J., & Sijpesteijn, A. K. (1973). Studies on the methylation of mercuric chloride by pure cultures of bacteria and fungi. *Antonie van Leeuwenhoek, 39*(1), 505–513. doi:10.1007/BF02578894 PMID:4201588

Walton, B. T., & Anderson, T. A. (1990). Microbial degradation of trichloroethylene in the rhizosphere: Potential application to biological remediation of waste sites. *Applied and Environmental Microbiology, 56*(4), 1012–1016. PMID:2339867

Wang, S., Wei, S., Ji, D., & Bai, J. (2015). Co-Planting Cd Contaminated Field Using Hyperaccumulator Solanum Nigrum L. Through Interplant with Low Accumulation Welsh Onion. *International Journal of Phytoremediation, 17*(9), 879–884. doi:10.1080/15226514.2014.981247 PMID:25581317

Whipps, J. M. (2001). Microbial interactions and biocontrol in the rhizosphere. *Journal of Experimental Botany, 52*(suppl 1), 487–511. doi:10.1093/jexbot/52.suppl_1.487 PMID:11326055

White, P. M. Jr, Wolf, D. C., Thoma, G. J., & Reynolds, C. M. (2006). Phytoremediation of alkylated polycyclic aromatic hydrocarbons in a crude oil-contaminated soil. *Water, Air, and Soil Pollution, 169*(1-4), 207–220. doi:10.1007/s11270-006-2194-0

Wiedemeier, T. H. (1999). *Natural attenuation of fuels and chlorinated solvents in the subsurface*. John Wiley & Sons. doi:10.1002/9780470172964

Wojtera-Kwiczor, J., Żukowska, W., Graj, W., Małecka, A., Piechalak, A., Ciszewska, L., & Barałkiewicz, D. et al. (2014). Rhizoremediation of Diesel-Contaminated Soil with Two Rapeseed Varieties and Petroleum degraders Reveals Different Responses of the Plant Defense Mechanisms. *International Journal of Phytoremediation*, *16*(7-8), 770–789. doi:10.1080/15226514.2013.856848 PMID:24933884

Wood, T. K. (2008). Molecular approaches in bioremediation. *Current Opinion in Biotechnology*, *19*(6), 572–578. doi:10.1016/j.copbio.2008.10.003 PMID:19000765

Wu, C. H., Wood, T. K., Mulchandani, A., & Chen, W. (2006). Engineering plant-microbe symbiosis for rhizoremediation of heavy metals. *Applied and Environmental Microbiology*, *72*(2), 1129–1134. doi:10.1128/AEM.72.2.1129-1134.2006 PMID:16461658

Xiao, W., Wang, H., Li, T., Zhu, Z., Zhang, J., He, Z., & Yang, X. (2013). Bioremediation of Cd and carbendazim co-contaminated soil by Cd-hyperaccumulator Sedum alfredii associated with carbendazim-degrading bacterial strains. *Environmental Science and Pollution Research International*, *20*(1), 380–389. doi:10.1007/s11356-012-0902-4 PMID:22529002

Xu, L., Teng, Y., Li, Z.-G., Norton, J. M., & Luo, Y.-M. (2010). Enhanced removal of polychlorinated biphenyls from alfalfa rhizosphere soil in a field study: The impact of a rhizobial inoculum. *The Science of the Total Environment*, *408*(5), 1007–1013. doi:10.1016/j.scitotenv.2009.11.031 PMID:19995667

Yang, M., Jawitz, J. W., & Lee, M. (2015). Uranium and cesium accumulation in bean (Phaseolus vulgaris L. var. vulgaris) and its potential for uranium rhizofiltration. *Journal of Environmental Radioactivity*, *140*, 42–49. doi:10.1016/j.jenvrad.2014.10.015 PMID:25461514

Yang, S., Liang, S., Yi, L., Xu, B., Cao, J., Guo, Y., & Zhou, Y. (2014). Heavy metal accumulation and phytostabilization potential of dominant plant species growing on manganese mine tailings. *Frontiers of Environmental Science & Engineering*, *8*(3), 394–404. doi:10.1007/s11783-013-0602-4

Yee, D. C., Maynard, J. A., & Wood, T. K. (1998). Rhizoremediation of Trichloroethylene by a Recombinant, Root-Colonizing Pseudomonas fluorescensStrain Expressing Toluene ortho-Monooxygenase Constitutively. *Applied and Environmental Microbiology*, *64*(1), 112–118. PMID:9435067

Yenn, R., Borah, M., Boruah, H. D., Roy, A. S., Baruah, R., Saikia, N., & Tamuli, A. et al. (2014). Phytoremediation of abandoned crude oil contaminated drill sites of Assam with the aid of a hydrocarbon-degrading bacterial formulation. *International Journal of Phytoremediation*, *16*(9), 909–925. doi:10.1080/15226514.2013.810573 PMID:24933892

Yoshitomi, K., & Shann, J. (2001). Corn (Zea mays L.) root exudates and their impact on 14 C-pyrene mineralization. *Soil Biology & Biochemistry*, *33*(12), 1769–1776. doi:10.1016/S0038-0717(01)00102-X

Zabłudowska, E., Kowalska, J., Wojas, S., Skłodowska, A., & Antosiewicz, D. (2009). Search for a plant for phytoremediation–What can we learn from field and hydroponic studies? *Chemosphere*, *77*(3), 301–307. doi:10.1016/j.chemosphere.2009.07.064 PMID:19733893

Zanaroli, G., Perez-Jimenez, J., Young, L., Marchetti, L., & Fava, F. (2006). Microbial reductive dechlorination of weathered and exogenous co-planar polychlorinated biphenyls (PCBs) in an anaerobic sediment of Venice Lagoon. *Biodegradation*, *17*(2), 19–27. doi:10.1007/s10532-005-3752-7 PMID:16477348

Zhu, Z., Yang, X.-e., Wang, K., Huang, H., Zhang, X., Fang, H., & He, Z. et al. (2012). Bioremediation of Cd-DDT co-contaminated soil using the Cd-hyperaccumulator Sedum alfredii and DDT-degrading microbes. *Journal of Hazardous Materials*, *235*, 144–151. doi:10.1016/j.jhazmat.2012.07.033 PMID:22868749

Zhuang, X., Chen, J., Shim, H., & Bai, Z. (2007). New advances in plant growth-promoting rhizobacteria for bioremediation. *Environment International*, *33*(3), 406–413. doi:10.1016/j.envint.2006.12.005 PMID:17275086

KEY TERMS AND DEFINITIONS

2, 4, 5-Trichlorophenol (TCP): Specific herbicide, insecticide or even fungicide with chlorinated structural.

Dichlorodiphenyltrichloroethane (DDT): A commercial insecticide with high environmental impact.

Hyperaccumulator Plant Species (HPS): The plants which can grow and absorb high concentration of pollutants (metallic elements) in extremely polluted media.

Plant Growth Promoting Rhizobacteria (PGPR): Soil bacteria living on/around the root of plants.

Poly Aromatic Hydrocarbons (PAH): Hazardous pollution mutagenic and carcinogenic properties.

Polychlorinated Biphenyl (PCB): A chlorine organic compound which used in electrical apparatus as coolant fluid.

Total Petroleum Hydrocarbon (TPH): Toxic organic compounds with hazardous effect on aquatic, air and terrestrial environment.

Trichloroethylene (TCE): It is a hazardous compound which could be found in hazardous waste sites and industrial section with toxic effect after breath, eat, drink or even touch the contaminated air, food or water.

Chapter 10
Molecular Overview of Heavy Metal Phytoremediation

Ved Prakash
College of Engineering and Technology, IILM, India

Sarika Saxena
Madhav Institute of Technology and Science, India

ABSTRACT

Metal toxification has remained one of the problems with the advent of industrial revolution. Plant based remediation are showing increasing promise for use in soils contaminated with organic and inorganic pollutants. A large number of plant families has been identified which has shown significant result in detoxification of heavy metals. Hyperaccumulator plant is capable of sequestering heavy metals in their shoot tissues. High tolerance to HM toxicity is dependent on a reduced metal uptake or increased internal sequestration, which depends on plant and environmental condition. Recent progresses on understanding cellular/molecular mechanisms of metal tolerance by plants are reviewed. This chapter aims to focus on molecular mechanism involved in heavy metal detoxification and tolerance by plants. A different method by which plant effectively converts toxic metal in less toxic compounds has been explained in this chapter. Further, mode of accumulation and sequestration of metals have been explained which are utilized by hyper accumulators.

INTRODUCTION

Phytoremediation can be defined as the process of utilizing plants to absorb, accumulate, detoxify and render harmless, contaminants in the growth substrate (soil, water and air) through physical, chemical or biological processes (Cunningham & Berti, 1993). Plant based remediation are showing increasing promise for use in soils contaminated with organic and inorganic pollutants. For removal of contaminants from soil, phytoremediation involves different processes, such as enzymatic degradation, that potentially lead to contaminant detoxification (Van Aken, 2008). However, despite great promise, rather slow removal rates and potential accumulation of toxic compounds within plants might have limited the application of phytoremediation (Eapen, Singh & D'souza 2007).

DOI: 10.4018/978-1-5225-2325-3.ch010

Phytoremediation is proposed as a cost-effective alternative for the treatment of contaminated soils. Topsoil would be preserved and the amount of hazardous material reduced significantly (Ensley, 2000). In recent years, public concerns relating to ecological threats caused by heavy metal led to intensive research of new economical plant based remediation of contaminated soil namely chemical, physical and microbiological methods are costly to install and operate (Danh, Truong, Mammucari, Tran & Foster, 2009). The ability to hyperaccumulate metals in above ground tissue without phytotoxic effects has evolved in almost 500 plant species mainly those in the Brassicaceae family (Kramer, 2010). At least three processes make a major contribution to the ability to hyperaccumulate/hypertolerate metals.

1. Enhanced root uptake and loading into xylem.
2. Superior root to shoot translocation and
3. Efficient detoxification via chelation and sequestration predominantly within leaf cell vacuoles (Clemens, kim, Neumann, & Schroeder, 2002).

WHAT ARE HYPERACCUMULATOR PLANTS?

The term "hyperaccumulator" describes a number of plants that belong to distantly related families, but share the ability to grow on metalliferous soils and to accumulate extraordinarily high amounts of heavy metals in the aerial organs, far in excess of the levels found in the majority of species, without suffering phytotoxic effects. Three basic hallmarks distinguish hyperaccumulators from related non-hyperaccumulating taxa: a strongly enhanced rate of heavy metal uptake, a faster root-to-shoot translocation and a greater ability to detoxify and sequester heavy metals in leaves (Rascio & Navari, 2011).The term "hyperaccumulator" was coined for plant that actively take up exceedingly large amount of one or more heavy metals from the soil. Moreover heavy metals are not retained in the roots but are translocated to the shoots and accumulated in above ground organs, especially leaves at conc. 100-1000 fold higher than those found in non hyperaccumulating species. They show no symptoms of phytotoxicity (Reeves

Table 1. Cost of soil treatment

Treatment	Cost ($/Ton)	Additional Factors
Vitrification	75-425	Long-term monitoring
Landfilling	100-500	Transport/excavation/monitoring
Chemical treatment	100-500	Recycling of contaminants
Electrokinetics	20-200	Monitoring
Soil washing	120-200	Transport/Monitoring
Low temp. thermal Desorption	45-200	Transport/Monitoring
Incineration	200-600	-
Vitrification	700	-
Pneumatic fracturing	8-12	-
Excavation/Retrieval Disposal	270-460	Transport/Monitoring
Disposal alone	35-60	-

Saxena, 1999.

& Baker, 2000). Over 400 plants species composing of 101 families have been reported including members of the *Asteraceae, Brassicaceae, Caryophyllaceae, Cyperaceae, Cunouniaceae, Fabaceae, Flacourtiaceae, Lamiaceae, Poaceae, Violaceae* and *Euphobiaceae*. Metal hyperaccumulation occur in approximately 0.2% of all angiosperm and is particularly well represented in Brassicaceae (Kramer 2010). Initially the term hyper accumulator referred to plants able to accumulate more than 1mg/g Ni (dry wt.) in the shoot, an exceptionally high heavy metal conc. Considering that in vegetative organ of most plant Ni toxicity starts from 10 to 15 ug/g. Threshold values were successively provided to define the hyperaccumulation of each other heavy metal based on its specific phytotoxicity. According to such a criterion hyperaccumulator are plants that when growing on native soils concentrate >10 mg/g (1%) Mn or Zn, >1mg/g (0.1%) As, Co, Cr, Cu, Ni, Pb, Sb, Se or Ti and >0.1 mg/g (0.01) Cd in the aerial organs without suffering phytotoxic damage (Rascio & Navari-Izzo, 2011).

To date, there are approximately 400 known hyperaccumulators metal in the world and the number is increasing. However, the remediation potential of many of these plants is limited because of their slow growth and low biomass.

MOLECULAR ASPECT OF PHYTOREMEDIATION

Phytoremediation involves molecular mechanistic aspects like cellular, biochemical, genomics and proteomic etc. in different types of remediation. Different ways that plants follow for metal tolerance have been detailed in Figure 1.

Figure 1. Different mechanism of action for remediating metal polluted soil, sediment, or water

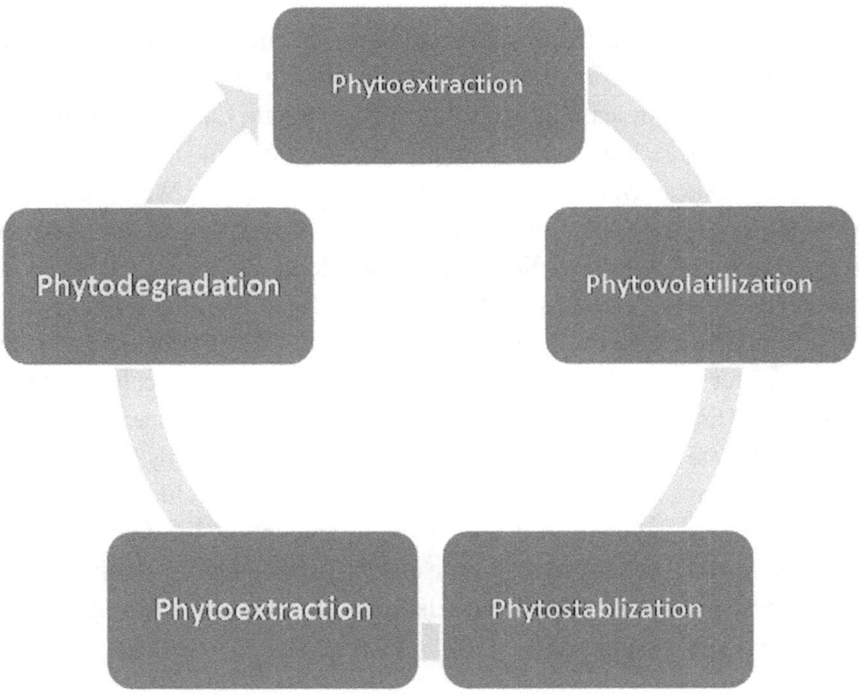

Phytoextraction

Phytoextraction is the removal of metals from contaminated soils into harvested plant tissues. The rate of phytoextraction is governed by both soil and plant characteristics. Most effort has focused on identifying appropriate plants for phytoextraction, but the benefits from this effort will be marginal unless the metals are in phytoavailable forms in the rhizosphere. The concentration of a metal in the rhizosphere can be estimated using solute transfer models that incorporate: the metal concentration in the bulk soil solution, the buffer power of the soil, diffusion coefficient of the metal, water movement, root size and morphology and rate of entry of metal into the roots (Whiting, Broadley & White, 2003). Plants absorb metals from soil through the root system and translocate them to harvestable shoots where they accumulate. Hyperaccumulator mostly used this process to extract metals from contaminated site (Sarma, 2011). Recent advances have been made in understanding the mechanisms responsible for hyperaccumulation of Zn, Cd, Ni and As by plants. Attempts to engineer metal tolerance and accumulation have so far been limited to Hg, As and Cd, and although promising results have been obtained that may be some way from practical application. More fundamental understanding of the traits and mechanism involved in hyperaccumulation are need so that phytoextraction can be optimized (McGrath & Zhao, 2003). Phytoextraction of meta-contaminated soil relies on the use of plants to extract and translocate metals to their harvestable parts. The current knowledge on how root exudates as well as organic acids and synthetic chelator amendments, might hasten both the phytoextraction of metals from soil and their translocation to shoots is explored ahead. As the ability of plants to accumulate metals is somewhat dependent on their capacity to tolerate high levels of metals in tissues, some mechanisms involved in metal accumulation by plants such as compartmentation in the vacuole and chelation in the cytoplasm are also examined (Nascimento & Xing, 2006).

Phytovolatilization

The use of plant to make volatile chemical species of soil element (Fulekar, Singh & Bhaduri, 2009). Phytovolatization is the uptake of contaminants by the roots followed by their conversion to volatile compounds and their subsequent release into the atmosphere (Singh, Labana, Pandey, Budhiraja & Jain, 2003). Indian mustard Brassicae Juncea can naturally extract selenium from the soil, where it is often present as the highly toxic as selenite or selenite (de Souza, Pickering, Walla & Terry, 2002). Using fast-growing tress such as yellow poplar and willow as phytovolatization systems could be an efficient and cost-effective option for phytoremediation. Trees have a long life-span, they have a deep and extensive root system that gives them a powerful hydraulic pull and stabilizes the soil and they produce large amount of litter which might increase the bioavailability of metal (Cluis, 2004). A promising development in this regard is the transfer of MerA to yellow poplar. Primary results indicate that transgenic poplars were able to volatilize 10 times more mercury than control plantlets (Kramer & Chardonnens, 2001). In a case study where the movement of volatile organics was monitored by fourier transform infrared spectrometry in hybrid poplars, tamarix parviflora and Medicago sativa chlorinated hydrocarbons were found to move readily through the plants did not. However, amounts of the contaminants transpired are in proportion to water flow and are relatively low, especially in the field. Selenium is a special case of a metal that is taken up by plants and volatilized, selenium can also be volatilized following conversion to dimethyl selenide by microbes and algae.

Phytostablization

In this process, the plant roots and microbial interaction can immobilized organic and some inorganic contaminant by binding them to soil particle and as a result reduce migration of contaminant to water. The process includes transpiration and root growth that immobilizes contaminants by reducing leaching, controlling erosion, creating an aerobic environment in the root zone and adding organic matter to the substrate that binds the contaminant. Microbial activity associated with the plants roots may accelerate the degradation of organic contaminants such as pesticides and hydrocarbons to nontoxic forms. Phytostabilization can be enhanced by using soil amendments that immobilize metals combined with plant species that are tolerant of high levels of contaminants and low-fertility soils or tailings. Soil amendments used to enhance immobilization may need to be periodically reapplied to maintain their effectiveness (Meena, 2013). The main goal of stabilization is to reduce the risk of an uncontrolled heavy metal transfer in groundwater and the biosphere. To reach this aim, the plant available heavy-metal fraction in soils has to be inactivated or stabilized. This can be achieved by increasing soil pH with limiting or by increasing metal binding capacity with the addition of clay minerals, iron oxides, or waste products such as gravel sludge (Krebs, Gupta, Furrer & Schulin, 1998).

Phytofiltration

Phytofiltration is the use of plants roots (rhizofiltration) or seedlings (blastofiltration) to absorb or adsorb pollutants mainly metals from water and aqueous waste streams (Prasad & Freitas 2003). Biosorption is a relatively new process that has proven very promising in the removal of contaminants from aqueous effluents. Biomaterial immobilization and chemical modification improves the adsorption capacity and stability of biosorbents. Mechanisms involved in the biosorption process include chemisorption, complexation, surface and pore adsorption-complexation, ion exchange, microprecipitation, hydroxide condensation onto the biosurface, and surface adsorption. Chemical modification and spectroscopic studies have shown that cellular components including carboxyl, hydroxyl, sulfate, sulfhydryl, phosphate, amino, amide, imine, and imidazol moieties have metal binding properties and are therefore the functional groups in the biomass. Column studies using support matrices for biomass immobilization such as silica, agar, polyacrilamide, polysulfone, alginates, cellulase, and different cross-linking agents have been performed to improve the biomass adsorption capacity and reusability (Gardea-Torresdey, Rosa la De & Peralta, 2004). P. vittata and P. creticahad similar arsenic phytofiltration efficiency and were able to rapidly remove arsenic from water to achieve arsenic levels below the new drinking water limit of 10 μg L-1. However, N. exaltata failed to reduce water arsenic to achieve the limit under the same experimental conditions. The significantly higher efficiency of arsenic phytofiltration by arsenic-hyperaccumulating fern species is associated with their ability to rapidly translocate absorbed arsenic from roots to shoots (Anawar, Garcia-Sanchez, Kul Alam & Rahman 2008). Micranthemum umbrosum commonly known as Water fern, Baby's tears, or Pearl grass, belongs to the family Linderniaceae, is one of them, that significantly absorbs both As and Cd from contaminated water. After culturing M. umbrosum for 7 days in a hydroponic experiment, the accumulation of about 1220 μg As g-1 and 800 μg Cd g-1 were observed in the leaves, from 1000 μg As L-1 and 1000 μg Cd L-1 of water, respectively and it can removed 79.3–89.5% As and 60–73.1% Cd from 200 to 1000 μg As L-1 and 300 to 3000 μg Cd L-1 solutions (Islam, 2014).

Figure 2. Destruction of organic contaminants by Phytodegradation

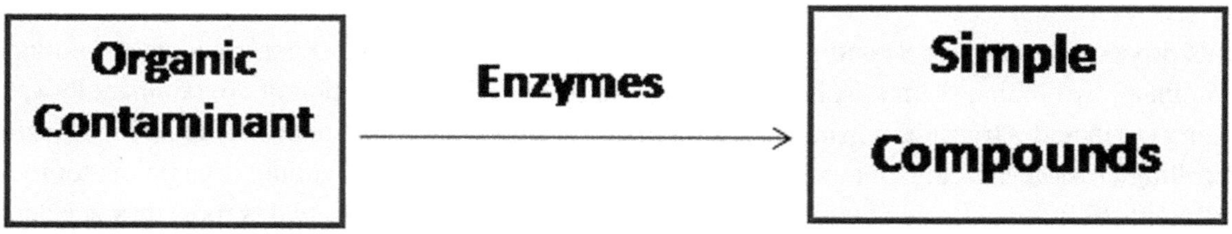

Phytodegradation

Phytodegradation which is also known as phyto-transformation is the breakdown of contaminants taken up by plants through metabolic processes within the plant, or the breakdown of contaminants surrounding the plant through the effect of enzymes produced by the plants (Figure 2). Plants are able to produce enzymes that catalyze and accelerate degradation. Hence, organic pollutants are broken down into simpler molecular forms and are incorporated into plant tissues to aid plant growth. The phytodegradation of organic compounds can take place inside the plant or within the rhizosphere of the plant. Many different compounds and classes of compounds can be removed from the environment by this method, including solvents in groundwater, petroleum and aromatic compounds in soils, and volatile compounds in the air (Newman & Reynolds, 2004). The overall metabolic process involved in phytodegradation is in some ways analogous to human metabolism of xenobiotic chemicals; thus, a 'green liver' conceptual model is often used to describe phytodegradation (Sandermann, 1994). The uptake of hydrophobic organic chemicals is very efficient while extremely hydrophilic or hydrophobic compounds are not very good candidates for phytoremediation. Such contaminants cannot be easily translocated within the plant, as they are either bound strongly to the surface of the roots or are not sorbed by roots and are actively transported through plant membranes (Burken, Shanks & Thompson, 2000).

MECHANISM OF ACCUMULATION

All plants possess a natural way to tolerate non-essential metals up to certain limits. Different families of plants have wide variety by which either they detoxify the metals or convert it to less toxic form. Few mechanisms have been focused below.

Cellular Mechanism

- **Mobilization (Uptake by Roots):** The movement of mineral elements to the root surface depends on the following: (1) Diffusion of elements along the concentration gradient formed due to uptake and thereby depletion of the element in the root vicinity. (2) Root interception, where soil volume is displaced by root volume due to root growth. (3) Mass flow, transport from bulk soil solution along the water potential. Metals are first taken into the apoplast of the roots. Then some of the total amount of the metal is transported further into the cells, some is transported further in the apoplast, and some becomes bound to cell wall substances. The movement of metals from the external solution into the cell walls is a non-metabolic, passive process, driven by diffusion or mass flow (Marschner, 2012). Metal ions are probably generally taken up into cells by membrane transport proteins designed for acquisition of nutrient metals. Transport of Fe across the root plasma membrane works by a Fe(II) transporter (IRT1), and this transporter might facilitate the transport of heavy-metal divalent cations (Cohen, Fox, Garvin & Kochian, 1998).

- **Uptake and Sequestration:** Metal sequestration in less bioactive cells such as epidermis and trichomes is one of the possible mechanisms for heavy metal detoxification in plants (Hall, 2002).
- **Xylem Transport:** The transport of heavy metals in the xylem is only partly understood. Whereas there seems to be general agreement that citrate and malate are the main transporters for Fe, Mn and Zn, Cu is perhaps chelated by histidine, glutamine, and asparagine in the xylem exudate of tomato (White, Baker, Chaney & Decker, 1981).

Translocation

Translocation from root to shoot requires a membrane transport step from root symplast into xylem apoplast. Heavy metals are largely transported apoplastically in plant tissue. Translocation of metals to the shoot is performed in the xylem and this translocation is promoted by transpiration of water via leaves. A young plant, however has a small ratio of shoot-to-root mass and in such plants the root pressure determines the translocation of xylem sap to the shoot. Translocation is also promoted for some metal ions by cation exchange at the negative charges of the xylem vessel walls (Wolterbeek, 1987). Organic pollutants pass the membrane between root symplast and xylem apoplast via simple diffusion. Transpiration stream concentration factor (TSCF) is the ratio of the concentration of a compound in the xylem fluid relative to the external solution, and is measure of uptake into the plant shoot. Entry of organic pollutants into the xylem depends on similar passive movement over membranes as their uptake into the plants. Mass flow in the xylem from the shoot creates negative tension in the xylem that pulls up water and solutes (Taiz, 1992). Plant transpiration depends on plant properties and environmental conditions. Plant species differ in transpiration rate, due to metabolic differences (for example C3/C4/CAM photosynthetic pathway) and anatomical differences (for example surface to volume ratio, stomatal density, rooting depth). Transpiration is maximal at high temperature, moderate wind, low relative air humidity as well as light intensity (Nwoko, 2013).

Sequestration

Heavy metal ions are disposed of from cytosol by sequestration. Vacuole plays a vital role in it. Mediators of sequestration are transporters in the respective membrane. Proteins like ZIP1-4, ZNT1, IRT1, COPT1, tVramp-1/3/4 and LCT1 on the plasma membrane-cytosol interface; ZAT, ABC type, AtMRP, HMT1, CAX2 seen in vacuoles; and RAN1 seen in golgi bodies. Manipulations of these transporters to achieve removal of metal ions from the cell hold great potential (Tong, Kneer & Zhu, 2004).

Role of ATP Type Transporter

While the concept of H+-coupling has dominated studies of energy-dependent organic solute transport in plants for over two decades, recent studies have demonstrated the existence of a group of organic solute transporters, belonging to the ATP-binding cassette (ABC) superfamily, that are directly energized by

MgATP rather than by a transmembrane H+-electrochemical potential difference. ABC transporters are constituted of one or two copies each of an integral membrane sector and cytosolically oriented ATP-binding domain. To date, two major subclasses, the multidrug resistance-associated proteins (MRPs) and multidrug resistance proteins (MDRs) (so named because of the phenotypes conferred by their animal prototypes), have been identified molecularly in plants. MRPs are considered to participate in the transport of exogenous and endogenous amphipathic anions and glutathionated compounds from the cytosol into the vacuole (Rea, drozdowicz & Martinoia, 1998). Hmt1 from S.pombe was the first gene identified that encodes a protein involved in vacuolar sequestration. Hmt1 was cloned on the basis of the complementation of a Cd-hypersensitive *S. pombe* strain lacking vacuolar sulphide containing PC-metal complexes. The corresponding protein belongs to the large family of ABC-type transporters. It localizes to the vacuolar membrane and transports PC-Cd complexes and apo-PCs (Ortiz, Ruscitti, McCue & Ow, 1995). Two yeast ATP-binding cassette (ABC) transporters sequester heavy metals into vacuoles. ScYCF1 is an ABC transporter of Saccharomyces cerevisae that contributes to cadmium resistance by pumping glutathione-conjugated cadmium into the vacuole (Li et al., 1997) and SpHMT1, a half size ABC transporter from Schizosaccharomyces pombe transports phytochelatin-Cd complexes into the vacuole (Kim, Maeshima, Martinoia & Lee, 2007). In *S. cerevisiae*, the transporter YCF1 is required for Cd tolerance of yeast cells (Szczypka, Wemmie, Moye-Rowley & Thiele, 1994) YCF1-deficient cells are Cd hypersensitive. From transport studies with vacuolar membrane vesicles purified from wild type and ycf- *S. cerevisiae* cells it is known that YCF1 mediates the Mg-ATP-dependent transport of bis (glutathionnato) cadmium into vacuoles (Li et al, 1997).

Cation Diffusion Facilitator (CDFs)

CDF family is a ubiquitous family of heavy metal transporters, these are group of proten which facilitates metal ion transport. Members of this family have been described in yeast, animals and plants (Nies 1995). ShMTP1 was found to confer Mn2+ tolerance to yeast by internal sequestration rather than by efflux of Mn2+. ShMTP1 conferred Mn2+ tolerance through internal sequestration. ShMTP1 proteins are members of the CDF family involved in conferring Mn2+ tolerance and that at least one of these proteins(ShMTP1) confers tolerance by sequestering Mn2+ into internal organelles (Delhaize, Kataoka, Hebb, White & Ryan, 2003). *S. cerevisiae* proteins COT1 and ZRC1 confer tolerance of cobalt (COT1). Expression of the Arabidopsis and poplar MTP11 cation diffusion facilitators in a manganese-hypersensitive yeast mutant restores manganese tolerance to wild type levels. Microsomes from yeast expressing AtMTP11 exhibit enhanced manganese uptake. Arabidopsis mtp11 mutants exhibit enhanced manganese concentrations in shoots and roots (Peiter *et al.*, 2007). The ZRC1 gene was cloned as a multicopy suppressor of zinc toxicity in Saccharomyces cerevisiae. Zrc1 belongs to CDF (cation diffusion facilitator) family. The transporters belonging to this family are thought to play an important role in metal detoxification (Miyabi, Izawa & Inoue, 2000). Ectopic overexpression of AtHMA3 improved plant tolerance to Cd, Cobalt, Pb and Zn, Cd accumulation increased by about 2 to 3 fold in plants overexpressing AtHMA3 compared with wild type plants. Thus AtHMA3 likely plays a role in the detoxification of biological (Zn) and nonbiological (Cd, cobalt and Pb) heavy metals by participating in their vacuolar sequestration (Morel *et al*, 2009).

Detoxification

Although micronutrients such as Zn, Mn, Ni and Cu are essential for plant growth and development, high intracellular concentration of these ions can be toxic. To deal with this potential stress, common non accumulator plants have evolved several mechanisms to control the homeostasis of intracellular ions. Such mechanism include regulation of ion influx (stimulation of transporter activity at low intracellularion supply and inhibition at high concentrations) and extrusion of intracellular ions back into the external solution metal hyperaccumulator species, capable of taking up metals in the thousands of ppm, possess additional detoxification mechanism.(Lasat, 2000).

Hyperaccumulator plants have developed detoxification mechanism which makes to withstand the high amount of heavy metal in their cellular parts without suffering any phytotoxic effect. Plants have developed their own mechanism to protect themselves from negative impact of high concentration of metal. The detoxifying/sequestrating mechanisms in aerial organs of hyperaccumulators consist mainly in heavy metal complexation with ligands and/or in their removal from metabolically active cytoplasm by moving them into inactive compartments, mainly vacuoles and cell walls (Revathi & Venugopal, 2013) Chelation of metals in the cytosol by peptides such as phytochelatin for the repair of stress-damaged proteins and compartmentation of metals in the vacuole by tonoplast-located transporters. Detoxification occurs in hyperaccumulators through vacuolar compartmentalization and volatilization mechanism.

Vacuolar Compartmentalization

Heavy metal is stored in vacuoles in plant cells. Vacuolar compartmentalization is quite effective in controlling the distribution and concentration of metal ions. Vacuoles restrict metal ions to limited space thus other parts of cell have no access to these metal ions.

Transport of heavy metal from cytosol to vacuole is potentially important for heavy metal tolerance in plant. Earlier studies showed that the vacuole is the site for the accumulation of a number of heavy metals including Zn and Cd (Ernst, 1992, De, 2000). Several vacuolar heavy metal transporters, such as HMA, and ABC transporter, HMT, and CAX, transport either free or conjugated heavy metals across biological membranes extruding from the cytosol and thus play an important role in the deposition, sequestration, and storage of excess or toxic metals in vacuoles (Martinoia, maeshima & Neuhaus, 2007, Schneider et. al., 2009). Detoxification of Cd in Paxillusinvolotus involves binding of Cd to the cell walls and accumulation in the vacuole (Blaudez, Bottom & Chalot, 2000). Recently it has been shown that the formation of a glutathione conjugate with cyanidin-3-glucoside is a critical step in the biosynthesis and vacuolar accumulation of anthocyaninin maize (Marrs, Alfenito, Lloyd & Walbot, 1995). The GSH-dependent detoxification pathway (including vacuolar compartmentalization) is one which is likely to be used for the detoxification of many xenobiotics and could be the basis for the selectivity and resistance mechanism of some herbicides. (Colemann, Blake-Kalff & Davies, 1997). MTP1, a gene encoding a protein localized at tonoplast.

Volatilization Mechanism

By converting metal ions into volatile state, some plant species avoid the lasting damage caused by accumulation and long-time stay of heavy metals. Transgenic tobacco plants engineered with bacterial merA and merB genes via the chloroplast genome were investigated to study the uptake, translocation of different forms of mercury (Hg) from roots to shoots, and their volatilization. The combined expression of merAB via the chloroplast genome enhanced conversion of Hg2+ into Hg0 conferred tolerance by rapid volatilization and increased uptake of different forms of mercury, surpassing the concentrations found in the soil (Hussein, Ruiz, Terry & Daniell, 2007).

Detoxification by Chelation of Trace Metals

Hyperaccumulated metals in plant are assumed to be bound to ligands, such as organic acids, amino acids, peptides and proteins. Metallothineins and phytochelatins play a prominent role in detoxification process.

Metallothineins

Higher plants contain two major types of cysteine-rich, metal binding peptides, the metallothioneins (MTs) and the phytochelatins. MTs are gene coded polypeptides that are usually classified into two groups (Hall, 2002). Class 1 MTs possess cysteine residues that align with a mammalian (equine) renal MT; class 2 MTs also possess similar cysteine clusters but these cannot be easily aligned with class 1 MTs (Robinson, Tommey, Kuske & Jackson, 1993; Prasad, 1999). MT3 and MT4 types have also been identified. Transgenic plants expressing metallothioneins have been created, and although these plants exhibited enhanced tolerance to high metal concentration, the uptake of metals was not enhanced. To enhance higher plant metal sequestration, the yeast metallothionein CUP1 was introduced into tobacco plants, and the cup1 gene expression and cu and cd phytoextraction were determined (Thomas et al, 2003).

Exposition of A. thaliana to cadmium ions lead to the increased expression of the genes for O-acetylserine(thiol)lyase (OASTL), gamma-glutamyl cysteine synthetase (GCS) and glutathione synthetase (GS). All these findings support importance of sulphur compounds in cadmium detoxification (Lee et al. 2003). External application of GSH does not lead to the increased tolerance to cadmium, but it is toxic also for plants without Cd exposition. This fact indicates the role of Gsh in biosynthesis of phytochelatins.

Phytochelatins

Phytochelatins are capable of binding various metals including Cd, Cu, Zn, Cd, or metalloid As. The structure of PCs is (y Glu-Cys)nX in which X is Gly, Y-Ala, Ser or Glu and n=2-11 depending on the organism. (Munir & Ahmad, 2010). The two standard characteristics attributed to the PCs are that (1) PC synthesis can be stimulated in cells exposed to various other metal ions such as CU2+, Zn2+, Pb2+, and Ag+, and (2) the formed PCs are capable of binding to multiple types of metals and metalloids (Gekeler, 1989). Phytochelatins are enzymatically synthesized Cys-rich peptides. After the structures of PCs had been elucidated and it was found that these peptides are distributed widely in the plant kingdom, it was proposed that PCs were the functional equivalent of MTs (Grill, Winnacker & Zenk, 1987). PCs are synthesized from glutathione (y-glutamyl-cysteinyl-glycine) in steps mediated by PC synthase. Genes (CAD1, PCS1) of the enzyme have been isolated from plants, fission yeast and some animals. Inhibition studies of PC biosynthesis via glutathione have demonstrated their fundamental roles in the metal detoxification in yeast and fungi, green algae and some aquatic plants, and also in the suspension-cultured cells and intact tissues in higher plants (Inouhe, 2005). Phytochelatins play major roles in metal detoxification in plants and fungi. However genes encoding phytochelatin synthases have not been identified. By screening for plant genes mediating metal tolerance wheat cDNA, TaPCS1 was identified whose expression in *Saccharomyces cerevisae* results in a dramatic increase in cadmium tolerance (Clemens, Kim, Neumann & Schroeder, 1999). The analysis of a *Schizosaccharomyces pombe* PCS knockout strain and of *S. cerevisae* cells expressing PCS genes from different sources showed that the formation of PCs confer Cd tolerance and to a limited degree Cu, arsenate and Hg tolerance. Based on these yeast studies, there is no indication for a role in for instance Zn, Co, or Ni detoxification. In the meantime PC-mediated as tolerance was confirmed for plants (Oksman & Barz, 2002). Agrobacterium mediated transformation, the induction and overexpression of a wheat gene encoding phytochelatin synthase (TaPCS1) in this particularly plant greatly increased its tolerance to metals such as Pb and Cd, developing seedling roots 160% longer than wild type plants. In addition, seedlings of transformed plants grown in mining soil containing high levels of Pb(1572 ppm) accumulated double concentration of this heavy metal than wild type (Gisbert et al, 2003). Arabidopsis thaliana AtPCS1 gene, encoding phytochelatin synthase, moderate expression levels showed significantly higher tolerance to Cd and Zn stress, but accumulated significantly less Cd and Zn than wild type plants in both shoot and root tissues (Gasic & Korban, 2007). A modified Arabidopsis thaliana PCS sequence (AtPCS1) was active in Escherichia coli. When AtPCS1 was overexpressed in Arabidopsis from a strong constitutive Arabidopsis actin regulatory sequence (A2), the A2::AtPCS1 plants were highly resistant to arsenic, accumulating 20-100 times more biomass on 250 and 300 uM arsenate than wild type (WT) (Li et al, 2004).

CONCLUSION

The chapter has focused on evidences that show molecular mechanism followed by plants to resist high concentration of metals. Hyperaccumlator can serve the solution for decontamination of heavy metal polluted soil. Although many microbial species has been identified for the bioremediation but they all are effected by edaphic characteristics. Phytoremediation have minimum impact on the environment and minimizing potential toxic by product. It is well established fact that phytoremediation has benefit has benefits to restore balance to a stressed environment. A large number of plants have already been identified which serve as hyperaccumulator, the same can be genetically modified introducing the genes responsible for remediation into other plants. The chapter has focused on recent evidence that identifies potential mechanism that may be involved in the resistance, tolerance as well as hyperaccumulation of heavy metals. The identification of individual function of metal transporters will help to develop knowledge of plant metal homeostasis. Transgenic plants can serve as a renaissance in field of phytoremediation. In depth research study is needed to find out which plant is maximum resistant and best adapted in particular metallic environment or region.

REFERENCES

Anawar, H. M., Garcia-Sanchez, A., Kul Alam, M. T., & Rahman, M. M. (2008). Phytofiltration of water polluted with arsenic and heavy metals. *International Journal of Environment and Pollution*, *33*(2-3), 292–312. doi:10.1504/IJEP.2008.019400

Ashraf, M., Ozturk, M., & Ahmad, M. S. A. (2010). *Plant adaptation and phytoremediation*. New York: Springer. doi:10.1007/978-90-481-9370-7

Blaudez, D., Botton, B., & Chalot, M. (2000). Cadmium uptake and subcellular compartmentation in the ectomycorrhizal fungus Paxillus involutus. *Microbiology*, *146*(5), 1109–1117. doi:10.1099/00221287-146-5-1109 PMID:10832638

Burken, J. G., Shanks, J. V., & Thompson, P. L. (2000). Metabolism of Explosives and Nitroaromatic Compounds. *Biodegradation of Nitroaromatic Compounds and Explosives*, 239.

Clemens, S., Kim, E. J., Neumann, D., & Schroeder, J. I. (1999). Tolerance to toxic metals by a gene family of phytochelatin synthases from plants and yeast. *The EMBO Journal*, *18*(12), 3325–3333. doi:10.1093/emboj/18.12.3325 PMID:10369673

Clemens, S., Palmgren, M. G., & Krämer, U. (2002). A long way ahead: Understanding and engineering plant metal accumulation. *Trends in Plant Science*, *7*(7), 309–315. doi:10.1016/S1360-1385(02)02295-1 PMID:12119168

Cluis, C. (2004). Junk-greedy greens: Phytoremediation as a new option for soil decontamination. *BioTeach Journal*, *2*(6), 1-67.

Cohen, C. K., Fox, T. C., Garvin, D. F., & Kochian, L. V. (1998). The role of iron-deficiency stress responses in stimulating heavy-metal transport in plants. *Plant Physiology*, *116*(3), 1063–1072. doi:10.1104/pp.116.3.1063 PMID:9501139

Coleman, J., Blake-Kalff, M., & Davies, E. (1997). Detoxification of xenobiotics by plants: Chemical modification and vacuolar compartmentation. *Trends in Plant Science*, *2*(4), 144–151. doi:10.1016/S1360-1385(97)01019-4

Cunningham, S. D., & Berti, W. R. (1993). Remediation of contaminated soils with green plants: An overview. *In Vitro Cellular & Developmental Biology. Plant*, *29*(4), 207–212. doi:10.1007/BF02632036

Danh, L. T., Truong, P., Mammucari, R., Tran, T., & Foster, N. (2009). Vetiver grass, Vetiveria zizanioides: A choice plant for phytoremediation of heavy metals and organic wastes. *International Journal of Phytoremediation*, *11*(8), 664–691. doi:10.1080/15226510902787302 PMID:19810597

De, D. N. (2000). *Plant cell vacuoles*. Academic Press.

De Souza, M. P., Pickering, I. J., Walla, M., & Terry, N. (2002). Selenium assimilation and volatilization from selenocyanate-treated Indian mustard and muskgrass. *Plant Physiology*, *128*(2), 625–633. doi:10.1104/pp.010686 PMID:11842165

Delhaize, E., Kataoka, T., Hebb, D. M., White, R. G., & Ryan, P. R. (2003). Genes encoding proteins of the cation diffusion facilitator family that confer manganese tolerance. *The Plant Cell*, *15*(5), 1131–1142. doi:10.1105/tpc.009134 PMID:12724539

Eapen, S., Singh, S., & Dsouza, S. F. (2007). Advances in development of transgenic plants for remediation of xenobiotic pollutants. *Biotechnology Advances*, *25*(5), 442–451. doi:10.1016/j.biotechadv.2007.05.001 PMID:17553651

Ensley, B. D. (2000). *Phytoremediation of toxic metals: using plants to clean up the environment*. John Wiley & s Sons, Incorporated.

Ernst, W. H. O., Verkleij, J. A. C., & Schat, H. (1992). Metal tolerance in plants. *Acta Botanica Neerlandica*, *41*(3), 229–248. doi:10.1111/j.1438-8677.1992.tb01332.x

Fulekar, M. H., Singh, A., & Bhaduri, A. M. (2009). Genetic engineering strategies for enhancing phytoremediation of heavy metals. *African Journal of Biotechnology*, *8*(4).

Gardea-Torresdey, J. L., De La Rosa, G., & Peralta-Videa, J. R. (2004). Use of phytofiltration technologies in the removal of heavy metals: A review. *Pure and Applied Chemistry*, *76*(4), 801–813. doi:10.1351/pac200476040801

Gasic, K., & Korban, S. S. (2007). Expression of Arabidopsis phytochelatin synthase in Indian mustard (Brassica juncea) plants enhances tolerance for Cd and Zn. *Planta*, *225*(5), 1277–1285. doi:10.1007/s00425-006-0421-y PMID:17086401

Gekeler, W., Grill, E., Winnacker, E. L., & Zenk, M. H. (1989). Survey of the plant kingdom for the ability to bind heavy metals through phytochelatins. *Zeitschrift für Naturforschung C*, *44*(5-6), 361–369.

Gisbert, C., Ros, R., De Haro, A., Walker, D. J., Bernal, M. P., Serrano, R., & Navarro-Aviñó, J. (2003). A plant genetically modified that accumulates Pb is especially promising for phytoremediation. *Biochemical and Biophysical Research Communications*, *303*(2), 440–445. doi:10.1016/S0006-291X(03)00349-8 PMID:12659836

Grill, E., Winnacker, E. L., & Zenk, M. H. (1987). Phytochelatins, a class of heavy-metal-binding peptides from plants, are functionally analogous to metallothioneins. *Proceedings of the National Academy of Sciences of the United States of America*, *84*(2), 439–443. doi:10.1073/pnas.84.2.439 PMID:16593801

Hall, J. L. (2002). Cellular mechanisms for heavy metal detoxification and tolerance. *Journal of Experimental Botany*, *53*(366), 1–11. doi:10.1093/jexbot/53.366.1 PMID:11741035

Hussein, H. S., Ruiz, O. N., Terry, N., & Daniell, H. (2007). Phytoremediation of mercury and organomercurials in chloroplast transgenic plants: Enhanced root uptake, translocation to shoots, and volatilization. *Environmental Science & Technology*, *41*(24), 8439–8446. doi:10.1021/es070908q PMID:18200876

Inouhe, M. (2005). Phytochelatins. *Brazilian Journal of Plant Physiology*, *17*(1), 65–78. doi:10.1590/S1677-04202005000100006

Islam, M. (2014). *Phytofiltration of arsenic and cadmium from the water environment using Micranthemum umbrosum*. Academic Press.

Kim, D. Y., Bovet, L., Maeshima, M., Martinoia, E., & Lee, Y. (2007). The ABC transporter AtPDR8 is a cadmium extrusion pump conferring heavy metal resistance. *The Plant Journal*, *50*(2), 207–218. doi:10.1111/j.1365-313X.2007.03044.x PMID:17355438

Krämer, U. (2010). Metal hyperaccumulation in plants. *Annual Review of Plant Biology*, *61*(1), 517–534. doi:10.1146/annurev-arplant-042809-112156 PMID:20192749

Krämer, U., & Chardonnens, A. (2001). The use of transgenic plants in the bioremediation of soils contaminated with trace elements. *Applied Microbiology and Biotechnology*, *55*(6), 661–672. doi:10.1007/s002530100631 PMID:11525612

Krebs, R., Gupta, S. K., Furrer, G., & Schulin, R. (1998). Solubility and plant uptake of metals with and without liming of sludge-amended soils. *Journal of Environmental Quality*, *27*(1), 18–23. doi:10.2134/jeq1998.00472425002700010004x

Lasat, M. M. (2000). Phytoextraction of metals from contaminated soil: A review of plant/soil/metal interaction and assessment of pertinent agronomic issues. *Journal of Hazardous Substance Research*, *2*(5), 1–25.

Lee, S., Moon, J. S., Ko, T. S., Petros, D., Goldsbrough, P. B., & Korban, S. S. (2003). Overexpression of Arabidopsis phytochelatin synthase paradoxically leads to hypersensitivity to cadmium stress. *Plant Physiology*, *131*(2), 656–663. doi:10.1104/pp.014118 PMID:12586889

Li, Y., Dhankher, O. P., Carreira, L., Lee, D., Chen, A., Schroeder, J. I., & Meagher, R. B. (2004). Overexpression of phytochelatin synthase in Arabidopsis leads to enhanced arsenic tolerance and cadmium hypersensitivity. *Plant & Cell Physiology*, *45*(12), 1787–1797. doi:10.1093/pcp/pch202 PMID:15653797

Li, Z. S., Lu, Y. P., Zhen, R. G., Szczypka, M., Thiele, D. J., & Rea, P. A. (1997). A new pathway for vacuolar cadmium sequestration in Saccharomyces cerevisiae: YCF1-catalyzed transport of bis (glutathionato) cadmium. *Proceedings of the National Academy of Sciences of the United States of America*, *94*(1), 42–47. doi:10.1073/pnas.94.1.42 PMID:8990158

Marrs, K. A., Alfenito, M. R., Lloyd, A. M., & Walbot, V. (1995). *A glutathione S-transferase involved in vacuolar transfer encoded by the maize gene Bronze-2*. Academic Press.

Marschner, H. (2012). *Marschner's mineral nutrition of higher plants* (Vol. 89). Academic press.

Martinoia, E., Maeshima, M., & Neuhaus, H. E. (2007). Vacuolar transporters and their essential role in plant metabolism. *Journal of Experimental Botany*, *58*(1), 83–102. doi:10.1093/jxb/erl183 PMID:17110589

McGrath, S. P., & Zhao, F. J. (2003). Phytoextraction of metals and metalloids from contaminated soils. *Current Opinion in Biotechnology*, *14*(3), 277–282. doi:10.1016/S0958-1669(03)00060-0 PMID:12849780

Meena, S. K. (2013). Bioterrorism is a dangerous to environment-A review. *Innovare Journal of Agricultural Sciences*, 5-7.

Miyabe, S., Izawa, S., & Inoue, Y. (2000). Expression of ZRC1 coding for suppressor of zinc toxicity is induced by zinc-starvation stress in Zap1-dependent fashion in Saccharomyces cerevisiae. *Biochemical and Biophysical Research Communications*, *276*(3), 879–884. doi:10.1006/bbrc.2000.3580 PMID:11027563

Morel, M., Crouzet, J., Gravot, A., Auroy, P., Leonhardt, N., Vavasseur, A., & Richaud, P. (2009). AtHMA3, a P1B-ATPase allowing Cd/Zn/Co/Pb vacuolar storage in Arabidopsis. *Plant Physiology*, *149*(2), 894–904. doi:10.1104/pp.108.130294 PMID:19036834

Nascimento, C. W. A. D., & Xing, B. (2006). Phytoextraction: a review on enhanced metal availability and plant accumulation. *Scientia Agricola, 63*(3), 299-311.

Newman, L. A., & Reynolds, C. M. (2004). Phytodegradation of organic compounds. *Current Opinion in Biotechnology*, *15*(3), 225–230. doi:10.1016/j.copbio.2004.04.006 PMID:15193330

Nies, D. H. (1995). The cobalt, zinc, and cadmium efflux system CzcABC from Alcaligenes eutrophus functions as a cation-proton antiporter in Escherichia coli. *Journal of Bacteriology*, *177*(10), 2707–2712. doi:10.1128/jb.177.10.2707-2712.1995 PMID:7751279

Nwoko, C. O. (2010). Trends in phytoremediation of toxic elemental and organic pollutants. *African Journal of Biotechnology*, *9*(37), 6010–6016.

Oksman-Caldentey, K. M., & Barz, W. H. (Eds.). (2002). *Plant biotechnology and transgenic plants* (Vol. 92). CRC press. doi:10.1201/9780203910849

Ortiz, D. F., Kreppel, L., Speiser, D. M., Scheel, G., McDonald, G., & Ow, D. (1992). Heavy metal tolerance in the fission yeast requires an ATP-binding cassette-type vacuolar membrane transporter. *The EMBO Journal*, *11*(10), 3491. PMID:1396551

Ortiz, D. F., Ruscitti, T., McCue, K. F., & Ow, D. W. (1995). Transport of metal-binding peptides by HMT1, fission yeast ABC-type vacuolar membrane protein. *The Journal of Biological Chemistry*, *270*(9), 4721–4728. doi:10.1074/jbc.270.9.4721 PMID:7876244

Peiter, E., Montanini, B., Gobert, A., Pedas, P., Husted, S., Maathuis, F. J., & Sanders, D. (2007). A secretory pathway-localized cation diffusion facilitator confers plant manganese tolerance. *Proceedings of the National Academy of Sciences of the United States of America*, *104*(20), 8532–8537. doi:10.1073/pnas.0609507104 PMID:17494768

Prasad, M. N. V. (1999). Metallothioneins and metal binding complexes in plants. In *Heavy metal stress in plants* (pp. 51–72). Springer Berlin Heidelberg. doi:10.1007/978-3-662-07745-0_3

Rascio, N., & Navari-Izzo, F. (2011). Heavy metal hyperaccumulating plants: How and why do they do it? And what makes them so interesting? *Plant Science*, *180*(2), 169–181. doi:10.1016/j.plantsci.2010.08.016 PMID:21421358

Rea, P. A., Li, Z. S., Lu, Y. P., Drozdowicz, Y. M., & Martinoia, E. (1998). From vacuolar GS-X pumps to multispecific ABC transporters. *Annual Review of Plant Biology*, *49*(1), 727–760. doi:10.1146/annurev.arplant.49.1.727 PMID:15012252

Reeves, R. D., & Baker, A. J. (2000). *Metal-accumulating plants. Phytoremediation of toxic metals: using plants to clean up the environment*. New York: Wiley.

Revathi, S., & Venugopal, S. (2013). Physiological and biochemical mechanisms of heavy metal tolerance. *International Journal of Environmental Sciences*, *3*(5), 1339.

Robinson, N. J., Tommey, A. M., Kuske, C., & Jackson, P. J. (1993). Plant metallothioneins. *The Biochemical Journal*, *295*(Pt 1), 1–10. doi:10.1042/bj2950001 PMID:8216201

Sandermann, H. Jr. (1994). Higher plant metabolism of xenobiotics: Thegreen liverconcept. *Pharmacogenetics and Genomics*, *4*(5), 225–241. doi:10.1097/00008571-199410000-00001 PMID:7894495

Sarma, H. (2011). Metal hyperaccumulation in plants: A review focusing on phytoremediation technology. *Journal of Environmental Science and Technology*, *4*(2), 118–138. doi:10.3923/jest.2011.118.138

Saxena, P. K. KrishnaRaj, S., Dan, T., Perras, M. R., & Vettakkorumakankav, N. N. (1999). Phytoremediation of heavy metal contaminated and polluted soils. In Heavy metal stress in plants (pp. 305-329). Springer Berlin Heidelberg.

Schneider, T., Schellenberg, M., Meyer, S., Keller, F., Gehrig, P., Riedel, K., & Martinoia, E. (2009). Quantitative detection of changes in the leaf-mesophyll tonoplast proteome in dependency of a cadmium exposure of barley (Hordeum vulgare L.) plants. *Proteomics*, *9*(10), 2668–2677. doi:10.1002/pmic.200800806 PMID:19391183

Singh, O. V., Labana, S., Pandey, G., Budhiraja, R., & Jain, R. K. (2003). Phytoremediation: An overview of metallic ion decontamination from soil. *Applied Microbiology and Biotechnology*, *61*(5-6), 405–412. doi:10.1007/s00253-003-1244-4 PMID:12764555

Szczypka, M. S., Wemmie, J. A., Moye-Rowley, W. S., & Thiele, D. J. (1994). A yeast metal resistance protein similar to human cystic fibrosis transmembrane conductance regulator (CFTR) and multidrug resistance-associated protein. *The Journal of Biological Chemistry*, *269*(36), 22853–22857. PMID:7521334

Taiz, L. (1992). The plant vacuole. *The Journal of Experimental Biology*, *172*(1), 113–122. PMID:9874729

Thomas, J. C., Davies, E. C., Malick, F. K., Endreszl, C., Williams, C. R., Abbas, M., & Ostenkowski, P. et al. (2003). Yeast metallothionein in transgenic tobacco promotes copper uptake from contaminated soils. *Biotechnology Progress*, *19*(2), 273–280. doi:10.1021/bp025623q PMID:12675559

Tong, Y. P., Kneer, R., & Zhu, Y. G. (2004). Vacuolar compartmentalization: A second-generation approach to engineering plants for phytoremediation. *Trends in Plant Science*, *9*(1), 7–9. doi:10.1016/j.tplants.2003.11.009 PMID:14729212

Van Aken, B. (2008). Transgenic plants for phytoremediation: Helping nature to clean up environmental pollution. *Trends in Biotechnology*, *26*(5), 225–227. doi:10.1016/j.tibtech.2008.02.001 PMID:18353473

Vara Prasad, M. N., & de Oliveira Freitas, H. M. (2003). Metal hyperaccumulation in plants: Biodiversity prospecting for phytoremediation technology. *Electronic Journal of Biotechnology*, *6*(3), 285–321.

White, M. C., Baker, F. D., Chaney, R. L., & Decker, A. M. (1981). Metal complexation in xylem fluid II. Theoretical equilibrium model and computational computer program. *Plant Physiology*, *67*(2), 301–310. doi:10.1104/pp.67.2.301 PMID:16661665

Whiting, S. N., Broadley, M. R., & White, P. J. (2003). Applying a solute transfer model to phytoextraction: Zinc acquisition by Thlaspi caerulescens. *Plant and Soil*, *249*(1), 45–56. doi:10.1023/A:1022542725880

Wolterbeek, H. T. (1987). Cation exchange in isolated xylem cell walls of tomato. I. Cd2+ and Rb+ exchange in adsorption experiments. *Plant, Cell & Environment*, *10*(1), 39–44. doi:10.1111/j.1365-3040.1987.tb02077.x

KEY TERMS AND DEFINITIONS

Accumulation Mechanism: Plants adapt different mechanism for transport and accumulation of heavy metals via xylem and phloem transport. Vacuoles play a vital role in accumulation of heavy metals.

Cellular Mechanism: It involves all those transport ways through which xylem transport heavy metal from soil to different parts of plant including uptake, sequestration and translocation methods.

Detoxification: The process by which contaminants are converted to less toxic or non-toxic form of compounds. Metallothineins and phytochelatins play a prominent role in detoxification process.

Heavy-Metals: Class of toxicants that cannot be broken down to non-toxic form easily. It has high density and high atomic number. These metals keep on accumulating in system posing serious threat to environment and organisms.

Hyperaccumulator: Plants which are capable of growing in high metal concentration and have capability to detoxify it by different mechanisms. A large number of plant families have been identified which can tolerate metal contaminants.

Phytoremediation: It is method of applying plants for in situ removal or degradation of contaminants into non-toxic form or less toxic form.

Sequestration: Heavy metal ions are disposed of from cytosol by sequestration. Vacuole plays a vital role in it. Mediators of sequestration are transporters in the respective membrane.

Translocation: The process by heavy metals or contaminants is Trans located to shoots after intake by roots. Translocation factor has been found prominent in estimating translocation.

Transporters: Heavy metals enter plant through specific metal transporters. Proteins like ZIP14, ZNT1, IRT1, COPT1, tVramp-1/3/4 and LCT1 on the plasma membrane-cytosol interface; ZAT, ABC type, AtMRP, HMT1, CAX2 seen in vacuoles; and RAN1 play a vital in transport phenomenon.

Chapter 11
Mycoremediation of Lignocelluloses

Saritha Vara
GITAM University, India

ABSTRACT

The most abundant aromatic biopolymer on earth Lignin is extremely recalcitrant to degradation. It creates a barrier to solutions or enzymes by linking to both hemicellulose and cellulose preventing the penetration of lignocellulolytic enzymes into the interior lignocellulosic structure. Global attention has been gained by fungi owing to the potential use of their versatile enzymes for agriculture, medicines, industries and bioremediation. The combination of extracellular ligninolytic enzymes, mediators, organic acids and accessory enzymes make some of the basidiomycete white-rot fungi to be able to degrade lignin efficiently. This review describes remediation of lignocelluloses by fungi, properties of fungi, their spatial distribution and the mechanisms of action which render them attractive candidates in biotechnological applications like biopulping, animal feed, genetic engineering and space exploration.

INTRODUCTION

Principal structural wood polymers are cellulose, hemicelluloses and lignin, which are the most abundant biopolymers in the earth's carbon cycle. These polymers form the lignocellulose complex in all woody tissues. This is a complex structure in which the cellulose is surrounded by a monolayer of hemicellulose and embedded in a matrix of hemicellulose and lignin. The highly ordered structure of cellulose microfibril aggregates which are embedded in a matrix of hemicelluloses and lignin provides the basis for the mechanical strength of the lignocelluloses (Salmén & Burgert, 2009). Furthermore lignin specifically creates a barrier to enzymatic attack (Daniel, 2003) the highly crystalline structure of cellulose is insoluble in water while hemicellulose and lignin create a protective sheath around cellulose (Karin & Manfred, 2012). Mycoremediation is a term that is applied to the use of fungi to clear the environment of contaminating substances. It can also be understood as bioremediation by fungi. Fungi perform mycoremediation by using extracellular enzymes to reduce/remove toxic wastes from the environment. Lignocellulose degradation apart from upgrading the quality of the degraded biomass

DOI: 10.4018/978-1-5225-2325-3.ch011

also helps in producing variety of commercial enzymes and many additional by products which are of interest simultaneously (Sharma & Arora, 2015).

Biochemical systems to degrade the lignocelluloses to perform conversion and mineralization of wood to carbon dioxide and water have been developed only by higher fungi. Most of these fungi belong to the Basidiomycetes. Though related phylogenetically (Binder & Hibbett, 2002; Floudas et al., 2012) strategies of their wood degradation are diverse: wood polysaccharides are primarily degraded by brown-rot fungi which leave behind a polymeric but highly modified lignin, while almost all polymeric wood constituents are degraded at similar rates by white-rot fungi. Selective white-rot fungi, which lack the ability to degrade cellulose efficiently, cause extensive delignification of wood. Soft-rot decay leading to softening of wet wood is performed by Ascomycetes and Deuteromycetes.

STRUCTURE OF LIGNOCELLULOSE

Three main polymeric constituents that form wood and other lignocellulosic materials are cellulose, lignin, and hemicelluloses (Higuchi, 1997). Cellulose, a linear and highly ordered (often crystalline) polymer of cellobiose (D-glucopyranosyl-β-1, 4-D-glucopyranose) represents over 50% of wood weight. On the other hand contrastingly lignin is a three-dimensional network buildup of dimethoxylated (syringyl, S), monomethoxylated (guaiacyl, G) and non-methoxylated (*p*-hydroxyphenyl, H) phenylpropanoid units which are derived from the corresponding *p*-hydroxycinnamyl alcohols, giving rise to variety of subunits including different ether and C—C bonds. Analytical pyrolysis has enabled identification of acetylated lignin units recently identified in non-woody plants (Del Río, Gutiérrez & Martínez, 2004).

PROPERTIES OF LIGNOCELLULOSES

The word lignin is derived from Latin word 'Lignum' meaning wood. It embodies three components of lignocellulosic biomass in addition to cellulose and hemicellulose. Lignin, the second most abundant natural substance in nature after cellulose is produced approximately 5 x 10^6 metric tons annually through industrial processes (Mai, Majcherczyk, & Huttermann, 2000). Lignin, most abundant renewable source of aromatic polymers owes it degradation mandatory for carbon recycling. Lignin chemically is a cross-linked macromolecule derived from the oxidative coupling of monolignols, mainly hydroxycinnamyl alcohols which are three main components comprising of p-coumaryl, coniferyl and synapyl alcohols. The composition of lignin is plant-specific, with molecular weight and linkage motifs varying according to plant species and environmental factors.

Mechanical resistance to wood confers to lignin's high resistance towards chemical and biological degradation. Middle lamella is where the highest concentration of this recalcitrant polymer is found, acting as a cement between wood fibers, it is also present in the layers of the cell wall (secondary cell-wall), with hemicelluloses forming an amorphous matrix embedded with cellulose fibrils which are protected against biodegradation (Loredano, Fabio, Elena, & Rosini, 2015). Highest lignin content made up with mostly G units is found in woody gymnosperms, which are contrasted by the presence of S and G units in the lignin of woody angiosperms. Lignin composition varies between wood tissues and cell-wall layers.

The third structural component, hemicelluloses (polyoses), has an intermediate degree of complexity and is made up of different pentose and hexose residues, which are often acetylated, and generally form branched chains. Typically, hemicelluloses in softwood are glucomannans, whereas those in hardwoods are mainly xylans together with variable percentages of galactose, arabinose, rhamnose and methylglucuronic acid units, and acetyl groups. Other non-structural components of wood include compounds extractable with organic solvents (the so-called extractives) which can be either polar (e.g. phenols and tannins) or apolar (e.g. fats and sterols), water-soluble compounds (e.g. sugars and starch), as well as proteins and ashes. These components together generally represent less than 5% of the dry weight of wood, but can reach 20% in some softwoods (e.g. in some *Cupressaceae*) (Fengel & Wegener, 1984; Ángel et al., 2005).

IMPORTANCE OF FUNGAL DEGRADATION

Fungi belong to the kingdom monophyletic eukaryotic lineage that consists of chemo-organotrophic organisms which have two distinct growth forms: Yeasts – with spherical cells and filamentous fungi which have threadlike structures called hyphae. Filamentous fungi have specific importance with reference to terrestrial ecosystems as they enable exploration of soils by bridging air-filled gaps (pores) and penetrating into solid material (Hoffland et al., 2004; Klein & Paschke, 2004; Money, 2007; Wurzbacher, Ba¨rlocher, & Grossart, 2010). Apart from this, hyphae possess the ability of trans locating nutrients through nutrient-poor patches and supplying growth-limiting elements to zones of metabolic activity (Frey, Elliott, Paustian, & Peterson, 2000). Hence fungi have been characterized as spatial integrators (Ritz, 2007). In addition mycelial growth form also enables biomass recycling, further increasing efficiency of nutrient use in patchy environments (Boddy, 1999; Falconer, Bown, White, & Crawford, 2007). Fungi have developed as significant components of terrestrial ecosystem functioning owing to their success through hyphal growth form (De Boer, Folman, Summerbell, & Boddy, 2005), specifically in respect to the decomposition of organic matter. Decomposition of organic matter manages the balance between soil carbon storage and CO_2 release into the atmosphere, releasing mineral nutrients, making available for plant growth.

The association between the ecology of fungi and carbon cycling has been acknowledged, but the dynamics and interactions of fungal species in the process of decomposition still need to be fully understood. Increased attention have been received during the last decade with reference to differentiation in fungal niche, connection between diversity of fungi and decay, role of decomposing fungi in rhizosphere, climate change impact on functioning of fungal communities, factors that are incorporated in fungal decomposition models, fungal community composition effected by fungal species and selection of fungal community composition which is specialized in decomposing litter of the local plant species or vegetation (Annemieke, Thomas, Geydan, Kuyper, & Wietse, 2013).

SPATIAL DISTRIBUTION OF DECOMPOSER FUNGI

Estimation of fungal biomass can be taken up through a number of methods like microscopically, biochemically, physiologically to biological molecular methods (Joergensen & Wichern, 2008; Strickland & Rousk, 2010; Pre´vost-Boure´ et al., 2011; Baldrian & Sˇtursova, 2012). Nevertheless microscopically and biochemical methods are capable of separating Glomeromycota from most other fungi, but they are unable to distinguish decomposing fungi from ectomycorrhizal and ericoid mycorrhizal fungi. Therefore, only the first clue of total decomposing fungal density is provided by most fungal biomass measurements (Strickland & Rousk, 2010). Hence such 'detection' measurements are to be complemented with 'activity' measurements, like the incorporation of 13C from labeled substrates (SIP) into fungal biomarkers (Malosso, English, Hopkins, & O'Donnell, 2004; Moore-Kucera & Dick, 2008; Hannula, Boschker, De Boer, Van, & Veen, 2012a).

All the above said measurements provide a generalized image that the availability of decomposing fungi is abundant in most terrestrial ecosystems, nonetheless their biomass can be poorly signified in fresh and marine water environments, along with their sediments (Jørgensen & Stepanauskas, 2009). The advantages associated with hyphal growth form above unicellular growth forms of terrestrial ecosystems cannot be validated for aquatic ecosystems, due to less spatial heterogeneity and the absence of air-filled pores (Wurzbacher et al., 2010). Moreover, most parts of aquatic ecosystems precisely sediments, are anaerobic in nature restricting contribution of fungi only to the processes of fermentation. Exclusively, degradation of lignin is highly sensitive to oxygen constraint (Ten Have & Teunissen, 2001). Besides, plants in aquatic habitat does not require to build similar physical support structure, the lignocellulose complexes, as required by plants in terrestrial habitat, consequently aquatic organic matter is much less recalcitrant for bacterial decomposers than terrestrial organic matter (Hedges & Oades, 1997).

Distribution of decomposer fungi is not uniform all through the soil. Abundance of saprotrophic fungi can be seen the upper layers (L and F), of temperate and boreal forest soils having well-developed organic layer where the rates of decomposition are considerably higher (Cairney, 2005; Lindahl et al., 2007; Osono, 2007). Also, organic-rich blotches of mineral soil are also observed to have high densities of fungi (Ritz, 2007). Different strategies have been developed by fungi that decompose in order to gain access to new patches, for example, forming cords of hyphae which can explore the environment (Boddy. Hynes, Bebber, & Fricker, 2009; Garbeva, Hol, Termorshuizen, Kowalchuk, & De, Boer, 2011). Presence of decomposer fungi in abundance is also seen in the rhizosphere zone that surrounds roots, where the activity of microbes is influenced by input of compounds derived from root. Previously this was considered to be dominated by bacteria owing to the rapid growth of bacteria on exudates of soluble root (Bue´e, De, Boer, Martin, Van, Overbeek, & Jurkevitch, 2009a). Presence of abundant decomposer fungi in the natural vegetation of rhizosphere can be related to the presence of roots with different ages, from which more recalcitrant compounds are released from senescent roots (Hegde & Fletcher, 1996).

DIVERSITY OF LIGNOCELLULOSE-DEGRADING ORGANISMS

Cellulose breakdown perhaps might have an ancient origin owing to the fact that it is generated by diverged group of marine organisms. At the end of the Permian age the evolution of lignin degradation, corresponded with the decline in the burial of organic carbon (Floudas et al., 2012). The appearance of land plants was seen after diverging evolution of chief branches of the tree of life; hence the capability of deconstruction of lignocellulose has numerous origins and has sustained to progress in varied smaller branches widely, but sparingly dispersed across the tree of life. To quote an example, ecologically important symbiosis of insect-protist that facilitates digestion of lignocellulose has developed in late Jurassic (Brune, 2014) and digestion of wood assisted by bacteria was the feature of the common last ancestor of the bivalve families of Teredinidae and Xylophagainae (Distel et al., 2011). Factors affecting deconstruction include wide range of environmental conditions, chiefly pH, temperature, redox potential and pressure.

Degradation of lignocellulose is the crucial step for recycling of carbon in terrestrial ecosystems. Besides decay of wood in service by fungi is the reason for loss of a billion-euro. The main wood rotters because of their capability of degrading or modifying lignin, the enzymatic process that has originated during the Upper Devonian period corresponding with the evolution of vascular plants are Basidiomycetes (Eriksson & Blanchette, 1990). There are two classes of wood-rotting basidiomycetes which are white-rot and brown-rot fungi classified based on their macroscopic aspects (Schwarze, Engels, & Mattheck, 2000). Basidiomycetes have unique properties of overcoming complications in wood decay like low nitrogen content and also the presence of toxin and antibiotic compounds.

Degradation of cell-wall components is mediated by the extracellular oxidative enzymes which are secreted by fungi. Characterization of the common wood-rotting fungi is based on their capability to degrade lignin, hemicelluloses and cellulose that give rise to white material enriched with cellulose. The ability of white-rot basidiomycetes in degradation of lignin concurrently along with cellulose presents two patterns of white- in different kinds of wood which are selective delignification called sequential decay, and simultaneous rot (Otjen & Blanchette, 1986). Only 7% of wood-rotting basidiomycetes growing principally on softwoods represent Brown-rot fungi. Basidiomycetes of this group degrade wood polysaccharides only after partial lignin modification, resulting in brown material which consists of oxidized lignin, which represents a potential source of aromatic compounds for the stable organic matter fraction in forest soils (Simon et al., 2015).

Degradation of wood is extensively done by only white-rot and brown-rot basidiomycetes. Besides some ascomycetes along with their asexual states, i.e., deuteromycetes, can also colonize wood that is in contact with soil. This kind of degradation results in a decrease of the mechanical properties of wood, giving rise to soft-rot, a process that frequently involves bacteria. Degradation of wood, even under extreme environmental conditions that would prohibit the activity of other fungi is also possible with Soft-rot fungi. Furthermore, soft-rot types of decay pattern are observed in some basidiomycetes also. As a final point, some limited numbers of ascomycetous fungi also called stain fungi are able to colonize wood across parenchymatic rays and resin channels leading to discoloration of softwood tissues, but very limited degradation, which mainly affects extractives and water-soluble materials (Simon et al., 2015).

MECHANISM OF FUNGAL DECOMPOSITION PROCESSES

A key contribution to terrestrial organic matter decomposition in particular to the more recalcitrant fractions is made by fungi (Dighton, 2003; De Boer, Kowalchuk, & Van Veen, 2006; Berg & McClaugherty, 2008). This ability of decomposing recalcitrant fractions of terrestrial organic matter is based on a combination of morphological characteristics which allows penetration of solid material and on physiological characteristics that allow degradation of the lignocellulose complex (Money, 2007; Baldrian & Vala´sˇkova´, 2008; Floudas et al., 2012).

Specifically the ability of decomposing lignin, heterologous aromatic polymer, seems to be mainly limited to Basidiomycota (Agaricomycotina) which are known as white-rot fungi (Baldrian, 2008; Floudas et al., 2012), even though lignin breakdown has also been reported by Xylariales, within the Ascomycota (Worrall, Anagnost, & Zabel, 1997; Osono, Hagiwara, & Masuya, 2011a; Osono, To-Anun, Hagiwara, & Hirose 2011b). Brown-rot fungi have the ability to modify lignin in order to gain access to cellulose, which together with hemi-cellulose forms the chief energy resource for litter and wood degrading fungi (Curling, Clausen, & Winandy, 2002; Yelle, Ralph, Lu, & Hammel, 2008; Eastwood, Floudas, & Binder,

2011; Martinez, Rencoret, Nieto, Jime´nez-Barbero, Gutie´rrez, & del Rı´o, 2011). Cellulolytic ascomycetes, even though are not capable of degrading or modifying lignin like white and brown rot fungi, can add significantly to the decomposition of lignin-rich organic matter, because the thin perforating hyphae of these fungi is able to reach cellulose-rich layers of woody cell walls (Schmidt, 2006).

Bacteria are also reported to cause a modification of lignin and decomposition of derivatives of lignin (Bugg, Ahmad, Hardiman, & Singh, 2011). These processes might prove to be important in the degradation of lignin in such environments where the growth of fungi is limited, for example, periodic anoxic conditions (De Angelis et al., 2011). But the direct involvement of bacteria in decomposition of natural lignocellulose complexes of terrestrial ecosystems seems to be very less (Kirby, 2005; Floudas et al., 2012; Schneider et al., 2012). Nonetheless, bacteria possibly play an important but an indirect role in the decomposition of organic material rich in lignocellulose and humus formation through metabolizing of intermediates that are released by fungal enzymes or through other interactions with fungi (De Boer & Van der Wal, 2008).

Degradation of organic matter containing lignocellulose is considered to be primarily fungal niche, but the fungi which decompose is not restricted only to these resources (De Boer, Kowalchuk, & Van Veen, 2006; Van der Wal, Van Veen, Pijl, Summerbell, & De Boer, 2006a; De Graaff, 2010). There are extensive assortments of organic compounds that are decomposed by relatively specific fungi (Baldrian et al., 2011). Most of these fungi also called molds possess characteristics like being opportunistic; fast-growing which can produce large numbers of spores or melanized resting structures of hyphae which help to survive conditions when degradable carbon sources are not available easily (Van der Wal, De, Boer, & Klein, Gunnewiek, 2009a). These organisms have to compete with bacteria for various resources as they degrade the same range of labile compounds (De Graaff, 2010). Factors that determine the relative importance of fungi and bacteria in the decomposition of easily degradable compounds are pH and soil moisture (De Boer, Kowalchuk, & Van Veen, 2006). On a general not performance of fungi is found to be better at lower pH and moderately dry conditions (Bapiri & Rousk, 2010; Rousk, Brookes, & Baath, 2010a; Yuste et al., 2011).

Unicellular fungi, Yeasts have a property to adapt and grow in environments which require tolerance against high concentrations of sugars, for example, rotting fruits and floral nectar (Gasch, 2007; Tekolo, Mckenzie, Botha, & Prior, 2010). Moreover, they can also grow anaerobically, utilizing fermentation as the chief process for generation of energy. Uncommon anaerobic decomposition in fungi is found among chytrids in the cattle rumen (Trinci et al., 1994). Even though some of the soil yeasts seem to be specialized in decomposition of small aromatic compounds, their role in soil organic matter decomposition still remains unclear (Botha, 2011).

Fungi degrading biomass depend on intricate degradative mechanisms that generally catalyse two kinds of processes: one, direct enzymatic depolymerization, for example, using cellobiohydrolases and two, oxidative species generation (e.g., radicals) which then acts on biomass. Terminology of categorization stays changing along with new genomic information of Basidiomycota signifying that fungal species conventionally classified as white rot or brown rot may no longer are apt to these categories because of progressions, both in the expression of metabolites and the resulting patterns of decay. Conventionally though, in typical white rot degradation, the mode of attack employed by fungi is primarily enzymatic.

The wood cell wall is attacked only from the surfaces of lignocellulose in white rot fungi since degradative enzymes are large enough enabling penetration into the intact cell wall. A whole suite of cellulases enzymes is employed by white rot fungi which also produce a suite of enzymes that can oxidise lignin include ligninase, manganese peroxidase, versatile peroxidase or laccase, or a combination of all these (Loredano, Fabio, Elena, & Rosini, 2015). Large numbers of LPMO genes have also been shown by some white rot fungi. Due to multiple time evolution from the precursors of current white rot fungi, brown rot fungi have lost lignolytic enzyme systems and crucial types of cellulases.

As a substitution for much of the cellulolytic enzyme mechanism a chelator-mediated Fenton (CMF) system has evolved in three orders of brown rot fungi (Gloeophyllales, Polyporales and Boletales), consequently creating an alternative effectual mechanism for biomodification and depolymerization of biomass (Eastwood, Floudas, & Binder, 2011). Uniqueness of CMF system among biological systems is that it is the only stated substrate deconstruction system which is based on oxygen radical chemistry permitting non-enzymatic deconstruction at a considerable distance (several microns) from the organism. Brown rot fungi gain advantages of this efficiency of CMF system in exploiting ecological niches, for example, these fungi have evacuated white rot precursors in the degradation of conifer wood. Few fungi belonging to ascomycete are also reported to degrade wood cell walls by forming chains of diamond-shaped cavities which generally follow orientation of the S2 elementary fibrils resulting in soft rot (Bjordal, 2012). Production of a full complement of cellulolytic enzymes is also reported by soft rot fungi, but their lignin degrading ability has been erratically reported to contain unspecified extracellular peroxidases and oxidases which appear to be limited in function than those isolated from white rot fungi.

FUNGAL ENZYMES IN BIODEGRADATION

Microbes that are capable of degrading cellulose and other plant cell wall fibres (French word for fibers in English) were many as reported including fungi and bacteria. Over 14,000 fungal species that were capable of expressing this ability were isolated, but only a few of them were studied in detail by 1976 (Mandelsv & Sternberg, 1976).

Production and secretion of several lignocellulolytic enzymes that act synergistically in the environment is seen in many fungi, which contribute significantly in the process of decay of lignocellulosic residues naturally. Lignocellulosic biomass breakdown comprises of formation of long-chain polysaccharides including mainly cellulose and hemicellulose and also resulting in the subsequent hydrolysis of these formed polysaccharides into their component 5- and 6- carbon chain sugars. These sugars can are further converted to bioethanol through fermentation processes in the production of biofuel (Zhou, & Ingram, 2000; Dashtban, Schraft, & Qin, 2009).

Basidiomycetes white-rot fungi are naturally efficient in lignin degradation in the process of wood decay. Although many white rot fungi attack lignin simultaneously along with hemicellulose and cellulose, some white-rot fungi preferentially work on lignin in a choosy manner. For example, while *Ceriporiopsis subvermispora* (Guerra, Mendonca, Ferraz, Lu, & Ralph, 2004), *Phlebia* sp. (Arora, & Sharma, 2009; Fackler, Gradinger, Hinterstoisser, Messner, & Schwanninger, 2006), *Physisporinus rivulosus* (Hilden, Hakala, Maijala, Lundell, & Hatakka, 2007) and *Dichomitus squalens* (Fackler et al., 2006) selectively attack lignin, *Trametes versicolor* (Tanaka, Itakura, Enoki, 1999), *Heterobasidium annosum* (Daniel, Asiegbu, & Johansson, 1998), *P.chrysosporium* (Sanchez, 2009) and *Irpex lacteus* (Xu, Ma, & Zhang, 2009) simultaneously degrade all cell wall components.

For many years laccases have been reported to play a diversity of roles like fruit-body morphogenesis, detoxification and synthesis of pigments in plants, fungi and insects (Blanchette et al., 1992). Production of laccases plate cultures of fungi was considered to be a unique characteristic of white-rot basidiomycetes (Fackler et al., 2006), even though some brown-rot fungi could produce laccase in liquid cultures

(Tanaka et al., 1999). The low redox potential of these phenoloxidases allows the direct oxidation of only phenolic lignin units, which are often less than 10% of the total polymer. With the discovery of laccases ability to oxidize high redox potential substrates in the presence of synthetic mediators have increased interest in laccases for biotechnological applications, this phenomenon allows the degradation of bleaching of paper pulp chlorine-free (Wang, Yan, Chen, Huang, & Gao, 2008) and xenobiotic compounds.

Although some of the lignin-derived phenols might act as efficient laccase mediators, natural mediators that are involved in biodegradation of lignin remain to be identified. Discovered in the mid-1980s lignin peroxidase (LiP) and manganese peroxidase (MnP) in *P. chrysosporium* were described as true ligninases owing to their high redox potential (Gold, Youngs, & Gelpke, 2000; Martínez, 2002). LiP are capable of degrading non-phenolic lignin units up to 90%, whereas Mn^{3+} generated by MnP acts as a diffusible oxidizer on phenolic or non-phenolic lignin units through lipid peroxidation reactions (Jensen, Bao, Kawai, Srebotnik, & Hammel, 1996). Versatile peroxidase (VP) are recently described as a third type of ligninolytic peroxidase that cartels the catalytic properties of LiP, MnP, and plant/microbial peroxidases oxidizing phenolic compounds which are found in *Pleurotus* (Martínez, Ruiz-Dueñas, Guillén, & Martínez, 1996; Ruiz-Dueñas, Martínez, & Martínez, 1999) and other fungi (Heinfling et al., 1998).

Oxidases that generate H_2O_2 and mycelium associated dehydrogenases which reduce lignin-derived compounds are the other extracellular enzymes that are involved in degradation of wood lignin. H_2O_2 generating oxidases comprise of aryl-alcohol oxidase (AAO) as described in *Pleurotus eryngii* and other fungi (Kersten, 1990). Aryl-alcohol dehydrogenases (AAD) and quinone reductases (QR) from fungi are also reported to be involved in the degradation of lignin (Guillén, Martínez, Muñoz, & Martínez, 1997; Gutiérrez, Caramelo, Prieto, Martínez, & Martínez, 1994).

Fungi producing lignocellulolytic enzymes are widespread and also include species from ascomycetes (e.g. *Trichoderma reesei*) and phyla of basidiomycetes such as white-rot (e.g. *Phanerochaete chrysosporium)* and brown-rot fungi (e.g. *Fomitopsis palustris*). Moreover, few anaerobic species (e.g. *Orpinomyces sp.*) are also found to be capable of degrading cellulose in ruminant animal (animal) gastrointestinal tracts (Guillén, Martínez, & Martínez, 1992). Degradation of biomass by these fungi is achieved through complex mixtures of cellulases (Guillén et al., 1997), hemicellulases (Guillén et al., 1992) and ligninases with reference to the complexity of the materials. Degradation of cellulosic biomass in nature is performed by mixtures of hydrolytic enzymes which are collectively known as cellulases. Cellulases are both endoglucanases (endo acting) and cellobiohydrolases (exo acting) enzymes, which perform synergistic way in biomass degrading microbes.

Lignin degrading fungi on a selective basis might have substantial prospective biotechnological applications, for example, in the removal of lignin to obtain intact cellulose in the processes of biopulping and also in techniques where the prime objective is the provision of an unprotected carbohydrate for following use (e.g. biofuel substrate and/or animal feed) (Martínez et al., 1996; Mayer & Staples, 2002). *C. subvermispora,* lacking cellulase, was selected for the processes of biopulping to be selective degrader of lignin for example (Martínez et al., 1996; Mayer & Staples, 2002).

Brown-rot fungi like *Postia placenta*, *Laetiporus portentosus*, *Piptoporus betulinus* and *Gloeophyllum trabeum* in contrast to white-rot fungi can degrade wood carbohydrates, but would not oxidized lignin, resulting in brown-colored rot (Fengel & Wegener, 1984). Even though ascomycetes are typically capable of degrading cellulose and hemicellulose, their ability in conversion of lignin is narrow (Fengel & Wegener, 1984). *Fusarium solani* f. sp. *glycines* plant pathogenic fungi are capable of lignin degradation through the production of lignin peroxidase and laccase. Fungal degradation of lignin degradation is advocated to have a key role in sudden death syndrome (SDS) of soybean (Otjen et al., 1986).

FUNGAL EXTRACELLULAR LIGNINASES

Fungal lignin degradation is performed by secretion of enzymes, which are termed "ligninases" collectively. Ligninases can be categorized as both phenol oxidases like laccase or heme peroxidases which include lignin peroxidase (LiP), manganese peroxidase (MnP) and versatile peroxidase (VP) (Martinez et al., 2005). On a general note the distinction between laccases are peroxidases is that the former use molecular oxygen as electron acceptors while the employ hydrogen peroxide as co-substrate (Mai, Kues, & Militz, 2004). Secretion of lignin-modifying enzymes (LMEs) by white-rot fungi is varied in addition to necessary other compounds for active degradation of lignin (Wesenberg, Kyriakides, & Agathos, 2003). Several LiP and MnP isoenzymes are produced by *P. chrysosporium* but not laccase (Singh & Chen, 2008). Likewise, ten LiP and five MnP genes are present in the genome of *P. chrysosporium* contains. Apart from that glyoxal oxidase (GLOX), H_2O_2-generating enzyme, has also been found in cultures of *P. chrysosporium* (Kersten, 1990). *T. versicolor*, *P. chrysosporium* and *Coriolus versicolor* (Moredo et al., 2003) of white-rot basidiomycetes, were found to possess greatest efficiency for degrading lignin. Even though LiP is capable of oxidizing non-phenolic part of lignin (that forms 80- 90% lignin composition), it is not found in many lignin degrading fungi. Studies through electron microscopy regarding the early stages of degradation of wood by fungi have revealed that oxidative ligninolytic enzymes are extremely large in order to penetrate into the wood cell wall micropores (Srebotnik, Messner, & Foisner, 1988). Hence, initiation of changes to the lignin structure by low-molecular weight diffusible reactive oxidative compounds prior to the enzymatic attack has been suggested (Srebotnik et al., 1988; Tanaka et al., 1999). To avoid the toxic substances released from plant cell that might inhibit the process of degradation cautious attention is to be paid in selecting lignocellulose material that is to be incorporated in the process of biotechnology. Inhibition of white rot fungi has been observed due to intrinsic substrate toxicity (Wolfaardt, Jacobs, Rabie, Smit, Taljaardt, & Wing-field, 1999).

Lignin-carbohydrate complexes (LCCs) toxicity to rumen microflora has been argued, but *in vivo* toxicity is supposed to be insignificant with reference to dilution and swift detoxification by the microflora (Cornu, Besle, Mosoni, & Grent, 1994; Wilson & Mertens, 1995). However LCCs toxic effects at the foci of biodegradation largely remain unknown. Aromatic lignin degradation intermediates result during lignocellulose degradation owing to its polyphenolic composition, which then enter diversity of man-made and natural environments. Wheat straw extracts have been detected with free phenolic acids and aldehydes (Galletti, Piccaglia, & Concialini, 1990; Milstein, Vered, Gressel, & Flowers, 1981). In particular release of intermediates that might enter anaerobic environments is observed in degradation of lignin by white-rot fungi. Under anaerobic conditions, degradation of lignin is very slow which in turn depends on a time scale and pressure that lead to the buildup of humus forming peat, organic soil matter followed by lignite and coal. Though polymeric lignin is stable under anaerobic conditions for longer durations products of smaller degradation are actively metabolized by bacteria consortia (Egland, Pelletier, Dispensa, Gibson, & Harwood, 1997; Elder & Kelly, 1994).

BIOTECHNOLOGICAL APPLICATION OF LIGNOCELLULOSE AND ITS BIODEGRADATION

Lignocellulose pretreatment is taken up by various industries with the prime objective of accessing potential cellulose and hemicellulose which are encrusted by lignin within the matrix of lignocellulose.

Capability of white-rot fungi in combination with solid-state fermentation (SSF) technology to degrade lignin selectively has made it possible for the implementation of industrial-scale biotechnologies based on lignocellulose. Attention of advantages and disadvantages with reference to SSF has been received by Mudgett (1986). SSF offering advantages of robust technology outperform conventional technologies of fermentation with respect to cost effectiveness, simplicity and maintenance requirements, making it an attractive technology for environmental problems associated with limited expertise and money. Nevertheless, there are also problems associated with SSF which are commonly heat build-up, scale-up, and biomass growth estimation, control of substrate moisture content and bacterial contamination (Lonsane, Ghildyal, fBudiatman, & Ramakrishna, 1985).

Lignocellulose-Based Technologies Using Unsterile Substrates

An example of working technology, which is based on unsterile lignocellulose substrates is silage manufacture. It is produced through controlled fermentation of plant material under moist conditions (McDonald, 1981). Silage effluent is considered as a serious threat to natural water supplies by water authorities (Haigh, 1994), on the other hand, from a farmer's perspective, loss of valuable nutrients is associated with silage effluent (McDonald, 1981).

- **Biopulping:** Cellulose-based wood processing faces problem with lignin as it should be separated from cellulose with vast energy, environmental and chemical expense. Hence, in solid-state fermentation processes wood chips are being treated with white-rot fungi in order to improve the process of delignification in Biopulping. This process reduces energy costs and environmental impact in comparison to relative traditional pulping operations (Breen & Singleton, 1999). Biopulping benefits were verified by Scott et al., (1998) by using a 40-ton scale experiment: tear, tensile and burst indexes of the subsequent paper were enhanced (which indicates higher degree conservation of cellulose during the process of pulping); pulp brightness were also increased (representing better removal of lignin) along with improved savings on energy with 30–38%. Problems that are prevalent with SSF still affect this concept. Key parameters that influence fungal activity is inoculation; heat removal and. Competition with microorganisms that are naturally occurring can be reason for the poor colonization of wood chips by white-rot (Wolfaardt et al., 1999). Secondary contamination by microorganisms that are airborne in nature must be prevented at additional costs along with sterilization of substrate which also is usually a major expense. Decades of dissatisfaction related to high principal costs for making SSF a viable option for the pulp and paper industry has been expressed by Breen and Singleton (1999), who summarized: "Overcoming these challenges will determine, in a large part, if biopulping becomes a reality".
- **Animal Feed:** Although animal itself does not produce cellulose-hydrolyzing enzymes, cellulose is the most important source of carbon and energy of ruminant's diet (Czerkowski, 1986). Cellulose and other plant carbohydrates are used as a source of carbon and energy by rumen microorganisms. Therefore, the conversion of carbohydrates in large amounts of propionic, acetic and butyric acids which can be used by higher animals as energy and carbon sources is done by the microorganisms (Colberg, 1988). Preferential delignification of lignocellulose materials by white-rot fungi has been utilized to enhance the nutritional value of forages (Agosin, & Odier, 1985; Akin et al., 1995; Chen, Fales, Varga, & Royse, 1995; Zadrazil & Isikhuemhen, 1997). This enhanced digestibility offers organic carbon, which can be fermented to organic acids under

anaerobic conditions like the rumen. But this concept of elevation of animal feed using white-rot fungi could not reach expectations of industrial proportions.

- **Applications of Genetic Engineering:** The possibility of lignocellulose-based applications are intensifying rapidly towards applications of genetic engineering. Repression of lignin biosynthesis was recently achieved in *Populus tremuloides* that has resulted in cellulose accumulation leading to healthy growth of these transgenic trees (Hu et al., 1999). Chemical modification of cellulose and lignocellulose fibers can be adopted to render it useful for various miscellaneous applications of the textile industry (Ghosh & Gangopadhyay, 2000). Actinomycetes which degrade lignocellulose have been used as biocontrol agents in order to preserve golf green turf-grass (Chamberlain & Crawford, 2000). Presently, metabolic engineering is being applied to enable concurrent fermentation of hexoses and pentoses to ethanol (Aristidou & Penttilä, 2000). The future may perceive applications of microorganisms which are genetically engineered (containing chiefly lignocellulases) for applications in biotechnology where wastes of lignocellulosic serve as on-site nutrient and carbon source.

The commercializing of lignin waste production might offset process costs as seen from the studies of (Dizhbite et al., 1999) where commercial byproducts of lignocellulose were converted to fuel ethanol and has found application as absorbents of organic pollutants and also as enterosorbents. Lignocellulose is also has potential value in medicine. Besides being essential as fiber for human diet, this can also be a source of compounds with biological activity. These compounds are potent stimulators of the human immune system and also as antiviral agents (Kiyohara, Matsumoto, & Yamada, 2000; Sakagami et al., 1999).

- **Potential of Lignocellulose in Space Exploration:** Research advances in lignocellulose enable scientists in contributing to space science. In near future space travel will benefit from such research, as the transport of lignocellulose to space can result in considerable cost savings. Lignocelluloses are considered to be a feedstock which can provide for all basic needs like feedstock chemicals, water, food, energy, and fuel. Therefore, lignocelluloses can be "super fuel" of future – owing to being compact natural polymer which contains enough potential energy to sustain man and machine in space.

CONCLUSION

A great deal of interest has been attracted towards lignocellulolytic microorganisms, especially fungi, as potential biomass degraders for large-scale applications owing to their ability to produce immense amounts of extracellular lignocellulolytic enzymes. A most recalcitrant component of lignocellulosic material, lignin acts as a barrier for a solution or enzymes by linking to both celluloses and hemicelluloses, preventing penetration of lignocellulolytic enzymes into the interior structure of lignocelluloses. Primarily white-rot fungi of basidiomycetes are responsible for efficient degradation of lignin in the processes of wood decay. Extracellular enzymes produced by them known as lignin-modifying enzymes (LMEs) are responsible for facilitating degradation process. Besides accessory enzymes like oxidases which produce H_2O_2 that is required by peroxidases have also been found to be involved in the degradation of lignin. White and brown-rot fungi mainly use Hydrogen peroxide to produce •OH in Fenton reaction. Eventually, these free radicals attack polysaccharides including lignin of plant cell walls in non-specific manner creating cleavage sites that allow easier penetration by lignocellulolytic enzymes.

Biorefineries can adopt these natural processes occurring in fungal wood degradation for industrial purposes to achieve greater potential for wood-processing and cellulose-producing industries along with high value-added conversion of lignocellulosic waste materials. In Particular, molecular mechanisms from selective white-rot fungi provide series of applications in the field of biotechnology of renewable resources (Bajpai, 2012). Hence, lignin might signify a renewable source of aromatic fine chemicals. Strategies for depolymerization of lignin for production of valuable chemicals are catalytic, reduction, cracking or hydrolysis reactions, enzymatic reactions and catalytic oxidation (Pandey & Kim, 2011; Zhou, Xia, Lin, Tong, & Beltramini, 2011; Nanayakkara, Patti, & Saito, 2014; Dashtban et al., 2010).

As a feedstock of the massive potential for production of low-molecular-weight chemicals, lignin is well represented. Owing to its complexity in structure, substantial efforts have been dedicated to understanding major natural pathways and enzymes involved in degradation of lignin. From this it is clear that lignin degradation in nature is a multi-enzymatic process that involves an array of accessory enzymes in addition to four major ligninolytic oxidases.

Enzymes from ligninolytic fungi have found applications in a vast number of industrial processes other than biorefinery. Development of innovative ligninolytic enzymes will pave the way for additional industrial uses. Hence, research in this direction should be focused on:

- Understanding role and interaction of enzyme combinations in lignin degradation
- Determination of the role of mediators of small molecule in the catalytic mechanism of the ligninolytic enzymes, including the overall process of lignin degradation;
- Studying enzyme treatment in combination with other pretreatment methods.

REFERENCES

Agger, J. W., Isaksen, T., Varnai, A., Vidal-Melgosa, S., Willats, W. G. T., Ludwig, R., & Westereng, B. et al. (2014). Discovery of LPMO activity on hemicelluloses shows the importance of oxidative processes in plant cell wall degradation. *Proceedings of the National Academy of Sciences of the United States of America*, *111*(17), 6287–6292. doi:10.1073/pnas.1323629111 PMID:24733907

Agosin, E., & Odier, E. (1985). Solid-state fermentation, lignin degradation and resulting digestibility of wheat straw fermented by selected whiterot fungi. *Applied Microbiology and Biotechnology*, *21*(6), 397–403. doi:10.1007/BF00249988

Akin, D. E., Rigsby, L. L., Sethuraman, A., Morrison, W. H. III, Gamble, G. R., & Eriksson, K. E. L. (1995). Alterations in structure, chemistry, and biodegradability of grass lignocellulose treated with the white rot fungi *Ceriporiopsis subvermispora* and *Cyathus stercoreus*. *Applied and Environmental Microbiology*, *61*, 1591–1598. PMID:7747973

Annemieke, V. (2013). A thready affair: Linking fungal diversity and community dynamics to terrestrial decomposition processes. *FEMS Microbiology Reviews*, *37*(4), 477–494. doi:10.1111/1574-6976.12001 PMID:22978352

Aristidou, A., & Penttilä, M. (2000). Metabolic engineering applications to renewable resource utilization. *Current Opinion in Biotechnology*, *11*(2), 187–198. doi:10.1016/S0958-1669(00)00085-9 PMID:10753763

Arora, D. S., & Sharma, R. K. (2009). Enhancement in in vitro digestibility of wheat straw obtained from different geographical regions during solid state fermentation by white rot fungi. *BioResour*, *4*, 909–920.

Bajpai, P. (2012). *Biotechnology for pulp and paper processing*. New York: Springer. doi:10.1007/978-1-4614-1409-4

Baldrian, P. (2008). Wood-inhabiting lignolytic basidiomycetes in soils: Ecology and constraints for application in bioremediation. *Fungal Ecology*, *1*(1), 4–12. doi:10.1016/j.funeco.2008.02.001

Baldrian, P., Kolarˇık, M., & Sˇtursova´, M. (2012). Active and total microbial communities in forest soil are largely different and highly stratified during decomposition. *The ISME Journal*, *6*(2), 248–258. doi:10.1038/ismej.2011.95 PMID:21776033

Baldrian, P., & Valášková, V. (2008). Degradation of cellulose by basidiomycetous fungi. *FEMS Microbiology Reviews*, *32*(3), 501–521. doi:10.1111/j.1574-6976.2008.00106.x PMID:18371173

Baldrian, P., Vorˇisˇkova´, J., Dobiasˇova´, P., & Merhautova, V. (2011). Production of extracellular enzymes and degradation of biopolymers by saprotrophic microfungi from the upper layers of forest soil. *Plant and Soil*, *338*(1-2), 111–125. doi:10.1007/s11104-010-0324-3

Bapiri, A., Bååth, E., & Rousk, J. (2010). Drying-rewetting cycles affect fungal and bacterial growth differently in an arable soil. *Microbial Ecology*, *60*(2), 419–428. doi:10.1007/s00248-010-9723-5 PMID:20635180

Berg, B., & McClaugherty, C. (2008). *Plant Litter: Decomposition, Humus Formation, Carbon Sequestration*. Heidelberg, Germany: Springer-Verlag. doi:10.1007/978-3-540-74923-3

Binder, M., & Hibbett, D. S. (2002). Higher-level phylogenetic relationships of homo basidiomycetes (mushroom-forming fungi) inferred from four rDNA regions. *Molecular Phylogenetics and Evolution*, *22*(1), 76–90. doi:10.1006/mpev.2001.1043 PMID:11796031

Martínez, Speranza, Ruiz-Dueñas, Ferreira, Camarero, Guillén, … Del Río. (2005). Biodegradation of lignocellulosics: Microbial, chemical, and enzymatic aspects of the fungal attack of lignin. *International Microbiology*, *8*, 195–204. PMID:16200498

Blanchette, R. A., Burnes, T. A., Eerdmans, M. M., & Akhtar, M. (1992). Evaluating Isolates of *Phanerochaete chrysosporium* and *Ceriporiopsis subvermispora* for Use in Biological Pulping Processes. *Holzforschung*, *46*(2), 109–115. doi:10.1515/hfsg.1992.46.2.109

Boddy, L. (1999). Saprotrophic cord-forming fungi: Meeting the challenge of heterogeneous environments. *Mycologia*, *91*(1), 13–32. doi:10.2307/3761190

Boddy, L., Hynes, J., Bebber, D. P., & Fricker, M. D. (2009). Saprotrophic cord systems: Dispersal mechanisms in space and time. *Mycoscience*, *50*(1), 9–19. doi:10.1007/S10267-008-0450-4

Botha, A. (2011). The importance and ecology of yeasts in soil. *Soil Biology & Biochemistry*, *43*(1), 1–8. doi:10.1016/j.soilbio.2010.10.001

Breen, A., & Singleton, F. L. (1999). Fungi in lignocellulose breakdown and biopulping. *Current Opinion in Biotechnology*, *10*(3), 252–258. doi:10.1016/S0958-1669(99)80044-5 PMID:10361070

Brune, A. (2014). Symbiotic digestion of lignocellulose in termite guts. Detailed review of the complex role microbial endosymbionts in termite digestion. *Nature Reviews. Microbiology*, *12*(3), 168–180. doi:10.1038/nrmicro3182 PMID:24487819

Buee, M., De Boer, W., Martin, F., van Overbeek, L., & Jurkevitch, E. (2009a). The rhizosphere zoo: An overview of plant associated communities of microorganisms, including phages, bacteria, archaea, and fungi, and some of their structuring factors. *Plant and Soil*, *321*(1-2), 189–212. doi:10.1007/s11104-009-9991-3

Bugg, T. D. H., Ahmad, M., Hardiman, E. M., & Singh, R. (2011). The emerging role for bacteria in lignin degradation and bioproduct formation. *Current Opinion in Biotechnology*, *22*(3), 394–400. doi:10.1016/j.copbio.2010.10.009 PMID:21071202

Busk, P. K., & Lange, L. (2015). Classification of fungal and bacterial lytic polysaccharide monooxygenases. *BMC Genomics*, *16*(1), 368. doi:10.1186/s12864-015-1601-6 PMID:25956378

Cairney, J. W. G. (2005). Basidiomycete mycelia in forest soils: Dimensions, dynamics and roles in nutrient distribution. *Mycological Research*, *109*(1), 7–20. doi:10.1017/S0953756204001753 PMID:15736859

Chamberlain, K., & Crawford, D. L. (2000). Thatch biodegradation and antifungal activities of two lignocellulolytic Streptomyces strains in laboratory cultures and in golf green turfgrass. *Canadian Journal of Microbiology*, *46*(6), 550–558. doi:10.1139/w00-025 PMID:10913977

Chen, J., Fales, S. L., Varga, G. A., & Royse, D. J. (1995). Biodegradation of cell wall components of maize stover colonized by white-rot fungi and resulting impact on in-vitro digestibility. *Journal of the Science of Food and Agriculture*, *68*(1), 91–98. doi:10.1002/jsfa.2740680115

Colberg, P. J. (1988). Anaerobic microbial degradation of cellulose, lignin, oligolignols, and monoaromatic lignin derivatives. In A. J. B. Zehnder (Ed.), *Biology of Anaerobic Microorganisms* (pp. 333–372). New York: John Wiley & Sons.

Cornu, A., Besle, J. M., Mosoni, P., & Grent, E. (1994). Lignin-carbohydrate complexes in forages: Structure and consequences in the ruminal degradation of cell-wall carbohydrates. *Reproduction, Nutrition, Development*, *24*(5), 385–398. doi:10.1051/rnd:19940501 PMID:7802932

Curling, S. F., Clausen, C. A., & Winandy, J. E. (2002). Experimental method to quantify progressive stages of decay of wood by basidiomycete fungi. *International Biodeterioration & Biodegradation*, *49*(1), 13–19. doi:10.1016/S0964-8305(01)00101-9

Daniel, G. (2003). Microview of wood under degradation by bacteria and fungi. *Advances in Our Changing World*, *845*, 34–72. doi:10.1021/bk-2003-0845.ch004

Daniel, G., Asiegbu, F., & Johansson, M. (1998). The saprotrophic wood-degrading abilities of *Heterobasidium annosum* intersterility groups P and S. *Mycological Research*, *102*(8), 991–997. doi:10.1017/S0953756297005935

Dashtban, M., Schraft, H., & Qin, W. (2009). Fungal bioconversion of lignocellulosic residues; opportunities & perspectives. *International Journal of Biological Sciences*, *5*, 578–595. doi:10.7150/ijbs.5.578 PMID:19774110

Dashtban, M., Schraft, H., Syed, T. A., & Qin, W. (2010). Fungal biodegradation and enzymatic modification of lignin. *Int J Biochem Mol Biol*, *1*, 36–50. PMID:21968746

De Angelis, K. M., Allgaier, M., Chavarria, Y., Fortney, J. L., Hugenholtz, P., Simmons, B., & Hazen, T. C. et al. (2011). Characterization of trapped lignin-degrading microbes in tropical forest soil. *PLoS ONE*, *6*(4), e19306. doi:10.1371/journal.pone.0019306 PMID:21559391

De Boer, W., Folman, L. B., Summerbell, R. C., & Boddy, L. (2005). Living in a fungal world: Impact of fungi on soil bacterial niche development. *FEMS Microbiology Reviews*, *29*(4), 795–811. doi:10.1016/j.femsre.2004.11.005 PMID:16102603

De Boer, W., Kowalchuk, G. A., & van Veen, J. A. (2006). Root-food and the rhizosphere microbial community composition. *The New Phytologist*, *170*(1), 3–6. doi:10.1111/j.1469-8137.2006.01674.x PMID:16539597

De Boer, W., & Vanderwal, A. (2008). Interactions between saprotrophic basidiomycetes and bacteria. In L. Boddy, J. C. Frankland, & P. Van West (Eds.), *Ecology of Saprotrophic Basidiomycetes* (pp. 142–15). Amsterdam: Academic Press. doi:10.1016/S0275-0287(08)80010-0

De Graaff, M. A., Classen, A. T., Castro, H. F., & Schadt, C. W. (2010). Labile soil carbon inputs mediate the soil microbial community composition and plant residue decomposition rates. *The New Phytologist*, *188*(4), 1055–1064. doi:10.1111/j.1469-8137.2010.03427.x PMID:21058948

Del Río, J. C., Gutiérrez, A., & Martínez, A.T. (2004). Identifying acetylated lignin units in non-wood fibers using pyrolysis-gas chromatography/ mass spectrometry. *Rapid Commun Mass Spectrom, 18,* 1181-1185.

Dighton, J. (2003). *The Role of Fungi in Ecosystem Processes*. New York, NY: Marcel Dekker. doi:10.1201/9780203911440

Distel, D. L., Amin, M., Burgoyne, A., Linton, E., Mamangkey, G., Morrill, W., & Yang, J. et al. (2011). Molecular phylogeny of Pholadoidea Lamarck, 1809 supports a single origin for xylotrophy (wood feeding) and xylotrophic bacterial endosymbiosis in Bivalvia. *Molecular Phylogenetics and Evolution*, *61*(2), 245–254. doi:10.1016/j.ympev.2011.05.019 PMID:21684342

Dizhbite, T., Zakis, G., Kizimia, A., Lazareva, E., Rossinskaya, G., Jurk-jane, V., & Viesturs, U. et al. (1999). Lignin – a useful bio-resource for the production of sorption-active materials. *Bioresource Technology*, *67*(3), 221–228. doi:10.1016/S0960-8524(98)80004-7

Eastwood, D., Floudas, D., Binder, M., Majcherczyk, A., Schneider, P., Aerts, A., & Watkinson, S. C. et al. (2011). The plant cell wall-decomposing machinery underlies the functional diversity of forest fungi. *Science*, *333*(6043), 762–765. doi:10.1126/science.1205411 PMID:21764756

Egland, P. G., Pelletier, D. A., Dispensa, M., Gibson, J., & Harwood, C. S. (1997). A cluster of bacterial genes for anaerobic benzene ring biodegradation. *Proceedings of the National Academy of Sciences of the United States of America*, *94*(12), 6484–6489. doi:10.1073/pnas.94.12.6484 PMID:9177244

Elder, D. J., & Kelly, D. J. (1994). The bacterial degradation of benzoic acid and benzenoid compounds under anaerobic conditions: Unifying trends and new perspectives. *FEMS Microbiology Reviews*, *13*(4), 441–468. doi:10.1111/j.1574-6976.1994.tb00061.x PMID:8011356

Eriksson, K. E. L., Blanchette, R. A., & Ander, P. (1990). *Microbial and enzymatic degradation of wood components*. Berlin: Springer-Verlag. doi:10.1007/978-3-642-46687-8

Fackler, K., Gradinger, C., Hinterstoisser, B., Messner, K., & Schwanninger, M. (2006). Lignin degradation by white rot fungi on spruce wood shavings during short-time solid-state fermentations monitored by near infrared spectroscopy. *Enzyme and Microbial Technology*, *39*(7), 1476–1483. doi:10.1016/j.enzmictec.2006.03.043

Falconer, R. E., Bown, J. L., White, N. A., & Crawford, J. W. (2007). Biomass recycling: A key to efficient foraging by fungal colonies. *Oikos*, *116*(9), 1558–1568. doi:10.1111/j.0030-1299.2007.15885.x

Fengel, D., & Wegener, G. (1984). *Wood: Chemistry, ultrastructure, reactions*. Berlin: De Gruyter.

Floudas, D., Binder, M., Riley, R., Barry, K., Blanchette, R. A., Henrissat, B., & Yadav, J. S. et al. (2012). The Paleozoic origin of enzymatic lignin decomposition reconstructed from 31 fungal genomes. *Science*, *336*, 1715–1719. doi:10.1126/science.1221748 PMID:22745431

Frey, S. D., Elliott, E. T., Paustian, K., & Peterson, G. A. (2000). Fungal translocation as a mechanism for soil nitrogen inputs to surface residue decomposition in a no-tillage agroecosystem. *Soil Biology & Biochemistry*, *32*(5), 689–698. doi:10.1016/S0038-0717(99)00205-9

Galletti, G. C., Piccaglia, R., & Concialini, V. (1990). Optimization of electrochemical detection in the high-performance liquid chromatography of lignin phenolics from lignocellulosic byproducts. *Journal of Chromatography. A*, *507*, 439–450. doi:10.1016/S0021-9673(01)84223-0 PMID:2380306

Garbeva, P., Hol, W. H. G., Termorshuizen, A. J., Kowalchuk, G. A., & de Boer, W. (2011). Fungistasis and general soil biostasis – a new synthesis. *Soil Biology & Biochemistry*, *43*(3), 469–477. doi:10.1016/j.soilbio.2010.11.020

Gasch, A. P. (2007). Comparative genomics of the environmental stress response in ascomycete fungi. *Yeast (Chichester, England)*, *24*(11), 961–976. doi:10.1002/yea.1512 PMID:17605132

Ghosh, P., & Gangopadyay, R. (2000). Photofunctionalization of cellulose and lignocellulose fibers using photoactive organic acids. *European Polymer Journal*, *36*(3), 625–634. doi:10.1016/S0014-3057(99)00093-2

Gold, M. H., Youngs, H. L., & Gelpke, M. D. (2000). Manganese peroxidase. *Metal Ions in Biological Systems*, *37*, 559–586. PMID:10693145

Guerra, A., Mendonca, R., Ferraz, A., Lu, F., & Ralph, J. (2004). Structural characterization of lignin during Pinus taeda wood treatment with *Ceriporiopsis subvermispora*. *Applied and Environmental Microbiology*, *70*(7), 4073–4078. doi:10.1128/AEM.70.7.4073-4078.2004 PMID:15240285

Guillén, F., Martínez, A. T., & Martínez, M. J. (1992). Substrate specificity and properties of the aryl-alcohol oxidase from the ligninolytic fungus *Pleurotus eryngii. European Journal of Biochemistry, 209*(2), 603–611. doi:10.1111/j.1432-1033.1992.tb17326.x PMID:1425667

Guillén, F., Martínez, M. J., Muñoz, C., & Martínez, A. T. (1997). Quinone redox cycling in the ligninolytic fungus *Pleurotus eryngii* leading to extracellular production of superoxide anion radical. *Archives of Biochemistry and Biophysics, 339*(1), 190–199. doi:10.1006/abbi.1996.9834 PMID:9056249

Gutiérrez, A., Caramelo, L., Prieto, A., Martínez, M. J., & Martínez, A. T. (1994). Anisaldehyde production and aryl-alcohol oxidase and dehydrogenase activities in ligninolytic fungi of the genus *Pleurotus. Applied and Environmental Microbiology, 60*, 1783–1788. PMID:8031078

Haigh, P. M. (1994). A review of agronomic factors influencing grass silage effluent production in England and Wales. *Journal of Agricultural Engineering Research, 57*(2), 73–87. doi:10.1006/jaer.1994.1007

Hannula, S. E., Boschker, H. T. S., De Boer, W., & Veen, J. A. (2012a). 13C pulse-labeling assessment of the community structure of active fungi in the rhizosphere of a genetically starch-modified potato culivar and its parental isoline. *New Phytol, 194*, 784–799.

Hedges, J. I., & Oades, J. M. (1997). Comparative organic geochemistries of soils and marine sediments. *Organic Geochemistry, 27*(7-8), 319–361. doi:10.1016/S0146-6380(97)00056-9

Hegde, R. S., & Fletcher, J. S. (1996). Influence of plant growth stage and season on the release of root phenolics by mulberry as related to development of phytoremediation technology. *Chemosphere, 32*(12), 2471–2479. doi:10.1016/0045-6535(96)00144-0

Heider, J., & Fuchs, G. (1996). Anaerobic metabolism of aromatic compounds. *European Journal of Biochemistry, 243*(3), 577–596. doi:10.1111/j.1432-1033.1997.00577.x PMID:9057820

Heinfling, A., Ruiz-Dueñas, F. J., Martínez, M. J., Bergbauer, M., Szewzyk, U., & Martínez, A. T. (1998). A study on reducing substrates of manganese-oxidizing peroxidases from *Pleurotus eryngii* and *Bjerkandera adusta. FEBS Letters, 428*(3), 141–146. doi:10.1016/S0014-5793(98)00512-2 PMID:9654123

Higuchi, T. (1997). *Biochemistry and molecular biology of wood.* London: Springer Verlag. doi:10.1007/978-3-642-60469-0

Hilden, K., Hakala, T. K., Maijala, P., Lundell, T. K., & Hatakka, A. (2007). Novel thermotolerant laccases produced by the white-rot fungus *Physisporinus rivulosus. Applied Microbiology and Biotechnology, 77*(2), 301–309. doi:10.1007/s00253-007-1155-x PMID:17805527

Hoffland, E., Kuyper, T. W., Wallander, H., Plassard, C., Gorbushina, A. A., Haselwandter, K., & van Breemen, N. et al. (2004). The role of fungi in weathering. *Frontiers in Ecology and the Environment, 2*(5), 258–264. doi:10.1890/1540-9295(2004)002[0258:TROFIW]2.0.CO;2

Hu, W. J., Harding, S. A., Lung, J., Popko, J. L., Ralph, J., Stokke, D. D., & Chiang, V. L. et al. (1999). Repression of lignin biosynthesis promotes cellulose accumulation and growth in transgenic trees. *Nature Biotechnology, 17*(8), 808–812. doi:10.1038/11758 PMID:10429249

Jensen, K. A. Jr, Bao, W., Kawai, S., Srebotnik, E., & Hammel, K. E. (1996). Manganese-dependent cleavage of nonphenolic lignin structures by *Ceriporiopsis subvermispora* in the absence of lignin peroxidase. *Applied and Environmental Microbiology*, *62*, 3679–3686. PMID:16535418

Joergensen, R. G., & Wichern, F. (2008). Quantitative assessment of the fungal contribution to microbial tissue in soil. *Soil Biology & Biochemistry*, *40*(12), 2977–2991. doi:10.1016/j.soilbio.2008.08.017

Jørgensen, N., & Stepanauskas, R. (2009). Biomass of pelagic fungi in Baltic rivers. *Hydrobiologia*, *623*(1), 105–112. doi:10.1007/s10750-008-9651-2

Karin, F. (2012). How spectroscopy and microspectroscopy of degraded wood contribute to understand fungal wood decay. *Applied Microbiology and Biotechnology*, *96*(3), 587–599. doi:10.1007/s00253-012-4369-5 PMID:22983562

Kersten, P. J. (1990). Glyoxal oxidase of *Phanerochaete chrysosporium*: Its characterization and activation by lignin peroxidase. *Proceedings of the National Academy of Sciences of the United States of America*, *87*(8), 2936–2940. doi:10.1073/pnas.87.8.2936 PMID:11607073

Kirby, R. (2005). Actinomycetes and lignin degradation. *Advances in Applied Microbiology*, *58*, 125–168. doi:10.1016/S0065-2164(05)58004-3 PMID:16543032

Kiyohara, H., Matsumoto, T., & Yamada, H. (2000). Lignin-carbohydrate complexes: Intestinal immune system modulating ingredients in kampo (Japanese herbal) medicine, juzen-taiho-to. *Planta Medica*, *66*, 20–24. doi:10.1055/s-2000-11116 PMID:10705728

Klein, D. A., & Paschke, M. W. (2004). Filamentous fungi: The indeterminate lifestyle and microbial ecology. *Microbial Ecology*, *47*(3), 224–235. doi:10.1007/s00248-003-1037-4 PMID:15037964

Kudanga, T., & Le Roes-Hill, M. (2014). Laccase applications in biofuels production: Current status and future prospects. *Applied Microbiology and Biotechnology*, *98*(15), 6525–6542. doi:10.1007/s00253-014-5810-8 PMID:24841120

Lindahl, B. D., Ihrmark, K., Boberg, J., Trumbore, S. E., Ho¨gberg, P., Stenlid, J., & Finlay, R. D. (2007). Spatial separation of litter decomposition and mycorrhizal nitrogen uptake in a boreal forest. *The New Phytologist*, *173*(3), 611–620. doi:10.1111/j.1469-8137.2006.01936.x PMID:17244056

Lonsane, B. K., Ghildyal, N. P., Budiatman, S., & Ramakrishna, S. V. (1985). Engineering aspects of solid state fermentation. *Enzyme and Microbial Technology*, *7*(6), 258–265. doi:10.1016/0141-0229(85)90083-3

Lonsane, B. K., Saucedo-Castaneda, G., Raimbault, M., Roussos, S., Viniegra-Gonzalez, G., Ghildyal, N. P., & Krish-naiah, M. M. et al. (1992). Scale-up strategies for solid state fermentation systems. *Process Biochemistry (Barking, London, England)*, *27*(5), 259–273. doi:10.1016/0032-9592(92)85011-P

Loredano, P. (2015). Lignin-degrading enzymes. *The FEBS Journal*, *282*(7), 1190–1213. doi:10.1111/febs.13224 PMID:25649492

Mai, C., Kues, U., & Militz, H. (2004). Biotechnology in the wood industry. *Applied Microbiology and Biotechnology*, *63*, 477–494. doi:10.1007/s00253-003-1411-7 PMID:12937955

Mai, C., Majcherczyk, A., & Huttermann, A. (2000). Chemo-enzymatic synthesis and characterization of graft copolymers from lignin and acrylic compounds. *Enzyme and Microbial Technology*, *27*(1-2), 167–175. doi:10.1016/S0141-0229(00)00214-3 PMID:10862917

Malosso, E., English, L., Hopkins, D. W., & ODonnell, A. G. (2004). Use of 13C-labelled plant materials and ergosterol, PLFA and NLFA analyses to investigate organic matter decomposition in Antarctic soil. *Soil Biology & Biochemistry*, *36*(1), 165–175. doi:10.1016/j.soilbio.2003.09.004

Mandels, M., & Sternberg, D. (1976). Recent advances in cellulase technology. *Ferment. Technol*, *54*, 267–286.

Martínez, A. T. (2002). Molecular biology and structure-function of lignin degrading heme peroxidases. *Enzyme and Microbial Technology*, *30*(4), 425–444. doi:10.1016/S0141-0229(01)00521-X

Martinez, A. T., Rencoret, J., Nieto, L., Jimenez-Barbero, J., Gutierrez, A., & del Rıo, J. C. (2011). Selective lignin and polysaccharide removal in fungal decay as evidenced by in situ structural analyses. *Environmental Microbiology*, *13*(1), 96–107. doi:10.1111/j.1462-2920.2010.02312.x PMID:21199251

Martinez, A. T., Speranza, M., Ruiz-Duenas, F. J., Ferreira, P., Camarero, S., Guillen, F., Martinez, M. J., & Gutierrez, A., & del, Rio, J. C. (2005). Biodegradation of lignocellulosics: Microbial, chemical, and enzymatic aspects of the fungal attack of lignin. *International Microbiology*, *8*, 195–204. PMID:16200498

Martínez, M. J., Ruiz-Dueñas, F. J., Guillén, F., & Martínez, A. T. (1996). Purification and catalytic properties of two manganese-peroxidase isoenzymes from *Pleurotus eryngii*. *European Journal of Biochemistry*, *237*(2), 424–432. doi:10.1111/j.1432-1033.1996.0424k.x PMID:8647081

Mayer, A. M., & Staples, R. C. (2002). Laccase: New functions for an old enzyme. *Phytochemistry*, *60*(6), 551–565. doi:10.1016/S0031-9422(02)00171-1 PMID:12126701

McDonald, P. (1981). The biochemistry of silage. John Wiley & Sons.

Milstein, O., Vered, Y., Gressel, J., & Flowers, H. M. (1981). Biodegradation of wheat straw lignocarbohydrate complexes (LCC) II. Fungal growth on aqueous hydrolysate liquors and particulate residues of wheat straw. *Eur. J. Appl. Microbiol. Biotechnol*, *13*(2), 117–127. doi:10.1007/BF00499699

Money, N. P. (2007). Biomechanics of invasive hyphal growth. In Biology of the Fungal Cell Wall. Springer-Verlag.

Moore-Kucera, J., & Dick, R. P. (2008). Application of 13C-labeled litter and root materials for in situ decomposition studies using phospholipid fatty acids. *Soil Biology & Biochemistry*, *40*(10), 2485–2493. doi:10.1016/j.soilbio.2008.06.002

Moredo, N., Lorenzo, M., Domínguez, A., Moldes, D., Cameselle, C., & Sanroman, A. (2003). Enhanced ligninolytic enzyme production and degrading capability of Phanerochaete chrysosporium and Trametes versicolor. *World Journal of Microbiology & Biotechnology*, *19*(7), 665–669. doi:10.1023/A:1025198917474

Mudgett, R. E. (1986). Solid-state fermentations. In A. L. Demain & N. A. Solomon (Eds.), *Manual of Industrial Microbiology and Biotechnology* (pp. 66–83). Washington, DC: American Society of Microbiology.

Nanayakkara, S., Patti, A. F., & Saito, K. (2014). Chemical depolymerization of lignin involving the redistribution mechanism with phenols and repolymerization of depolymerized products. *Green Chemistry*, *16*(4), 1897–1903. doi:10.1039/C3GC41708E

Osono, T. (2007). Ecology of ligninolytic fungi associated with leaf litter decomposition. *Ecological Research*, *22*(6), 955–974. doi:10.1007/s11284-007-0390-z

Osono, T., Hagiwara, Y., & Masuya, H. (2011a). Effects of temperature and litter type on fungal growth and decomposition of leaf litter. *Mycoscience*, *52*(5), 327–332. doi:10.1007/S10267-011-0112-9

Osono, T., To-Anun, C., Hagiwara, Y., & Hirose, D. (2011b). Decomposition of wood, petiole and leaf litter by Xylaria species from northern Thailand. *Fungal Ecology*, *4*(3), 210–218. doi:10.1016/j.funeco.2010.11.003

Otjen, L., & Blanchette, R. A. (1986). A discussion of microstructural changes in wood during decomposition by white rot basidiomycetes. *Canadian Journal of Botany*, *64*(5), 905–911. doi:10.1139/b86-121

Pandey, M. P., & Kim, C. S. (2011). Lignin depolymerization and conversion: A review of thermochemical methods. *Chemical Engineering & Technology*, *34*(1), 29–41. doi:10.1002/ceat.201000270

Prevost-Boure, N. C., Christen, R., Dequiedt, S., Mougel, C., Lelievre, M., Jolivet, C., & Ranjard, L. et al. (2011). Validation and application of a PCR primer set to quantify fungal communities in the soil environment by real-time quantitative PCR. *PLoS ONE*, *6*(9), e24166. doi:10.1371/journal.pone.0024166 PMID:21931659

Rousk, J., Brookes, P. C., & Baath, E. (2010a). Investigating the mechanisms for opposing pH relationships of fungal and bacterial growth in soil. *Soil Biology & Biochemistry*, *42*(6), 926–934. doi:10.1016/j.soilbio.2010.02.009

Ruiz-Dueñas, F. J., Martínez, M. J., & Martínez, A. T. (1999). Molecular characterization of a novel peroxidase isolated from the ligninolytic fungus *Pleurotus eryngii*. *Molecular Microbiology*, *31*(1), 223–235. doi:10.1046/j.1365-2958.1999.01164.x PMID:9987124

Sakagami, H., Satoh, K., Ida, Y., Koyama, N., Premanathan, M., Arakaki, R., & Yoshida, T. et al. (1999). Induction of apoptosis and anti-HIV activity by tannin- and lignin-related substances. *Basic Life Sciences*, *66*, 595–611. PMID:10800465

Salmén, L., & Burgert, I. (2009). Cell wall features with regard to mechanical performance. A review. *Holzforschung*, *63*, 121–129.

Sanchez, C. (2009). Lignocellulosic residues: Biodegradation and bioconversion by fungi. *Biotechnology Advances*, *27*(2), 185–194. doi:10.1016/j.biotechadv.2008.11.001 PMID:19100826

Sarikaya, A., & Ladisch, M. R. (1997). Mechanism and potential applications of bio-ligninolytic systems in a CELSS. *Applied Biochemistry and Biotechnology*, *62*(2-3), 131–149. doi:10.1007/BF02787990 PMID:9170251

Schmidt, O. (2006). *Wood and Tree Fungi. Biology, Damage, Protection, and Use*. Heidelberg, Germany: Springer.

Schneider, T., Keiblinger, K. M., Schmid, E., Streflinger-Gleixner, K., Ellersdorfer, G., Roschitzki, B., & Riedel, K. et al. (2012). Who is who in litter decomposition? Metaproteomics reveals major microbial players and their biogeochemical functions. *The ISME Journal*, *6*(9), 1749–1762. doi:10.1038/ismej.2012.11 PMID:22402400

Schwarze, F. W. M. R., Engels, J., & Mattheck, C. (2000). *Fungal strategies of decay in trees*. Berlin: Springer. doi:10.1007/978-3-642-57302-6

Scott, G. M., Akhtar, M., Lentz, M. J., Kirk, T. K., & Swaney, R. (1998). New technology for paper making: Commercializing biopulping. *Tappi J*, *81*, 220–225.

Sharma, R. K., & Arora, D. S. (2015). Fungal degradation of lignocellulosic residues: An aspect of improved nutritive quality. *Critical Reviews in Microbiology*, *41*(1), 52–60. doi:10.3109/1040841X.2013.791247 PMID:23855359

Simon, M. (2015). Lignocellulose degradation mechanisms across the Tree of Life. *Current Opinion in Chemical Biology*, *29*, 108–119. doi:10.1016/j.cbpa.2015.10.018 PMID:26583519

Singh, D., & Chen, S. (2008). The white-rot fungus Phanerochaete chrysosporium: Conditions for the production of lignin-degrading enzymes. *Applied Microbiology and Biotechnology*, *81*(3), 399–417. doi:10.1007/s00253-008-1706-9 PMID:18810426

Srebotnik, E., Messner, K., & Foisner, R. (1988). Penetrability of white rot-degraded pine wood by the lignin peroxidase of *Phanerochaete chrysosporium. Applied and Environmental Microbiology*, *54*, 2608–2614. PMID:16347765

Strickland, M. S., & Rousk, J. (2010). Considering fungal: Bacterial dominance in soils – methods, controls, and ecosystem implications. *Soil Biology & Biochemistry*, *42*(9), 1385–1395. doi:10.1016/j.soilbio.2010.05.007

Tanaka, H., Itakura, S., & Enoki, A. (1999). Hydroxyl radical generation by an extracellular low-molecular-weight substance and phenol oxidase activity during wood degradation by the white-rot basidiomycete Trametes versicolor. *Journal of Biotechnology*, *75*(1), 57–70. doi:10.1016/S0168-1656(99)00138-8 PMID:10704993

Tekolo, O. M., Mckenzie, J., Botha, A., & Prior, B. A. (2010). The osmotic stress tolerance of basidiomycetous yeasts. *FEMS Yeast Research*, *10*(4), 482–491. doi:10.1111/j.1567-1364.2010.00612.x PMID:20214685

Ten, H. (2001). Oxidative mechanisms involved in lignin degradation by white-rot fungi. *Chemical Reviews*, *101*(11), 3397–3414. doi:10.1021/cr000115l PMID:11749405

Trinci, A. P. J., Davies, D. R., Gull, K., Lawrence, M. I., Bonde Nielsen, B., Rickers, A., & Theodorou, M. K. (1994). Anaerobic fungi in herbivorous animals. *Mycological Research*, *98*(2), 129–152. doi:10.1016/S0953-7562(09)80178-0

Van der Wal, A., De Boer, W., Klein, Gunnewiek, P. J. A., & Van Veen, J. A. (2009b). Possible mechanism for spontaneous establishment of Calluna vulgaris in a recently abandoned agricultural field. *Restor Ecol, 17,* 308–313.

Van der Wal, A., Van Veen, J. A., Pijl, A. S., Summerbell, R. C., & De Boer, W. (2006a). Constraints on development of fungal biomass and decomposition processes during restoration of arable sandy soils. *Soil Biol Biochem, 38,* 2890–2902.

Vered, Y., Milstein, O., Flowers, H. M., & Gressel, J. (1981). Biodegradation of wheat straw lignocarbohydrate complexes (LCC). I. Dynamics of liberation of hot aqueous LCCs from wheat straw and partial characterization of the products. *Eur. J. Applied Microbiology and Biotechnology*, *12*(3), 183–188. doi:10.1007/BF01008341

Wang, L., Yan, W., Chen, J., Huang, F., & Gao, P. (2008). Function of the iron-binding chelator produced by Coriolus versicolor in lignin biodegradation. *Science in China. Series C, Life Sciences*, *51*(3), 214–221. doi:10.1007/s11427-008-0033-9 PMID:18246309

Wesenberg, D., Kyriakides, I., & Agathos, S. N. (2003). White-rot fungi and their enzymes for the treatment of industrial dye effluents. *Biotechnology Advances*, *22*(1-2), 161–187. doi:10.1016/j.biotechadv.2003.08.011 PMID:14623049

Wignarajah, K., Pisharody, S., & Fisher, J. W. (2000). Can incineration technology convert CELSS wastes to resources for crop production? A working hypothesis and some preliminary findings. *Advances in Space Research*, *26*(2), 327–333. doi:10.1016/S0273-1177(99)01072-8 PMID:11543171

Wilson, J. R., & Mertens, D. R. (1995). Cell wall accessibility and cell structure limitations to microbial digestion of forage. *Crop Science*, *35*(1), 251–259. doi:10.2135/cropsci1995.0011183X003500010046x

Wolfaardt, J. F., Jacobs, A., Rabie, C. J., Smit, S., Taljaardt, J. L., & Wing-field, M. J. (1999). Factors affecting colonization of freshly chipped softwood by white-rot fungi. Poster presented at an International Energy Agency Bioenergy Workshop on the Bioconversion of Lignocellulose, Itala Game Reserve, South Africa.

Wong, D. W. S. (2009). Structure and action mechanism of ligninolytic enzymes. *Applied Biochemistry and Biotechnology*, *157*(2), 174–209. doi:10.1007/s12010-008-8279-z PMID:18581264

Worrall, J. J., Anagnost, S. E., & Zabel, R. A. (1997). Comparison of wood decay among diverse lignicolous fungi. *Mycologia*, *89*(2), 199–219. doi:10.2307/3761073

Wurzbacher, C. M., Ba¨rlocher, F., & Grossart, H. P. (2010). Fungi in lake ecosystems. *Aquatic Microbial Ecology*, *59*, 125–149. doi:10.3354/ame01385

Xu, C., Ma, F., & Zhang, X. (2009). Lignocellulose degradation and enzyme production by Irpex lacteus CD2 during solid-state fermentation of cornstover. *Journal of Bioscience and Bioengineering*, *108*(5), 372–375. doi:10.1016/j.jbiosc.2009.04.023 PMID:19804859

Yelle, D. J., Ralph, J., Lu, F., & Hammel, K. E. (2008). Evidence for cleavage of lignin by a brown rot basidiomycete. *Environmental Microbiology*, *10*(7), 1844–1849. doi:10.1111/j.1462-2920.2008.01605.x PMID:18363712

Yuste, J. C., Penuelas, J., Estiarte, M., Garcia-Mas, J., Mattana, S., Ogaya, R., & Sardans, J. et al. (2011). Drought-resistant fungi control soil organic matter decomposition and its response to temperature. *Global Change Biology*, *17*(3), 1475–1486. doi:10.1111/j.1365-2486.2010.02300.x

Zadrazil, F., & Isikhuemhen, O. (1997). Solid state fermentation of lignocellulosics into animal feed with white rot fungi. In S. Roussos, B. K. Lonsane, M. Raimbault, & G. Viniegra-Gonzalez (Eds.), *Advances in Solid State Fermentation* (pp. 23–38). Dordrecht, The Netherlands: Kluwer Academic Publishers. doi:10.1007/978-94-017-0661-2_22

Zhou, C. H., Xia, X., Lin, C. X., Tong, D. S., & Beltramini, J. (2011). Catalytic conversion of lignocellulosic biomass to fine chemicals and fuels. *Chemical Society Reviews*, *40*(11), 5588–5617. doi:10.1039/c1cs15124j PMID:21863197

Zhou, S., & Ingram, L. O. (2000). Synergistic hydrolysis of carboxymethyl cellulose and acid-swollen cellulose by two endoglucanases (CelZ and CelY) from *Erwinia chrysanthemi*. *Journal of Bacteriology*, *182*(20), 5676–5682. doi:10.1128/JB.182.20.5676-5682.2000 PMID:11004164

KEY TERMS AND DEFINITIONS

Biopulping: It is a type of industrial biotechnology. It uses natural fungus to convert wood chips to paper pulp.

Biorefineries: Co-production of a variety of biologically-based products such as food, feed, materials, chemicals and energy i.e fuels, power, heat from biomass.

Biotechnological Applications: It is the use of living systems and organisms to develop or make products. Applications include the production of certain drugs, synthetic hormones, and bulk foodstuffs as well as the bioconversion of organic waste and the use of genetically altered bacteria in the cleanup of oil spills.

Extracellular Enzymes: These are enzymes that are secreted by a cell which function outside of that cell, where their function is to break down complex macromolecules into smaller units to be taken up by the cell for growth and assimilation. These enzymes degrade complex organic matter such as cellulose and hemicellulose into simple sugars that enzyme-producing organisms use as a source of carbon, energy, and nutrients.

Fungal Degradation: It is a beneficial activity of fungi in carrying out biodegradation using chemical substances as carbon and energy source for metabolism, thereby breaking down larger molecules to smaller ones.

Fungi: Plural for fungus which is a group of unicellular, multicellular, or syncytial spore-producing organisms including yeasts and molds, as well as multicellular fungi that produce familiar fruiting forms known as mushrooms.

Genetic Engineering: A set of technologies used to change the genetic makeup of cells, including the transfer of genes within and across species boundaries to produce improved or novel organisms.

Polymers: Compounds with high molecular weight consisting of repeated linked units, each a relatively light, similar and simple molecule. e.g. natural compounds like lignin and cellulose and many synthetic organic materials used as plastics and resins.

Chapter 12
Extraction of Preformed Mixed Phase Graphene Sheets from Graphitized Coal by Fungal Leaching

Manoj Balachandran
Christ University, India

ABSTRACT

The potential use of coal as source of carbon nano structure is seldom investigated. Herein we report a facile fungal solubilization method to extract mixed phase carbon structure from low grade coal. Coal had been used as a primary source for the production of carbon nanostructure with novel property, in addition to its main utility as a fuel. The major hurdle in its application is the inherent mineral embedded in it. An environmentally benign demineralization procedure make coal as a widely accepted precursor for the novel carbon materials. With Aspergiilus niger leaching, the randomly oriented preformed crystalline mixed phase nanocarbon in coal can be extracted. Raman studies revealed the presence of E_2g *scattering mode of graphite. The* sp^3 *domains at ~1355* cm^{-1} *(D band) is an indication of diamond like structure with disorder or defect. In the 2D region, multilayer stacking of graphene layers is noticed. The ratio of the defect to graphitic bands was found to be decreasing with increasing rank of coal. Bio leaching of coal enhances the carbon content in coal while eliminating the associated minerals in it. These defected carbon is an ideal material for graphene quantum dots and carbon dots, which are useful in drug delivery and bio imaging applications.*

INTRODUCTION

Coal is extensively used as solid fuel across the world and has sizeable share in the global fossil fuel reserve. Utilization of this resource generate excess spoil and low grade waste to the environment. Solubilization of minerals in coal has the potential to elevate this inherently dirty solid fuel to a source of value-added products like graphene, graphene quantum dots or carbon nanotubes (Manoj, 2014; Ruquan et al, 2013). Microorganisms have the ability to breakdown while beneficiating the low grade coal.

DOI: 10.4018/978-1-5225-2325-3.ch012

Bioremediation technologies have unfortunately not been realized for obtaining value added products from coal. Currently the application of bioremediation in controlling and eradicating mineral pollution has generated great interest. Alternatively bio-sorption, which exploits resources of biological origin, like bacteria, fungi, yeast, algae, are also gaining momentum. These microorganisms have the property to sequester metal-and decrease them to *ppb* level. It can effectively and quickly sequester metal ions out of complex molecule and is an ideal candidate for the treatment of industrial waste (Wang, & Chen, 2006; Volesky, 2001; Lin et al., 2006). The bio-sorbent materials like algae, fungi and bacteria have the ability to bind on metal elements and is considered from the toxicological point of view. This biomass can passively bind on metallic ions or organic compounds through various physico-chemical mechanisms (Wang, & Chen, 2006). The challenge is the proper selection of biomass from a large pool of readily available biomaterial, which has high mineral/metal binding property. In recent past, extensive research is being carried out on the bio-sorption phenomena, particularly on the demineralization or leaching of minerals to enhance the quality of ore (Wang, & Chen, 2006; Volesky, 2001, 2007).

Graphene is sp^2 hybridized form of carbon, appear in many forms like graphite, nanotubes or fullerenes (Manoj, 2014; Manoj, & Kunjomana, 2014). With the recent advances in nano carbon research, its synthesis plentiful natural source like coal has gained momentum (Elcey, & Manoj, 2016; Raquan et al., 2013). Coal or black carbon is a solid product formed by the partial combustion or anaerobic digestion of vegetation. It comprises of crystalline carbon with turbostratic structure with highly disordered amorphous addends on the edges (Elcey, & Manoj, 2010; 2013; 2014; Dong et al., 2014). Nano domain in coal consist of about 3-4 stacked carbon layers, with lateral dimension and interlayer spacing of 3nm and ~ 0.35 nm respectively. The amorphous realms constituted of polycyclic aromatic compounds, which are graphene sheets having irregular onion-like arrangement along with other organic constituents. The main hindrance in the structural characterization of coal is the inherent minerals present in it. Being inherently low quality due to the mineral matter and ash, leaching of coal is need for any application. Generally chemical leaching with mineral acids or alkalis are used for the demineralization of coal. Even though mineral matter could be removed, the disposal of the used chemical in safe manner is a serious concern. These chemicals generally corrode the environment and generates serious environmental pollution. In this scenario, a facile, ecofriendly demineralization technique of coal is worthwhile problem to be investigated.

Direct application of organic acids may affect mineral weathering rates by at least 3 mechanisms: by changing the dissolution rate far from equilibrium through decreasing solution pH or through forming complexes with cations at the mineral surface or affecting the saturation state of the solution with respect to the mineral (Elcey, & Manoj, 2010; 2013). Use of mineral acids in demineralization not only modifies the surface morphology and deteriorates the carbon structure, but also reduces the calorific value. These acids have strong oxidizing power and the safe disposal of the spent liquid is a major environmental concern. For commercial utility of coal bio-demineralization, fungal leaching is an ideal eco-friendly method. These mild organic acid partially removes the minerals in coal without much modification to the stacking structure. Coal being amorphous in nature, the ordering of carbon in it happens only with a strong acid.

In the present work, efficacy of the fungus *Aspergillus niger* in leaching coal and its effect on nano carbon layer stacking is discussed. It is a green extraction of preformed carbon structure from low rank coal. Changes in the stacking structure, nature of defect, degree of graphitization and change surface morphology with bioleaching is also highlighted. The mechanism of demineralization with carboxylic acid secreted by *Aspergillus niger* was elaborated.

FUNGAL CHARACTERISTICS

Fungi are large groups of eukaryotic microorganisms, of which three groups have paramount importance: molds, yeast and mushrooms. They are able to bind metallic elements and can alter the rate of fermentation process. Fungi *A. niger* has the ability to eliminate metal ions and radio-nuclides from aqueous solutions. Filamentous fungi have widely used to produce a variety of extracellular enzymes such as gluco amylase, pectinase, acidic lipase, along with organic acids like gluconic acid and citric acid (Celin, Acharya, Sukla, & Misra, 2005; Ghorbani et al., 2007; Ghorbani, Oliazadeh, & Shavedi, 2008; Sharma, & Wadhwa, 1997). These enzymes and acids are generally regarded as safe for food and drug administration. *A. niger* is of prime importance in biodegradation of toxic chemicals and conversion of waste water sludge. As it secretes organic acids, it can be used to leach out metals from mining ores (Kubicek, Scherferl, & Wohres, 1988; Jaklitsch, Kubicek, & Scrutten, 1991).

About 10 g selected coal samples were dried, ground to a particular size and inoculated into 200 ml modified Czepek Dox medium with low sucrose. Cultures of *A. niger* was grown on potato dextrose agar medium till sporulation. The spores was inoculated to modified Czepek Dox medium containing coal sample and *A. niger* in triplicates. The cultures were incubated for 10 days at room temperature under stationary conditions. A control experiment was set up with the same amount of coal samples in the medium without test organism. The cultures were grown to mycelial mass on the surface of medium. Filtrates (1 ml each) were drawn at regular intervals for the analysis of pH and metabolites. After 10 days of incubation, the cultures as well as the control were decanted to obtain the residual coal particles, which were further centrifuged and separated into culture and filtrate. The residual coal samples were thoroughly washed with de-ionized water, dried, powdered and subjected to ultimate and proximate analysis. The culture filtrates were placed on nutrient and potato dextrose agar plates to establish their purity. The organic acids produced by the fungi were determined by HPLC with the help of a UV detector at 210 nm. An Agilent Technologies 1200 series high performance liquid chromatography with C18 column was used for the purpose.

The presence minerals like silicate, aluminates, quartz and calcites in coal samples were determined using a scanning electron microscopy (SEM-JEOL model JED-2300). The X-ray profile was recorded by a Bruker AXS D8 Advance X-ray powder diffractometer. Powdered samples were scanned from 4-70° in 2θ range with 0.020°step intervals and 2 s/step counter time. The spectroscopic properties of the samples were analyzed by Micro Raman Spectrometer Horiba Jobin Lab RAM HR system with an excitation wavelength of 514.5 cm^{-1}. Spectral analysis and curve fitting was carried out with the aid of origin 8 software.

CHARACTERIZATION OF ORGANIC ACIDS IN FERMENTED MEDIA

HPLC elution profiles of cultural filtrate of the sample treated with *A. niger* revealed the presence of various organic acids like oxalic acid, citric acid and gluconic acid (secreted by *A. niger*) along with enzymes, acetyl-CoA and oxalo acetate (Kubicek, Scherferl, & Wohres, 1988; Jaklitsch, Kubicek, & Scrutten, 1991; Ta-Fa Lin et al., 2006; Manoj, & Elcey, 2010). Protons of organic acid, produced as a result of acidolysis are free to cause reduction of pH in the culture of *A. niger*.

The critical parameters for citric acid production by *A. niger* include: high carbohydrate concentration, low but, finite manganese concentrations, high dissolved oxygen, constant agitation and low pH

(Wang, & Chen, 2006; Ghorbani et al., 2007). The production of citrate from glucose or sucrose involves a large number of enzymatic steps occurring in two different membrane bound cellular compartments namely, the cytosol and mitochondrion. Glucose is taken into cell and converted to pyruvate, a three-carbon acid via the glycolytic pathway in the cytosol. One of the molecules of pyruvate is decarboxylated with the formation of acetyl-CoA by the mitochondrial pyruvate dehydrogenase complex and another is decarboxylated to oxaloacetate in the cytosol by pyruvate carboxylase. Oxaloacetate is transferred into mitochondrion (via malate) and condensed with acetyl-CoA to form citrate. The product is then transported out of the mitochondrion and cell. High yields in citric acid are possible, because all the six carbons of the substrate are conserved in the six-carbon product, citric acid, through the glycolytic pathway (Jaklitsch, Kubicek, & Scrutten, 1991).

$C_6H_{12}O_6 + 1.5\ O_2 \rightarrow C_6H_8O_7 + 2\ H_2O$ (Citric acid).

In *A. niger*, biosynthesis of oxalate is exclusively due to the action of oxalo-acetase, which catalyses the hydrolysis of oxalo-acetate to oxalate and acetate. The enzyme is located in the cytosol and expansion is induced at pH value greater than 4.

$C_6H_{12}O_6 + 4.5\ O_2 \rightarrow 3\ C_2H_2O_4 + 3\ H_2O$ (Oxalic acid).

Gluconic acid production is exceptional among the organic acid fermentations and it occurs completely outside the cytoplasmic membrane. Glucose oxidase is catalyzed in steps, which oxidizes β-D-glucopyranose to D-glucono-1,5-lactone. During the process of conversion, glucose oxidase present in *A. niger* undergoes self-reduction by the removal of two hydrogen atoms. The reduced form of enzyme is further oxidized by molecular oxygen, which results in the formation of hydrogen peroxide, a by-product in the reaction. It produces catalase, which acts on hydrogen peroxide releasing water and oxygen. Hydrolysis of glucono-δ-lactone to gluconic acid is facilitated by lactonase. The two key parameters which influence the gluconic acid production, are oxygen availability and pH of the culture medium.

MECHANISM OF BIO-DEMINERALIZATION

The mechanism and role of carboxylic acid in demineralization is discussed elsewhere (Manoj, & Elecey, 2013; Ghorbani et al., 2007). During bio-solubilization, the alumina cation (Al^{+3}) present in coal is fully dissociated in aqueous solution and aluminum citrate is formed.

$C_6H_8O_7 \rightarrow (C_6H_5O_7)^{3-} + 3\ H^+$

$(C_6H_5O_7)^{3-} + Al^{3+} \rightarrow Al(C_6H_5O_7)$

Similarly, oxalic acid contains two carboxylic groups, which forms possible complexes of alumina cation with oxalate ions are as follows:

$C_2H_2O_4 \rightarrow (C_2HO_4)^{1-} + H^+$

$3\ (C_2HO_4)^{1-} + Al^{3+} \rightarrow Al(C_2HO_4)_3$

$C_2H_2O_4 \rightarrow (C_2O_4)^{2-} + 2\ H^+\ (pKa_1 = 4.2)$

$3\ (C_2O_4)^{2-} + 2\ Al^{3+} \rightarrow Al_2(C_2O_4)_3$

$C_2H_2O_4 \rightarrow (C_2O_4)^{2-} + 2\ H^+$

$(C_2O_4)^{2-} + Ca^{2+} \rightarrow CaC_2O_4$

The citrate complexes thus formed are precipitated and removed from the sample by washing with water and filtration. Silicate solubilization involves a combination of proton attack and complex reaction with organic acids produced by the fungi. The large scale production of gluconic acid is attributed to the silicate solubilizing ability of fungi (Ramachandran, Pierre, & Pandey, 2006). Gluconic acid can cause substantial reduction of calcium minerals with the possible formation of calcium gluconate. The carboxylic acids secreted during fungal solubilization react with the mineral content by the following mechanism:

$2\ (C_6H_{12}O_7) \rightarrow (C_{12}H_{22}O_{14})^{2-} + 2\ H^+$

$C_{12}H_{22}O_{14} + Ca^{2+} \rightarrow C_{22}H_{22}CaO_{14}$ (Calcium gluconate)

The metal complexes were removed from the solution by filtration and the purified coal was recovered. The demineralized coal sample was further subjected to X-ray and Raman analysis to understand the change in stacking.

INFLUENCE OF BIOLEACHING ON STACKING STRUCTURE

The X-ray profile of the bio solubilized coal samples are presented in Figure 1 and Figure 2. A broadened Π band is observed at 25.5°, due to the formation of poly crystalline graphitic carbon. Higher intensity of this band indicates higher degree of crystallinity/graphitization. This band found to be broadened owing to the presence of nanocrysatlline carbon in the coal matrix. To understand the different types of carbonaceous species present in the coal, the spectral deconvolution of X-ray profile is carried out. There are four gaussian peaks formed in the bituminous coal profile (Figure 2). They are due to the presence of nanocrystalline carbon (n-C), sp^3 type carbon (diamond like), sp^2 type carbon (Graphitc type) and amorphous carbon (a-C). In the case of sub-bituminous caol, five peaks are formed. Being immature coal, the presence of nano carbon (n-C) or defective carbon are more in sub-bituiminous coal. The intensity ratio of disorder carbon to ordered carbon (I_{20}/I_{26}) is estimated to be 1.1 and 0.8 for the sub-bituminous coal and bituminous coal respectively. A higher value in defect indicate a higher degree of disorder in the graphitic structure in sub-bituminous coal.

Raman spectra of nano carbon from biosolubilized coal are presented in Figure 3 and Figure 4. Two overlapping peaks are found at 1345 cm^{-1} and 1585 cm^{-1} which are due to defect (D) and graphitic and (G) in the coal (Manoj, 2016; 2013; 2012; Elcey, & Manoj, 2016; Xiao, Liu, & Yang, 2015). The

Figure 1. X-ray analysis of nano carbon in sub-bituminous coal

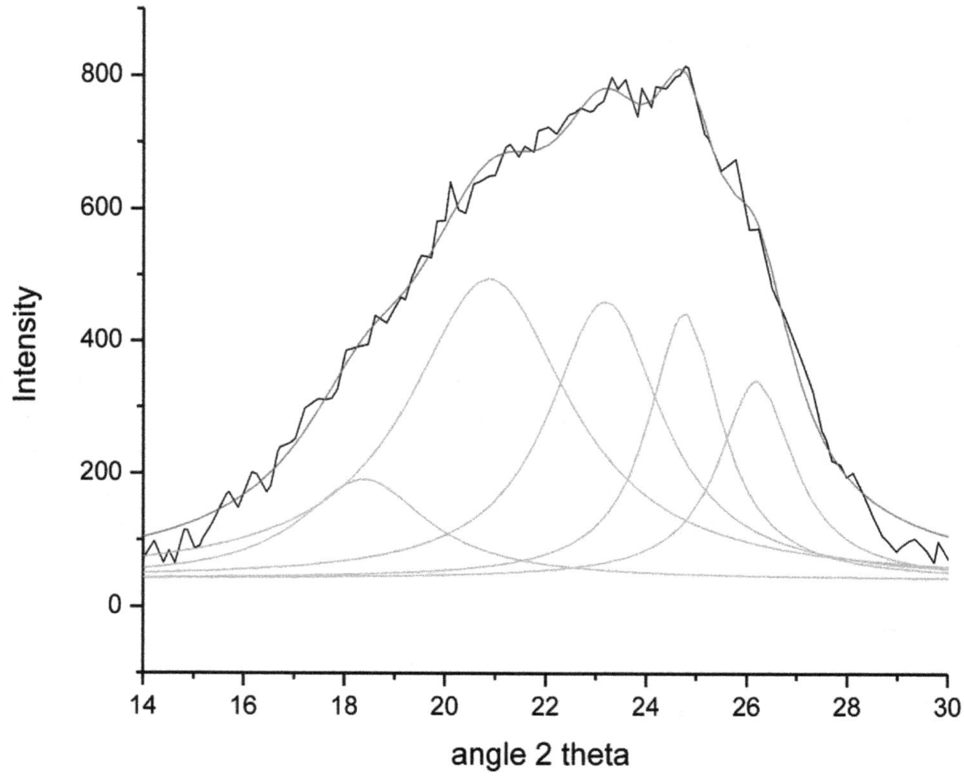

Figure 2. X-ray analysis of nano carbon in bituminous coal

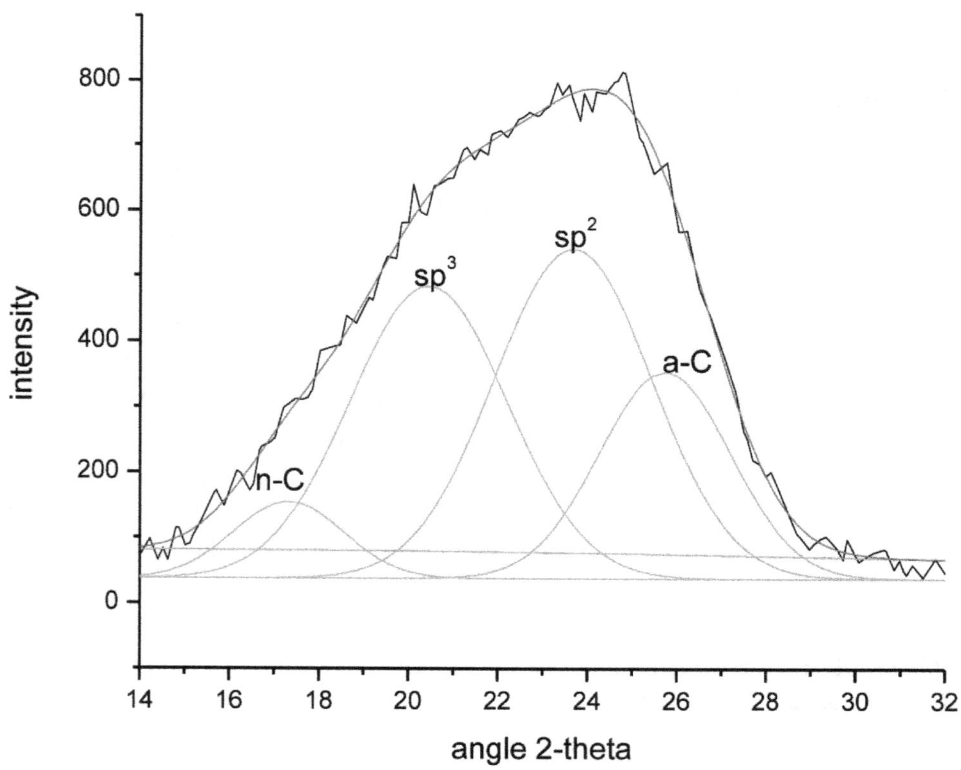

intensity ratio of these two bands are estimated to be 0.84 and 0.86 respectively for the bituminous coal and sub-bituminous coal. This result confirms the lesser degree of diorder in the graphitic domain of the bituminous coal compared to sub-bituminous coal. This result is in support with the X-ray analysis.

A small shoulder peak is observed near 1600 cm^{-1} and is designated as D' peak. With the rise in the concentration of defect, this peak gets resolved (up to 1/3rd intensity of G peak). The ratio of G peak to D' peak ($I_G/I_{D'}$) is found to be 3.67 and 5.78 while the $I_D/I_{D'}$ ratio is calculated to be ~2.9 and ~4.85 for the two samples respectively. This defect in the layers of carbon in the bituminous coal is due to on-site defect while that in the sub-bituminous coal is attributed to hopping defect (Elcey, & Manoj, 2016; Eckmann et al., 2012). These defects are mainly due to the out-of-plane bending of sp^3 carbon atom in the graphene layers of coal. The lateral size of the aromatic lamellae was calculated by the modified Scherrer formula and is found to be 5.24 and 5.11 nm for the bituminous and sub-bituminous coal respectively.

The 2D region of the Raman spectrum of the biosolubilized coal samples shows a broad bump confirming the stacking of graphene layers in the sample. This region is deconvoluted into constituent peaks to understand the nature of defects in the synthesized graphene layers (Figure 5). The 2D band is fitted with multiple Lorentian profile has 4 peaks at 2445, 2720, 2925 and 3160 cm^{-1} designated as G*, G', D+D' and 2D' band. This is primarily due to splitting of π electron dispersion energies caused by the interaction between adjacent carbon layers (Rama, Prabhu, & Adarsh, 2012, Ferrari, & Basko, 2013; Manoj, 2014). Another reason could be the first overtone of the D band, which induces defect in the layers of graphene. The position of the G' band depends on the stacking layers and order of the graphite sample. In a well stacked multi layer graphene (less than 10 layers), this peak exhibits two peak profile (Rama Prabhu, & Adarsh, 2012; Ferrari, & Robertson, 2001). With increase in disorder and randomness, this peak shift upwards and finally merges into a single 2D band.

Figure 3. Raman analysis of bituminous coal

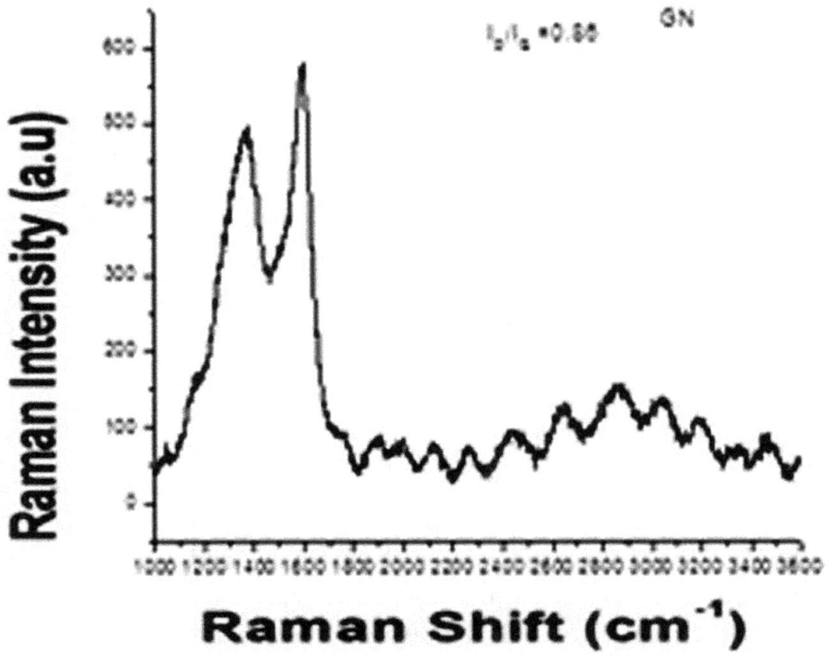

Figure 4. Raman analysis of sub-bituminous coal

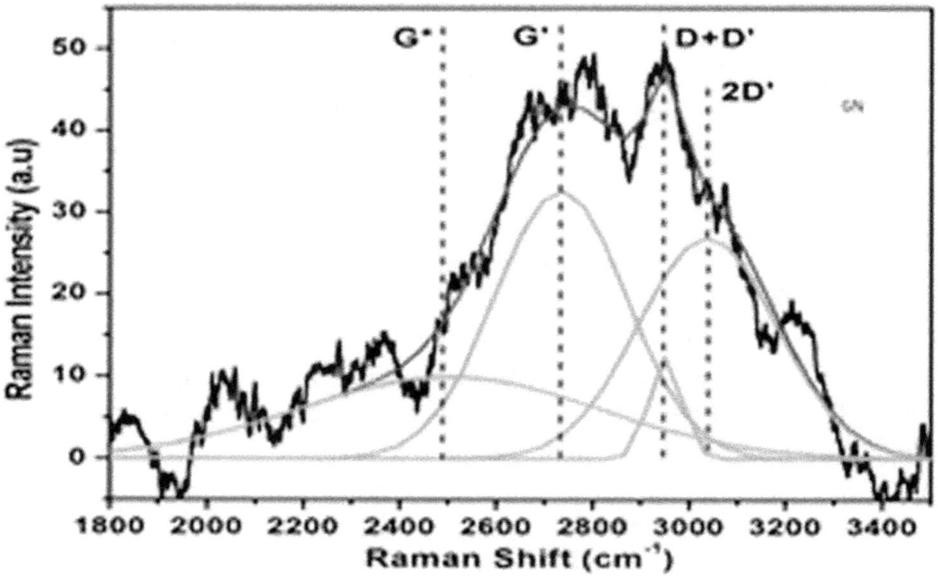

Figure 5. Curve fit for the second order Raman spectrum of sub-bituminous coal (GN)

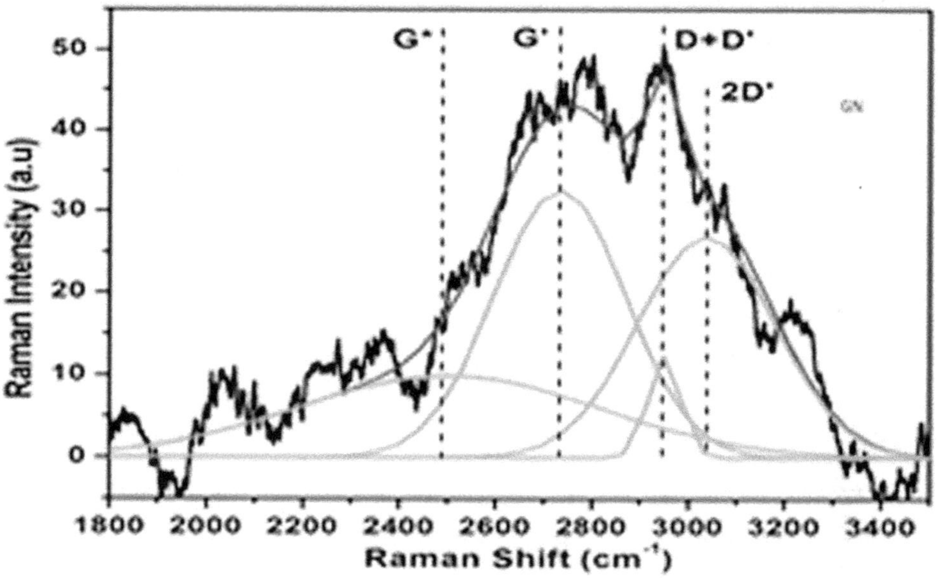

Raman analysis reveals that during bioleaching, the arrangement of carbon in coal was modified into a mixed phase graphene nanostructure with sp^2 and sp^3 components. The broad graphitic band and defect band confirms that formed nano carbon has both amorphous and graphene domain in it. The Raman investigation confirms the presence of multi-layer mixed phase graphene nanosheets (mixed phase sp^2 - sp^3) than a pure sp^2 graphene layer. This findings are agreement with the XRD analysis.

MORPHOLOGY OF MINERALS

The surface morphology of the virgin and bio-solubilized products were investigated in order to understand the change in mineral content and stacking. The images (Figures 6 and 7) revealed that the micro surface is formed by homogeneous network of mineral crystallites embedded in the carbon matrix (Wu, & Steel, 2007). The elemental analysis by EDS (Si-0.76 wt%; Al-0.66 wt% and Ca-traces) revealed that Si and Al as major elements in the raw coal. *A. niger* (Figure 6) leaching has modified the morphology of coal. The minerals were eliminated and large flakes of carbon were formed. The EDS analysis confirmed that only carbon and oxygen are present in the product. It support the finding that carbonization is happened with leaching. The graphite oxide layer formed was folded and stacked. The proximate analysis showed a remarkable reduction of 98% in ash content (not shown). This drastic loss in ash was reported for the first time in single stage leaching and is attributed to the solubilization of minerals by secreted organic acids during *A. niger* treatment.

CONCLUSION

The present research work is focused on the extraction of graphene sheets from coal by leaching it with *Aspergillus niger*. The investigations helped to elucidate a novel mechanism and biocatalyst involved in coal bioconversion process. The significant results obtained are summarized as follows. The presence of nanocrystalline carbon (n-C), sp^3 type carbon (diamond like), sp^2 type carbon (Graphitc type) and amorphous carbon (a-C) are evident in the bioleached coal. The defect formed on the graphene system is on-site defect in bituminous coal while that in the sub-bituminous coal is hopping defect. After bioleaching, the nano carbon in coal has a mixed phase graphene nanostructure containing both sp^2 and sp^3

Figure 6. SEM/EDS analysis of virgin coal

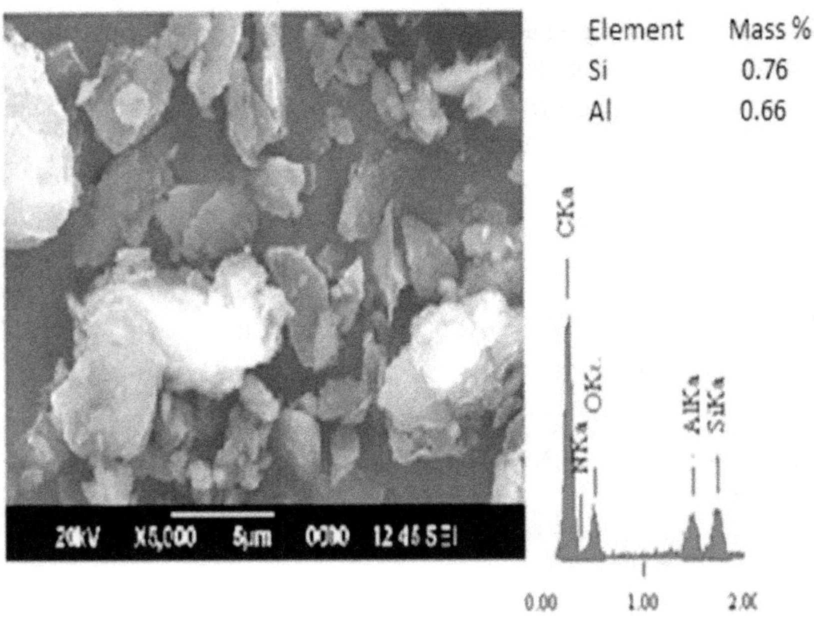

Figure 7. SEM/EDS analysis of Aspergillus niger leached coal

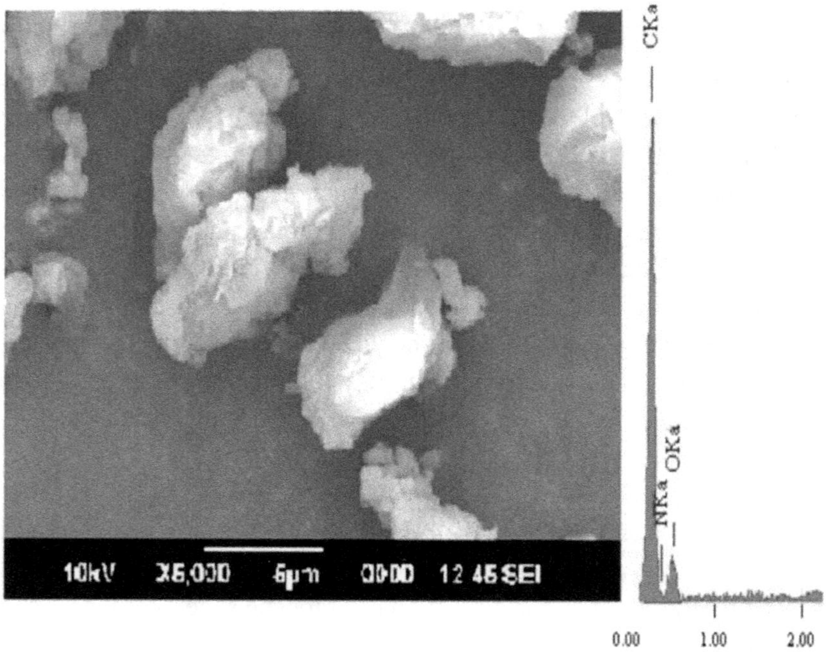

components. The Raman analysis confirms the formation of stacking of graphene nanosheets with sp^2 - sp^3 mixed phase. This findings are well supported the XRD analysis. The HPLC analysis confirmed secretion of oxalic acid, gluconic acid and citric acid. Secretion of these organic acid lower the pH and fasten the demineralization process by loosening the mineral bond with the carbon matrix. SEM-EDS analysis confirmed the removal of silicates, aluminates and calcites during leaching and stacking and folding of carbon layers as that of bulk graphite. The synthesized nano crystalline carbon from coal have wide application in synthesizing fluorescent carbon dots which are biocompatible. These materials have the potential application in bio imaging as they have tunable band gap and photoluminescence. They can encapsulate the pharmaceutical molecule which will be helpful in selective drug delivery. They are useful precursor for the large scale synthesis of cheap and good quality graphene quantum dots and nano diamonds.

REFERENCES

Celin, A., Sukla, L. B., & Misra, V. N. (2005). Biological elimination of sulphur from high sulphur coal by Aspergillus-like fungi. *Fuel*, *84*, 1597–1600.

Dong, Y., Lin, J., Chen, Y., Fu, F., Chi, Y., & Chen, G. (2014). Graphene quantum dots, graphene oxide, carbon quantum dots and graphite nanocrystals in coals. *Nanscale*, *6*(13), 7410–7415. doi:10.1039/c4nr01482k PMID:24875280

Eckmann, A., Felten, A., Mishchenko, A., Britnell, L., Krupke, R., Novoselov, K. S., & Casiraghi, C. (2012). Probing the nature of defects in graphene by Raman spectroscopy. *Nano Letters*, *12*(8), 3925–3930. doi:10.1021/nl300901a PMID:22764888

Elcey, C. D., & Manoj, B. (2010). Demineralization of coal by stepwise leaching: A study of sub- bituminous Indian coal by FTIR and SEM, *Journal of the University Chemical Technology and metallurgy*, *45*(4), 385-390.

Elcey, C. D., & Manoj, B. (2013). Demineralization of sub-bituminous Coal by fungal leaching: structural characterization by X-ray and FTIR analysis. *Research Journal of Chemistry and Environment, 17*(8), 11-15.

Elcey, C. D., & Manoj, B. (2014). An investigation on bio-solubilization of coal by *Aspergillus niger*. *Chemical Science Transactions*, *3*(3), 1200–1206.

Elcey, C. D., & Manoj, B. (2016). Graphitization of coal by Bio-solubilization: Structure probe by Raman spectroscopy. *Asian Journal of Chemistry*, *28*(7), 1557–1560. doi:10.14233/ajchem.2016.19750

Ferrari, A. C., & Basko, M. (2013). Raman Spectroscopy as a versatile tool for studying the properties of graphene. *Nature Nanotechnology*, *8*(4), 235–246. doi:10.1038/nnano.2013.46 PMID:23552117

Ferrari, A. C., & Robertson, J. (2001). Resonant Raman spectroscopy of disordered, amorphous and diamond like carbon. *Phys Rev B., 64,* 075414-1–075414.

Ghorbani, Y., Oliazadeh, M., & Shavedi, A. (2008). Aluminum solubilization from red mud by some indigenous fungi in Iran. *Journal of Applied Biosciences*, *7*, 207–213.

Ghorbani, Y., Oliazadeh, M., Shavedi, A., Roohi, R., & Pirayehgar, A. (2007). Use of some isolated fungi in biological Leaching of Aluminium from low grade bauxite. *African Journal of Biotechnology*, *6*(11), 284–1288.

Jaklitsch, W. M., Kubicek, C. P., & Scrutten, M. C. (1991). Intracellular location of enzymes involved in citrate production by *Aspergillus niger. Canadian Journal of Microbiology*, *37*(11), 823–827. doi:10.1139/m91-142 PMID:1777859

Kubicek, C. P., Scherferl, K. G., & Wöhres, W. (1988). Evidence for a cytoplasmic pathway of oxalate biosynthesis in *Aspergillus niger. Applied and Environmental Microbiology*, *54*(3), 633–637. PMID:3132096

Lin, T. F., Huang, H. I., Shen, F. T., & Young, C.-C. (2006). The protons of gluconic acid are the major factor responsible for the dissolution of tricalcium phosphate by *Burkholderia cepacia* CC-A174. *Bioresource Technology*, *97*(7), 957–960. doi:10.1016/j.biortech.2005.02.017 PMID:16356711

Manoj, B. (2012). Chemical demineralization of high volatile Indian bituminous coal by carboxylic acid and characterization of the products by SEM/EDS. *Journal of Environmental Research and Development, 6*(3A), 653-659.

Manoj, B. (2013). Bio-demineralization of Indian bituminous coal by *Aspergillus niger* and characterization of the products. *Research Journal of Biotechnology*, *8*(3), 49–54.

Manoj, B. (2014). Investigation of nano-crystalline structure in selected carbonaceous materials. *International Journal of Minerals Materials and Metallurgy, 21*(9), 940-946.

Manoj, B. (2016). A Comprehensive analysis of various structural parameters of Indian coals with the aid of advanced analytical tools. *International Journal of Coal Science and Technology*, *3*(2), 123–132. doi:10.1007/s40789-016-0134-1

Manoj, B., & Kunjomana, A. G. (2014). Systematic Investigations of graphene layers in sub-bituminous coal. *Russian Journal of Applied Chemistry*, *87*(11), 1726–1733. doi:10.1134/S1070427214110251

Ramachandran, S., Pierre, F., & Pandey, A. (2006). Gluconic acid: Properties, applications and microbial production. *Food Technology and Biotechnology*, *44*(2), 185–195.

Ramaprabhu, S. A., & Kaniyoor, A. (2012). Raman spectroscopic investigation of graphite oxide derived graphene. *AIP Advances*, *2*(3), 032183, 032183–13. doi:10.1063/1.4756995

Ruquan, Y. C., Xiang, J., Lin, Z., & Peng, K., Huang, & Torr. (2013). Coal as an abundant source of graphene quantum dots. *Nature Communications*, *4*, 2943–2947. PMID:24309588

Sharma, D. K., & Wadhwa, G. (1997). Demineralization of coal by stepwise bioleaching: A comparative study of three Indian coals by Fourier transform infrared and X-ray diffraction techniques. *World Journal of Microbiology & Biotechnology*, *13*(1), 29–36. doi:10.1007/BF02770804

Volesky, B. (2001). Detoxification of metal-bearing effluents: Biosorption for the next century. *Hydrometallurgy*, *59*(2-3), 203–216. doi:10.1016/S0304-386X(00)00160-2

Volesky, B. (2007). Biosorption and me. *Water Research*, *41*(18), 4017–4029. doi:10.1016/j.watres.2007.05.062 PMID:17632204

Wang, J., & Chen, C. (2006). Biosorption of heavy metals by *Saccharomyces cerevisiac*: A review. *Biotechnology Advances*, *4*, 417–451.

Wu, Z., & Steel, K. M. (2007). Demineralization of a UK bituminous coal using HF and ferric ions. *Fuel*, *86*(14), 2194–2200. doi:10.1016/j.fuel.2007.02.038

Xiao, J., Liu, P., & Yang, W. (2015). Nanodiamonds from coal under ambient conditions. *Nanoscale*, *7*(14), 6114–6125. doi:10.1039/C4NR06186A PMID:25774000

KEY TERMS AND DEFINITIONS

Crystalline Carbon: Carbon lattice with crystal structure or periodicity throughout.

Defect Band (D-Band): Is attributed to the defect or disorder in graphitic lattice which arises due to the breathing mode of graphitic structure. It also originates from the diamond/dimond like structure of carbon.

Eukaryotic Microorganisms: Aspergillus niger is an Eukaryotic microorganisms which is capable of secreting large amount of proteins and metabolites. These microorganisms have cells with nucleous and other organelle enclosed within membrane.

Fungal Leaching: Solubilization of minerals associated with coal matrix to remove unwanted contamination using fungus. During this process fungus secrete organic acid which removes the minerals by loosening the bond strength.

Graphene: Graphene is a single layer of carbon atoms. Few layer is also considered as graphene (3-4 layers).

Graphitization (G-Peak): Graphitic peak of carbon- This peak is due to the bond stretching of all pairs of sp^2 carbon atoms in both rings and chains. Degree of graphitization is an indication of presence of graphene like structure present in the carbon sample.

Low grade Coal: Coal whose rank or quality is low. Normally with low carbon content and high content of ash and moisture.

Mixed Phase Structure: A carbon structure with both diamond and graph.

Nano Carbon: Carbon stacking in the nano meter range. Can be carbon dots, graphene quantum dots or nano diamond.

Turbostratic Structure: In Turbostratic structure carbon layers are arranged in a haphazardly folded or crumbled manner.

Chapter 13
Application of Potential Biological Agents in Green Bioremediation Technology: Case Studies

Debajyoti Kundu
University of Kalyani, India

Deblina Dutta
University of Kalyani, India

Subinoy Mondal
The University of Burdwan, India

Smaranya Haque
The University of Burdwan, India

Jatindra Nath Bhakta
University of Kalyani, India

Bana Behari Jana
University of Kalyani, India

ABSTRACT

Upgradation and advancement in every field related to mankind leads to the origin of a contaminated environment. Development in science and technology enabled humans to combat the rate of contaminants by using biological agents, commonly known as bioremediation. The chapter deals with the different species of bioremediation agents viz. bacteria, fungi, algae, plants, animals and organic wastes to treat diverse environmental pollution. The extent of environmental bioremediation encompasses inorganic viz. arsenic, chromium, mercury, cyanide etc. and organics viz. Hydrocarbons, petroleum, pesticides etc. Thus, the reasons for the control of water and soil by considering bioremediation are concern on public health, protection of environment, and cost reduction of decontamination. Different case studies have been demonstrated herein to understand the enigmatic process and evaluate practical efficacy of the environment to decontaminate itself by the presence of various biological organisms.

INTRODUCTION

Expeditious growth of population and their anthropogenic activities in industry, agriculture and domestic sectors leads to the generation of huge hazardous toxic materials. These toxic elements can persist in the environment for a long time. Thus, continuous generation along with long persistence make a

DOI: 10.4018/978-1-5225-2325-3.ch013

favorable condition for pollutants to accumulate in different matrices (air, water and soil). This filthy condition is gradually capturing our world with its adverse effect on the human health. Hence there is an utmost requirement to combat with the pollution load for our better health and environment. There are several processes to deal with the contaminants, among them biological process is recently gaining popularity for its eco-friendly nature. Nature teaches us different remedies to manage any type of condition. Development in scientific research, scientists are able to develop efficient and economic processes for remediation of contaminants (Elekwachi et al., 2014). Use of biological agents like plants, animals, algae, fungi, microbes, etc. in pollution abatement is commonly known as bioremediation. Some species of plants, animals, algae, fungi, microbes, etc. also increases in certain places so as to make them highlighted by being polluted.

The principle of bioremediation involves the biological organisms to wipe out the contaminants or convert them into less harmful variety by formation of porous fence, intra and extracellular confiscation, active transport, enzyme mediated detoxification, chelation of metals, precipitation through the production of organic harmless substances and biotransformation reactions (methylation, oxidation, reduction and volatilization). They multiply themselves by generating energy and nutrients from the atmosphere. In very few cases, the nature provides all the essential materials for cleaning up the contaminated sites, known as intrinsic bioremediation. But in most of the cases, essential materials are regularly supplied from an outer source to the organisms, this process is known as engineered bioremediation. These two types can further be classified under different techniques viz. biostimulation i.e., by stimulating viable native microbial population; bioaugmentation i.e., artificially introducing viable population; bioaccumulation i.e., applying live cells; biosorption i.e., using dead microbial biomass; phytoremediation i.e., by using various species of plants and rhizoremediation i.e., by using plant and microbes at the same time (Chowdhury et al, 2012). Different species of these agents are used to identify pollution as well as treat diverse environmental pollution. Inorganic, organic as well as air pollutants can be treated through bioremediation. Aquatic plants also act as biosensor and help in the remediation of polluted water. It is an interesting fact that besides plants and microbes, soil animals like protozoans, nematodes, earthworms, mites, insect larvae, arachnids, millipedes and centipedes are also involved in bioremediation. They usually play a very regulative role in the decomposition rate of the contaminants. The process of bioremediation is very simple and very cost effective usually driven by the solar energy (Prasad, 2011). The only part for effective bioremediation includes the proper selection of biological species for a definite problem of contamination in polluted sites may it be land or water. Progress in various paths for the destruction and degradation of pollutants and contaminants has been developed for ensuring a sustainable bioremediation strategy. Following proper techniques of bioremediation can result in productive and non-deleterious quality of the environment. Till now several researches are going on to find the best agent for providing best green remediation. This chapter explains the causes, processes and mechanisms used in bioremediation emphasizing efficiency of different living organisms (tools) and case studies to cleanup the environmental contaminants.

A BRIEF HISTORICAL OVERVIEW

In bioremediation, different microorganisms, fungi, algae and plant species has been extensively used to decrease, degrade, transform and detoxify the pollutants from the contaminated sites. Potential living organisms present in the contaminated sites spontaneously may involve in the remediation process, mimicking this natural process, researchers are trying to make the process more efficient.

The concept of bioremediation is not new; Romanian in 600 B.C established a sewer system for biological wastewater treatment. Before the 1940s, scientists hypothesized the degradation of petroleum hydrocarbons by microbes but their mechanism and limitation was unknown. However, in 1960 George M Robinson has invented a bioremediation process during experiments with polluted containers. His pioneer concept for commercial utilization of dried microbes gained acceptance throughout the 1960s (Sonawdekar, 2012). In 1970 Dr. Ananda Mohan Chakrabarty discovered crude oil degrading bacterial strain. First commercially, bioremediation techniques were used in 1972 to clean up the Sun oil pipeline spill in Ambler, Pennsylvania. In this meantime, research had progressed in the field of genetic modification of microbes to enhance its stability, substrate specificity and efficiency and therefore, in 1975 Dr. Chakrabarty once again brought revolution by discovering oil-eating superbug (*Pseudomonas putida*) using genetic cross linking. Burning, mechanical cleanup and chemical dispersants methods were used to remediate the contaminated shoreline after Exxon Valdez oil spills in 1989. By 1990s, bioremediation got its significance through the commercial application of this process for the treatment of contaminated soils and groundwater. According to the Environment Protection Agency (EPA) survey report in 1992, 240 bioremediation cases were identified. In the year of 2005, UK researchers identified uranium degrading bacteria from a decomposed state in an insoluble forms (Evan, 2005). In 2014, Norwegian researchers in Trondheim have gotten remarkable results of nature's ability in oil spills clean up (Doan, & Domingo, 2015; Pal et al., 2010).

MECHANISM OF BIOREMEDIATION

For the bioremediation purpose environmental factors and the properties of the biological agent involved in this process plays a crucial role. A proper uniform contact of agents with pollutants is also an important factor. Activity and growth of microorganism, degradation mechanism of contaminants and efficiency of the process are broadly depended on various factors like moisture, pH, temperature, types of matrix, redox potential, availability of oxygen, mass transfer characteristics etc (Uqab et al., 2016). Mechanism of bioremediation basically depend on the unique property/tricks of living organisms involve in this process by the means of absorption, adsorption, uptake, conversion to simple no/less toxic materials, degradation and decomposition (Figure 1).

Adsorption

Adsorption is carried out by the binding of pollutants in the binding sites of microbial cellular structure. Extracellular polymeric substances (EPS) which is a reactive compound have a significant effect on acid-base properties and metal adsorption through the mechanism of proton exchange and micro-precipitation of metals (Dixit et al., 2015).

For the plant species surface of the roots binds many pollutants. Interaction of the plant root and microbes resulted into secretion of proton, organic acids, enzymes and phytochelatins increase the metal dissolution in rhizosphere zone (Revathi et al., 2013).

Figure 1. Mechanism of various bioremediation pathways driven by different living machines (biological components) in decontaminating and detoxifying the hazardous substances in environment (Up take- Up, Accumulation- Ac, Conversion- Cn, Enzymatic hydrolysis- Eh, Degradation-Deg, Decomposition-Dc, Absorption-Ab., Adsorption-Ad, Precipitation-Pt, Detoxification-Dt).

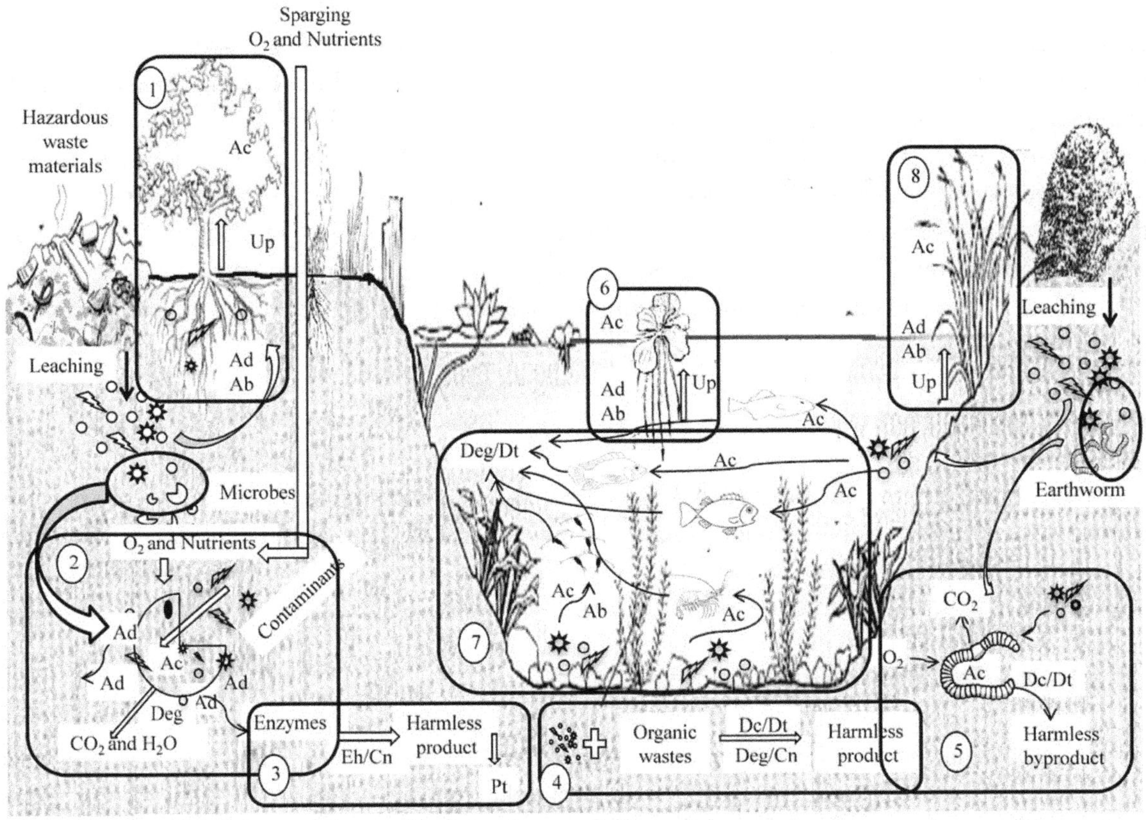

Plant tools- system 1, 6, 8 (phytoremediation); Microbial tools- system 2, 7 (microbial remediation); Enzyme tools- system 3 (enzymatic remediation); Animal tools- system 5,7 (zooremediation); Organic waste- system 4 (biosorption)

Biosorption

By the biosorption process positively charged heavy metal ions get absorbed by the negatively charged microbial cell membranes and polysachharides. In this process contaminants first entered in to the cell cytoplasm through cell membrane by the help of transporter proteins. From the surface, the metals are transported into the cell cytoplasm through the cell membrane with help of transporter proteins. Specific binding sites present in the transporter proteins binds and transports ions from outer side to inside of the cell (Sinha et al., 2013).

Precipitation

Precipitation of metals occurs by compartmentalizing and/or converting to more innocuous form. Micro-organism precipitate metals in different forms like metal sulphides, metal oxides, metal protein aggregates or elemental metal crystals. These forms of metals form particulates which are closely associated with cytoplasmic membrane (Bhakta, 2016).

Detoxification

Possible mechanisms of the detoxification of contaminants by microorganisms are 'valence transformation', by 'extracellular chemical precipitation' or by 'volatilization'. In general by redox conversion of inorganic pollutants or conversion of inorganic pollutants to organic forms, detoxification occurs by microbial population. Aerobic bacteria produce harmless CO_2 and water by the degradation of natural and synthetic hazardous organic substance on the other hand anaerobic bacteria are responsible for the degradation of halogen and nitrosamine, reduction of epoxides to olefins, reduction of nitro groups. Microbial methylation of metals is another aspect of detoxification. Living organisms are used for biomethylation of arsenic to gaseous arsines, selenium to volatile dimethyle selenide and lead to dimethyl lead (Sinha et al., 2013).

Enzymatic Bioconversion

Microbial enzymes act as a biocatalysis which facilitate the bio-conversion of substrates into products. Different types of enzymes like monooxygenase, dioxygenase, laccase, lignin peroxidase, manganese peroxidase, versatile peroxidase, lipase, cellulase and protease are involved in oxidoreductase and hydrolysis of toxic materials (Karigar and Rao, 2011).

Degradation/Decomposition

Several microbes has been identified which are involved in the breakdown of toxic organic substances and xenobiotics in the environment. Heterotrophic microbes which cannot produce their own food are involved in the decomposition process to gain their growth nutrients. These microbes utilize Oxygen (O_2), ammonia (NH_3), and phosphate (PO_4^{3-}) from the organic material as a nutrient source and produce harmless materials. Degradation process may be carried out in the absence (Anaerobic process) or presence (Aerobic process) of oxygen. Anaerobic process produces carbon dioxide (CO_2) to methane (CH_4) in the presence of the available hydrogen (H_2).

Hydrocarbons of waste organics (*Acetogens*) → Acetate + Organic Acids + CO_2 + H_2O

Organic Acids + CO_2 + H_2 (*Methanogens*) →Byproducts + Energy

On the other hand aerobic process produces CO_2 and water.

Organics + Molecular Oxygen (Aerobes) → CO_2 + H_2O + Other products + Energy (Sinha et al., 2013).

LIVING MACHINE (TOOL) OF BIOREMEDIATION

A vast array of biological components in ecosystem play important role in self-cleaning process of ecosystem as living machines or tools. Several studies determine and demonstrated their services and efficacy in remediation of polluted/contaminated environment. Therefore, it is well established that biological agents have potent capacity in bioremediation by removing and/or transforming from higher to lower toxic forms. Biological components act a very important role in biotransformation and biodegradation of contaminants to less toxic substances. Biotransformation deals with the amendment of the structure of a substrate by microorganisms, where biodegradation deals with the breakdown of organic and biotransformation of inorganic compounds into non-toxic compounds. Bioremediation utilize different types of biological agents like bacteria, fungi, algae, plant, etc. to clean up the contaminated sites. This technology promotes the growth of the microorganism in contaminated sites and can be enhanced by adding nutrients, adding terminal electron acceptor; maintaining temperature and moisture content (Singh et al., 2014).

Bacteria

Microbes have been widely used for environmental clean-up. Microbial degradation process involved in metal reduction, methylation, demethylation, enzymatic reduction, metal organic complexion, metal ligand degradation (Okeke et al., 2014). Different types of species of which are extensively utilized for the management of contaminated sites like *Streptomyces*, *Pseudomonas* and *Bacillus* for metal biosorptions. Few others commonly use microorganisms are the species of *Achromobacter, Citrobacter, Escherichia, Cinetobacter, Flavobacterium, Micrococcus, Mycobacterium, Norcardia, Pseudomonas* and *Rhodococcus* etc. (Bhatnagar, & Kumari, 2013; Kumar et al., 2011). Versatile properties of these microorganisms can change in metal mobility, speciation, toxicity and mineralization and thus helps in the degradation and transformation of the huge variety of compounds viz., metals, hydrocarbons, chlorinated solvents, chlorinated phenol, BITEX, polychlorinated biphenyls (PCBs), polyaromatic hydrocarbons (PAHs), pesticides, radionuclide (Joutey et al., 2013). Bioremediation can be classified into 3 phases i) by natural attenuation ii) by supplying nutrients and iii) by adding microbes to the system (Seo et al., 2009). Many factors like pH, temperature, moisture, and availability of nitrogen, phosphorous, carbon, production of toxic metabolites, solubility of contamination and concentration of pollutants influence the existence of microorganism population and the capability of degradation (Naik and Duraphe, 2012). Hence, there is an urge for the preparation of high potential microorganism to work a wide range of environmental condition and thus genetically engineered microorganisms (GEM) came into the scenario. Four principle approaches for the development of GEMs for bioremediation application are i) identification of the microorganism which are suitable for modification ii) pathway construction and regulation iii) adaptation of enzyme specificity and affinity and iv) bioaffinity, bioreporter sensor applications in chemical sensing, toxicity reduction and end point analysis (Kulshreshtha, 2013). Some examples of GEMs are *Alcaligenes eutrophus* 2050, *A. eutrophus* H850Lr, *Burkholderia. cepacia* BRI6001, Comamonas testosteroni VP44, *E. coli* FM5/pKY287, *Pseudomonas* sp. B13, *P. putida* KT2442, *Pseudomonas* sp. FR1, *Pseudomonas* sp. LB400, *P. pseudoalcaligenes* KF707-D2, *P. putida* TVA8, *P. fluorescens* HK44, *P. fluorescens* 10586s/pUCD607 and *P. fluorescens* 10586s/pUCD607. These microorganisms are used in the process monitoring, end point analysis, stress response and toxicity assessment (Menn et al., 2008).

Fungi

Morphological, physiological and genetic features of fungi are important to make them unique in bioremediation process. They can easily persist and colonize in all matrices in the environment and thus play an important role in the reduction, elimination and transformation of contaminants. Air is the main vehicle for fungal propagules, though aquatic environment is steadily colonized by fungi but the major habitat is soil. Fungi can be found in different ecological niche extended from arctic tundra to the desert sand dunes. All fungi are heterotrophic in nature and utilize organic substance in their growth by three ways: saprophytism, parasitism and mutualistic symbiosis. One or more fungal species can degrade almost all natural organic compounds by producing a variety of enzymes viz., amylases, lipases, and proteases. They degrade natural polymers like keratin, chitin, and lignin, which are mainly complex and resistant to microbial attack. Wood fungi can degrade numerous aromatic and aliphatic, xenobiotic compounds, different pollutants like polycyclic aromatic hydrocarbons (PAHs), polychlorinated biphenyls (PCBs), pesticides, and herbicides. White-Rot Fungi, Brown-Rot Fungi, Litter Fungi, Soil Fungi and Mycorrhizal Fungi are some examples which are extensively used in the bioremediation process (Anastasi et al., 2013). Some examples of fungi and their application in pollution remediation are listed in Table 1.

Yeasts show the capability to resist under adverse environmental condition. The major application of yeast is dye decolourization of food industry effluents. Different types of yeast species like *Clavispora, Candida, Rhodosporidium, Pichia* and *Rhodotorula* etc. are used in the bioremediation process (Bhatnagar, & Kumari, 2013).

Algae

Algae are important bioremediation agents, used in pollutant abatement in waste water. The major advantage of algae as a bioremediation agent is its high biomass yield capability which leads to high absorption and accumulation of metals, tolerance to heavy metals, capability to grow both autotrophically and heterotrophically, large surface area/volume ratios and potential for genetic manipulation (Chekroun, & Mourad, 2013). Algal species are used broadly for the bioremediation of heavy metal, oil, pesticides, toxins, radioactive and several organic compounds (Prabha et al., 2016). A list of some algal species and their bioremediation of specific pollutants has been enlisted in Table 2.

Table 1. List of some potential fungi used in bioremediation

Fungi	Application
Phanerochaete chrysosporium	Degrade lignin macro molecules, organopollutants (PAH's, PCBs) dioxins, synthetic dyes, pesticides, nitrocranditics, chlorolignins, chlorophenols
Phanerochaete flavido-alba	Decolorize olive oil mill wastewater
Trametes versicolor	Degrade lignin, PAHs, polychlorinated biphenyl mixture and synthetic dyes. Delignify, dechlorinate and decolorize bleach kraft pulp
Pleurotus ostreatus	Degradation of PAHs
Pleurotus tuber-regium	Ameliorate crude oil polluted soil
Lentinus squarrosulus	Mineralization of soil tainted with crude oil and thus at the end nutrient contents increased in the soil
Pleurotus pulmonarius	Management of cement and battery polluted soils

Adenipekun, & Lawal, 2012.

Table 2. Potential algae involved in pollutant removal from diverse field

	Algal species	Elements/Contaminants
Heavy metal removal	*Anabena doliolum*	Cu^{2+} and Fe
	Sargassum fluitans	Pb^{2+}, Cd^{2+}, Cu^{2+}, Ni^{2+} and Zn^{2+}
	Cladophora fascicularis	Cu^{2+}
	Padina sps	Cu^{2+}
	Spirulina	Pb^{2+}
	Chlorella vulgaris	Cu^{2+}, Cd^{2+} and Pb
	Spirogyra	Pb^{2+}
	Spirogyra hyaline	Cd^{2+}, Hg^{2+}, Pb^{2+}, As and Co^{2+}
	Scenedesmus obliques	Oil effluents
	Scenedesmus obliquus	Cd^{2+}
	Sargassum sinicola	Cd^{2+} and Cu^{2+}
	Dunaliella	Hg^{2+}, Cd^{2+} and Pb^{2+}
	Enteromorpha algae and its silicates bonded material	Pb
	Spirulina platensis	Cu^{+} and Cu^{2+}
	Tetraselmis chuii	Cu^{2+}
	Spirogyra sp., *Nostoc* commune	Se
	Anabaena variabilis, *Aulosira* sp., *Nostoc muscorum*, *Oscillatoria* sp. and *Westiellopsis* sp.	Cr^{6+}, Ni^{2+}
	Chlorella minutissima	Cr^{6+}
	Chlorella pyrenoidosa	Zn^{2+}, Cu^{2+}, As, Pb^{2+}, Cd^{2+}, Cr, Ni, Hg^{2+}
	Ascophyllum nodosum	Au^{+}, Co^{2+}, Ni^{2+}, Pb^{2+}
	Caulerpa racemosa	B, As, Zn^{2+}, Ni^{2+}
	Laminaria japonica	Zn^{2+}
	Micrasterias denticulata	Cd^{2+}
	Phormedium bohner	Cr
	Platymonas subcordiformis	Sr
	Sargassum filipendula	Cu^{2+}
	Sargassum natans	Pb^{2+}
	Sargassum vulgare	Pb^{2+}
Wastewater treatment	*Chlorella vulgaris* and *Scenedesmus dimorphus*	NH_3 and P
	Oscillatoria	Chemical Oxygen Demand (COD) and Biological Oxygen Demand (BOD)
	Chlorella vulgaris, Spirulina platensis and *Dunaliella* saline	COD and Phenolic Compounds
	Chlorella vulgaris and *Scenedesmus quadricauda*	PO_4^{3-}, NO_3^-, COD and BOD
	Chlorella minutissima	P and NO_3^-
	Oscillatoria, Synechococcus, Nodularia, Nostoc and *Cyanothece*	BOD, COD, Dissolved Oxygen, NH_3, NO_3^-, Mg, Organic and Inorganic Phosphates
	Isochrysis zhanjiangenesis	P, NH_3, N, NO_3^- and NO_2^-
	Spirogyra sps. and *Oscillatoria* sps.	Cd^{2+} and Ni^{2+}
	Aphanocapsa sps., *Chlorella autotrophica, Coccochloris elabens, Dunaliella tertiolecta, Oscillatoria* sps., *Scenedesmus obliques, Synechococcus elongates* and *Volvox*	Crude Oil
	Chlorella vulgaris	Textile waste effluents (Color and COD)

continued on following page

Table 2. Continued

	Algal species	Elements/Contaminants
Bioremediation of oil	Cyanobacteria (Blue-green algae)	Benzene, toluene, naphthalene, phenanthrene, pyrene
	Mycobacterium sp.	n-Undecane
	Strains of *Pseudomonas putida*	Phenol cresols
	Trichosporon pullulans	Paraffins
	Acinetobacter calcoaceticus	Petroleum derivatives
	Microcystis aeruginosa	Acrylonitrile
	Aeromonium sp.	Total petroleum hydrocarbons
Pesticides	*Chlamydomonas* sp.	Mirex, Lindane, naphthalene, phenol,
	Selenastrum capricornutum	Benzene, toluene, chlorobenzene, 1,2-dichlorobenzene,nitro benzene naphthalene, 2,6- dinitrotoluene, phenanthrene, di-nbutylphthalate, pyrene
	Cylindrotheca sp	Dichlorodiphenyltrichloroethane (DDT)
	Euglena gracilis	DDT, parathion
	Scenedesmus obliquus	DDT, parathion, Naphthalene sulfonic acid
	Chlorella sp.	Toxaphene, methoxychlor, Lindane, chlordimeform
	Chlorococcum sp.	Mirex
	Dunaliella sp.	Mirex, DDT, naphthalene

Prabha et al., 2016; Chekroun, & Mourad, 2013; Priyadarshani et al., 2011.

Plants

Use of plants is a low cost and eco-friendly tool for effective and efficient remediation process of pollutants. It plays a significant role in the removal of toxic materials from contaminated soil and water. Plant species provide advantages by its unique and selective uptake capabilities of root, accumulation and degradation of pollutants by whole plant. Pollutants accumulation capacity varies from plant to plant and thus high accumulator plants known as 'hyperaccumulator' which can take up toxic materials up to the levels of thousand ppm (Tangahu et al., 2011). Examples of some hyper accumulator plants are *Armeria maririma* (Seapink thrift), *Ambrosia artemisiifolia* (Ragweed), *Brassica juncea* (Indian mustard), *Brassica napus* (Rape, Rutabaga, Turnip), *Brassica oleracea* (Flowering/ornamental kale and cabbage, Broccoli), *Festucaovina* (Blue/sheep fescue), *Helianthus annuus* (Sunflower), *Thalspi rotundifolium* (Pennycress), *Triticum aestivum* (Wheat), *Zea mays* (Corn) (Paz-Alberto and Sigua, 2013). Plants have a very efficient mechanism and highly specific to attain essential micronutrients when present in low concentration. Plant root produces chelating agent and thus make pH change and redox reaction which helps to solubilize and uptake the nutrients. Not only for uptaking, but these same mechanisms are involved in the translocation of storage of pollutants in different parts of the plant. There are several mechanisms of pollutant uptake by plants, these are phytoremediation phytoextraction, phytostabilisation, rhizofiltration, and phytovolatilization (Tangahu et al., 2011). A list of some plants involved in phytoremediation process presented in Table 3.

Not only for water and soil, plant species are extensively used to remediate pollutants from the air. These types of plants should have certain properties like evergreen, big leaved, coarse bark, native, low water

Table 3. Examples of some plant species used in remediation process

Plant Species	Toxic Compounds
Ambrosia artemisifolia and *Apocynum cannabinum*	Pb^{2+}
Brassica juncea	Se, Pb^{2+} and Cu^{2+}
Helianthus annus	As and U
Medicago sativa	Benzopyreneand PAHs
Melastoma malabathricum	Al
Nephrolepis exaltata	Hg^{2+}
Pteridium esculentum	As
Pteris vitata	As, Hg^{2+}, Cs and Sr
Salix viminlais	Cd^{2+}, Zn^{2+} and Cu^{2+}
Raphanus sativus	Cu^{2+}
Silene vulgaris and *Thlaspi caerulescens*	Cd^{2+} and Zn^{2+}
Brassica juncea, hybrid poplars grasses	As, Cd^{2+}, Cr, Cu^{2+}, Pb^{2+} and Zn^{2+}
Brassica juncea, pennycress, alyssum sunflowers, hybrid poplars	Ag^{2+}, Cd^{2+}, Co^{2+}, Cr^{6+}, Cr^{3+}, Cu^{2+}, Hg^{2+}, Mn^{2+}, Mo, Ni^{2+}, Pb^{2+}, Zn^{2+} 90Sr, 137Cs, 239Pu, 238,234U
Helianthus annus, *Brassica juncea*, water hyacinth	Metals and radionuclides
Poplars, alfalfa black locust, *Brassica juncea*	Chlorinated solvents
Stonewort black willow, bald cypress hybrid poplar	Organic compounds, chlorinated solvents phenols, herbicides
Red mulberry, grasses, hybrid poplar, cattail, rice	Organic compounds (TPH, PAHs, pesticides chlorinated solvents PCBs)
Vetiveria zizanioides L.	As^{3+} As^{5+}, Cd^{2+}, Cr^{3+}, Cr^{6+}, Cu^{2+}, Pb^{2+} and Hg^{2+}

Bhatnagar, & Kumari, 2013; Pandey, & Bagga, 2013; Paz-Alberto, & Sigua, 2013.

requisite, high absorption of pollutants, agro-climatic suitability, pollution tolerance and dust scavenging capacity etc. *Tamarindus indicus, Mangifera indica, Derris indica,Thespepsia populnea,Polyalthea longifolia, Azadirchta indica, Bombax ceiba, Delonix regia, Ficus religiosa, Jacaranda mimosifolia, Lagerstroemia indica, Plumeria rubra, Syzygium cumini* are suitable for pollutants elimination from air (Kumar et al., 2013).

Animals

Animal induced bioremediation capacities are not widely known and are rarely used for bioremediation for its ethical concern. Animal mediated bioremediation can be achieved by the introduction, culture, and harvest of animals from the contaminated sites. Some examples of animal involved in aquatic system bioremediation are oysters, mussels, clams, fish, polychaetes and sponges. This process is termed as 'zooremediation' (Gifford et al., 2006). The mechanisms employed are zooextraction, zoostabilisation, zootansformation and animal metal-hyperaccumulation. This remediation process occurs in two ways i) by taking part in the process, increasing in number and soil metabolic activity. ii) as a good indicator of contaminated soil, to evaluate the toxicity and risk of the end product (Haimi, 2000). Aquatic animals

Table 4. Name and groups of some animals involved in bioremediation process

Animal Group	Species
Earthworms	*Eisenia fetida/Andrei* *Aporrectodea caliginosa* *Lumbricus terrestris*
Enchytraeids	*Enchytraeus albidus*, also other species
Collembola	*Folsomia candida* several species, e.g. *Folsomia fimetaria* and *Orchesella cincta*
Mites	*Platynothrus peltifer*
Isopods	*Porcellio scaber* *Trichoniscus pusillus*
Nematodes	*Plectus parietinus*
Protozoans	*Colpoda cucullus*
Animal-Hyperacculators	*Crassostrea virginaca* *Mytilus edulis* *Pinetada albino albina* *Crassotrea rhizophorea*
Molluscs	*Pera viridis.* *Chione fluctifraga* *Mytilus edulis*
Sponges	*Chondrilla nucula* *Hymeniacidon perleve* *Halichondria panicea* *Spongia officinalis*
Arthropods	Carabid beetle *Poecilus cupreus* *Pterostchus oblogopunctatus*

Haimi, 2000; Okeke et al., 2014.

accumulate contaminants in several orders of magnitude higher than the surrounding water. Accumulation is depend on several factors like temperature, pH and salinity of the matrix and biological factors like age, sex, sexual maturity stage of premeditators (Gupta and Singh, 2010). Earthworms also found as a potential species for the removal of contaminants. It involves in the remediation of pesticides, heavy metals and lipophilic organic micropollutants like the polycyclic aromatic hydrocarbons (PAH) from the soil. In the soil the area of earthworm and pollutants interactions is known as "drilosphere system" and the process of remediation called as "Vermiremediation" (Sinha et al., 2009). Some names of decomposer animal are listed in Table 4.

Organic Wastes

Organic manures/wastes have been extensively used throughout the world for the improvement of soil fertility by adding organic matter and nutrients (Anyanwu, & Okorie, 2013; Umar et al., 2013). Like other agents application of organic waste is a promising tool in bioremediation process (Bahadure et al., 2013). Basically addition or amendment of the soil by organic waste increases the microbial population and activity by providing nutrients (Chijioke-Osuji et al., 2014).

SOME CASE STUDIES

Bacterial Bioremediation

Case 1

Singh and Fulekar in 2010 isolated *Pseudomonas putida* strain MHF 7109 for biodegradation of benzene, toluene and o-xylene (BTX). By this process, they successfully degrade BTX (50mg/L) within a time period of 48, 96, and 168 h, respectively, but higher concentration of BTX (250 and 500mg/L) repressed the action of microorganisms. 100 mg/L benzene and toluene was completely degraded within 120 and 168 h respectively where same concentration of o-xylene was degraded within 168 h by 62%.

Case 2

A pilot scale study was conducted by Mukherjee, & Bordoloi (2011) for the reclamation of petroleum oil hydrocarbons contaminated soil. Bacterial consortium (20% v/w) consisting three strains (*Bacillus subtilis* DM-04 and *Pseudomonas aeruginosa* M and NM) were used for these remediation processes for a period of 180 days. After the experimental period, results showed a significant degradation of contaminants in polluted soil by 76%. An ecotoxicity study by observing seed growth of Bengal gram (*Cicer aretinum*) and green gram (*Phaseolus mungo*) in contaminated and remediated soil was carried out where 72–74% seeds were germinated in the remediated soil (after 180 days experiment) which is very much similar to uncontaminated soil.

Case 3

In the year of 2013, Ahluwalia, & Goyal reported 100% removal of chromium (VI) from the metal processing industry effluent. In this study a non living biomass (*Aspergillus terreus*) was used as a biosorbent to remove chromium. Removal efficiency up to 10 cycles and 100% removal of metal up to 5 weeks made this process more cost-effective and commercially viable.

Case 4

Pilot scale studies were carried out at different oil refineries in India for the purpose of the removal of petroleum hydrocarbon. Four species of bacterial consortium were used for remediation of 48,914 tons of different types of oily waste. Initially, total petroleum hydrocarbon (TPH) content varied 83.50 to 531.30 gm/kg of oily waste in 44 contaminated sites. This process successfully degraded 10 gm/kg of oily waste within the time period of 2-12 months (Mandal et al., 2012).

Case 5

A new metal resistance lactic acid bacterial (LAB) strain was isolated (*E. faecium* Pb12) for the purpose of the removal of metal contaminants. This strain shows minimum inhibitory concentration of 120 mg/L for cadmium and 800 mg/L for lead. The researchers successfully removed cadmium and lead by 0.0377 mg/h/g and 0.0460 mg/h/g of cells (wet weight) respectively. Thus, a potential metal resistant probiotic strain has been discovered which can be used for heavy metal removal from fish intestinal milieu (Bhakta et al., 2012).

Case 6

A mixture of 6 microbial consortiums was used for the remediation of oil contaminated area in upper Assam, India. From 2009 to 2010, they were able to remediate about 1210 metric ton (MT) and 5045 m^3 of contaminated soil and water respectively. After 7-13 months 74-90% degradation was occurred in soil and in case of water, 90% contaminants were degraded within 4-6 months (Agrawal et al., 2010).

Case 7

Mandal et al. (2012) utilized a four species of bacterial consortium for the degradation of TPH of the oily waste. After the period of 2-12 months of successful application of the microbial consortium, 95% degradation of TPH was occurred.

Case 8

Jové et al. (2016) studied on the degradation potentiality of anthracene in cork by three ligninolytic white-rot fungi and three non-ligninolytic fungi. *Aspergillus niger, Penicillium simplicissimum and Mucor racemosus* were selected as ligninolytic white-rot fungi and *Phanerochaete chrysosporium, Irpe lacteus and Pleurotus ostreatus* were selected as non-ligninolytic fungi. After 61 days of experimental period, *P. simplicissimum* was capable to degrade the highest amount of contaminant (86%) followed by *P. chrysosporium* (40%) and *I. lacteus* (38%) in cork.

Case 9

Mouhamadou et al. (2013) studied bioremediation of polychlorinated biphenyl (PCB) using different autochthonous filamentous fungal strains. From this research *Myceliophthora thermophila, Thermoascus crustaceus, Aspergillus fumigates, Doratomyces nanus, Phoma eupyrena, Doratomyces purpureofuscus, Doratomyces verrucisporus* strains showed >70% degradation ability of PCBs. Among all these strains *Aspergillus fumigatus* S2-9 has degraded 84.99% of PCB.

Case 10

Lee et al. (2015) carried out the bioremediation of Polycyclic Aromatic Hydrocarbons (PAHs) in Creosote contaminated Soil by *Peniophora incarnata* KUC8836 strain. After 2 weeks, concentration of phenanthrene, fluoranthene, and pyrene were degraded by 86.5%, 77.4%, and 82.6% respectively. Total initial PAHs concentration (229.49 mg/kg) was reduced by 85.7% after 4 weeks of incubation period.

Algal Bioremediation

Case 1

El-Sheekh et al. (2016) treated a mixture of wastewater (sewage, sea and well) by fresh water (*Chlorella vulgaris*) and marine (*Chlorella salina*) algae. Both alga species were successfully reduced total dissolved solids (TDS), biological oxygen demand (BOD), chemical oxygen demand (COD), nutrients concentration (nitrate, ammonia, phosphate, sulphate, calcium, magnesium, sodium, potassium) and heavy metals concentration (Zn, Cu, Mn, Ni, Co, Fe and Cr) after 10 days of experiment. Overall *C. vulgaris* was more efficient in pollutants removal than *C. salina*.

Case 2

Alcohol distillery wastewater treatment has been done by using a new strain of *Chlorella sorokiniana* using a high-density photobioreactor. 92.5% removal of COD, 95%, 77%, 35% removal of nitrate, phosphate and sulfate, respectively were achieved by this process (Solovchenko et al., 2014).

Case 3

Degradation of Triclosan, a widely used antimicrobial agent, was carried out by *Chlorella pyrenoidosa,* a ubiquitous green alga. After 96 h of experiment, 77.2% degradation was achieved. Degradation of Triclosan leads to the disruption of the chloroplast and the release of organic material into the aquatic body and thus affects the membrane metabolism.

Case 4

Spirulina platensis was successfully used as a biosorbent for the removal of azo dye, C.I. Basic Red 46 (C.I. BR 46). After 1 h of wastewater treatment, for the particular algal species maximum dye biosorption was accomplished as 25.46 mg/g (Deniza, & Kepekcib, 2015).

Case 5

An experiment on the bioremediation rate of deepwater horizon crude oil was carried out using *Aphanocapsa* sp., *Chlorella autotrophica*, *Coccochloris elabens*, *Dunaliella tertiolecta*, *Oscillatoria* sp., *Scenedesmus obliquus, Synechococcus elongatus* and *Volvox aureus*. Maximum 34.99% crude oil degradation was observed by *Coccochloris elabens*. Hence it can be concluded that algae is potential degrader of crude oil and can be flourish in the high crude oil environment (Cao et al., 2013).

Case 6

A pilot scale study was conducted for the treatment of chrome-sludge generated from electroplating industry. *Desmococcus olivaceus* (Persoon et Acharius) J.R. Laundon VIAT029 was utilized for the treatment process. In this open pond system TDS, total suspended solid (TSS), ammonia, nitrate, phosphate and chromium were successfully removed by the algal species (Sivasubramanian et al., 2010). Another pilot scale study as well as World's first phytoremediation plant has been established at Ranipet, India. From 2006 this pond is in operation. About 30-40 kiloliters of effluent has been generated from alginate industry every day and 100% sludge has been achieved by this process. A blue green alga *Chroococcus turgidus* has been used for this treatment process (Sivasubramanian, 2011).

Chlorella vulgaris was used for the treatment of raw effluent of a Confectionery industry in Tamil Nadu, India, which generates 50-70 kiloliters of effluent per day. Algal pretreatment has been employed at the flow rate of 3500 L/h before anaerobic digestion, because this pretreatment process helps to maintain pH, reduce BOD, COD and sugar level (Sivasubramanian, 2011).

Plant Mediated Bioremediation

Case 1

Reclamation of a 1600000 m^2 (approx.) metal-contaminated site known as Ahangaran lead and zinc mine of Malayer city, southwest of Hamedan, was successfully done by 12 plant species (*Euphorbia macroclada Boiss, Centaurea virgata Lam, Reseda lutea* L, *Echinophora platyloba DC, Cichorium intybus* L, *Tamarix ramosissima Ledeb, Cirsium congestum Fisch and C. A. Scrophularia scoparia Pennell, Chaerophyllum macropodum Boiss, Cardaria draba* (L.) *Desv, Scariola orientalis (Boiss.) Sojak, Reseda alba* L.). Mainly from the top 20 cm of the soil was contaminated with heavy metals such as Zn, Mn, Cu, Pb and Fe. These hyper accumulator plants showed suitable remediation efficiency for the reclamation of this contaminated site (Cheraghi et al., 2011).

Case 2

Phytoremediation of heavy metals was successfully done by *Typha latifolia, Eichhornia crassipes, Salvinia molesta and Pistia stratiotes* through constructed wetland. The source of effluent was a rare earth separating industry, Kerala, India, which contain high concentration of lead, arsenic, cadmium and copper. 100 x 45 x 45 (cm) plastic tubs were constructed for the remediation purpose. Lead, copper, arsenic and cadmium were significantly removed by *Typha latifolia, Eichhornia crassipes, Salvinia molesta and Pistia stratiotes* from the effluents of Travancore Titanium Products (TTP). Initial concentration of lead, arsenic, copper and cadmium in TTP effluent were 2.984 ± 0.021 ppm, 0.016 ± 0.004 ppm, 0.096 ± 0.012 ppm and 0.253 ± 0.019 ppm, respectively. (Sukumaran, 2013). Maximum lead was removed by *Eichhornia* sp. (1.012 ± 0.11 ppm residual), arsenic, copper and cadmium by *Typha* sp. (0.002 ± 0.0003, 0.024 ± 0.002 and 0.102 ± 0.015 ppm residual).

Case 3

A highly sulfonated textile dye Remazol Red (RR) was completely degraded by a macrophyte (*Alternanthera philoxeroides Griseb*). Initial concentration of dye was 70 mg/L which was completely degraded within 72 h. Several enzyme activities like azoreductase and riboflavin reductase was found in the root and stem tissues respectively. On the other hand, lignin peroxidase, laccase and dichlorophenolindophenol (DCIP) reductase activity were present in leaf tissues (Rane, 2015).

Case 4

Amaranthus dubus which is leafy high biomass content vegetable crop, effectively applied for decontamination of metallic elements such as Fe, Cd, Co, Pb, and Cr in Patancheru, Hyderabad, India. Its rapid growth and huge biomass content make it suitable for phytoremediation. *Amaranthus tricolor*, *Amaranthus dubius* wild I and *Amaranthus dubius* wild II were used in combination with Phosphate Solubilising Bacteria (PSB) and solely for the remediation purpose. *Amaranthus dubius* wild II accumulate 2.54 fold and 3.56 fold elevated Cr and Cd concentrations, respectively than control one. In the case of Co *Amaranthus dubius* wild I show highest accumulation than other treatment (2.70 fold than control). *Amaranthus tricolor* with the combination of PSB shows the highest removal of Pb (26.8 mg/kg) (Shankar, 2011).

Case 5

A heavy metal (Zn, Cu, Cr, Ni, Pb, Cd) contaminated area (6,667 m^2) in Chang-Hwa City, Taiwan, was selected for the decontamination of pollutants through plant species. *Jatropha curcas* was used for the remediation purpose. About 800 plants were planted with the density of 0.12 plant/m^2. From 0-30 cm depth, soil sample was collected. After 2 years of plantation, plant species were successfully removed Cd, Cr, Cu, Ni, Pb and Zn by 51.16, 12.80, 38.88, 14.10, 17.61, 17.88%, respectively (Chang, et al., 2014).

Animal Assisted Bioremediation

Case 1

Kaur & Hundal (2015) successfully established the metal accumulation potentiality of earth worm (*Eisenia fetida*). Organism was able to removal of 42.3%, 24.5% and 8.5% for lead, chromium and nickel respectively from contaminated soil. This experiment was carried out in the contaminated soil collected from the vicinity of Buddha Nullah at Ludhiana.

Case 2

Cadmium was removed by the *Anodonta woodiana* from cadmium contaminated synthetic water. Cd was accumulated in different tissues of the mussels. Maximum Cd accumulation was found in the digestive gland (0.142 mg/g) followed by gills and mantle of the mussels. Though it is an experiment but make a print of evidence of pollutants accumulation by animal species (Bhakta & Munekage, 2008).

Case 3

Kundu et al. (2016) observed the accumulation of lead (Pb) in East Kolkata Wetland (22°27' N 88°27' E), India through fish (*Oreochromis niloticus*). Lead concentration in fish *was* highest in intestine (0.0182 µg/mg) and lowest in fin (0.0004 µg/mg) with an order as intestine > liver > skin > muscle > scale > bone > fin.

Case 4

Removal of phenol was successfully carried out by Mohan Krishna et al, 2011 using *Eisenia foetida*. Different concentration of phenol contaminated (20 to 100 ppm) were taken for the experiment. As a result *Eisenia foetida* established as a potential tolerant species for phenol removal than other two earthworm species (*Eudrillus eugenia* and *Anantapur* speciesand can uptake upto 100 ppm phenol completely with in 72hours.

Case 5

Eisenia fetida and green waste compost (GWC) were to desalinate a coastal saline soil (38° 46' N, 117°13' E) in Tianjin, North China. Mean annual temperature of this area is 14°C, where the mean monthly temperature ranges from -2 °C in January to 28 °C in July. Maximum soil salinization occurs during March-May moths in the topsoil. 60 number of *E. fetida* were used in per m^2 area and municipal waste was used as green waste compost which moisture content was 60% (w/w) and the C/N ratio of 25-30. Results revealed increased contents of humic acid, available N, and available P in the GWC-amended soil in reclaimed soil. Modeling from this study predict that total salt content of the saline soil would decrease by > 2 g/kg when 29-90 individuals per m^2 of *E. fetida* (Zhang et al., 2015).

Case 6

A successful bioremediation of polluted water was carried out in off the coastline of the urban center of Zanzibar-Stone Town by the implementation of sponge and pearl oyster mariculture. The experiment reveals the strong pollution filtration abilities of both species. Sponges and oysters help to decrease the phosphate concentration of polluted water from 3.93 ug/L to 1.33 and 1.73 ug/L, respectively. Marine sponges also reduce the turbidity level within 36 h (Oakland, 2013).

Waste Based Bioremediation

Case 1

Hara et al. (2013) used food waste compost in order to remediate oil contaminated site in Japan. Bioremediation was carried out in an oil factory site in Japan from the month of August to November, 2006. Climatic condition such as temperature was recorded as 9.5 to 31.5 °C and precipitation was 0 to 50.5 mm per day during the experiment time. The experiments were carried out taking 2,000 kg of contaminated soil in 4 sections in such as control section, compost section (3.8% [w/w] compost), chemical nutrient section (0.5% [w/w] ammonium sulphate and 0.1% [w/w] phosphorus fertilizer), mushroom section (3% [w/w] *Merulius tremellosus*) and biofertilizer section (0.2% [w/w] bionutrients and TerraXyme). In the food waste compost treatment maximum degradation was occurred from the concentration 4,700 ppm to 2,300 ppm of TPH after 74 days followed by other treatments.

Case 2

In the Southeast of Mexico, drilling mud-polluted site (contaminated with petroleum hydrocarbons) was reclaimed by composting in biopiles with addition of urea and K_2HPO_4 were added as nitrogen and phosphorous sources to achieve optimum C/N/P ratio. Biopiles were amended with the addition of nutrients to get a C/N/P ratio of 100/3/0.5 and straw as a bulking agent (30 kg of bulking agent per each ton of soil). A polluted mudpit site near an oil well drilled in the early 1970s, at the Mexican State of Tabasco was chosen for the remediation purpose. The average atmospheric temperature was in the range of 26 °C to 35 °C during the experimentation period. Soil was clay loam and slightly alkaline (pH 7.9) in nature with the water content of 25% and an existing C/N/P ratio of 6089/76/1. Average TPH concentration of contaminated soil was 99300 ± 23000 mg per kg soil. After 180 days TPH concentration was reduced by 77% (22900 ± 7800 mg per kg soil as residue) in the soil (Rojas-Avelizapa et al, 2007).

Case 3

Three types of organic wastes such as goat manure, poultry droppings and cow dung, were used for the treatment of oil contaminated soil due to pipeline vandalization and local refinery activity, collected from Isaka Mangrove, in Okirika Local Government Area of Rivers State, Nigeria. Organic wastes were mixed with the soil in the ratio of 10/1 (soil/organic waste). Total count of heterotrophic bacteria, hydrocarbon utilizing bacteria and heterotrophic fungi were increased gradually during the experimental tenure while TPH concentration were decreased by 87.1%, 76.6%, 70.7%, and 32.1% after treated with goat manure, poultry droppings and cow dung, respectively after the 28 days experiment (Obiakalaije et al., 2015).

CONCLUSION AND FUTURE PROSPECTS

Bioremediation is a natural way to clean the environment without having any adverse effect on the society. As the process involves all the biological organisms, such as bacteria, fungi, algae, plant and animals, it is considered as a safe technology, environmental friendly and aesthetically pleasant and also of low cost as compared to other physical or chemical treatment methods. These agents effectively remediate pollutants from contaminated matrix by the diverse mechanism of absorption, adsorption, precipitation, enzymatic hydrolysis, detoxification and degradation. Each and every agent has particular niche and unique mechanism to deals with the contaminants. It is very easy to implement and maintain and also can be done on-site or off-site. Though there several benefits but the process is not without its drawbacks, long time taking, limitation in contaminants degradation, difficult to extrapolate from bench and pilot scale study, lack of potential biological agents, lack of proper knowledge are the main setback of bioremediation process.

Prior to go for bioremediation, meticulous knowledge of the environment must be taken and proper selection and characterization of the bioremediation agents should be done. Time taking nature of the process can be overcome by the development of modified versions of the species by the help of biotechnological techniques which may serve very effective, less time consuming process. It will also enhance the capacity of the agents for biotransformation of most toxic contaminants present in the environment. Recently, application of potential biological tools in designing the engineered ecosystem for the benefit of both human and environment is an emerging field. On the basis of decontamination efficacy of various useful biological tools/component, the ecosystem can be designed following the ecological engineering technology, which would be an important environmental friendly solution for cleaning up the contaminated domain of the environment.

Future research in the area of protocol evaluation, innovative site characterization techniques, model improvement and the capabilities of bioremediation to deep will open the new frontiers in bioremediation. Special focus on understanding the biological agent's communities, their responses towards natural environment and pollutants, technology development for the degradation of complex mixture of contaminants may open the new door in bioremediation history. On the other hand a wide variety of potential bioremediation agents are waiting to be explored. However with the existing laboratory scale process development, field trials of the technology for the remediation of organic compounds, metals, metalloids, radionuclides and complex polycyclic hydrocarbons should be employed immensely. Improvement of the microbial genomics studies will be providing several robust technologies for the bioremediation of contaminated waters and land contaminated with metals. Though, the concept of potential biological tool based-engineered ecosystem is in very infancy stage, it has significant future prospect in sustainable development of pollutants free environment.

REFERENCES

Adenipekun, C. O., & Lawal, R. (2012). Uses of mushrooms in bioremediation: A review. *Biotechnology and Molecular Biology Review*, *7*(3), 62–68.

Agrawal, S. P., Subramanian, R., Tripathi, R., Rautela, M. S., & Kumar, S. (2010). A case study of bioremediation of contaminated soil and waste water in upper Assam. Petrotech-2010, New Delhi, India.

Ahluwalia, S. S., & Goyal, D. (2013). Microbial waste biomass for removal of chromium (VI) from chrome effluent. *Bioremediation Journal*, *17*(3), 190–199. doi:10.1080/10889868.2013.807770

Anastasi, A., Tigini, V., & Varese, G. C. (2013). The bioremediation potential of different ecophysiological groups of fungi. In E. M. Goltapeh, Y. R. Danesh, & A. Varma (Eds.), Fungi as Bioremediators, Soil Biology 32 (pp. 29-49). Springer. doi:10.1007/978-3-642-33811-3_2

Bahadure, S., Kalia, R., & Chavan, R. (2013). Comparative study of bioremediation of hydrocarbon fuels. *International Journal of Biotechnology and Bioengineering Research.*, *4*(7), 677–686.

Bhakta, J. N. (2016). Microbial response against metal toxicity. In A. K. Rathoure & V. K. Dhatwalia (Eds.), *Toxicity and Waste Management Using Bioremediation* (pp. 75–96). IGI Global.

Bhakta, J. N., & Munekage, Y. (2008). Role of ecosystem components in Cd removal process of aquatic ecosystem. *Ecological Engineering*, *32*(3), 274–280. doi:10.1016/j.ecoleng.2007.12.004

Bhakta, J. N., Munekage, Y., Ohnishi, K., & Jana, B. B. (2012). Isolation and identification of cadmium and lead-resistant lactic acid bacteria for application as metal removing probiotic. *International Journal of Environmental Science and Technology*, *9*(3), 433–440. doi:10.1007/s13762-012-0049-3

Bhatnagar, S., & Kumari, R. (2013). Bioremediation: A sustainable tool for environmental management-A review. *Annual Review & Research in Biology*, *3*(4), 974–993.

Cao, X., Ying, X., & Janelle, L., (2013). The effects of micro-algae characteristics on the bioremediation rate of deepwater horizon crude oil. *Journal of Emerging Investigators,* 1-7.

Chang, F. C., Ko, C. H., Tsai, M. J., Wang, Y. N., & Chung, C. Y. (2014). Phytoremediation of heavy metal contaminated soil by *Jatrophacurcas. Ecotoxicology (London, England), 23*(10), 1969–1978. doi:10.1007/s10646-014-1343-2 PMID:25236867

Chekroun, K. B., & Mourad, B. (2013). The role of algae in phytoremediation of heavy metals: A review. *J Mater Environ Sci, 4*(6), 873–880.

Cheraghi, M., Lorestani, B., & Yousefi, N. (2011). Introduction of hyper accumulator plants with phytoremediation potential of a lead-zinc mine in Iran. *International Journal of Environmental, Chemical, Ecological. Geological and Geophysical Engineering, 5*(5), 289–294.

Chijioke-Osuji, C. C., Ibegbulam-Njoku, P. N., & Belford, E. J. D. (2014). Biodegradation of crude oil polluted soil by co-composting with agricultural wastes and inorganic fertilizer. *Journal of Natural Sciences Research., 4*(6), 28–39.

Chowdhury, S., Bala, N. N., & Dhauria, P. (2012). Bioremediation- A Natural way for cleaner environment. *Int J Pharm Chem Biol Sci, 2*(4), 600–611.

Deniza, F. (2015). Bioremediation of contaminated water with unnatural dye using blue-green alga *Spirulina Platensis. Environ Prog Sustain Energy, 34*(5), 1414–1419. doi:10.1002/ep.12137

Dixit, R., Wasiullah, E. Y. N. Y., Malaviya, D., Pandiyan, K., Singh, U., Sahu, A., & Paul, D. et al. (2015). Bioremediation of heavy metals from soil and aquatic environment: An overview of principles and criteria of fundamental processes. *Sustainability, 7*(2), 2189–2212. doi:10.3390/su7022189

Doan, J., & Domingo, D. (2015). *Bioremediation: Microorganisms cleaning the environment.* Retrieved July 5, 2016, from: http://jdbioremed.wix.com/bioremediation#!timeline/c16s3

El-Sheekh, M. M., Farghl, A. A., Galal, H. R., & Bayoumi, H. S. (2016). Bioremediation of different types of polluted water using microalgae. *Rendiconti Lincei, 27*(2), 401–410. doi:10.1007/s12210-015-0495-1

Elekwachi, C. O., Andresen, J., & Hodgman, T. C. (2014). Global use of bioremediation technologies for decontamination of ecosystems. *J Bioremediat Biodegrad, 5*(4), 1–9. doi:10.4172/2155-6199.1000225

Evans, J. (2005). *Bioremediation mechanism discovered.* Royal Society of Chemistry. Retrieved October 1, 2016, from https://www.chemistryworld.com/news/6-july-2005-bioremediation-mechanism-discovered/1013110.article

Gifford, S., Dunstan, R. H., O'Connor, W., Koller, C. E., & MacFarlan, G. R. (2006). Aquatic zooremediation: Deploying animals to remediate contaminated aquatic environments. *Trends Biotechnol, 25*(2), 0-65.

Gupta, S. K., & Singh, J. (2011). Evaluation of mollusc as sensitive indicator of heavy metal pollution in aquatic system: A review. *IIOAB-India, 2*(1), 49–57.

Haimi, J. (2000). Decomposer animals and bioremediation of soils. *Environmental Pollution, 107*(2), 233–238. doi:10.1016/S0269-7491(99)00142-6 PMID:15093000

Hara, E., Kurihara, M., Nomura, N., Nakajima, T., & Uchiyama, H. (2013). Bioremediation field trial of oil-contaminated soil with food-waste compost. *Journal of JSCE, 1*(1), 125–132. doi:10.2208/journalofjsce.1.1_125

Joutey, N. T., Bahafid, W., Sayel, H., & Ghachtouli, N. E. (2013). Biodegradation: Involved microorganisms and genetically engineered microorganisms. In R. Chamy & F. Rosenkranz (Eds.), *Biodegradation-Life of Science* (pp. 289–318). InTech.

Jové, P., Olivella, M. À., Camarero, S., Caixach, J., Planas, C., Cano, L., & Heras, F. X. De L. (2016). Fungal biodegradation of anthracene-polluted cork: A comparative study. *J Environ Sci Health A Tox Hazard Subst Environ Eng, 51*(1), 70-77.

Karigar, C. S., & Rao, S. S. (2011). Role of microbial enzymes in the bioremediation of pollutants: A review. *Enzyme Research, 2011*, 805187. doi:10.4061/2011/805187 PMID:21912739

Kaur, G., & Hundal, S. S. (2015). Bioremediation of heavy metal contaminated soil using earthworm *Eisenia fetida. Journal of Environment, 4*(2), 25–29.

Kulshreshtha, S. (2013). Genetically engineered microorganisms: A problem solving approach for bioremediation. *Bioremediation & Biodegradation, 4*(4), 1–2.

Kumar, A., Bisht, B. S., Joshi, V. D., & Dhewa, T. (2011). Review on bioremediation of polluted environment: A management tool. *International Journal of Environmental Sciences., 1*(6), 1079–1093.

Kumar, S. R., Arumugam, T., Anandakumar, C. R., Balakrishnan, S., & Rajavel, D. S. (2013). Use of plant species in controlling environmental pollution-A review. *Bulletin of Environment. Pharmacology and Life Sciences, 2*(2), 52–63.

Kundu, D., Mondal, S., Dutta, D., Haque, S., & Ghosh, A. R. (2016). Accumulation and contamination of lead in different trophic levels of food chain in sewage-fed East Kolkata Wetland, West Bengal, India. *Int J Env Tech Sci, 2*, 61–68.

Lee, H., Yun, S. Y., Jang, S., Kim, G. H., & Kim, J. J. (2015). Bioremediation of polycyclic aromatic hydrocarbons in creosote contaminated soil by *Peniophora incarnata* KUC8836. *Bioremediation Journal, 19*(1), 1–8. doi:10.1080/10889868.2014.939136

Li, Y., Yang, H., Liu, N., Luo, J., Wang, Q., & Wang, L. (2015). Cadmium accumulation and metallothionein biosynthesis in cadmium-treated freshwater mussel *Anodonta woodiana. PLoS ONE, 10*(2), e0117037. doi:10.1371/journal.pone.0117037 PMID:25647043

Mandal, A. K., Sarma, P. M., & Jeyaseelan, C. Paul., Channashettar, V. A., Singh, B., Lal, B., & Datta, J. (2012). Large scale bioremediation of petroleum hydrocarbon contaminated waste at Indian oil refineries: Case studies. *Int J Life Sci Pharma Res, 2*(4), L-114–L-128.

Mandal, A. K., Sarma, P. M., Singh, B., & Jeyaseelan, C. P. (2012). Bioremediation: An environment friendly sustainable biotechnological solution for remediation of petroleum hydrocarbon contaminated waste. *ARPN Journal of Science and Technology, 2*(Special Issue), 1–12.

Menn, F.-M., James, P. E., & Gary, S. S. (2008). Genetically engineered microorganisms and bioremediation. In H. J. Rehmand, & G. Reed (Eds.), Bio technology (2nd ed.; pp. 441-463). New York: VCH Publishers Inc.

Mohan Krishna, V., Hrushikesh, N., & Sreehari, K., Aravind Kumar, T., Vidyavathi, N., & Pallavi, A. (2011). Studies on bioremediation of phenol by earthworm. *International Journal of Environmental Sciences*, *1*(6), 1268–1273.

Mouhamadou, B., Faure, M., Sage, L., Marcais, J., Souard, F., & Geremia, R. A. (2013). Potential of autochthonous fungal strains isolated from contaminated soils for degradation of polychlorinated biphenyls. *Fungal Biology*, *117*(4), 268–274. doi:10.1016/j.funbio.2013.02.004 PMID:23622721

Mukherjee, A. K., & Bordoloi, N. K. (2011). Bioremediation and reclamation of soil contaminated with petroleum oil hydrocarbons by exogenously seeded bacterial consortium: A pilot-scale study. *Environl Sci Pollut Res*, *18*(3), 471–478. doi:10.1007/s11356-010-0391-2 PMID:20835890

Naik, M. G., & Duraphe, M. D. (2012). Review paper on-Parameters affecting bioremediation. *Int J Life Sci Pharma Res*, *2*(3), L-77–L-80.

Oakland, H. (2013). *Bioremediation mariculture in Zanzibar, Tanzania: A viability assessment of using bath sponge and pearl oyster farms to filter highly polluted waters in the Zanzibar channel.* Independent Study Project (ISP) Collection. Paper 1526. Retrieved October 5, 2016, from http://digitalcollections.sit.edu/isp_collection/1526

Obasi, N. A., Eze, E., Anyanwu, D. I., & Okorie, U. C. (2013). Effects of organic manures on the physicochemical properties of crude oil polluted soils. *Afr J Biochem Res*, *7*(6), 67–75.

Obiakalaije, U. M., Makinde, O. A., & Amakoromo, E. R. (2015). Bioremediation of crude oil polluted soil using animal waste. *International Journal of Environmental Bioremediation & Biodegradation*, *3*(3), 79–85.

Okeke, J. J., Isreal, O. I., & Akubukor, F. C. (2014). Role of invertebrates as bioremediators: A Review. *Am J Life Sci Res*, *3*(1), 13–20.

Pal, S., Patra, A. K., Reza, S. K., Wildi, W., & Poté, J. (2010). Use of bio-resources for remediation of soil pollution. *Nat Resour*, *1*, 110–125.

Pandey, S. S., & Bagga, D. (2013). Phytoremediation an alternative. *International Journal of Environmental Engineering and Management*, *4*(5), 483–488.

Paz-Alberto, A. M., & Sigua, G. C. (2013). Phytoremediation: A green technology to remove environmental pollutants. *American Journal of Climate Change*, *2*(01), 71–86. doi:10.4236/ajcc.2013.21008

Prabha, Y., Soni, S. K., Gupta, S., & Sonal, . (2016). Potential of algae in bioremediation of wastewater: Current Research. *Int J Curr Microbiol Appl Sci*, *5*(2), 693–700. doi:10.20546/ijcmas.2016.502.076

Prasad, M. N. V. (2011). *A state-of-the-art report on bioremediation, its applications to contaminated sites in India*. Ministry of Environment & Forests, Government of India.

Priyadarshani, I., Sahu, D., & Rath, B. (2011). Microalgal bioremediation: Current practices and perspectives. *J Biochem Technol*, *3*(3), 299–304.

Rane, N. R., Chandanshive, V. V., Watharkar, A. D., Khandare, R. V., Patil, T. S., Pawar, P. K., & Govindwar, S. P. (2015). Phytoremediation of sulfonated remazol red dye and textile effluents by *Alternanthera philoxeroides*: An anatomical, enzymatic and pilot scale study. *Water Research*, *83*, 271–281. doi:10.1016/j.watres.2015.06.046 PMID:26164661

Revathi, S., & Venugopal, S. (2013). Physiological and biochemical mechanisms of heavy metal tolerance. *International Journal of Environmental Sciences*, *3*(5), 1339–1354.

Rojas-Avelizapa, N. G., Roldan-Carrillo, T., Zegarra-Martınez, H., Muñoz-Colunga, A. M., & Fernández-Linares, L. C. (2007). A field trial for an ex-situ bioremediation of a drilling mud-polluted site. *Chemosphere*, *66*(9), 1595–1600. doi:10.1016/j.chemosphere.2006.08.011 PMID:16997351

Seo, J. S., Keum, Y. S., & Li, Q. X. (2009). Bacterial degradation of aromatic compounds. *International Journal of Environmental Research and Public Health*, *6*(1), 278–309. doi:10.3390/ijerph6010278 PMID:19440284

Shankar, K. S., Devi, Ch. A., & Rao, C. S. V. R. (2011). Phytoremediation of heavy metals with *Amaranthus dubius* in semi-arid soils of Patancheru, Andhra Pradesh. *Indian Journal of Dryland Agricultural Research and Development*, *26*(2), 71–76.

Singh, D., & Fulekar, M. H. (2010). Biodegradation of petroleum hydrocarbons by Pseudomonas putida strain MHF 7109 isolated from cow dung microbial consortium. *Clean (Weinh)*, *38*(8), 781–786.

Singh, R., Singh, P., & Sharma, R. (2014). Microorganism as a tool of bioremediation technology for cleaning environment: A review. *Proceedings of the International Academy of Ecology and Environmental Sciences*, *4*(1), 1–6.

Sinha, R. K., Valani, D., Sinha, S., Singh, S., & Heart, S. (2009). Bioremediation of contaminated sites: A low-cost nature's biotechnology for environmental clean up by versatile microbes, plants & earthworms. In T. Faerber & J. Herzog (Eds.), *Solid Waste Management and Environmental Remediation* (pp. 1–72). Nova Science.

Sivasubramanian, V. (2011). Microalgae for potential phycoremediation of Industrial effluent - A case study. *ENVIS CENTRE Newsletter, 9*(4).

Sivasubramanian, V., Subramanian, V. V., & Muthukumaran, M. (2010). Bioremediation of chrome-sludge from an electroplating industry using the micro alga *Desmococcus olivaceus* - A pilot study. *J Algal Biomass Util*, *1*(3), 104–128.

Solovchenko, A., Pogosyan, S., Chivkunova, O., Selyakh, I., Semenova, L., Voronova, E., & Lobakova, E. et al. (2014). Phycoremediation of alcohol distillery wastewater with a novel *chlorella sorokiniana* strain cultivated in a photobioreactor monitored on-line via chlorophyll fluorescence. *Algal Res*, *6*, 234–241. doi:10.1016/j.algal.2014.01.002

Sonawdekar, S. (2012). Bioremediation: A boon to hydrocarbon degradation. *International Journal of Environmental Sciences*, *2*(4), 2408–2424.

Sukumaran, D. (2013). Phytoremediation of heavy metals from industrial effluent using constructed wetland technology. *Applied Ecology and Environmental Sciences*, *1*(5), 92–97. doi:10.12691/aees-1-5-4

Tangahu, B. V., Abdullah, S. R. S., Basri, H., Idris, M., Anuar, N., & Mukhlisin, M. (2011). A review on heavy metals (As, Pb and Hg) uptake by plants through phytoremediation. *International Journal of Chemical Engineering*, *2011*, 1–31. doi:10.1155/2011/939161

Umar, A. F., Tahir, F., Larkin, M., Oyawoye, O. M., & Musa, B. (2012). In-Situ biostimulatory effect of selected organic wastes on bacterial atrazine biodegradation. *Adv Microbiol*, *2*(04), 587–592. doi:10.4236/aim.2012.24076

Uqab, B., Mudasir, S., Sheikh, A. Q., & Nazir, R. (2016). Bioremediation: A management tool. *J Bioremediat Biodegrad*, *7*(2), 1–5.

Wang, S., Wang, X., Poon, K., Wang, Y., Li, S., Liu, H., & Cai, Z. et al. (2013). Removal and reductive dechlorination of triclosan by *Chlorella pyrenoidosa*. *Chemosphere*, *92*(11), 1498–1505. doi:10.1016/j.chemosphere.2013.03.067 PMID:23648333

Zhang, T., Li, S., Sun, X., Zhang, Y., Gong, X., Fu, Y., & Jia, L. (2015). The earthworm *Eisenia fetida* can help desalinate a coastal saline soil in Tianjin, North China. *PLoS ONE*, *10*(12), e0144709. doi:10.1371/journal.pone.0144709 PMID:26699869

KEY TERMS AND DEFINITIONS

Biological Tools: It refers to organisms or biological substances used in any systems.

Bioremediation: It is an eco-friendly biological process employed in remediation of environmental problems (such as, pollution, contamination, etc.) using biological system or organisms.

Contaminants: Presence of unwanted substances in any matrix which cause several physicochemical changes and made the substrate polluted, toxic, unsafe, harmful to environment as well as causes health hazards.

Decontamination: The process involved in the elimination or neutralization of contaminants.

Eco-Friendly: It refers to method or product or system having no negative impacts on environment.

Ecological Engineering: It refers to an emerging technology for designing the sustainable ecosystem by integrating human society with its natural environment for the benefit of both or the fusion of engineering and ecology for the development of human and environment by the design, construction and management of ecosystems which are beneficial for both human and environment.

Ecosystem Services: Services provided by the ecosystem by the means of provisioning, regulating, cultural and supporting services. This is a self-regulating mechanism driven by the ecosystem components which fulfill the basic needs of human beings.

Living Machine: Biotic system or organisms used in bioremediation process to protection environment from degradation/destruction.

Phytoremediation: Use of different types of plants in the bioremediation, such as purification of wastewater, air and reclamation of contaminated soil are known as phytoremediation.

Zooremediation: It is a kind of bioremediation process driven by the animal systems using the mechanism of uptake, accumulation, and transformation of contaminants.

Chapter 14
Contribution of Earthworm to Bioremediation as a Living Machine:
Bioremediation

Shweta Yadav
Dr. H. S. Gour Vishwavidalaya (A Central University), India

ABSTRACT

The literature regarding the benefits of earthworms is fairly ancient. Aristotle, the Greek philosopher, referred to them as 'intestines of earth' because of their habit of ingesting and ejecting the soil. In the subtropical regions of Egypt and India, the success of the ancient civilization of the Nile and the Indus Valley was partly due to the fertile soils created by the activity of the earthworms and by the continual renewal of the land by the alluvium process. During the Cleopatra era (69-30 BC), the earthworm was declared a sacred animal in the ancient Egypt. Later Darwin remarked the earthworms have played more roles throughout the history of the world than any other animal. Besides to contribute in physical structure and nutritive value of the soil by burrowing and feeding they can also be a potential pollution hazard. They are useful tool in environment monitoring and are good indicators of condition of soils. This chapter reviews soil contamination that influences earthworms and how they cope-up in contaminated environment.

INTRODUCTION

The literature regarding the benefits of earthworms is fairly ancient. Aristotle, the Greek philosopher, referred to them as '*intestines of earth*' because of their habit of ingesting and ejecting the soil. In the subtropical regions of Egypt and India, the success of the ancient civilization of the Nile and the Indus Valley was partly due to the fertile soils created by the activity of the earthworms and by the continual renewal of the land by the alluvium process. During the Cleopatra era (69-30 BC), the earthworm was declared a sacred animal in the ancient Egypt. Later Darwin remarked the earthworms have played

DOI: 10.4018/978-1-5225-2325-3.ch014

more roles throughout the history of the world than any other animal. Besides to contribute in physical structure and nutritive value of the soil by burrowing and feeding they can also be a potential pollution hazard. Earthworms (Annelida: Oligochaeta) ranges in length from 2 centimeters to over 2 meters, and found in boreal to tropical, sea level to 5000 meters elevation, semi-arid to extremely humid, aerobic to nearly anoxic and even aquatic. Some are predatory but majority feed on forms of organic matter in or above soil (Yadav, 2016a). Beyer (1981) have reported that earthworms can take up and accumulate heavy metals such as cadmium, mercury and gold in their tissues, when living both in non-contaminated and contaminated environment (Helmkeet al., 1979). They have potential to bioremediate soils by reducing the pollutants concentration through a bioaccumulation mechanism within the body of earthworm and the chemical changes in their alimentary tract that may render various metals more available to plants (Oberdoster et al., 2005).

As earthworms occupy major invertebrate biomass (>80%) in terrestrial ecosystem and have over 600 million years of experience as environmental managers in the ecosystem as ‘waste managers’ as ‘soil managers’, ‘fertility improvers’ and ‘plant growth promoters’ for long time (Sinha et al., 2010). But relatively new discoveries about their role in bioremediation or detoxification mechanism of industrial wastes, chemically contaminated soil, dairy industry waste material, and detergent industries have revolutionized the understanding of functioning of this unheralded soldier of mankind. Earthworms are the consumers of many detritus food chains while, carnivorous animals (depends on their assimilation efficiency), may accumulates these toxicants by preying on contaminated earthworms. The digestive system of earthworms is capable of detaching heavy metal ions from the complex aggregates between ions and humic substances (Dominguez et al., 2005. Various enzyme- driven processes seem to lead assimilation of the metal ions by the worms so they are locked up in organisms ‘tissues rather than released back to soil (Yadav 2016a). Mineralization of dead earthworms’ releases less toxic form of metals –accumulated metals in the environment. Chemical changes in alimentary tract of earthworms may render various metals more available to plants, and mineralization of dead earthworms’ releases metals–accumulated, metals in the environment. They have also been reported to host microbes in their gut which can biodegrade chemicals (Edwards &Bohlen, 1996). They can tolerate high environmental concentrations of toxic heavy metals anddo not absorb metal, accumulate it in a non-toxic form or excrete it efficiently (Ireland, 1979). Yet, the mechanism of detoxification in earthworms is poorly understood. It varies with the metals concerned, but metal accumulation moderately high levels reasonably short periods may appears to have little deleterious effects on their biomass or growth. Long-term sub lethal concentrations of heavy metals may reduce their fecundity.

ENVIRONMENTAL BENEFIT OF EARTHWORM

Soil biodiversity has both direct (the organisms themselves and/or their metabolic products) and indirect (the long-term outcome of their activities) benefit on the economy basis. It is estimated that the value of ecosystem services provided each year by soil biota in agricultural systems worldwide (e.g., organic waste disposal, soil formation, N_2 fixation, bioremediation and biocontrol) may exceed US$1.5 trillion (by AGL/FAO as modified from Pimentel et al., 1997). Earthworms are the main governors of soil biology and as such they rightly deserve special attention for without them the soil diversity is poor and the soil-food-web interactions may be weak. Earthworms are ubiquitous in soils capable of supporting higher plants. They are found only in refugee in arid and frozen regions, but in most other terrestrial ecosystems they dominate soil macrofaunal communities with densities and biomass bulk as high as one million per hectare and 2 tons per hectare (Lavelle, 1983). They play a vital role in the formation and maintenance of fertile soil and are of paramount importance to primary production. Recent resurgence of interest in earthworms is driven by urgent environmental and economic concerns, particularly the need to understand and utilize their function in sustainable agriculture and bioremediation of soil toxicants; to appreciate their potential for restoring damaged soils; and to exploit their recycling of organic 'wastes' [Lee (1985); Sims & Gerard (1999); Edwards & Bohlen (1996)]. The importance of earthworms to the survival of terrestrial life relates to their intimate relationship with pedogenesis, the formation of soil, and the effect this has *via* recycling of nutrients and the maintenance of soil fertility and soil structure on plant productivity. Dr Bill Mollison, co-founder of Permaculture, points out that by sliding through their tunnels (Blakemore, 2010), earthworms *act as an innumerable army of pistons pumping air in and out of the soils on a 24 hour cycle (more rapidly at night).*

SOIL CONTAMINATION AND IMPACT

Soil Contamination

Modern industries generate enormous quantities of wastes especially heavy metals. That finds its way into the soil organic layer, making it a sink for contaminants. These contaminants effects negatively to soil organisms and inhibits microbial process in soil. Metals from anthropogenic sources are widely spread in the atmosphere including aquatic and terrestrial environments. And, are major group of compounds involved in soil contamination and causes damage to ecosystems close to emission sources by affecting the abundance, diversity and distribution of organisms (Lobersli & Steinnes 1988; Kozlov et al., 1993; Haimi& Siira-Pietikainen, 1996; Yaron et al., 1996). Indiscriminate disposal of these wastes lead inevitably to ecological problems because of wide dispersal of metals. The heavy metals, which occur naturally in all soils, waters and living organisms, can have severe toxic impact on living system in excessive concentrations. Main sources of metal contaminants are mining, metal pigment and chemical manufacturing as well as combustion of fossil fuels including the exhaust emission of motor vehicles. Metals *viz.,* arsenic, cadmium, chromium, copper, mercury and zinc play a significant role in soil contamination. Some metals, arsenic, cadmium, chromium, copper, mercury and zinc have potential risk of toxicity to the environment even in low concentrations (Yaron et al., 1996). However, Zinc and copper are essential trace elements for organisms but becomes toxic above certain concentrations and exposure times (Hopkin, 1989).

Heavy metals (atomic density greater than $5g/cm^3$) can divided into two groups, elements essential to life and non–essential to life. The elements essential to life like copper and zinc can become toxic if present at high concentrations. While, elements like arsenic, cadmium, chromium, copper, mercury, nickel, lead and zinc are considered as threat to environment. They can enter the soil environment *via* atmospheric pollution (including motor vehicle emissions), combustion of fossil fuel, agricultural chemicals, organic manures, municipal sludge, industrial waste and mining. Among them disposal of industrial effluent into river and lakes has led to a series of well documented ecological disasters. The emitted metals finally end up in surface soil layers and bind especially to organic matter. This binding reduces metal mobility and bioavailability but on the other hand, metals are bound in those parts of ecosystems where essential biological processes, (decomposition and nutrient mineralization) take place. Ultimately, this metal does not degrade and causes long-term effects on soil organisms.

Exposure of Heavy Metals to Soil Organism

The ecological consequences of heavy metals pollution of soils are related largely to mobility and solubility of heavy metal. These chemically related factors determine mobility through the soil to the water table, availability to soil microbes, soil animals, agricultural crops and ultimately to the human population. And, these metals are subjected to various reactions including soil interactions, cation exchange, specific absorption, co-precepitation, organic complexation, plant uptake, volatilization and bioaccumulation (Wuana et al., 2014) in organisms. Their entry into biological system depends on specific receptor to organism, the route of entry, time of exposure and the matrix containing the contaminant (Adriano, 2003). Soil has varying degree of spatial and temporal gradients of physical and chemical characteristics. These characteristics together with environmental conditions and physiology, behavior of organisms determine the bioavailability of metals or other substances in soils to the organisms. Langdon et al. (2001) observed the metal-spiked soil in laboratory experiments is more toxic to soil organisms than the equivalent metal concentrations in the field soils. Several methods for extraction of metals and their bioavailability have been proposed but none of them have been accepted as its absolute measure. Luo et al.(2003) suggested sequential extraction (using strength of extracting solution) fractionates metals into forms of different solubility & mobility and may be improved tool for predicting metal availability to soil biota, movement in the soil profile and transformation between different forms in soils. However, the fate of metals in environment and their biological impacts in changing environmental conditions are difficult to predict.

Measuring Effects of Heavy Metals on Soil Organism

The heavy metals may accumulate and remains in the soil for years (Mattson et al., 2002). At specific concentration these metals are potentially toxic to livestock and to humans and have been shown to interfere with soil microbiological processes. The effect of heavy metals and other anthropogenic substances to soil organisms can be assessed with chemical analyses to different kinds of biotests and field monitoring, mainly measuring mortality or reproduction of selected species or estimating numbers and diversity of organisms in the contaminated field. Some of the tests have been standardized and some are undergoing standardization processes in the Organization for Economic Cooperation and Development (OECD), the International Organization for Standardization (ISO) and European Committee for Standardization (CEN). Standardized acute toxicity (ISO-11268-1, 1993) and reproduction (ISO-11268-2, 1997) test on earthworms have proven to be useful tools in assessing the toxicity and possible risk of soil contaminants (Lokke & Van Gestel, 1998). However, test don't provide information on how individuals may react when facing contaminants in their environments. Still, early warning signals (sub lethal responses) remain overlooked that may provide very useful information when assessing the environment risk of contamination (Yadav, 2016b). It may be suggested that biological tests and bioindicators together with chemical analyses may give a natural reliable and repeatable view of the condition of the environment contamination. Analysis of microcosms may be also an important tool to decrease the gap between simple laboratory studies and the data derived from field conditions (Sheppard, 1997). Soil organism may respond to contamination either by acclimatization or enough genetic adaptation over several generation (Bengtsson et al., 1992).They can survive in contaminated environments by increasing resistance or by mechanisms that can minimize their exposure. The increase in detoxification or excretion processes and decrease in cuticular or intestinal uptake helps to develop resistance in soil organism against harmful contaminants. But, acquired ability of resistance may have negative impacts on other fitness components like reproduction or growth. This may also be reflected in life history-traits and activity of organisms, which affect overall properties and efficacy of soil organism in contaminated ecosystem.

Effect of Heavy Metals on Earthworms

Earthworms are important link in the natural food chain, in addition to their role in soil processes. They can accumulate and transfer beneficial and hazardous materials. The metabolic activity of earthworm increases the solubility of heavy metals (like Pb and Zn) in soils contaminated with heavy metals and their activity may be important in reactions controlling the availability of heavy metals to plants (Cernansky et al., 2009). They have been considered to be a valuable animal for assessment of soil quality and contamination as well useful tools in environment monitoring. They also reflect site-specific contamination levels because of mobility capacity and close relationships to different soil conditions. While burrowing, earthworms ingest large amount of soil and are therefore exposed to contaminants through their intestine as well as through their skin. As a result substances are concentrated from the soil into their body, making them ideally suited for assessing the bioavailability of many harmful substances. Pathirana et al., (1994) and Hobbelen et al., (2004) reported *Lumbricus rubellus* and *Aporrectodea caliginosa*can accumulate higher amounts of contaminants in comparison to other invertebrates such as isopods and millipedes. Responses of earthworms to soil contamination may vary in different species and the exposure history of the individuals (Reincke et al., 2001). Species behavior such as, burrowing and feeding activities also affected by metal contamination (Morgan &Morgan, 1999). In heterogeneously contaminated habitats, earthworms avoid the most contaminated soil patches (Stephenson et al., 1998). Earthworms can perceive a diverse range of chemical stimuli to help them in selection of food and in detecting adverse environmental conditions such as changed soil acidity and harmful contamination (Bolen et al., 2003). Acute chemical sensitivity of chemoreceptors and the ability for dispersal make the earthworms capable of avoiding highly contaminated soil (Naidu et al., 2010). The characteristics of earthworms make them potentially attractive organisms to analyze for monitoring the biological availability of natural and anthropogenic metals in soils. Their feeding behavior expose them to the soil fraction that often contains high concentrations of heavy metals and their size, ubiquity and ease of collection simplifies analysis. Earthworms are also capable of detecting and avoiding contaminated soils. First they try to avoid metal contaminated soil when they face contaminated soil patches in their habitat. If they don't have possibilities to avoid metal contaminated soil induces biochemical responses. If contamination continues and biochemical detoxification systems against metal exposure are not adequate to reduce the harmful effects of metals, changes in individual activity and biomass appears. If an exposure continues or earthworms are not able to tolerate the contamination level, the individual may dies. Yet, is not possible to draw any firm conclusions concerning metal toxicity to earthworms in soil from the available data base. However, it is known that the very high concentrations of metals that can be accumulated by earthworms pose potential hazards to themselves and their predators. Bio-concentration of heavy metals may impact the environment as they are important food resources for several bird and small mammals. The concentration of metals present in the tissues of earthworms and other animals including birds does not necessarily reflect the extent of soil contamination with metals, presumably because bioavailability is also a function of other components in the diet which modulate absorption of toxic metals by animals. In addition other soil chemical properties such as pH also exert a strong influence on bioavailability of metals in body tissue.

HEAVY METALS BIOACCUMULATION POTENTIAL OF EARTHWORM

Several researches showed that effects of metals on soil organisms depend on exposure to concentrations that were available to uptake. Soil organism including earthworms may be exposed to metals by direct dermal contact in soil solution or by ingestion of pore water, polluted food and/or soil particles (Yadav, 2016a). Now-a-days contaminated soils have become a primordial problem since they lead to ground water contamination and biomagnifications of chemical compounds through food webs and sometimes affect human health (Loureiro et al., 2005). The re-use of these contaminated sites is big challenge and could be only possible by realistic assessment of their level of harm at the contaminated site (Davies et al., 2002). Earthworms have specificity and capacity to accumulate metals in their bodies which is species–specific (Hopkin, 1989) and depends on seasonal changes (Morgan & Morgan, 1993). In general, earthworm consume a great amount of soil to achieve their daily nutrition and their digestive process liberates metals in their free forms in the gut lumen. These freely available metals are then absorbed by the gut epithelial lining. By this enzymatic process, earthworms tend to accumulate considerable amount of metals within cells of alimentary canal. They accumulate metal within their issue by binding with metallothioneins (metalloproteins). However, the degree of accumulation may be species as well as site specific and also depends on biomass & abundance of metals present in the soil. Earthworms' tissue heavy metal contaminations may be an indicator of exposure to heavy metals contaminated soil. Accumulation of heavy metals in earthworms could reduce the biotoxicity or phototoxicity of metals and therefore, this could be exploited for detoxification of pollutant compounds in the environment. It has high affinity for accumulating Cd and may absorb 100mg/kg from sludge soil. Zinc is also accumulated by earthworms relative to soil concentrations, whereas Cu, Pb and Ni are not accumulated in earthworms (Nakashima et al., 2008).It has been observed that concentrations of Cd, Cu and Zn in earthworms increased, whereas Se content decreased with increasing rate of sludge application. The effect on Se may be due to antagonism between Se and heavy metals. In contrast, Hg and Cr may not bio-available to earthworms (Olaniran, 2013). Zinc status within animal tissue appears to be effectively regulated and a homeostatic situation was maintained irrespective of dietary intake even in severely polluted ecosystem. The results of tracer studies revealed that Cd, Cu, Hg and Zn are readily accumulated by earthworms when these are supplied to soluble forms (Wuana et al.,2014).The relatively long, biological half-life of Cd and Hg compared to Zn indicates that relatively higher concentrations of these toxic metals can be accumulated by earthworms. Cadmium pollution has clear ecotoxicological consequences as it is readily bio-available and has reported tendency to bioaccumulate in soil-biota (Ireland, 1976). In terms of relationship between metals and earthworms, cadmium has been extensively studied and is known to affect earthworm's physiological state in term within biostimulation. As consequences, some metals are regulated dynamically after uptake by earthworms whereas other such as Cd are accumulated continuously and become tightly bound and immobilized. This accumulation is reinforced because of the low toxicity of Cd to earthworms. The continuous accumulation of metals within the earthworms after transfer to clean soil was observed by Van Gestel et al., (1997). Wielgus-Serafnska & Kawka (1976) reported lead accumulation in the glandular epithelial cells of epidermis and intestine of *Eisenia fetida*, deposition of lead and zinc in posterior part of alimentary tract and active deposition of calcium in anterior part of alimentary tract in *Dendrobaena rubida.*Ireland (1976) detected lowest concentrations of lead, zinc and calcium in their body wall and small amount of lead in urine, mucus and coelomic fluid. Thus dorsal pores are involved in regulation of intra-coelomic pressure and also playing an important role in excretory function.

Nieboer & Richardson(1980) distinguished two groups of metals: one that tends to bind metals (Li, Na, Mg,Ca,Sr) to oxygen sites in biological ligands and act as macro-nutrients and a second (Cu, Zn, Cd, Hg Pb) that forms strong bond with N or S and exerts toxicity on earthworms by affecting protein structure and enzyme function. However, the metals in the group which act as macronutrients may become toxic under certain doses, conditions and end points. Toxic metals can interact with trace metals as well as metal macronutrients. Substitution between these forms is an important process in this context, as demonstrated for Ca-Cd and Ca-Pb interactions by Ma (1988). Janssen et al., (1998) showed the potassium affects the uptake of Cs but not of Na, whereas Morgan& O'Reilly (2002) reported a competitive inhibition of heavy metal accumulation by Mn.

Ireland (1979) examined the distribution of metals in *Lumbricus rubellus* at three sites in Wales containing high concentrations of lead and manganese, zinc and copper. At the predominantly, lead and manganese–polluted site, the highest concentrations of lead, manganese and calcium were found in earthworms. Despite, the high soil concentrations of zinc at one site and copper at the other, the burden of these metals in earthworm body was not significantly different at all studied sites and it was suggested that *L.rubellus* might regulate zinc and copper. Higher concentrations of zinc and manganese and the lowest concentrations of lead were also recorded in *L.rubellus*. *Dendrobaena veneta* contained the higher concentration of cadmium and *Eiseniella tetraedra* contained the highest concentration of lead. The difference in metal content of various species could be result of difference in uptake mechanisms. Piearce (1978) reported that intestinal uptake of calcium was much higher in *L.rubellus* than *A. caliginosa.* However, interactions between metals can also influence the amount of specific metal assimilated (Kirchgressner, 1978). The heavy metal accumulation in earthworms may be effected by soil pH, concentration and chemical speciation of metals, intrinsic rate of bio-accumulation, tolerance of an individual, properties of digestive system and immobilization /storage mechanism in the tissues. Removal of toxicant from sensitive tissues of earthworms can be regulated by excretion from the body; bound within the matrix of inorganic granules; attached to proteins or other ligands. If detoxified primarily by excretion, body concentrations should decrease when previously exposed individuals are transferred to a clean environment (Yadav, 2016 b). However, if an element is bound in an organic matrix or to organic ligands, metal level may remain constant. Spurgeon & Hopkin (1999) recorded copper and zinc metals primarily detoxified by excretion however for xenobiotic metals (cadmium and lead) excretion was slow or absent in earthworms. It may be suggested main detoxification pathways for xenobiotics may be sequestration within inorganic matrices or binding to organic ligands.

BIOREMEDIATION MECHANISM OF EARTHWORM

Earthworm coelomic cells (non-movable scavenger); amebocytes (mainly immune function) and eleocytes (mainly nutritive function exists in various functional states and stages of maturation. It has been assumed that amebocytes derived from mesenchymal lining of coelom whereas, eleocytes originate by detachment of chloragogen cells covering intestinal tract (Affar et al., 1998; Hamed et al., 2002). Amebocytes participate in the transport and storage of nutritive substances (Valembois et al., 1988), phagocytosis (Stein & Cooper, 1981; Bilej et al., 1990; Dales &Kalae, 1992; Ranzelli-Cain & Kaloustian, 1995; Cossarizza et al., 1996), while; eleocytes play an important role in immune responses producing bacterial substances (Valembois et al., 1982; Ville et al., 1995; Milochau et al., 1997) and also participates in reaction of encapsulation and formation of brown bodies (Cooper & Stein,1981). Fischer & Molnar (1993) suggested chloragogenous tissues are main site of storage of metals. The cells of these tissue contain two types of granules (Chloragosomes);

1. Phosphate rich complexes containing calcium and to a lesser extent, zinc (Prento,1979), these granules are involved in the binding of type A or borderline metals such as lead by a process involving exchange with the matrix-associated calcium (Morgan & Morgan,1999);
2. Second granule type contains sulphur-donating ligands probably in the form of residues of metallothionein –like proteins, these proteins bind type B metals, such as cadmium. Chloragosomes may act as specific electron acceptors, substituting molecular oxygen in the metabolism when the environment becomes hypoxic.

A more important function of chloragosomes, regarding metal contamination is their cation-exchange properties. It also contains phosphoric acid, carboxyl, phenolic hydroxyl and sulphonic acid groups, all of which are constituents of ion-exchange compounds. The addition of lead to a chloragosomal suspension caused a drop in pH (Ireland, 1978) and release of mainly calcium, indicating that cation-exchange was taking place. Thus, chloragosomes of earthworms may therefore function as cation exchanger for heavy metals, thus reduce their toxic effects.

Figure 1. Mechanism of uptake and biotransformation of toxic substances by earthworms

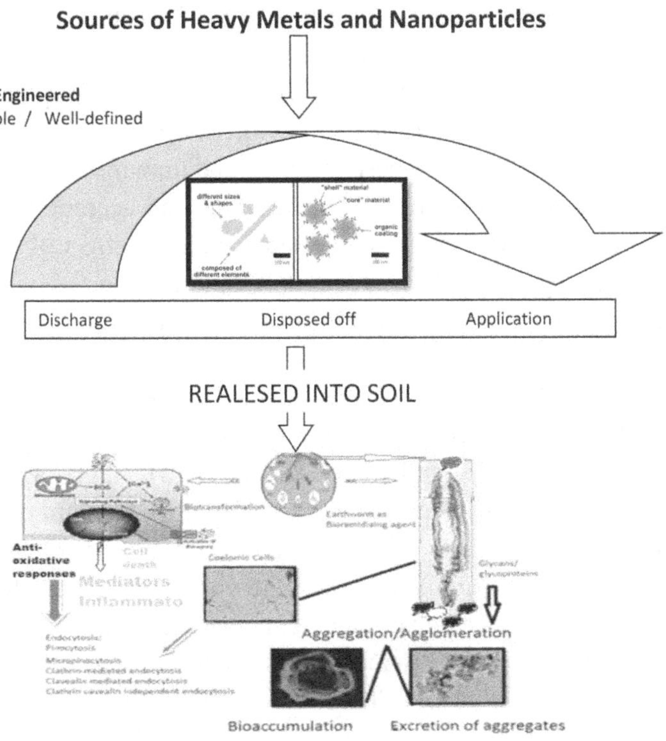

Earthworms to Nullify Toxic Effect of Nanoparticles

Nanoparticles may be released from various products through normal use and then enter into wastewater stream. A major portion of these nanoparticles may release into sewage sludge those are disposed of in landfills, incinerated or applied to agriculture lands. Thus, soil system is an alternative sink for large portion of nanoparticles (Gottschalk et al., 2009). The cytotoxic molecules increase the intracellular calcium concentration that participates in exocytosis. Exhausted phagocytes with NPs (nanoparticles) can then be eliminated through the dorsal pores. The internalized NPs can also be excreted by nephridia, while agglomerated /hetero aggregated NPs can be eliminated by process of encapsulation. The internalization of NPs in coelomic cells occurs due to TLR (toll like receptors) and their recognition is mediated by pattern-recognition receptors (PRRs, which leads into various inflammatory cytokines and antimicrobial peptides. It has been reported that intact nanoparticles absorbed by earthworms and, bio-distributed to tissues remote from portal of entry (Gupta et al., 2014; Yadav et al., 2016b) and in soil have potential to enter in terrestrial food webs and also (Unrine et al., 2010). Earthworms can uptake NPs from soil in two forms *i.e.* dietary pathways (~ 95% of total uptake) and dermal pathways (~ 5% of total uptake, Gupta et al., 2014). However, their skin with mucus is the first-line defense barrier against foreign endeavor, but once they enter the coelom, they are exposed to cellular and humoral responses.

Coelomic cytolytic factor (CCF) is a well-characterized 42-k Da lytic protein (secreted into coelomic fluid in a stable form) in earthworms that act as pattern-recognition molecule and is present on the cells of the mesenchymal lining of coelomic cavity as well as on free coelomocytes. CCF is formed by two spatially distinct lectin like domain located in the central part of the molecule which interacts with lipo-polysaccharide and β-1,3, glucans and second domain is located in C-terminal part interacts with peptidoglycan constituents. However, coelomic cytolytic factor 1(CCF1), a pattern recognition receptor was found suppressed in lifelong experiment. Clathrin-mediated endocytosis, caveolae-mediated phagocytosis and micro-pinocytosis may involve in uptake of NPs by coelomocytes. Scavenger receptor class A (a pattern recognition receptors) of coelomocytes is potential pathway to phagocytosis the NPs by amebocytes of earthworms (Cooper& Stein, 1981). Macrophage receptor with collagenous structure (MACRO) of amebocytes recognizes and associates with nanoparticles for phagocytic clearance. Activation of calcium signaling on exposure NPs reflects stress and immune-responses like in highly conserved signal transduction cascade, MARK pathways. On the other hand, secretion/modulation of cytokine from phagocytes is also effected by NPs may be suggested as indirect effect of intracellular communication of coelomocytes during exocytosis of NPs in coelomic fluid as reported in our earlier studies (Gupta et al., 2014). Thus phagocytic population of coelomocytes seems to have potential to uptake and excreted out in the form as aggregates to scavenging nanaomaterials from the soil system.

The aggregates of ZnO-NPs (50-100nm) were recorded in coelomic fluid and tissues of *Eisenia fetida* in our earlier studies (Gupta et al., 2015). Accumulation of gold and quantum dots has been recently reported in mucus materials of jelly fish by Patwa et al. (2015). Aggregation of Au-NPs in gut tissues of *E. fetida* were also recorded by Unrine et al. (2010). Deposits of 50nm in He La cells (Chitrani et al., 2006) and aggregates of 10-58nm NPs in mice (Hillyer & Albrechet, 2001) were recorded by earlier workers. Aggregation of NPs in coelomic fluid of earthworms may be due to presence of glycoproteins and glycans(high manse type N-linked glycans, mucin type O-linked glycans and high levels of pentose-containing oligomeres. This may be due to interactions of electric charges, which allow particles to link together resulting in a zero point of charge. Thus, the size and density of aggregates of particles in coelomic fluid may be due to interactions between NPs surfaces and glycoproteins/glycans. The formation of aggregates may be regulated by distribution of charges in polymers, molecular weight (Mabire et al., 1984), polymer conformation (Zeng et al., 2015), hydrogen bonding; ionic interactions and dehydration of polar groups (Ulrich, 2006). The stabilization of these aggregates might be to obtain the mixture from intestinal and body wall mucus secretions and to measure the extent to which they interact with soil component and produce stable aggregates. Gum like nitrogenous mucus materials probably muco-protein left behind when earthworm pass over the soil surface, readily adhere to the surfaces with which they come into contact. In case where aggregate forming secretions are resistant to microbial attack, the micro-aggregates and aggregates formed could be expected to be stable. However, the soil particles in passing through the earthworm gizzard; aggregates wouldn't survive in this process, although domains could, and some very small micro- aggregates might exist. A re-association of aggregate forming components must take place as the materials pass through the earthworm intestine.

Hayes (1983) suggested that calciferous glands provides appropriate amount of calcium to allow divalent cations bridging to take place between the humic substances and the negatively charged inorganic colloids. This bridging is actually resulted into formation of aggregates, which takes place in the intestine in presence of excess water in the intestine. Drying of the casts after ejection brings the particles into close proximity and allow to polymers secreted in the intestine, as well as the indigenous humic substances and polysaccharides to form the necessary bridges (aggregates). The humus substance provides appropriate polymers for binding of these particles and stabilizing aggregates. These polymers may have to be of appropriate size and in abundance, to bridge the distances between the components of aggregates/hetero-aggregates (Yadav, 2014). The larger the aggregates, the longer and the more numerous of these bridges must be to confer stability of aggregates (non-toxic) forms of nanoparticles.

Figure 2. Layout of potential base technology for nanoparticles using earthworms as living machine (NP, nanoparticle; EW, earthworm)

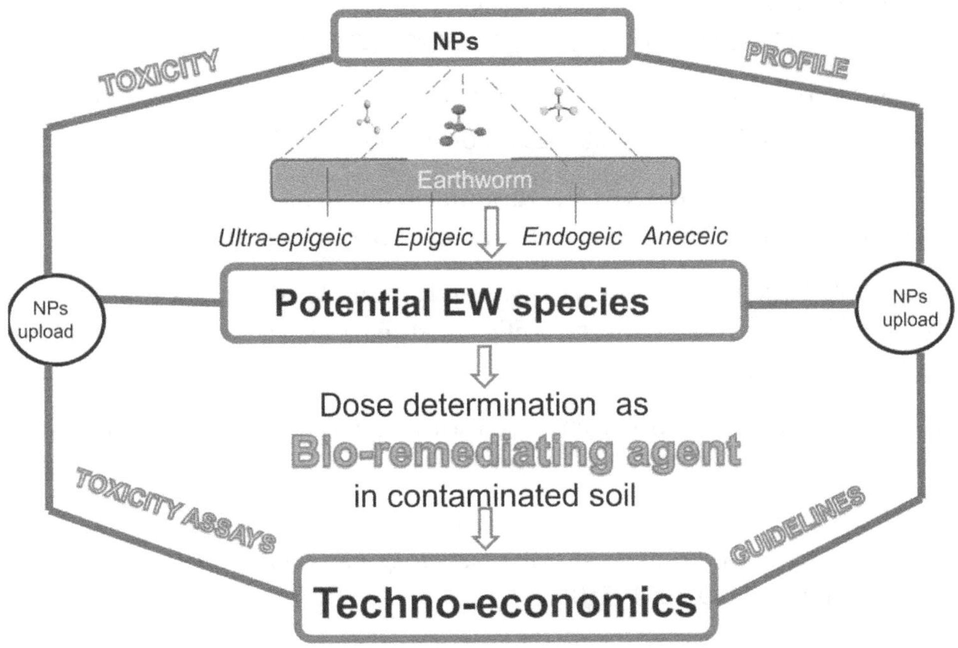

FUTURE PROSPECTIVES

Earthworms are potential bioaccumulator to mitigate toxicity of industrial and municipal wastes. They either "biotransform' or 'biodegrade' the chemical contaminants turning them harmless in their gut before entering of the metals in their tissues. Those entering in gut metabolized, immobilized and excreted or sequestered in tissues or vacuoles.

Thus earthworms might contribute to:

1. The bioaccumulation and bio-flocculation of metals & nanomaterials in their coelomic fluid using bacteria and fungus to deposit homo and hetero-aggregates of nanoparticles to reduce their toxic impact in soil system;
2. Facilitate trapping properties to capture toxicants from contaminated soil using mucus glycoproteins/ and or glycans with intestinal mucosa;
3. Earthworm gut is immunity booster that allow earthworms to live in the "hostile environments" with full contaminants including synthetics nanoparticles-polysaccharide;
4. The presence of glycoproteins and naturally organic materials may play significant role in formation of aggregates or hetero-aggregates of nanoparticles in their body cavity to nullify their toxicity and released less toxic forms into soil system.

REFERENCES

Adriano, D. C. (2003). *Trace elements in the terrestrial environment biogeochemistry, bioavailability and risks of heavy metals*. New York: Springer-Verlag.

Affar, E. B., Dufour, M., Poirier, G. G., & Nadeau, D. (1998). Isolation, purification and partial characterization of chloragocytes from the earthworm species *Lumbricus terrestris. Molecular and Cellular Biochemistry*, *185*(1/2), 123–133. doi:10.1023/A:1006882207581 PMID:9746218

Bengtsson, G., Ek, H., Rundgren, S., & Bengtsson, G. (1992). Evolutionary response of earthworms to long-term metal exposure. *Oikos*, *63*(2), 289–297. doi:10.2307/3545390

Beyer, W. N. (1981). Metals and terrestrial earthworms (Annelida:Oligochaeta). In *Workshop on the role of earthworms in the stabilization of organic residues*. Beech Leaf Press.

Bilej, M., Vetvicka, V., Tuckova, L., Trebichavsky, I., & Koukal,, M. M., & Ima, P. (1990). Phagocytosis of synthetic particles in earthworms: Effect of antigenic stimulation and opsonisation. *Folia Biologica*, *36*, 273–275. PMID:2279582

Blakemore, R. J. (2010). Saga of Herr Hilgendorfs worms. *Zoology in the Middle East*, *49*(sup2), 7–22. doi:10.1080/09397140.2010.10638453

Bolen, N. S., Adriano, D. C., Natesan, R., & Koo, B. J. (2003). Effects of organic amendments on the phyoavalability of chromate in mineral soil. *Journal of Environmental Quality*, *32*(1), 120–128. doi:10.2134/jeq2003.1200 PMID:12549550

Cernansky, S., Kolenck, M., Sevc, J., Unk, M., & Hiller, E. (2009). Fungal votalization of trivalent and pentavalent arsenic under laboratoryconditions. *Bioresource Technology*, *100*(30), 1037–1043. doi:10.1016/j.biortech.2008.07.030 PMID:18774290

Chitrani, B. D., Ghazani, A. A., & Chan, W. C. D. (2006). Determining the size and shape dependence of gold nanoparticles uptake into mammalian cells. *Nano Letters*, *6*(4), 662–668. doi:10.1021/nl052396o PMID:16608261

Cooper, E. L., & Stein, E. A. (1981). *Invertebrate blood cells*. London: Academic Press.

Cossarizza, A., Cooper, E. L., Suzuki, M. M., Salvioli, S., Capri, M., Gri, G., & Franceschi, C. et al. (1996). Earthworm leucocytes that are not phagocytic and cross-react with several human epitopes can kill human tumor cell lines. *Experimental Cell Research*, *224*(1), 174–182. doi:10.1006/excr.1996.0125 PMID:8612683

Dales, R. P., & Kalac, Y. (1992). Phagocytic defense by the earthworm *Eisenia fetida* against certain pathogenic bacteria. *Comparative Biochemistry and Physiology*, *101*(3), 487–490. doi:10.1016/0300-9629(92)90499-G

Davies, N. A., Hodson, M. E., & Black, S. (2002). Changes in toxicity and bioavailability of lead in contaminated soils to the earthworm *Eisenia fetida* (Savigny, 1826) after bone meal amendments to the soil. *Environmental Toxicology and Chemistry*, *21*(12), 2685–2691. doi:10.1002/etc.5620211222 PMID:12463565

Dominguez, J., Velando, A., & Ferreiro, A. (2005). Are *Eisenia fetida* (Savigny, 1926) and *Eisenia andrei* Bouché (1972) (Oligochaeta, Lumbricidae) different biological species?[. *Pedobiologia*, *49*(1), 81–87. http://webs.uvigo.es/avelando/pdfs_archivos/Eisenias.pdf doi:10.1016/j.pedobi.2004.08.005

Edwards, C. A., & Bohlen, P. J. (1996). *Biology and ecology of earthworms* (3rd ed.). London: Chapman & Hall.

Fischer, E., & Molnar, L. (1993). Environmental aspects of the chloragogenous tissue of earthworms. *Soil Biology & Biochemistry*, *24*(12), 1723–1727. doi:10.1016/0038-0717(92)90177-Y

Gottschalk, F., Sonderer, T., Scholz, R. W., & Nowack, B. (2009). Modeled environmental concentrations of engineered nanomaterials (TiO2, ZnO, Ag, CNT, Fullerenes) for different regions. *Environmental Science & Technology*, *43*(24), 9216–9222. doi:10.1021/es9015553 PMID:20000512

Gupta, S., & Kushwah, T. (2015). Exposure of ZnO-NPs enhanced gut-associated microbial activity in *Eisenia foetida*. *Journal of Toxicology and Environmental Health Sciences*, *7*(2), 9–17. doi:10.5897/JTEHS2015.0330

Gupta, S., Kushwah, T., & Yadav, S. (2014). Earthworm coelomocytes as a nanoscavanger to ZnO-NPs. *Nanoscale Research Letters*, *9*(1), 259. doi:10.1186/1556-276X-9-259 PMID:24959107

Haimi, J., & Siira-Pietikainen, A. (1996). Decomposer animal communities in forest soil along heavy metal pollution gradient. *Fresenius' Journal of Analytical Chemistry*, *354*, 672–675. PMID:15067469

Hamed, S. S., Kauchke, E., & Cooper, E. L. (2002). Cytochemical properties of earthworm coelomocytes enriched by Percoll. In New model for analyzing antimicrobial properties with biomedical applications (pp. 29-37). IOS Press.

Hayes, M. H. B. (1983). Darwin's "vegetable mould" and some modern concepts of humus structure and soil aggregation. In J. E. Satchell (Ed.), *Earthworm Ecology: From Darwin to Vermiculture*. London: Chapman & Hall. doi:10.1007/978-94-009-5965-1_3

Helmke, P.A., Robarge, W.P., Korotev, R.L., & Schomber, P.J. (1979).Effects of soil-applied sewage sludge on concentrations of elements in eartworms. *J.Environ. Qual., 8,* 322-327.

Hillyer, J. F., & Albrecht, R. M. (2001). Gastrointestinal persorption and tissue distribution of differently sized colloidal gold nanoparticles. *Journal of Pharmaceutical Sciences, 90*(12), 1927–1936. doi:10.1002/jps.1143 PMID:11745751

Hobbelen, P. H. F., Koolhaas, J. E., & Van Gestel, C. A. M. (2004). Risk assessment of heavy metals for detritivores in floodplain soils in the Biesbosch, the Netherlands, taking bioavailability into account. *EnvironmentPollution, 129,* 409–419. PMID:15016462

Hopkin, S. P. (1989). *Ecophysiology of Metals in Terrestrial Invertebrates*. London: Elsevier.

Ireland, M. P. (1976). Excretion of lead, zinc and calcium by the earthworms by the earthworm *Dendrobaena rubida* living in soil contaminated with zinc and lead. *Soil Biology & Biochemistry, 8*(5), 347–350. doi:10.1016/0038-0717(76)90031-6

Ireland, M. P. (1978). Heavy metal binding properties of earthworm chloragosomes. *Acta Biol.Sci.Hung, 29,* 385–394. PMID:757448

Ireland, M. P. (1979). Metal accumulation by the earthworm *Lumbricus rubellus, Dendrobaena veneta* and *Eiseniella tetraedra* living in heavy metal polluted sites. *Environmental Pollution, 19*(3), 201–206. doi:10.1016/0013-9327(79)90041-7

Janssen, C. (1998). Alternative assays for routine toxicity assessments: a review. In G. Schüürmann & B. Markert (Eds.), *Ecotoxicology: Ecological Fundamentals, Chemical Exposure, and Biological Effects* (pp. 813–839). New York: John Wiley & Sons, Inc. and Spektrum AkademischerVerlag.

Kirchgessner, M. (1978). *Trace element metabolism in man and animals*. Freising, Weihenstephan: Technische Universitat Munchen.

Kozlov, M. V., Haukioja, E. E., & Yarmishko, V. T. (1993). Aerial pollution in kola peninsula. *International Workshop*, St.Petersburg, Russia.

Langdon, C. J., Piearce, T. G., Meharg, A. A., & Emple, K. T. (2001). Survival and behavior of behavior of the earthworms *Lumbricus rubellus* and *Dendrodrilus rubidus* from arsenate-contaminated and non-contaminated sites. *Soil Biology & Biochemistry, 33*(9), 1239–1244. doi:10.1016/S0038-0717(01)00029-3

Lavelle, P. (1983). The structure of earthworm communities. In J. E. Satchell (Ed.), *Earthworm Ecology*. London: Chapman & Hall. doi:10.1007/978-94-009-5965-1_39

Lee, K. E. (1985). *Earthworms - Their Ecology and Relationships with Soils and Land Use*. Sydney: Academic Press.

Lobersli, E. M., & Steinnes, E. (1988). Metal uptake in plants from a birch forest area near a copper smelter in Norway. *Water Air Pollution*, *37*, 25–39.

Lokke, H., & Van Gestel, C. A. M. (1998). *Handbook of soil invertebrate toxicity tests*. Chichester, UK: John Wiley & Sons.

Loureiro, S., Soares, A. M. V. M., & Nogueira, A. J. A. (2005). Terrestrial avoidance behaviour tests as screening tool to assess soil contamination. *Environmental Pollution*, *138*(1), 121–131. doi:10.1016/j.envpol.2005.02.013 PMID:15885863

Luo, Y., Jiang, X., Wu, L., Song, J., Wu, S., Lu, R., & Christie, P. (2003). Accumulation and chemical fractionation of Cu in a paddy soil irrigated with Cu-enriched organic waste materials. *Ecological Bulletins*, *39*, 53–56.

Ma, W. C. (1988). Toxicity of copper to lumbricid earthworms in sandy agricultural soils amended with Cu-enriched organic waste material. *Ecological Bulletins*, *39*, 53–56.

Mabire, F., Audebert, R., & Quivoron, C. (1984). Flocculation properties of some water-soluble cationic copolymers toward silica suspensions: A semi-quantative interpretation of the role of molecular weight and cationicity through a patchwork model. *Journal of Colloid and Interface Science*, *97*(1), 120–136. doi:10.1016/0021-9797(84)90280-7

Mattson, D. J., French, M. G., & French, S. P. (2002). Consumption of earthworms by yellow stone grizzly bears. *Ursus*, *13*, 153–158.

Milochau, A., Lassegues, M., & Valembois, P. (1997). Purificaton,characterization and activities of two hemolytic and antibacterial proteins from coelomic fluid of the annelid *Eisenia fetida andrei*. *Biochimica et Biophysica Acta*, *1337*(1), 123–132. doi:10.1016/S0167-4838(96)00160-4 PMID:9003444

Moragan, J. E., & Morgan, A. J. (1999). The accumulation of metals (Cd,Cu,Pb,Zn and Ca) by two ecologically contrasting earthworm species (*Lumbricus rubellus* and *Aporrectodea caliginosa*): Implications for ecotoxicological testing. *Applied Soil Ecology*, *13*(1), 9–20. doi:10.1016/S0929-1393(99)00012-8

Morgan, A. J., & O'Reilly, M. (2002). The manganese relationships of epigeic and endogeic earthworm species living in soils associated with disused manganese mines.*Proceedings Sixth International Earthworm Ecology Symposium*.

Morgan, J. E., & Morgan, A. J. (1993). Seasonal changes in the tissue-metal (Cd, Zn and Pb) concentrations in two eco-physiologically dissimilar earthworm species: Pollution-monitoring implications. *Environmental Pollution*, *82*(1), 1–17. doi:10.1016/0269-7491(93)90155-H PMID:15091792

Naidu, R., Bolan, N., Megharaj, M., Juhanz, A., Lombi, E., & Smith, E. (2010).Human and ecological risk assessment of contaminated sites-Key knowledge gaps. *19th Congress of Soil Science. Soil Solutions for a changing World.*

Nakashima, T., Okada, T., Asahi, J., Yamashita, A., Kawa, K., Kasai, H., & Hirano, T. et al. (2008). 8-Hydroxydeoxyguanosine generated in the earthworm *Eisenia fetida* grown in metal-containing soil. *Mutation Research, 654*(2), 138–144. doi:10.1016/j.mrgentox.2008.05.011 PMID:18585474

Nieboer, E., & Richardson, D. H. S. (1980). The replacement of the nondescript term "heavy metals" by a biologically and chemically significant classification of metal ions. *Environmental Pollution*, 3–26.

Oberdorster, G., Oberdorster, E., & Oberdorster, J. (2005). Nanotoxicology: An emerging discipline evolving from studies of ultrafine particles. *Environmental Health Perspectives, 113*(7), 825–839. doi:10.1289/ehp.7339 PMID:16002369

Olaniran, A. O., Balgobind, A., & Pillay, B. (2013). Bioavailability of heavy metals in soil impact on microbial biodegradati on of organic compounds and possible improvement strategies. *International Journal of Molecular Sciences, 14*(5), 10197–102228. doi:10.3390/ijms140510197 PMID:23676353

Parry, M. J. (1975). Evidence of mitotic divisions of coelomocytes in the normal, wounded and grafted earthworm *Eisenia fetida. Experimentia, 32*(4), 449–451. doi:10.1007/BF01920788

Pathirana, S., Connell, D., & Vowles, P. (1994). Distribution of polycylic aromatic hydrocarbons (PAHs) in an urban roadway system. *Ecotoxicology and Environmental Safety, 28*(3), 256–269. doi:10.1006/eesa.1994.1051 PMID:7525221

Patwa, A., Thiery, A., Fabien, L., & Martin, L. K.S., Boisset Claire, Bramard, Jean-Francois, Bottero Jean-Yves, &Barthelemy Philippe. (2015). Accumulation of nanoparticles in jellyfish mucus: A bio-inspired route to decontamination of nano-waste. *Scientific Reports*. doi:10.1038/screp11138

Piearce, T. G. (1978). The calcium relations of selected Lumbricidae. *Journal of Animal Ecology, 41*(1), 167–188. doi:10.2307/3511

Pimentel, D., Wilson, C., McCullum, C., Huang, R., Dwen, P., Flack, J., & Cliff, B. et al. (1997). Economic and Environmental Benefits of Biodiversity. *Bioscience, 47*(11), 747–757. doi:10.2307/1313097

Prento, P. (1979). Metal and phosphate in the chloragosomes of *Lumbricus terrestris* and their possible physiological significance. *Cell and Tissue Research, 196*, 123–134. doi:10.1007/BF00236353 PMID:421244

Ranzelli-Cain, R., & Kaloustian, K. V. (1995). Evidence for the involvement of opioid peptides in phagocytosis, conformation, granulation and aggregation of immunocompetent *Lumbricus terrestris* amoebocytes. *Comparative Biochemistry and Physiology, 111*, 205–211. PMID:8521243

Reincke, A. J., Reinecke, S. A., & Maboeta, M. S. (2001). Cocoon production and viability as endpoints in toxicity testing of heavy metals with three earthworm species. *Pedobiologia, 45*(1), 61–68. doi:10.1078/0031-4056-00068

Sheppard, S. (1997). Toxicity testing using microcosms. In Soil Toxicology (pp. 345-374). Lewis Publishers.

Sims, R. W., & Gerard, B. M. (1999). *Earthworms: Notes for the identification of British species* (4th ed.). Shrewsbury, UK: Published for The Linnean Society of London and The Estuarine and Coastal Sciences Association by Field Studies Council, Montford Bridge.

Sinha, R. K., Agarwal, S., Chauhan, K., & Valani, D. (2010). The wonders of earthworms &its vermicompost to farm production:Charles Darwins friends of farmers with potential to replace destructive chemical fertilizers. *Agricultural Sciences*, *1*(02), 76–94. doi:10.4236/as.2010.12011

Spurgeon, D. J., & Hopkin, S. P. (1999). Tolerance to zinc in populations of earthworm Lumbricusrubellus from uncontaminated and metal- contaminated ecosystem. *Archives of Environmental Contamination and Toxicology*, *37*(3), 332–337. doi:10.1007/s002449900522 PMID:10473789

Stein, E. A., & Cooper, E. L. (1981). Cytochemical observations of coelomocytes the earthworm *Lumbricusterrestris*. *Developmental and Comparative Immunology*, *5*, 15–25. PMID:7009243

Stephenson, G. L., Kaushik, A., Kaushik, N. K., Solmon, K. R., Steele, T., & Scroggins, R. F. (1998). Use of an avoidance-response test to assess the toxicity of contaminated soils to earthworms. In Advances in earthworm toxicology (pp. 67-81). Setac Press.

Ulrich, S., Seijo, M., Laguecir, A., & Stoll, S. (2006). Nanoparticle adsorption on a week polyelectrolyte. Stiffness, pH, charge mobility and ionic concentration effects investigated by Monte Carlo simulations. *Journal of Physical Chemistry*, *110*(42), 20954–20964. doi:10.1021/jp063671d PMID:17048913

Unrine, J. M., Hunyadi, S. E., Tsyusko, O. V., Rao, W., Shoults-Wilson, W. A., & Bertsch, P. M. (2010). Evidence for bioavailability of an nanoparticles from soil and biodistribution with in earthworms (*Eisenia fetida*). *Environmental Science & Technology*, *44*(21), 8308–8313. doi:10.1021/es101885w PMID:20879765

Valembois, P., Roch, P., & Lassegues, M. (1988). Evidence of plasma clotting system in earthworms. *Journal of Invertebrate Pathology*, *51*(3), 221–228. doi:10.1016/0022-2011(88)90029-8

Valembois, P., Roch, P., Lassegues, M., & Cassand, P. (1982). Antibacterial activity of the hemolytic system from the earthworm *Eisenia fetida andrei*. *Journal of Invertebrate Pathology*, *40*(1), 21–27. doi:10.1016/0022-2011(82)90032-5

Van Gestel, C. A. M., Leon, C. D., & Van Straalen, N. M. (1997). Evaluation of soil fauna ecotoxicity tests regarding their use in risk assessment. In J. Tarradellas, G. Bitton, & D. Rossel (Eds.), *Soil Ecotoxicology* (pp. 291–318). New York: Lewis Publishers, New Publishers.

Ville, P., Roch, P., Cooper, E., Masson, P., & Narrbomnne, J. (1995). PCBs increase molecular-related activities (lysosome, antibacterial, hemolysis,proteases) but inhibit macrophage-related functions (phagocytosis, wound healing) in earthworms. *Journal of Invertebrate Pathology*, *65*(3), 217–224. doi:10.1006/jipa.1995.1033 PMID:7745278

Wielgus-Serafinska, E., & Kawa, E. (1976). Accumulation and localization of lead in *Eisenia fetida* (Oligochaeta) tissues. *Folia Histochemica et Cytochemica*, *14*, 315–320. PMID:1030682

Wuana, A., Raymond, O., & Felix, E. (2014). Mixed interactions in soil: Implications for bioavailability,risk assessment and remediation. *African J. Environmental Science & Technology*, *8*(12), 691–708.

Yadav, S. (2014). Glycoprotein of coelomic fluid triggers formation of aggregates/hetero-aggregates of ZnO-NPs in Earthworms. *Int. J. Curr. Microbiol. App. Sci.*, *3*(11), 1040–1045.

Yadav, S. (2016a). *Earthworm Biology: Recent Advances in techniques to study earthworms*. Lambert Academic Publishing.

Yadav, S. (2016b). Screening of Immuno-competent coelomic cells in earthworms. *International J of Sciences*, *5*(4), 43–51. doi:10.18483/ijSci.999

Yaron, B., Calvet, R., & Prost, R. (1996). *Soil pollution*. Berlin: Springer-Verlag. doi:10.1007/978-3-642-61147-6

Zeng, Z., Patel, J., Lee, S.-H., McCallum, M., Tyagi, A., Yan, M., & Shea, K. J. (2015). Synthetic polymer nanoparticle-polysaccharide interactions: A systematic study. *Journal of the American Chemical Society*, *134*(5), 2681–2690. doi:10.1021/ja209959t PMID:22229911

KEY TERMS AND DEFINITIONS

Aggregation: Collection or gathering of particles.

Bioaugmentation: Addition of living cells capable of degradation.

Biodegradation: Disintegration of materials by biological means.

Bioflocculation: Clumping together of fine, dispersed organic particles by the action of organisms.

Bioremediation: Biological process to breakdown contaminants and biosorption in the presence of optimum environmental conditions and sufficient nutrients.

Biostimulation: It involves the modification of the environment to stimulate organism those are capable of bioremediation.

Biotransformation: It is chemical alternation of chemicals or substrates by biological system.

Chloragogen Cells: These are excretory cells similar to the liver in vertebrates. They store glycogen and neutralize toxins, are yellowish in color due to the presence of yellow granules called chloragosomes.

Earthworm Coelomic Cells: Theseare non-movable scavenger and are of two types, amebocytes (mainly immune function) and eleocytes (mainly nutritive function exists in various functional states and stages of maturation).

Nanoparticles: Particles between 1 and 100 nanometers in size.

Nanotoxicology: The study of the toxicity of nanomaterials.

Particle Agglomeration: This refers to formation of assemblages in a suspension and, particles dispersed in the liquid phase stick to each other, and spontaneously form irregular particle clusters, flocs or aggregates.

Chapter 15
Bioremediation of Agricultural, Municipal, and Industrial Wastes

Shivani Garg
Kurukshetra University, India

ABSTRACT

Growth of agriculture and manufacturing industries has resulted in increased a wide range of complex and hazardous compounds to the environment. Excess growth of hazardous waste has led to reduce availability of clean water and disturbances of soil thus limiting crop production. Waste generated from different sources like Industrial, domestic and agricultural etc. having different kinds of chemical compound i.e. organic or inorganic. Traditional methods are not able to deal with some of these chemical compounds. Bioremediation process is good option in such environmental problems. Bioremediation provides a technique for cleaning up pollution by enhancing the natural biodegradation processes. It treats such waste with the help of microorganism. Number of microbes including aerobes, anaerobic and fungi are involved in bioremediation process. Specific types of microbes are used to treat specific type of chemical contaminant. The chapter include all the techniques of bioremediation used to treat different kinds of contaminant.

INTRODUCTION

Ecosystems are threatened with natural environmental changes and disturbances that remove biomass from a community, such as fire, flood, drought, or predation over time and geographic space. (Levin, 1992) These disturbances needs new techniques emerge out of the patchwork of natural experimentation and opportunity implying a good measure of ecological resilience are a cornerstone theory in an ecosystem (Folke et al., 2004). The most common types of waste can be categorized into four types: agricultural, industrial, municipal and nuclear (Alloway, 1995). Agricultural waste, have both natural and non-natural wastes, is a general term used to describe waste produced from the fields through various activities. Agricultural wastes are the by-products of various agricultural activities such as crop production, harvesting of crops, saw milling, agro-industrial processing, and others. In India sugar industry alone produces about 90 MT of bagasse per year and being used in the manufacturing of insulation boards, wall panels, printing paper and corrugating medium (Sengupta, 2002). Generally, agricultural waste is produced in two forms- solid and liquid waste. The food harvesting and production industry generate crop residuals,

DOI: 10.4018/978-1-5225-2325-3.ch015

as well as pre- and post- consumer food wastes. Some of the farmer burn agro- wastes after harvesting time; this causes a huge amount of air and land pollution. Municipal waste (MSW) composition can vary considerably with places and time depending on many factors, including socio-economic, climatic conditions, living standards, waste collection, and disposal methods, sampling and sorting procedures. MSW is heterogeneous in nature and have a number of different materials derived from various types of activities. Municipal waste generated from residential and domestic sources. This type of waste is commonly called trash or garbage and includes everyday items, spoiled food, broken things, or simply any item a person no longer needs or wants. The most commonly disposed of items as municipal solid waste are paper, yard trimmings, food, plastics, metals, rubbers, and textiles. In recent years, the amount of electronic waste, also known as e-waste, has increased drastically as people become more dependent on electronics, such as cell phones and computers that are replaced and disposed of frequently. Based on the definition of municipal solid waste, the waste that you dispose of everyday would fall into this category. Municipal solid waste categorized into two types: garbage and rubbish. Garbage or food waste is the animal and vegetable residue resulting from the preparation, cooking and serving of food. These wastes contain organic matter and moisture. Kitchens, restaurants, and markets are sources of garbage. Rubbish consists of old tin cans, newspaper, tires, packaging materials, bottles, plastics etc. Both combustible and non-combustible solid wastes are included in this category but do not include garbage. Trash is the combustible portion of rubbish. The rise in the number of industries producing glass, textile, leather, plastic and metal products food, and electronics, has significantly contributed to waste production. Rapid industrialization has resulted in the generation of a huge quantity of wastes, both solid and liquid, in industrial sectors such as fruit and food processing, pulp and paper, sugar, sago / starch, distilleries, dairies, tanneries, poultries, etc. These wastes are generally dumped on land or discharged into water bodies, without adequate treatment, so pollute environment and cause health hazard. In a broad sense, industrial wastes could be classified into two types i.e. Hazardous wastes and non- hazardous wastes (Table 1). Hazardous industrial wastes in India can be categorized broadly into two categories: Hazardous wastes generated from various industries in India and Hazardous industrial wastes imported into

Table 1. Different types of industrial waste classified on the basis of their composition

	Hazardous Waste Component	Sources
Heavy Metals	Arsenic	Mining, non-anthropogenic geo-chemical formation
	Cadmium	Mining, fertilizer industry, battery waste
	Chromium	Mining areas, Tanneries
	Lead	Lead acid battery smelters
	Manganese	Mining areas
	Mercury	Chloro-alkali industries, healthcare institutes
	Nickel	Mining, metal refining
Hydrocarbons	Benzene	Petrochemical industries, solvents
	Vinyl chloride	Plastics
Pesticides		Insecticides
Organic chemicals	Dioxins	Waste incineration, herbicides
	PCBs	Fluorescent lights, e-waste, Hydraulic fluid

India from Western Countries for re-processing and recycling. Hazardous waste, in particular, includes products that are explosive, flammable, irritant, harmful, toxic, carcinogenic, corrosive, infectious, or toxic to reproduction. The other one is Non-hazardous or ordinary industrial waste that generated by industrial or commercial activities but is similar to household waste by its nature and composition. It is not toxic, hazardous and requires no special treatment. In particular, it includes ordinary waste produced by companies, shopkeepers and tradespeople (paper, cardboard, wood, textiles, pack, etc.). Due to its non-hazardous nature, this waste is often sorted and treated in the same facilities as household waste.

Bioremediation involves use of microbes to detoxify and degrade environmental pollutants. Recently, research is being focused in the development of genetically modified microbes or consortia for the detoxification of environmental pollutants. Microbes degrade the complex substances to simpler one that can be reuse or recycle in environmental processes. Microbes involved in the degradation naturally are aerobic or anaerobic bacteria, fungi or algae. Most of the techniques in bioremediation are aerobic in nature, but anaerobic processes are also being developed to help degrade pollutants in oxygen deficit areas (Franchi et al., 2016). In bioremediation processes, microorganisms use the contaminants as nutrient or energy sources (Hess *et al.*1997; Agarwal, 1998; Tang et al., 2007). Bioremediation is a process by which microorganisms are stimulated to rapidly degrade hazardous organic pollutants to environmentally safe levels in soils, sediments, substances, materials and ground water. Environmental parameters should be favourable for microorganisms to grow and remediate the pollutants at a rapid rate (de la Cueva et al., 2016). Bioremediation is currently being used to reduce the organic chemical waste content of ground water, soils, effluent from food industry and chemical plants, sludge from petroleum refineries and oil spills in the ocean. Bioremediation techniques fall into two categories — in situ and ex situ bioremediation. Using bioremediation technology TERI, India, has treated 48,914 tons of waste oily sludge in batches at various oil refineries in India (Mandal et al., 2012). Bioremediation is concerned with the biological restoration and rehabilitation of historically contaminated sites and with the clean-up of areas contaminated in more recent times, either accidentally or incidentally, as a result of the production, storage, transport, and use of organic and inorganic chemicals. The uses of plants can improve bioremediation rates as plants provide habitats to microbes; improve soil porosity and help in cycling of micro nutrients with their microbial metabolism. In return, microorganisms improve plant growth by reducing soil toxicity, producing plant growth promoting metabolites, provide plant nutrients by sequestration from soil etc. (Gkorezis et al., 2016)Some examples of bioremediation technologies are phytoremediation, bioleaching, bioventing, bioreactor, land farming, composting, bio augmentation, rhizofiltration, and biostimulation. The biological approach of remediating the environment with the help of bioremediation, defined as *'the process that use biological agents to reclaim waters and soils contaminated by hazardous substances to human health and the environment'* (Atlas and Pramer, 1990). Bioremediation is an option that offers the possibility to destroy or render harmless various contaminants using natural biological activity. Industrial waste contains much kind of organic and inorganic components. Wastes produced from different sources may be in solid, liquid or gaseous form. Treatment of different types of wastes is done with the help of processes listed in Figure 1. All the techniques require specific type of conditions such as, microbial profile, energy source and electron acceptor, moisture and pH, nutrients and temperature, absence of toxicity, removal of metabolites, absence of competitive microorganisms, etc. for digestion and degradation. Some of the processes are explained below which are used for the treatment of agricultural, municipal or industrial waste. The main objective of this chapter to study and find out the different bioremediation processes which is effective in dealing with agricultural, municipal and industrial waste disposed or present in the environment and cannot be degraded by conventional processes or naturally without harming the environment.

Figure 1. Different strategies used for management of wastes with bioremediation

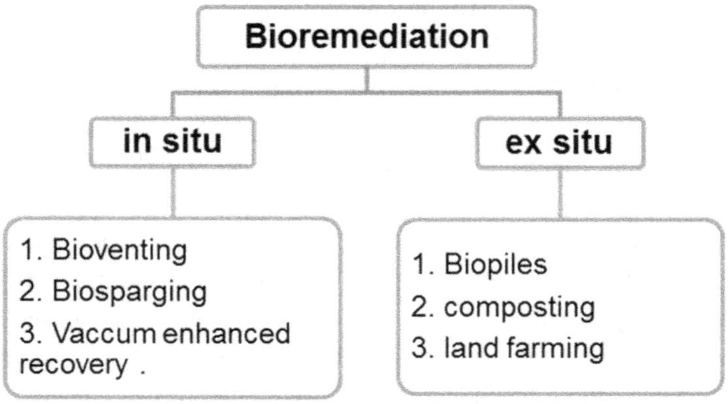

BIOREMEDIATION OF AGRICULTURAL WASTE

There are two main methods of bio remediating agricultural waste. The first method is anaerobic digestion. The other method is composting, an aerobic process where microorganisms break down the organic material in the waste when air and water are added to the waste product. Following processes are involved in the management of agricultural waste with the help of bioremediation process which are commonly used:

- Anaerobic digestion,
- Composting,
- Vermicomposting,
- Biogas production,
- Biopile,
- Land farming, etc.

Digester systems consist of a vessel in which the waste is heated to a certain temperature, to allow the microorganisms to consume at an efficient rate. The method in which this is done varies widely, from using simple composting piles to composting vessels. Yet another option is to combine one or more composting methods to produce the desired composition of compost. All the techniques are used for bio remediating agricultural waste explained in detail below.

Anaerobic Digestion

In this process about 95% of biodegradable carbon is decomposed into biogas and the rest 5% into biomass. Three main steps are involved in the breakdown of organic waste under anaerobic conditions. These steps essentially hydrolysis of biopolymers to monomers, fermentation of monomers to volatile acids and Methanogenesis. The microbial species responsible for anaerobic degradation are *Actinomyces, Arthrobacter, Citrobacter, Escherichia, Lactobacillus, Micrococcus* etc. Anaerobic treatment is a biological process carried out in the absence oxygen for the stabilization of organic materials by conversion into methane and inorganic end-products such as CO_2 and NH_3. Anaerobic process is of two types: anaerobic

Figure 2. Process of anaerobic digestion of organic agricultural waste

$$\text{Organic waste} \xrightarrow[\text{Steps}]{\text{Hydrolysis}} \text{Organic acids, Alcohol} \xrightarrow[35^{o}C,\ pH,\ 5\text{-}6]{\text{Acetogenesis}} \text{Acetic Acid, } CO_2,$$

$$H_2 \xrightarrow[45^{o}C,\ pH,\ 6\text{-}8]{\text{Methanogenesis}} CH_4,\ CO_2$$

fermentation and anaerobic respiration. In anaerobic fermentation, the product obtained by accepting the electrons released during breakdown of organic matter and ethanol is also produced along with energy. Anaerobic respiration is used mainly by prokaryotes that live in environments devoid of oxygen. In this process, energy is used by the reaction for degradation of organic material. General process of anaerobic digestion of organic agricultural waste is as follows

There are four different types of anaerobic wastewater treatment bioreactors which are anaerobic filter reactor, anaerobic contact process reactor, fluidized bed reactor and up flow anaerobic sludge blanket (UASB). Most commonly used bioreactor is up flow anaerobic sludge blanket (UASB).

Up Flow Anaerobic Sludge Blanket Reactor (UASB)

UASB reactors can treat all type of high-strength wastewater. It can be used at large-scale (e.g. agro-industrial wastes) or as decentralised treatment systems for domestic wastewaters; yet domestic treatment is still relatively new and not always successful as domestic wastewater generally lower strength. Anaerobic treatment systems such as UASB do not require an energy consuming aeration system and can be built simply than aerobic process. This process converts the organic matter into biogas, which can be recovered. The nutrient-rich effluent can be used for agricultural irrigation. Sludge, even is partly stabilised (mineralised) and can be used as an organic soil fertiliser after composting or drying. The UASB reactor is a methanogens (methane-producing) digester that evolved from the clarifier. UASB uses an anaerobic process whilst forming a blanket of granular sludge which suspends in the tank. Wastewater flows upwards through the blanket and is processed (degraded) by microorganisms. The upward flow combined with the settling action of gravity suspends the blanket with the aid of flocculants. The blanket begins to reach maturity at around 3 months. In the absence of any support matrix, the flow conditions create a selective environment in which only those microorganisms, capable of attaching to each other, survive and proliferate (Figure 2).

Eventually the aggregates form into dense compact biofilms referred to as "granules". The government of India has made a major commitment to anaerobic treatment technology in its national river basin improvement program. As of 1996, 13 new anaerobic treatment plants, with an aggregate treatment capacity of over 306 MLD are under construction in India. Anaerobic waste water treatment is the most popular method, cheaper for treatment of high strength waste water; produce energy is in the form of biogas and required very small space for installation than aerobic treatment. Biogas produced form this technique can be used for domestic or commercial purposes in household industries. This technique usually used for the treatment of liquid waste i.e. wastewater generated from agricultural land or domestic area dependent on agriculture for their occupation, sewage waste and industrial effluent.

Figure 3. General model of up flow anaerobic sludge bioreactor for treatment of wastewater

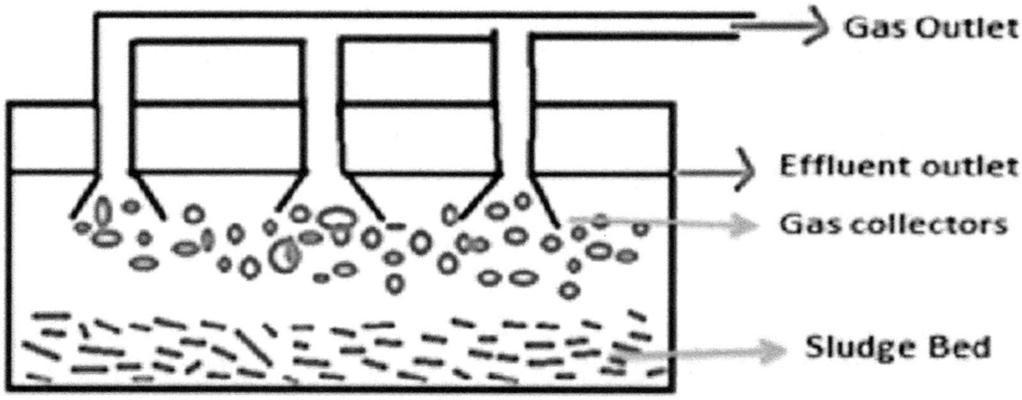

It could be recognised by some newly developed European waste treatment systems that some microbial degradation or transformation occur only in the absence of oxygen. Under anaerobic conditions, microorganisms growing on wastes can produce valuable fuels. The dry anaerobic composting process converts the organic fraction of biodegradable organic waste and refuse into energy in the form of biogas (methane and carbon dioxide) and humus- like material (Figure 3).

Composting

Composting of manure is a valuable management tool that increases the flexibility for land application and storage, while significantly reducing the potential for contamination to soils and water that might occur if raw manure were to be improperly stored or land spread. Composting of carcasses should be restricted to poultry and small animals. Composting systems should be located at least 90 metres from a water course or a water body. Proper management of the composting system is required to ensure optimum conditions are maintained for decomposition. It is a self-heating, substrate-dense, managed

Figure 4. Natural method of Biogas production and uses of biogas in electricity production

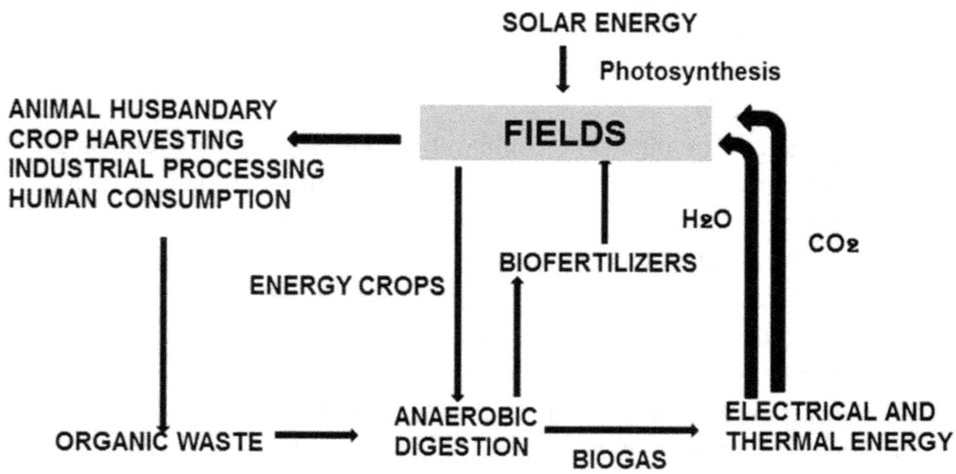

microbial system and one solid phase biological treatment technology which are suitable to treatment of large amount of contaminated solid materials. It is a biological process where fresh organic wastes are allowed to be decomposed into humus like substance. The process is conducted by a complete automatic system which consists of several steps: dumping of refuse into a container, metallic particles removed by magnetic separator or mechanically, and then material is transferred to wet condition to a rotatory cylinder, analogous to a rotatory drier. The cylinder rotates slowly on large tyres and the wastes move from one end to the other. They are thoroughly mixed and pulverized by abrasion. Air is introduced at low pressure throughout the length of the cylinder. The hazardous compound reported to disappear through composting include aliphatic and aromatic hydrocarbons and certain halogenated compound. Composting can be done in- open system i.e. land treatment and closed system. The open land system is inexpensive treatment method but the temperature fluctuates from summer to winter. Therefore, rate of biodegradation of waste material declines. Secondly, land treatment system may become oxygen limited, depending on amount of substrates, depth of waste application etc. However efficiency of open treatment system can be increased by passing air. This approach is referred to as engineered soil piles and forced aeration treatment. Closed treatment system is preferred over the open land treatment system because controlled air is supplied to maintain the microbial activity. As a result of microbial growth and volatilization of hazardous compounds, increase internal temperature gradually. Therefore, use of blower for air circulation and exhaust for removal of toxic volatile are set up in closed treatment system. Ventilators supply O_2 and remove heat through evaporation of water. The main products of the biogas plants are fuel gas and organic manure (Figure 4). Biogas is a flammable gas. Composition of biogas is methane, carbon dioxide, hydrogen, nitrogen and hydrogen sulphide.

Figure 5. Flowchart of composting treatment sequence
Hogan,1998.

In several parts of India, compost plants are in operation for example, in Ahmedabad, Kolkata, Mumbai, Chennai, Delhi, Pune and Hyderabad. Compost has been used by Indian Agricultural Research Institute (IARI) to produce blue green algae- coated granulated compost.

Vermicomposting

Vermiculture or use of earthworm in biotechnology, where the earthworm feed on and degrade a variety of organic waste, eliminate toxic elements and convert the waste into vermicompost which is high- grade, nutrient- rich compost and a very significant bio fertilizer and soil conditioners. Vermicomposting enhances soil productivity and increases crop yield. Over 3000 species of earthworm have been identified from different parts of the world out of which 385 species have been found in India. The useful species found in India are *Lampito mauritii, Dichogaster boloui, Drawida willsi I etc.* The most important effect of earthworms in agro-ecosystems is the increase in nutrient cycling, particularly nitrogen. Lorimor et al. (2001) concluded that, in comparison with conventional composting system (as commonly used in India to manage crop residues and other related wastes), the vermicomposting often results in mass reduction, in shorter time for processing, and produce more humus with reduced phytotoxicity in ready material, i.e., vermicompost (Figure 5).

The crop residues amended with cattle dung in different ratios in order to produce three feed mixtures. The mixed substrate material served as bedding as well as feed for the earthworms. Earthworms alter the chemical properties of the substrates and indirectly promote microbial propagation and colonization, resulting in higher mineralization rates in vermicomposting beds than general composting (Suthar, 2008). It was concluded that the vermicomposting process accelerates the mineralization rate and subsequently enriches the end product with more available forms of plant nutrients. Earthworms keep on turning the material through different layers and decompose the organic material. It takes about 45 days to make vermicompost, which is a black granular material.

Figure 6. General model of vermicomposting pit for treatment of municipal waste

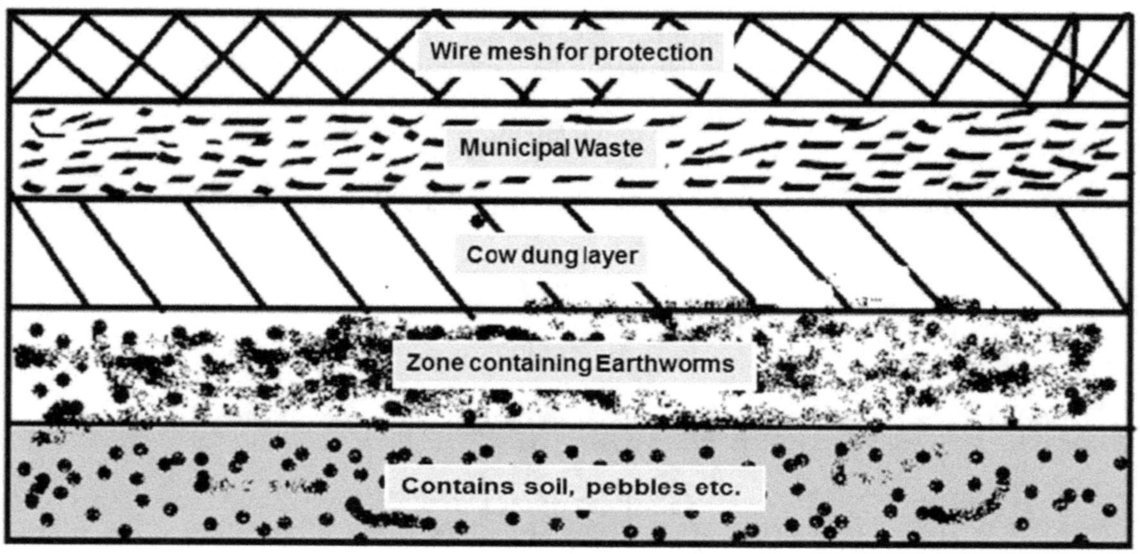

Biopiling

It is a hybrid form of composting and land farming. Bio pile treatment is also known as bio cells/ bio heaps/ bio mounds/ compost piles which are used to reduce the concentration of petroleum hydrocarbon constituents in soil through the use of biodegradation. This technology involves heaping the contaminants soil into piles and stimulates the aerobic microbial activity within soil through aeration and addition of mineral nutrients and moisture. Soil bio piles are a process where piles of soil are placed over top of a bug vacuum pump. The vacuum pump pulls air through the pile of soil to allow oxygen to get though the soil to the microorganisms. Contaminants turn into gas forms is easily recovered. The basic bio pile system includes a treatment bed, an aeration system, an irrigation/nutrient system and a leach ate collection system. For proper treatment, there should be control of moisture, heat, nutrients, oxygen, and pH. The irrigation system is buried under the soil and provides air high temperature. It is mainly used for the removal of gasoline, oil, petroleum etc. The rate removal of these substances is varied from one site to another site. Bio pile treatment is a full-scale technology in which excavated soils are mixed with soil amendments, placed on a treatment area, and bio remediated using forced aeration. The contaminants are reduced to carbon dioxide and water. Soil piles can be up to 20 feet high and may be covered with plastic to control runoff, evaporation and volatilization, and to promote solar heating. If volatile organic compounds (VOCs) in the soil volatilize into the air stream, the air leaving the soil may be treated before they are discharged into the atmosphere. Treatment time is typically 3 to 6 months.

Landfarming

Land farming is a bioremediation treatment process that is performed in the upper soil zone or in bio treatment cells. In this technique the waste is carefully mixed with surface soil on a suitable tract of land. Microbes that can metabolize the waste may be added, along with nutrients. The soil is spread out over a pad which has a system that is used to collect any of the residual liquids from the soil which may be toxic. The soil is mixed regularly to allow aeration of excavated soil. So that the micro-organisms present in the soil are more able to break down the contaminants in the soil. Contaminated soils, sediments, or sludge are incorporated into the soil surface and periodically turned over or tilled to aerate the mixture. In land forming make a sandwich layer of excavated soil between a clean soil and a clay and concrete. Firstly, the clean soil at bottom and uppermost layers should be concrete layer. After this allow it for natural degradation. Proper maintenance of oxygen, nutrition and moisture and pH should also require with near the pH 7 by using lime. Land forming is useful mainly for pesticides. Biological treatment of certain organic wastes, such as those from the petroleum industry, is also an option (Figure 6).

BIOREMEDIATION OF MUNICIPAL WASTE

According to Ministry of Urban Development (MoUD, 2000), Government of India 1 lakh MT of Municipal Solid Waste was generated daily in the country (Figure 7, Table 2). During the year 2004-05, Central Pollution Control Board (CPCB) through National Environmental Engineering Research Institute (NEERI), Nagpur, A survey conducted in 59 cities and estimated that 39,031 Tons per day MSW generated in these cities. The survey conducted by the central institute of Plastics Engineering and technology (CIPET) at the instance of CPCB has reported generation of 50,592 tonnes of MSW per day in the year

Figure 7. On the basis of nature, different types of municipal solid waste

2010-11 in same 59 cities (Keisham and Paul, 2015). Annually, Asia alone generates 4.4 billion tonnes of solid wastes and MSW comprise 790 million tonnes (MT) of which about 48 (6%) MT is generated in India (Yoshizawa, Tanaka, Shekdar, 2004). Despite the best efforts to reduce, reuse and recycle, there will always be residual waste requiring disposal. The alternative treatment and disposal technologies are:

1. Composting,
2. Anaerobic digestion,
3. Pelletisation / Refuse Derived Fuel (RDF),
4. Landfills - Sanitary Landfill / Bioreactor landfill / Secured landfill (for inert waste),
5. Incineration include gasification and pyrolysis.

Among the various treatment and disposal processes as mentioned above, MoEF (Ministry of Environment and Forest) has notified composting (windrow composting, vermicomposting), anaerobic digestion, incineration, pelletization and landfill technologies and has given relevant standards for compost quality, leachate disposal, incineration operations and emissions and landfill specifications.

Bioremediation process is a natural process so MSW management include composting, anaerobic digestion, and landfill and palletisation process. These all the techniques are explained in above part as bioremediation of agricultural wastes excluding palletisation and incineration but these both the process are chemical based process not natural and bioremediation process is a natural phenomenon.

Table 2. Physical characteristics of municipal solid waste in Indian cities

Population Range (in Millions)	No. of Cities Surveyed	Paper	Rubber, Leather, Synthetics	Glass	Metal	Total Compostable Matter	Inert Material
0.1- 0.5	12	2.91	0.78	0.56	0.33	44.57	43.59
0.5- 1.0	15	2.95	0.73	0.56	0.32	40.04	48.38
1.0-2.0	09	4.71	0.71	0.48	0.49	38.95	44.73
2.0-5.0	03	3.78	0.48	0.48	0.59	56.67	40.07
>5.0	04	6.43	0.28	0.94	0.80	30.84	53.90

Source: Sharholy et al., 2008.

BIOREMEDIATION OF INDUSTRIAL WASTE

The degradation and bioremediation of industrial wastes are a challenging task because there is no reliable technology till date which is effective in complete removal/reduction of these pollutants. Ultimately, diverse groups of microorganism that are already present in the nature may provide solution for the degradation and bioremediation of toxic industrial wastes. Microorganisms that are being used for the bioremediation and transformation of pollutants from long times (Okpokwasili, 2007). For the management of industrial waste, bioremediation processes are classified according to composition of waste to be managed.

Bioremediation of Hydrocarbons or Toxic Organic Pollutant

Petroleum and its products are hydrocarbons. Oil constitutes a variety of hydrocarbons like xylenes, naphthalene, octane's, camphor etc. The pollutants can be degraded by a consortium of microorganisms, e.g. *Corynebacterium, Pseudomoas, Mycobacterium, Arthrobacter,* and *Nocardia*. Genetically engineered bacterial strains are used to enhance bioremediation.In 1990, the USA government allowed him to use this superbug for cleaning up of an oil spill in water of State of Texas (Figure 8). It was produced on a large scale in laboratory, mixed with straw and dried.

Bioremediation of Pesticides

Effluent discharge from pesticide manufacturing or formulating industry and agricultural runoff are major sources of pesticides in the environment and the wastewater discharged from these industries cause a high hazard for waterbodies (Rajeswari and Kanmani, 2009). Excess fertilizer, herbicides and pesticides when washed by rain into rivers causes serious danger to life. However, improper handling and disposal may cause pollution. The immediate consequence is drop in dissolved oxygen and the Biological Oxygen Demand (BOD). Water pollution is also caused by hot water discharge from cooling

Figure 8. Process involved in bioremediation of hydrocarbons

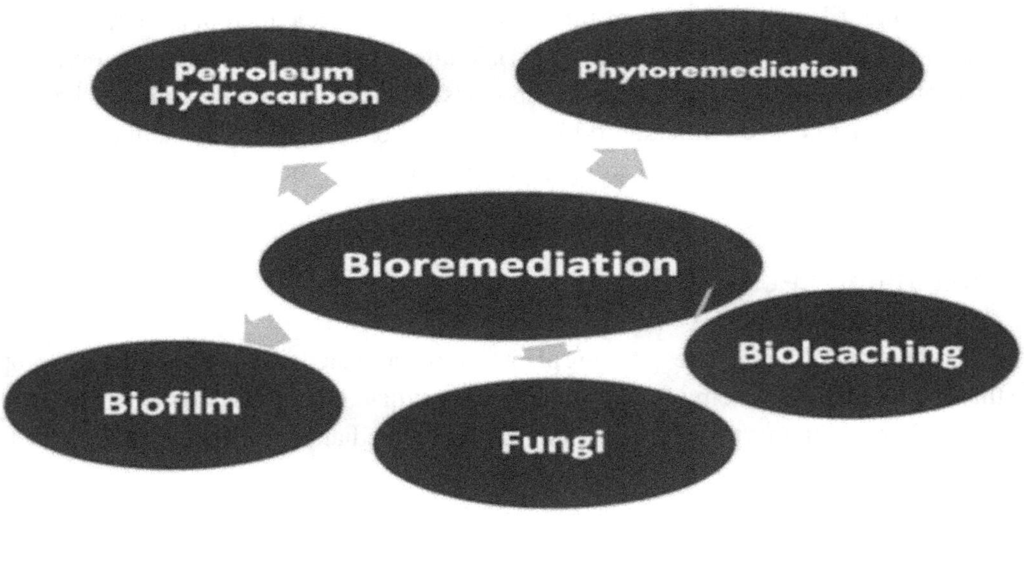

engines in the industries. This increases water temperature and lowers the metabolic rate of organisms. The constituents of gases that are produced in wastewater are carbon dioxide (CO_2), carbon monoxide (CO), nitrogen dioxide (NO_2) and sulphur dioxide (SO_2). There are numerous example of bioaccumulation of different organosynthetic insecticides and pesticides due to their uncontrolled use in the past decades. Bioremediation of pesticides from wastewater can be enhanced by the use of lignolytic enzymes such as Laccases. Removal of phenolic or xenobiotic pollutants from wastewater using Laccase can be occured in several ways: 1) using purified immobilised enzyme; 2) using purified free enzyme; 3) using enzyme obtained directly from culture broths; and 4) bioremediation in reactors with immobilized or free cells (Majeau, Satinder and Tyagi, 2010). Advanced oxidation processes (AOPs) in which hydroxyl radical (OH–) is produced by different reactions like chemical, photochemical, photocatalytic or electrochemical reactions, are very effective for treatment of wastewaters containing highly recalcitrant organic pollutants (Gogate and Pandit, 2004). Rajeswari and Kanmani (2009) used TiO_2 for photocatalytic oxidation of pesticide wastewater. The degradation revealed 99% removal of active ingredient of pesticide and 76% mineralization. Complete mineralization could be achieved by further increasing the reaction time or by adding some additives, which would speed up reaction.

Microbial Degradation of Pesticides

It has been observed that a significant portion of pesticides residues in the environment is degraded in the soil and aquatic environment by microbes. The majority of pesticides can be degraded by fungi and bacteria. The transportation of parent pesticide molecules to less complex products is often achieved by concerted efforts of microbial communities. It is estimated that 1 g of soil contains more than 5000–7000 different species of bacteria and more than ten thousand fungal colonies (Dindal 1990; Melling 1993). The use of microbial metabolic potential for eliminating soil pollutants offers a safe and inexpensive substitute to other commonly used physico-chemical strategies (Vidali, 2001). Indigenous microorganisms (natural attenuation) can be used for detoxification of contaminants in the environment (Siddique, Okeke, Arshad and Frankenberger 2003; Kumar and Philip 2006). Soil conditions include warm temperatures, favourable pH levels, adequate soil moisture, aeration (oxygen) and fertility favours microbial growth. The amount of adsorption also influences microbial degradation. The soil fungi for e.g., *Fusarium, Penicillium* are now known to degrade pesticides with greater efficacy (Twigg and Socha, 2001). The branching, filamentous mode of fungus allows more efficient colonization and exploration of contaminated soil (Aust, Swaner and Stahl, 2004). White-rot fungi are filamentous organisms offer advantages over bacteria able to oxidise many compounds (Pointing, 2001). In addition, they are robust organisms and are generally more tolerant to high concentrations of chemical pollutants than bacteria (Evans and Hedger, 2001) and are a powerful prospective tool in soil bioremediation and some species have already been patented.

Phytoremediation of Pesticides

Phytoremediation technology comprises growing plants on contaminated sites so that polluting components percolate through the radical system of the plants and accumulate in various parts (roots, stems, leaves, etc.). There are other constraints to the use of plants alone in remediation; plant growth is dependent on a number of environmental factors like nutrients and water availability, soil texture and pH, etc. The maximum benefits of phytoremediation may therefore, be achieved in long-term applications, or when used in conjunction with other immediate remedial actions. Excluding these limitations, plants are identified to absorb a wide range of air-borne chemicals through the foliage surface. The in situ (on-site treatment) phytoremediation has widely used in Europe and the United States (Meharg and Cairney, 2000; Gaur and Adholeya 2004). This technology is beneficial when the roots of the plant are in contact with the pollutants. It could also be necessary to proceed through several cycles of culture and harvest to restore a site completely. And harvested plants must be disposed carefully (Mulligan, Yong and Gibbs, 2001).

Some Other Bioremediation Processes for Degradation of Pesticides

Bio-Stimulation

This refers to addition of specialized nutrients and suitable physiological conditions for the growth of the indigenous microbial populations. This promotes increased metabolic activity, which then degrades the contaminants (Trindade, Sobral, Rizzo, Leite and Soriano, 2005). Bio stimulation has been used to remove pesticides contamination in the environment. In this process, the activity of the indigenous microorganisms is stimulated by adding the organic and/or inorganic additives such as N or P, etc. The modifications added would be used by the indigenous microorganisms for cell growth to increase cell number as well as their activities to degrade the pesticides. In addition, the amendments could be necessary as the enzyme-inducers and/or the co-metabolic substrates in the pesticide degradation pathways (Robles-Gonzalez, Fava, Poggi- Varaldo, 2008).

Bio-Augmentations

Bio-augmentation is the introduction of exogenous microorganisms with specific catabolic abilities into the contaminated environment or into a bioreactor to initiate the bioremediation process. This can be an in situ or ex situ treatment process in which naturally occurring microbes are added to contaminated sites to eliminate toxic contaminants (Head and Oleszkiewicz, 2004; Perelo, 2010). The technique is reported to be an effective bioremediation approach for improving pesticide degradation in contaminated soils and water that lack indigenous microbial activity (Dams et al. 2007; Plangklang and Reungsang, 2010). It has been employed to degrade a wide range of chemical contaminants such as ammonia, hydrogen sulphide, insecticides, petroleum compounds, and a growing number of toxic organic chemicals present in soil and water (Hwang and Cutright, 2002; Jianlong, Xiangchun, Libo, Yi and Hegemann, 2002; Quan, Shi, Liu, Wang and Qian,2004).

Both bio-stimulation and bio-augmentation possess the ultimate goal of decontamination through biotransformation. The toxic substance is generally converted into other nontoxic forms (transformed) by a chemical reaction using microbes, aerobically or anaerobically. Anaerobic biotransformation is essentially used for degradation of organic compounds such as chlorinated hydrocarbons, polychlorinated phenols, and nitro aromatics which are recalcitrant to aerobic treatment processes (Ferguson and Pietari 2000; Marschner, Yang, Lieberei and Crowley, 2001; Somsamak, Cowan and Haggblom, 2001; Bhushan et al. 2003).

Bioremediation of Metal Contaminated Site

Accumulation of heavy metals by emissions from industrial areas, mine tailings, disposal of high metal wastes, leaded gasoline and paints, spillage of petrochemicals, sewage sludge, pesticides, wastewater irrigation, coal combustion residues, land application of fertilizers, animal manures, and atmospheric deposition contaminated soil extremely and rapidly. Soils are the major sink for heavy metals released into the environment and also their total concentration in soils persists for a long time after their introduction. To overcome the problem of soil or land pollution and restore the fertility of the soil, it is necessary to remediate it. Bacteria, algae, fungi, actinomycetes and higher plants accumulate high amount of heavy metals in their cells and bio remediate the contaminate site.

Algae

The species of *Chlorella, Anabaena inaequalis, Westiellopsisprolifica, Stigeocloniumtenue, Synechococcus sp.* tolerate heavy metals. However, several species of *Chlorella, Anabaena,* marine algae have been used for the removal of heavy metals. But the operational condition limits the practical application of these organisms. The naturally occurring cells showed higher efficiency for Cd^{++} and Ni^{++} as compared to laboratory cells.

Fungi

Fungi also are capable of accumulating heavy metals in their cells. However, several mechanisms operate in them for removal of heavy metals from the solution; a few of these have been discussed below:

Metabolism-Independent Accumulation

The positively charged ions in the solution are attracted to negatively charged ligands in cell materials. Biosorption of metal ion occurs on microbial cell surface but composition of biomass and other factors affect biosorption. For example, in Rhizopusarrhizus adsorption depends on ionic radius of Li^{3+}, Mn^{2+}, Cu^{2+}, Zn^{2+}, Cd^{2+}, Ba^{2+}, Hg^{2+} and Pb^{2+}. However, binding of Hg^{2+}, Ag^{2+}, Cd^{2+}, Al^{3+}, Ni^{2+}, Cu^{2+} and Pb^{2+}strongly depends on concentration of yeast cells.

Metabolism-Dependent Accumulation

In fungi and yeast, heavy metal ions are transported into the cells through cell membrane. However, as a result of metabolic processes ions are precipitated around the cells, and synthesized intracellularly as metal-binding proteins. Energy-dependent uptake of Cu^{2+}, Cd^{2+}, Co^{2+}, Ni^{2+}, $Zn^{2+}$by fungi has been demonstrated. Moreover, intracellular uptake is influenced by certain external factors such as pH, anions, cations and organic materials, growth phase, etc. Microbes like algae, bacteria and fungi as well as higher plants have capabilities to uptake these pollutants. After uptake, these either accumulates or is assimilated by them. Accumulated heavy metals are recovered by recycling or disposal.

The following bioremediation processes are used for metal contaminated sites:

- Biosorption,
- Bioprecipitation/ Biomineralization,
- Biomining:
  - Biooxidation,
  - Bioleaching.

Biosorption

It is the physiochemical process that occurs naturally in certain biomass which allows it to passively concentrate and bind the contaminant on its cellular structure. Microorganisms responsible for this process are called biosorbents. It involves the microbial immobilization of free metal ions and their changed species and is also used for the remediation of mine waste soil. Biosorbents which are used for biosorption of metals from solutions could be classified into three broad categories i.e. exo-polysaccharides, living cultures and non- living biomass and preparations. Living and non- living microbial both cells are active in metal sorption. Non- living microbial biomass is usually preferable for biosorption technology as those precludes the necessity of adding nutrients required for microbial development, maintaining sterility of the process and adjusting parameters to values favourable for microbial growth. The biosorption process involves a solid phase and liquid phase containing a dissolved species to be sorbed. Due to the higher affinity of the absorbent for the absorbate species, the latter is attracted and bound there by different mechanisms.

The process continues till equilibrium is established between the amount of solid- bound absorbate species and its portion remaining in the solution. The degree of absorbent affinity for the absorbate determines its distribution between the solid and liquid phases. The metallic species could be deposited

Figure 9. Processes of bio sorption mechanism

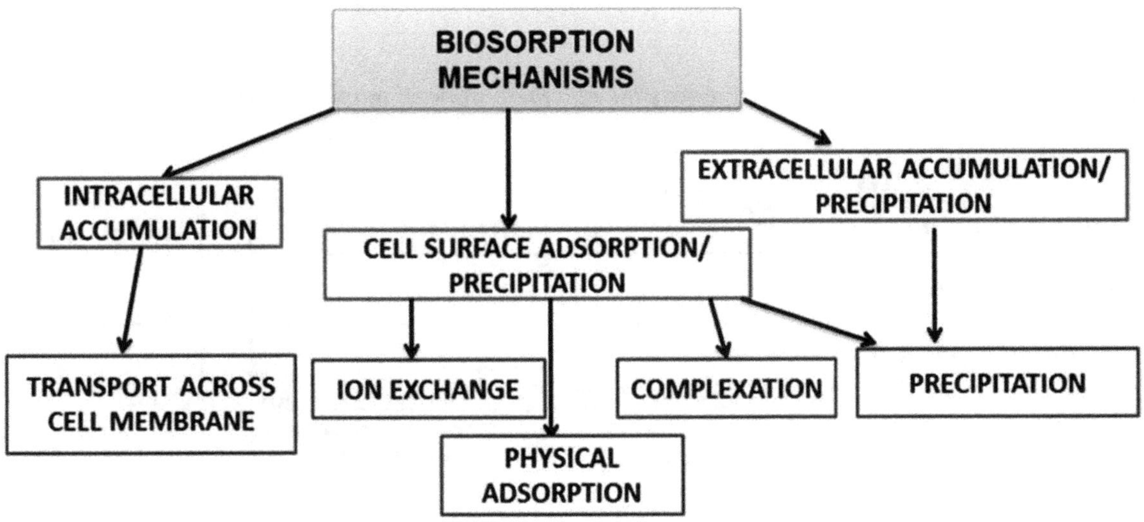

in the solid biosorbent through various sorption processes such as transport across membrane, physical adsorption, precipitation and chemisorption by ion exchange, coordination, chelation, and complexion.

1. **Transport Across Membrane:** Heavy metal transport across microbial cell membranes may be mediated by the same mechanism used to convey metabolically important ions such potassium, magnesium and sodium. The metal transport system may become difficult by the presence of heavy metals ions of the same charge and ionic radius associated with metabolic activity.
2. **Physical Adsorption:** it includes the Vander Waal's forces, Electrostatic interactions between the metal ions in the solution and cell wall of microbial cells (Figure 9 & 10).
3. **Ion Exchange:** it is reversible chemical reaction wherein an ion in the solution is exchanged for a similarly charged ion attached to an immobile solid particles. Cell walls of microorganisms contain polysaccharides and bivalent metal ions exchange with the counter ions of polysaccharides.
4. **Complexion and Precipitation:** The metal removal from solution may also take place by complex formation on the cell surface after the interaction between the metal and the active groups. Complexion found to be the only mechanism responsible for calcium, magnesium, cadmium, zinc, copper and mercury accumulation by *Pseudomonas suringae*. Fungi produce several extracellular products which can complex or precipitate heavy metals. For example, many fungi and yeast release high affinity Fe-binding compounds that chelate iron. It is called siderophores. The Fe^{3+}chelates which are formed outside the cell wall are taken up into the cell. In *Saccharomyces cerevisiae* removal of metals is done by their precipitation as sulphides e.g. Cu^{2+} is precipitated as CuS (Beveridges, 1989).

Bioprecipitation

It is the concept of rainmaking bacteria in which metal contaminants get precipitated by microorganisms in liquid medium. It is also called biomineralization. It removes metals by forming insoluble precipitates that either move into deeper layer osf soil or remain in soil without imposing any harmful effect on crops or vegetation.

Figure 10. Mechanism of bio sorption across the cell membrane of microorganisms

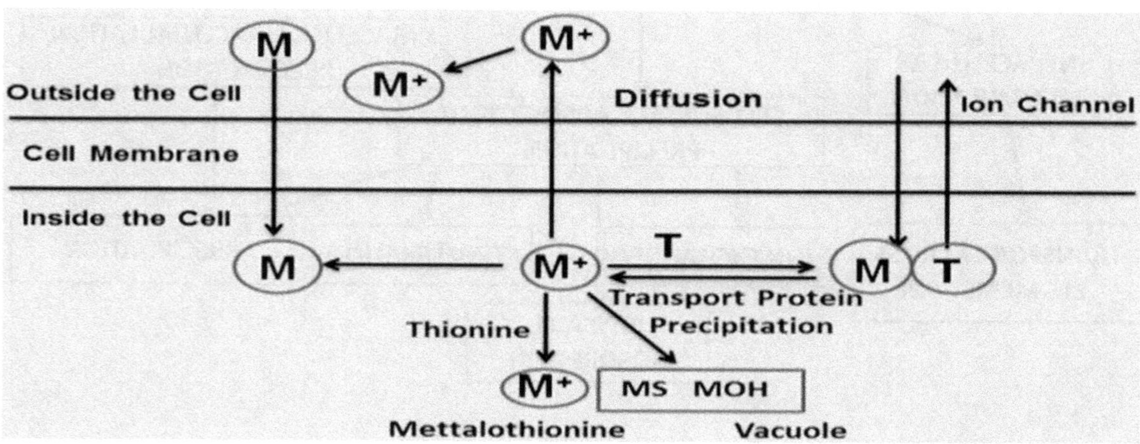

Biomining

It is the extraction of desired minerals from their ores through the use of living microorganisms. It is used to aid the recovery of metal form low grade ores. It is economically sound hydrometallurgical process with lesser environmental problems than conventional commercial application. It has tremendous practical application. It consists of two processes:

Bio Oxidation

It is the process which is applied to enhance the extraction of gold and silver. An oxidation process caused by microbes where the valuable metal remains (but becomes enriched) in the solid phase. Biooxidation leaves the metal values in the solid phase and the solution may be discarded. This contrasts with bioleaching where the valuable metal is solubilized. (Both bioleaching and biooxidation are oxidation processes. If the valuable metal in the ore is hosted by, but not bound to, a surrounding undesired mineral matrix which is blocking the valuable metal from further chemical treatment the matrix can be degraded by selective dissolution of the undesired mineral matrix. Here the microbes causing the selective dissolution may enrich the valuable metal and facilitate metal extraction by selective dissolution of the undesired mineral matrix.

Bioleaching

Bioleaching is the process by which metals are dissolved from ore bearing rocks using microorganisms. For the last 10 centuries, microorganisms have assisted in the recovery of copper dissolved in drainage from water. Microbial technology renders helps in case of recovery of ores which cannot be economically processed with chemical methods, because they contain low grade metals. The low grade ores are discarded in waste heaps which enter in the environment. The low grade ores contain significant amount of nickel, lead, and zinc ores which could be processed by microbial leaching. Bioleaching of uranium and copper has been widely commercialized. But large scale leaching process may cause environmental problems when dump is not managed properly. These results in seepage of leach fluids containing large quantity of metals and low pH into nearby natural water supplies and ground water. Thus, bio mining is economically sound hydrometallurgical process with lesser environmental problem than conventional commercial application. However, it is an inter-disciplinary field involving metallurgy, chemical engineering, microbiology and molecular biology. It has tremendous practical application. In a country like India bio mining has great national significance where there is vast unexploited mineral potential (Mogal and Desai, 1992). The most commonly used microorganisms for bioleaching are *Thiobacillus thiooxidans* and *T.ferrooxidans*. The other microorganisms may also be used in bioleaching, *Bacillus licheniformis, B. luteus, B. megaterium, B. polymyxa, Leptospirillum ferrooxidans, Pseudomonas fluorescens, Sulfolobusacido caldarius, Thermothrix thioparus, Thiobacillus thermophilica, etc.*

Bioremediation of Dyes

Synthetic dyes are widely used in a number of industries, such as textile dyeing or printing, paper printing, cosmetics, pharmaceuticals and leather industries. These industries uses large amounts of energy and water, and they release a substantial amount of waste water with many contaminants. With the increased demand for textile products, the waste waters generated from this industry have also increased proportionally and polluting the receiving water worldwide. In India, mill discharges is an average about 1.5 million litres of effluent per day, which causes chronic and acute toxicity (Sandhya et al., 2005 and Arami et al., 2006). The organic dye stuffs, chrome dyes and other chemicals during various operations and produce a large quantity of solid and liquid waste containing hexavalent chromium [chromium (VI)], salts of zinc, sulphates, copper, sodium and potassium etc., which causes severe toxicity to aquatic organisms. There is limited study on microbial degradation of azo and reactive dyes. Maximum number of dyes undergoes degradation through reduction. Kulla (1981) discussed azo- reductase of Pseudomonas strains in the chemostat culture. These enzyme catalyses azo-linkage of the dye. During degradation process of azo, NAD(P) acts as electron donor. Srivastava et al. (1995) observed degradation of black liquor pulp mill effluents by the strains of *Pseudomonas putida.* Some anaerobic bacteria, Streptomyces and fungi (e.g. *Phaenerochaete chrysosporium*) have been characterized for decoloration of chromogenic dyes. The enzymes involved in dye degradation are lignases (lignin peroxidase), Mn (II) dependent peroxidase and glyoxal oxidase. These enzymes are well associated with lignin degrading system.

Bioremediation of Xenobiotic Compounds

Man-made chemicals present in the nature at high concentrations polluting the environment are known as Xenobiotic compounds. Some microbes are capable in breaking down of xenobiotic compounds to some extent. But most of the xenobiotic compounds are non-degradable in nature and remain in the environment for many years leading to bioaccumulation or bio-magnification. Such compounds are known to be recalcitrant in nature. The recalcitrant xenobiotic compounds can be divided into different groups depending on their chemical composition like halocarbons, synthetic polymers, alkyl benzyl sulphonates, oil mixture etc. These compounds are highly toxic in nature and can affect survival of lower as well as higher eukaryotes. It also poses health hazards to humans like various skin problems, reproductively and even known as a trigger for causing cancer. They also find a way into the food chains.

For breaking down xenobiotics compounds, the enzymes act on certain groups present in the compound. For e.g. in the halocarbons, the halogen group is targeted. Enzymes like oxygenases play a major role. The bonds like ester-, amide-, or ether bonds present in the compounds are first attacked leading to breaking down of compounds. In some cases the aliphatic chains and in aromatic compounds the aromatic components may be targeted. Often it is seen that the xenobiotic do not act as a source of energy to microbes and as a result they are not degraded. The presence of a suitable substrate induces its breakdown. This substrate is known as co–metabolite and the process of degradation are known as co-metabolism. In another process, the xenobiotic serve as substrates and are acted upon to release energy. This is called gratuitous metabolism. Certain microbes on continuous exposure to xenobiotic develop the ability to degrade the same as a result of mutations. Certain mutations also resulted in developing new enzymatic pathway for xenobiotic degradation. Use of mixed population of microbes is usually recommended as it has been seen to yield faster results as the two different microbes attack different parts through different mechanisms resulting in effective break down. It also creates a condition of co metabolism. The modification of certain genes of microbes to break down xenobiotic is also recommended and is seen to produce high level of accuracy.

CONCLUSION AND FUTURE PROSPECTS OF BIOREMEDIATION

The compounds or chemical which are produced from different sources are needed to be reused or recycle if possible so that their direct impact can be reduced on environment. It is dynamic to increase understanding of microbial communities and their activities to the environment and pollutants, expanding the knowledge of the genetics of the microbes to increase capabilities to degrade pollutants, conducting field studies of new bioremediation techniques which are cost effective, and dedicating sites which are set aside for long term research purpose. These opportunities offer potential for significant advances. *Trichoderma viridie, Beauveria bassiana, Verticillium lecanii, Metarhizium anisopliae* are tested for remediation of waste materials like temple waste, vegetable waste, fruit waste and saw dust waste. Despite its short-comings, its importance to the world is unquestionable considering present day environmental hazards. There is no doubt that bioremediation is in the process of paving a way to greener pastures and national safety in general. Bioremediation technology has also some limitations like several microorganisms cannot break toxic metals into harmless metabolites and have inhibitory effects on microbial activity. Modification in the outer membrane proteins of bacteria with potential bioremediation properties for improving metal binding abilities is to enhance their capacity for biotransformation of toxic metals. Future studies should focus on the factors involved in improved by using genetically engineered microorganisms (GEM) in situ bioremediation strategies and also the applicability and adaptability of these GEMs in all the possible adverse conditions and multiple heavy metal polluted conditions. Bioremediation and phytoremediation can be good treatment strategies for PHC pollutants but more research information and techniques is required to prepare methods for treatment.

REFERENCES

Agarwal, S. K. (1998). *Environmental Biotechnology*. New Delhi, India: APH Publishing Corporation.

Arami, M., Limaee, N. Y., Mahmoodi, N. M., & Tabrizi, N. S. (2006). Equilibrium and kinetics studies for the adsorption of direct and acid dyes from aqueous solution by soy meal hull. *Journal of Hazardous Materials*, *135*(1-3), 171–179. doi:10.1016/j.jhazmat.2005.11.044 PMID:16442216

Atlas, R. M., & Pramer, D. (1990). Focus on bioremediation. *ASM News*, *56*, 352–354.

Aust, S. D., Swaner, P. R., & Stahl, J. D. (2004). Detoxification and metabolism of chemicals by white-rot fungi. In Pesticide decontamination and detoxification. Oxford.

Beveridge, T. J. (1989). Role of cellular design in bacterial metal accumulation and mineralization. *Annual Review of Microbiology*, *43*(1), 147–171. doi:10.1146/annurev.mi.43.100189.001051 PMID:2679352

Bhushan, B., Paquet, L., Halasz, A., Spain, J. C., & Hawari, J. (2003). Mechanism of xanthine oxidase catalyzed biotransformation of HMX under anaerobic conditions. *Biochemical and Biophysical Research Communications*, *306*(2), 509–515. doi:10.1016/S0006-291X(03)01001-5 PMID:12804594

Dams, R. I., Paton, G. I., & Killham, K. (2007). Rhizo-remediation of pentachlorophenol by *Sphingobium chlorophenolicum*, ATCC 39723. *Chemosphere*, *68*(5), 864–870. doi:10.1016/j.chemosphere.2007.02.014 PMID:17376504 de la Cueva, S. C., Rodríguez, C. H., Cruz, N. O. S., Contreras, J. A. R., & Miranda, J. L. (2016). Changes in Bacterial Populations During Bioremediation of Soil Contaminated with Petroleum Hydrocarbons. *Water, Air, and Soil Pollution*, *227*(3), 1–12. doi:10.1007/s11270-016-2789-z

Deeb, R. A., & Cohen, L. (2000). Aerobic biotransformation of gasoline aromatics in multicomponent mixtures. *Bioremediation Journal*, *4*(1), 1–9.

Dindal, D. L. (1990). *Soil biology guide*. New York: Wiley.

Elba. (2014). Environmental hazards associated with pesticide import into Costa Rica, 1977 2009. *J. Environ. Biol,35*.

Evans, C., & Hedger, J. (2001). Degradation of cell wall polymers. In G. Gadd (Ed.), *Fungi in Bioremediation*. doi:10.1017/CBO9780511541780.002

Ferguson, J. F., & Pietari, J. M. H. (2000). Anaerobic transformations and bioremediation of chlorinated solvents. *Environmental Pollution*, *107*(2), 209–215. doi:10.1016/S0269-7491(99)00139-6 PMID:15092997

Folke, C., Carpenter, S., Walker, B., Scheffer, M., Elmqvist, T., Gunderson, L., & Holling, C. S. (2004). Regime shifts, resilience, and biodiversity in ecosystem management. *Annual Review of Ecology and Systematics*, *35*(1), 557–581. doi:10.1146/annurev.ecolsys.35.021103.105711

Franchi, E., Agazzi, G., Rolli, E., Borin, S., Marasco, R., Chiaberge, S., & Barbafieri, M. (2016). Exploiting hydrocarbon-degrader indigenous bacteria for bioremediation and phytoremediation of a multi-contaminated soil. *Chemical Engineering & Technology*, *39*(9), 1676–1684. doi:10.1002/ceat.201500573

Frankenberger, D. G., & Fotopoulou, A. (2005). Non-specific biodegradation of the organophosphorus pesticides, cadusafos and ethoprophos, by two bacterial isolates. *FEMS Microbiology Ecology*, *53*(3), 369–378. doi:10.1016/j.femsec.2005.01.012 PMID:16329956

Gaur, A., & Adholeya, A. (2004). Prospects of arbuscularmycorrhizal fungi in phytoremediation of heavy metal contaminated soils. *Current Science*, *86*(4), 528–534.

Gkorezis, P., Daghio, M., Franzetti, A., Van Hamme, J. D., Sillen, W., & Vangronsveld, J. (2016). The Interaction between Plants and Bacteria in the Remediation of Petroleum Hydrocarbons: An Environmental Perspective. *Frontiers in Microbiology,7*.

Gogate, P. R., & Pandit, A. R. (2004). A Review of imperative technologies for wastewater treatment I: Oxidation technologies at ambient conditions. *Advances in Environmental Research*, *8*(3-4), 501–551. doi:10.1016/S1093-0191(03)00032-7

Head, M. A., & Oleszkiewicz, J. A. (2004). Bio-augmentation for nitrification at cold temperatures. *Water Research*, *38*(3), 523–530. doi:10.1016/j.watres.2003.11.003 PMID:14723920

Hess, A., Zarda, B., & Hahn, D. (1997). In situ analysis of denitrifying toluene and m-xylene degrading bacteria in a diesel fuel contaminated laboratory aquifer column. Applied and Environmental Microbiology. *J. App. Enviro. Micro*, *63*, 21362141.

Hwang, S., & Cutright, T. J. (2002). Biodegradability of Aged Pyrene and Phenenthrene in a Natural Soil. *Chemosphere*, *47*(9), 891–899. doi:10.1016/S0045-6535(02)00016-4 PMID:12108695

Jianlong, W., Xiangchun, Q., Libo, W., Yi, Q., & Hegemann, W. (2002). Bioaugmentation as a Tool to Enhance the Removal of Refractory Compound in Coke Plant Wastewater. *Process Biochem, Oxford, UK, 38*(5), 777–781. doi:10.1016/S0032-9592(02)00227-3

Keisham, S., & Paul, B. (2015). A Review on the Recent Scenario of Municipal Solid Waste Management in India. *International Journal of Engineering Research and General Science.*

Kumar, M., & Philip, L. (2006). Enrichment and Isolation of a Mixed Bacterial Culture for Complete Mineralization of Endosulfan. *J Environ Sci Health, 41*(1), 81–96. doi:10.1080/03601230500234935 PMID:16393897

Levin, S. A. (1992). The Problem of Pattern and Scale in Ecology: The Robert H. Mac Arthur Award. *Ecology, 73*(6), 1943–1967. doi:10.2307/1941447

Lorimor, J., Fulhage, C., Zhang, R., Funk, T., Sheffield, R., Sheppard, C., & Newton, G. L. (2001). *Manure Management Strategies/ Technologies.* White Paper on Animal Agriculture and the Environment for National Centre for Manure and Animal Waste Management.

Majeau, J. A., Satinder, K. B., & Tyagi, R. D. (2010). Laccases for removal of recalcitrant and emerging pollutants. *BiresourTechnol, 101*(7), 2331–2350. doi:10.1016/j.biortech.2009.10.087 PMID:19948398

Mandal, A. K., Sarma, P. M., Jeyaseelan, C. P., Channashettar, V. A., Singh, B., Lal, B., & Datta, J. (2012). Large-scale bioremediation of petroleum hydrocarbon contaminated waste at Indian oil refineries: Case studies. *Int. J. Life Sci. Pharm. Res, 2*(4), L114–L128.

Marschner, P., Yang, C. H., Lieberei, R., & Crowley, D. E. (2001). Soil and plant specific effects on bacteria community composition in the rhizosphere. *Soil Biology & Biochemistry, 33*(11), 1437–1445. doi:10.1016/S0038-0717(01)00052-9

Meharg, A. A., & Cairney, J. W. G. (2000). Ecto-mycorrhizae -Extending the capabilities of rhizosphere remediation. *Soil Biology & Biochemistry, 32*(11-12), 1475–1484. doi:10.1016/S0038-0717(00)00076-6

Melling, F. B. Jr. (1993). *Soil microbial ecology: Applications in agricultural and environmental management.* New York: Marcel Dekker.

Mulligan, C. N., Yong, R. N., & Gibbs, B. F. (2001). Remediation alternative treatment option for heavy metal bearing wastewaters: A review. *Bioresource Technology, 53*, 195–206.

Okpokwasili, G. C. (2007). Biotechnology and clean environment. In *Proceedings of the 20th Annual conference of the Biotechnology Society of Nigeria (BSN).* Ebonyi State University.

Perelo, L. W. (2010). Review: In situ and bioremediation of organic pollutants in aquatic sediments. *Journal of Hazardous Materials, 177*(1-3), 81–89. doi:10.1016/j.jhazmat.2009.12.090 PMID:20138425

Plangklang, P., & Reungsang, A. (2010). Bio-augmentation of carbo furan by *Burkholderia cepacia* PCL_3 in a bio slurry phase sequencing batch reactor. *Process Biochemistry, 45*(2), 230–238. doi:10.1016/j.procbio.2009.09.013

Pointing, S. B. (2001). Feasibility of bioremediation by white-rot fungi. *Applied Microbiology and Biotechnology, 57*(1-2), 20–33. doi:10.1007/s002530100745 PMID:11693920

Quan, X., Shi, H., Liu, H., Wang, J., & Qian, Y. (2004). Removal of 2,4-dichlorophenol in a conventional activated sludge system through bioaugmentation. *Process Biochem, Oxford, UK, 39*(11), 1701–1707. doi:10.1016/S0032-9592(03)00307-8

Rajeswari, R., & Kanmani, S. (2009). A study on degradation of pesticide wastewater by TiO_2 catalysis. *Journal of Scientific and Industrial Research, 68*, 1063–1067.

Robles-Gonzalez, I. V., Fava, F., & Poggi-Varaldo, H. M. (2008). A review on slurry bioreactors for bioremediation of soils and sediments. *Microbial Cell Factories, 7*(1), 1–16. doi:10.1186/1475-2859-7-5 PMID:18312630

Sandhya, S., Padmavathy, S., Swaminathan, K., Subrahmanyam, Y. V., & Kaul, S. N. (2005). Micro aerophilic- aerobic sequential batch reactor for treatment of azo dyes containing simulated wastewater. *Process Biochemistry, 40*(2), 885–890. doi:10.1016/j.procbio.2004.02.015

Sengupta, J. (2002). Recycling of Agro- industrial wastes for manufacturing of building materials and components in India: An overview. *Civil Engineering & Construction Review, 15*(2), 23–33.

Sharholy, M., Ahmad, K., Mahmood, G., & Trivedi, R. C. (2008). Municipal solid waste management in Indian cities. A review. *Journal of Waste Management, 28*(2), 459–467. doi:10.1016/j.wasman.2007.02.008 PMID:17433664

Sharma, A., Mishra, M., Sheet, S., & Thite, M. (2013). Role of microbes as cleaning degrading industrial wastes for environmental sustainability- a review. *Recent Res. Sci. Technol, 5*, 21–25.

Shen, Y., West, C., & Hutchins, S. R. (2000). In vitro cytotoxicity of aromatic aerobic biotransformation products in bluegill sunfish BF-2 cells. *Ecotoxicology and Environmental Safety, 45*(1), 27–32. doi:10.1006/eesa.1999.1843 PMID:10677264

Siddique, T., Okeke, B. C., Arshad, M., & Frankenberger, W. T. (2003). Enrichment and isolation of endosulfan-degrading microorganisms. *Journal of Environmental Quality, 32*(1), 47–54. doi:10.2134/jeq2003.4700 PMID:12549541

Somsamak, P., Cowan, R. M., & Haggblom, M. M. (2001). Anaerobic biotransformation of fuel oxygenates under sulfate-reducing conditions. *FEMS Microbiology Ecology, 37*, 259–264. doi:10.1111/j.1574-6941.2001.tb00873.x

Tang, C. Y., Criddle, Q. S., Fu, C. S., & Leckie, J. O. (2007). Effect of flux and technique. *Biologie Medicale, 1*(3), 1–6.

Trindade, P. V. O., Sobral, L. G., Rizzo, A. C. L., Leite, S. G. F., & Soriano, A. U. (2005). Bioremediation of a weathered and recently oil-contaminated soil from Brazil: A comparison study. *Chemosphere, 58*(4), 515–522. doi:10.1016/j.chemosphere.2004.09.021 PMID:15620743

Twiggy, L. E., & Socha, L. V. (2001). Defluorination of sodium mono fluoroacetate by soil microorganisms from central Australia. *Soil Biology & Biochemistry, 33*(2), 227–234. doi:10.1016/S0038-0717(00)00134-6

Vidali, M. (2001). Bioremediation: An overview. *Pure and Applied Chemistry, 73*(7), 1163–1172. doi:10.1351/pac200173071163

Yoshizawa, S., Tanaka, M., & Shekdar, A. V. (2004). Global trends in waste generation. In Recycling waste treatment and clean technology. TMS Mineral, Metals and Materials Publishers.

Zhang, P., Sheng, G. Y., Feng, Y. C., & Miller, D. M. (2005). Role of wheat residue-derived char in the biodegradation of benzonitrile in soil: Nutritional stimulation versus adsorptive inhibition. *Environmental Science & Technology*, *39*(14), 5442–5448. doi:10.1021/es0480670 PMID:16082978

KEY TERMS AND DEFINITIONS

Agricultural Waste: Waste in the form of the crop residues in the farm, manure from livestock operations, including dairy and piggery effluent, and poultry litter.

Anaerobic Digestion: The conversion of biodegradable waste matter into compost in the absence of oxygen.

Biogas: Biogas is the gaseous emissions from anaerobic degradation of organic matter (from plants or animals) by a consortium of bacteria. Biogas is principally a mixture of methane (CH_4) and carbon dioxide (CO_2) along with other trace gases.

Bioremediation: Bioremediation is a waste management technique that involves the microorganisms to remove or neutralize pollutants from a contaminated site.

Industrial Waste: Industrial waste is the waste produced by industrial activity which includes any material that is rendered useless during a manufacturing process such as that of factories, industries, mills, and mining operations.

Municipal Waste: Commonly known as trash or garbage—consists of everyday items we use and then throw away, such as product packaging, grass clippings, furniture, clothing, bottles, food scraps, newspapers, appliances, paint, and batteries. This comes from our homes, schools, hospitals, and businesses.

Pesticides: A pesticide is any substance or mixture of substances used to destroy, suppress or alter the life cycle of any pest. Pesticides include bactericides, baits, fungicides, herbicides, insecticides, lures, rodenticides and repellents.

Vermicomposting: The processing of organic wastes through earthworms. It is a natural, odourless, aerobic process, much different from traditional composting.

Xenobiotic Compounds: Man-made chemicals present in the nature at high concentrations polluting the environment are known as Xenobiotic compounds. These compounds are not commonly produced by nature.

Chapter 16
Bioremediation of Pharmaceutical Wastes

Alka Bali
Panjab University, India

ABSTRACT

The astounding increase in the use of pharmaceuticals in the last decade has raised concerns over their occurrence in the soils and wastewaters posing potential dangers to the general public health and environment. Considering the limitations of chemical remediation measures in treatment of recalcitrant xenobiotics, several bioremediation measures are being currently investigated and proposed for removal of pharmaceutical contaminants from the environment. Several bacterial, fungal and plant species have shown promising bioremediation potential with regard to the removal of pharmaceuticals. Varying configurations of anaerobic and aerobic reactors have been utilized for pharmaceutical wastewater treatment. This chapter is intended to give a compilation and overview of the various types of bioremediation measures currently being employed or investigated to remove the pharmaceutical pollutants.

INTRODUCTION

Pharmaceuticals comprise the broad group of synthetic as well as naturally obtained chemicals, intended for human or veterinary use as medicines. The use of pharmaceuticals and personal care products (PPCPs) has shown an astonishing increase in the last decade. Pharmaceutical industry production involves a wide range of raw materials, solvents, antibiotics, drugs and cosmetic products and hence, pharmaceutical effluents contain a wide variety of potentially harmful constituents (Dixit & Parmar, 2013). Pharmaceuticals enter the waste water, ground water, surface waters and soil through various means (Deziel, 2014). These include the sewage carrying excreta of patients on prescription or over the counter medicines, uncontrolled and unaccountable disposal of drugs by individuals as well as organizations, errant production houses lacking proper facilities for the disposal of drug waste, etc. The most frequently detected classes of pharmaceuticals in wastewater are NSAIDs, antibiotics, beta-blockers, lipid-regulators, steroids, antiepileptics, X-ray contrast media and tranquilizers (Drewes, 2007). The sewage treatment plants employed for treating wastewaters have been found to act upon the previously conjugated pharmaceutical residues (bound to a bile acid) excreted from the patients. The latter undergo

DOI: 10.4018/978-1-5225-2325-3.ch016

de-conjugation, yielding higher levels of free pharmaceutical substance in the plant outlet than in its incoming water. Wastewater reuse in areas with water scarcity leads to contamination of drinking water with pharmaceuticals like sulphamethoxazole, ibuprofen and carbamazepine in low levels (Luo, Guo, Ngo, Nghiem & Hai, 2014). Limited studies have shown that presence of these low level pharmaceutical contaminants in wastewaters can be potential dangers to human health (Pomati et al, 2006).

Further, the continual input of antibiotics into the environment is leading to emergence of newer antibiotic-resistant strains of microorganisms (Larsson, 2014). Currently, the antibiotic contamination is being recognized as a global environment issue. The recalcitrant nature of antibiotics sprouting from their inherent resistance to biodegradation due to their antimicrobial properties further complicates the issue in comparison with other classes of pharmaceuticals. This antibiotic contamination can promote the mobilization as well as the fixation of resistance genes among human and animal commensal bacteria along with other environmental and clinically-relevant species (Baquero, Tedim, & Coque, 2013; Kristiansson et al, 2011)). Several studies have shown the harmful effects of wastewater contamination on various animal species. Hormonal drugs have demonstrated capability of feminizing or masculinizing fish (Gilbert, 2012). Further, the non-steroidal anti-inflammatory drug (NSAID) diclofenac, can cause damage to the gills and lungs in fish populations (Gilbert, 2012).

The concerns regarding the presence of pharmaceuticals in the environment are many (Patneedi & Prasadu, 2015), and concrete remedial measures are yet to gain momentum but, this relatively recent field is rapidly acquiring eminence with several bioremediation measures being investigated and proposed for removal of pharmaceutical contaminants from the environment. Chemical methods can possibly lead to the generation of newer xenobiotics which can persist in the environment and may not be amenable to biological transformation. Biological processes utilizing microorganisms, plants and enzymes represent an eco-friendly approach towards the removal of pollutants from the environment. This book chapter is intended to explore and compile the various types of bioremediation approaches currently applicable (or holding considerable future promise) to successfully treat pharmaceutical industry effluents.

PHARMACEUTICAL WASTES AND THEIR ENVIRONMENTAL IMPACT

In recent times, concerns are being raised over the occurrence of pharmaceuticals in the environment and their potential adverse effects on general public health, flora and fauna and the aquatic environment. The pharmaceuticals may be categorized into four classes based on their occurrence in environment and their risk effects as:

1. Hormonal drugs and their derivatives, found in effluents at nanogram levels but pose a health risk due to their high potency and potential effects as endocrine disruptors,
2. Antineoplastics (anticancer) agents, with cytotoxic, carcinogenic, mutagenic and embryotoxic potential on account of their high reactivity,
3. Anti-infective agents, including antibiotics, and
4. Halogenated compounds such as X-ray contrast media, which are highly resistant to biodegradation and hence, highly persistent in the environment and food chain.

Several studies have highlighted the toxicity concerns of pharmaceuticals in wastewater, particularly, their potential to induce DNA damage and to cause lipid peroxidation effects (Mansour, Mosrati, Barillier,

Figure 1. Antimicrobials (antibiotics and sulphonamides) as pharmaceutical wastes

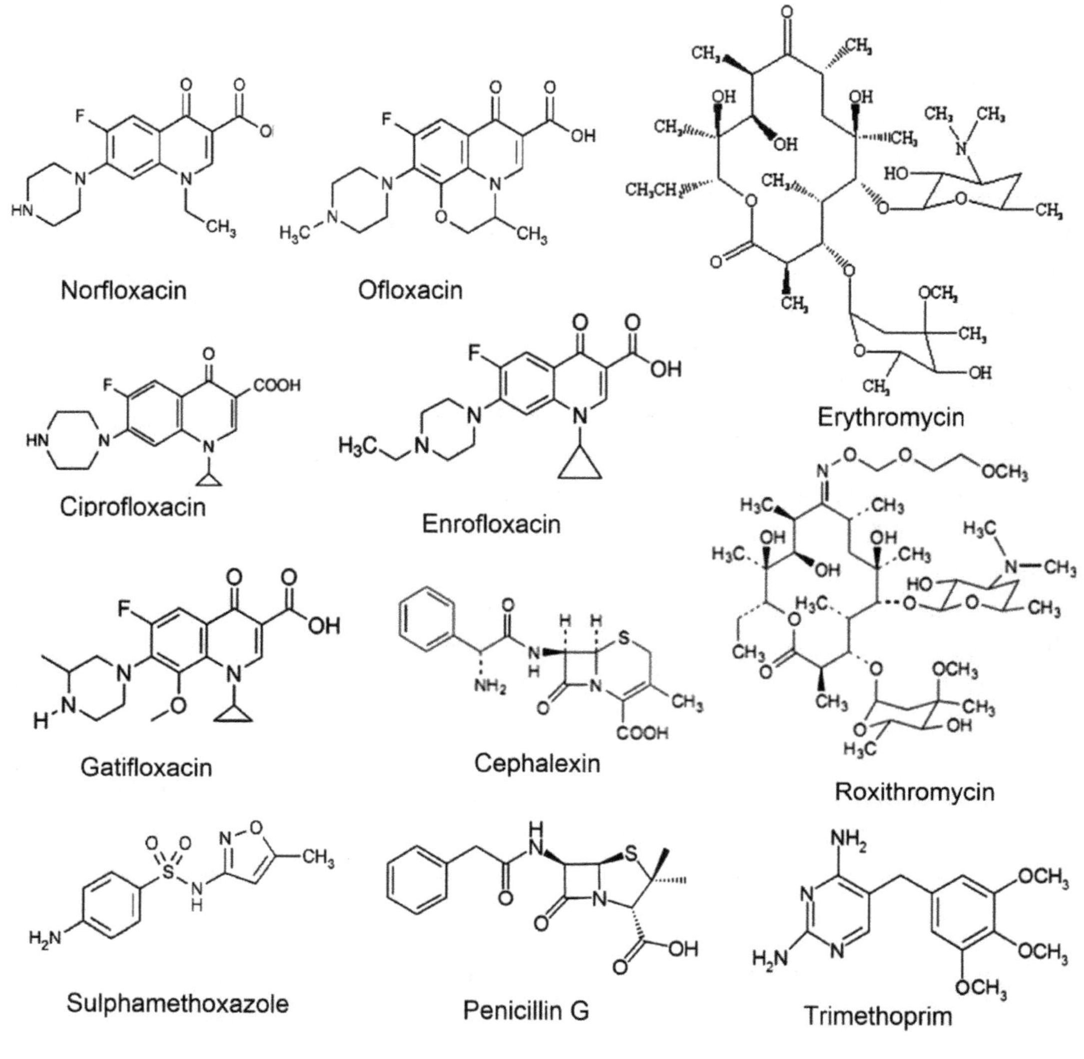

Ghedira, & Chekir-Ghedira, 2012). The US-EPA and other international agencies have listed the antibiotics (Figure 1), analgesics and other drugstore products (Figure 2) as top priority pollutants. The large scale production and use of antibiotics both in clinical and veterinary medicine, aquaculture, horticulture and agriculture has led to significant discharges of antibiotics into the environment. The bacterial isolates in the pharmaceutical effluents get exposure to sub-optimal dose of antibiotics produced by the factories and these constitute a major threat to public health. Many antibiotics are known to be persistent in the environment (Zuccato, Calamari, Natangelo, & Fanelli, 2000). A three year study conducted on seventy two pharmaceuticals and personal care products concluded that the half-lives for antibiotics when they reach agricultural soils can be phenomenally high, e.g., 55 to 578 days for tetracycline and 120 to 2310 days for ciprofloxacin (Walters, McClellan, & Halden, 2010). There have been several reports on isolation and characterization of newer antibiotic-resistant bacterial strains from pharmaceutical industrial wastewaters demonstrating varying patterns of multi-drug resistance (Tahrani et al., 2015).

A wide array of chemical as well as biological treatment methods are being employed by pharmaceutical industries for disposal of wastewater (Deegan et al., 2011). These include the use of ozone, Fenton's method, coagulation, electrocoagulation (Dixit & Parmar, 2013), pH correction, oxidation, sand filtration (Saleem, 2007), sedimentation, membrane separation, UV irradiation, photelectrocoagulation, peroxi-photelectrocoagulation, peroxi-electrocoagulation, adsorption, chlorination, distillation, reverse osmosis, bacterial treatments (Madukasi, Dai, He, & Zhao, 2010), phytoremediation methods including algal treatment, methods using membrane bioreactor (MBR) (C. Chang, J. Chang, Vigneswaran, & Kandasamy, 2008), fungal treatment (Spina, Anastasi, Prigione, Tigini, & Varese, 2012), anaerobic fixed film reactor (Rao, Naidu, Prasad, Rao, Mohan, Jetty, & Sharma, 2004), aerobic sequencing batch reactor (ASBR) (Patil, Kulkarni, S. Kore, & V. Kore, 2013), membrane-aerated biofilm reactor (MABFR) (Wei et al., 2012) and activated sludge method (Mayabhate, Gupta, & Joshi, 1988).

Chemical/physicochemical purification methods utilize costly chemicals and treatment units, which are difficult to manage at the industrial unit level and they can also possibly generate additional by-products with complex pharmacological/toxicity profiles, e.g., chlorination of phenols resulting in more toxic chlorinated products (Krug, Ziegler & Straube, 1985; Lee, Lu, & Chuang, 1994). Microorganisms play a critical role in the degradation of xenobiotics and in maintaining the steady-state concentration of chemicals in the environment (Misal, Lingojwar, Shinde, & Gawai, 2011). Their incredible versatility to catabolize, degrade or transform various contaminants has been increasingly investigated for removal of different types of non-pharmaceutical and pharmaceutical contaminants. The development of various types of aerobic and anaerobic biological reactors employing these microorganisms has revolutionized this field and this has become a prime area of research today. Further, phytoremediation and enzyme based approaches are being explored and encouraging results are generating novel leads in this area which can open up newer avenues in future.

BIOREMEDIATION APPROACHES OF PHARMACEUTICAL WASTES

Different types of bioremediation approaches for treatment of pharmaceutical wastewaters rely on specific strains of bacteria and fungi which have the capability to degrade or transform the pharmaceutical pollutants under anaerobic or aerobic conditions. Various types of anaerobic and aerobic reactors have been designed employing bacterial and fungal consortia as bioremediation measures. Besides these, plants and certain enzymes have also been explored as a means of removal of pharmaceuticals from industrial effluents and wastewaters.

BACTERIAL BIOREMEDIATION

Several bacterial strains like *Pseudomonas, Enterobacter, Streptomonas, Aeromonas, Acinetobacter* and *Klebsiella* have been shown to demonstrate upto 44% reduction in the chemical oxygen demand (COD) in anaerobically treated spent wash (Ghosh, Verma, Mengoni, & Tripathi, 2004). Soil bacteria such as *Arthrobacter* and *Pseudomonas* predominantly present in rhizospheric soil, possess a special phenolic biodegradation potential. Numerous bacteria from the genera *Comamonas, Rhodococcus and Ralstonia* have been shown to degrade phenolic and other complex organic compounds and hence, have potential utility in degrading the pharmaceutical pollutants as well. Some *Clostridium* species (fermenting bacteria)

Figure 2. Selected non-antibiotic pharmaceutical contaminants in wastewaters

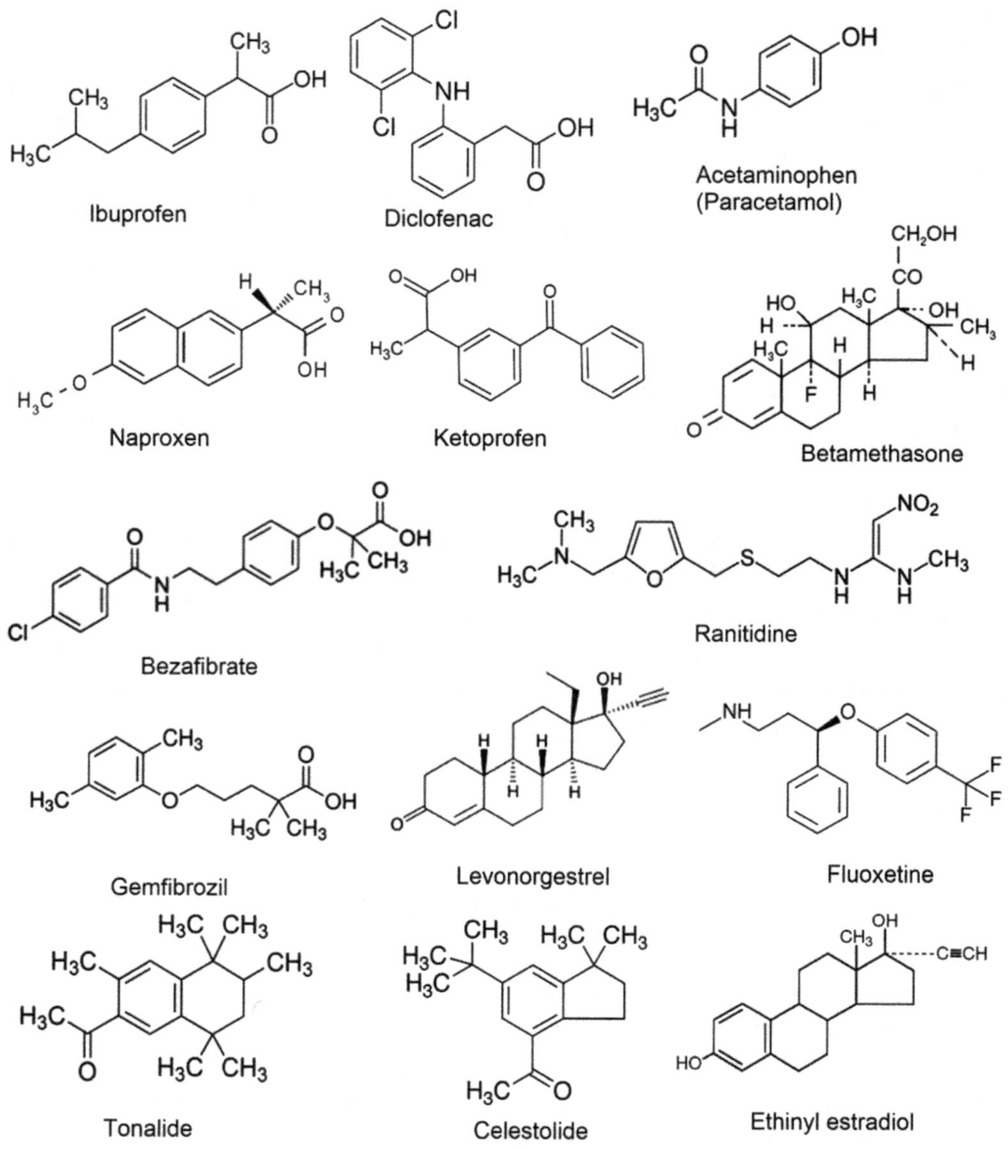

can degrade resorcinol (Kavitha & Beebi, 2003). Thermophilic bacterium *Bacillus thermoglucosidasius* A7 has been shown to degrade phenols/cresol at 65°C *via* the meta-cleavage pathway (Duffner, Kirchner, Bauer, & Muller, 2000). The bacterium *Pseudomonas fluorescence* has also been reported to demonstrate the capability to degrade phenols and chlorophenols (Agarry & Solomon, 2008). Biodegradation kinetics of phenol and catechol has been studied using *Pseudomonas putida* MTCC 1194 (A. Kumar, S. Kumar, & S. Kumar, 2005). Phototropic bacterium *Rhodobacter sphaeroides* has been used for treatment of pharmaceutical wastewater and 80% COD removal was achieved (Madukasi et al., 2010).

The specific bacterial species acclimatized in presence of the compounds needing remedial methods can be assets in bioremediation approaches and several of these have been investigated. Many researchers have worked on the isolation and identification of species from the sludge of biological treatment plants in pharmaceutical industries and have used those bacteria for utilization of other pharmaceutical ingredients. Bacteria highly resistant to a large number of antibiotics and non-antibiotics have been identified in such studies. The same bacteria, however, have also been seen to demonstrate a good potential to degrade and bioremediate some other non-antibiotic pharmaceutical effluents (Basu, Dastidar, Mukhopadhyay, & Gangopadhyay, 2012). Effective bioremediation of pharmaceutical waste water by *Pseudomonas putida* mt-2 has been recently reported (Mansour et al., 2012; Mansour, Dellai, & Ayed, 2012). *P. putida* was found to be very efficient in reducing the chemical oxygen demand (COD), total dissolved solids (TDS), and turbidity of solution (85.5, 89.1, and 81.5%, respectively). Further, genotoxicity evaluation studies on the effluent showed that the chromosomal aberrations and lipid peroxidation effects induced by the pharmaceutical wastewater (PW) were significantly reduced on bioremediation with *P. putida*. *Pseudomonas putida* IFO 14671 has also been isolated, and identified from the *Gomeya*/cow dung microbial consortium as a high-potential phenol degrader (Rao et al., 2004; Singh & Fulekar, 2009). *P. putida* isolated from cow dung microflora has been demonstrated a potential to degrade benzene (Singh & Fulekar, 2010). Bioremediation of effluents with *P. putida* has been shown to completely remove the murine chronotoxicity to pharmaceutical wastewater (Dridi, Bchir, Zouiten, Tahrani, & Mansour, 2013). Recently, biodegradation of the pharmaceutical waste, 4-nitrophenol has been shown with *P. putida* (Guleria, 2014).

Antibiotics, on account of their inherent properties, pose problems to bioremediation measures. However, there are several reports where these have been biodegraded in presence of consortium of microbes under specific conditions such as the absence of alternative sources of carbon and nitrogen, presence of moisture, etc. (Drillia et al., 2005; Gartiser, Elke, Radka, & Kümmerer, 2007). Streptomycin and cycloheximide have been reported to be biodegraded by soil bacteria and fungi (Badalucco, Pomaré, Grego, Landi, & Nannipieri, 1994). In a recent study, the bacteria and fungi from aquatic environment have been shown to degrade a mixture of two broad spectrum antibiotics, ciprofloxacin and erythromycin (Nnenna, Lekiah, & Obemeata, 2011).

Gomeya/cow dung has been used traditionally as an excellent, cheap and effective bioremediation measure against harmful pollutants. Randhawa and Kullar (2011) have reviewed the available studies on the bioremediation of a vast range of pharmaceuticals, pesticides, and petrochemicals with gomeya/cow dung. Manure (cattle dung) has been shown to effectively degrade the antibiotic enrofloxacin due to the presence of the known strains of coprophilous basidiomycetes, particularly, *Cyathus stercoreus* (Freer & Detroy, 1982; Wicklow, 1992). Reports are there in literature showing degradation of the antibiotics such as erythromycin, streptomycin, penicillin and bacitracin present in manure within 30 days (Gavalchin & Katz, 1994; Wang, Bradford, Zheng, & Yates, 2006). The sulphonamide drug sulfadimethoxine is shown to undergo biodegradation in manure and its degradation kinetics has also been studied (Wang et al., 2006).

FUNGAL BIOREMEDIATION (MYCOREMEDIATION)

Several fungal strains play an important role in the bioremediation of industrial effluents. Fungi, especially, the white-rot fungi are ubiquitously present in nature and these have adaptability to extreme conditions, thus, making them good biodegraders. These produce enzymes like lignin peroxidase, laccase and manganese peroxidase which are involved in degradation of lignin in their natural lignocellulosic substrates. This ligninolytic system of white rot fungi is directly involved in the degradation of various xenobiotics including pharmaceutical drugs and dyes by causing one-electron oxidation of aromatic compounds, with subsequent formation of aryl cation radicals, which spontaneously decompose by various pathways. (Christian, Shrivastava, Shukla, Modi, & Vyas, 2005; Tripathi, Harsh, & Gupta, 2007). Manganese peroxidase catalyzes the oxidation of Mn^{2+} to Mn^{3+}, which in turn can cause oxidation of several phenolic substrates (Sayad and Ellouz, 1995). However, fungal bioremediation (mycoremediation) measures suffer from certain drawbacks in terms of long growth cycle and spore formation.

Studies have shown that some fungal strains like *Bjerkandera adusta* MUT 2295, *Aspergillus niger, Aspergillus fumigatus, Penicillium decumbens* and *Penicillium lignorum* are effective in causing COD reduction in industrial effluents (Jayaraman, Muthuswamy, & Swaminathan, 2003; Mohammad et al. 2006). A comparative study was conducted by Spina et al. (2012) in which effectiveness of fungal treatment (Bjerkandera adusta MUT 2295) was compared with the activated sludge for treatment of pharmaceutical industrial wastewater. A 91% reduction in COD was noted with the fungal method, which was significantly higher than that observed from the activated sludge (78%). The degradation of enrofloxacin, a fluoroquinolone antibacterial drug used in veterinary medicine, has been investigated with the brown rot fungus *Gloeophyllum striatum* presumably by an oxidative mechanism. The findings suggest that the brown rot fungi are capable of producing highly reactive hydroxyl radicals, which can transform the xenobiotics by oxidative metabolism (Wetzstein, Schmeer, & Karl, 1997). Azo dyes are used extensively in the pharmaceutical and other industries and these are known to possess potentially mutagenic and/or carcinogenic properties. *Phanerochaete chyrosporium,* a major wood rotting fungus is capable of degrading a wide range of recalcitrant xenobiotics including azo-dyes through its non-specific hydrogen peroxide dependent extracellular lignin degradation enzyme system (Aust, 1990; Cripps, Bumpus, & Aust, 1990). *Aspergillus niveus,* a litter degrading fungus has been found useful in the biotreatment of distillery effluent (Angayarkanni, Palaniswamy, & Swaminathan, 2003). Recently, degradation of pharmaceuticals in non-sterile urban wastewater by white-rot fungus *Trametes versicolor* was reported in a fluidized bed bioreactor (Cruz-Morató et al., 2013). In a similar study, a fungal bioreactor based on *Trametes versicolor* pellets was reported as a new alternative for pre-treatment of hospital wastewater resulting in efficient removal of pharmaceuticals and endocrine disruptor compounds (Cruz-Morató et al., 2014). Further, the degradation of pharmaceuticals by *Trametes versicolor* has been found to be enhanced by addition of external nutrients (Badia-Fabregat et al., 2015). Very recently, *Trametes versicolor* has been applied for the removal of pharmaceuticals from a veterinary hospital wastewater and sixty-six per cent removal was achieved in a non-sterile batch bioreactor inoculated with *T. versicolor* pellets (Badia-Fabregat et al., 2016). Biodegradation by the fungus *Pycnoporus sanguineus* is another promising method for the mineralization of recalcitrant compounds present in pharmaceutical effluents (Watanabeet al., 2012).

Periconiella species of fungus isolated from cow dung has been reported to be an excellent degrader of biomedical waste and thus, represents a cheap, safe, and environment friendly method of biomedical waste disposal (Pandey & Gundevia, 2008). The efficiency of two white-rot fungi, *Trametes versicolor* and *Ganoderma lucidum,* to eliminate thirteen pharmaceutical pollutants has been recently investigated. Total removal of the drugs diclofenac, gemfibrozil, ibuprofen, progesterone and ranitidine was noted with the method whereas, lower removals were achieved for 4-acetamidoantipyrin, clofibric acid, atenolol, caffeine, carbamazepine, hydrochlorothiazide, sulfamethoxazole and sulpiride (Vasiliadou et al., 2016).

ENZYMATIC BIOREMEDIATION

In the last two decades, the application of enzymes as biocatalysts for environmental remediation has been widely studied as these offer advantages over living systems like microbes or fungi with respect to high reaction kinetics of degradation under subtle conditions of pH or temperature. Further, these are selective and are not affected by the biological activity of the targeted compounds. Enzymes can act on specific recalcitrant pollutants to remove them by precipitation or transformation to other products. The enzymes like oxidoreductases EC1 (*i.e.*, peroxidases, polyphenol oxidases (PPO), hydrolases EC3 (*i.e.*, proteases, esterases, lipases and cellulases) and lyases EC4 have been recommended for various wastewater treatment applications including pharmaceutical wastewaters and their effectiveness extends to wide range of pharmaceutical substances like phenolic drugs, antibiotics and other drugs. The application of oxidoreductive enzymes to catalyze the removal of phenol from wastewater has been investigated extensively. An example of such a method employs hydrogen peroxide and an immobilized derivative of soybean peroxidase, covalently bound to a glass support in a continuous tank reactor to remove phenol from aqueous solutions with 80% removal efficiency (Gómez et al. 2008). A recent article discusses the effectiveness of peroxidase based enzymatic treatment for continuous removal of phenol from wastewater (Chiong, Lau, Khor, & Danquah, 2014). A pharmaceutical effluent has been bio-remediated using yeast and amylase as the active agents (Uwadiae, Yerima, & Azike, 2011). Based on this technology, several types of enzymatic reactors have been designed (EMRs).

PHYTOREMEDIATION

Phytoremediation of wastewater is an emerging low-cost technique which utilizes the natural or transgenic green plants to bio-accumulate hazardous substances, particularly, metal ions such as lead and cadmium, and other xenobiotics (Campos, Merino, Casado, Pacios, & Gómez, 2008). The method has an advantage over microbial techniques as the microorganisms don't absorb heavy metals easily. This technique has a great potential for commercialization particularly in Europe and United states (Campos et al., 2008; McIntyre & Lewis, 1997; Watanabe et al., 1997). The method is still in preliminary stage and some reports on usage of aquatic plants as bioremediation measures are there. A recent review has addressed phytotoxicity of common pharmaceuticals and bioremediation potential of plants (Carvalho, Basto, Almeida, & Brix, 2014).

- **Aquatic Plants:** Phytoremediation through aquatic macrophytes treatment system for the removal of pollutants and contaminants from various natural sources is an accepted environmental protection technique. The biosorption capacity of the water hyacinth (*Eichhornia crassipes*), a highly invasive aquatic weed, in minimising various contaminants present in the industrial wastewater has been extensively studied (Priya & Selvan, 2014). Water hyacinth and the plants *Typha latipholia* and *Phragmitis karka* have been used in some studies to demonstrate their usefulness in reducing COD, BOD levels in pharmaceutical wastewaters (Billore et al., 2001). The algae are effective in hyper-accumulation of heavy metals and can bring about bio-degradation of xenobiotics (Suresh & Ravishankar, 2004). *Dunaliella* algae have been found effective in causing biological removal of heavy metals from wastewater at concentrations of about 85 mg/l (Qari & Hassan, 2014). Biodegradable adsorbents are being developed for the treatment of contaminated waters due to their low cost and greater efficiency in the removal of organic and inorganic pollutants. The application of marine algae and derivatives as potential adsorbents for pharmaceutical products is being studied as an inexpensive and eco-friendly alternative. Different types of marine algae and hydrogel beads of alginate and chitosan have been reported for the bio-removal of antibiotics such as amoxicillin, enrofloxacin, penicillin G and other non-antibiotic drugs such as caffeine.

- **Aerial Plants**: Several plants are being investigated for their potential to remove xenobiotics from soils. The phytoremediation potential of *Cicer arietinum* (black chick pea) has been evaluated for aspirin (Gahlawat, Makhijani, Chauhan, Valsangkar, & Gauba, 2014) and tetracycline (Makhijani, Gahlawat, Chauhan, Valsangkar, & Gauba, 2014) under hydroponic conditions. Recently, Gahlawat & Gauba (2016) have evaluated the phytoremediation of aspirin and tetracycline by *Brassica juncea.*
- **Plant Root Remediation:** Peroxidases (EC 1.11.1.7) are versatile heme-containing biological catalysts that have been used for wide spectrum applications in a number of enzyme based industrial processes to preserve water quality (Ashraf & Husain 2010). Hairy roots in various plants offer an attractive cost-effective alternative to enzyme-based methods in this regard, as their exudates contain enzymes like peroxidase, and some metal chelating compounds to sequester harmful inorganic or organic pollutants, including phenol (Angelini, Orejas, Medina, & Agostini, 2011). Plant root remediation of contaminated soils and groundwater is being recognized as having a great potential for future development due to its environmental compatibility and cost-effectiveness (Zhou, Tang, & Wu, 2013). Phytoremediation of 2,4-dichlorophenol and phenol by *Brassica napus* hairy roots cultures has been reported (Agostini, Coniglio, Milrad, Tigier, & Giulietti, 2003; Angelini et al., 2011; Coniglio et al., 2008). Phytoremediation of phenol at pilot scale has been carried out by tobacco hairy roots and various parameters suggested that these could provide adequate matrix for the successful removal of phenol (Sosa Alderete, Ibáñez, Agostini, & Medina, 2012).
- **Constructed Wetlands:** Constructed wetlands have been viewed as a promising approach for the removal of pharmaceuticals from waters and wastewaters (Li et al., 2014). Constructed wetlands are engineered wetlands that utilize natural processes involving wetland soil, vegetation, and consorted microbial aggregations to assist in treating polluted waters (Helt, Weber, Legge, & Slawson, 2012). Five different wetland plant species have been evaluated for their efficiency in scavenging commonly used recalcitrant fluoroquinolone antibiotics ciprofloxacin, gemifloxacin mesylate, ofloxacin and gatifloxacin from aqueous medium. *Taxodium distichum* was found to be the most suitable for the removal of ofloxacin, gatifloxacin and ciprofloxacin and *Canna indica* was the most effective in the removal of gemifloxacin mesylate (Tandon et al., 2013).

A particular concern in viewing the phytoremediation methods as viable alternative bioremediation techniques is that several of the reported studies have not clearly addressed the relation between the xenobiotic removal and plant uptake of pharmaceuticals. In such studies, photo-degradation and oxidative degradation could also be additional possible contributors and their contribution needs to be ruled out as a possible cause of drug removal. Results from phytoremediation plants show that in many cases, the plant uptake usually represents a low percentage of the mass depleted from the systems. However, it has been debated that phytoremediation depends both on the plant uptake and the rhizospheric microbial communities; therefore, removal efficiencies must be analysed in terms of the overall processes involved (Carvalho et al., 2014).

IMPORTANT BIOREMEDIATION TECHNOLOGIES

Biological treatment technologies in pharmaceutical industrial wastewater treatment are broadly based on aerobic and anaerobic methods (Deegan et al., 2011). Several high-rate reactors have been designed for the biological treatment of pharmaceutical effluents at full-scale operation.

Aerobic Methods

Activated Sludge Process

The conventional activated sludge process (CAS) uses a microbial biomass to treat the influent wastewater. The typical configuration consists of a primary clarifier, aeration tank, a secondary clarifier and a process for returning a portion of the secondary sludge. In the aeration tank, the wastewater is mixed and aerated with the microbial suspension. Under well-controlled operating conditions, the classical aerobic AS reactors can be relatively efficient for the removal of poorly biodegradable persistent drug pollutants diclofenac, sulfophenylcarboxylates and antibiotics, like tetracycline or erythromycin, and hardly-degradable micropollutants, such as carbamazepine.

Several researchers have reported the aerobic biological treatment of pharmaceutical wastewater (Altaf and Ali, 2010; Khan and Mostafa, 2011; Lapara, Nakatsu, Pantea, & Alleman, 2001) with significant reductions in the chemical oxygen demand (COD) and biochemical oxygen demand (BOD) amounting to 75-98%. The use of activated sludge process (ASP) for treatment of wastewater from herbal pharmaceutical industry resulted in 68.98–91.02% reduction in COD, 76.90–97.26% reduction in BOD and 74.61–95.54% reduction in suspended solids (Vanerkar, Satyanarayan, & Dharmadhikari, 2013). The **aerobic** sequencing batch reactor (ASBR) and activated sludge process are the most commonly employed aerobic methods for the treatment of industrial wastewater.

Sequencing batch reactors (SBRs) carry out the activated sludge process in a sequential, time-oriented manner employing a single vessel for all the phases of the process. These offer advantages over the ASP in being more cost-effective due to lower operating costs, amenability to online control, a reduced site space, and adaptability to suit variable wastewater compositions and flow rates (Al-Rekabi et al., 2007). Sequencing batch reactor (SBR) has been used to treat the mixed pharmaceutical industrial resulting in significant removal in BOD (approximately 85%), COD (approximately 87%), ammonia (approximately 96%), suspended solids (approximately 98%), and phosphate (approximately 5%) (Ileri, Sengil, Kulac, & Damar, 2003). Another similar study with sequential batch reactor (SBR) has been reported by Altaf and Ali (2010). Optimization of SBR operating conditions for treatment of high-strength mixed pharmaceutical wastewater with reduction in BOD (94%) and COD (83%) has been carried out by Elmolla et al. (2012).

Rotating Biological Contactor (RBC) Biofilm Process

The rotating biological contactor (RBC) biofilm process is an important biofilm process (attached growth process) in existing treatment facilities for municipal and industrial wastewater treatment with medium strength (COD concentration is 600–2000mg/l) of organic substances but has not been applied yet to pharmaceutical effluents (Akunna & Jefferies, 2000; Vidal, Ba˜nares-Alc´antara, Rodr´ıguez-Roda, & Poch, 2002). The RBC process has the advantages of a high microbial concentrations in good condition, high suitability for fluctuations in flow rate and organic concentration and high efficiency for organic degradation (Chen, Wen, Wang, & Li, 2006; Wang, Liu, Zhang, & He, 2009). RBC also has the advantages of simultaneous nitrification and denitrification accompanied by carbon removal (Hiras, Manariotis, & Grigoropoulos, 2004; Tawfik, Klapwijk, El-Gohary, & Lettinga, 2005). Further, low water content in the sludge causes it to easily settle, allows easy operation and its lower energy consumption makes it a economically viable option as well (Taseli, G¨okc¸ay, & G¨urol, 2008; Torkian, Yazdani, & Alinejad, 2003). Recently, high-strength antibiotic pharmaceutical residual liquid mixed with ordinary wastewater has been treated by a simple three-stage RBC system with resultant COD removal efficiency of 45% and BOD removal efficiency of 85% at low ambient temperatures thus, showing the effectiveness of RBC for such an inherently bio-recalcitrant wastewater (Su et al., 2015). Another study has been reported involving a two-stage RBC for biological removal of the pharmaceuticals belonging to different therapeutic classes including caffeine, sulfamethoxazole, ranitidine and carbamazepine and the maximum removals achieved for all the drugs were greater than 85% (Vasiliadou, Molina, Martínez, & Melero, 2014).

Nitrifying Activated Sludge (NAS) Method

An alternative procedure employing nitrifying bacteria (e.g., *nitrosomonas europaea*) is the nitrifying activated sludge (NAS) method which has been demonstrated to be more effective in the removal of pharmaceuticals. This method has been successfully employed to remove ethinyl estradiol (Vader et al., 2000) and other estrogens (Anderson, Siegrist, Halling-Sorensen, & Ternes, 2003; Kanda & Churchley, 2008) in sewage treatment plants. Batt et al. (2006) have reported higher removal of the drugs iopromide and trimethoprim in the NAS compared to conventional AS method in lab batch reactors as well as full-scale plants. Another comparative study has shown the complete removal of trimethoprim (poorly removed by AS and primary sewage treatments), sulfamethazine, sulfamethoxazole and sulfathiazole in NAS batch experiments on laboratory scale (Perez, Eichhorn, & Aga, 2005). Tran et al (2009) have also reported higher laboratory scale removal of ten APIs (clofibric acid, gemfibrozil, ibuprofen, fenoprofen, ketoprofen, naproxen, diclofenac, indomethacin, propyphenazone and carbamazepine) by the NAS method compared to CAS method. Good removal of the drugs fluoxetine, citalopram, ethinyl estradiol, ibuprofen, naproxen, diclofenac, carbamazepine, trimethoprim, roxithromycin, sulfamethoxazole, erythromycin, galaxolide, tonalide, celestolide and diazepam has been reported with NAS system (Suarez, Lema, & Omil, 2010).

Anaerobic Methods

Anaerobic treatment of the pharmaceutical industrial wastewater (PIWW) involves the step-wise decomposition of the organic material into gaseous methane and CO_2 with the help of microorganisms in the absence of molecular oxygen. Anaerobic treatment systems do not require an energy consuming aeration system and offer several additional advantages over aerobic methods such as good removal efficiency even at high loading rates and low temperature, slow sludge production, need of low nutrient amount and production of biogas (>50% methane) which can be utilized as a renewable energy source for energy production (Nandy & Kaul, 2001; Seghezzo, Zeeman, van Lier, Hamelers, & Lettinga, 1998). Typically, pharmaceutical wastewater is characterized by high chemical oxygen demand (COD), and some pharmaceutical wastewaters can have COD as high as 80,000 mg/l. Due to their high organic content, anaerobic technology is being viewed as a promising alternative for treatment of pharmaceutical effluents. Various reactor configurations are available such as up-flow anaerobic stage reactor (UASR), upflow anaerobic sludge blanket reactor (UASB), anaerobic filter, anaerobic contact reactor (ACR), fluidized bed reactor (FBR) and anaerobic fixed film reactor (AFFR) to treat wastewaters from different industries.

Up-Flow Anaerobic Stage Reactor (UASR)

An up-flow anaerobic stage reactor (UASR) has been investigated as a pre-treatment system at an existing pharmaceutical production plant to reduce the antibiotics in the trade effluent. The pharmaceutical wastewater originally containing macrolide antibiotics (Tylosin and Avilamycin) showed 95% reduction in Tylosin concentrations on UASR treatment (Chelliapan, Wilby, & Sallis, 2006).

Up-Flow Anaerobic Packed Bed Reactor

A fully packed up-flow anaerobic packed bed provides a quiescent inlet region for large dense biomass aggregates to develop, which are not prone to washout. Furthermore, the reactor provides a surface to facilitate biofilm accumulation. This assures a shorter start-up period due to greater amount of retained inoculum. An anaerobic packed bed reactor has been recently designed and employed to treat highly polluted of pharmaceutical wastewater of Samarra Drug Factory (SDI) in Iraq showing best hydraulic retention time (HRT) of 24 hours for anaerobic stage and removal efficiencies of COD and BOD as 87% and 90% (Hashimi, Al-Ali, & Abbas, 2011). A similar treatment of highly polluted pharmaceutical wastewater employing an anaerobic packed bed reactor has also been reported by Chelliapan et al., (2011). An Up-Flow Anaerobic Packed Bed reactor has been employed to treat pharmaceutical wastewater containing recalcitrant antibiotics. Results from the study showed a COD reduction of 60–70% at an organic loading rate (OLR) of 1.5–4.6 kg $COD.m^{-3}.d^{-1}$, suggesting that the biomass had acclimated to the antibiotics (Chelliapan & Golar, 2011). Antibiotic wastewaters containing a high proportion of spent fermentation broths usually require long HRT for efficient treatment (Lapara et al., 2001), presumably, on account of their complex organic carbon content. Hence, this factor can limit the performance of the packed bed reactor at lower HRT (2 d).

Wastewaters from pharmaceutical industry usually contain a high concentration of sulphate as a result of sulphate use during down-stream processing of fermentation broth. The production of sulphide during anaerobic treatment of sulphate containing wastewaters can inhibit methanogens and can precipitate the nutrients required by the methanogens (Ozdemir, Cirik, Akman, Sahinkaya, & Cinar, 2013). An anaerobic reactor system has been developed employing phase separation (especially at high OLRs) by accomodating different microbial populations in different stages according to their favourable substrate conditions. This two-stage anaerobic digestion system has been recently utilized for anaerobic treatment pharmaceutical wastewaters with high sulphate content (Chelliapan & Sallis, 2015).

Upflow Anaerobic Sludge Blanket (UASB) Reactor

A hybrid upflow anaerobic sludge blanket (UASB) reactor is a single tank process in an anaerobic centralised or decentralised industrial wastewater or blackwater treatment system achieving high removal of organic pollutants. A UASB reactor operates as a suspended growth system where microorganisms attach themselves to each other or to small particles of suspended matter to form agglomerates of granules which settle easily. The latter form an active suspended sludge blanket that filters and treats the wastewater as it flows upwards through it. The process combines the advantages of the UASB and anaerobic filter (AF) concepts. This method has been widely used for the treatment of wastewater from the beverage, brewery, food and tannery industries (To´th, Gergely F, & Mizsey, 2011) and is reported as a preferred option for certain chemical synthesis-based pharmaceutical wastewaters due to several operational advantages over other reactor configurations. These advantages include high organic loadings, greater efficiency, low energy requirements, shorter hydraulic retention time (HRT) and easy reactor construction. (Oktem, O. Ince, Sallis, Donnelly, & B. Ince, 2007). It is a popular anaerobic reactor for both high and low temperatures. Among the recently developed high-rate processes, it has attracted most commercial and research interests (Lettinga, van Velsen, Hohma, de Zeeuw, & Klapwijk, 1980). A study by Rodriguez-Martinez et al (2005) in an Up-Flow Anaerobic Sludge Bed (UASB) treating pharmaceutical wastewater containing Penicillin G antibiotics has shown a COD removal efficiency of 90% at an OLR of 1.5 kg COD.m^{-3}.d^{-1} and HRT 11 d. The performance of a pilot-scale UASB reactor, treating an antibiotic chemical synthesis-based pharmaceutical wastewater, has been recently evaluated under different operating conditions with good results (Rao, Habiulla, Prasad, & Anand, 2014).

The hybrid UASB reactor has been evaluated for efficacy in reduction of chemical oxygen demand (COD) and biochemical oxygen demand (BOD) of bulk drug pharmaceutical wastewater containing hazardous compounds like phenol, 1,2-methoxy phenol, 2,4,6-trichloro phenol, dibutyl phthalate, 1-bromo naphthalene, carbamazepine and antipyrine. After the treatment, all the pharmaceutical compounds degraded almost completely except carbamazepine. Thermophilic *methanothrix* and *methanosaetae* like bacteria were present in the granular sludge (Sreekanth, Sivaramakrishna, Himabindu, & Anjaneyulu, 2009). Herbal pharmaceutical wastewater treatment has been reported by Satyanarayan et al (2009) by a pilot scale UASB reactor resulting in COD, BOD and suspended solids (SS) removal in the range of 86.2%-91.6%, 90.0%-95.2% and 62.6%-68.0% respectively.

Low-temperature or psychrophilic ($<20^0$C) anaerobic digestion (PAD) has recently been demonstrated as a cost-effective option for the treatment of solvent containing pharmaceutical wastewater (Enright, McHugh, Collins, & O'Flaherty, 2005). Two UASB reactors with different operating temperatures, mesophilic (35 ± 1°C) and thermophilic (54 ± 1°C) have been studied for the treatment of a non-penicillin based product factory waste water (ISA, Kutty & Bin Abd Rahim, 2010) affording the highest percentage of COD removal for the mesophilic and thermophilic reactors as 95% and 93% respectively.

Structured Packing Rotating Biological Contactor (spRBC)

A removal process of the antibiotic enrofloxacin has utilized a biological treatment unit consisting of a structured packing rotating biological contactor (spRBC). The removal rate reached a maximum of 70% in this biological unit and removal efficiency for a combined system including spRBC and an ultraviolet radiation unit was 94% (Zamanpour & Mehrabani-Zeinabad, 2014).

Anaerobic Fixed Film Reactor (AFFR)

The anaerobic fixed film reactor (AFFR) possesses a biofilm support structure for biomass attachment. Utilization of the anaerobic fixed film reactor (AFFR) is reported to be more effective due to simplicity of construction, stability, higher threshold to absorb toxic shock load and faster recovery. An AFFR may employ glass beads, red drain clay, polyurethane foam, waste tyre rubber, splintered glass, polyacryl nitrile acrylamide, corrugated plastic, coconut coir, and charcoal or nylon fibre as a support media (Borja, Sanchez, Marin, & Jimenez, 1996; Garcia, Rivas, Figueroa, & Monroe, 1995; Perez, Romero,

Nebot, & Sales, 1997). The laboratory scale bio-methanation of high strength herbal pharmaceutical wastewater has been achieved with anaerobic fixed film fixed bed reactor system (Nandy & Kaul, 2001) with COD removal efficiencies ranging from 76 to 98%. Studies have been carried out on the treatment of wastewater from a bulk drug industry employing an AFFR designed and fabricated in the laboratory The AFFR has been shown to be very efficient in lowering COD, total dissolved solids (TDS) and total suspended solids (TSS) (Rao et al., 2004).

Anaerobic Fluidized Bed (AFB) Reactor

The anaerobic fluidized bed (AFB) reactor comprises small media, such as granular activated carbon or sand to which the bacteria attach. High flow rate around the particles results in decreased clogging. The large pore spaces formed through bed expansion and high specific surface area of the carriers significantly improves the efficiency of the fluidized bed reactors. The expanded granular sludge bed (EGSB) reactor is a modification of the AFB reactor with a difference that the fluid's upward flow velocity is not as high which results in partial bed fluidization (Rittmann & McCarty, 2001). Bioaugmentation and anaerobic treatment of pharmaceutical effluent has been carried out in fluidized bed reactor affording maximum COD removal of 91.2% (Saravanane, Murthy & Krishnaiah, 2001a). The treatment of cephalexin drug-based pharmaceutical effluent in an upflow anaerobic fluidized bed system coupled with bio-augmentation also resulted in 88.5% COD removal (Saravanane, Murthy & Krishnaiah, 2001b). A large quantity of concentrated effluent resulting from the production of cephradine, a cephalosporin antibiotic (COD of 14000-18000 mg/l) has been treated using a fluidized bed reactor under anaerobic conditions with bioaugmentation (Saravanane, Murthy & Krishnaiah, 2001c). Degradation of phenol has been studied in anaerobic fluidized bed reactors (AFBR) packed with polymeric particulate supports including polystyrene, polyvinyl chloride (PVC) and polyethylene terephthalate (PET). Excellent removal efficiencies of ranging from 90-100% have been demonstrated (Sancinetti et al., 2012).

Anaerobic Suspended Film Contact Reactor

Bio-treatability studies have been reported on the effluents from a bulk drug industry using an anaerobic suspended film contact reactor (Mohan, Prakasham, Satyavathi, Annapurna, & Ramakrishna, 2001). The COD reduction was found to be in the range of 60 to 80% for organic loading rates varying from 0.25 kg/m^3/day to 2.5 kg/m^3/day.

Membrane Bioreactors (MBR)

MBR technology is a combination of the activated sludge (AS) process with a micro- or ultrafiltration membrane system (Mutamim, Noor, Hassan, & Olsson, 2012). The activated sludge treatment is responsible for the biodegradation of the organic waste compounds, and the membrane system (pore diameter 0.01-0.1 μm) physically separates the solids from the liquid (Hoinkis et al., 2012). The membrane bioreactors (MBR) have been reported to be much more efficient than classical AS systems (Kimura, Hara, & Watanabe, 2007; Radjenović, Petrovic, & Barcelo, 2009). At worst, these systems perform as well as activated sludge treatment (Clara et al., 2005; Joss et al., 2005). Recently, several reports have come up emphasizing the relevance of this technique in the current bioremediation scenario. Membrane bioreactor (MBR) has been used for the large-scale wastewater treatment of industrial wastewater in the pharmaceutical industry (C. Chang, J. Chang, Vigneswaran, & Kandasamy 2008). MBRs can operate at high suspended solid concentrations because the membrane separation step is not limited by sedimentation capacity, which will result in longer solid retention times (SRTs) (Cirja, Ivashechkin, Schäffer, & Corvini, 2008). These conditions can be favourable for API (active pharmaceutical ingredient) biodegradation as more diverse microbial population can flourish at longer SRTs resulting in diverse biodegradation pathways (Clara, Kreuzinger, Strenn, Gans, & Kroiss, 2005). Moreover, the extended contact time gives the bacterial biomass an opportunity to acclimate to certain xenobiotics by broadening their enzymatic (Suarez et al., 2010) or metabolic capabilities (Ternes, Joss, & Siegrist, 2004). An external loop airlift membrane bioreactor (ELAMBR) has been applied for treatment of a synthetic pharmaceutical wastewater and the performance of this system was evaluated in removal of acetaminophen as the main pollutant of the pharmaceutical wastewater. The removal efficiency was found to be significantly higher in the ELAMBR system compared to the AS process and 100% of the acetaminophen was removed within two days (Shariati et al., 2010). Another comparative study has been conducted with a laboratory-scale MBR and the AS process on pharmaceutical wastewater containing several drugs like diclofenac, ketoprofen, ranitidine, gemfibrozil, bezafibrate, pravastatin, ofloxacin, ibuprofen, naproxen, acetaminophen, paroxetine, and hydrochlorothiazide. Pharmaceuticals were, in general, removed to a greater extent by the MBR integrated system than by the AS process (Radjenović et al., 2009). Treatment of pharmaceutical wastewater containing three NSAIDs (acetaminophen, ketoprofen and naproxen) and three antibiotics (roxithromycin, sulfamethoxazole and trimethoprim) has been reported in two MBRs with sludge retention times (SRTs) of 15 (MBR-15) and 30 (MBR-30) days over a period of 4 weeks. The NSAIDs were removed with higher efficiency compared to the antibiotics, and the MBR-30 afforded better removal efficiencies than the MBR-15 for all the drugs (Tambosi et al., 2010). A very similar study with similar results has been reported by Schröder et al. (2012) with wastewater containing three NSAIDs (acetaminophen, ketoprofen and naproxen) and three antibiotics (roxithromycin, sulfamethoxazole and trimethoprim) performed MBR-15 and MBR-30. Removal rates ranged from 55% (sulfamethoxazole) up to 100% (acetaminophen, ketoprofen). An efficient removal of ammonia, total organic matter and four corticosteroids, levonorgestrel, betamethasone valerate, betamethasone dipropionate (Betha-D) and medroxyprogesterone acetate, has been achieved in a laboratory-scale submerged membrane bioreactor (MBR) treating wastewater from a pharmaceutical industry. Chemical analysis of MBR effluent revealed that all the corticosteroids were removed at high percentages (> 93%) in all the cases (López-Fernández, Martínez, & Villaverde, 2012). The application of membrane separation technology (MBR) on pharmaceutical wastewater has been recently reviewed (Cazes, Abejón, Belleville, & Sanchez-Marcano, 2014; Fazal, Zhang, Zhong, Gao & Lu, 2015).

Combined Anaerobic-Aerobic Processes

Although, innumerable reports have come up relating to the successful organic removal from pharmaceutical wastewater employing varying configurations of anaerobic reactors, yet, there have been differences in anaerobic treatment performances possibly explained by the differing wastewater characteristics including salinity/toxicity and presence of refractory compounds poorly biodegraded under anaerobic conditions. In light of the tightening legislations on effluent water quality standards, a subsequent aerobic process is required for further removal of organic residues from the anaerobic effluents (Chen, Ren, Wang, Zhang, & Shi, 2008). Application of activated sludge system as an aerobic post-treatment process has been shown to further enhance the COD removal rate from 62% to 96.4% in the treatment of purified terephthalic acid wastewater (Pophali, Khan, Dhodapkar, Nandy, & Devotta, 2007). This sequential two stage process has been widely employed for the treatment of paper and pulp industry effluents (Tezel, Guven, Erguder, & Demirer, 2001), textile wastewaters, food solid waste leachates, etc. Agdag and Sponza (2005) have employed the combination of an anaerobic sludge blanket reactor (UASB) and completely stirred tank reactor (CSTR) systems for the treatment of municipal landfill leachate. Isik and Sponza (2008) have reported the anaerobic/aerobic treatment of a simulated textile wastewater. Recently, granular sludge was successfully cultivated employing an intermittent anaerobic and aerobic reaction phase strategy using synthetic textile wastewater containing mixed dyes with 93% removal of COD and ammonia (Muda, Aris, Salim, Ibrahim, & Yahya, 2012). The application of sequential anaerobic/aerobic treatment to pharmaceutical wastewater treatment has been fairly recent and relatively fewer applications have been reported. In a recent study, pharmaceutical wastewater with high total dissolved solids (TDSs) and chemical oxygen demand (COD) content has been treated through a sequential anaerobic–aerobic treatment process. For the anaerobic process, application of up-flow anaerobic sludge blanket (UASB) resulted in COD removal efficiency of 41.3 ± 2.2% achieved with an organic loading rate of 8.11 ± 0.31 g COD/l/d. Significant improvement was achieved by treating the UASB effluent by a membrane bioreactor (MBR) and a sequencing batch reactor (SBR). Both the UASB + MBR and UASB + SBR systems have demonstrated an excellent organic removal efficiency, with COD removal of 94.7% and 91.8% respectively. Best performance was noted with UASB + MBR system in terms of organic removal as well as nitrification (Shi, Lefebvre, K. Ng, & H. Ng, 2014) The continuous sequential biological anaerobic/aerobic treatment of Samarra Drugs Factory wastewater has included a pilot plant of upflow anaerobic filter (UAF) (with mesophilic bacterial growth at a temperature range of 35-37 °C) for the anaerobic stage followed by an air diffuser for the aerobic stage. However, there was no significant advantage in terms of the removal efficiency of COD and BOD in anaerobic/aerobic stage (92% and 93% respectively) over the anaerobic stage alone (87% and 90% respectively) (Hashimi & Abbas, 2007). A pilot-scale system composed of an anaerobic baffled reactor followed by a biofilm airlift suspension reactor has been designed for removing organic matter from the high-strength pharmaceutical wastewater containing antibiotics ampicillin and aureomycin (Zhou, Su, Li, Qian, 2006). Recently, ozone and thermal hydrolysis pre-treatment has been employed to improve anaerobic digestion (AD) efficiency to reduce tetracycline resistance genes in municipal and pharmaceutical waste sludge. Thermal hydrolysis pre-treatment was found to be more efficient than ozone for reduction after AD (Pej, Yao, Ren & Yu, 2016).

Coupling Bioremediation with Alternative Techniques

Bioremediation measures have also coupled with other alternative or conventional wastewater management processes to ensure more effective removal of the pollutants. An example is provided by a recent report wherein, distinct laboratory scale biological processes: suspended activated sludge and attached-growth biomass or moving bed biofilm flow-through reactors have been investigated for wastewater treatment to remove pharmaceuticals including ibuprofen, naproxen, ketoprofen, carbamazepine and diclofenac and an active metabolite of the lipid regulating agent clofibric acid. The suspended activated sludge process resulted in 74% removal of ibuprofen, naproxen and ketoprofen but removal of clofibric acid, carbamazepine and diclofenac was inconsistent. However, the coupling of the attached-growth biomass biological treatment with UV treatment and hydrodynamic cavitation/hydrogen peroxide process afforded excellent removal efficiencies for clofibric acid (>90%) and carbamazepine and diclofenac (>98%), while the remaining compounds were reduced to levels below their limits of detection (LOD). A combined treatment process (solar photo-Fenton + immobilized biomass reactor) was utilized for the treatment of pharmaceutical industrial wastewater containing nalidixic acid (an antibiotic pertaining to the quinolone group), which is not easily biodegraded. This process resulted in 95% reduction in COD out of which, 62% reduction was accounted for by the immobilized biomass reactor (Sirtori et al., 2009). The coupling of solar photo-Fenton process with aerobic sequencing batch reactor attained COD removal of 98% in pharmaceutical wastewater (Adishkumar, Sivajothi, & Banu, 2012).

FUTURE RESEARCH DIRECTIONS

The emerging field of bioremediation of pharmaceuticals is quite challenging due to the plethora of information still waiting to be explored not only in relation to their removal from to the environment but also for gaining insight into their chemical characteristics and susceptibilities. The structural diversity of the pharmaceuticals encompasses a vast organic chemical space, hence, diverse range of approaches are required to be explored for their effective and complete removal from the environment. There is a need to have comprehensive information on the most common pharmaceutical contaminants in the environment. Further, there is limited information on the deleterious impact of low level pharmaceutical contaminants on human and animal health. As drug selection in different geographical regions varies depending upon economic and social factors, the potential bioremediation measures need to be customized according to the contaminants present. Generation of comprehensive database of common pharmaceutical pollutants in different geographical regions of the world should be the first step in this direction. In-depth and accurate analysis of industrial effluents and wastewater and ground water needs to be carried out in order to assess the most common pharmaceutical pollutants present. Particularly recalcitrant and stubborn drugs and pharmaceuticals need to be identified in terms of their prevalence in wastewater, groundwater and soils to chalk out a future course of action for devising specific bioremediation measures. Pharmaceutical research is a fast evolving field and with the introduction of newer pharmaceuticals in therapeutic arena, newer challenges will arise for their safe disposal.

CONCLUSION

Presence of wide variety of pharmaceuticals in the environment is a cause of concern on account of their harmful effects to the humans and also due to their deleterious effects on terrestrial as well as aquatic plant/animal life. Bioremediation of pharmaceuticals is still an emerging field. In recent times, there has been a growing awareness about this global issue but the real gravity of the problem is yet to be recognized and understood completely. The chemical diversity of pharmaceuticals, frequent introduction of newer pharmaceuticals into the market and poor existing information about the susceptibility of different types of plant and animal species to their toxic effects adds up to the complexity of the problem. Several bioremediation measures have been proposed in the recent times utilizing the versatility of various types of microorganisms such as bacteria and fungi to catabolize or degrade different types of pharmaceutical contaminants. Several configurations of anaerobic and aerobic reactors have been devised which have demonstrated their potential to detoxify pharmaceutical industrial effluents by removing the various drugs, solvents and other non-pharmaceutical raw materials used in the pharmaceutical manufacturing.

Enzyme based studies have also yielded encouraging results but their costs involved could preclude their wide applicability. Phytoremediation has been utilized for removal of a number of non-pharmaceutical industry effluents but recently, this is gaining prominence for its potential to degrade pharmaceuticals as well. All these represent green or eco-friendly approaches towards conservation of the environment and in the coming times, this field in expected to grow tremendously because the scientific community as well as the general population is getting increasingly sensitized towards alleviation of this problem.

REFERENCES

Adishkumar, S., Sivajothi, S., & Banu, J. R. (2012). Coupled solar photofenton process with aerobic sequential batch reactor for treatment of pharmaceutical wastewater. *Desalination and Water Treatment*, *48*(1-3), 89–95. doi:10.1080/19443994.2012.698799

Agarry, S. E., & Solomon, B. O. (2008). Kinetics of batch microbial degradation of phenols by indigenous *Pseudomonas fluorescence. International Journal of Environmental Science and Technology*, *5*(2), 223–232. doi:10.1007/BF03326016

Agdag, O. N., & Sponza, D. T. (2005). Anaerobic/aerobic treatment of municipal landfill leachate in sequential two-stage up-flow anaerobic sludge blanket reactor (UASB)/completely stirred tank reactor (CSTR) systems. *Process Biochemistry*, *40*(2), 895–902. doi:10.1016/j.procbio.2004.02.021

Akunna, J. C., & Jefferies, C. (2000). Performance of family-size sequencing batch reactor and rotating biological contactor units treating sewage at various operating conditions. *Water Science and Technology*, *41*(1), 97–104.

Altaf, M. S., & Ali, T. A. (2010). Wastewater treatment using sequential batch reactor and development of microbiological method for the analysis of relative toxicity. *Pakistan Journal of Nutrition*, *9*(6), 574–576. doi:10.3923/pjn.2010.574.576

Andersen, H., Siegrist, H., Halling-Sorensen, B., & Ternes, T. A. (2003). Fate of estrogens in a municipal sewage treatment plant. *Environmental Science & Technology*, *37*(18), 4021–4026. and doi:10.1021/es026192a

Angayarkanni, J., Palaniswamy, M., & Swaminathan, K. (2003). Biotreatment of distillery effluent using *Aspergillus niveus. Bulletin of Environmental Contamination and Toxicology*, *70*(2), 268–277. and doi:10.1007/s00128-002-0187-2

Angelini, V. A., Orejas, J., Medina, M. I., & Agostini, E. (2011). Scale up of 2,4-dichlorophenol removal from aqueous solutions using *Brassica napus* hairy roots. *Journal of Hazardous Materials*, *85*(1), 269–274. and doi:10.1016/j.jhazmat.2010.09.028

Ashraf, H., & Husain, Q. (2010). Studies on bitter gourd peroxidase catalyzed removal of p-bromophenol from wastewater. *Desalination*, *262*(1-3), 267–272. doi:10.1016/j.desal.2010.05.044

Aust, S. D. (1990). Degradation of environmental pollutants by *Phanerochaete chyrosporium. Microbial Ecology*, *20*(1), 197–209. and doi:10.1007/BF02543877

Badia-Fabregat, M., Lucas, D., Gros, M., Rodríguez-Mozaz, S., Barceló, D., Caminal, G., & Vicent, T. (2015). Identification of some factors affecting pharmaceutical active compounds (PhACs) removal in real wastewater. Case study of fungal treatment of reverse osmosis concentrate. *Journal of Hazardous Materials*, *283*, 663–671. and doi:10.1016/j.jhazmat.2014.10.007

Badia-Fabregat, M., Lucas, D., Pereira, M. A., Alves, M., Pennanen, T., Fritze, H., & Caminal, G. et al. (2016). Continuous fungal treatment of non-sterile veterinary hospital effluent: Pharmaceuticals removal and microbial community assessment. *Applied Microbial Biotechnology*, *100*(5), 2401–2415. and doi:10.1007/s00253-015-7105-0

Baquero, F., Tedim, A. P., & Coque, T. M. (2013). Antibiotic resistance shaping multi-level population biology of bacteria. Frontiers in Microbiology, 4, 15.10.3389/fmicb.2013.00015

Basu, S., Dastidar, S. G., Mukhopadhyay, S., & Gangopadhyay, A. (2012). Utilization of bulk drugs by a highly antibiotic resistant microorganism isolated from waste water of a bulk drug industry. *Journal of Pharmaceutical Science and Technology*, *2*(1), 41–46.

Batt, A. L., Kim, S., & Aga, D. S. (2006). Enhanced biodegradation of iopromide and trimethoprim in nitrifying activated sludge. *Environmental Science & Technology*, *40*(23), 7367–7373. and doi:10.1021/es060835v

Billore, S. K., Singh, N., Ram, H. K., Sharma, J. K., Singh, V. P., Nelson, R. M., & Dass, P. (2001). Treatment of a molasses based distillery effluent in a constructed wetland in central India. *Water Science and Technology*, *44*(11–12), 441–448.

Borja, R., Sanchez, E., Marin, A., & Jimenez, A. M. (1996). Kinetic behaviour of waste tyre rubber as microorganism support in an anaerobic digester treating cane molasses distillery slops. *Bioprocess and Biosystems Engineering*, *16*(1), 17–23. doi:10.1007/s004490050282

Campos, V. M., Merino, I., Casado, R., Pacios, L. F., & Gómez, L. (2008). Phytoremediation of organic pollutants. *Spanish Journal of Agricultural Research*, *6*(S1), 38–47. doi:10.5424/sjar/200806S1-372

Carvalho, P. N., Basto, M. C. P., Almeida, C. M. R., & Brix, H. (2014). A review of plant-pharmaceutical interactions: From uptake and effects in crop plants to phytoremediation in constructed wetlands. *Environmental Science and Pollution Research International*, *21*(20), 11729–11763. and doi:10.1007/s11356-014-2550-3

Cazes, de M., Abejón, R., Belleville, M. & Sanchez-Marcano, J. (2014). Membrane bioprocesses for pharmaceutical micropollutant removal from waters. *Membranes (Basel)*, *4*(4), 692–729.

Chang, C. Y., Chang, J. S., Vigneswaran, S., & Kandasamy, J. (2008). Pharmaceutical wastewater treatment by membrane bioreactor process—a case study in southern Taiwan. *Desalination*, *234*(1-3), 393–401. doi:10.1016/j.desal.2007.09.109

Chelliapan, S., & Golar, S. (2011). Pre-treatment of antibiotic wastewater using an anaerobic reactor. *Research Journal of Chemical Sciences*, *1*(6), 59–63.

Chelliapan, S., & Sallis, P. J. (2015). Anaerobic treatment of high sulphate containing pharmaceutical wastewater. *Journal of Scientific and Industrial Research*, *74*, 526–530.

Chelliapan, S., Wilby, T., & Sallis, P. J. (2006). Performance of an up-flow anaerobic stage reactor (UASR) in the treatment of pharmaceutical wastewater containing macrolide antibiotics. *Water Research*, *40*(3), 507–516. and doi:10.1016/j.watres.2005.11.020

Chelliapan, S., Yuzir, A., Din, M. F., & Sallis, P. J. (2011). Anaerobic pre-treatment of pharmaceutical wastewater using packed bed reactor. *International Journal of Chemical Engineering and Applications*, *2*(1), 32–37. doi:10.7763/IJCEA.2011.V2.71

Chen, Z., Ren, N., Wang, A., Zhang, Z., & Shi, Y. (2008). A novel application of TPAD–MBR system to the pilot treatment of chemical synthesis-based pharmaceutical wastewater. *Water Research*, *42*(13), 3385–3392. and doi:10.1016/j.watres.2008.04.020

Chen, Z., Wen, Q., Wang, J., & Li, F. (2006). Simultaneous removal of carbon and nitrogen from municipal-type synthetic wastewater using net-like rotating biological contactor (NRBC). *Process Biochemistry*, *41*(12), 2468–2472. doi:10.1016/j.procbio.2006.06.003

Chiong, T., Lau, S., Khor, E., & Danquah, M. (2014). Enzymatic approach to phenol removal from wastewater using peroxidases. *Biochemical Engineering & Bioprocess Engineering. OA Biotechnology*, *3*(1), 9.

Cirja, M., Ivashechkin, P., Schäffer, A., & Corvini, P. (2008). Factors affecting the removal of organic micropollutants from wastewater in conventional treatment plants (CTP) and membrane bioreactors (MBR). *Reviews in Environmental Science and Biotechnology*, *7*(1), 61–78. doi:10.1007/s11157-007-9121-8

Clara, M., Kreuzinger, N., Strenn, B., Gans, O., & Kroiss, H. (2005). The solids retention time - a suitable design parameter to evaluate the capacity of wastewater treatment plants to remove micropollutants. *Water Research*, *39*(1), 97–106. and doi:10.1016/j.watres.2004.08.036

Clara, M., Strenn, B., Gans, O., Martinez, E., Kreuzinger, N., & Kroiss, H. (2005). Removal of selected pharmaceuticals, fragrances and endocrine disrupting compounds in a membrane bioreactor and conventional wastewater treatment plants. *Water Research*, *39*(19), 4797–4807. and doi:10.1016/j.watres.2005.09.015

Coniglio, M. S., Busto, V. D., González, P. S., Medina, M. I., Milrad, S., & Agostini, E. (2008). Application of *Brassica napus* hairy root cultures for phenol removal from aqueous solutions. *Chemosphere*, *72*(7), 1035–1042. and doi:10.1016/j.chemosphere.2008.04.003

Cripps, C., Bumpus, J. A., & Aust, S. D. (1990). Biodegradation of azo and heterocyclic dyes by *Phanerochaete chyrosporium. Applied and Environmental Microbiology*, *56*, 1114–1118.

Cruz-Morató, C., Ferrando-Climent, L., Rodriguez-Mozaz, S., Barceló, D., Marco-Urrea, E., Vicent, T., & Sarrà, M. (2013). Degradation of pharmaceuticals in non-sterile urban wastewater by *Trametes versicolor* in a fluidized bed bioreactor. *Water Research*, *47*(14), 5200–5210. and doi:10.1016/j.watres.2013.06.007

Cruz-Morató, C., Lucas, D., Llorca, M., Rodriguez-Mozaz, S., Gorga, M., Petrovic, M., & Marco-Urrea, E. et al. (2014). Hospital wastewater treatment by fungal bioreactor: Removal efficiency for pharmaceuticals and endocrine disruptor compounds. *The Science of the Total Environment*, *493*, 365–376. and doi:10.1016/j.scitotenv.2014.05.117

Das, M. P., Bashwant, M., Kumar, K., & Das, J. (2012). Control of pharmaceutical effluent parameters through bioremediation. *Journal of Chemical and Pharmaceutical Research*, *4*(2), 1061–1065.

Deegan, A. M., Shaik, B., Nolan, K., Urell, K., Oelgemo¨ller, M., Tobin, J., & Morrissey, A. (2011). Treatment options for wastewater effluents from pharmaceutical companies. *International Journal of Environmental Science and Technology*, *8*(3), 649–666. doi:10.1007/BF03326250

Deziel, N. (2014). Pharmaceuticals in wastewater treatment plant effluent waters. *Scholarly Horizons: University of Minnesota. Morris Undergraduate Journal*, *1*(2), 12.

Dixit, D., & Parmar, N. (2013). Treatment of pharmaceutical waste water by electro-coagulation and natural coagulation process[review]. *VSRD International Journal of Environmental Science and Technology*, *4*(5), 79–88.

Drewes, J. E. (2007). Removal of pharmaceutical residues during wastewater treatment. In M. Petrovic & D. Barcelo (Eds.), *Comprehensive analytical chemistry* (pp. 427–446). Oxford, UK: Elsevier.

Dridi, D., Bchir, F., Zouiten, A., Tahrani, L., & Mansour, H. B. (2013). Murine chronotoxicity to pharmaceutical wastewater. *Biological Rhythm Research*, *45*(2), 167–181. doi:10.1080/09291016.2013.787684

Drillia, P., Dokianakis, S. N., Fountoulakis, M. S., Kornaros, M., Stamatelatou, K., & Lyberatos, G. (2005). On the occasional biodegradation of pharmaceuticals in the activated sludge process: The example of the antibiotic sulfamethoxazole. *Journal of Hazardous Materials*, *122*(3), 259–265. and doi:10.1016/j.jhazmat.2005.03.009

Duffner, F. M., Kirchner, U., Bauer, M. P., & Muller, R. (2000). Phenol/cresol degradation by the thermophillic *Bacillus thermoglucosidasius* A7: Cloning and sequence analysis of five genes involved in the pathway. *Gene*, *256*(1-2), 215–221S. and doi:10.1016/S0378-1119(00)00352-8

Elmolla, E. S., Ramdass, N., & Choudhari, M. (2012). Optimization of Sequencing Batch Reactor operating conditions for treatment of high strength pharmaceutical wastewater. *Journal of Environmental Science and Technology*, *5*(6), 452–459. doi:10.3923/jest.2012.452.459

Enright, A.-M., McHugh, S., Collins, G., & OFlaherty, V. (2005). Low-temperature anaerobic biological treatment of solvent containing pharmaceutical wastewater. Water Research, 39(19), 4587–4596.10.1016/j.watres.2005.08.037

Fang, H. H. P., Chui, H. K., & Li, Y. Y. (1994). Microbial structure and activity of UASB granules treating different wastewaters. *Water Science and Technology*, *30*, 87–96.

Fazal, S., Zhang, B., Zhong, Z., Gao, L., & Lu, X. (2015). Membrane separation technology on pharmaceutical wastewater by using MBR (Membrane Bioreactor). *Journal of Environmental Protection*, *6*(04), 299–307. doi:10.4236/jep.2015.64030

Freer, N., & Detroy, R. W. (1982). Biological delignification of C-labeled lignocelluloses by basidiomycetes: Degradation and solubilization of the lignin and cellulose components. *Mycologia*, *74*(6), 943–951. doi:10.2307/3792724

Gahlawat, S., & Gauba, P. (2016). Phytoremediation of aspirin and tetracycline by *Brassica juncea*. *International Journal of Phytoremediation*, *18*(9), 929–935. and doi:10.1080/15226514.2015.1131230

Gahlawat, S., Makhijani, M., Chauhan, K., Valsangkar, S., & Gauba, P. Accessing the phytoremediation potential of *Cicer arietinum* for aspirin. *International Journal of Genetic Engineering and Biotechnology, 5*(2), 161-168.

Garcia, A., Rivas, H. M., Figueroa, J. L., & Monroe, A. L. (1995). Case history: Pharmaceutical wastewater treatment plant upgrade, Smith Kline Beecham Pharmaceuticals Company. *Desalination, 102*(1–3), 255–263. doi:10.1016/0011-9164(95)00061-6

Gartiser, S., Elke, U., Radka, A., & Kümmerer, K. (2007). Ultimate biodegradation and elimination of antibiotics in inherent tests. *Chemosphere, 67*(3), 604–613. and doi:10.1016/j.chemosphere.2006.08.038

Ghosh, M., Verma, S. C., Mengoni, A., & Tripathi, A. K. (2004). Enrichment and identification of bacteria capable of reducing chemical oxygen demand of anaerobically treated spent wash. *Journal of Applied Microbiology, 6*, 241–278.

Gilbert, N. (2012). Drug-pollution law all washed up. Nature. *International Weekly Journal of Science., 491*(7425), 503–504. and doi:10.1038/491503a

Gómez, J. L., Bódalo, A., Gómez, E., Hidalgo, A. M., Gómez, M., & Murcia, M. D. (2008). A transient design model of a continuous tank reactor for removing phenol with immobilized soybean peroxidase and hydrogen peroxide. *Chemical Engineering Journal, 145*(1), 142–148. doi:10.1016/j.cej.2008.07.050

Guleria, A. (2014). Biodegradation of pharmaceutical waste, 4-nitrophenol by *Pseudumonas putida. Life Sciences, 3*(5), 127–130.

Hashimi, M. A. I., & Abbas, A. H. (2007). Sequential anaerobic/aerobic treatment of pharmaceutical wastewater. *Tikrit Journal of Eng. Sciences, 14*(2), 1–31.

Hashimi, M. A. I., Al-Ali, M. I., & Abbas, A. H. (2011). Anaerobic treatment of pharmaceutical wastewater using packed bed reactor. *Journal of Environmental Studies, 7*, 33–40.

Helt, C. D., Weber, K. P., Legge, R. L., & Slawson, R. M. (2012). Antibiotic resistance profiles of representative wetland bacteria and faecal indicators following ciprofloxacin exposure in lab-scale constructed mesocosms. *Ecological Engineering, 39*, 113–122. doi:10.1016/j.ecoleng.2011.11.007

Hiras, D. N., Manariotis, I. D., & Grigoropoulos, S. G. (2004). Organic and nitrogen removal in a two-stage rotating biological contactor treating municipal wastewater. *Bioresource Technology, 93*(1), 91–98. and doi:10.1016/j.biortech.2003.06.005

Hoinkis, J., Deowan, S. A., Panten, V., Figoli, A., Huang, R. R., & Drioli, E. (2012). Membrane bioreactor (MBR) technology – a promising approach for industrial water reuse. Procedia Engineering, 33, 234–241. 10.1016/j.proeng.2012.01.1199

Ileri, R., Sengil, I. A., Kulac, S., & Damar, Y. (2003). Treatment of mixed pharmaceutical industry and domestic wastewater by sequencing batch reactor. *Journal of Environmental Science and Health. Part A, Toxic/Hazardous Substances & Environmental Engineering, 38*(10), 2101–2111. and doi:10.1081/ESE-120023336

Isa, M. H., Kutty, S. M., & Bin Abd Rahim, M. F. (2010). *Anaerobic treatment of pharmaceutical wastewater under mesophilic and thermophilic conditions.International Conference on Environment.*

Isik, M., & Sponza, D. T. (2008). Anaerobic/aerobic treatment of a simulated textile wastewater. *Separation and Purification Technology*, *60*(1), 64–72. doi:10.1016/j.seppur.2007.07.043

Jayaraman, A., Muthuswamy, P., & Swaminathan, K. (2003). Biotreatment of Distillery Effluent Using *Aspergillus niveus. Bulletin of Environmental Contamination and Toxicology*, *70*(2), 268–277. and doi:10.1007/s00128-002-0187-2

Joss, A., Keller, E., Alder, A. C., Gobel, A., McArdell, C. S., Ternes, T. A., & Siegrist, H. (2005). Removal of pharmaceuticals and fragrances in biological wastewater treatment. *Water Research*, *39*(14), 3139–3152. and doi:10.1016/j.watres.2005.05.031

Kavitha, G. V., & Beebi, S. K. (2003). Biodegradation of phenol in a packed bed reactor using peat media. *Asian Journal of Microbiology, Biotechnology &. Environmental Sciences*, *5*(2), 157–159.

Khan, M. Z. H., & Mostafa, M. G. (2011). Aerobic treatment of pharmaceutical wastewater in a biological reactor. *International Journal of Environmental Sciences*, *1*(7), 1797–1805.

Kimura, K., Hara, H., & Watanabe, Y. (2007). Elimination of selected acidic pharmaceuticals from municipal wastewater by an activated sludge system and membrane bioreactors. *Environmental Science & Technology*, *41*(10), 3708–3714. and doi:10.1021/es061684z

Kristiansson, E., Fick, J., Janzon, A., Grabic, R., Rutgersson, C., Weidegård, B., & Larsson, D. G. J. et al. (2011). Pyrosequencing of antibiotic-contaminated river sediments reveals high levels of resistance and gene transfer elements. *PLoS ONE*, *6*(2), e17038. and doi:10.1371/journal.pone.0017038

Krug, M., Ziegler, H., & Straube, G. (1985). Degradation of phenolic compounds by the yeast Candida tropicalis HP15.I. Physiology of growth and substrate utilization. *Journal of Basic Microbiology*, *25*(2), 103–110. and doi:10.1002/jobm.3620250206

Kumar, A., Kumar, S., & Kumar, S. (2005). Biodegradation kinetics of phenol and catechol using *Pseudomonas putida* MTCC 1194. *Biochemical Engineering Journal*, *22*(2), 151–159. doi:10.1016/j.bej.2004.09.006

Lapara, T. M., Nakatsu, C. H., Pantea, L. M., & Alleman, J. E. (2001). Aerobic biological treatment of a pharmaceutical wastewater: Effect of temperature on COD removal and bacterial community development. *Water Research*, *35*(18), 4417–4425. and doi:10.1016/S0043-1354(01)00178-6

Larsson, D. G. J. (2014). Antibiotics in the environment. *Upsala Journal of Medical Sciences*, *119*(2), 108–112. and doi:10.3109/03009734.2014.896438

Lee, C. M., Lu, C. J., & Chuang, M. S. (1994). Effects of immobilized cells on the biodegradation of chlorinated phenols. *Water Science and Technology*, *30*, 87–90.

Lettinga, G., van Velsen, A. F., Hohma, S. M., de Zeeuw, W., & Klapwijk, A. (1980). Use of upflow sludge blanket (UASB) reactor concept for biological wastewater treatment. *Biotechnology and Bioengineering*, *22*, 699–734. doi:10.1002/bit.260220402

Li, Y., Zhu, G., Ng, J. W., & Tan, S. K. (2014). A review on removing pharmaceutical contaminants from wastewater by constructed wetlands: Design, performance and mechanism. *The Science of the Total Environment*, *468-469*, 908–932. and doi:10.1016/j.scitotenv.2013.09.018

López-Fernández, R., Martínez, L., & Villaverde, S. (2012). Membrane bioreactor for the treatment of pharmaceutical wastewater containing corticosteroids. *Desalination*, *300*, 19–23. doi:10.1016/j.desal.2012.05.032

Luo, Y., Guo, W., Ngo, H. H., Nghiem, L. D., Hai, F. I., Zhang, J., & Wang, X. C. et al. (2014). A review on the occurrence of micropollutants in the aquatic environment and their fate and removal during wastewater treatment. *The Science of the Total Environment*, *473-474*, 619–641. and doi:10.1016/j.scitotenv.2013.12.065

Madukasi, E. I., Dai, X., He, C., & Zhou, J. (2010). Potentials of phototrophic bacteria in treating pharmaceutical wastewater. *International Journal of Environmental Science and Technology*, *7*(1), 165–174. doi:10.1007/BF03326128

Makhijani, M., Gahlawat, S., Chauhan, K., Valsangkar, S., & Gauba, P. (2014). Phytoremediation potential of *Cicer arietinum* for Tetracycline. *International Journal of Genetic Engineering and Biotechnology*, *5*(2), 153–160.

Mansour, H. B., Dellai, A., & Ayed, Y. (2012). Toxicities effects of pharmaceutical, olive mill and textile wastewaters before and after degradation by *Pseudomonas putida* mt-2. *Cancer Cell International*, *12*(1), 4. and doi:10.1186/1475-2867-12-4

Mansour, H. B., Mosrati, R., Barillier, D., Ghedira, K., & Chekir-Ghedira, L. (2012). Bioremediation of industrial pharmaceutical drugs. *Drug and Chemical Toxicology*, *35*(3), 235–240. and doi:10.3109/01480545.2011.591799

Mayabhate, S. P., Gupta, S. K., & Joshi, S. G. (1988). Biological treatment of pharmaceutical wastewater. *Water, Air, and Soil Pollution*, *38*(1–2), 189–197.

McIntyre, T., & Lewis, G. M. (1997). Advancement of phytoremediation as an innovative environmental technology for stabilization, remediation, or restoration of contaminated sites in Canada: A discussion paper. *Journal of Soil Contamination*, *6*(3), 227–224. doi:10.1080/15320389709383562

Misal, S. A., Lingojwar, D. P., Shinde, R. M., & Gawai, K. R. (2011). Purification and characterization of azoreductase from alkaliphilic strains *Bacillus badius*. *Process Biochemistry*, *46*(6), 264–269. doi:10.1016/j.procbio.2011.02.013

Mohan, S. V., Prakasham, R. S., Satyavathi, B., Annapurna, J., & Ramakrishna, S. V. (2001). Biotreatability studies of pharmaceutical wastewater using an anaerobic suspended film contact reactor. *Water Science and Technology*, *43*(2), 271–276.

Muda, K., Aris, A., Salim, M. R., Ibrahim, Z., & Yahya, A. (2012). Textile wastewater treatment using biogranules under intermittent anaerobic/aerobic reaction phase. *Journal of Water and Environment Technology*, *10*(3), 303–315. doi:10.2965/jwet.2012.303

Mutamim, N. S. A., Noor, Z. Z., Hassan, M. A. A., & Olsson, G. (2012). Application of membrane bioreactor technology in treating high strength industrial wastewater: A performance review. *Desalination*, *305*, 1–11. doi:10.1016/j.desal.2012.07.033

Nandy, T., & Kaul, S. N. (2001). Anaerobic pre-treatment of herbal-based wastewater using fixed-film reactor with recourse to energy recovery. *Water Research*, *35*(2), 351–362. and doi:10.1016/S0043-1354(00)00263-3

Nnenna, F.-P., Lekiah, P., & Obemeata, O. (2011). Degradation of antibiotics by bacteria and fungi from the aquatic environment. *Journal of Toxicology and Environmental Health Sciences*, *3*(10), 275–285.

Oktem, Y. A., Ince, O., Sallis, P., Donnelly, T., & Ince, B. K. (2007). Anaerobic treatment of a chemical synthesis-based pharmaceutical wastewater in a hybrid upflow anaerobic sludge blanket reactor. *Bioresource Technology*, *99*(5), 1089–1096. and doi:10.1016/j.biortech.2007.02.036

Ozdemir, S., Cirik, K., Akman, D., Sahinkaya, E., & Cinar, O. (2013). Treatment of azo-dye containing synthetic textile dye effluent using sulfidogenic anaerobic baffled reactor. *Bioresource Technology*, *146*, 135–143. and doi:10.1016/j.biortech.2013.07.066

Pandey, A., & Gundevia, H. S. (2008). Role of the fungus—*Periconiella* sp. in destruction of biomedical waste. *Journal of Environmental Science and Engineering*, *50*(3), 239–240.

Patil, P. G., Kulkarni, G. S., Kore, S. V., & Kore, V. S. (2013). Aerobic sequencing batch reactor for wastewater treatment: A review. *International Journal of Engineering Research and Technology*, *2*(10), 534–550.

Patneedi, C. B., & Prasadu, K. D. (2015). Impact of pharmaceutical wastes on human life and environment. *Rasayan Journal of Chemistry*, *8*(1), 67–70.

Perez, M., Romero, L. I., Nebot, E., & Sales, D. (1997). Colonisation of a porous sintered glass support in anaerobic thermophilic bioreactors. *Bioresource Technology*, *59*(2-3), 177–183. doi:10.1016/S0960-8524(96)00154-X

Perez, S., Eichhorn, P., & Aga, D. S. (2005). Evaluating the biodegradability of sulfamethazine, sulfamethoxazole, sulfathiazole, and trimethoprim at different stages of sewage treatment. *Environmental Toxicology and Chemistry*, *24*(6), 1361–1367. and doi:10.1897/04-211R.1

Pomati, F., Castiglioni, S., Zuccato, E., Fanelli, R., Vigetti, D., Rossetti, C., & Calamari, D. (2006). Effects of a complex mixture of therapeutic drugs at environmental levels on human embryonic cells. *Environmental Science & Technology*, *40*(7), 2442–2447. and doi:10.1021/es051715a

Pophali, G. R., Khan, R., Dhodapkar, R. S., Nandy, T., & Devotta, S. (2007). Anaerobic–aerobic treatment of purified terephthalic acid (PTA) effluent: A technoeconomic alternative to two-stage aerobic process. *Journal of Environmental Management*, *85*(4), 1024–1033. and doi:10.1016/j.jenvman.2006.11.016

Priya, E. S., & Selvan, P. S. (2014). *Water hyacinth (Eichhornia crassipes) – An efficient and economic adsorbent for textile effluent treatment – A review*. Arabian Journal of Chemistry; doi:10.1016/j.arabjc.2014.03.002

Qari, H. A., & Hassan, I. A. (2014). Removal of pollutants from waste water using *Dunaliella* algae. Biomedical & Pharmacology Journal, 7(1), 147–151. 10.13005/bpj/465

Radjenović, J., Petrovic, M., & Barcelo, D. (2009). Fate and distribution of pharmaceuticals in wastewater and sewage sludge of the conventional activated sludge (CAS) and advanced membrane bioreactor (MBR) treatment. *Water Research, 43*(3), 831–841. and doi:10.1016/j.watres.2008.11.043

Rana, R. S., Singh, P., Kandari, V., Singh, R., Dobhal, R., & Gupta, S. (2014). A review on characterization and bioremediation of pharmaceutical industries' wastewater: An Indian perspective. *Applied Water Science*, 1-12. 10.1007/s13201-014-0225-3

Randhawa, G. K., & Kullar, J. K. (2011). Bioremediation of pharmaceuticals, pesticides, and petrochemicals with gomeya/cow dung. *ISRN Pharmacology, 2011*, 362459. and doi:10.5402/2011/362459

Rao, A. G., Naidu, G. V., Prasad, K. K., Rao, N. C., Mohan, S. V., Jetty, A., & Sharma, P. N. (2004). Anaerobic treatment of wastewater with high suspended solids from a bulk drug industry using fixed film reactor (AFFR). *Bioresource Technology, 93*(3), 241–247. and doi:10.1016/j.biortech.2003.11.004

Rao, P. N., Habiulla, S., Prasad, K. V., & Anand, Ch. (2014). A treatment study of pharmaceutical industrial wastewater using UASB. *Environmental Sciences, 3*(10), 195–197.

Rittmann, B. E., & McCarty, P. L. (2001). *Environmental biotechnology: Principles and applications.* New York: McGraw-Hill.

Saleem, M. (2007). Pharmaceutical wastewater treatment: A physicochemical study. *Journal of Research (Science), 18*(2), 125–134.

Sallis, P. J., Christian, V., Shrivastava, R., Shukla, D., Modi, H. A., & Vyas, B. R. M. (2005). Degradation of xenobiotic compounds by lignin-degradibg white-rot fungi: Enzymology and mechanisma involved. *Indian Journal of Experimental Biology, 43*, 301–312.

Sancinetti, G. P., Sader, L. T., Varesche, M. B. A., Amorim, E. L. C., Omena, S. P. F., & Silva, E. L. (2012). Phenol degradation in an anaerobic fluidized bed reactor packed with low density support materials. *Brazilian Journal of Chemical Engineering, 29*(1), 87–98. doi:10.1590/S0104-66322012000100010

Saravanane, R., Murthy, D. V., & Krishnaiah, K. (2001a). Bioaugmentation and anaerobic treatment of pharmaceutical effluent in fluidized bed reactor. *Journal of Environmental Science and Health. Part A, Toxic/Hazardous Substances & Environmental Engineering, 36*(5), 779–791. and doi:10.1081/ESE-100103760

Saravanane, R., Murthy, D. V., & Krishnaiah, K. (2001b). Bioaugmentation and treatment of cephalexin drug-based pharmaceutical effluent in an upflow anaerobic fluidized bed system. *Bioresource Technology, 76*(3), 279–281. and doi:10.1016/S0960-8524(00)00121-8

Saravanane, R., Murthy, D. V., & Krishnaiah, K. (2001c). Treatment of anti-osmotic drug based pharmaceutical effluent in an upflow anaerobic fluidized bed system. *Waste Management (New York, N.Y.), 21*(6), 563–568. and doi:10.1016/S0956-053X(00)00109-4

Satyanarayan, S., Karambe, A., & Vanerkar, A. P. (2009). Herbal pharmaceutical wastewater treatment by a pilot scale upflow anaerobic sludge blanket (UASB) reactor. *Water Science and Technology, 59*(11), 2265–2272. and doi:10.2166/wst.2009.273

Sayad, S., & Ellouz, R. (1995). Roles of lignin peroxidase and manganese peroxidase from phanerochaete chrysosporium in the decolorization of olive mill wastewaters. *Applied and Environmental Microbiology*, *61*(3), 1098–1103.

Schröder, H. F., Tambosi, J. L., Sena, R. F., Moreira, R. F., José, H. J., & Pinnekamp, J. (2012). The removal and degradation of pharmaceutical compounds during membrane bioreactor treatment. *Water Science and Technology*, *65*(5), 833–839. and doi:10.2166/wst.2012.828

Seghezzo, L., Zeeman, G., van Lier, J. B., Hamelers, H. V. M., & Lettinga, G. (1998). A review: The anaerobic treatment of sewage in UASB and EGSB reactors. *Bioresource Technology*, *65*(3), 175–190. doi:10.1016/S0960-8524(98)00046-7

Shariati, F. P., Mehrnia, M. R., Salmasi, B. M., Heran, M., Wisniewski, C., & Sarrafzadeh, M. H. (2010). Membrane bioreactor for treatment of pharmaceutical wastewater containing acetaminophen. *Desalination*, *250*(2), 798–800. doi:10.1016/j.desal.2008.11.044

Shi, X., Lefebvre, O., Ng, K. K., & Ng, H. W. (2014). Sequential anaerobic–aerobic treatment of pharmaceutical wastewater with high salinity. *Bioresource Technology*, *153*, 79–86. and doi:10.1016/j.biortech.2013.11.045

Singh, D., & Fulekar, M. H. (2009). Bioremediation of phenol by a novel partitioning bioreactor using cow dung microbial consortia. *Biotechnology Journal*, *4*(3), 423–431. and doi:10.1002/biot.200800204

Singh, D., & Fulekar, M. H. (2010). Benzene bioremediation using cow dung microflora in two phase partitioning bioreactor. *Journal of Hazardous Materials*, *175*(1-3), 336–343. and doi:10.1016/j.jhazmat.2009.10.008

Sirtori, C., Zapata, A., Oller, I., Gernjak, W., Aguera, A., & Malato, S. (2009). Decontamination industrial pharmaceutical wastewater by combining solar photo-Fenton and biological treatment. *Water Research*, *43*(3), 661–668. and doi:10.1016/j.watres.2008.11.013

Sosa Alderete, L. G., Ibáñez, S. G., Agostini, E., & Medina, M. I. (2012). Phytoremediation of phenol at pilot scale by tobacco hairy roots. *International Journal of Environmental Sciences*, *3*(1), 398–407.

Spina, F., Anastasi, A., Prigione, V., Tigini, V., & Varese, G. C. (2012). Biological treatment of industrial wastewaters: A fungal approach. *Chemical Engineering Transactions*, *27*, 175–180.

Sreekanth, D., Sivaramakrishna, D., Himabindu, V., & Anjaneyulu, Y. (2009). Thermophilic treatment of bulk drug pharmaceutical industrial wastewaters by using hybrid up flow anaerobic sludge blanket reactor. *Bioresource Technology*, *100*(9), 2534–2539. and doi:10.1016/j.biortech.2008.11.028

Su, R., Zhang, G., Wang, P., Li, S., Ravenelle, R. M., & Crittenden, J. C. (2015). Treatment of antibiotic pharmaceutical wastewater using a rotating biological contactor. *Journal of Chemistry*, *2015*, 705275. doi:10.1155/2015/705275

Su, R., Zhang, G., Wang, P., Li, S., Ravenelle, R. M., & Crittenden, J. C. (2015). Treatment of antibiotic pharmaceutical wastewater using a rotating biological contactor. *Journal of Chemistry*, *2015*, 705275. doi:10.1155/2015/705275

Suarez, S., Lema, J. M., & Omil, F. (2010). Removal of pharmaceutical and personal care products (PPCPs) under nitrifying and denitrifying conditions. *Water Research*, *44*(10), 3214–3224. and doi:10.1016/j.watres.2010.02.040

Suresh, B., & Ravishankar, G. A. (2004). Phytoremediationa novel and promising approach for environmental clean-up. *Critical Reviews in Biotechnology*, *24*(2-3), 97–124. and doi:10.1080/07388550490493627

Tahrani, L., Soufi, L., Mehri, I., Najjari, A., Hassan, A., Van Loco, J., & Mansour, H. B. et al. (2015). Isolation and characterization of antibiotic-resistant bacteria from pharmaceutical industrial wastewaters. *Microbial Pathogenesis*, *89*, 54–61. and doi:10.1016/j.micpath.2015.09.001

Tambosi, J. L., de Sena, R. F., Favier, M., Gebhardt, W., Joséa, H. J., Schröderb, H. F., & Moreiraa, R. P. M. (2010). Removal of pharmaceutical compounds in membrane bioreactors (MBR) applying submerged membrane. *Desalination*, *261*(1-2), 148–156. doi:10.1016/j.desal.2010.05.014

Tandon, S. A., Kumar, R., Bajwa, R. K., & Yadav, S. A. (2013). Phytoremediation of fluoroquinolone group of antibiotics from waste water. Natural Science, 5(1212a), 21–27. 10.4236/ns.2013.512A004

Taseli, B. K., G¨okc¸ay, C. F., & G¨urol, A. (2008). Influence of nickel (II) and chromium (VI) on the laboratory scale rotating biological contactor. *Journal of Industrial Microbiology & Biotechnology*, *35*(9), 1033–1039. and doi:10.1007/s10295-008-0379-9

Tawfik, A., Klapwijk, A., El-Gohary, F., & Lettinga, G. (2005). Potentials of using a rotating biological contactor (RBC) for posttreatment of anaerobically pre-treated domestic wastewater. *Biochemical Engineering Journal*, *25*(1), 89–98. doi:10.1016/j.bej.2005.03.013

Ternes, T. A., Joss, A., & Siegrist, H. (2004). Scrutinizing pharmaceuticals and personal care products in wastewater treatment. *Environmental Science & Technology*, *38*(20), 392A–399A. and doi:10.1021/es040639t

Tezel, U., Guven, E., Erguder, T. H., & Demirer, G. N. (2001). Sequential (anaerobic/aerobic) biological treatment of dalaman SEKA pulp and paper industry effluent. *Waste Management (New York, N.Y.)*, *21*(8), 717–724. and doi:10.1016/S0956-053X(01)00013-7

To'th, A. J., Gergely, F., & Mizsey, P. (2011). Physicochemical treatment of pharmaceutical process wastewater: Distillation and membrane processes. *Chemical Engineering (Albany, N.Y.)*, *55*(2), 59–67.

Torkian, A., Yazdani, O., & Alinejad, K. (2003). Treatability evaluation of municipal wastewater and anaerobically-treated industrial effluent in a rotating biological contactor. *Industrial Journal of Engineering*, *16*(6), 16–22.

Tran, N. H., Urase, T., & Kusakabe, O. (2009). The characteristics of enriched nitrifier culture in the degradation of selected pharmaceutically active compounds. *Journal of Hazardous Materials*, *171*(1-3), 1051–1057. and doi:10.1016/j.jhazmat.2009.06.114

Tripathi, A. K., Harsh, N. S. K., & Gupta, N. (2007). Fungal treatment of industrial effluents: A mini-review. *Life Science Journal*, *4*(2), 78–81.

Uwadiae, S. E., Yerima, Y., & Azike, R. U. (2011). Enzymatic biodegradation of pharmaceutical wastewater. *International Journal of Energy and Environment*, *2*(4), 683–690.

Vader, J. S., van Ginkel, C. G., Sperling, F., de Jong, J., de Boer, W., de Graaf, J. S., & Stokman, P. G. W. et al. (2000). Degradation of ethinyl estradiol by nitrifying activated sludge. *Chemosphere*, *41*(8), 1239–1243. and doi:10.1016/S0045-6535(99)00556-1

Vanerkar, A. P., Satyanarayan, S., & Dharmadhikari, D. M. (2013). Full scale treatment of herbal pharmaceutical industry wastewater. *International Journal of Chemical & Physical Sciences*, *2*, 52–62.

Vasiliadou, I. A., Molina, R., Martínez, F., & Melero, J. A. (2014). Experimental and modeling study on removal of pharmaceutically active compounds in rotating biological contactors. *Journal of Hazardous Materials*, *274*, 473–482. and doi:10.1016/j.jhazmat.2014.04.034

Vasiliadou, I. A., Sánchez-Vázquez, R., Molina, R., Martínez, F., Melero, J. A., Bautista, L. F., & Morales, G. et al. (2016). Biological removal of pharmaceutical compounds using white-rot fungi with concomitant FAME production of the residual biomass. *Journal of Environmental Management*, *180*, 228–237. and doi:10.1016/j.jenvman.2016.05.035

Vidal, N., Bañares-Alcántara, R., Rodríguez-Roda, I., & Poch, M. (2002). Design of wastewater treatment plants using a conceptual design methodology. *Industrial & Engineering Chemistry Research*, *41*(20), 4993–5005. doi:10.1021/ie010652b

Walters, E., McClellan, K., & Halden, R. U. (2010). Occurrence and loss over three years of 72 pharmaceuticals and personal care products from biosolids-soil mixtures in outdoor mesocosms. *Water Research*, *44*(20), 6011–6020. and doi:10.1016/j.watres.2010.07.051

Wang, J., & Katz, S. E. (1994). The persistence of fecal-borne antibiotics in soil. *Journal - Association of Official Analytical Chemists*, *77*, 481–484.

Wang, K., Liu, S., Zhang, Q., & He, Y. (2009). Pharmaceutical wastewater treatment by internal micro-electrolysis-coagulation, biological treatment and activated carbon adsorption. *Environmental Technology*, *30*(13), 1469–1474. and doi:10.1080/09593330903229164

Wang, Q.-Q., Bradford, S. A., Zheng, W., & Yates, S. R. (2006). Sulfadimethoxine degradation kinetics in manure as affected by initial concentration, moisture, and temperature. *Journal of Environmental Quality*, *35*(6), 2162–2169. and doi:10.2134/jeq2006.0178

Watanabe, M. E. (1997). Phytoremediation on the brink of commercialization. *Environmental Science & Technology*, *31*(4), 182A–186A. and doi:10.1021/es972219s

Watanabe, R. A., Sales, P. T. F., Campos, L. C., Garcia, T. A., Valadares, M. C., Schimidt, F., & Santiago, M. F. (2012). Evaluation of the use of *Pycnoporus sanguineus* fungus for phenolics and genotoxicity decay of a pharmaceutical effluent treatment. *Ambi-Agua. Taubaté*, *7*(3), 41–50.

Wei, X., Li, B., Zhao, S., Wang, L., Zhang, H., Li, C., & Wang, S. (2012). Mixed pharmaceutical wastewater treatment by integrated membraneaerated biofilm reactor (MABR) system—a pilot-scale study. *Bioresource Technology*, *122*, 189–195. and doi:10.1016/j.biortech.2012.06.041

Wetzstein, H. G., Schmeer, N., & Karl, W. (1997). Degradation of the fluoroquinolone enrofloxacin by the brown rot fungus Gloeophyllum striatum: Identification of metabolites. *Applied and Environmental Microbiology*, *63*(11), 4272–4281.

Wicklow, D. T. (1992). The coprophilous fungal community: an experimental system. In G. C. Carrol & D. T. Wicklow (Eds.), *The Fungal Community. Its Organisation and Role in the Ecosystem* (2nd ed., pp. 715–728). New York: Marcel Dekker.

Zamanpour, G., & Mehrabani-Zeinabad, A. (2014). An experimental study on bioremediation and photolysis of enrofloxacin. *Water Science and Technology*, *70*(5), 932–938. and doi:10.2166/wst.2014.320

Zhou, M., Tang, Y., & Wu, Y. (2013). Plant hairy roots for remediation of aqueous pollutants. *Plant Molecular Biology Reporter*, *31*(1), 1–8. doi:10.1007/s11105-012-0465-z

Zhou, P., Su, C., Li, B., & Qian, Y. (2006). Treatment of high-strength pharmaceutical wastewater and removal of antibiotics in anaerobic and aerobic biological treatment processes. *Journal of Environmental Engineering*, *132*(1), 129–136. doi:10.1061/(ASCE)0733-9372(2006)132:1(129)

Zuccato, E., Calamari, D., Natangelo, M., & Fanelli, R. (2000). Presence of therapeutic drugs in the environment. *Lancet*, *355*(9217), 1789–1790. and doi:10.1016/S0140-6736(00)02270-4

KEY TERMS AND DEFINITIONS

Aerobic: A process requiring the presence of oxygen.

Anaerobic: A process occurring in the absence of oxygen.

Bacterial Bioremediation: Treatment of wastewater with bacteria having the capability to consume or digest the chemical contaminants including drugs and pharmaceuticals.

BOD: Biochemical oxygen demand or biological oxygen demand. It is an important water quality parameter representing the amount of dissolved oxygen needed for biodegradation of organic material present in a given water sample at certain temperature over a specific time period.

COD: Chemical oxygen demand is a water quality parameter (unit mg/L) based on chemical oxidation methods employing strong oxidizing agents.

Membrane Bioreactor: Wastewater treatment process in which a membrane filter process like microfiltration or ultrafiltration is integrated with a biological process like suspended growth bioreactor yielding high quality treated effluent.

Mycoremediation: Bioremediation method employing fungi to degrade or accumulate organic and heavy metal contaminants in the environment.

Phytoremediation: Use of green plants and solar energy for degradation and removal of environmental contaminants.

Chapter 17
Heavy Metal(loid) Remediation Using Bio-Waste:
A Potential Low-Cost Green Technology for Cleaning Environment

Sukanta Rana
University of Kalyani, India

Jatindra Nath Bhakta
University of Kalyani, India

ABSTRACT

Heavy metal(loid)s are hazardous, biologically non-essential, non-biodegradable and persistent in nature, which can accumulate in plants and animals as well as in environment especially agri- and aqua- culture ecosystems. It is severely responsible for causing several health hazards problems in human, such as, cardiovascular, pulmonary, hepatic, nephrological, dermatological, neurological disorders as well as carcinogenic effects. Removal of these heavy metals from living systems is extensively expensive and also unsuccessful in sent percent removal. Therefore, in order to protect the environment, the removal of heavy metal(loid)s from polluted effluents is essential before discharging into environment. Besides various treatment technologies, sorption of metal(loid)s using bio-wastes are highly potent alternatives in recent years. The present chapter deals with the removal efficiencies of various bio-wastes, orange peels, waste tea leaves, rice husk, wheat stalk, sugar cane bagasse, coconut husk, sun flower stalk, corn cob, nut shell, water hyacinth, crab shell particle, activated carbons etc. The present discussion has also revealed that bio-waste could be a low-cost eco-friendly and green emerging alternative technology in treating the metal(loid)s contaminated environment without posing any further adverse environmental impacts.

INTRODUCTION

Industrialization, urbanization and population explosion are the prime reason for tremendous environmental pollution in last few decades (Bhakta, Son, & Munekage, 2009; Bhakta & Munekage, 2010). Massive amount of pollutants generating from anthropogenic activities are indiscriminately discharging

DOI: 10.4018/978-1-5225-2325-3.ch017

into the environment which contaminating the life system through the food chain (Jillian, Robert, & Rajakaruna, 2015). Effluents from industries like chlor-alkali, electrical, electronic, electroplating, glass manufacture, photography, oil refinery, pesticide, paint-pigment, plastic, paper and pulp, pharmaceutical, gun, steel etc. contain heavy metals such as, Hg, Cd, Pb, Cr, Cu, As, Ni etc (Tack, Vanhaesebroeck, Verloo, Rompaey, & Ranst, 2005). in extensively high concentration and are disposed into sea, river, canal or pond directly and contaminate them. These metals percolate into ground water and contaminate. Irrigated water is also not safe from heavy metals. Aquatic plants and animals and ultimately human beings are being affected greatly. Domestic wastes (such as, batteries, paints, electrical and electronics goods etc.) have also similar contribution.

Heavy metal(loid)s are hazardous, biologically non-essential, non-biodegradable and persistent in nature and have severe several toxic impacts on lives (Bhakta, Ohnishi, Munekage, Iwasaki, & Wei, 2012). They enter into the body of plants and animals through absorption of water in plants and drinking of water and ingestion of food in animal and get accumulated within tissue. These metals bind to functional groups such as amino ($-NH_2$), imino (=NH), sulfhydryl (-SH), carboxyl (-COOH), phenolic (C_6H_5-), phosphoryl ($-P_2O_5$), imidazole ($C_3H_4N_2$) and phosphate ($-PO_4$) of bio-molecules and form stable coordination complexes within the tissues (Perez Marín, Aguilar Meseguer, Ortũno, & Śaez, 2008; Gönen & Serin, 2012). They tend to form more stable complexes with sulphur and nitrogen ligands than oxygen ligand. Thus, heavy metals are accumulated in liver, kidney, brain, bones, teeth, lungs and blood and damage and impair various systems in human. This accumulation when crosses the critical level causes death of the living machine.

Therefore, decontamination of environment from the severe pollution of heavy metal(loid)s is an urgent issue to scientists. To overcome this problem, many technologies have been developed by researches, sorption, precipitation, coagulation, membrane filtration, etc. so far. Sorption is one of the important technologies in treating the polluted water using various inorganic (chemical) and organic based sorbent media in this respect. However, most of them are expensive and having adverse impact to environment. There is always significant demand in developing the low-cost green eco-friendly technology in this respect. In recent years, biomass based sorbents have drawn much attention to the scientist for applying as a potential sorbent in treating various pollutants of polluted water especially heavy metal(loid)s. Most common domestic biomass wastes (such as - banana peel, orange peel, cotton fiber, cellulose paper, *Eichhornia* plant, coconut coir, waste tea leaf etc.) and different industrial and agricultural bio-wastes (such as, sugarcane baggasse, rice husk, wheat and rice straw, pineapple crown top, sun flower stalk, olive stone, sea weeds, crab shell etc.) have the capability in binding of heavy metal by the process of adsorption and /or absorption, and thus, they can remove heavy metals from polluted water significantly.

The objective of this chapter is to evaluate the bio-sorbent potentiality of biomass waste as a potent low-cost agent of eco-technology in the heavy metal(loid)s remediation of environment. The chapter also concisely discussed applicability of various bio-wastes for remediation of heavy metal(loid)s from contaminated aquatic ecosystem or aqueous solution without harming environment.

CHEMICAL PROPERTIES OF BIO-WASTE

Bio-wastes possess many functional groups such as -COOH, -OH which play important role to bind metal ions coordinately and form a stable complexes resulting sorption. According to Perez Marín,

Aguilar Meseguer, Ortũno, Śaez. (2008) and Gönen and Serin (2012), orange peel is made up of cellulose, hemicellulose, pectin, limonene ($C_{10}H_{16}$) and many other low molecular weight compounds which contain hydroxyl (-OH) and carboxyl group (-COOH) as important functional group for heavy metal binding. Banana peel contains pectin such as galacturonic acid which provides –COOH as functional group (Arunakumar, Walpola, & Yoon, 2013). Study also revealed that lignin contains –COOH and –OH as functional groups (Todorcius et al., 2015) those help to remove metals from aqueous media. FTIR analysis showed that activated carbon possessed nitrile (C≡N), alcohol (-OH) and carboxyl group (-COOH) (Rajesh, Pujari, & Uppaluri, 2014), tea wastes possess phenolic hydroxyl group (-OH), alcoholic hydroxyl group (-OH), carboxyl group (-COOH) and amino group ($-NH_2$) (Dwivedi & Rajput, 2014). It is apparent that the binding of metals with these functional groups and formation of complex is the basic principle of bio materials in removing toxic pollutants especially metals from ambient environment.

COMMON BIO-SORBENT PREPARATION PROCESS

Although, there are variations in developing the bio-sorbent from biomass, basic processes followed in different cases are almost same. Bio-wastes are processed to develop adsorbent in the following subsequent ways:

- **Collection:** Bio-wastes such as, orange or banana peel, cotton fiber, cellulose paper, *Eichhornia* plant, coconut coir, waste tea leaf, sugarcane baggasse, rice husk, wheat and rice straw, pineapple crown top, sun flower stalk, olive stone, sea weeds, crab shell etc. are collected from local market or houses.
- **Washing:** The collected wastes are washed thoroughly under tap water to remove any particulate matter and are sprayed with distilled water thereafter.
- **Chopping:** The large biomass can be chapped into small pieces.
- **Drying:** Chopped wastes can be dried by air or hot air oven at specific temperature (generally 70 - 80^{0}C) for certain period.
- **Pyrolysis:** This step is followed only to prepare activated carbon (AC). A certain amount of dried sample is taken and heated at high temperature (generally 500^{0}C) using micro-oven or other device for a period to carbonize fully and then it is cooled.
- **Crushing:** Dried materials or pyrolysed biomass are crushed to powder by mortar and pestle or mix grinder machine.
- **Sieving:** Powdery substances can be sieved by sieve of various mesh size. After sieving, the dried powder or carbonized powder can be used as sorbent.
- **Depigmentation:** This process is followed in some cases for producing the bio-sorbent. Biosobent substrates are soaked with alkaline-ethanol solution made of 95% C_2H_5OH and 1% NaOH at 1:1 ratio by volume at 25^{0}C for 24 hours and pigments like chlorophyll, β-carotene and other low molecular weight compounds are removed. Or, substances are boiled in distilled water at 85^{0}C for 30 minutes to remove colors. Now, the de-pigmented substances are hot air oven dried at specific temperature for specific periods.

CHEMICAL ACTIVATION OF BIO-SORBENT

The metal removal activity of bio-sorbent or other sorbent can be improved by means of activation using various chemical agents or heating process. Some activation processes generally followed in this regard are discussed hereunder:

- **Amination(-NH_2):** According to Shrestha, Kour, Homagai, Pokhrel, & Ghimire (2013), samples are charred with concentrated H_2SO_4 and then is washed with distilled water till neutrality. It is oven dried at 80^{0}C for 24 hours. The charred samples are chlorinated with $SOCl_2$ in the presence of pyridine at 0^{0}C and then heated at 70^{0}C for 30 minutes. Now, it is washed with distilled water. The chlorinated sample is mixed with ethylene-di-amine (NH_2-CH_2–CH_2-NH_2) and Na_2CO_3 in the presence of di-methyl-sulphonide ($DMSO_3$) and heated at 70^{0}C for 6 hours. Then, the aminated adsorbent is washed with 0.1M HCl and followed by distilled water till neutrality. Finally, the sample is oven dried at 70^{0}C for over night.
- **Xanthate:** It is described by Liang et al. (2012) as follows. About 50 ml of 4M NaOH are added in dried sample (about 15 g) taken in an Erlenmeyer flask, and is stirred magnetically for 3 hours at 25^{0}C. Then, 10 ml of carbondisulfide (CS_2) is mixed and allowed to settle it for one hour. The precipitated mixture is washed with acetone and then is dried in desiccator. Finally, the product is xanthate that can be used as adsorbent.
- **H_2SO_4 Activation (Liang et al., 2012):** Sizable and dried sample is mixed with concentrated 1N H_2SO_4 in a ratio of 1:2 (w/v) at a temperature of 150^{0}C for 24 hours. Then the mixture is washed with distilled water, kept in 1% $NaHCO_3$ solution and then is filtered. Finally, the solid matter is oven dried at 105^{0}C till to obtain a stable weight which acts as strong adsorbent.
- **NaOH Activation (Liang et al., 2012**): About 200 g of dried matter is mixed with 1L of 0.1M NaOH solution and is left for 24 hrs. Then it is neutralized with distilled water wash. Now, it is oven dried at 105^{0}C and is ready to use as potential sorbent.
- **CH_3OH Activation (Liang et al., 2012):** A mass of raw material after chopping is mixed with methanol (purity: 99.9%) as per as 9 g/633ml and a small quantity of 0.1N HCl is added. The mixture is heated at 600^{0}C for 48 hrs. Then it is washed thoroughly with distilled water and filtered. Finally, the sample is oven dried at 50 – 60^{0}C till to gain a constant weight.
- **Methylation (Liang et al., 2012):** Dried sample is mixed with a formaldehyde-acetic acid solution as per as 9 g/180ml and is shaken for 6 hrs. The blended solution is prepared by mixing of HCHO with CH_3COOH in a ratio of 1:1 (v/v). It is then washed with distilled water, filtered and oven dried at 50 – 60^{0}C up to a constant weight.
- **Acetylation (Liang et al., 2012):** Refluxing of chopped raw matter in acetic acid as per as 12g in 64 ml at 80^{0}C for 10 hrs adds acetyl group (CH_3-CO-) to the adsorbent. The acetylated substance is washed with distilled water filtered and oven dried at 50 -60^{0}C up to a constant weight.

SORPTION MECHANISM OF BIO-SORBENT

The mechanism of bio-sorbent is mainly based on the high affinity of functional groups to metal ions to attached to solid sorbent to make a stable complex till the saturation of all functional groups. Bio-sorption follows many mechanisms which may vary bio-mass to bio-mass and also may adopt many mechanisms simultaneously in a sorption process. Some mechanisms described by a number of scientists are summarized here in the following:

Trans-Membrane Transport

This mechanism is mediated by trans-membrane protein across the membrane from extracellular medium into cytosol (Ma, Jacobsen, & Giedroc, 2010). Cu^{2+}, Zn^{2+}, Ni^{2+}, Co^{+}, Mn^{2+}, Fe^{3+} and Cd^{2+} can be transported by this mechanism (Ma, Jacobsen, & Giedroc, 2010). Metal binding to cell wall is essential prior to transportation (Ahalya et al., 2003). This mechanism is applicable only for living cells like bacteria (*Escherichia coli)*, fungi (*Saccharomyces cerevisiae)* etc.

Physical Adsorption

The van der Waal`s force causes the adsorption of divalent ions like Zn^{2+}, Cd^{2+}, Cu^{2+}, Co^{2+} etc. to the cell wall of dead plant biomass. The components (cellulose, hemicellulose, pectin, limonene, etc.) of cell wall act as potent adsorbent in this process sorption.

Ion Exchange

Cell wall polysaccharides (such as alginates in marine algae) of living microorganisms uptake divalent metals (Cu^{2+}, Cd^{2+}, Zn^{2+} etc.) with exchanging of Na^{+}, K^{+}, Ca^{2+} and Mg^{2+}. This mechanism has been described in *Aspergillus niger* (Ahalya, Ramachandra, & Kanamadi, 2003).

Complexation

Complex formation between di-metalic ions with amino ($-NH_2$) or carboxyl groups (-COOH) of polysaccharide components of cell wall forming coordination bond is another mechanism for removal of Ca^{2+}, Mg^{2+}, Cd^{2+}, Zn^{2+}, Cu^{2+} and Hg^{2+} from extracellular solution and these metals get accumulated within cell. Most of the microorganisms (*Pseudomonas syringae, Zoogloea ramigera* and alga *Chiarella vulgaris etc.)* are found to uptake metals by this mechanism.

Precipitation

It is associated with mechanism of formation of metallic compound which precipitates in the solution. This is one of the active defense mechanism of microorganism in the presence of toxic heavy metals like Cd^{2+}, Hg^{2+}, Pb^{2+}, Ni^{2+} etc. Here the metal interacts with cell surface and a metallic compound is formed.

BIOSORPTION INFLUENCING FACTORS

Metal sorption capacity of bio-sorbent is significantly influenced by a number of factors, such as, initial concentration of metal ions, pH, temperature, quantity of adsorbent used and number of functional group. The highest metals sorption capacity of bio-sorbent is found at specific conditions or states of aforesaid factors, i.e., optimum condition of influencing factors. It is varied from sorbent to sorbent on the basis of their chemical properties and functional groups responsible for sorption mechanism. The per cent of removal was greater with decrease in initial concentration of metal ions (10 – 200 mg/L) and with increase in amount of adsorbent used at pH 5 and 25^{0}C (Gönen and Serin, 2012). Most of the experiments showed that maximum removal was recorded at the pH range 4 – 7 (Yasuda, Ozaki, Nohata, Ohshiro, & Sonda, 2010; Gönen and Serin, 2012; Mohammed, 2012; Ahn & Kim, 2016). The amount of adsorption increased with increasing pH vales up to 7 (Ahn & Kim, 2016). Under optimum condition, the amount of adsorption was followed the order Pb^{2+} > Cu^{2+} > Cd >Zn^{2+} >Ni^{2+} (Shrestha et al., 2013; Dwivedi & Rajput, 2014) and Co > Cd > Zn (Mohammed, 2012). Temperature was a great influencing factor. Lower temperature was conducive for better adsorption whereas higher temperature inhibited the adsorption (Rakshit, 2004). Metal – sorbent interaction studied in all experiment between 25^0 – 30^{0}C reported that the adsorption capacity was not influenced by the temperature because adsorptions were isothermal (Bhakta, Rana, Lahiri, & Munekage, 2014; Dwivedi & Rajput, 2014; Ahn & Kim, 2016). Presence of undesirable metal ions – K^+, Ca^{++}, Na^+, Mg^{++} etc. and surfactans in the aqueous solution reduced adsorption capacity but sulfonated (-HSO_3) and carboxylated (-COOH) adsorbent showed better adsorption in the presence of surfactants (Ahn & Kim, 2016). Adsorption capacity also depended on porosity and fineness of adsorbent (Rakshit, 2004) and was proportional to porosity, surface area (Figure 1) and to some extent to incubation period (Figure 2) (Bhakta et al., 2014). It is also found that sorption mechanism is also significantly depended on the amount of sorbent and concentration of metals in the solution (Bhakta & Munekage, 2013).

Figure 1. Porosity, surface area and Hg binding in porous matrix of activated carbon. 1- outer surface carbon-metal complex formation, 2- inner surface complex formation, 3 – lateral diffusion, 4 – ion particle growth and diffusion back to aqueous solution after saturation
After Bhakta et al., 2014.

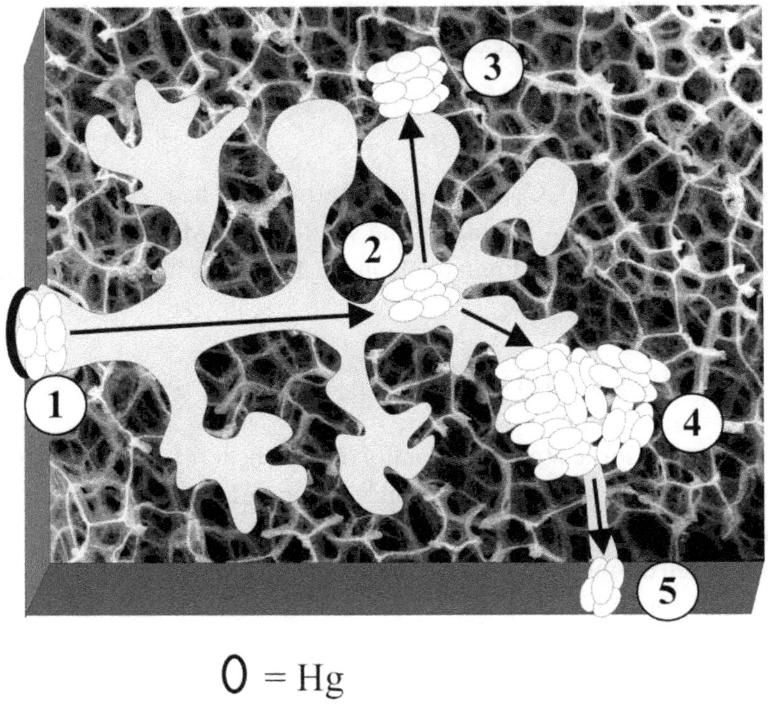

Figure 2. Hg sorption capacity of activated carbon developed from banana peel ($ACWM_B$), orange peel ($ACWM_O$), cotton fiber ($ACWM_C$) and paper waste ($ACWS_P$) in aqueous phase
After Bhakta et al., 2014.

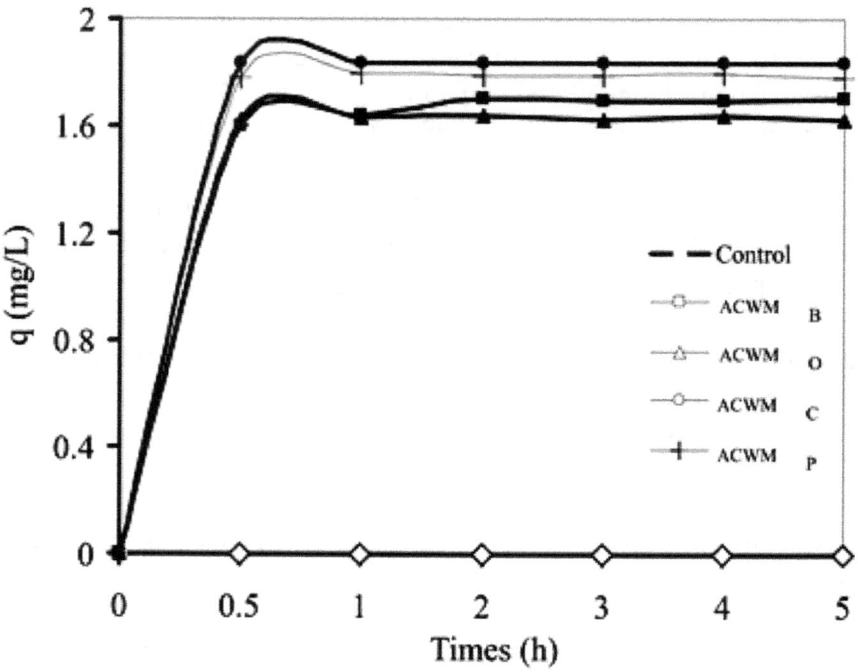

SORPTION KINETICS

The metal sorption capacity of adsorbent can be estimated using mass balance relationship at room temperature in following (Bhakta & Munekage, 2013).

$$q_e = \{(C_0 - C_t) \text{ x } V\}/M \tag{1}$$

where, q_e – sorption capacity (mg/g), C_0 and C_t – concentration of heavy metal (mg/L) in aqueous medium at time 0 and t respectively, V – volume of the solution and M – quantity of adsorbent Sorption isotherm capacity can be calculated by Freundlich and Langmuir isotherm equations. The Freundlich isotherm equation is as foloows:

$$x/m = kc^{1/n} \text{ or, } \text{Log}(x/m) = \text{Log } k + 1/n \text{ Log } c \tag{2}$$

where, x – quantity of metal adsorbed (g), m – quantity of adsorbent (g), x/m – amount of metal adsorbed per unit mass of the adsorbent, c – equilibrium concentration of metal, k and n – empirical constants depending on the nature of adsorbent and adsorbate and the temperature.

The relationship between the number of active sites of the surface undergoing adsorption and pressure is explained by Langmuir equation –

$$\theta = kp/(1+kp) \tag{3}$$

where, θ – number of sites of the surface which are covered with gaseous molecules, p – pressure, k – equilibrium constant for distribution of adsorbate between surface and gaseous phase.

Langmuir isotherm adsorption equation is valid at low pressure because, (1+kp) is negligible and θ = kp ; but at high pressure (1+kp) equals to kp. Then, θ = kp/kp = 1 that is not found in reality.

Adsorbent sorption capacity for metals can also be calculated by the alternative Langmuir equation –

$$q = q_{max}\, bc_{eq} / (1 + bc_{eq}) \tag{4}$$

where, q - metal accumulated per adsorbent (mg/g), q_{max} – maximum uptake of specific metal corresponding to site saturation, ceq – metal residual concentration at equilibrium and b – ratio of rate of adsorption and rate of desorption

Monolayer adsorbate of Langmuir isotherm was replaced by BET equation which was on the basis of poly layer adsorbate developed by couple action of van der Walls force (physical adsorption) and valence forces (chemisorption) with adsorbent. The surface area of adsorbent can be quantified by the following BET equation -

$$p / v(p^0\text{-}p) = 1 / v_m.c + (c\text{-}1)/v_m.c \text{ x } p/p^0 \tag{5}$$

where, v – volume of gas adsorbed under pressure p, p^0 – saturated vapor pressure at same temperature, V_m – volume of gas adsorbed when the surface is covered with a unimolecular layer and c – a constant for a given adsorbate. When, $c = e^{(E1 - EL)/RT}$ in which, E_1 – heat of absorption in first layer and E_L – heat of liquefaction of the gas

The metal removal percentage from aqueous media can be quantified following the equation:

$$Re = \{(C_0 - C_t) / C_0\} \text{ x } 100 \tag{6}$$

where Re – removal (%), C_0 and C_t – concentration of heavy metal (mg/L) in aqueous medium at time 0 and t respectively.

SOME CASE STUDIES

Although, varieties of waste biomass have been employed as bio-sorbent to remediate the contaminated water using the metal bio-soorption mechanism, some of them are summarizing in following:

Orange Peel

The adsorption property of a xanthate produced from mixing dechlorophyllated and dried orange peel with carbon disulfide (CS_2) was investigated by Liang et al. (2010) and was shown that the maximum adsorption capacity was 77.60 mg/g for Cu^{2+}, 76.57 mg/g for Cd^{2+}, 218.34 mg/g for Pb^{2+}, 49.85 mg/g for Zn^{2+} and 15.45 mg/g for Ni^{2+}. Orange peel after proper washing with double distilled water was oven dried at 70°C for 24 hours and was then crushed to powder. About 50 g of the crushed matter was socked with a mixed solution of 250 ml of C_2H_5OH and 250 ml of 1% NaOH at room temperature for 24 hours to remove chlorophyll pigments and other low molecular weight compounds.

Khalfaoui & Meniai (2012) used crushed and dried orange peel and chemically activated orange peel as sorbent in an experiment of Cu^{2+} removal from aqueous solution and reported that the removal efficiency was 75% in case of orange peel and more than 99% in activated peel. The orange peel was activated with concentrated H_2SO_4, NaOH, CH_3OH and acetic anhydride ($C_4H_6O_3$).

An adsorption study carrying out mixing 0.2 g of orange peel with 100 ml of $Ni(NO_3)_2$ of 10 – 200 ppm using an orbital shaker operating at 200 rpm revealed that greater percentage of Ni^{2+} removal was achieved with decrease in initial concentration Ni^{2+} and maximum removal was observed at pH 5 at room temperature (Gönen & Serin, 2012). Solutions were adjusted to the optimum pH prior to the addition of adsorbent.

Rajput et al. (2015) also conducted similar experiment using 0.5 g of finely crushed (particulate diameter: < 3.35 mm) and dried orange peel and maximum 78% Pb^{2+} ions was removed at optimum condition at room temperature from a solution of $Pb(NO_3)$ of 1000 ppm.

Banana Peel

Arunakumar et al. (2013) used banana peel as adsorbent to remove heavy metals from their aqueous solution and showed that the adsorption capacity of banana peel was 2.55 mg/g for Co^{2+}, 5.71 mg/g for Cd^{2+}, 5.80 mg/g for Zn^{2+}, 6.88 mg/g for Ni^{2+}, 7.97 mg/g for Pb^{2+} and 28.0 mg/g for Cu^{2+}. They also reported that more than 100 millions tone of banana are produced annually world wide and about 40 million tones of peel remains unused.

Waste Tea Leaves

Mohammed (2012) used black tea waste as an adsorbent to remove Co^{++}, Cd^{++} and Zn^{++} from wastewater. In an experiment, he allowed to interact 0.5 g of black tea waste with the aqueous solution of these heavy metals for 30 minutes at pH 6 and observed that the adsorption capacity of the black tea waste was 15.39 mg/g for Co, 13.77 mg/g for Cd and 12.24 mg/g for Zn. Tea waste collected from tea stall and restaurant was used to adsorb Cd^{++} and Cu^{++} from the aqueous solution of $Cd(NO_3)_2$ of 1 g/L and $Cu(NO_3)_2$ of 1 g/L respectively and a removal efficiency of 87% for Cd and 89% for Cu was observed (Dwivedi and Rajput, 2014). The crude tea waste was decolorized boiling in distilled water at 85^0C and was then dried in hot air oven at 105^0C for 12 hours. These dried leaves were pounded to powder of about 120 μm in size and that was used as sorbent as per 0.5 g for 100 ml solution.

Shrestas et al. (2013) removed Cd^{++} and Pb^{++} from wastewater using aminated tea leave as sorbent. The tea leaves were first oven dried at 80^0C for 24 hours and then they were pounded and sieved to pass through 212 μm mesh. Then the sizable leaves charred with concentrated H_2SO_4 followed by distilled water wash and oven drying at 80^0C for 24 hours. The charred sample was chlorinated with $SOCl_2$ in the presence of pyridine followed by amination with ethylene diamine $(CH_2NH_2)_2$ in the presence of Na_2CO_3 and dimethylsulphonide ($DMSO_3$) at 70^0C (Figure 3). The dried aminated leaves were active absorber for Hg^{++} and the adsorption capacity were 83.04 mg/g for Pb^{++} and 57 mg/g for Cd^{++}.

Green tea leaf is also a good adsorbent for heavy metals. Catechin extracted green tea leaves was used in removal of Cr^{6+}, Fe^{3+}, Cd^{2+}, Co^{2+}, Cu^{2+}, Mn^{2+}, Ni^{2+} and Zn^{2+} from their aqueous solution and the removal was increased with increasing the catechin extraction time i.e., a negative correlation between Catechin content in leaf and % of removal (Yasuda et al., 2010). The extraction at pH 4 accompanied

Figure 3. Scanning electron microscope images of waste tea leaves before (A) and after amination with ethyle diamine (B)

(a)

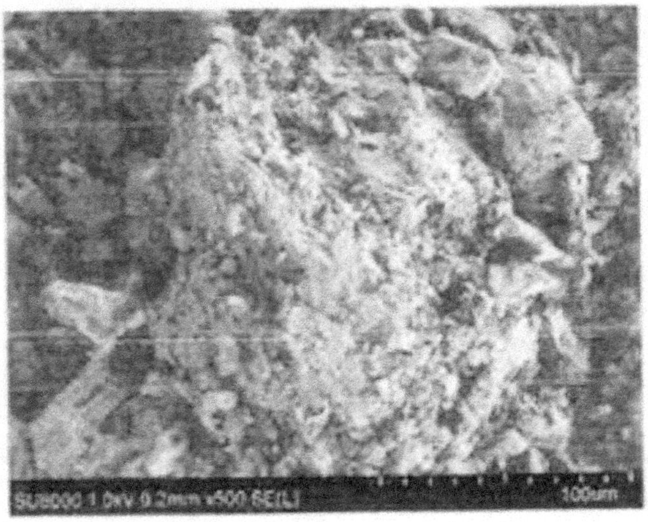

(b)

with stirring for 30 minutes at room temperature was best for optimum adsorption for Cr^{6+}, Fe^{3+} and Pb^{2+} whereas, at pH 6 for Cd^{2+}, Co^{2+}, Cu^{2+}, Mn^{2+}, Ni^{2+} and Zn^{2+}(Yasuda et al., 2010). Chemically modified used tea leaves i.e., thiolated (-SH), sulfonated (-HSO_3) and carboxylated (-COOH) tea leaves can be used as Cu^{2+} sorbent. Cu^{2+} were removed up to 53%, 22% and 19% by thiolated, carboxylated and sulfonated used tea leaves respectively at pH 7 (Ahn and Kim, 2016). They observed that, the adsorption efficiency of thiolated leaves was decreased in the presence of salt and surfactants whereas, the sulfonated and carboxylated leaves showed better adsorption in the presence of surfactants.

Rice Husk

Rice hush collected from rice mill is also a good adsorbent. Ye, Zhang, Zhang, Wu, and Du (2012) estimated its physical properties - surface area - 438.05 m^2/g, bulk density – 0.3086 g cm^3, porosity – 0.38 by fraction and chemical composition (by weight) – cellulose – 34.4%, hemicellulose-29.3%, lignin – 19.2% and ash – 17.1%. Ngah and Hanafiah (2008) used rice husk as an adsorbent for removal of Pb, Cd, Zn, Ni and As from both ground and surface water and found that treated hush showed better performance than untreated ones because, it created problems like – low adsorption capacity, high BOD, COD and Total Organic Carbon (TOC). Dada, Ojediran, and Olalekan (2013) treated the rice husk collected from mill to soak H_3PO_4 from its solution and used the treated husk as adsorbent in his experiment for heavy metal removal and found that the adsorption capacity of the adsorbent for Pb^{2+} was 138.89 mg /g at pH 6.3 – 6.5. Lashkenari, Davodo, and Eisazadeh (2011) reported that the adsorption capacity of polyanilinated rice husk for As (III) was 34.48 mg /g at pH 10 and at a dosage of 10 g/L.

Rice Straw

Raw rice straw is very poor or no metal sorbent because it lacks of any functional group for metal binding. Rice straw treated with NaOH was a good adsorbent for removal of Cr^{6+} and Ni^{2+} from their aqueous solution and its adsorption followed Lagergren second order kinetics (Brahmainah et al., 2015):

$$t/q_t = 1/k_2 q_e^2 + (1/q_e)t$$

where, k_2 – rate constant (g/mg-min.), q_e and q_t – amount of adsorbate adsorbed (mg/g) at time t and time at equilibrium time e. Washed, Dried and powdered straw samples was used for the study. About 16 g of powdered sample was mixed with 240 ml of NaOH solution and was kept for 24 hrs at room temperature. It was then oven dried at 100^0C for 24 hrs which was treated as potential sorbent.

Sugar Cane Bagasse

Sugar cane bagasse procured from food canning processing industry could remove 96.4% of Cd^{2+} from an aqueous solution of 20 mg/L. The bagasse was first washed with distilled water, then oven dried at 60^0C for 24 hrs. The dried substance was milled to powder and sieved into a particle size of 0.5 – 4 mm which was used as adsorbent (Kumar & Sahu, 2013). Sugar cane bagasse polymerized with acrylic acid and acrylamide using N,N-methylene bis-acrylamide as a cross linker used as an highly effective adsorbent as much as 268 mg/g for Cu^{2+}, 300 mg/g for Cd^{2+} and 700 mg/g for Pb^{2+} (Kong, Ren, Wang, & Chen, 2014). About 98% removal of arsenic (As) was also reported using sugar cane bagasse with adsorption capacity 11.9 mg/g at pH 9.0 (Tagermina, Ebadi, Nasernejad, & Ghafori, 2014).

Wheat Straw

Partially de-lignified Ca^{++}-Mg^{++} ion exchanged product was developed from ligno-cellulostic wheat straw to remove Pb^{2+}, Cd^{2+}, Hg^{2+}, Co^{2+}, Ni^{2+}, Mn^{2+}, Zn^{2+} and to de-toxicate Cr^{6+}. A strong affinity to Pb^{2+}, Cd^{2+}, Cu^{2+}, Zn^{2+} and Hg^{2+} and a weak affinity to Co^{2+}, Ni^{2+} and Mn^{2+}, were observed (Krishnami, 2016). Lignin is a polymer and possesses many functional groups like – (-COOH) and (-OH). Lignin extracted from wheat straw by boiling in alkaline solution acted as strong adsorbent for Cu ions (Todorcius, Bulgariu, & Popa, 2015). The adsorbent was prepared by addition of 1.25 g of lignin powder to 25ml of Cu $(NO_3)_2$ solution of 10^{-2} mol/L and was shaken for 24 hrs to obtain equilibrium condition and then the lignin was separated by filtration.

Sunflower Stalk

Jalali & Aboulghazi (2013) reported that fine particles derived from sun flower stalk was a potent sorbent that showed maximum removal efficiency 97% for Pb^{2+} and 87% for Cd^{2+} from a solution of 1g/L at pH 5.0 and adsorption capacity 182 mg/g and 70 mg/g for Pb^{2+} and Cd^{2+} respectively. The adsorbent was prepared by washing with de-ionized water until the effluent was colourless. The stalk then was oven dried at 60^0C for 24 hrs, ground to power and passed through 2.0 mm sieve.

Coconut Husk and Shell

Coconut husk rich in tanin is a good adsorbent. It removed 90% of both Cu^{2+} and Pb^{2+} at pH 5.0 (Abdulrasad & Baseru, 2010). Coconut shell treated with acid and coated with chitosan resulted 93% removal performance for Zn ions at pH 6.0 and temperature 25^0C during 3 hrs of interaction of adsorbents of 30 g/L with metal solution (Lim & Aris, 2014).

Palm Fruit Fiber

Palm fruit fiber was an excellent bio-sorbent which was proven in an experiment done by Ideriah, David, and Ogbonna (2012) that resulted 100% removal of Pb and Cu ions at pH 10 and room temperature (Figure 4).

Nut Shell

Chemically treated cashew nut shell showed a great removing potential of heavy metal ions from their contaminated aqueous solution with maximum adsorption capacities - 406.6 mg/g for Cu^{2+}, 436.7 mg/g for Cd^{2+}, 455.7 mg/g for Zn^{2+} and 456.3 mg/g for Ni^{2+} following Langmuir isotherm (Kumar et al., 2012). They reported that Cu ions were removed by 100% at pH 5.0 and at 30^0C. Chest nut shell treated with 4% NaOH solution was more efficient sorbent for removal of Cd^{++}, Pb^{++}, Pb^{++} and Zn^{++} s compared to chest nut shell treated with acid formaldehyde because alkaline treatment caused swelling of adsorbent with increased surface area of adsorbent coupled with more negative charge on particle surface (Vàzquez et al., 2012).

Corn Cob

Excellent metal removal efficiency by corn cob was observed as much as 95.05% for Cr^{6+} at pH 5.4 and 4.18 hrs of interaction, 99.9% for Ni at pH 6.5 and 2.38 hrs and 91.35% for Zn at pH 6.5 and 0.15hrs from electric industrial effluents by fixed bed column technique (Ali, Khalid, & Majid, 2014). Corn cob collected from market was crushed by grinder and mixed with de-ionized water in a shaker for 24 hrs and then washed with fresh de-ionized water. These were put in a large glass baking dish and dried

Figure 4. Removal efficiency of metal ions by palm fibre with respect to different concentrations of metal ions (A), interacting time (B) and pH values of aqueous solution of metal ions. 2.0 g of dried fibre was used in 50 ml of metallic solution.
After Ideriah et al. 2012.

A.

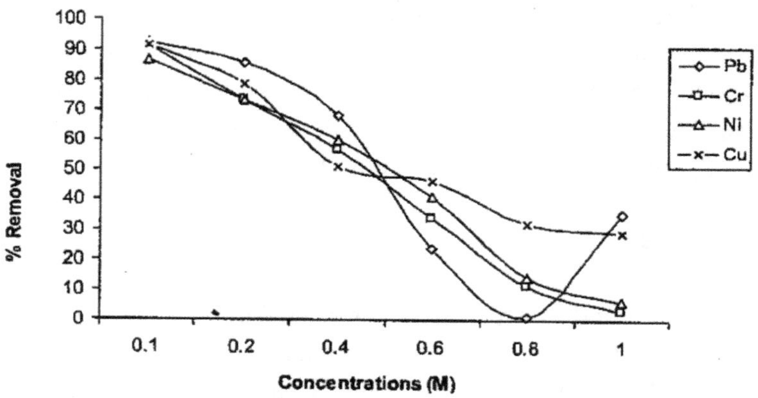

B.

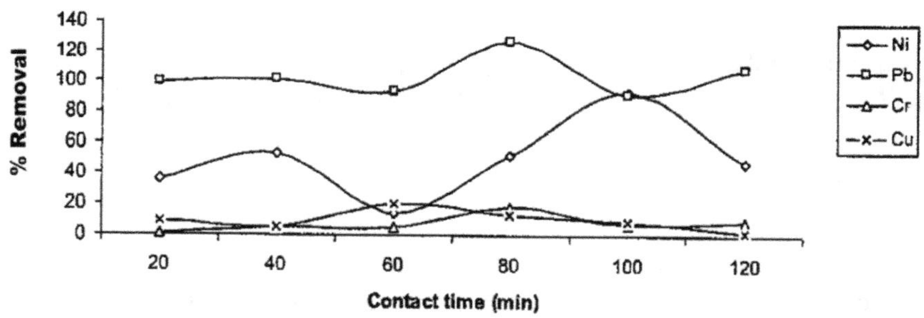

C.

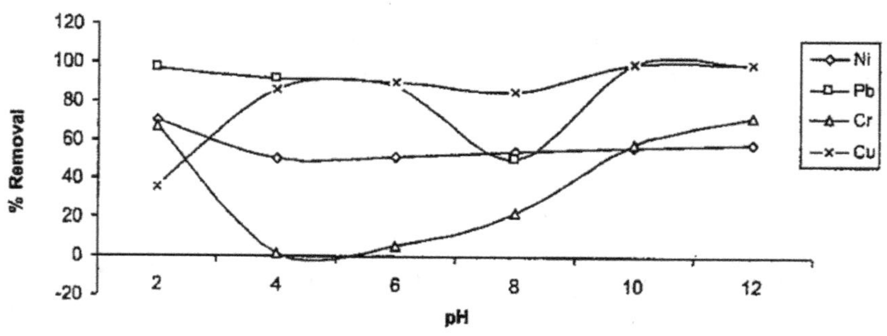

at room temperature. Now, the sorbent was treated with 10% HCl solution for 6 hrs at 30^0C followed by wash with double distilled water and oven dried at 125^0C for 24 hrs. Then the samples were sieved to 4 mm and stored in plastic bottle for usages as potential sorbent.

Weeds

The most irritating aquatic weed water hyacinth (*Eichhornia crassipes*) has tremendous adsorption capacity for metal ions. An experiment carried out by Dixit, Dixit, and Goswami (2015) using pulverized matter from dried water hyacinth as sorbent revealed that the removal efficiency were 98 - 99% for Pb, Cd, Cr and Zn (Table-1).

The marine algae (*Sargassum myriocystum*) has been reported to be one of the potential adsorbent for Pb ions when 2g of dried algal biomass was mixed with a solution of Pb of 1000 mg/L with pH 5 and allowed to interact for 60 minutes at room temperature and a maximal removal of Pb^{2+}s was observed by 89.75% (Sweetly et al., 2014). Fresh algae were washed with distilled water to remove salt and sand particles and oven dried at 60^0C overnight. It was then acid washed to remove sorbent inhibitors – Ca^{2+}, K^+, Mg^{2+} and Na^+s. Solid samples then was the dried at room temperature for 3 - 4 days until stable weight. The adsorptive performance of sea weed (*Caulerpa serrulata*) for copper, lead and cadmium ions from an aqueous solution was investigated by Mwangi and Ngila (2012) before and after ethelenediamination of the adsorbent and was resulted that the adsorption capacities of aminated sea weed were 5.27 mg/g for Cu, 2.12 mg/g for Cd and 2.16 mg/g for Pb ions whereas, the capacities for untreated sorbent were 3.29 mg/g, 4.57 mg/g and 1.06 mg/g for Cu, Cd and Pb ions respectively at pH 4 – 6. About 10g of green weeds was suspended in 200 ml of N,N-dimethyl formamide (DMF) followed by drop-wise addition of 35 ml of thionyl chloride ($SOCl_2$) at 80^0C under mechanical stirring. The chlorinated weeds were then washed NH_4OH solution to neutralize and the sample was vacuum dried at room temperature. Thereafter, the sample was reacted with 25 ml of ethylene 1-2-diamine under reflux with mechanical stirring for 3 hours. The solution was filtered through a sintered glass filter and vacuum dried at room temperature for 24 hrs.

Table 1. Removal efficiencies of water hyacinth of several heavy metals from aqueous solution of different concentration at room temperature

Heavy Metal	Initial Concentration (mg/L)	Quantity of Adsorbent Used (g/L)	Removal Efficiency (%)
Pb	10	2.5	93
Cd	10	3.0	82
Cr	10	1.5	82
Zn	10	4.0	75
Pb	50	3 – 4	98 – 99
Cd	50	3 – 4	98 – 99
Cr	50	3 - 4	98 – 99
Zn	50	3 - 4	98 - 99

After Dixit et al., 2015.

Crab Shell Particle

Crab particles have strong affinity to heavy metals and chemical treatment causes more and stronger affinity. Budin, Subhramanium, Tair, and Ali (2014) produced particle from shell of crab (*Scylla serata*) and cockle (*Anadara granosa*) and used as sorbent as per as 0.5 g/L to remove lead and chromium ions from an aqueous solution of Pb and Cr of 1 g/L respectively and observed that the maximum removal efficiency of shell particles was 97.45% for Cr^{6+}s and to some extent low for Pb^{2+}s at pH 4.0 and room temperature. They also reported that sorption capacity of crab shell particle was better than cockle shell. The crab was first boiled and its shell was separated. It was then washed with distilled water and sun dried. The dried shell was pulverized and sieved of 0.5mm in size that was used as sorbent.

The adsorption capacity of pretreated crab shell particle (*Portunus sanguinolentus*) – 71.46 mg/g at pH 4.5 and 25⁰C for zinc ions was also reported from an experiment of Zn^{2+}s removal from an aqueous of 100 mg/L performed by Senthil and Parthiban (2011). Waste shells of the crab were collected from Marina sea beach, washed with distilled water and then sun dried. It was now crushed to desired size by ball mill. The particle was washed with 0.1M HCl solution for 4 hrs to remove $CaCO_3$. It was then washed with de-ionized water and sun dried. Thereafter, these particles were soaked in 1%NaOH solution overnight at room temperature to remove bulk protein. The residues (chitin) were then rinsed with de-ionized water and oven dried at 100^0C for 6-7 hrs. The chitin was soaked in 50% NaOH solution for 2 hrs to remove some or all acetyl residues (CH_3CO-) and all chitin molecules were converted to chitosan molecules which were finally sun dried and acted as potential sorbent.

Activated Carbon (AC)

Activated carbon is also a highly potent sorbent for heavy metal due to its numerous pore, large active surface area and good mechanical properties (Cai and Jia, 2010). Heavy metal ions – Cr^{6+}, Cd^{2+} and Cu^{2+} were removed from aqueous solution by the sorbent – activated carbon developed from peel of Colombian orange through pyrolysis treatment at 823 K and the adsorption capacity of activated carbon was exhibited greatest for Cu^{2+} ions because of its size and pH (Moreno-Pirajan & Giraldo, 2012).

Bhakta et al. (2014) experimentally showed that activated carbon generated from microwave mediated pyrolysis of banana peel, orange peel, cotton fiber and cellulose paper waste, was a potent mercury (Hg^{2+}) sorbent. He reported that the Hg sorption efficiency or capacity was directly proportional to the surface area, porosity and carbon content (C) in that activated carbon. He estimated the following parameters (Table-2) and calculated the Hg removal efficiency of activated carbon used as sorbent as per 0.05 g/L for removal of Hg from an aqueous solution of $HgCl_2$ as concentration as 100 µg/L and concluded that the removal performance was better in cotton fiber and paper waste than other banana and orange peel (Table-2). He also cited that other components (such as, K_2O, CaO, SiO_2 and Cl) in activated carbon (Table-3) inhibited sorption process and thus reduced sorption capacity of activated carbon to Hg ions.

Rice husk carbonized at 400^0C showed the adsorption capacity as high as 2.69 mg/ g for Ni^{2+}, 5.28mg/g for Zn^{2+} and 6.01mg/g for Pb^{2+} (Taha et al., 2011). An equilibrium and kinetic study of Ni^{2+} using activated carbon developed from bamboo (*Bambuseae*) as potential adsorbent carried out by (Rajesh et al., 2014) resulted that the BET surface area, metal uptake and metal removal efficiency of the adsorbent were 116 m^2/g, 121.72 mg/g and 92.47% respectively from a an aqueous solution of Ni of 50 – 300 mg/L at pH 2-10, adsorption dosages 0.02 – 0.1g/50ml and interaction time 15 – 300 minutes. The bamboo stem was cut into small pieces and then pulverized and sieved to 78µm. About 10 – 20 g of

Table 2. Surface area, porosity, carbon content and removal efficiency of activated carbon generated from orange and banana peel, cotton fiber and cellulose paper waste

Activated carbon generated from	BET surface area (m^2/g)	Micropore surface area (m^2/g)	Micropore volume (cc/g)	Hg Removal efficiency (%)
Banana peel	115	51	0.0628	85
Orange peel	109	43	0.0519	84
Cotton fiber	138	65	0.0910	95
Cellulose paper waste	119	54	0.0791	94

After Bhakta et al., 2014.
BET: Brunauer Emmett and Teller

Table 3. Constituents of activated carbon developed from different waste employed for Hg sorption study

Constituents (wt %)	Activated Carbon Derived from			
	Banana Peel	Orange Peel	Cotton Fiber	Paper Waste
C	64.29	89.0	95.04	94.4
P_2O_5	5.29	5.24	5.57	5.59
K_2O	27.15	2.59	-	-
SiO_2	4.59	-	-	-
Cl	1.29	-	-	-
CaO	0.08	3.16	-	-

After Bhakta et al., 2014.

the powder was mixed with 14 ml of H_3PO_4 and heated at 150^{0}C in muffle furnace operated with N_2 gas flow at atmospheric pressure for 2 hrs. The mixture was sintered at a temperature range of 600 – 700^{0}C for 4 hrs. It was then cooled to normal temperature in flowing N_2 gas. Finally, the sample was washed with distilled water till pH 7 and dried at 105^{0}C for the use as adsorbent.

COST EFFECTIVENESS AND ENVIRONMENTAL BENEFIT OF BIO-WASTE TECHNOLOGY

The generation of bio-wastes is unlimited throughout the world. They are available and easily collected from houses, markets and industries without any cost. The preparation of bio-sorbent is very simple and follow some physical processes such as washing, drying, powdering and sieving but some sorbent undergo chemical treatment for addition of more functional groups such as ethylenediamination, methylation, acetylation etc. for binding metal. Carbonization process takes place in micro oven. There is no mechanical complexity no such costly infrastructure is required. Application of adsorbent is direct mixing with contaminated water. Small quantity of sorbent is well efficient in removing metals approximately 90%. All mechanical set up are cost reducing. Technology does not demand expert engineers and personnel; rather, skilled labours are enough to carry out the process involved in technology. The spent sorbent of

Figure 5. Bio-sorbent of bio-waste is a wealth of environment

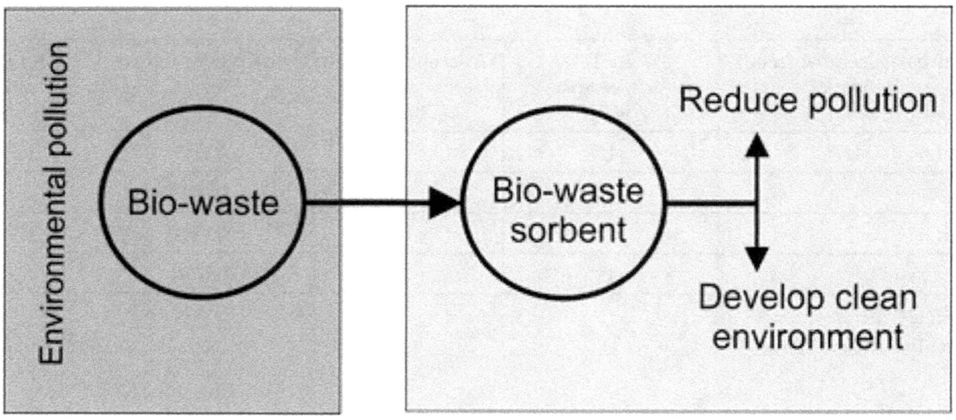

bio-waste has no further adverse environmental impacts since; it can be recycled and can be converted into building materials (such as ingredient of brick). Besides, this technology converts the bio-waste into wealth (Figure 5). Therefore, it is absolutely a low-cost economically feasible technology for the present scenario of environmental pollution.

CONCLUSION AND FUTURE PROSPECTS

Summarily, it can be concluded that bio-waste are capable to remediate the heavy metal(loid)s from their concerned solution at significantly higher rate (75% - 100%) at specific pH and at around room temperature. Additionally, chemical modifications can increase its removal efficiency. The removal efficiencies of various bio-wastes, orange peels, waste tea leaves, rice husk, wheat stalk, sugar cane bagasse, coconut husk, sun flower stalk, corn cob, nut shell, water hyacinth, crab shell particle, activated carbons etc. are extensively high and can be recommended as a low-cost media for treating the metal(loid)s contaminated environment. It would be a low-cost, environment friendly, feasible metal(loid)s treatment green technologies in developing and under developed countries.

Recently, researches also reported that algae are promising for heavy metal remediation due to its low cost, non polluting and high sorptive capacity (Dwivedi, 2012). The capacity depends on the interaction between heavy metals and alginate compounds located within cell wall of algae. So, further research will be focused on binding mechanism and some chemical modification for better sorption. These algae can also provide clue for development of more effective sorbent. Culture of microalgae in heavy metal contaminated medium can be a great effective green technology in heavy metal remediation. Because, extracellular polysachharides, lipid and fatty acid components in cell membrane may bind heavy metal and thus microalgae can be a potential sink for heavy metal. Filamentous fungi have a great sorptive potential for Pb, Zn, Cd, Cr, As and Ni because they can produce large amount biomass within short time (Siddiquee et al., 2015). So it can be a potential sorbent.

Development of potential sorbent from bio-waste and its application in removal of heavy metal(loid)s from wastewater or their contaminated aqueous solution are a low cost, eco-friendly and green technology which is available and feasible in underdeveloped countries. Further research is needed to increase its perfection and performance. Further study is also needed to search alternative high performance low cost biomass waste and green technology. In conclusion, it can be summarized that bio-waste has significant prospect as a low-cost and green metal(loid)s removal technology for leaning the contaminated environment.

REFERENCES

Abdulrasaq, O. O., & Basiru, O. G. (2010). Removal of Copper (II), Iron (III) and Lead (II) ions from mono component stimulated waste effluent by adsorption on coconut husk. *African Journal of Environmental Science and Technology*, *4*(6), 382–387. doi:10.5897/AJEST09.224

Ahalya, N., Ramachandra, T. V., & Kanamadi, R. D. (2003). Biosorption of heavy metals. *Research Journal of Chemistry and Environment*, *7*(4), 71–79.

Ahn, P., & Kim, B. (2016). Adsorption of heavy metals using chemically modified tea leaves. *International Journal of Biotechnology and Bioengineering*, *3*(1), 342.

Ali, S. M., Khalid, A. R., & Majid, R. M. (2014). The removal of Zinc, Chromium and Nickel from industrial waste water using corn cob. *Iraqi Journal of Science*, *55*(1), 123–131.

Arunakumar, K. K. I. U., & Walpola, B. C. (2013). Banana peel: A green revolution for metal removal from contaminated waters. *Korean Journal of Environmental Agriculture*, *32*(2), 108–116. doi:10.5338/KJEA.2013.32.2.108

Bhakta, J. N., & Munekage, Y. (2010). Spatial distribution and contamination status of arsenic, cadmium and lead in some coastal shrimp (*Macrobrachium rosenbergii*) farming ponds of Viet Nam. *Pacific Journal of Science and Technology*, *11*(1), 606–615.

Bhakta, J. N., & Munekage, Y. (2013). Identification of potential soil adsorbent for the removal of hazardous metals from aqueous phase. *International Journal of Environmental Science and Technology*, *10*(2), 315–324. doi:10.1007/s13762-012-0116-9

Bhakta, J. N., Ohnishi, K., Munekage, Y., Iwasaki, K., & Wei, M. (2012). Characterization of lactic acid bacteria-based probiotics as heavy metals sorbents. *Journal of Applied Microbiology*, *112*(6), 1193–1206. doi:10.1111/j.1365-2672.2012.05284.x PMID:22404232

Bhakta, J. N., Rana, S., Lahiri, S., & Munekage, Y. (2014). *Sorption of mercury onto waste material derived low-cost activated carbon*. Applied Water Science. doi:10.1007/s13201-014-0236-0

Bhakta, J. N., Son, L. T., & Munekage, Y. (2009). Distribution and potential impact of metal pollutants in the coastal environment: A case study with special reference to coastal aquaculture in red river delta of Viet Nam. *Electronic Journal of Biology, 5*(2), 22–27. Retrieved from http://www.ejbio.com

Brahmaiah, T., Spurthi, L., Chandrika, K., Ramanaiah, S., & Prasad, K. S. Sai. (2015). Kinetics of heavy metal (Cr & Ni) removal for the wastewater by using low cost adsorbent. *World Journal of Pharmacy and Pharmaceutical Sciences*, *4*(11), 1600–1610.

Budin, K., Subhramanium, Y., Tair, R., & Ali, S. A. M. (2014). The ability of crab and cockle shell to adsorb lead and chromium from industrial effluent. *IOSR Journal of Environmental Science. Toxicology and Food Technology*, *8*(12), 4–6. doi:10.9790/2402-081210406

Cai, J. H., & Jia, C. Q. (2010). Mercury removal from aqueous solution using coke-derived sulfur impregnated activated carbons. *Industrial & Engineering Chemistry Research*, *49*(6), 2716–2721. doi:10.1021/ie901194r

Dada, A. O., Ojediran, J. O., & Olalekan, A. P. (2013). Sorption of Pb2+ from aqueous solution unto modified rice husk: Isotherm studies. Advances in Physical Chemistry: Vol. 2013. (*Article ID: 842425,6*). Cairo, Egypt: Hindawi Publishing Corporation.

De Filipps, P., Di Palma, L., Petrucci, E., Scarsella, M., & Verdone, N. (2013). Production and characterization of adsorbent materials from sewage sludge by pyrolysis. *Chem Eng Trans*, *32*, 205–210.

Dixit, A., Dixit, S., & Goswami, C. S. (2015). Eco-friend alternatives for the removal of heavy metal using dry biomass of weeds and study the mechanism involved. *Journal of Bioremediation and Biodegradation*, *6*(3), 290–297.

Dwivedi, A. K., & Rajput, D. P. S. (2014). Studies on adsorption removal of heavy metal (Cu and Cd) from aqueous solution by tea waste adsorbent. *Journal of Industrial Pollution Control*, *30*(1), 85–90.

Dwivedi, S. (2012). Bioremediation of heavy metal by algae: Current and future perspective. *Journal of Advanced Laboratory Research in Biology*, *3*(3), 195–199.

Gall, J. E., Boyd, R. S., & Rajakaruna, N. (2015). Transfer of heavy metals through terrestrial food webs: A review. *Environmental Monitoring and Assessment*, *187*(4), 201. doi:10.1007/s10661-015-4436-3 PMID:25800370

Gönen, F., & Serin, D. S. (2012). Adsorption study on orange peel: Removal of Ni(II) ions from aqueous solution. *African Journal of Biotechnology*, *11*(5), 1250–1258.

Gönen, F., & Serin, D. S. (2012). Adsorption study on orange peel: Removal of Ni (II) ions from aqueous solution. *African Journal of Biotechnology*, *11*(5), 1250–1258.

Ideriah, T. J. K., David, O. D., & Ogbonna, D. N. (2012). Removal of heavy metal ions in aqueous solutions using palm fruit fibre as adsorbent. *Journal of Environmental Chemistry and Exotoxicology*, *4*(4), 82–90.

Jalali, M., & Aboulghazi, F. (2013). Sunflower stalk, an agricultural waste, as an adsorbent for the removal of lead and cadmium from aqueous solutions. *Journal of Material Cycles and Waste Management*, *15*(4), 548–555. doi:10.1007/s10163-012-0096-3

Khalfaoui, A., & Meniai, A. H. (2012). Application of chemically modified orange peels for removal of Copper (II) from aqueous solution. *Theoretical Foundations of Chemical Engineering*, *46*(6), 732–739. doi:10.1134/S0040579512060103

Kong, W., Ren, J., Wang, S., & Chen, Q. (2014). Removal of heavy metals from aqueous solution using acrylic-modified sugar cane bagasse – based adsorbents: Equilibrium and kinetic studies. *BioResources*, *9*(2), 3184–3196. doi:10.15376/biores.9.2.3184-3196

Krishnani, K. K. (2016). Lignocellulostic wheat straw – derived ion exchange adsorbent for heavy metals removal. *Applied Biochemistry and Biotechnology*, *178*(4), 670–686. doi:10.1007/s12010-015-1901-y PMID:26494139

Kumar, A., & Sahu, O. (2013). Sugar industry wastes removal of toxic metals from waste water. *World Journal of Chemical Education*, *1*(1), 17–20.

Kumar, P. S., Ramlingam, S., Abhinaya, R. V., Kiruphe, S. D., Murugesan, M., & Sevanesan, S. (2012). Adsorption of metal ions onto the chemically modified agricultural waste. *Clean-Soil, Air. Water (Basel)*, *40*(2), 188–197.

Lashkenari, M. S., Davodo, B., & Eisazadeh, H. (2011). Removal of Arsenic from aqueous solution using poly aniline rice husk nano composite. *Korean Journal of Chemical Engineering*, *28*(7), 1532–1538. doi:10.1007/s11814-011-0014-8

Liang, Y. J., Chi, L. Y., Min, X. B., Tang, C. J., & Zhang, H.-J. (2012). Hydrothermal sulfidation and floatationtreatment of heavy-metal-containing sludge for recovery and stabilization. *Journal of Hazardous Materials*, *217*, 307–310. doi:10.1016/j.jhazmat.2012.03.025 PMID:22487137

Lim Ai, P., & Aris, A. Z. (2014). A review on economically adsorbents on heavy metals removal in water and wastewater. *Reviews in Environmental Science and Biotechnology*, *13*(2), 163–181. doi:10.1007/s11157-013-9330-2

Ma, Z., Jacobsen, F. E., & Giedroc, D. P. (2010). Metal transporters and metal sensors: How coordination chemistry controls bacterial metal homeostasis. *Chemical Reviews*, *109*(10), 4644–4681. doi:10.1021/cr900077w PMID:19788177

Mahammed, R. R. (2012). Removal of heavy metals from wastewater using black tea waste. *Arabian Journal for Science and Engineering*, *37*(6), 1505–1520. doi:10.1007/s13369-012-0264-8

Moreno-Pirajan, J. C., & Giraldo, L. (2012). Heavy metal ions adsorption from wastewater using activated carbon from orange peel. *E-Journal of Chemistry*, *9*(2), 926–937. doi:10.1155/2012/383742

Mwangi, I.W., & Ngile, J.C. (2012). Removal of heavy metals from contaminated water using ethylene diamine modified green sea weed (*Caulerpa serrulata*). *Physics and Chemistry of the Earth, 50 – 52*, 111 – 120.

Ngah, W. S. W., & Hanafiah, M. A. K. M. (2008). Removal of heavy metal ions from wastewater by chemically modified plant wastes as adsorbent: A review. *Bioresource Technology*, *99*(10), 3935–3948. doi:10.1016/j.biortech.2007.06.011 PMID:17681755

Perez Marín, A. B., Aguilar, M. I., Meseguer, V. F., Ortũno, J. F., Sáez, J., & Lloréns, M. (2008). Biosorption of chromium (III) by orange (Citrus cinensis) waste: Batch and continuous studies. *Chemical Engineering Journal*, *155*(1-2), 199–206. doi:10.1016/j.cej.2009.07.034

Rajesh, Y., Pujari, M., & Uppaluri, R. (2014). Equilibrium and kinetic studies of Ni(II) adsorption using pine apple and bamboo stem based adsorbents. *Separation Science and Technology*, *49*(4), 533–544. do i:10.1080/01496395.2013.845426

Rajput, M. S., Sharma, A. K., Sharma, S., & Verma, S. (2015). Removal of Pb (II) from aqueous solution by orange peel. *International Journal of Applied Research, 1*(9), 411–413.

Rakshit, P. C. (2004). *Physical Chemistry* (7th ed.). Kolkata, India: Sarat Book Distributors.

Senthil, K. R., & Parthiban, R. (2011). Removal of zinc (Zn++) ions with crab shell particles from aqueous solution. *International Journal of Environmental Sciences, 1*(7), 1965–1977.

Shrestha, B., Kour, J., Homagai, P. L., Pokhrel, M. R., & Ghimire, K. N. (2013). Surface modification of the biowaste for purification of wastewater contaminated with toxic heavy metals – lead and cadmium. *Advances in Chemical Engineering and Science, 3*(03), 178–184. doi:10.4236/aces.2013.33022

Siddiquee, S., Rovina, K., Azad, S. A., Naher, L., Suryani, S., & Chaikaew, P. (2015). Heavy metal contaminants removal from wastewater using the potential filamentous fungi biomass: A review. *Journal of Microbial and Biochemical Technology, 7*(6), 384–393. doi:10.4172/1948-5948.1000243

Sweetly, D. J., Sangeetha, K., & Suganthi, B. (2014). Biosorption of heavy metal lead from aqueous solution by non-living biomass of *Sargassum myriocystum. International Journal of Application and Innovation in Engineering and Management, 3*(4), 39–45.

Tack, F. M. G., Vanhaesebroeck, T., Verloo, M. G., VanRompaey, K., & VanRanst, E. (2005). Mercury baseline levels in Flemish soils (Belgium). *Environmental Pollution, 134*(1), 173–179. doi:10.1016/j.envpol.2004.05.031 PMID:15572235

Tagermina, H., Ebadi, T., Nasernejad, B., & Ghafori, M. (2014). Arsenic removal from water by sugar cane bagasse: An application of response surface methodology (RSM). *Water, Air, and Soil Pollution, 225*(7), 2028–2050. doi:10.1007/s11270-014-2028-4

Taha, M. F., Keat, C. F., Shaharum, M. S., & Raml, A. (2011). *Removal of Ni(II), Zn(II) and Pb (II) ions from single metal aqueous solution using activated carbon prepared from rice husk.* World Academy of Science, Engineering and Technology.

Todorcius, T., Bulgariu, L., & Popa, V. I. (2015). Adsorption of Cu (II) from aqueous solution on wheat straw lignin: Equilibrium and kinetic studies. *Cellulose Chemistry and Technology, 49*(5-6), 439–447.

Vàzquez, G., Mosquera, O., Freire, M. S., Antorrena, G., Gouzàalez, A., & Levrez, J. (2012). Alkaline pre-treatment of waste chest nut shell from a food industry to enhance Cadmium, Copper, Lead and Zinc ions removal. *Chemical Engineering Journal, 184*, 147–155. doi:10.1016/j.cej.2012.01.019

Yasuda, M., Ozaki, K., Nohata, Y., Ohshiro, A., & Sonda, T. (2010). Removal of heavy metal ions from water by used green tea leaves. *Journal of Home Economics of Japan, 61*(6), 349–354.

Ye, H., Zhang, L., Zhang, B., Wu, G., & Du, D. (2012). Adsorption removal of Cu (II) from aqueous solution using modified rice husk. *International Journal of Engineering Research and Applications, 2*(2), 855–863.

KEY TERMS AND DEFINITIONS

Bioaccumulation: Formation of stable complexes between bio-molecules and non bio-degradable inorganic substances within tissue.

Bio-Waste: Unnecessary biomass produced from living body of animals and plants.

Chemical Modification: Covalent addition or alteration of functional group or groups of bio-molecules by chemical treatments for better efficiency.

Green Technology: Biomass based technology which has no harmful effect on environment.

Heavy Metal: Metal and metalloid with high density (> 5g cm^{-3}) in earth`s crust.

Metal Binding Groups: Functional groups of bio-molecules which can bind metal ions covalently, non-covalently or ionically and form a complex.

Remediation: Removal of hazardous substances from an environment.

Sorption Kinetics: Rate of adsorption or absorption process or reaction of heavy metal to bio-molecule.

Sorption: Adherence or entrance of metallic ion within cell or tissue.

Chapter 18
Bacterial Remediation of Phenolic Compounds

Veena Gayathri Krishnaswamy
Stella Maris College, India

ABSTRACT

Environmental pollution has been an irrefutable fact of life for many centuries; but it has become a real problem, since the start of the industrial revolution. Discharge of these toxic compounds without treatment results in serious health risks to humans and the marine ecosystem. Several physical, chemical and biological methods have been employed for the remediation of the phenolics. Bioremediation is identified as the most efficient, cost effective and eco-friendly ways for treatment of phenolic compounds. This article is a comprehensive review on the sources of phenolic compounds, their hazards, and their fate once released into the environment; the treatment technologies employed and bioremediation of these compounds using both non-extremophlic and extremophilic organisms. The review, throws light on the enzymes involved in the remediation of phenolic compounds, highlights the importance of extremophilic organisms and biological treatment of phenol containing industrial wastewaters. Such comprehensive information on the research work performed for the remediation of phenolic compounds provide ways to explore the role played by micro organisms in the remediation of phenolic compounds, which could be applied in the remediation of phenol /contaminated sites even under extreme conditions.

INTRODUCTION

Environmental pollution is a by-product of modern industrial society. With the immense growth of industries, a major problem is encountered with regards to the contamination of the environment with hazardous chemicals (Lakshmi & Sridevi, 2009).

The release of man-made (anthropogenic) organic compounds into the environment poses a very serious risk to public health. Phenolics, are one among the major pollutants, are discharged into the wastewater from the various industries such as phenol resin and pharmaceutical, oil refineries, petrochemical plants, ceramic plants, steel plants, herbicide manufacturing, fibre glass manufacturing and petrochemicals (El-Ashtoukhy et al., 2013; Veeresh et al., 2004; Jadhav & Vanjara, 2004) and coal conversion processes.

DOI: 10.4018/978-1-5225-2325-3.ch018

Phenol is the basic structural unit in a wide variety of synthetic organic compounds. Due to their toxic nature, these molecules tend to accumulate in water and soil after being discharged without an adequate treatment. Phenol at a concentration as low as 5 mg/L, imparts typical smell upon chlorination. The World Health Organization (WHO) prescribes 1 mg/L as the permissible concentration of phenol in drinking water (Saravanan et al., 2008).

Phenol and its derivatives are some of the major hazardous compounds in industrial wastewater (Watanabe et al., 1996b; Peters et al., 1997). Carbolic acid (hydroxy benzene) and cresols are antiseptic agents used in disinfection, which indicates that they are toxic to microbes. For instance, phenol is released into water from industrial effluent discharges such as petroleum refinery wastewater (Pfeiffer, 1979) and from other industrial discharges (Hawthorne & Sievers, 1984). Phenol has been detected in groundwater due to leaching of phenolic compounds through soil after leakage of phenol (Delfino & Dube, 1976), from landfill sites (Clark & Piskin, 1977), and from hazardous waste sites (Plumb, 1987). Therefore, phenol needs to be disposed-off in a safe and environmentally acceptable way to preserve the environmental quality.

Wastewater generated from various industries like pharmaceutical, textile, tannery which have high pH, salinity, Total dissolved Solids and recalcitrant compound which posses' potential hazards to the environment. Thus the wastewater becomes difficult to treat with conventional microorganisms as high salt and pH content induces cell lysis; hence there is need for microorganisms which can withstand extreme levels of pH, salt content and will degrade the recalcitrant phenolic compounds present in the wastewater and contaminated sites.

A goal of bioremediation is to transform organic pollutants into harmless metabolites or mineralize the pollutants into carbon dioxide and water. A feasible remedial technology requires microorganisms being capable of quick adaptation to and efficiently uses the pollutants of our interest in particular case in reasonable period of time.

This chapter would be a comprehensive review that outlines the sources of phenolic compounds, hazards, their treatment technologies and biological degradation using both non-extremophlic and extremophilic organisms. Therefore the present review focuses on types of pollution caused by the phenolic compounds and the health problems associated with the release of such xenobiotic compounds. The review also focuses on the different ways of remediation of the phenolic compounds and the importance of using extremophiles in the removal of the recalcitrant compounds underlining the importance of extremophilic microorganisms in the treatment of phenol laden- wastewaters and phenolic compounds contaminated sites.

Sources of Phenolic Compounds

The phenol originates in the environment either form natural, man-made and endogenous habitats .Phenol primarily enter the waters from industrial effluent discharges. Phenol is a mono-hydroxy derivative of benzene and the substituent with chloro, nitro, methyl etc that are named as derivatives of phenols or collectively referred to as "Phenolics" in wastewater. The sources of phenol in the environment are shown in Figure 1.

Figure 1. Sources of phenol contamination in the environment

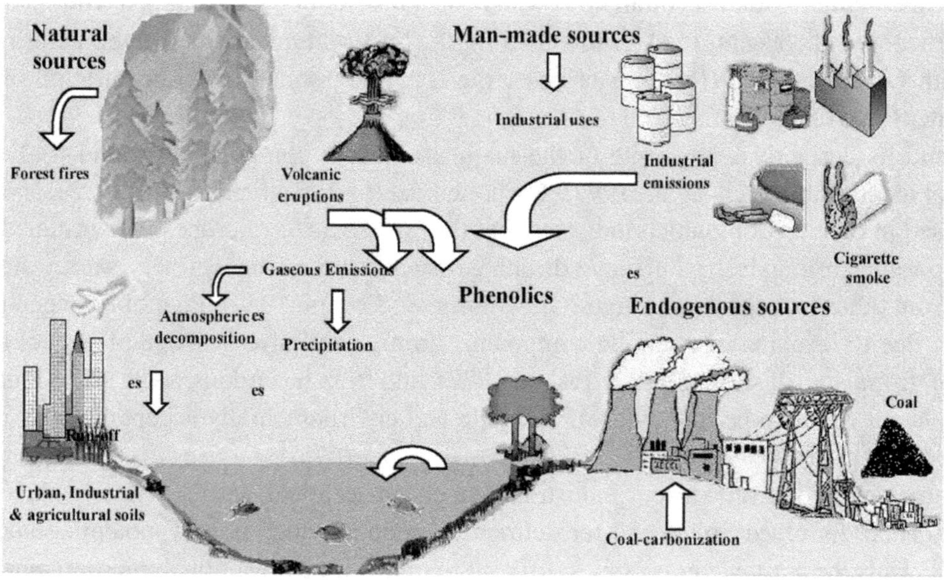

Natural Sources

Phenol is a constituent of coal tar, and is formed during decomposition of organic materials. Increased environmental levels of phenol may result from forest fires (Hubble et al., 1981). It has been detected among the volatile components from liquid manure at concentrations of 7-55 μg/kg dry weight and has an average concentration in manure of 30 μg/kg dry weight (RIVM, 1986).

Anthropogenic Sources

Man-made sources include industrial wastes from fossil fuel extraction, chemical manufacturing processes such as phenol manufacturing plants, pharmaceutical industry, wood processing industry and pesticide manufacturing plants. Industrial sources of phenols and other related aromatics from petroleum refinery, petrochemicals, basic organic chemical manufacture, coal refining, pharmaceuticals, tannery and pulp and paper mills (Table 1) (Kumaran & Parachuri 1997).

Endogenous Sources

An important additional source of phenol may be the formation from various xenobiotics such as benzene (Pekari et al., 1992) under the influence of light (Hoshino & Akimoto, 1978). Phenol and cresols (methyl substituted) are used extensively in the formulation of novolak resins, bakelite, caprolactum (intermediates in the synthesis of nylon), bisphenol and also in the preparation of drugs like aspirin. Shivaraman and Pandey (2000) reported that the low temperature coal carbonization waste contains 2500 mg/L of phenol, 200 mg/L 3-methylphenol, 480 mg/L 2-methylphenol, 470 mg/L 4-Methylphenol and 400 mg/L 2,4 dimethylphenol. Preis et al. (1997) indicated that the ash dump wastewater contains phenoilcs like phenol and cresol. Chloro substituted phenols have been used widely as a starting material or an

Table 1. Sources of phenols and other related aromatic compounds in wastewater

Sources	Significant Phenolic Compounds
Petroleum refining	Hydrocarbons (alkanes, cycloalkanes, polyaromatic hydrocarbons), benzenes, substituted benzenes, toluenes, *n*-octanes, *n*-decanes, naphthalenes, biphenyls, phenol, cyanide, sulphide and ammonia
Petrochemicals	Naphthalene, hepatanes, benzenes, butadiene, C-4 alcohols, phenol and resorcinol.
Basic organic chemical Manufacturing	*m*-amino phenol, resorcinol, dinitrophenol, *p*-nitrophenol, trinitrophenol, benzene sulphonic acids, aniline, chlorobenzenes, toluene and resorcinol.
Coal refining	*Phenol, catechol, o-, m-, p-cresols, resorcinol, hydroquinone, pyrogallol, polyaromatic hydrocarbons, pyridine, pycolines, lutidines, xylenes, toluenes, benzoicacid.*
Pharmaceuticals	*Toluenes, benzyl alcohols, phenyl acetic acid, chlorinated products of benzene, chloroform, ether, ethyl alcohol.*
Tannery	*Tannin, catechin, phenol, chlorophenol, nitrophenols.*
Pulp and Paper mills	*Lignin, vanillin, vanillic acid, dehydrodivanillin, ferulic acid, cinnamic acid, synringic acid, vieratric acid, protocatechuic acid, gentisic acid, benzoic acid, guadiachols, catechol, coniferyl alcohol, dehydrodihydroconiferyl alcohol, phenyl propionic acid, phenols and chlorophenols.*

Kumaran & Paruchuri, 1997.

intermediate in the manufacture of phenoxyaceticacid (2,4-D, 2,4,5-T), lindane wood preservatives and so forth. The wastewater from bleaching kraft mill contains 2,4-dichlorophenol, 2,4,6-trichlorophenol along with guaiacol (Perez et al., 2002). Cresols (mixed isomers) are widespread in nature, occurring, for instance, in petroleum, coal tar and crude oil. They are emitted from municipal incinerators, during coal and wood combustion, with vehicle exhaust, from oil refineries and cigarette smoke. Cresols are also products of the photo oxidation of toluene. *p*-Cresol is an endogenous metabolite of the aminoacid tyrosine in humans and warm-blooded animals.

The annual industrial production of chlorophenols was estimated at 0.2 x 10^6 metric tonnes in 1989 (WHO, 1989; Xun et al., 1992a; Stuart & Woods, 1998; Tiirola et al., 2002). PCP along with TCP and TeCP isomers were used historically as fungicides in wood preservative formulations (McAllister et al., 1996). Chlorophenols are observed as by-products of chlorine bleaching in the pulp and paper industry (Annachhatre & Gheewala 1996; Tartakovsky et al., 2001).

RELEASE OF PHENOL INTO THE ENVIRONMENT

Man-made phenolic compounds are found in the environment in abundance, due to agricultural and industrial activities. It has been reported that an estimated total of 23.5 million pounds (10.6 million kg) of phenol is released to the environment from 689 large processing facilities (TRI, 1998). Phenol has been found in surface water, ground water, soil and sediments.

- **Water:** It has been estimated that 99,546 pounds (~45 metric tonnes) of phenol was released to surface water from 677 domestic manufacturing and processing facilities in 2005, and it accounted for about 0.1% of the estimated total environmental releases from facilities required to report to the TRI, (TRI05, 2007).

Table 2. Most common levels of phenol concentration in the wastewater of selected industries

Industrial Activity	Phenol Concentration (mg/L)	Reference
Phenol production	3,000-4000	Godjevargova et al. 2003
Pulp and paper	33.1-40	Peralta-Zamora et al. 1998, Minussi et al. 1998
Textile	12.3	Kunz et al. 2001
Olive oil mill	3000-10,000	Klibanov et al. 1983, Borja et al. 1992, Hamdi 1992, Martinez-Neito et al. 1992, Knupp et al. 1996, 1998
Coal conversion plant	4-4780	Parkhurst et al. 1979
Oil- shale wastewater	4.5	Hawthorne and Sievers, 1984
Ash-heap water (oil shale)	500	Kahru et al. 1998
Phenolic resins production	1200->10,000	Patterson 1985, Kavitha and Palanivelu 2004
Methyl violet and cumene phenol production	310-660	Kanekar et al. 1999
Chemical specialities manufacturing	0.01-0.30	Jungclaus et al. 1978
Petroleum oil refinery	33.5	Pfeffer 1979

The most common anthropogenic sources of phenol in natural water include coal tar (Thurman, 1985) and wastewater from manufacturing industries such as resins, plastics, fibers, adhesives, iron, steel, aluminum, leather, rubber, textile, dying, phenolic resin manufacturing, glass fiber units, varnish industries and smelting related to metallurgical operations (Juang & Tsai, 2006; Yan et al., 2005; Bai et al., 2007) and influents from synthetic fuel manufacturing (Parkhurst et al., 1979). Phenol is also released from paper pulp mills (Keith, 1976) and wood treatment facilities (Goerlitz et al., 1985). Phenol has been detected in the effluent discharges of a variety of industries. Levels of phenol concentration in wastewater from selected industries are shown in Table 2. Other releases of phenol result from commercial use of phenol and phenol-containing products, including slimicides, general disinfectants (Budavari et al., 1989; Hawley, 1981), and medicinal preparations such as throat lozenges, mouthwashes, gargles, and antiseptic lotions.

Two natural sources of phenol in aquatic media are animal wastes and decomposition of organic wastes (EPA, 1989). As a metabolite of benzene, phenol may be released from publicly owned treatment works (POTWs) and sewage over flow. For example, it has been estimated that 3.8 kg/day of phenol are released to Newark Bay, in New Jersey, from municipal treatment facilities (Crawford et al., 1995).

- **Air:** The estimated releases of phenol of 7.6 million pounds (~3,480 metric tonnes) of phenol to the atmosphere from 677 domestic manufacturing and processing facilities in 2004, accounted for about 64% of the estimated total environmental releases from facilities required to report to the TRI (TRI05, 2007). During manufacturing, phenol is released to the atmosphere from storage tank vents during transport loading (Delaney & Hughes, 1979). Other major sources of release to the atmosphere are from residential wood burning and automobile exhaust. Phenol has been detected in the exhaust gases of private cars at concentration of 0.3 ppm (approximately 1.2 mg/m3) to 1.4-2.0 ppm (5.4-7.7 mg/m3) (Kuwata et al., 1980; Verschueren, 1983). Phenol has been detected from other sources such as emissions from waste incinerator plant at 0.36 ppb (Jay &

Stieglitz, 1995), in cigarette smoke and plastics (Graedel, 1978). It has been identified in cigarette smoke, in quantities that are comparable to an average emission of 0.4 mg/cigarette (Groenen, 1978). Emission of gases from all material incinerators and home fires, especially wood-burning, may contain substantial quantities of phenol (Den Boeft et al., 1984). Volatilization from environmental waters and soil has been shown to be a slow process and not expected to be a significant source of phenol in the atmosphere.

- **Soil:** The estimated releases of phenols of 1 million pounds (~472 metric tons) to soils from 689 domestic manufacturing and processing facilities in 2005, accounted for about 14% of the estimated total environmental releases from facilities required to report to the TRI (TRI05, 2007). An additional 1.5 million pounds (~670 metric tons), constituting about 19% of the total environmental emissions, were released via underground injection (TRI05, 2007).

Phenols are released to the soil during its manufacturing process, loading and transport when spills occur, and when it leaches from hazardous wastes sites and landfills (Xing et al., 1994). According to ATSDR (2006), generally the data on concentrations of phenol found in soil at sites other than hazardous sites are lacking. This may be due to a rapid rate of biodegradation and leaching. Phenol can be expected to be found in soils that receive continuous or consistent releases from a point source. Phenol that leaches through soil to groundwater spends at least some time in that soil as it travels to the groundwater. Phenol has been found in groundwater, mainly at or near hazardous wastes sites.

Fate of Phenol in the Environment

- **Water:** Phenol is highly soluble in water and relatively low in vapour pressure at room temperature. With these properties, phenol is expected to end up largely in the water phase upon distribution between air and water.

Phenol absorbs light in the region of 290-330 nm (Sadtler, 1960), thus it might photo-degrade directly in surface waters. Phenols react relatively rapid in sunlight natural water via reaction with photochemically produced hydroxyl radicals and peroxy radicals. According to Mill and Mabey (1985) the typical half-lives for hydroxyl and peroxyl radical reactions are on the order of 100 and 19.2 hours of sunlight, respectively. The rate constant for the reaction of phenol with ozone in water has been reported to range from 1.5 x 10-5 to 6 x 10-5 / milliseconds (Beltran & Alvarez, 1996). Phenol has been detected in ground water as a result of leaching through soil from a spill of phenol (Delfino & Dube, 1976), from landfill sites (Clark & Piskin, 1977), and from hazardous waste sites (Plumb, 1987). Phenol is readily biodegradable in natural water, provided its concentration is not high enough to cause inhibition. A complete degradation of phenol takes less than 1 day in for river water from lakes (Rubin & Alexander, 1983) and river after 2-4 days depending on the temperature (Ludzack & Ettinger, 1960). However, degradation of phenol was reported to be slower in salt water, and a half-life of 9 days has been reported in an estuarine river (Lee & Ryan, 1979). While the evidence cited above suggest that phenol rapidly gets degraded in natural water, but it may be still present in the environment because the exact conditions under which phenol is rapidly degraded are not known in all instances. One method of phenol breakdown is accomplished by the bacterium *Pseudomonas* sp. CF600, which uses a set of enzymes encoded by the plasmid dmp operon (Powlowski & Shingler, 1994).

The use of sequence batch reactors (SBR) in treating sludge contaminated with phenolic compounds has proven effective in breaking down the compounds biologically with no evidence of phenol volatility (Al-Harazin et al., 1991). Concentration as high as 1,600 mg/L can be broken down by 75% within 1 day of retention time. In some situation, phenol concentration may be too high or the population of microorganisms may not be present in sufficient numbers for significant biodegradation to occur. Phenol was found not to bioconcentrate in aquatic organisms. Reported log bioconcentration factors (BCF) in fish for phenol was estimated to be 0.28 for goldfish (Kobayashi et al., 1989) and 1.3 for golden orfe (Freitag et al., 1984).

- **Air:** The gas-phase reaction of phenol with photochemically produced hydroxyl radicals is probably the major removal mechanism in the atmosphere. An estimated half-life for phenol for this reaction is 0.61 days (EPA, 1989). The reaction of phenol with nitrate radicals during the night may constitute a significant removal process. This is based on a rate constant of 3.8 x 10-12 cm3/molecule second for this reaction, corresponding to a half-life of 15 minutes at an atmospheric concentration of 2 x 108 nitrate radicals per cm3 (Atkinson et al., 1987). The reaction of phenol with nitrate radicals present in the atmosphere during smog episodes may decrease the half-life of phenol in polluted atmospheres. The above data indicate that phenol has a short half-life in the atmosphere, probably <1 day. Phenol does not absorb light in the region of 290–330 nm (Lide, 1993) therefore; it should not photodegrade directly in the atmosphere. It has been concluded that deposition may contribute to the disappearance of phenol from the atmosphere.
- **Soil and Sediment:** Partition coefficient (KOC) values of phenol for two silt loams were reported to be 39 and 91 dm3/kg. Based on this KOC values, phenol would be expected to be highly mobile in soil, and therefore may leach to groundwater (Howard, 1989). This was confirmed by Scott et al. (1982) who found that low adsorption of phenol to two sterile 14 silt loams (pH 5.4, organic matter content 1.1 and 3.6, respectively). Ehrlich et al. (1982) reported that there was very little sorption of phenol onto aquifers materials which suggests that phenol sorption to sediments may be minimal. Based on the soil sorption coefficient, phenol released to soil is expected to leach to ground water. However, the rate of phenol degradation in the soil may be rapid, except in cases of large releases such as spills or continuous releases such as leaching from landfill sites, the probability of groundwater contamination may be low (Ehrlich et al., 1982). According to Xing et al. (1994), sorption coefficient for phenol by soil increases with increasing soil organic matter which may indicate that soil organic matter may act as a primary phenol sorbent in soil. Volatilization from dry-near surface soil should be relatively rapid (Howard, 1989). Phenol is degradable in soil under both aerobic and anaerobic conditions. The half-life of phenol in soil is generally less than 5 days (Baker & Mayfield, 1980; HSDB, 1998) but acidic soils and some surface soils may have half-life between 20 and 25 days (HSDB, 1998). Haider et al. (1974) found that mineralization in alkaline soil under aerobic conditions was 45.5 after 3 days, 48% after 7 days and 65% after 70 days. The half-lives for degradation of low concentration of phenol in silt loam soils were between 2.7 and 3.5 hours (Scott et al., 1982). Phenol degradation under anaerobic conditions is slower. However, phenol can be degraded completely in soil under both aerobic and anaerobic conditions, and phenol is not expected to be absorbed to sediment (HSDB 1998). Plants can readily uptake phenol (Cataldo et al., 1987) however, bioaccumulation does not take place due to high rate respiratory decomposition of phenol to CO2 (ATSDR, 2000). Figure 2 shows the Chemical structures of common phenolic compounds.

Figure 2. Chemical structure of common phenolic compounds
Taghreed Al-Khalid & Muftah H. El-Naas 2012.

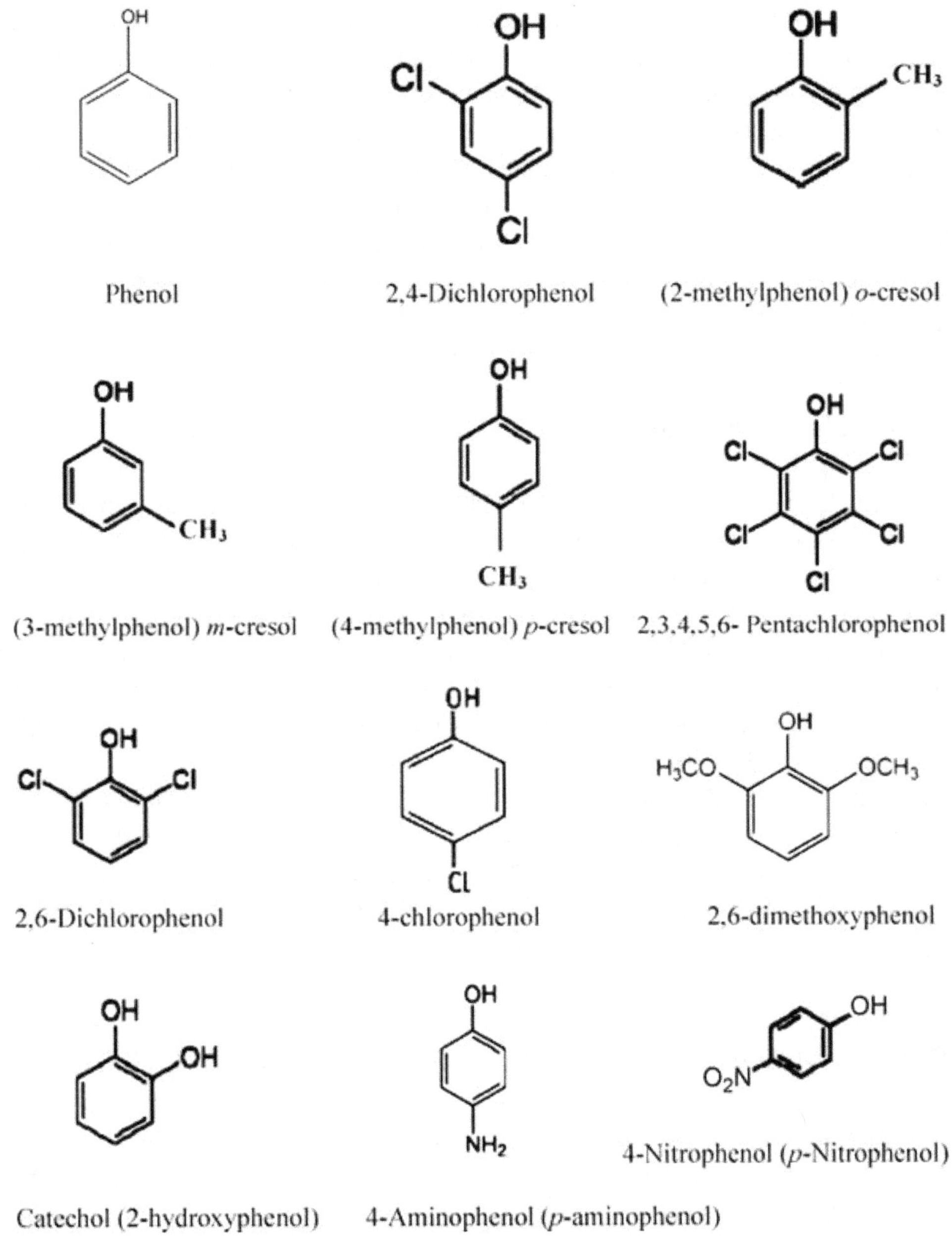

HAZARDS OF PHENOLIC COMPOUNDS AND ITS IMPACT

Aromatic hydrocarbons are not as readily biodegradable as the normal and branched alkanes, they are somewhat more easily degradable than the alicyclic hydrocarbons (Perry, 1984; Leahy & Colwell, 1990). Many of these compounds are toxic and some are known or suspected carcinogens (Verschueren, 1977; Klibanov, 1982; Sheeja & Murugesan, 2002). The presence of phenol in drinking water and irrigation water represents a serious health hazards to humans, animals, plants and microorganisms (Shailubhai, 1986; Sharma et al., 1997).

As compared to phenol, chlorophenolic compounds are more persistent in the environment. Toxicity and bioaccumulative potential of chlorophenols increases with the degree of chlorination (Loehr & Krishnamoorthy, 1988) and with chlorophenol lipophilicity (Sierra-Alvarez et al., 1994). Bioaccumulation potential, which is quantified by the bioconcentration factor (BCF), is positively correlated to the octanol water partition coefficient (Kow) (Loehr & Krishnamoorthy, 1988). On the basis of log Kow values, bioaccumulation followed the following order: 2-chlorophenol (CP) < 4-CP < 2,4,5-TCP (TCP) < 2,4,6-TCP < 2,3,4,6-TeCP (tetraCP) < PCP. This clearly indicates that the bioaccumulation potential increases with the number of chlorine substituent's on the phenolic ring. Phenol at a concentration as low as 5µg/L imparts typical smell upon chlorination and that is why WHO has prescribed a concentration of 1µg as the guideline concentration for drinking water (WHO, 1989).

Effects of Phenolic Compounds

Phenols have relatively high water solubility and widely known to be acutely toxic to a range of organisms. It produces undesirable taste, odour, colour to water and is considered toxic (Klibanov, 1983).

Continuous ingestion of phenol for prolonged period of time causes mouth sore, diarrhoea, excretion of dark urine and impaired vision at concentrations levels ranging between 10 and 240 mg/L (Barker et al., 1978). Phenols are toxic to several biochemical functions (Kumaran, 1986) and to fish life. It acts as substrate inhibitor in the biotransformation (Hill & Robinson, 1975). Lethal blood concentration for phenol is around 4.7 to 130 mg/100 ml. Phenol affects the nervous system and key organs, i.e. spleen, pancreas and kidneys (Manahan, 1994).

Phenol also contributes to off-flavours in drinking and food-processing waters (Hill & Robinson, 1975). Although not found to be bio-accumulative (Loehr & Krishnamoorthy, 1988) humans exposed to phenol in well water at concentrations of 1300 mg/L exhibited a statistically significant increase in diarrhoea, mouth sores, dark urine and burning of the mouth (USEPA, 1980). The toxic effects of phenol on humans, animals and fishes are presented in the Table 3.

METHODS INVOLVED IN THE REMOVAL OF PHENOLIC COMPOUNDS

Based on the mode of action, the removal of phenolic compounds is categorized into physical, chemical and biological which are represented in Figure 3. Physical methods involve the removal of phenols from aqueous stream, which includes adsorption, solvent extraction, stripping etc. Chemical methods mostly destroy the phenolic compounds either by oxidation or reduction process. Each treatment process and their application are specific to phenol and their substituted phenolic compounds are discussed below.

Physical and Chemical Methods

Physical treatment is mainly done through adsorption with charcoal, chitosan etc, while the chemical processes include oxidation reactions using ozone, ozone /uv, wet air oxidation, supercritical water oxidation and electro chemical oxidation (Annadurai et al., 2000 a & b).

Table 3. Summary of toxic effects of phenolic compounds

Route of Exposure	Toxic Effects	Reference
On Human Acute Exposure		
Route –Inhalation	Anorexia, weight loss, headache, vertigo, salivation and dark urine indicative of nephrotoxicity	IPCS 1994,1999
Route-Ingestion	Cardiovascular effects, respiratory distress, metabolic acidosis, renal failure, neurological effects, shock, coma and death	IPCS 1994,1999
Dermal exposure	Dermal inflammation, erythema, anaesthetic, painless blanching, burns, corrosion and necrosis	IPCS 1994,ATSDR 1998
Chronic Exposure		
Route –Inhalation	Gastrointestinal effects such as anorexia, progressive weight loss and excess production of saliva, Muscle pain and weakness have also been reported, enlarged liver, and elevated concentrations of liver enzymes and dark urine.	IPCS 1994,1999; ATSDR 1998
Route-Ingestion	Gastrointestinal irritation, cardiovascular, CNS and respiratory effects and decreased body weight	IPCS 1994,1999 ATSDR 1998
Dermal exposure	Onychronosis (yellowing of the skin), skin irritation skin eruption, as well as dermal inflammation and necrosis	IPCS 1994,1999 ATSDR 1998, DEFRA 2003
On Animals		
Route –Inhalation	Acute effects are generally attributed to the depression of the CNS, neuromuscular hyperexcitability (twitching and convulsions), increased heart rate followed by slow and irregular heart rate), hypertension followed by hypotension, salivation, dyspnoea and hypothermia.	IPCS 1999
Route-Ingestion	Cardiovascular effects, respiratory distress, metabolic acidosis, renal failure, neurological effects, shock, coma and death.	IPCS 1999, ATSDR 1998
Dermal exposure	Erythema, inflammation, oedema, skin irritation, discolouration, eczema and necrosis, congenital malformations	IPCS 1999, NTIS 2005
Effects on fishes, Route- general	Phenol at 5-25 mg/L are toxic to fishes causing genotoxic, carcinogenic, pytotoxic, immunotoxic, haematological, physiological effects, antibiosis, ovipositional deterrence and reduced growth rate	Saha et al. 1999, gonzalez et al. 1990, Rodriguez 1988, Girolami et al. 1981, Capsso et al. 1992, Nahed and Amal 2008

Adsorption

Adsorption is regarded as a physical process in which certain solid materials have the capacity to adsorb or extract specific substances out of the solution. Activated carbon is one of the most widely used adsorbent for removing organic compounds from water and wastewater. Activated carbon and chitosan had been reported to adsorb phenolics from various effluents (Annaduari et al., 2000 a & b). The adsorption by chemicals depended on the mode of contact with a sorbate in the reactor. With the high cost of activated carbon, charcoal was found to be a better alternative. Charcoal treatment was suitable for small scale industries.

Liquid Membrane

It is one of the recovery techniques employed for the extraction of organics and metal ions from aqueous wastewater by using immiscible solvent and are simultaneously stripped. Chemical processes included oxidation reactions using ozone; ozone/UV, ozone/H_2O_2, The method was disadvantages with initial

Figure 3. Methods involved in the treatment of phenolic compounds

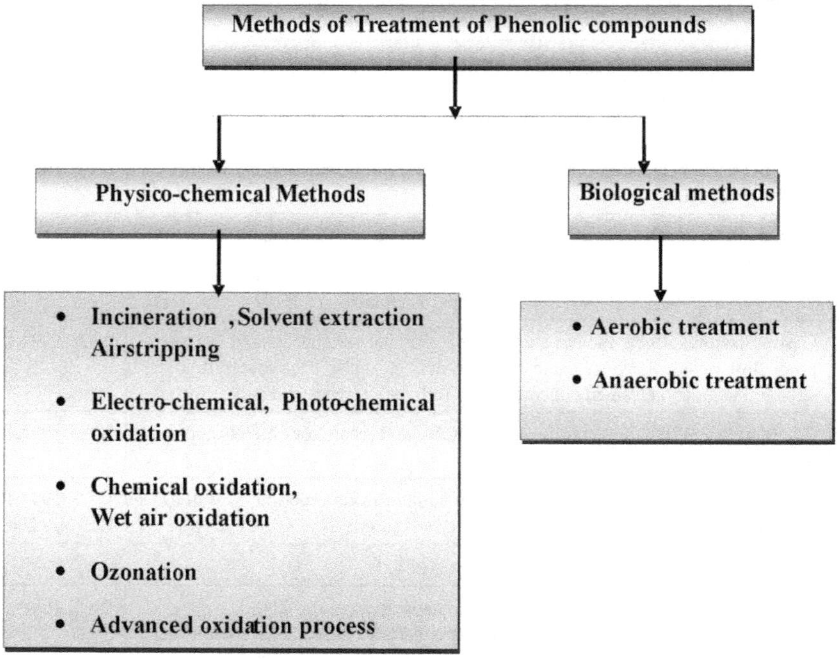

high cost of ozone. Air was an inexpensive oxidizing agent, but the reaction was very slow. Kadhim et al. (1999) suggested that the same cannot be adopted because of high cost, incomplete purification and formation of toxic intermediates.

Chemical Oxidation

This method is employed in chemical oxidation of organic compounds present in water and wastewater treatment because of its inherent capability of converting organic compounds to innocuous compounds of water and CO_2. Various oxidants like hydrogen peroxide, ozone and chlorine are used to oxidize hazardous organic compounds, particularly phenolics. Oxygen is generally not a strong oxidant at room conditions to oxidize phenols. However, Wet air oxidation and supercritical water oxidation utilized air and water at elevated pressure and temperatures to oxidize these compounds. The use of chlorine in the treatment of phenols was avoided due to the formation of chlorinated by-products, which are extremely toxic compared to phenol (Kinney, 1960).

Ozonation

The ozone is a powerful oxidizing agent; it is able to participate in a great number of reactions with organic and inorganic compound. Among the most common oxidizers, only hydroxyl and fluorine have oxidation potential more than the ozone. In the last twenty years, this chemical agent has acquired a great importance in wastewater treatments. Thus, the ozonation can constitute a advanced oxidation processes for itself, as hydroxyl radical coming from the decomposition of ozone which is catalyzed by hydroxyl ion or the initiated by the presence of traces of other substances, like transition metal cations (Beltran et al., 1997). The ozone treatment process can be supplemented with the use of ultraviolet radiation and/ or hydrogen peroxide.

Biological Methods

The main objective of the biological treatment is to remove soluble or colloidal or suspended biologically transformable complex substances to its simpler forms by the action of microorganisms and to stabilize the organic matter (Metcalf & Eddy, 2005). Compared with other treatment processes available, biological treatment methods are attractive because they are able to degrade complex organic compounds present in the wastewater resulting in lower concentrations of organics (Nakamura & Sawada, 2000) and they are inexpensive. Reported research on the biological degradation of phenolics ranges from laboratory scale studies with single culture to in-situ or field level applications. In future technologies of microbial systems might be potent tools to deal with environmental pollutants.

Aerobic microbes are more efficient to transfer the recalcitrant compounds because they grow faster than anaerobes and usually transform organic compounds to inorganic compounds and produce end products CO_2, H_2O_2 (Ucun, 2010; Dash, 2009; Mrozik, 2010). Aerobic processes are also preferred due to the low costs associated with this method. Based on these aspects utilization of phenols by anaerobic bacteria is limited. Several reports have been reported in the literature highlighting these points. (Hussain, 2010; Leven, 2006; Lin, 2009; Gali, 2007; Shibata, 2006; Boll, 2005). Since most biological treatments methods on phenolic compounds have used aerobic processes, this review will focus mainly on aerobic biodegradation of phenolic compounds by mesophiles and extermophlies (Figure 4).

Bacterial Bioremediation of Phenol by Mesophilies

Phenol and benzoic acids were converted to non-recalcitrant catechol by some soil bacteria. Phenol degrading thermophilic *Bacillus* sp. converted catechol to hydroxymuconic semialdehyde by meta-cleavage routes (Ceniglia, 1984). *Bacillus sterothermophilus* BR219 degraded phenol at levels of 15 mM (Gurujeyalakshmi & Oriel, 1988). It was a river sediment isolate and showed growth inhibition at higher concentrations of phenol.

Phenols are metabolized by microorganisms from a variety of different genera and species of bacteria, fungi and algae. The biodegradation of phenol and its derivatives by bacteria has been extensively studied and a large number of phenol-degrading bacteria have been isolated and characterized at the physiological and genetic levels (Wang et al., 2007; El-Sayed et al., 2003; Shen et al., 2004 ;Polymenakou & Stephanou, 2005; Wu et al., 2005; Arutchelvan et al., 2005)

The key components of bacterial communities responsible for degradation of phenolic wastes are *Pseudomonads*. Pure and mixed cultures of the Pseudomonas genus are the most commonly utilized biomass for the biodegradation of phenols (Stoilova et al., 2007). Specifically, *Pseudomonas putida* has been commonly used for the biodegradation of phenol due to its high removal efficiency (El-Naas et al., 2008). As shown in Table 4, *Pseudomonas putida* has been extensively investigated and has been reported to be capable of high rates of phenol degradation (Hutchinson & Robinson, 1988).

In *Pseudomonads*, many of its induced enzymes are non-specific and its metabolic pathway contains a high degree of convergence. The convergence of catabolic pathways allow for the efficient utilisation

Figure 4. Main pathways of phenol degradation under aerobic condition (ortho-and meta fission of the benzene ring) (Krug et al., 1985; Nagi et al., 1990)

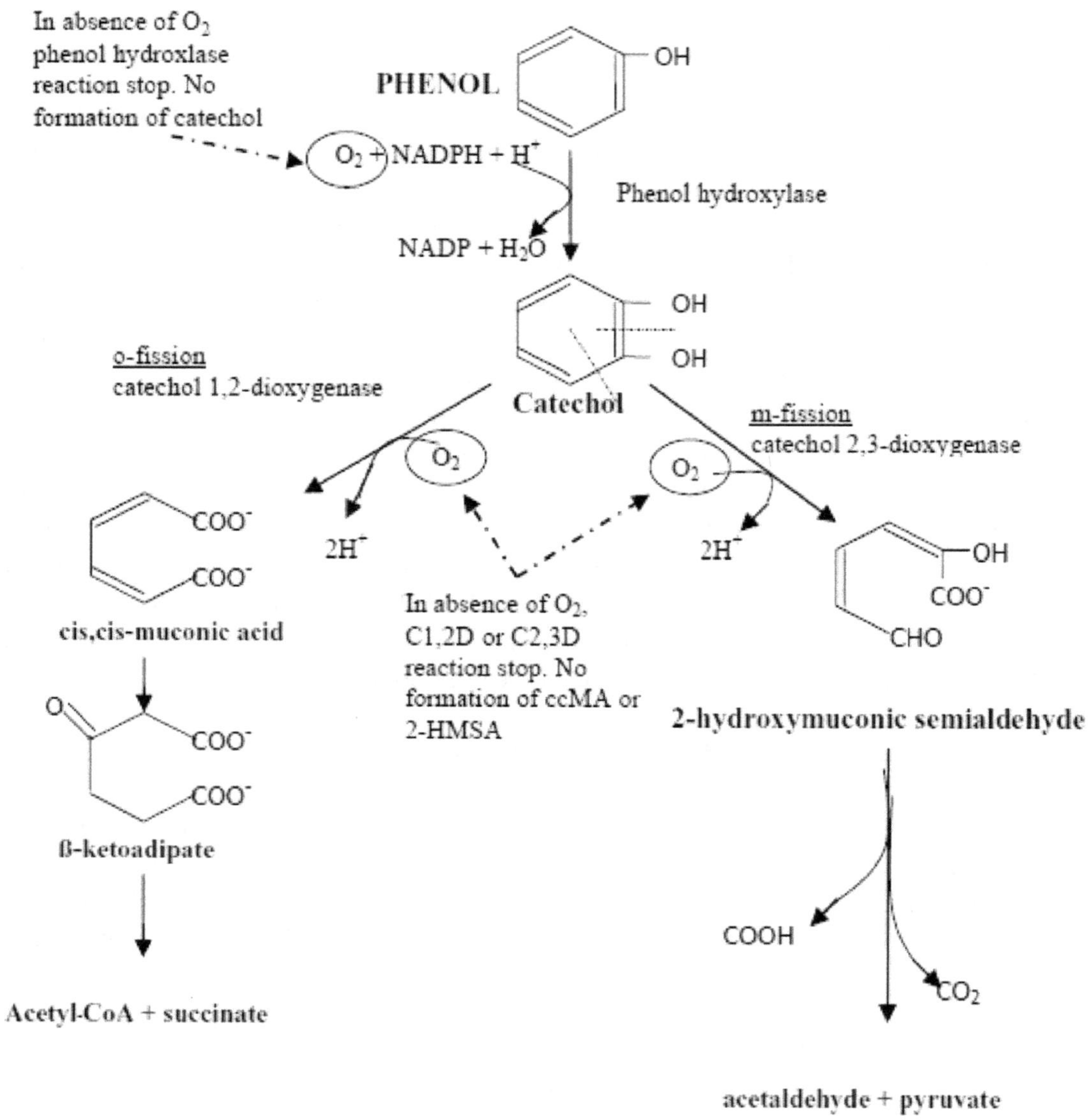

of a wide range of growth substrates while the non specificity of the induced enzymes allows for the simultaneous utilisation of several similar substrates without an excess of redundant genetic coding for enzyme induction (Hutchinson & Robinson, 1988).

Responses of *P. putida* to chemical stresses have indicated that its cells could use diverse protective mechanisms for survival in various extreme environments. These studies could help in synthesizing new bacterial strains with enhanced degradation capability and improved tolerance to toxic pollutants (Loh & Cao, 2008).

Table 4. Phenol- degradation by different species of bacteria

Microorganisms	Reference
Acinetobacter sp. *Acinetobacter* sp. W-17 *Acinetobacter calcoaceticus* *Acinetobacter johnsonii*	Tibbles and Baecker 1989 Abd-EL-Haleem et al. 2003 Nakamura and Sawada 2000 Heilbuth et al. 2003
Achromobacter sp.E1	Watanabe et al. 1996a
Alcaligenes faecalis	Bastos et al. 2000b
Alcaligenes sp. E2	Watanabe et al. 1996a
Arthrobacter sp.	Kar et al. 1996
Azoarcus sp.	Shinoda et al. 2000
Bacillus thermoleovorans-A2	Mutzel et al. 1996
Burkholderia cepacia G4	Moustafa El-Sayed 2003
Halomonas sp.	Hinteregger and Streichsbier, 1997
Comamonastestosteroni -P15, -E23 (mutant)	Yap et al. 1999
Halophilic bacteria CA00, CA08, SL03, SL08, SP04	Peyton et al. 2002
Magnetospirillum sp.	Shinoda et al. 2000
Micrococcus sp.	Tibbles and Baecker 1989
Nocardia sp.	Tibbles and Baecker 1989
Ochrobactrum tritici	El-Sayed et al. 2003
Phormidium valderianum BDU 30501	Shashirekha et al. 1997
Pseudomonas sp.	Kang and Park 1997
Pseudomonas pictorum-NICM-2077	Sheeja and Murugesan 2002
Pseudomonas putida ATCC 11172	Loh and Liu 2001
Pseudomonas putida ATCC 12633	Hughes and Cooper 1996
Pseudomonas putida – ATCC 17484	Gonzalez et al. 2001
Pseudomonas putida F1 -ATCC 700007	Abuhamed et al. 2003
Pseudomonas putida	Abuhamed et al. 2004
Pseudomonas putida BH	Reardon et al. 2000 Reardon et al. 2002
Pseudomonas putida BH	Soda et al. 1998
Pseudomonas putida – CCRC14365	Chung et al. 2003
Pseudomonas pictorium	Annadurai et al. 2000 a
Pseudomonas putida DSM 548	Monteiro et al. 2000
Pseudomonas putida EKII	Hinteregger et al. 1992
Arthrobacter sp, *Bacillus cereus*, *Citrobacter freundii*, *Micrococcus agilis*, *Pseudomonas putida* biovar B	Kanekar et al. 1999

Bioremediation Mechanisms of Phenolic Compound

The presence or absence of molecular oxygen plays a crucial role in determining the fate and biodegradation mechanisms of aromatic compounds. In general, phenol can be transformed both under aerobic and anaerobic conditions.

Microbial degradation of chemicals in the environment is a route for their removal. The microbial degradation of these compounds is a complex series of biochemical reactions and often different when different microorganisms are involved. The interdependence of biodegradation, biotransformation and biocatalysis has been reviewed by Parales et al. (2002). Microbial degradation of pollutants is crucial in order to predict their longevity and long term effects and also important in the actual remediation process (Landis & Yu, 2003).

Aerobic and anaerobic biodegradation of aromatic compounds have several common features. Structurally diverse compounds are degraded through many different peripheral pathways to a few intermediates that are further channelled via a few central pathways to the central metabolism of the cell. In the aerobic catabolic funnel, most peripheral pathways involve oxygenation reactions carried out by monooxygenases and /or hydroxylating dioxygenases that generate dihydroxy aromatic compounds (catechol, protocatechuate, gentisate, homoprotocatechuate, homogentisate, hydroquinone, hydroxyquinol). These intermediate compounds are the substrates of ring-cleavage enzymes that use molecular oxygen to open the aromatic ring between the two hydroxyl groups (ortho cleavage, catalyzed by intradiol dioxygenases) or proximal to one of the two hydroxyl groups (meta cleavage, catalyzed by extradiol dioxygenases) (Diaz, 2004).

The biodegradation process requires the presence of molecular oxygen to initiate enzymatic attack on the aromatic rings. A typical pathway for metabolizing phenol is to hydroxylate the ring, by the enzyme phenol hydroxylase, form catechol, and then open the ring through ortho- (also termed β-ketoadipate pathway) or meta-oxidation. (Melo, 2005; Lika, 2009)

Phenol hydroxylase represents the first enzyme in the metabolic pathway of phenol degradation. Both o- and m-pathways are distinguishable by measuring their characteristic enzyme activities. In the o-pathway, the aromatic ring is cleaved by the enzyme catechol 1,2-dioxygenase (C12O). In the m-pathway, the ring is cleaved by the enzyme catechol 2,3-dioxygenase (C23O). Thus, the ring is opened and the degraded (Agarry, 2008).

The ring cleavage can occur in two different orientations and this difference in cleavage site is used to classify catechol dioxygenases in two groups; the intradiol (such as C12O) and extradiol (such as C23O) cleaving enzymes. The genes of ring-cleavage dioxygenases may serve as good targets for monitoring the biodegrading biomass, thus providing a rapid method for monitoring the microbial community during the treatment process. (Heinaru et al., 2005). Biodegradation of phenol and their prediction of pathways has been studied and reported by various literature (Wang et al., 2010; Veena Gayathri & Vasudevan, 2010, 2011, 2015).

Bacterial Bioremediation of Phenol by Extremophiles

"Extremophiles" are organisms with the ability to thrive in extreme environments such as hydrothermal vents. Since they live in "extreme environments" (under different pH, high pressure and temperature), they can survive under a different range of conditions. The extreme environments vary in some aspects from those which humans consider as 'normal', where moderate conditions exist with pH around neutral, temperatures ranges between 20 °C and 40 °C, pressures around 0.1 MPA (1 atm), and adequate levels of nutrients and salts (Satyanarayana et al., 2005; Thiel, 2011). Extreme environments harbour well adapted organisms, called extremophiles, which can survive in the presence of heavy metals, acids, petroleum and natural gas; these are toxic to human being.

The extermophiles are classified according to the conditions in which they grow: As thermophiles and hyperthermophiles (organisms growing at high or very high temperatures, respectively), psychrophiles (organisms that grow best at low temperatures), acidophiles and alkaliphiles (organisms optimally adapted to acidic or basic pH values, respectively), barophiles (organisms that grow best under pressure), and halophiles (organisms that require NaCl for growth). In addition, these organisms are normally polyextremophiles, being adapted to live in habitats where various physicochemical parameters reach extreme values (Rampelotto, 2013).

Novel species are continuously being isolated from environments previously thought unsuitable for growth because of extraordinary conditions. While extremophiles have been shown to degrade some problematic substances such as aromatic, polycyclic, chlorinated, and nitrated hydrocarbons, the broader biodegradation potential for extremophiles is still considered to be limited Margesin & Schinner, (2001). It is common for extreme habitats to contain some hydrocarbons, naturally or by pollution, so that indigenous populations of extremophiles have developed resistance and degradative capabilities with respect to those contaminants, and may be successfully targeted for environmental clean-up given optimal nutrient conditions (Margesin & Schinner, 2001). Extremophiles in Soap Lake have been shown to degrade some aromatic compounds including phenol and coumaric acid (Alva & Peyton, 2003; Dimitriu et al., 2005).

Halophiles

Industrial and municipal effluents are often discharged into saline and hypersaline depression and intertidal zones, especially in developing countries (Lefebvre et al., 2004). Several industrial processes, such as pesticide, chemical and pharmaceutical production as well as gas and oil extraction, generate thousands of millions of liters of saline to highly saline wastewaters (Lefebvre et al., 2006). Saline and hypersaline environments are frequently contaminated with organic compounds as a result of industrial activities (Oren et al., 1992; Margesin, 2001). Contamination of these habitats constitutes a serious environmental problem mainly due to the high toxicity exhibited by aromatic hydrocarbons.

In most cases, biodegradation constitutes the primary mechanism for contaminant removal. However, biodegradation processes are difficult to perform under saline conditions (Oren et al., 1992; Margesin & Schinner, 2001). In addition, it is known that traditional pollutant biodegradation is less efficient or does not function when salinity increases above sea levels (Oren et al., 1992) and the conditions can kill or inhibit the functionality of non-extremophilic species. Phenol biodegradation by mesophilic microorganisms has been extensively evaluated as can be seen in the review by Nair et al. (2008). However, relative to mesophilic phenol degrading microorganisms only a small number of reports have been published describing the biodegradation of phenol and other aromatic compounds under hypersaline conditions and considerably less information exists on the associated genes, pathways, and mechanisms of their degradation (Woolard & Irvine 1994, 1995; Hinteregger & Streichsbier, 1997; Bastos et al., 2000; Munoz et al., 2001; Alva, & Peyton, 2003; Garcia, et al., 2005; Wang et al., 2009; Veena Gayathri &Vasudevan, 2010; Bonfa et al., 2011; Veena Gayathri & Vasudevan, 2014, Veena Gayathri & Vasudevan, 2015).

Hence, it is a challenging issue in the treatment of wastewater in petrochemical, pesticide, pharmaceutical, mining and other industries for effective biological treatment. Conventional methods of reverse osmosis, ion exchange, electrodialysis, and sizeable freshwater dilution are not always feasible for polluted environmental sites and industrial effluents, using halophiles is an attractive alternative to remediation with non-halophilic organisms for contaminant biodegradation of hypersaline wastewaters (Margesin & Schinner, 2001; Peyton et al., 2002).

The physiology of non-halophilic bacteria does not allow them to function properly in high saline solutions, as these conditions tend to induce plasmolysis (recession of cytoplasm) Kargi and Uygur (1996). Plasmolysis is induced by osmotic differences across the cell wall and causes an outward flow of intracellular water, resulting in cell dehydration. Halophilic bacteria have adapted to the osmotic stress of high salinity environments by actively accumulating K+, Glycerol, betaine and /or other compatible solutes within the cytoplasm. These solutes increase inonic strength within the cell, thus allowing the halophile to balance intra – and extra-cellular ionic strength to maintain cytoplasmic hydration. Halophilic organisms could play an important role in the treatment of such industrial wastewaters. Typical non-halophilic bacteria grow well in media containing NaCl concentrations below 1% w/v (10 g/L). Bacteria with growth optima in NaCl concentrations in excess of 1% are defined as halophilic, and are further categorized into three groups according to the salinity of their growth optima: slightly halophilic (1–3% w/v), moderately halophilic (3–15% w/v), or extremely halophilic (>15% w/v). The degradation of phenol by both pure and mixed cultures under saline conditions is given in the Table 5.

Table 5. Summary of degradation of phenol by microorganisms under saline condition

Microorganism	Method	Concentration of Phenol	Time	Concentration of NaCl (w/v)	Reference
Mixed culture	Sequential batch reactor	100 mg/L	150 h	1-15%	Woolard and Irvine 1995
Halomonas sp.	Bubble reactor	100 mg/L	13 h	3-5%	Hinteregger and Schreibier 1997
Mixed culture	Batch study	100 mg/L	-	2.6%	Kanekar et al. 1999
Candida tropicalis Alcaligenes faecalis	Batch study	1130-1600 mg/L	200 h	5.6%	Bastos et al. 2000
Deleya venusta	Batch study	1600 mg/L	10 d	8%	Munoz et al. 2001
Halomonas campisalis	Batch study – Kinetics	320 mg/L	18 h	10%	Peyton et al. 2002
Halomonas campisalis	Batch study	130 mg/L	100 h	15%	Alva and Peyton 2003
Halomonas sp. EF11	Batch study-Carbon-Nitrogen ratio with metabolic pathway	2mM	50 h	10%	Maskow and Kleinsteuber 2004
Halomonas organivorans	Batch study	Phenol-5mM *p*-cresol-2 mM	-	10%	Garcia et al. 2004
Thalassobacillus devorans	Isolation study	50 mg/L	96 h	7.5-10%	Garcia et al. 2005
Arthrobacter sp.	Batch study-Mixture of phenolics	Phenol- 200mg/L *p*-cresol-100mg/L	70 h	0%-10%	Wang et al. 2009

Chlorinated hydrocarbons are widely used because of their chemical and thermal stability as well as their fungicidal, herbicidal, and insecticidal properties. The ability of halophiles to degrade halogenated aromatic compounds frequently in pesticides is also known. The dechlorination of chlorophenols and chlorophenoxyphenols has been observed in anaerobic groundwater next to a hypersaline alkaline lake in the USA (Boone et al., 1989). The halophilic Archaea *Haloarcula, Halobacterium* and *Haloferax* also degrade hydrocarbons such as halogenated trichlorophenols and insecticides lindane and DDT.

Chlorinated phenols have been widely used as wood preservatives and agricultural biocides. Maltseva and Oriel (1997) reported the isolation of the alkaliphilic and slightly halophilic *Nocardioides* sp. strain M6, which was able to grow on 2,4-dichlorophenol (2,4-DCP) as well as on 2,4,5- and 2,4,6-trichlorophenol (2,4,5-TCP and 2,4,6- TCP) as the sole source of carbon and energy. Strain M6 degraded 0.7 g/l of 2,4,5-TCP in 120 h and could grow at concentrations up to 1.6 g/l of 2,4,6-TCP; however, the length of the lag phase increased on increasing the concentration of this substrate. Optimum growth occurred at pH-9-9.4 and 1-2.4% NaCl. Recently another halophilic phylogenetically close and *Pseudomonas beijerinckii, Chromohalobacter marismortui* have used the phosphonoacetate, 2-aminoethyl-, 3--aminopropyl, 4-aminobutyl-, methyl-and ethyl-phosphonate as a source of phosphorus for growth at 10% NaCl (Hayes et al., 2000).

A novel *Bacillus* sp., strain YAS1, could transform tyrosol (p-hydroxyphenylethanol) a phenolic compound from olive mill wastewater into p-hydroxyphenylacetic acid under saline conditions (50 g/L NaCl) in the presence of 1 g/l of yeast extract (Abdelkafi et al., 2005). Tabib et al. (2014) isolated phenol degrading bacterium, *Ralstonia* sp. Strain PHS1, which was isolated from oil-contaminated soil, was able to degrade up to 1100 mg/L of phenol. The strain PHS1strain grew well in the pH range of 4 to 9 and temperature ranging from 30 to 40 °C at different concentrations of NaCl ranging from 10 to 20% on the growth of bacteria was studied and it was found that this strain was able to show optimum degradation at 10% NaCl in 120 h. Haddadi and Shavandi (2013) isolated a moderately halophilic strain of *Halomonas* sp. strain PH2-2 from petroleum-contaminated soil in Iran that could degrade 400 mg/L phenol with a removal efficiency of 95% at 7% NaCl. Jiang et al. (2015) investigated the halophilic phenol-degrading fungus belongs to *Candida* genus which was isolated from activated sludge collected from a pharmaceutical plant. The strain utilized phenol as a sole source of carbon up to 800 mg/L of phenol after 2 days of incubation having the optimal pH value of 6.0. Lu et al. (2015) developed a high-throughput screening system for moderately halophilic phenol-degrading bacteria. The enrichments were cultivated in 48 deep well microplates, based on two characteristics, salt tolerance and phenol degradation, for environmental samples. Using this process, they screened halophilic bacterial communities from 43 bacterial enrichments and harvested a bacterial community that could degrade phenol 500 mg/L efficiently in a broad range of salinities up to 15% NaCl.

Bioremediation of Phenol by Physcrophiles and Thermophiles

Mesophilic microorganisms have been reported to degrade phenol under aerobic conditions at 30°C. There have been very few reports on the degradation of phenol under high temperature conditions using thermophilic bacteria (Mutzel et al., 1996; Feitkenhaur et al., 2001). Phenol degrading thermophilic *Bacillus* sp. converted catechol to hydroxymuconic semialdehyde by meta- cleavage routes (Ceniglia, 1984). *Bacillus sterothermophilus* BR219 degraded phenol at levels of 15 mM (Gurujeyalakshmi & Oriel, 1988). It was a river sediment isolate and showed growth inhibition at higher concentrations of phenol.

Phenolic compounds containing wastewaters are hot, only a few reports have dealt with phenol biodegradation by thermophilic (Mutzel et al., 1996; Ali et al., 1998; Feitkenhauer et al., 2001) or hyperthermophilic (Izzo et al., 2005; Christen et al., 2011) microorganisms. However, hyperthermophilic microorganisms can be a source of rare and robust biocatalysts with potential biotechnological applications (Niehaus, 1999), and their advantages compared with mesophilic ones have been described elsewhere. For example, heat- and acid-stable amylase, cyclomaltodextrinase and endoglucanase from *Alicyclobacillus acidocaldarius* and glycosyl hydrolases from *Sulfolobus solfataricus* have been purified and characterized for further possible expression of the corresponding genes in mesophilic hosts (Niehaus, 1999; Haki, 2003). In the field of thermostable enzymes, if many research works – and even applications – have dealt with heat-stable DNA polymerases or hydrolases such as proteases, amylases, cellulases or lipases, very few works on oxygenases have been reported up till now (Bertoldo, 2004).

Margesin and Bergauer (2004) compared the Phenol degradation efficiency of cold-tolerant *Arthrobacter* sp. AG31 and mesophilic *Pseudomonas putida* DSM6414. The cold-tolerant strain was cultivated at 10°C degraded 200 mg and 400 mg phenol/l within 48–72 h of cultivation could produced more biomass than the mesophile. Both strains oxidized catechol by the ortho type of ring fission. Bergauer et al. (2005) reported the degradation of phenol and phenol related compounds by Basidiomycetes yeast strains isolated from alpine soil at low temperature. Krallish et al. (2006) reported on the ability of three phenol degrading cold-tolerant yeast strains isolated from hydrocarbon-polluted alpine habitats to accumulate Extra polysaccharides and to form biofilm under salt stress conditions. Phenol degradation was reported up to 200 mg/L in four days by the bacterial strains *Cryptococcus terreus* (strain PB4) and *Rhodotorula creatinivora* (strains PB7, PB12) and they could accumulate the expolysaccharides at 25ºC.

Christen et al. (2012) reported the kinetics on degradation of phenol by the acidophilic and hyperthermophilic archaeon *Sulfolobus solfataricus* 98/2 which could degrade phenol at 80ºC at pH 3.2 which could help in the significant bioremediation of phenol contaminated in hot acidic sites. John et al. (2016) reported the degradation of different types of phenolic compounds like phenol, ferulic acid, ρ-coumaric acid and sinapinic acid by a thermo-acidophilic bacterium *Alicyclobacillus acidocaldarius*, and the laccase enzyme assays of cellular protein fractions suggested significant activity on a lignin analog between the temperatures of 45 and 90 °C.

Bioremediation of Phenol by Alkaliphiles

Non-conventional microorganisms such as Alkaliphiles and alkali-tolerant strains have been widely accepted that could specifically thrive in the alkaline conditions with pH range of 9–12, which was attributed to their ability of adjusting high pH to neutral value (Bull, 2010). There has been relatively limited information about the application of alkali-tolerant strains for aromatics-containing wastewater treatment (Al-Awadhi et al., 2007; Kanekar et al., 1999). Therefore, it is potential to use the Alkaliphiles or other alkali-tolerant microorganisms as bioaugmentation preparation for the treatment of alkaline wastewater containing aromatics such as phenol, etc. by combining the adjusting tank with the biochemical reaction pool in one unit (Fahy et al., 2008; Heinaru et al.. 2005; Qu et al., 2006).

Maltseva et al. (1996) reported the isolation of haloalkaliphilic bacteria related to the family Halomonadaceae (91% identity after analysis of 303 nucleotides of the 16S rRNA gene). These isolates were able to degrade 2,4-D as the sole source of carbon and energy, up to 3g 2,4 D/L was degraded in 3 days growing optimally at pH 8.4–9.4 and NaCl concentrations of 0.6–1.0 M. One of the isolates was further characterized revealing that it utilizes the same 2,4-D degradation pathway as most of the non-extremophilic 2,4-D degrading bacteria.

Qu et al. (2011) studied the degradation of phenol by a novel alkali-tolerant strain *Pseudomonas* JY-2, which could utilize phenol as sole source of carbon and energy up to 750 mg/L the degradation efficiencies was up to 48.2% within 40 h at pH 10.0. The alkaline phenol-containing wastewater treatment augmented with strain JY-2 in sequencing batch reactor (SBR) system was investigated, which suggested that the bioaugmented system exhibited the better performance for adjusting high pH to neutral value than the non-bioaugmented one.

Sarwade and Gawai (2014) investigated the degradation of phenol using alkaliphilic *Bacillus badius* D1 with various concentrations ranging from 0.42 to 1.68 g/L in alkaline broth of pH-9, in shaking incubator at110 rpm, 370C, higher concentration the rate of degradation was found higher than 70%. They concluded that the *Bacillus badius* follows ortho and meta-catechol pathway for degradation.

CONCLUSION AND FUTURE PROSPECT

This review gives a comprehensive outline of sources of phenolic compounds, their hazards, various treatment technologies and their degradation using both mesophilic bacteria and extremophiles. As phenolic compounds are considered priority pollutants must be eliminated from the environment to preserve the environmental quality. Among various methods available, biological treatment is environmentally friendly, cost-effective, and most promising compared with other methods for the prevention of pollution. Phenolic compound containing wastewaters are generated from different industries and even from extreme environments. They are to be remediated with microorganisms not only with the mesophilic groups but also with extremophilic microbes, which have the capability to utilize the recalcitrant compounds at extreme conditions.

Saline wastewater is generated from various industries, and this has salinity, greater than that of sea water. It is very difficult to treat these waters with a high salinity, with conventional microorganisms, as the high salt content induces cell lysis; hence, there is a need for microorganisms, which can withstand this high salt content and still degrade the phenolic contaminants. Environments with extreme pH can be treated with acidophiles or alkaliphiles and extreme temperature with psychrophiles and thermophiles. Hence, the present review focuses on remediation of phenolic compounds including both mesophilic and extremophilic bacterial community. In extreme environments, due to reduced microbial diversity, when compared with non-extreme environments, pollutant degradation becomes critical. There are numerous literature reports available on the remediation of phenolic compounds by mesophilic group of bacteria. While the bacterial remediation of phenolic compounds using extremophiles has started since recent years.

Nevertheless, significant research efforts could be made for the potential use of extreme environments to recover novel microbes to treat the pollutants; and efforts should also be focused on the study of different degradation mechanisms and to identify the enzymes involved in the remediation process and their metabolism regulation. Further study is required in genetic modification of microorganisms from extreme environments to meet the required recombinant strain to remove the recalcitrant compound. Research work should be directed to the isolation and identification of new microorganisms and their phenol-degrading ability. Further, advanced treatment technologies such as proteomics and genomics should be investigated and applied to provide more knowledge regarding the enzymatic mechanisms and intermediates involved in metabolic activities during biodegradation of phenolics. To conclude performance of various microorganisms from different ecological diversity and various interdisciplinary technologies could be considered as a research topic of high priority for future research work.

The data summarized in this review paper gives the research work carried out for the remediation of phenolic compounds by both the mesophilic and extremophilic bacteria used in the removal of phenolic compounds. Degradation of phenol and its derivatives using various bacterial strains has been the topic of scientific interest for a number of decades. Enzymatic methods have been widely studied in the degradation of phenolic compounds. These enzymes are either used as free forms or immobilized forms. Immobilized enzymes can be reused which will reduce the operating costs. Microbes used in the removal of phenolic compounds are both are and mixed cultures of microorganisms and enzymes can serve the purpose, mainly employed for treatment of industrial wastewaters containing phenols and on the contaminated sites. The findings of this review suggest that both microbial and enzymes could be involved in the remediation, further there is lot of scope on the extremophilic bacteria which could perform the similar function as mesophiles but under extreme conditions. Application of halophiles for the treatment of industrial wastewaters should provide the necessary knowledge on the development of robust wastewater treatment systems that can accept and provide valuable treatment option for hypersaline industrial wastewater containing phenolic compounds. In conjunction with functional genomics, proteomics and metabolomics are becoming more important for the elucidation of mechanisms of biodegradation and biotransformation of these recalcitrant compounds in the environment, particularly for the situations of multiple chemicals and microbial consortia.

REFERENCES

Abd-El-Haleem, D., Beshay, U., Abdelhamid Abdu, O., Moawad, H., & Zaki, S. (2003). Effects of mixed nitrogen sources on biodegradation of phenol by immobilized *Acinetobacter* sp. strain W-17. *African Journal of Biotechnology*, *2*(1), 8–12. doi:10.5897/AJB2003.000-1002

Abdelkafi, S., Chambkha, M., Casalot, L., Sayadi, S., & Labat, M. (2005). Isolation and characterization of a novel *Bacillus* sp., strain YAS1, capable of transforming tyrosol under hypersaline conditions. *FEMS Microbiology Letters*, *252*(1), 79–84. doi:10.1016/j.femsle.2005.08.032 PMID:16165329

Agarry, S. E., Durojaiye, A. O., & Solomon, B. O. (2008). Microbial degradation of phenols: A review. *International Journal of Environment and Pollution*, *32*(1), 12–28. doi:10.1504/IJEP.2008.016895

Agency for Toxic Substances and Disease Registry. (2006). *Case studies in environmental medicine*. Academic Press.

Al-Awadhi, H., Sulaiman, R. H., Mahmoud, H. M., & Radwan, S. S. (2007). Alkaliphilic and halophilic hydrocarbon-utilizing bacteria from Kuwaiti coasts of the Arabian Gulf. *Applied Microbiology and Biotechnology*, *77*(1), 183–186. doi:10.1007/s00253-007-1127-1 PMID:17710391

Al-Harazin, I. M., Nakhla, G. F., & Farooq, S. (1991). Start-up of sequencing batch reactors for toxic wastewater treatment. *Journal of Environmental Science and Health. Part A, Environmental Science and Engineering & Toxic and Hazardous Substance Control*, *26*(5), 673–687. doi:10.1080/10934529109375662

Alva, V., & Peyton, B. M. (2003). Phenol and catechol biodegradation by the haloalkaliphile *Halomonas campisalis*: Influence of pH and salinity. *Environmental Science & Technology*, *37*(19), 4397–4402. doi:10.1021/es0341844 PMID:14572091

Annachhatre, A. P., & Gheewala, S. H. (1996). Biodegradation of chlorinated phenolic compounds. *Biotechnology Advances*, *14*(1), 35–56. doi:10.1016/0734-9750(96)00002-X PMID:14536923

Annaduari, G., Rajesh, B. S., Mahesh, K. P. O., & Murugesan, T. (2000b). Absorption and biodegradation of phenol of Chitosan immobilised *Pseudomonas putida* (Mcm. 2174). *Bioprocess Engineering*, *22*(6), 493–501. doi:10.1007/s004499900092

Annadurai, G., Balan, M. S., & Murugesan, T. (2000a). Design of experiments in the biodegradation of phenol using immobilized *Pseudomonas pictorium* (NICM- 2077) on activated carbon. *Bioprocess Engineering*, *22*, 101–107.

Arutchelvan, V., Kanakasabai, S., Nagarajan, S., & Muralikrishnan, V. (2005). Isolation and identification of novel high strength phenol degrading bacterial strains from phenol-formaldehyde resin manufacturing industrial wastewater. *Journal of Hazardous Materials*, *B127*(1-3), 238–243. doi:10.1016/j.jhazmat.2005.04.043 PMID:16154261

Atkinson, R., Aschmann, S. M., & Winer, A. M. (1987). Kinetics of the reactions of NO3 radicals with a series of aromatic compounds. *Environmental Science & Technology*, *21*(11), 1123–1126. doi:10.1021/es00164a015

Bai, J., Wen, J. P., Li, H. M., & Jiang, Y. (2007). Kinetic modelling of growth and biodegradation of phenol and *m*-cresol using *Alcaligenes faecalis*. *Process Biochemistry*, *42*(4), 510–517. doi:10.1016/j.procbio.2006.10.004

Bastos, A. E. R., Moon, D. H., Rossi, A., Trevors, J. T., & Tsai, S. M. (2000b). Salt tolerant phenol-degrading microorganisms isolated from Amazon soil samples. *Archives of Microbiology*, *174*(5), 346–352. doi:10.1007/s002030000216 PMID:11131025

Beltran, F. J., & Alvarez, P. (1996). Rate constant determination of ozone-organic fast reaction in water using agitated cell. *Environ. Sci. Health*, *31*, 1159–1178.

Bergauer, P., Fonteyne, P. A., Nolard, N., Schinner, F., & Margesin, R. (2005). Biodegradation of phenol and phenol-related compounds by psychrophilic and cold- tolerant alpine yeasts. *Chemosphere*, *59*(7), 909–918. doi:10.1016/j.chemosphere.2004.11.011 PMID:15823324

Bertoldo, C., Dock, C., & Antranikian, G. (2004). Thermoacidophilic microorganisms and their novel biocatalysts. *Engineering in Life Sciences*, *4*(4), 521–532. doi:10.1002/elsc.200402155

Boll, M., & Fuchs, G. (2005). Unusual reactions involved in anaerobic metabolism of phenolic compounds. *Biological Chemistry*, *386*(10), 989–997. doi:10.1515/BC.2005.115 PMID:16218871

Bonfa, M. R. L., Grossman, M. J., Mellado, E., & Durrant, L. R. (2011). Biodegradation of aromatic hydrocarbons by Haloarchaea and their use for the reduction of the chemical oxygen demand of hypersaline petroleum produced water. *Chemosphere*, *84*(11), 1671–1676. doi:10.1016/j.chemosphere.2011.05.005 PMID:21621813

Boone, D. R., Johnson, R. L., Chen, D. C., Mathrani, I. M., & Mah, R. A. (1989). Methanogenesis and reductive dechlorinations in an alkaline, hypersaline sediment and groundwater. In M. S. Da Costa, J. C. Duarte, & R. A. D. Williams (Eds.), *Microbiology of Extreme Environments and its potential for Biotechnology* (pp. 205–215). London: Elsevier Applied Science.

Borja, R., Martin, A., Maestro, R., Alba, J., & Fiestas, J. A. (1992). Enhancement of the anaerobic digestion of olive mill wastewaters by removal of phenolic inhibitors. *Proc. Biochem.*, *27*(4), 231–237. doi:10.1016/0032-9592(92)80023-V

Budavari, S. (1996). *The Merck Index: An encyclopedia of chemicals, drugs and biochemicals, Whitehouse station*. Merck.

Bull, A. T. (2010). Actinobacteria of the extremobiosphere. In Extremophiles Handbook. Berlin: Springer-Verlag GmbH.

Cataldo, D. A., Bean, R. M., & Fellow, R. J. (1987). Uptake and fate of phenol aniline and quinoline in terrestrial plants. In R. H. Gray, . . . (Eds.), Health and environmental research on complex organic mixtures. Academic Press.

Christen, P., Davidson, S., Combet-Blanc, Y., & Auria, R. (2011). Phenol biodegradation by the thermoacidophilic archaeon *Sulfolobus solfataricus* 98/2 in a fed batch bioreactor. *Biodegradation*, *22*(3), 475–484. doi:10.1007/s10532-010-9420-6 PMID:20886261

Christen, P., Vega, A., Casalot, L., Simon, G., & Auria, R. (2012). Kinetics of aerobic phenol biodegradation by the acidophilic and hyperthermophilic archaeon *Sulfolobus solfataricus* 98/2. *Biochemical Engineering Journal*, *62*, 56–62. doi:10.1016/j.bej.2011.12.012

Clark, T. P., & Piskin, R. (1977). Chemical quality and indicator parameters for monitoring leachate in Illinois. *Environ. Geol.*, *1*(6), 329–340. doi:10.1007/BF02380501

Crawford, D. W., Bonnevie, N. C., & Wenning, R. J. (1995). Sources of pollution and sediment contamination in Newark Bay. *Exotoxicology and Environmental Safety, New Jersey*, *35*(1), 85–100. doi:10.1006/eesa.1995.1010 PMID:7540541

Dash, R. R., Gaur, A., & Balomajumder, C. (2009). Cyanide in industrial wastewaters and its removal: A review on biotreatment. *Journal of Hazardous Materials*, *163*(1), 1–11. doi:10.1016/j.jhazmat.2008.06.051 PMID:18657360

Delaney, J. L., & Hughes, T. W. (1979). Source assessment: Manufacture of acetone and phenol from cumene. Mosanto Research Corp.

Delfino, J.J., & Dube, D.J. (1976). Persistent contamination of ground water by phenol. *Environmental Science and Health, A11*, 345-355.

Den Boeft, J., Kruiswijk, F. J., & Schulting, F. L. (1984). Air pollution by combustion of solid fuels. The Hague: Ministry of Housing, Physical Planning and Environment (Publication Lucht No.37).

Diaz, E. (2004). Bacterial degradation of aromatic pollutants: A paradigm of metabolic versatility. *International Microbiology*, *7*, 173–180. PMID:15492931

Dimitriu, P. A., Shukla, S. K., Conradt, J., Marquez, M. C., Ventosa, A., Maglia, A., & Mormile, M. R. et al. (2005). *Nitrincola lacisaponensis* gen. nov., sp. nov., a novel alkaliphilic bacterium isolated from an alkaline, saline lake. Int. J. Syst. *Evol. Micro.*, *55*(6), 2273–2278. doi:10.1099/ijs.0.63647-0 PMID:16280482

Ehrlich, G. G., Goelitz, D. F., & Godsy, E. M. (1982). Degradation of phenolic contaminants in ground water by anaerobic bacteria, St. Louis Park, MN. *Ground Water*, *20*(6), 703–710. doi:10.1111/j.1745-6584.1982.tb01390.x

El-Ashtoukhy, S. Z., El-Taweel, Y. A., Abdelwahab, O., & Nassef, E. M. (2013). Treatment of petrochemical wastewater containing phenolic compounds by electrocoagulation using a fixed bed electrochemical reactor. *International Journal of Electrochemical Science*, *8*, 1534–1550.

El-Naas, M., Al-Zuhair, S., & Makhlouf, S. (2008). Continuous biodegradation of phenol in a Spouted Bed Bioreactor (SBBR). *Chemical Engineering Journal*, *160*(2), 565–570. doi:10.1016/j.cej.2010.03.068

El-Sayed, W. S., Ibrahim, M. K., Abu-Shady, M., El-Beih, F., Ohmura, N., Saiki, H., & Ando, A. (2003). Isolation and characterization of phenol-catabolizing bacteria from a coking plant. *Bioscience, Biotechnology, and Biochemistry*, *67*(9), 2026–2029. doi:10.1271/bbb.67.2026 PMID:14519997

Fahy, A., Ball, A. S., Lethbridge, G., Timmis, K. N., & McGenity, T. J. (2008). Isolation of alkali-tolerant benzene-degrading bacteria from a contaminated aquifer. *Letters in Applied Microbiology*, *47*(1), 60–66. doi:10.1111/j.1472-765X.2008.02386.x PMID:18544140

Feitkenhauer, H., Schnicke, S., Müller, R., & Märkl, H. (2001). Determination of the kinetic parameters of the phenol-degrading thermophile Bacillus thermoleovorans sp. A2. *Applied Microbiology and Biotechnology*, *57*, 744–750. doi:10.1007/s002530100823 PMID:11778888

Freitag, D., Lay, J. P., & Korte, F. (1984). Environmental hazard- the results as related to structures & transplantation into the environment. In QSAR Environ. Toxicol. Proc. Work. Quant. Struct. Act. Relat. D. Reidel Publishing Co.

Gali, V. S., Kumar, P., & Mehrotra, I. (2006). Biodegradation of phenol with wastewater as a cosubstrate in upflow anaerobic sludge blanket. *Journal of Environmental Engineering*, *132*(11), 1539–1542. doi:10.1061/(ASCE)0733-9372(2006)132:11(1539)

Garcia, E., Mellado, J. C. O., & Ventosa, A. (2004). *Halomonas organivorans* sp. nov., a moderate halophile able to degrade aromatic compounds. *International Journal of Systematic and Evolutionary Microbiology*, *54*(5), 1723–1728. doi:10.1099/ijs.0.63114-0 PMID:15388735

Garcia, M. T., Ventosa, A., & Mellado, E. (2005). Catabolic versatility of aromatic compound-degrading halophilic bacteria. *FEMS Microbiology Ecology*, *54*, 97–109. doi:10.1016/j.femsec.2005.03.009 PMID:16329976

Godjevargova, T., Ivanova, D., Aleksieva, Z., & Dimova, N. (2003). Biodegradation of toxic organic components from industrial phenol production wastewaters by free and immobilized *Trichosporon cutaneum* R57. *Proc. Biochem.*, *38*(6), 915–920. doi:10.1016/S0032-9592(02)00170-X

Goerlitz, D. F., Troutman, D. E., & Gody, E. M. (1985). Migration of wood processing chemical in contaminated ground water in a sand aquifer at Pensacola, Florida. *Environmental Science & Technology*, *19*(10), 955–961. doi:10.1021/es00140a012

Gonzalez, G., Herrera, M. G., Garcia, M. T., & Peña, M. (2001). Biodegradation of phenol in a continuous process: Comparative study of stirred tank and fluidized-bed bioreactors. *Bioresource Technology*, *76*(3), 245–251. doi:10.1016/S0960-8524(00)00092-4 PMID:11198177

Graedel, T. E. (1978). *Chemical compounds in the atmosphere*. New York: Academic Press.

Groenen, P. J. (1978). Components of tobacco smoke. In Nature and quantity; potential influence on health. CIVO-TNO Institute (Rep. No. R/5787).

Gurujeyalakshmi, G., & Oreil, P. (1988). Isolation of phenol degrading *Bacillus stearothermophilus* and partial characterization of the phenol hydroxylase. *Applied and Environmental Microbiology*, *55*, 500–502. PMID:2719481

Haddadi, A., & Shavandi, M. (2013). Biodegradation of phenol in hypersaline conditions by *Halomonas* sp. Strain PH2–2 isolated from saline soil. *International Biodeterioration & Biodegradation*, *85*, 29–34. doi:10.1016/j.ibiod.2013.06.005

Haider, K., Jagnow, G., Kohnen, R., & Lim, S. U. (1974). Degradation of chlorinated benzenes, phenols and chlorinated derivatives by benzene phenol utilizing bacteria under aerobic conditions. *Archives of Microbiology*, *96*(1), 183–200. doi:10.1007/BF00590175 PMID:4134769

Haki, G. D., & Rakshit, S. K. (2003). Developments in industrial important thermostable enzymes: A review. *Bioresource Technology*, *89*(1), 17–34. doi:10.1016/S0960-8524(03)00033-6 PMID:12676497

Hamdi, M. (1992). Toxicity and biodegradability of olive mill wastewaters in batch anaerobic digestion. *Applied Biochemistry and Biotechnology*, *37*(2), 155–163. doi:10.1007/BF02921667

Hawley, G. G. (1981). The condensed chemical dictionary (10th ed.). New York: Van Nostrand Reinh8old Co.

Hawthorne, S. B., & Sievers, R. E. (1984). Emission of organic air pollutants from shale oil wastewaters. *Environmental Science & Technology*, *18*(6), 483–490. doi:10.1021/es00124a016 PMID:22247953

Hayes, V. E., Ternan, N. G., & McMillan, G. (2000). Organophosphonate metabolism by a moderately halophilic bacterial isolate. *FEMS Microbiol.*, *186*(2), 171–175. doi:10.1111/j.1574-6968.2000.tb09099.x PMID:10802166

Heilbuth, N. M., Linardi, V. R., & Santos, V. L. (2003). Phenol biodegradation by free and immobilized cells of *Acinetobacter johnsoni* (M.Sc. Thesis). Instituto De Ciencias Biologicas/Pos-Graduacao Em. Microbiologia.

Heinaru, E., Merimaa, M., Viggor, S., Lehiste, I., Truu, J., & Heinaru, A. (2005). Biodegradation efficiency of functionally important populations selected for bioaugmentation in phenol- and oil-polluted area. *FEMS Microbiology Ecology*, *51*(3), 363–373. doi:10.1016/j.femsec.2004.09.009 PMID:16329884

Hill, G.A., & Robinson, C.W. (1972). Substrate inhibition kinetics, phenol degradation by *P.putida. Biotechnol. Bioeng., 17,* 1599-1615.

Hinteregger, C., & Streichsbier, F. (1997). *Halomonas* sp., a moderately halophilic strain, for biotreatment of saline phenolic wastewater. *Biotechnology Letters, 19*(11), 1099–1102. doi:10.1023/A:1018488410102

Hoshino, M., Akimoto, H., & Okuda, M. (1978). Photochemical oxidation of benzene, toluene and ethylbenzene initiated by OH radicals in the gas phase. *Bulletin of the Chemical Society of Japan, 51*(3), 718–724. doi:10.1246/bcsj.51.718

Howard, P. H. (1989). Handbook of environmental fate and exposure data for organic chemicals. Lewis Publishers.

Hubble, B. R., Stetter, J. R., Gebert, E., Harkness, J. B. L., & Flotard, R. D. (1981). Experimental measurements from residential wood-burning stoves. *Proc. Int. Conf. Resid. Solid Fuels: Environ. Impacts Solut.*

Hughes, S. M., & Cooper, D. G. (1996). Biodegradation of phenol using the selfcycling fermentation (SCF) process. *Biotechnology and Bioengineering, 51*(1), 112–119. doi:10.1002/(SICI)1097-0290(19960705)51:1<112::AID-BIT13>3.0.CO;2-S PMID:18627094

Hussain, A., Kumar, P., & Mehrotra, I. (2010). Nitrogen biotransformation in anaerobic treatment of phenolic wastewater. *Desalination, 250*(1), 35–41. doi:10.1016/j.desal.2009.09.018

Hutchinson, D. H., & Robinson, C. W. (1988). Kinetics of the simultaneous batch degradation of *p*-Cresol and Phenol by *Pseudomonas putida. Applied Microbiology and Biotechnology, 29*(6), 599–604. doi:10.1007/BF00260991

Izzo, V., Notomista, E., Picardi, A., Pennacchio, F., & Di Donato, A. (2005). The thermophilic archaeon Sulfolobus solfataricus is able to grow on phenol. *Research in Microbiology, 156*(5-6), 677–689. doi:10.1016/j.resmic.2005.04.001 PMID:15921893

Jadhav, D. N., & Vanjara, A. K. (2004). Removal of phenol from wastewater using sawdust, polymerized sawdust and sawdust carbon. *Indian J Chem Technol, 11,* 35–41.

Jay, K., & Steiglitz, L. (1995). Identification and quantification of volatile organic components in emissions of waste incineration plants. *Chemosphere, 30*(7), 1249–1260. doi:10.1016/0045-6535(95)00021-Y

Jiang, Y., Yang, K., Wang, H., Shang, Y., & Yang, X. (2015). Characteristics of phenol degradation in saline conditions of a halophilic strain JS3 isolated from industrial activated sludge. *Marine Pollution Bulletin, 99*(1-2), 1–2, 230–234. doi:10.1016/j.marpolbul.2015.07.021 PMID:26187399

John, E. A., William, A. A., Brady, D. L., David, N. T., Jeffrey, A. L., Deborah, T. N., & Vicki, S. et al. (2016). Degradation of phenolic compounds by the lignocelluloses deconstructing thermoacidophilic bacterium *Alicyclobacillus acidocaldarius. Journal of Industrial Microbiology & Biotechnology, 43*(1), 13–23. doi:10.1007/s10295-015-1700-z PMID:26542284

Juang, R.S., & Tsai, S.Y. (2006). Growth kinetics of *Pseudomonas putida* in the Biodegradation of single and mixed phenol and sodium salicylate. *Biochem. Eng., 2*(1), 133-140.

Jungclaus, G. A., Lopez-Avila, V., & Hites, R. A. (1978). Organic compounds in an industrial wastewater: A case study of their environmental impact. *Environmental Science & Technology*, *12*(1), 88–96. doi:10.1021/es60137a015

Kadhim, H., Graham, C., Baratt, P., Evane, C. S., & Rastall, R. A. (1999). Removal of phenolic compounds by *Coriolus versicolor* grown on wheat bran. *Enzyme and Microbial Technology*, *24*(5-6), 303–307. doi:10.1016/S0141-0229(98)00123-9

Kahru, A., Reiman, R., & Rätseep, A. (1998). The efficiency of different degrading bacteria and activated sludges in detoxification of phenolic leachates. *Chemosphere*, *37*(2), 301–318. doi:10.1016/S0045-6535(98)00048-4 PMID:9650267

Kanekar, P. P., Sarnaik, S. S., & Kelkar, A. S. (1999). Bioremediation of phenol by alkaline bacteria isolated from alkaline lake of Lonar. *Journal of Applied Microbiology*, *85*(S1), 128S–133S. doi:10.1111/j.1365-2672.1998.tb05291.x PMID:21182701

Kang, M. H., & Park, J. M. (1997). Sequential degradation of phenol and cyanide by a commensal interaction between two microorganisms. *Chem. Technol. Biotechnol.*, *69*(2), 226–230. doi:10.1002/(SICI)1097-4660(199706)69:2<226::AID-JCTB677>3.0.CO;2-N

Kar, S., Swaminathan, T., & Baradarajan, A. (1996). Studies on biodegradation of a mixture of toxic and non-toxic pollutant using *Arthrobacter* species. *Bioprocess Engineering*, *15*(4), 195–199. doi:10.1007/BF00369481

Kargi, F., & Dinçer, A. R. (1996). Enhancement of biological treatment performance of saline wastewater by halophilic bacteria. *Bioprocess Engineering*, *15*, 1. doi:10.1007/BF00435529

Kargi, F., & Uygur, A. (1996). Biological treatment of saline wastewater in an aerated percolator unit utilizing halophilic bacteria. *Environmental Technology*, *17*(3), 325–330. doi:10.1080/09593331708616391

Kavitha, V., & Palanivelu, K. (2004). The role of ferrous ion in Fenton and photo-Fenton process for the degradation of phenol. *Chemosphere*, *55*(9), 1235–1243. doi:10.1016/j.chemosphere.2003.12.022 PMID:15081764

Keith, L. H. (1976). Identification of organic compounds in unbleached treated Kraft paper mill wastewaters. *Environmental Science & Technology*, *10*(6), 555–564. doi:10.1021/es60117a009

Kinney, J. E. (1960). Evaluating the taste and odour control problems. *Journal - American Water Works Association*, *52*, 505–516.

Klibanov, A. M. (1983). Enzymatic removal of hazardous pollutants from industrial aqueous effluents. *Enzy. Eng.*, *6*, 319–323.

Klibanov, A. M., Tu, T. M., & Scott, K. P. (1983). Peroxidase-catalyzed removal of phenols from coal-conversion waste waters. *Science*, *221*(4607), 259–260. doi:10.1126/science.221.4607.259-a PMID:17815191

Knupp, G., Rücker, G., Ramos-Cormenzana, A., Garrido Hoyos, S., Neugebauer, M., & Ossenkop, T. (1996). Problems of identifying phenolic compounds during the microbial degradation of olive oil mill wastewater, Inter. *Biodeterio. Biodeg.*, *38*, 277–282.

Kobayashi, T., Hasinaga, T., Mikami, E., & Suzuki, T. (1989). Methanogenic degradation of phenol and benzoate in acclimated sludges. *Water Science and Technology*, *21*, 55–56.

Krallish, I., Svetlana, G., Ludmila, S., Phillip, B., & Margesin, R. (2006). Phenol degradation by immobilized cold-adapted yeast strains of *Cryptococcus terreus* and *Rhodotorula creatinivora*. *Extremophiles*, *10*(5), 441–449. doi:10.1007/s00792-006-0517-0 PMID:16601913

Kumaran, P. (1986). Development of biotechnology of toxic waste treatment. *Proc. Symp. Environmental Biology of Coastal Ecosystems, 11.*

Kumaran, P., & Paruchuri, Y. L. (1997). Kinetics of phenol biotransformation. *Water Research*, *31*(1), 11–22. doi:10.1016/S0043-1354(99)80001-3

Kuwata, K., Uebori, M., & Yamazaki, Y. (1980). Determination of phenol in polluted air as p-nitrobenzene azophenol derivatives by reversed phase high performance liquid chromatography. *Analytical Chemistry*, *52*(6), 857–860. doi:10.1021/ac50056a020

Lakshmi, C., & Sridevi, V. (2009). A review on biodegradation of phenol from Industrial effluents. *Jr. of Industrial Pollution Control*, *25*(1), 13–27.

Landis, W. G., & Yu, M. H. (2003). *Introduction to environmental toxicology- Impacts of chemicals upon ecological systems* (3rd ed.). New York: Lewis Publishers.

Leahy, J. G., & Colwell, R. R. (1990). Microbial degradation of hydrocarbons in the environment. *Microbiological Reviews*, *54*, 305–315. PMID:2215423

Lee, R. F., & Ryan, C. (1979). Microbial degradation of oraganochlorine compounds in estuarine waters and sediments. *Proc. Workshop: Microbe degradation Pollut Mar Environ.*

Lefebvre, O., Habouzit, F., Bru, V., Delgenes, J. P., Godon, J. J., & Moletta, R. (2004). Treatment of hypersaline industrial wastewater by a microbial consortium in a sequencing batch reactor. *Environmental Technology*, *25*(5), 543–553. doi:10.1080/09593330.2004.9619345 PMID:15242230

Lefebvre, O., Vasudevan, N., Thanasekaran, K., Moletta, R., & Godon, J. J. (2006). Microbial diversity in hypersaline wastewater: The example of tanneries. *Extremophiles*, *10*(6), 505–513. doi:10.1007/s00792-006-0524-1 PMID:16738814

Leven, L., Nyberg, K., Korkea-aho, L., & Schnurer, A. (2006). Phenols in anaerobic digestion processes and inhibition of ammonia oxidising bacteria (AOB) in soil. *The Science of the Total Environment*, *364*(1-3), 229–238. doi:10.1016/j.scitotenv.2005.06.003 PMID:16125214

Lide, D. R. (1993). *CRC handbook of chemistry and physics.* CRC Press.

Lika, K., & Papadakis, I. A. (2009). Modeling the biodegradation of phenolic compounds by microalgae. *Journal of Sea Research*, *62*(2-3), 135–146. doi:10.1016/j.seares.2009.02.005

Lin, Y. H., Wu, C. L., Hsu, C. H., & Li, H. L. (2009). Biodegradation of phenol with chromium(VI) reduction in an anaerobic fixed-biofilm process—kinetic model and reactor performance. *Journal of Hazardous Materials*, *172*(2-3), 1394–1401. doi:10.1016/j.jhazmat.2009.08.005 PMID:19726129

Loehr, R. C., & Krishnamoorthy, R. (1988). Terrestrial bioaccumulation potential of phenolic compounds. *Hazardous Waste & Hazardous Materials*, *5*(2), 109–119. doi:10.1089/hwm.1988.5.109

Loh, K. C., & Cao, B. (2008). Paradigm in biodegradation using *Pseudomonas putida:* A review of proteomics studies. *Enzyme and Microbial Technology*, *43*(1), 1–12. doi:10.1016/j.enzmictec.2008.03.004

Lu, Z. Y., & Xiao-Jue, G. (2015). High-Throughput Screening for a Moderately Halophilic Phenol-Degrading Strain and Its Salt Tolerance Response. *International Journal of Molecular Sciences*, *16*(6), 11834–11848. doi:10.3390/ijms160611834 PMID:26020478

Ludzack, F. J., & Ettinger, M. B. (1960). Chemical structures resistant to aerobic biochemical stabilization. *Wat. Poll.Cont. Fed.*, *32*, 1173–1200.

Maltseva, O., & Oriel, P. (1997). Monitoring of an Alkaline 2,4, 6-Trichlorophenol-Degrading Enrichment Culture by DNA Fingerprinting Methods and Isolation of the Responsible Organism, Haloalkaliphilic *Nocardioides* sp. Strain M6. *Applied and Environmental Microbiology*, *63*, 4145–4149. PMID:16535723

Margesin, R., & Schinner, F. (2001). Biodegradation and bioremediation of hydrocarbons in extreme environments. *Applied Microbiology and Biotechnology*, *56*(5-6), 650–663. doi:10.1007/s002530100701 PMID:11601610

Margesin, R. P., Bergauer, S. G., & Gander, S. (2004). Degradation of phenol and toxicity of phenolic compounds: A comparison of cold-tolerant *Arthrobacter* sp. and mesophilic *Pseudomonas putida. Extremophiles*, *8*(3), 201–207. doi:10.1007/s00792-004-0378-3 PMID:14872323

Martinez-Neito, L., Ramos Cormenzana, A., Garcia Pareja, M. P., & Garrido Hoyos, S. E. (1992). Phenolic compounds biodegradation of olive mill wastewater with *Aspergillus terreus*. *Grasas y Aceites*, *43*, 75–81.

McAllister, K. A., Lee, H., & Trevors, J. T. (1996). Microbial degradation of pentachlorophenol. *Biodegradation*, *7*(1), 1–40. doi:10.1007/BF00056556

Melo, J. S., Kholi, S., Patwardhan, A. W., & DSouza, S. F. (2005). Effect of oxygen transfer limitations in phenol biodegradation. *Process Biochemistry (Barking, London, England)*, *40*(2), 625–628. doi:10.1016/j.procbio.2004.01.049

Metcalf & Eddy. (2005). *Wastewater Engineering-Treatment, disposal and reuse* (4th ed.). Tata McGraw Hill.

Mill, T., & Mabey, W. (1985). Photochemical transformations. In Environmental exposure from chemicals. Boca Raton, FL: CRC Press.

Mrozik, A., Cycon, M., & Piotrowska-Seget, Z. (2010). Changes of FAME profiles as a marker of phenol degradation in different soils inoculated with *Pseudomonas* sp. CF600. *International Biodeterioration & Biodegradation*, *64*(1), 86–96. doi:10.1016/j.ibiod.2009.11.002

Munoz, J. A., Perez-Esteban, B., Esteban, M., de la Escalera, S., Gomez, M. A., Martinez-Toledo, M. V., & Gonzalez-Lopez, J. (2001). Growth of moderately halophilic bacteria isolated from sea water using phenol as the sole carbon source. *Folia Microbiologica*, *46*(4), 297–302. doi:10.1007/BF02815617 PMID:11830940

Mutzel, A., Reinscheid, U. M., Antranikian, G., & Müller, R. (1996). Isolation and characterization of a thermophilic bacillus strain that degrades phenol and cresols as sole carbon source at 70 oC. *Applied Microbiology and Biotechnology*, *46*(5-6), 593–596. doi:10.1007/s002530050866

Nair, I., Jayachandran, C. K., & Shankar, S. (2008). Biodegradation of phenol. *African Journal of Biotechnology*, *7*(25), 4951–4958.

Nair, I. C., & Shashidhar, S. (2004). Microbial degradation of phenol by a species of *Alcaligenes* isolated from a tropical soil. *Soil Science*, *T-5*, 3, 4, 47–51.

Nakamura, Y., & Sawada, T. (2000). Biodegradation of phenol in presence of heavy metals. *Journal of Chemical Technology and Biotechnology (Oxford, Oxfordshire)*, *75*(2), 137–142. doi:10.1002/(SICI)1097-4660(200002)75:2<137::AID-JCTB194>3.0.CO;2-0

Niehaus, F., Bertoldo, C., Kähler, M., & Antranikian, G. (1999). Extremophiles as a source of novel enzymes for industrial application. *Applied Microbiology and Biotechnology*, *51*(6), 711–729. doi:10.1007/s002530051456 PMID:10422220

Oren, A. (2002). Diversity of halophilic microorganisms: Environments, phylogeny, physiology, and applications. *Journal of Industrial Microbiology & Biotechnology*, *28*(1), 56–63. doi:10.1038/sj/jim/7000176 PMID:11938472

Parales, R. E., Bruce, N. C., Schmid, A., & Wackett, L. P. (2002). Biodegradation, biotransformation, and biocatalysis (B3). *Applied and Environmental Microbiology*, *68*(10), 2699–4709. doi:10.1128/AEM.68.10.4699-4709.2002 PMID:12324310

Patterson, J. W. (1985). *Industrial wastewater treatment technology*. Boston: Butterworths.

Pekari, K., Vainotalo, S., Heikkila, P., Palotie, A., Luotamo, M., & Riihimaki, V. (1992). Biological monitoring of occupational exposure to low levels of benzene. *Scandinavian Journal of Work, Environment & Health*, *18*(5), 317–322. doi:10.5271/sjweh.1570 PMID:1439659

Peralta-Zamora, P., Gimenes, L. F., Cordi, L., Reyes, J., Alves, O. L., & Durán, N. (1998). Remediation of effluents from pulp and paper industry using horseradish peroxidase immobilized on ceramic materials. *Proc. 3rd Lat. Am. Biodegradation and Biodeterioration Symposium*.

Perez, M., Torrades, F., Garcia, H. J. A., Domennech, X., & Peral, J. (2002). Removal of organic contaminants in paper pulp treatment effluents under Fenton and photofenton conditions. *App. Cat. B. Environment*, *36*, 63–74.

Perry, J. J. (1984). Perry's chemical engineer handbook (6th ed.). McGraw Hill.

Peters, M., Heinaru, E., Talpsep, H., Ward, H., Stottmeister, U., Heinaru, A., & Nurk, A. (1997). Acquisition of a deliberately introduced phenol degradation operon, *pheAB*, by different indigenous *Pseudomonas* species. *Applied and Environmental Microbiology*, *63*, 4899–4906. PMID:9406411

Peyton, B. M., Wilson, T., & Yonge, D. R. (2002). Kinetics of phenol biodegradation in high salt solutions. *Water Research*, *36*(19), 4811–4820. doi:10.1016/S0043-1354(02)00200-2 PMID:12448524

Pfeiffer, F. M. (1979). The 1977 screening survey for measurement of organic priority pollutants in petroleum refinery wastewaters. *ASTM Special Technical Publication*, 181–190.

Plumb, R. H. (1987). A comparison of ground water monitoring data from CERCLA and RCRA sites. *Ground Water Monitoring Review*, *7*(4), 94–100. doi:10.1111/j.1745-6592.1987.tb00968.x

Polymenakou, P. N., & Stephanou, E. G. (2005). Effect of temperature and additional carbon sources on phenol degradation by an indigenous soil *Pseudomonas*. *Biodegradation*, *16*(5), 403–413. doi:10.1007/s10532-004-3333-1 PMID:15865154

Powlowski, J., & Shingler, V. (1994). Genetics and biochemistry of phenol degradation by *Pseudomonas* sp. CF600. *Biodegradation*, *5*(3-4), 219–236. doi:10.1007/BF00696461 PMID:7765834

Preis, S., Terentyeva, Y., & Rozkov, A. (1997). Photocatalytic oxidation of phenolic compounds in wastewater from oil shale treatment. *Water Science and Technology*, *35*(4), 165–174. doi:10.1016/S0273-1223(97)88394-1

Qu, Y. Y., Zhang, R., Fang, M., Zhou, J., & Yan, B. (1926). Bioaugmentation with a novel alkali-tolerant Pseudomonas strain for alkaline phenol wastewater treatment in sequencing batch Reactor. *World Journal of Microbiology & Biotechnology*, *27*(8), 1919–1926. doi:10.1007/s11274-011-0653-2

Qu, Y. Y., Zhou, J. T., Wang, J., Song, Z. Y., Xing, L. L., & Fu, X. (2006). Bioaugmentation of bromoamine acid degradation with *Sphingomonas xenophaga* QYY & DNA fingerprint analysis of augmented systems. *Biodegradation*, *17*(1), 83–91. doi:10.1007/s10532-005-3544-0 PMID:16453174

Rampelotto, P. H. (2013). Extremophiles and Extreme Environments. *Life (Basel)*, *3*(3), 482–485. PMID:25369817

RIVM. (1986). Bilthoven, The Netherlands: National Institute of Public Health and Environment Protection Document No. 738513002.

Rubin, H. E., & Alexander, M. (1983). Effects of nutrients on the rates of mineralization of trace concentrations of phenol and *p*-nitrophenol. *Environmental Science & Technology*, *17*(2), 104–107. doi:10.1021/es00108a008 PMID:22295962

Sadtler, N. A. (1960). *Sadtler standard spectra*. Sadtler Research Lab.

Said, M., Ahmad, A., & Abdul Wahab, M. (2013). Removal of phenol during ultrafiltration of Palm oil mill effluent (POME): Effect of pH, ionic strength, pressure and temperature. *Der Pharma Chemical*, *5*(3), 190–196.

Saravanan, P., Pakshirajan, K., & Saha, P. (2008). Growth kinetics of an indigenous mixed microbial consortium during phenol degradation in a batch reactor. *Bioresource Technology*, *99*(1), 205–209. doi:10.1016/j.biortech.2006.11.045 PMID:17236761

Sarwade, V.D., & Gawai, K.R. (2014). Biodegradation of phenol by alkaliphilic *Bacillus badius* D1IOSR. *Journal of Environmental Science, Toxicology and Food Technology*, *8*(2), 28-35.

Satyanarayana, T., Raghukumar, C., & Shivaji, S. (2005). Extremophilic microbes: Diversity and perspectives. *Current Science*, *89*, 78–90.

Scott, H. D., Wolf, D. C., & Lavy, T. L. (1982). Adsorption and degradation of phenol at low concentration in soil. *Environ. Qual.*, *11*, 107–111. doi:10.2134/jeq1982.00472425001100010025x

Shailubhai, K. (1986). *Treatment of Petroleum industry oil sludge in soil, Tibtech August 1986*. Amsterdam: Elsevier Science Publishers B.V.

Sharma, H., Barber, J. T., Ensley, H. E., & Polito, M. A. (1997). A comparison of the toxicity of phenol and chlorinated phenols by *Lemna gibba*, with reference to 2,4,5-trichlorophenol. *Environmental Toxicology and Chemistry*, *16*, 346–350. doi:10.1002/etc.5620160233

Shashirekha, S., Uma, L., & Subramaniam, G. (1997). Phenol degradation by the marine cyanobacterium *Phormidium valderianum* BDU 30501, Ind. *Microbial Biotechnology*, *19*(2), 130–133. doi:10.1038/sj.jim.2900438

Sheeja, R. Y., & Murugesan, T. (2002). Mass transfer studies on the biodegradation of phenols in up-flow packed bed reactors. *Journal of Hazardous Materials*, *B89*(2-3), 287–301. doi:10.1016/S0304-3894(01)00319-3 PMID:11744212

Shen, X., Liu, Z., & Wang, B. (2004). Isolation, identification of phenol degrading *Rhadococcus* sp. strain PNAN5 and characterization of its ring cleavage dioxygenases. *Acta Scientiae Circumstantiae*, *24*, 482–486.

Shibata, A., Inoue, Y., & Katayama, A. (2006). Aerobic and anaerobic biodegradation of phenol derivatives in various paddy soils. *The Science of the Total Environment*, *367*(2-3), 979–987. doi:10.1016/j.scitotenv.2006.01.031 PMID:16530810

Shinoda, T., Sakai, Y., Ue, M., Hiraishi, A., & Kato, N. (2000). Isolation and characterization of a new denitrifying *spirillum* capable of anaerobic degradation of phenol. *Applied and Environmental Microbiology*, *66*(4), 1286–1291. doi:10.1128/AEM.66.4.1286-1291.2000 PMID:10742201

Shivaraman, N., & Pandey, R. A. (2000). Characterization and biodegradation of phenolic wastewater, J. Ind. Assos. *Environmental Management*, *27*, 12.

Sierra-Alvarez, R., Field, J. A., Kortekaas, S., & Lettinga, G. (1994). Overview of the anaerobic toxicity caused by organic forest industry wastewater pollutants. *Water Science and Technology*, *29*, 353–363.

Stoilova, I., Krastanov, A., Yanakieva, I., Kratchanova, M., & Yemendjiev, H. (2007). Biodegradation of mixed phenolic compounds by *Aspergillus awamori* NRRL 3112. *International Biodeterioration & Biodegradation*, *60*(4), 342–346. doi:10.1016/j.ibiod.2007.05.011

Stuart, S. L., & Woods, S. L. (1998). Kinetic evidence for pentachlorophenol-dependent growth of a dehalogenating population in a pentachlorophenol- and acetate-fed methanogenic culture. *Biotechnology and Bioengineering*, *57*(4), 420–429. doi:10.1002/(SICI)1097-0290(19980220)57:4<420::AID-BIT5>3.0.CO;2-O PMID:10099218

Tabib, A., Haddadi, M., Shavandi, M. S., & Azizinikoo, P. (2012). Biodegradation of Phenol by Newly Isolated Phenol-degrading Bacterium *Ralstonia* sp. Strain PH-S1. *Journal of Petroleum Science and Technology*, *2*, 58–62.

Taghreed, A., & Muftah, H. (2012). Aerobic Biodegradation of Phenols: A Comprehensive Review. *Critical Reviews in Environmental Science and Technology*, *42*(16), 1631–1690. doi:10.1080/10643389.2011.569872

Tartakovsky, B., Manuel, M. F., Beaumier, D., Greer, C. W., & Guiot, S. R. (2001). Enhanced selection of an anaerobic pentachlorophenol- degrading consortium. *Biotechnology and Bioengineering*, *73*(6), 476–483. doi:10.1002/bit.1082 PMID:11344452

Thiel, V. (2011). Extreme environments. In Encyclopedia of Geobiology (pp. 362–366). Academic Press.

Thurman, E. M. (1985). Phenol. In Kirk-Othmer encyclopedia of chemical technology (3rd ed.). New York, NY: John Wiley and Sons.

Tibbles, B. J., & Baecker, A. A. W. (1989). Effect of pH and inoculum size on phenol degradation by bacteria isolated from from landfill waste. *Environmental Pollution*, *59*(3), 227–239. doi:10.1016/0269-7491(89)90228-5 PMID:15092404

Tiirola, M.A., Wang, H., Paulin, L., & Kulomaa, M.S. (2002). Evidence for natural horizontal transfer of the pcpB gene in the evolution of polychlorophenol-degrading sphingomonads. *Appl. Environ. Microbiol., 68*, 4495–4501.

TRI05 TRI explorer: Providing access to EPA's toxics release inventory data. (2007). Retrieved from http://www.epa.gov/triexplorer/

TRI. (1998). *Toxic Chemical Release Inventory*. National Library of Medicine, National Toxicology Information Program.

Ucun, H., Yildiz, E., & Nuhoglu, A. (2010). Phenol biodegradation in a batch jet loop bioreactor (JLB): Kinetics study and pH variation. *Bioresource Technology*, *101*(9), 2965–2971. doi:10.1016/j.biortech.2009.12.005 PMID:20053552

US EPA Toxics release inventory. (1989). Washington, DC: US Environmental Protection Agency, Office of Pesticides and Toxic Substances.

Veena Gayathri, K., & Vasudevan, N. (2010). Aerobic degradation of Catechol and Resorcinol by Bacterial strains and isolated from wastewater. *Int. J. Biotechnol. Res.*, *3*(2), 35–38.

Veena Gayathri, K., & Vasudevan, N. (2010). Effect of pH, Nitrogen sources and salts on the degradation of Phenol by the Bacterial Consortium under Saline Conditions. *Int. J. Biotech. Biochem*, *6*, 783–791.

Veena Gayathri, K., & Vasudevan, N. (2010). Enrichment and Isolation of Phenol degrading Moderately Halophilic Bacterial Consortium from Saline Environment *J. Biorem. Biodeg.*, *1*(1), 104.

Veena Gayathri, K., & Vasudevan, N. (2011). Ortho and meta cleavage dioxygenases detected during the degradation of phenolic compounds by a moderately halophilic bacterial consortium. *Int. Res. J. Microbiol.*, *2*(10), 406–414.

Veena Gayathri, K., & Vasudevan, N. (2013). Degradation of 4-Chlorophenol by a Moderately Halophilic Bacterial Consortium under Saline Conditions. *British Microbiology Research Journal*, *3*(4), 513–524. doi:10.9734/BMRJ/2013/3171

Veena Gayathri, K., & Vasudevan, N. (2015). Degradation of Chlorophenolic compounds by a moderately halotolerant bacterial consortium isolated from saline environment. *International Journal of Bioassays*, *4*(7), 4102–4108.

Veena Gayathri, K., & Vasudevan, N. (2015). Degradation of dual phenolics by a moderately halophilic bacterial consortium and its degradation products. International. *Journal of Current Microbiology and Applied Sciences.*, *4*(5), 1083–1095.

Veeresh, G. S., Kumar, P., & Mehrotra, I. (2004). Treatment of phenol and cresols in up-flow anaerobic sludge blanket (UASB) process: A review. *Water Research*, *39*(1), 154–170. doi:10.1016/j.watres.2004.07.028 PMID:15607175

Verschueren, K. (1983). *Handbook of Environmental Data on Organic Chemicals* (2nd ed.). New York: Van Nostrand Reinhold Co.

Wang, C. C., Lee, C. M., & Kuan, C. H. (2000). Removal of dichlorophenol by suspended and immobilized *Bacillus insolitus*. *Chemosphere*, *41*(3), 447–452. doi:10.1016/S0045-6535(99)00263-5 PMID:11057607

Wang, L., Li, Y., Yu, P., Xie, Z., Luo, Y., & Lin, Y. (2010). Biodegradation of phenol at high concentration by a novel fungal strain *Paecilomyces variotii* JH6. *Journal of Hazardous Materials*, *183*(1-3), 366–371. doi:10.1016/j.jhazmat.2010.07.033 PMID:20685040

Wang, Y., Tian, Y., Han, B., Zhaw, H. B., Bi, J. N., & Cai, B. L. (2007). Biodegradation of phenol by free and immobilized *Acinetobacter* sp. Strain PD12. *Journal of Environmental Sciences (China)*, *19*(2), 222–225. doi:10.1016/S1001-0742(07)60036-9 PMID:17915733

Watanabe, K. (2001). Microorganisms relevant to bioremediation. *Current Opinion in Biotechnology*, *12*(3), 237–241. doi:10.1016/S0958-1669(00)00205-6 PMID:11404100

Watanabe, K., Hino, S., & Takahashi, N. (1996). Effects of exogenous phenol degrading bacteria performance on ecosystem of activated sludge. *Ferm. Bioeng.*, *82*(3), 291–298. doi:10.1016/0922-338X(96)88822-2

WHO. (1989). *Chlorophenols other than Pentachlorophenol*. Geneva: World Health Organization.

Woolard, C. R., & Irvine, R. L. (1994). Biological treatment of hypersaline wastewater by a biofilm of halophilic bacteria. *Water Environment Research*, *66*(3), 230–235. doi:10.2175/WER.66.3.8

Woolard, C.R., & Irvine, R.L. (1995). Treatment of hypersaline wastewater in the sequencing batch reactor. *Wat. Res., 29*(4), 1159-68.

Wu, Z. G., Wang, Y. M., Xing, Z. H., Wu, X. W., & Song, S. S. (2005). Study on degrading phenol by immobilized *Ralstonia metallidurans* CH34. *Microbiology*, *32*, 31–36.

Xing, B., McGill, W. B., Dudas, M. J., Maham, Y., & Hepler, L. (1994). Sorption of phenol by selected polymers: Isotherms, energetics, and polarity. *Environmental Science & Technology*, *28*(3), 466–473. doi:10.1021/es00052a019 PMID:22165882

Xun, L. Y., Topp, E., & Orser, C. S. (1992). Confirmation of oxidative dehalogenation of pentachlorophenol by a *Flavobacterium* pentachlorophenol hydroxylase. *Journal of Bacteriology*, *174*(17), 5745–5747. doi:10.1128/jb.174.17.5745-5747.1992 PMID:1512208

Yan, J., Jianping, W., Hongmei, L., Suliang, Y., & Zongding, H. (2005). The biodegradation of phenol at high initial concentration by yeast *Candida tropicalis*. *Biochem. Eng.*, *24*(3), 243–247. doi:10.1016/j.bej.2005.02.016

KEY TERMS AND DEFINITIONS

Anthropogenic Compounds: Human impact on the environment.

ATSDR: Agency for Toxic Substances and Disease Registry.

Bioaccumulation: It refers to the accumulation of substances, such as pesticides, or other chemicals in an organism.

Bioremediation: It is a waste management technique that involves the use of organisms to remove or neutralize pollutants.

EPA: Environmental Protection Agency.

Extremophiles: An extremophile is an organism that thrives in physically or geochemically extreme conditions that are detrimental to most life on Earth.

Recalcitrance Compounds: resisting control for degradation.

RIVM: National Institute for Public Health & the Environment.

WHO: World Health Organisation.

Chapter 19
Advancement in Bioremediation of Pharmaceutical and Personal Care Products

Vasudha Agnihotri
G. B. Pant National Institute of Himalayan Environment and Sustainable Development, India

ABSTRACT

Pharmaceutical and personal care products (PPCPs) are regularly used by human being for their day to day life. These products contains various chemical compounds which are regularly added in our surface and ground water sources either through untreated or partially treated domestic and industrial wastes or through agricultural runoffs etc. These are present at the concentration of µg liter-1, and at such low concentration also these have shown harmful affect to living being. A lot of studies have been carried out for studying their effects on animal biodiversity, which can directly or indirectly affect human being also. So their removal from waste water is essential. Various technologies are being tested for removing these compounds from polluted water; bioremediation is also one of them. Present chapter will briefly give the description of various PPCPs present in waste water, their impact and removal technologies available for their removal with special emphasis on bio remediation.

INTRODUCTION

Water is one of the prime element and necessary for every living being on earth. Two third of the earth is covered by water and major fraction of our body is also made up of water. Further in the body, it regulates the activities of fluids, tissues, cells, lymph, blood and glandular secretions. So its contamination can easily affect our metabolism as well as that of all the living creatures. Fresh water present in the earth's surface is put to multifarious uses. It is used for drinking, domestic and municipal uses, agricultural irrigation, industries, navigation, recreation, etc. The used water becomes contaminated and is called waste water. Water passing through industries, agricultural lands and the urban settlements gets spoiled up and is rendered harmful to plants, animals and human being. This water gets loaded with organic and inorganic chemicals, pathogenic and non- pathogenic organisms including viruses and

DOI: 10.4018/978-1-5225-2325-3.ch019

cysts. Usually these waters do not remain fit for human consumption and for life of plants and animals as well. There are different types of pollutants polluting the water e.g. Sewage and fertilizers, chlorinated hydrocarbons and pesticides, heavy metals, oil and petroleum products, radioactive substances, plastics, pharmaceutical and personal care products (PPCPs) etc. Role of PPCPs as environmental pollutant is recent in comparison to other pollutants.

Now the question is what are pharmaceutical and personal care products? Different types of medications both for human as well as for livestock, cosmetic products, and antibiotics for agriculture etc are being used. Excess of these flush out through domestic effluents and become a part of water or sediments/ soil if not treated properly, known as PPCPs. Places where the waste water from hospitals, residential areas and industries are directly released into surface water without pretreatment, chances of presence of PPCPs becomes more prominent. As these group of compounds are still emerging therefore also known as a subgroup of "contaminants of emerging concern," as well as "emerging contaminants" or "trace organic compounds" (Sedlak et. al., 2000). Contaminants of emerging concern as a whole are compounds that are not commonly monitored due to their "emerging" nature. A wide range of PPCPs may be present in wastewater and the environment receiving it (Daughton & Ternes, 1999; Kolpin et. al., 2002). These compounds are mostly less biodegradable, high persistent in the environment, poorly removed in wastewater treatment plants (WWTPs), and often detected in the micro to nano grams per liter concentrations. Leaching of these compounds in groundwater is the potential risk as due to its usage for drinking purposes in most o the areas. The present chapter is providing different aspects of PPCPs such as type of PPCPs present in environment, their impacts on environment and available removal technologies for PPCPs from waste water with the emphasis on bioremediation.

BACKGROUND

Bioremediation was first used by the Romans, at around 600 BC, to clean their waste water. This process was officially invented in the 1960's by George Robinson. He had experimented with microbes inside of polluted glass jars. In the 1970s, Ananda Chakrabarty and his colleagues at General Electric discovered a strain of bacteria that is able to degrade some components in crude oil. He obtained this strain by isolating a Pseudomonas from a soil filled with. First commercial in situ bioremediation system was installed in 1972 to cleanup a Sun Oil pipeline spill in Ambler, Pennsylvania. The Exxon Valdez oil spill in 1989 in Prince William Sound, Alaska, was the genesis of global attention to this process. Importance of bioremediation is that it can completely destroy contaminants, converting them to carbon dioxide, water, and new cell mass. So this can be a good technique for removal of PPCPs from waste water.

Now days many PPCPs are detected at wide concentration ranges in a variety of environmental samples (Halling-Sørensen et al., 1998, Snyder et al., 2003) and in animal tissues, human blood, and breast milk samples (Rimkus et al., 1999; Adolfsson-Erici et al., 2002). Study was conducted by researchers at the U.S. Geological Survey in 1999 and 2000 to provide baseline information on the occurrences of pharmaceuticals, hormones and other wastewater contaminants in water resources. Researchers found organic wastewater contaminants (OWCs) in 80% of the streams e.g. coprostanol (a fecal steroid), cholesterol (a plant and animal steroid), tridosan (an antimicrobial disinfectant), 4-nonylphenol (a non-ionic detergent metabolite) and caffeine as the most frequently detected compounds (Nakada, 2007). The highest concentration reported was 40 ng/l for meprobamate (Benotti et al., 2009). Several pharmaceuticals were found in tap water at different concentrations ranging from nanograms to low micrograms per litre in several countries in Europe, including Germany, the Netherlands and Italy (Huerta-Fontela et al, 2011). 20 different drugs were found in rivers and streams of Germany having water coming out of sewage treatment plants (STP). Some of them were lipid regulator bezafibrate, antiphlogistics diclofenac, ibuprofen, naproxen, metabolites like clofibric acid, salicylic acid, betablocker metroprolol, propranolol, antiepileptic carbamazepine etc (Ternes, 1998). Two separate studies in Germany (Reddersen et al., 2002; Zühlke et al., 2004) found phenazone and propylphenazone (an analgesic and an antipyretic drug, respectively) in Berlin drinking-water, with the highest concentration being 400 ng/l for phenazone. This high value was largely attributed to groundwater, used as a drinking-water source, contaminated with sewage (Jones et al., 2005).

In the Netherlands, traces of antibiotics, antiepileptic and beta blockers were detected in the drinking-water supply at concentrations below 100 ng/l, with most concentrations below 50 ng/l (Mons et al., 2003). Pain reliever ibuprofen was detected in the influents of several Switzerland WWTPs at concentration levels of 1 to 3.3 μg L^{-1} (Buser et al., 1999). As medical substances are developed with the intention of performing some sort of biological function, they have a tendency to bioaccumulate and induce effects in aquatic and terrestrial ecosystems (Halling-Sorensen et al., 1998). These compounds can easily come out from water treatment plants in original or modified forms, so their removal techniques from waste or drinking water need to be modified.

PPCP: SOURCES IN WASTE WATER, IMPACT ON ENVIRONMENT AND APPROACHES FOR REMOVAL

Sources of PPCPs in Waste Water

Millions of dollars are spent each year on PPCPs, many of these products are never used or only partially used and end up thrown out in the trash or dumped down the drain (Daughton & Ternes, 1999; Meador et al., 2016). The pervasive use of biocides in consumer products to improve hygiene and destroy bacteria creates another set of chemical contaminants that pass readily into water supplies as they are usually associated with cleaning and are dumped directly into household or industrial drains. These can affect human health due to their potential biomagnifications through the food chain in aquatic organisms (Crane et al., 2006). Generally the water is examined in terms of conventional physico-chemical parameters i.e., major nutrients, bulk organic matter in terms of BOD, COD, TOC etc. But with development, problems related to PPCPs have increased as these are harmful even at ng/L levels. The tendency for these compounds to persist or be only partially degraded during treatment or to bypass treatment altogether via sewage overflows will, therefore, contribute to their load in receiving waters, many of which serve as recreational and drinking water sources. These include PAHs, PCBs, pesticides, detergent residues (alkylphenols), plasticisers (DEHP), linear alkylbenzene sulfonates (LASs), natural and synthetic estrogens, poly brominated diphenyl ethers (PBDEs), disinfection by-products and others (Figure 1, Figure 2, and Figure 3). High percentages of many pharmaceuticals can be excreted from the body unmetabolized and enter wastewater as biologically active substances (Buhner, 2002) e.g. 90% of the drug, propofol found in anesthesia, is excreted un-metabolized (Kummerer, 2001). Aquaculture wastes are also the source of antibiotics in the waste water as it contains fish feed, dead fish, faecal matter, egg shells and chemicals used for treatment of fishes (Antibiotic overflow. Down to Earth, 1-15 August, 2016).

Figure 1. Some important PPCPs reported in waste water throughout the world

S.No.	PPCPs	Structural formula	Log K_{ow}	Nature of solubility	Usage	Adverse affects on bioaccumulation in water or sediments	References
1	Triclosan	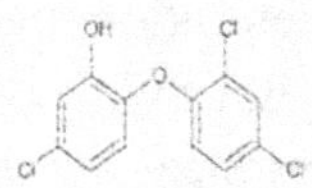	2.39 - 4.54	hydrophobic	bactericide in personal care products such as toothpaste, shampoos, and soaps as a stabilizing agent in a multitude of detergents and cosmetics	skin irritation, allergy susceptibility, ecological toxicity to the aquatic and terrestrial environment (effect on earth worms (Eisenia fetida) and on Japanese medaka fish	Adolfsson-Erici et al., 2012 Coogan et al., 2007 Lin et al., 2010 Nassef et al., 2010 Chen et al, 2012 Bhargava & Leonard, 1996
2	Ibuprofen	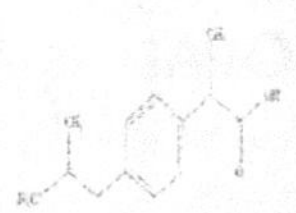	0.82 - 3.40	hydrophobic	Analgesic, anti-inflammatory , cyclooxygenase inhibitor	Risk to aquatic environment with chronic toxic effect (such as inhibited polyp regeneration and reduced reproduction)	Stumpf et al. (1999)
3	Carbamazepine	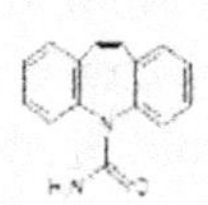	1.51	hydrophobic	anticonvulsant drug, and also as a specific analgesic for trigeminal neuralgia	Immune disorder	Tina et al., 2009 Deniz et al., 2000

Impact of PPCPs on Environment

PPCPs potentially impact both the ecology and environment as these compounds may be active at very low concentrations. These are wide spread and released in large quantities in water bodies through waste water. In an aquatic system, the residence time of a pollutant and its distribution between water, sediments, and biota depends on its capacity to bound to suspended particles and accumulate in sediments. Thus, the affinity of hydrophobic organic compounds for biotic and abiotic phases is an important determinant of both the rate of a detoxification of water system and its response time to changing loadings. Polarity of these compounds determines their fate in potential matrices of environment. Non polar compounds

Figure 2. Some important PPCPs reported in waste water throughout the world

4	Diclofenac	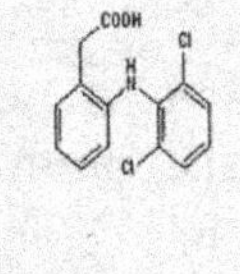	1.12	hydrophobic	anti-inflammatory	Risk to aquatic environment with chronic toxic effect (such as inhibited polyp regeneration and reduced reproduction)	Stumpf et al. (1999), Heberer et al. (2001), Heberer et al. (2002) Buser et al., 1998a
5	Estradiol	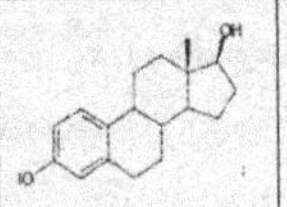	2.89	hydrophobic	Hormone	Cancer, endometriosis, birth defects etc.	Baronti et al., 2000, Johnson et al. (2000)
6	Codeine	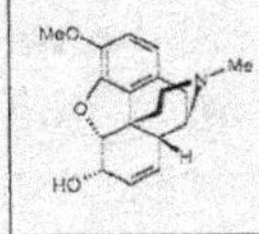	1.14	hydrophobic	Analgesic		Thai et al., 2016
7	Sulfamethoxazole	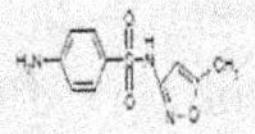	0.79	hydrophobic	Antibiotic	alterations to the balance of naturally-occurring microorganisms, the nitrogen cycle, genetic mutations, and aquatic toxicity.	Laville et al., 2004
8	Caffeine	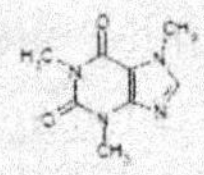	-0.07	hydrophobic	Stimulant	Affect respiration and function of heterotrophic microbial biofilms present in aquatic environment so impact the	Potera, 2012

Figure 3. Some important PPCPs reported in waste water throughout the world

						ecological environment of water sources	
9	Clofibric acid	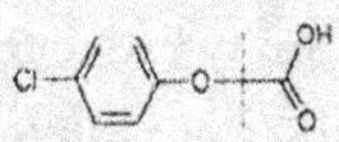	2.88	hydrophobic	Lipid regulator		Buser, et al., 1998b.
10	Alkylphenol polyethoxylates	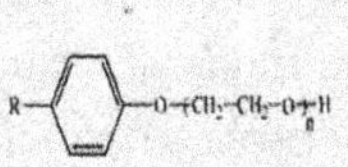	>4.5		Surfactant	Nonionic surfactants added in detergents	La Guardia et al., 2001

generally more attracted towards soil colloids and sediments as also accumulate in biota (Heberer, 2002), while the polar PPCPs are more labile in the environment.

These compounds can affect human health due to their potential biomagnifications through the food chain in aquatic organisms (Crane et al, 2006). These are found to be associated with various health hazards as presence of PPCPs in water can potentially leads to carcinogenic, reproductive developmental effects by interfering with various hormonal pathways, disruption in general metabolism, producing acute toxicity, immune system effects (Kotz et al., 2012). These compounds often exert their effects at relatively low concentrations and through specific mode of action. Reduction of vulture population by consuming declofenac contaminated carcasses (Oaks et al., 2004), feminization of fish due to ethinylestradiol contaminated water obtained after sewage treatment are the examples of effect of PPCPs on living creatures (Shanmugam et al., 2014, Desbrow et al., 1998, Larsson et al., 1999, Thorpe et al., 2003, Kidd et al., 2007). Most pharmaceuticals are non-bioaccumulative and eliminated from human or animal body after administration. After passing through body metabolism, hydrophilic functions generally added in the molecule for their successful excretion with urine/ feces (Katzung, 2001). These compounds often exert their effects at relatively low concentrations and through specific mode of action.

PPCPs are used as antibacterial agents, these when released in waterways and sediments allowed bacterial flora to develop antibacterial resistance to those particular agents. The development of resistance to antimicrobial agents makes treatment of infections very difficult to cure; therefore this issue is an important consideration for the treatment of wastewater, especially that which is discharged from hospitals, veterinary clinics, or other locations where large amounts of antibiotics are used. Interplay of bacteria with antibiotics present in waste water make them resistant and increases their populations in waste water. Quaternary ammonium compounds (QACs) are cationic microbial compounds that are important ingredients in disinfectants, which are used in hospitals as well as households. They are known to inhibit the proper functioning of aquatic microorganisms, and they have a low biodegradability. Inhibitory effects have also been found against denitrifying bacteria in very low concentrations of QACs. In India, high resistance is observed in treatment of prevalent infectious diseases like pneumonia, tuberculosis, urinary tract infection, and infection of bloodstream and gut ('Hatching Superbugs', Down to earth, 1-15 August 2014).

The tendency for these compounds to persist or be only partially degraded during treatment or to bypass treatment altogether via sewage overflows will, therefore, contribute to their load in receiving waters, many of which serve as recreational and drinking water sources.

Some Advanced PPCP Removal Technologies

PPCPs include a wide range of compounds with different chemical and physical proprieties that affect their removal. Due to the low PPCP concentrations reported in wastewaters (ppb or ppt) and their complex chemical structure, common technologies used in sewage and drinking water treatment plants may not be efficient enough to accomplish their complete removal. Various bench- scale studies have been conducted for removal of PPCPs from polluted water. Coagulation, lime softening, activated carbon adsorption, biofiltration, chlorination, membrane bioreactor, photocatalytic treatment, ultraviolet photodegradation, ozonationation and advanced oxidation (AOP) are some of the methods tried in batch scale studies for removal of PPCPs from waste water. Information about physico-chemical characteristics such as acidity, lipophilicity, volatility and sorption potential is a useful tool to understand the different removal patterns observed. In the removal efficiency of biological treatment processes for PPCPs is related to hydrophobicity and to consequent adsorption in activated sludge, whereas more polar compounds are released in the effluent (Heidler & Halden, 2008). In the conventional physical-chemical processes, surface water conventional treatments (i.e. coagulation, sand filtration, softening) frequently used as tertiary wastewater treatments, and are not efficient for the complete removal of organic micropollutants. Particularly, the use of aluminium sulphate, ferric chloride or lime removes less than 20% of many EDCs and PPCPs (Westerhoff et al., 2005).

Advanced treatments with high removal efficiencies of emerging contaminants are adsorption on activated carbon, nanofiltration (NF) and reverse osmosis (RO) (Snyder et al., 2007). However, the NF/RO processes produce a large amount of concentrate whose treatment and disposal is very expensive, therefore, their application is less suitable in inland areas. The typical doses used in disinfection by UV irradiation (< 40 mJ/cm^2) are unable to remove EDCs and PPCPs (removal efficiencies are often less of 20%) (Rosario-Ortiz et al., 2010). Ozone and chlorine oxidation are selective for the chemical structure and functional groups of PPCPs and EDCs, whose removal range between 90% with a prominent efficiency of ozonation (Westerhoff et al., 2005). However, chlorination and ozonation are known to lead to the formation of by-products that account for largely unidentified compounds to date. This issue raises concerns regarding their potential impact on the environment and human health. The AOPs, including ozonation which produces OH radicals in wastewater application, are able to degrade a wide range of emerging contaminants. Membrane processes are an effective mean to reduce the concentrations of several EDCs and PPCPs from drinking water and wastewater. The removal rate is related to membrane characteristics and to molecular properties of particular contaminant.

The performance of activated carbon depends on the properties of the activated carbon sorbent (surface area, pore size distribution, surface charge, oxygen content) and on the properties of the solute (shape, size, charge and hydrophobicity). Hydrophobic interactions are the dominant mechanisms of removal for most organic compounds in activated carbon adsorption systems. However, ion exchange interactions can result in removal of polar solutes (Crittenden et al., 1999; Matsumura et al., 1985; Youssef et al., 1982). As a result of the hydrophobic interactions, activated carbon efficiently removes most non-polar organic compounds. The ability of activated carbon to remove most polar compounds will depend upon the strength of the polar interactions. Tanghe & Verstraete (2001) showed that the sorption capacity of crushed granular activated carbon for nonyl phenol (NP) was at least 100 mg/g. Based on the results of their static adsorption tests, they concluded that a full-scale GAC filter unit would be adequate to remove environmentally relevant NP concentrations of 10 μg/L. Many other studies have been carried out for removing organic pollutants from contaminated water (Knappe et al, 1999; Rodriguez et al, 2016).

Biological activated carbon (BAC) consists of a pre-ozonation followed by a GAC filter which supports the growth of bacteria. This technology has been used for drinking water treatment and has proven to be able to significantly remove natural organic matter, ozonation by-products, disinfection by-products precursors as well as odour and taste compounds. Recently, it has been reported that BAC systems are very effective to remove EDCs and PPCPs (Gerrity et al., 2011; Reungoat et al., 2011). However, more studies are needed for a better evaluation of BAC removal efficiencies taking into account several operating conditions through full-scale applications. During IBAC treatment, the process of microorganisms immobilizing is of great complexity, which is not only related to various acting forces, but also involves microorganism growth as well as the ability for producing extracellular and external appendages (Collins et al., 2003). Although there is a relatively high repulsive force during the process when the bacteria approaching to the activated carbon, the bacteria will ultimately contact with the activated carbon under the bridging effect of EPSs (Takeuchi et al., 1996). There are various ways to immobilize microorganisms, and any method with a limitation for free flowing of microorganisms can be used to produce and immobilize microorganisms, an ideal and universal application way for immobilizing microorganisms is still not available (Jin et al., 2013).

MICROBIAL POTENTIAL IN PPCP BIOREMEDIATION

Bioremediation is the technology of using biological agents particularly microorganisms, to remove toxic pollutants from the environment, particularly from soil and water. In case of bacteria, it can grow in both the aerobic as well as anaerobic conditions. Under aerobic conditions, bacteria use molecular oxygen as their terminal electron acceptor and in anaerobic conditions other compounds such as carbon dioxide is used (Figure 4).

This is a continuous method as it cannot be turned on or off because it is a living system, so constant source of nutrients are required for microbes. Aerobic bacteria is suitable for degrading organic compounds for the range of 50 to 4000 mg/l with some exception up to lower range while anaerobic bacteria can work in the range of 4000 to 50000 mg/l concentration (Nyer, 1992). It was also found that organic contaminants are mineralized and recycled more efficiently when a diverse microbial population is present (Torsvik & Øvreås, 2002). However the relationship between microbial diversity and biodegradation is not linear as many metabolic processes are carried out by a consortium rather than individual microbial species (Jjemba, 2004). Mixed microbial communities have the most powerful biodegradable potential because the genetic information of more than one organism is necessary to degrade the complex mixture of organic compounds present in contaminated sites.

Pollutants are decomposed to non-toxic substances through microbial metabolism (Pelczar, et al., 1993). In-general biodegradation process is based on two processes: growth and co-metabolism (Figure 5).

Microbes use naturally occurring as well as synthetic chemicals for growth by using these molecules as source of carbon, nitrogen, phosphorus, sulfur, energy and any other elements needed by the cells (Riser-Roberts, 1998). For all these processes, optimum conditions of pH, temperature, nutrient concentrations and target organic concentration below toxic limit for microbes should be maintained (Figure 6).

Figure 4. Electron exchange during bioremediation by microbial cell

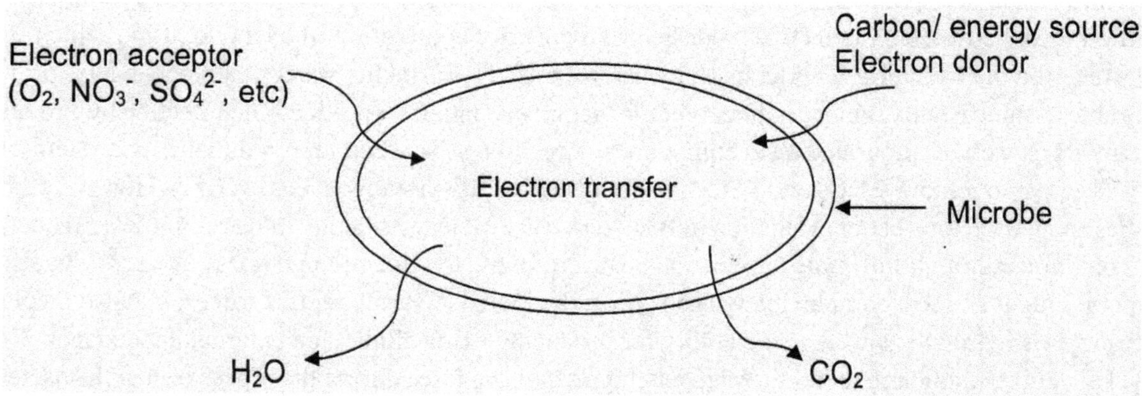

Figure 5. Biodegradation process in case of microbes

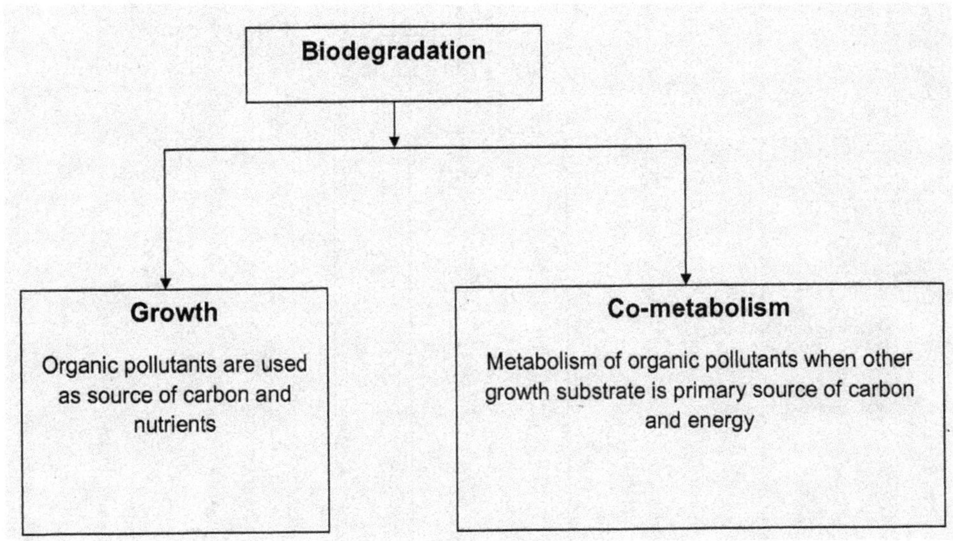

The target organic pollutant should be compatible with microbes. Initial intracellular attack of PPCPs is an enzymatic reaction. These are oxidation reactions catalyzed by oxygenases and peroxidases. Oxygen molecule is also required for respiration process. Peripheral degradation process converts organic pollutants into the intermediates of converts' organic pollutants into intermediates of tri-carboxylic acid (TCA) cycle, then biosynthesis of cell biomass takes place. In case of cometabolism, organic pollutants are transformed or degraded by microbes but they are unable to use the substrate as a source of energy or of one of its constituent elements.

In an aquatic system, the residence time of a pollutant and its distribution between water, sediments, and biota depends on its capacity to bound to suspended particles and accumulate in sediments. Thus, the affinity of hydrophobic organic compounds for biotic and abiotic phases is an important determinant of both the rate of a detoxification of water system and its response time to changing loadings. Polarity of these compounds determines their fate in potential matrices of environment. Non polar compounds generally more attracted towards soil colloids and sediments as also accumulate in biota (Heberer, 2002), while the polar PPCPs are more labile in the environment. Microbial degradation of any compound starts with low molecular weight compounds and proceeds to higher molecular weight compounds. Degradation of molecules proceeds through introduction of molecular oxygen into the structure of target molecule. Normal paraffin proceeds via terminal oxidation and subsequently β oxidation to yield C-3 units and carboxylic acid. In case of aromatic hydrocarbons, ring cleavage required de-hydroxylation reaction, introduction of both the oxygen atoms of the molecule in the aromatic ring and subsequent formation of cis- diol. Small molecules generate CO_2 and water after degradation but aromatic hydrocarbon yields phenolics and benzoic acid intermediates, which are then readily metabolized by β-oxidation and incorporated in TCA cycle with the uptake of some ATP (Figure 7) (Fritsche & Holrichter, 2005).

The distribution of non-polar organic compounds between water and natural solids (e.g., soils, sediments and suspended particles) or organisms, can be viewed in many cases as a partitioning process between the aqueous phase and the bulk organic matter present in natural solids or in biota (Schwarzenbach et al., 2004). Fate of their attachment to different environmental matrices can be determined

Figure 6. Bioremediation of PPCPs contaminated water

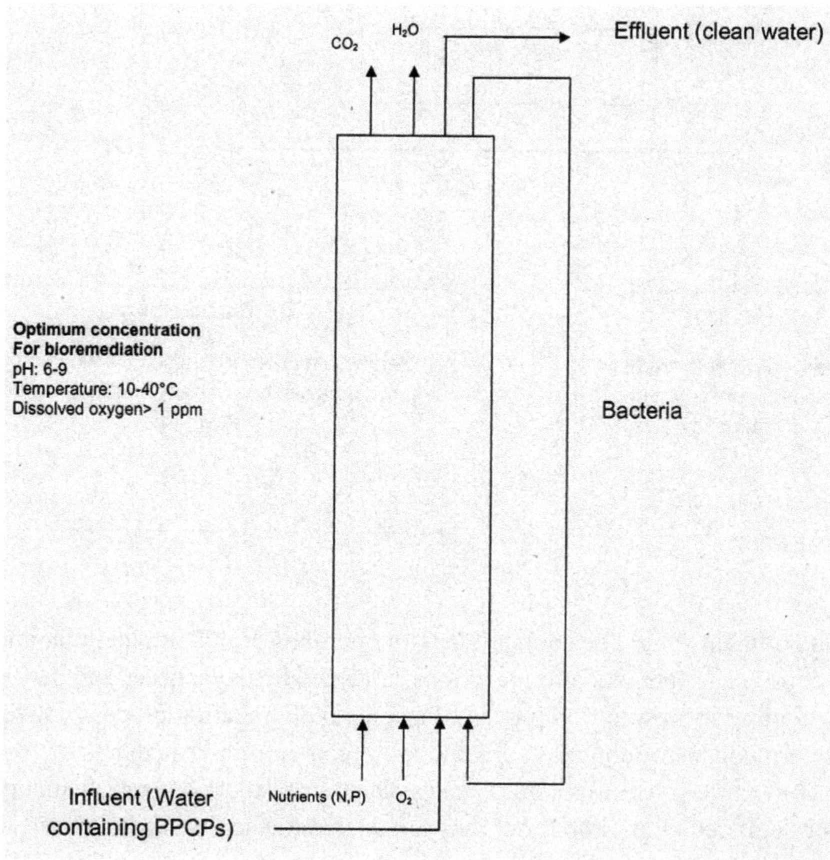

Figure 7. Biodegradation of benzene to carboxylic acid

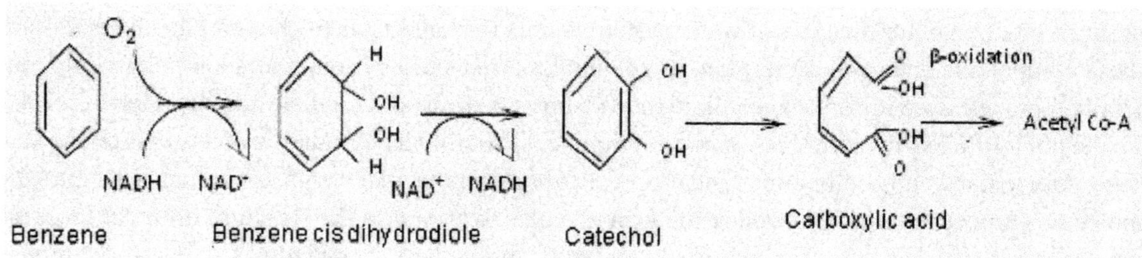

through analyzing log of K_{ow} value of PPCPs. The octanol/water partition coefficient (K_{ow}) is defined as the ratio of a chemical's concentration in the octanol phase (C_o) to its concentration in the aqueous phase (C_{aq}) of a two-phase octanol/water system (Equation 1).

$$K_{aw} = \frac{C_o}{C_{aq}} \tag{1}$$

Values of K_{ow} represent the tendency of the chemical to partition itself between an organic phase (e.g., a fish, a soil) and an aqueous phase. Biological activity of selected organic pollutant can be correlated with log K_{ow} through the equation (2)

$$BR = a + b \log K_{ow} \quad (2)$$

BR may be one of the large numbers of reactions of organisms after introduction of organic pollutants, a and b are the constants for particular system. When organic pollutants come in contact with the cell, then partitioning of the compound takes place from aqueous like phase to organic like phase, which is controlled by the partition coefficient of that compound. Hydrophobic compounds tend to accumulate in organisms because they are metabolized slowly and are effectively stores in tissues (Sangster J., 1997).

Compounds which are resistant to microbial attack can be rendered biodegradable by chemical treatment to yield low molecular weight carboxylic acid (Garg et al., 1994). Mode of degradation of compounds by microorganisms depends upon the nature of chain length of hydrocarbons along with its stereochemistry as well as species of microorganisms. Aliphatic organic compounds are more readily attacked than aromatic compounds by microbes. These are potential carbon and energy source for microbes so these are biodegradable (Bartha & Atlas, 1977). *Pseudomonads* (aerobic gram negative rods) were found to have immense potential of degrading organic compounds due to their catabolic enzymes along with their capability of metabolic regulations (Haughtan & Shanley, 1994). *Rhodococci* and *Corynaform* bacteria have also shown degradation capacity of organic compounds. Some bacteria under this group generate mycolic acid at the external surface of the cells which are long chain alcoholic and fatty acids, esterified to peptidoglycans of the cell wall. So these can work against the lipophilic/ hydrophobic organic compounds. Pharmaceutical and Personal care products also occur either as aliphatic or aromatic moieties with either polar or nonpolar nature. But major differences with other general organic compounds are that PPCPs are mainly designed to harm microorganisms responsible for infections. So selections of microbes from nature, which are not harmful, are essential.

Adsorption processes using activated carbons are widely used to remove pollutants from wastewaters and occurrence of large numbers of bacteria on GAC filters is well documented (American Water Works Association. 1981, Eberhardt, 1976). The microorganisms attached to the GAC then convert the biodegradable portion to biomass, carbon dioxide, and waste products before that material can occupy adsorption sites on the GAC. Biological Activated Carbon (BAC) process is developed on the basis of activated carbon technology, which uses the synergistic effect of adsorption on activated carbon and biodegradation to purify raw water. Activated carbon has a high specific surface area and a highly developed pore structure, so it is characterized by its great effect on absorbing dissolved oxygen and organics in raw water. For BAC technology, activated carbon is used as a carrier, by accumulating or artificially immobilizing microorganisms under proper temperature and nutrition conditions. The microorganisms will reproduce on the surface of the activated carbon and finally form BAC, which can exert the adsorption and biodegradable roles simultaneously (Qing & Jihua, 2006). Activated carbon is suited as supporting material by its large adsorptive surface (Andrew & Tien, 1981; Shimp & Pfaender, 1982). Microbes can take part in the degradation of harmful aromatic compounds as these are the growth substrate for microorganisms. The crucial step in their degradation is overcoming the resonance energy that stabilizes the ring structure (Fuchs et al., 2011). Inoculation and maintenance of bacteria with special degradation capabilities resulted in a lower consumption of GAC and thus to extended shelf life (Stucki & Thuer,

1994). Microorganisms which can carry out biodegradation in different environments are identified as active members of microbial consortiums. These microorganisms include: *Acinethobacter, Actinobacter, Acaligenes, Arthrobacter, Bacillins, Berijerinckia, Flavobacterium, Methylosinus, Mycrobacterium, Mycococcus, Nitrosomonas, Nocardia, Penicillium, Phanerochaete, Pseudomonas, Rhizoctomia, Serratio, Trametes* and *Xanthofacter* (Singh et al., 2014). Species of *Candida* and *Pseudomonas* have been used for their immobilization on activated carbon. Phenol adsorbed on carbon surface was 90% degraded by microbes (Ehrhardt & Rehm, 1985).

The relationship between the activated carbon and contaminants is simply the effect of adsorption on activated carbon, and the reaction depends on the properties of the activated carbon and contaminants. Meanwhile, the activated carbon can absorb dissolved oxygen (DO) and microorganisms which were adsorbed on the surface of activated carbon, feed on DO will biodegrade contaminants. The activated carbon used in BAC process, should be highly developed in the pore structure, especially for the filter pores. The basic principle of immobilization biological activated carbon (IBAC) is screening and acclimation dominant bio-community from nature, followed by immobilizing the community on activated carbon, to enhance the efficiency and rate of degradation. Meanwhile, dominant biocommunity should be nonpathogenic and has strong antioxidant capacity and enzyme activity which enable the growing and reproducing under poor nutrition environment. Therefore, the effect of biodegradation after activated carbon adsorption saturation is highly improved by Immobilization BAC (Scholz & Martin, 1997; Wang & Luo, 2003). Due to large surface area available for degradation of pollutants by microbe (Bancroft et al., 1983), biological activated carbon can be a better approach in combination with other conventional or advanced methods to solve the problem of PPCPs present in different matrices of environment.

FUTURE RESEARCH DIRECTIONS

PPCPs are a group of diverse chemicals consumed as drugs, cosmetics, as well as in agriculture etc. These compounds can easily become a part of fresh water resources either through agricultural runoff or through the addition of untreated or partially treated sewage water containing the waste water of household and industries. Various technologies are available for the treatment of waste water, and a lot of work on removal of PPCPs from waste water is also going on. Still in many areas of the world, even the preliminary screening of PPCPs in water resources is lacking, so the first requirement is to assess the concentration of these compounds in water resources (both the surface and ground water) used for the purpose of drinking. After getting the exact concentration of these compounds, the removal technologies should be used accordingly.

So, future area of research on PPCPs removal may be classified in two parts. First will be their assessment and quantification in existing water resources and their possible point sources and then the development of suitable removal technology for PPCPs based on their concentration in the area. As these compounds are harmful at very low concentration and varies in their physico-chemical properties, so no single removal technologies will be able to solve the problem. Combination of existing removal technologies with bioremediation process will be helpful for removal of these harmful compounds from water. In Bioremediation, microbes use these compounds as source of energy or as secondary substrate. So selection of suitable microbes for biodegradation of PPCP compounds present in waste water will be required. This will be helpful for utilization of microbes for bioremediation directly in the sludge or in the form of biological activated carbon.

CONCLUSION

Chemicals used as PPCPs are being released into the environment in extremely large quantities on a regular basis. Exact effect of these compounds on human system is not yet completely known, still based on some studies their harmful effects cannot be ignored. These compounds are tremendously harmful in respect to environmental and human health at very low concentration even at microgram level, therefore, effective and efficient removal technologies need to be developed to treat the contaminated effluents in order solve this dreaded problems. As bioremediation using microbes is based on biodegradation of pollutants and these living cells of microorganisms are able to degrade organic compounds at very low concentration, this could be play important role in this respect as an ecofriendly efficient technology for removing PPCPs from waste water.

REFERENCES

Adolfsson-Erici, M., Pettersson, M., Parkkonen, J., & Sturve, J. (2002). Triclosan, a commonly used bactericide found in human milk and in the aquatic environment in Sweden. *Chemosphere*, *46*(9-10), 1485–1489. doi:10.1016/S0045-6535(01)00255-7 PMID:12002480

American Water Works Association Research and Technical Practice Committee on Organic Contaminant. (1981). An assessment of microbial activity on GAC. *Journal - American Water Works Association*, *73*, 447–454.

Andrews, G. F., & Tien, C. (1981). Bacterial film growth in adsorbent surfaces. *AIChE Journal. American Institute of Chemical Engineers*, *27*(3), 396–403. doi:10.1002/aic.690270309

Bancroft, K., Maloney, S. W., Mcelhaney, J., Suffet, I. H., & Pipes, W. O. (1983). Assessment of bacterial growth and total organic carbon removal on granular activated carbon contactors. *Applied and Environmental Microbiology*, *46*(3), 683–688. PMID:6639023

Baronti, C., Curini, R., DAscenzo, G., Di Corcia, A., Gentili, A., & Samperi, R. (2000). Monitoring natural and synthetic estrogens at activated sludge sewage treatment plants and in a receiving river water. *Environmental Science & Technology*, *34*(24), 5059–5066. doi:10.1021/es001359q

Bartha, R., & Atlas, R. M. (1977). The microbiology of aquatic oil spills. *Advances in Applied Microbiology*, *22*, 225–266. doi:10.1016/S0065-2164(08)70164-3 PMID:337769

Bhargava, H. N., & Leonard, P. A. (1996). Triclosan: Applications and safety. *American Journal of Infection Control*, *24*(3), 209–218. doi:10.1016/S0196-6553(96)90017-6 PMID:8807001

Buhner, S. H. (2002). *The Lost Language of Plants- The Ecological Importance of Plant Medicines to Life on Earth.* White River Junction, Vermont: Chelsea Green Publishing.

Buser, H. R., Müller, M. D., & Theobald, N. (1998b). Occurrence of the pharmaceutical drug clofibric acid and the herbicide mecoprop in various Swiss lakes and in the North Sea. *Environmental Science & Technology*, *32*(1), 188–192. doi:10.1021/es9705811

Buser, H. R., Poiger, T., & Müller, M. D. (1998a). Occurrence and face of the pharmaceutical drug diclofenac in surface waters: Rapid photodegradation in a lake. *Environmental Science & Technology*, *32*(22), 3449–3456. doi:10.1021/es980301x

Buser, H. R., Poiger, T., & Müller, M. D. (1999). Occurrence and environmental behaviour of chiral pharmaceutical drug ibuprofen in surface waters and in wastewater. *Environmental Science & Technology*, *33*(15), 2529–2535. doi:10.1021/es981014w

Carol, P. (2012). Caffeine in Wastewater Is a Tracer for Human Fecal Contamination. *Environmental Health Perspectives*, *120*(3), a108–a109. doi:10.1289/ehp.120-a108a PMID:22382130

Chen, X., Nielsen, J. L., Furgal, K., Liu, Y., Lolas, I. B. Y., & Bester, K. (2011). Biodegradation of triclosan and formation of methyl-triclosan in activated sludge under aerobic conditions. *Chemosphere*, *84*(4), 452–456. doi:10.1016/j.chemosphere.2011.03.042 PMID:21507452

Collins, G., Woods, A., & Sharon, M. H. (2003). Microbial Community structure and Methanogenic Activity during Start-up of Psychrophilic Anaerobic Digesters Treating Synthetic Industrial Wastewaters. *FEMS Microbiology Ecology*, *46*(1), 159–170. doi:10.1016/S0168-6496(03)00217-4 PMID:19719569

Coogan, M. A., Edziyie, R. E., La Point, T. W., & Venables, B. J. (2007). Algal bioaccumulation of triclocarban, triclosan, and methyl-triclosan in a North Texas wastewater treatment plant receiving stream. *Chemosphere*, *67*(10), 1911–1918. doi:10.1016/j.chemosphere.2006.12.027 PMID:17275881

Crane, M., Watts, C., & Boucard, T. (2006). Chronic aquatic environmental risks from exposure to human pharmaceuticals. *The Science of the Total Environment*, *367*(1), 23–41. doi:10.1016/j.scitotenv.2006.04.010 PMID:16762401

Crittenden, J. C., Reddy, P. S., Arora, H., Trynoski, J. H., Perram, D. L., & Summers, R. S. (1991). Predicting gas performance with rapid small-scale column tests. *Journal - American Water Works Association*, *83*, 77–87.

Daughton, C. G., & Ternes, T. A. (1999). Pharmaceuticals and personal care products in the environment: Agents of subtle change? *Environmental Health Perspectives*, *107*(supplement 6), 907–942. doi:10.1289/ehp.99107s6907 PMID:10592150

Deniz, O., Tozkoparan, E., Ìlvan, A., Çìftçì, F., Bozkanat, E., & Öztürk, S. (2000). A Case of Carbamazepine-Induced Lymphocytic Interstitial Pneumonitis. *Turkish Journal of Medical Science*, *30*, 509–510.

Desbrow, C., Routledge, E. J., Brighty, C. G., Sumpter, J. P., & Waldock, M. (1998). Identification of estrogenic chemicals in STW effluent. 1. Chemical fractionation and in vitro biological screening. *Environmental Science & Technology*, *32*(11), 1549–1558. doi:10.1021/es9707973

Eberhardt, M. (1976). Experience with the use of biologically effective activated carbon. In Reports on special problems in water technology (vol. 9, pp. 331-347). U.S. Environmental Protection Agency, Municipal Environmental Research Laboratory.

Ehrhardt, H. M., & Rehm, H. J. (1985). Phenol degradation by microorganisms adsorbed on activated carbon. *Applied Microbiology and Biotechnology*, *21*, 32–36.

Fritsche, W., & Holrichter, M. (2005). Aerobic degradation of recalcitrant organic compounds by Microorganisms. In J. Joderning & J. Winter (Eds.), *Environmental biotechnology, Concepts and applications. Wiley-VCH Verlag GmbH & Co. KGaA*. doi:10.1002/3527604286.ch7

Fuchs, G., Boll, M., & Heider, J. (2011). Microbial degradation of aromatic compounds- from one strategy to four. *Nature Reviews. Microbiology*, *9*(11), 803–816. doi:10.1038/nrmicro2652 PMID:21963803

Gerrity, D., Gamage, S., Holady, J. C., Mawhinney, D. B., Quinones, O., Trenholm, R. A., & Snyder, S. A. (2011). Pilot-scale evaluation of ozone and biological activated carbon for trace organic contaminant mitigation and disinfection. *Water Research*, *45*(5), 2155–2165. doi:10.1016/j.watres.2010.12.031 PMID:21288550

Govindaraj, S., Srimurali, S., & Kumar, S. K. (2014). Non-steroidal anti-inflammatory drugs in Indian rivers. *Environmental Science and Pollution Research International*, *21*(2), 921–931. doi:10.1007/s11356-013-1957-6 PMID:23832803

Halling-Sørensen, B., Nielsen, S. N., Lanzky, P. F., & Ingerslev, F. (1998). Occurrence, fate and effects of pharmaceutical substances in the environment—A review. *Chemosphere*, *36*(2), 357–393. doi:10.1016/S0045-6535(97)00354-8 PMID:9569937

Heberer, T. (2002). Occurrence, fate, and removal of pharmaceutical residues in the aquatic environment: A review of recent research data. *Toxicology Letters*, *131*(1-2), 5–17. doi:10.1016/S0378-4274(02)00041-3 PMID:11988354

Heberer, T., Fuhrmann, B., Schmidt-Ba¨umler, K., Tsipi, D. V., Kout souba, & Hiskia. (2001). Occurrence of pharmaceutical residues in sewage, river, ground and drinking water in Greece and Ger many. In *Pharmaceuticals and personal care products in the environment: Scientific and regulatory issues* (pp. 70–83). American Chemical Society.

Heidler, J., & Halden, R. U. (2008). Meta-analysis of mass balances examining chemical fate during wastewater treatment. *Environmental Science & Technology*, *42*(17), 6324–6332. doi:10.1021/es703008y PMID:18800497

Houghton, J. E., & Shanley, M. S. (1994). Catabolic potential of pseudomonads: a regulatory perspective. In R. G. Chaudhry (Ed.), *Biological Degradation and Bioremediation of Toxic Chemicals* (pp. 11–32). London: Chapman & Hall.

Huerta-Fontela, M., Galceran, M. T., & Ventura, F. (2011). Occurrence and removal of pharmaceuticals and hormones through drinking water treatment. *Water Research*, *45*(3), 1432–1442. doi:10.1016/j.watres.2010.10.036 PMID:21122885

Jin, P., Jin, X., Wang, X., Feng, Y., & Wang, X. C. (2013). Biological Activated Carbon Treatment Process for Advanced Water and Wastewater Treatment. In Biomass Now-Cultivation and Utilization (pp. 153-192). InTech. doi:10.5772/52021

Jjemna, P. K. (2004). *Environmental microbiology: Principles and applications*. Enfield, NH: Science Publishers. Inc.

Johnson, A. C., Belfroid, A., & Di Corcia, A. (2000). Estimating steroid oestrogen inputs into activated sludge treatment works and observations on their removal from the effluent. *The Science of the Total Environment*, *256*(2-3), 163–173. doi:10.1016/S0048-9697(00)00481-2 PMID:10902843

Jones, O. A., Lester, J. N., & Voulvoulis, N. (2005). Pharamceuticals: A threat to drinking water? *Trends in Biology*, *23*(4), 163–167. doi:10.1016/j.tibtech.2005.02.001 PMID:15780706

Katzung, B. G. (2001). *Basic and clinical pharmacology* (8th ed.). New York: McGraw- Hill.

Kidd, K., Blanchfield, P., Mills, K., Palace, V., Evans, R., Lazorchak, J., & Flick, R. (2007). Collapse of a fish population after exposure to a synthetic estrogen. *Proceedings of the National Academy of Sciences of the United States of America*, *104*(21), 8897–8901. doi:10.1073/pnas.0609568104 PMID:17517636

Knappe, D. R. U., Snoeyink, V. L., Roche, P., Prados, M. J., & Bourbigot, M. (1999). Atrazine removal by preloaded GAC. *Journal - American Water Works Association*, (10): 97–109.

Kolpin, D. W., Furlong, E. T., Meyer, M. T., Thurman, E. M., Zaugg, S. D., Barber, L. B., & Buxton, H. T. (2002). Pharmaceuticals, Hormones and other organic wastewater contaminants in U.S. streams, 19992000: A national reconnaissance. *Environmental Science & Technology*, *36*(6), 1202–1211. doi:10.1021/es011055j PMID:11944670

Kotz, J. C. (2012). Chemistry and Chemical reactivity (8th ed.). Brookes/Cole, Cengage Learning.

Kummerer, K. (2001). Drugs in the Environment: Emission of Drugs, Diagnostic Aids and Disinfectants into Wastewater by Hospitals in Relation to Other Sources- a Review. *Chemosphere*, *45*(6-7), 957–969. doi:10.1016/S0045-6535(01)00144-8 PMID:11695619

La Guardia, M. J., Hale, R. C., Harvey, E., & Mainor, T. M. (2001). Alkylphenol ethoxylate degradation products in land-applied sew age sludge (biosolids). *Environmental Science & Technology*, *35*(24), 4798–4804. doi:10.1021/es0109040 PMID:11775155

Larsson, D. G. J., Adolfsson-Erici, M., Parkkonen, J., Pettersson, M., Berg, A. H., Olsson, P. E., & Forlin, L. (1999). Ethinyloestradiol - an undesired fish contraceptive? *Aquatic Toxicology (Amsterdam, Netherlands)*, *45*(2-3), 2–3, 91–97. doi:10.1016/S0166-445X(98)00112-X

Laville, N., Ait-Aissa, S., Gomez, E., Casellas, C., & Porcher, J. M. (2004). Effects of humanpharmaceuticals on cytotoxicity, EROD activity and ROS production in fish hepatocytes. *Toxicology*, *196*(1-2), 41–55. doi:10.1016/j.tox.2003.11.002 PMID:15036755

Lin, D., Zhou, Q., Xie, X., & Liu, Y. (2010). Potential biochemical and genetic toxicity of triclosan as an emerging pollutant on earthworms (Eisenia fetida). *Chemosphere*, *81*(10), 1328–1333. doi:10.1016/j.chemosphere.2010.08.027 PMID:20825966

Matsumura, Y., Yamabe, K., & Takahashi, H. (1985). The effects of hydrophilic structures of active-carbon on the adsorption of benzene and methanol vapors. *Carbon*, *23*(3), 263–271. doi:10.1016/0008-6223(85)90111-3

Meador, J., Yeh, A., Young, G., & Gallagher, E. P. (2016). Contaminants of emerging concern in a large temperate estuary. *Environmental Pollution*, *213*, 254–267. doi:10.1016/j.envpol.2016.01.088 PMID:26907702

Mons M.N., Hoogenboom A.C., & Noij T.H.M., (2003). *Pharmaceuticals and drinking water supply in the Netherlands*. Nieuwegein, Kiwa Water Research (Kiwa Report No. BTO 2003.040).

Nakada, N., Shinohara, H., Murata, A., Kiri, K., Managaki, S., Sato, N., & Takada, H. (2007). Removal of selected pharmaceuticals and personal care products (PPCPs) and endocrine-disrupting chemicals (EDCs) during sand filtration and ozonation at a municipal sewage treatment plant. *Water Research*, *41*(19), 4373–4382. doi:10.1016/j.watres.2007.06.038 PMID:17632207

Nassef, M., Matsumoto, S., Seki, M., Khalil, F., Kang, I. J., Shimasaki, Y., & Honjo, T. et al. (2010). Acute effects of triclosan, diclofenac and carbamazepine on feeding performance of Japanese medaka fish (Oryzias latipes). *Chemosphere*, *80*(9), 1095–1100. doi:10.1016/j.chemosphere.2010.04.073 PMID:20537681

Neelima, G., Garg, K. L., & Mukerji, K. G. (1994). Microbial degradation of environmental pollutants. In Recent Advances in Bio-deterioration and biodegradation (vol. 2, pp. 329-362). Naya Prokash.

Nyer, E. K. (1992). *Treatment for organic contaminants: Biological methods. In Groundwater Treatment Technology* (pp. 129–214). New York: Van Nostrand Reinhold.

Oaks J.L., Gilbert M., Virani M.Z., Watson R.T., Meteyer C.U., Rideout B.A., … Khan A.A. (2004). Diclofenac residues as the cause of vulture population decline in Pakistan. *Nature*, *427*(6975), 630-633.

Pelczar, M. J., Chan, E. C. S., & Kreig, N. K. (1993). *Microbiology concepts and application*. McGraw Hill, Inc.

Reddersen, K., Heberer, T., & Dünnbier, U. (2002). Identification and significance of phenazone drugs and their metabolites in ground- and drinking water. *Chemosphere*, *49*(6), 539–544. doi:10.1016/S0045-6535(02)00387-9 PMID:12430641

Reungoat, J., Escher, B. I., Macova, M., & Keller, J. (2011). Biofiltration of wastewater treatment plant effluent: Effective removal of pharmaceuticals and personal care products and reduction of toxicity. *Water Research*, *45*(9), 2751–2762. doi:10.1016/j.watres.2011.02.013 PMID:21450327

Rimkus, G. G., Gatermann, R., & Hühnerfuss, H. (1999). Musk xylene and musk ketone amino metabolites in the aquatic environment. *Toxicology Letters*, *111*(1-2), 5–15. doi:10.1016/S0378-4274(99)00190-3 PMID:10630699

Riser-Roberts, E. (1998). *Remediation of Petroleum Contaminated Soil: Biological, Physical, and Chemical Processes*. Boca Raton, FL: Lewis Publishers. doi:10.1201/9781420050578

Rodriguez, E., Campinas, M., Acero, J. L., & Rosa, M. J. (2016). Investigating PPCP Removal from Wastewater by Powdered Activated Carbon/ Ultrafiltration. *Water, Air, and Soil Pollution*, *227*(6), 177. doi:10.1007/s11270-016-2870-7

Rosario-Ortiz, F. L., Wert, E. C., & Snyder, S. A. (2010). Evaluation of UV/H_2O_2 treatment for the oxidation of pharmaceuticals in wastewater. *Water Research*, *44*(5), 1440–1448. doi:10.1016/j.watres.2009.10.031 PMID:19931113

Sangster, J. (1997). *Octanol-water partition coefficients: Fundamentals and physical chemistry*. John Wiley and Sons.

Scholz, M., & Martin, R. J. (1997). Ecological Equilibrium on Biological Activated Carbon. *Water Research*, *31*(12), 2959–2968. doi:10.1016/S0043-1354(97)00155-3

Schwarzenbach, R. P., Gschwend, P. M., & Imoden, D. M. (2004). Environmental Organic Chemistry (2nd ed.). J. Wiley and Sons.

Sedlak, D. L., Gray, J. L., & Pinkston, K. E. (2000). Understanding microcontaminants in recycled water. *Environmental Science & Technology*, *34*(23), 509A–515A. doi:10.1021/es003513e PMID:21662317

Shimp, R. J., & Pfaender, F. K. (1982). Effects of surface area and flow rate on marinal bacterial growth in activated carbon columns. *Applied and Environmental Microbiology*, *44*, 471–477. PMID:16346080

Singh, R., Singh, P., & Sharma, R. (2014). Microorganism as a tool of bioremediation technology for cleaning environment. *RE:view*, *4*(1), 1–6.

Snyder, S. A., Adham, S., Redding, A. M., Cannon, F. S., De Carolis, J., Oppenheimer, J., & Yoon, Y. et al. (2007). Role of membranes and activated carbon in the removal of endocrine disruptors and pharmaceuticals. *Desalination*, *202*(1-3), 156–181. doi:10.1016/j.desal.2005.12.052

Snyder, S. A., Westerhoff, P., Yoon, Y., & Sedlak, D. L. (2003). Pharmaceuticals, personal care products and endocrine disruptors in water:implications for the water industry. *Environmental Engineering Science*, *20*(5), 449–469. doi:10.1089/109287503768335931

Stucki, G., & Thuer, M. (1994). Increased removal capacity for 1,2- dichloromethane by biological modification of the granular modification of the granular activated carbon process. *Applied Microbiology and Biotechnology*, *42*(1), 167–172. doi:10.1007/BF00170241 PMID:7765815

Stumpf, M., Ternes, T. A., Wilken, R. D., & Rodrigues, S. V. (1999). Polar drug residues in sewage and natural waters in state of Rio de Janeiro, Brazil. *The Science of the Total Environment*, *225*(1-2), 135–141. doi:10.1016/S0048-9697(98)00339-8 PMID:10028710

Takeuchi, Y., Suzuki, Y., Mochidzuki, K., Yagishita, Y., Fukuda, T., Amakusa, H., & Abe, H. (1996). Biological activated carbon treatment wastewater containing organics and heavy metal ions. In M. D. LeVan (Ed.), *Proceedings of the Fifth International Conference on Fundamentals of Adsorption* (pp. 929-236). Kluwer Academic Publishers. doi:10.1007/978-1-4613-1375-5_116

Tanghe, T., & Verstraete, W. (2001). Adsorption of nonylphenol onto granular activated carbon. *Water, Air, and Soil Pollution*, *131*(1/4), 61–72. doi:10.1023/A:1011966914827

Ternes, T. A. (1998). Occurrence of drugs in German sewage treatment plants and rivers. *Water Research*, *32*(11), 3245–3260. doi:10.1016/S0043-1354(98)00099-2

Thai, P. K., Lai, F. Y., Bruno, R., van Dyken, E., Hall, W., OBrien, J., & Mueller, J. F. et al. (2016). Refining the excretion factors of methadone and codeine for wastewater analysis- Combining data from pharmacokinetic and wastewater studies. *Environment International*, *94*, 307–314. doi:10.1016/j.envint.2016.05.033 PMID:27295047

Thorpe, K. L., Cummings, R. I., Tutchinson, T. H., Scholz, M., Brighty, G., Sumpter, J. P., & Tyler, C. R. (2003). Relative potencies and combination effects of steroidal estrogens in fish. *Environmental Science & Technology*, *37*(6), 1142–1149. doi:10.1021/es0201348 PMID:12680667

Tian, Q., & Chen, J. (2006). Application of bioactivated carbon (BAC) Process in Water and Wastewater Treatment.[In Chinese]. *Environmental Engineering*, *24*(1), 84–86.

Torsvik, V., & Øvreås, L. (2002). Microbial diversity and function in soil: From genes to ecosystems. *Current Opinion in Microbiology*, *5*(3), 240–245. doi:10.1016/S1369-5274(02)00324-7 PMID:12057676

Wang, L., & Luo, Q. (2003). Study on Degradation of Immobilized Microorganism on Endocrine Disruptor Di-n-butyl Phthalate. *Chinese Journal of Public Health*, *19*(11), 1302–1303.

Westerhoff, P., Yoon, Y., Snyder, S. A., & Wert, E. (2005). Fate of endocrine-disruptor, pharmaceutical, and personal care product chemicals during simulated drinking water treatment processes. *Environmental Science & Technology*, *39*(17), 6649–6663. doi:10.1021/es0484799 PMID:16190224

Youssef, A. M., Ghazy, T. M., & Elnabarawy, T. (1982). Moisture sorption by modified-activated carbons. *Carbon*, *20*(2), 113–116. doi:10.1016/0008-6223(82)90415-8

Zühlke, S., Dunnbier, U., & Heberer, T. (2004). Detection and identification of phenazone-type drugs and their microbial metabolites in ground and drinking water applying solid-phase extraction and gas chromatography with mass spectrometric detection. *Journal of Chromatography. A*, *1050*(2), 201–209. doi:10.1016/S0021-9673(04)01401-3 PMID:15508313

KEY TERMS AND DEFINITIONS

Activated Carbon: It is a highly porous carbonaceous substances composed by carbon atoms having potential adsorption properties.

Biodegradation: Degradation of materials using biological systems e.g. enzymes generated by microbes etc.

Bio-Magnification: Accumulation of any toxic substance in living cells.

Bioremediation: Removal of pollutants using biological materials e.g. microbes, plants etc.

Hydrophilic: Material which attract water.

Hydrophobic: Material which repel water.

Partition Coefficient: Ratio of distribution of any compound between two immiscible mediums.

Water Partition Coefficient: It refers to ratio of concentrations of a compound in a mixture of two immiscible phases at equilibrium.

Chapter 20
Testing and Monitoring of Biodegradable Contaminants in Bioremediation Technique

Ajay Kumar
ITM University, India

Pragati Saini
KRG College, India

ABSTRACT

Environmental pollution is one of the most acute global problems. Pollution is present at the every corners of the globe. Several methods are used for monitoring the pollutants in the environment. The present chapter includes the wide variety of testing such as physical, chemical, biological and as well molecular to determine the level of biodegradable contaminated substances in the environment and also check the feasibility of different bioremediation techniques. The monitoring and testing are essential and have an enormous impact on the cost of full-scale remediation. Successful remediation techniques completely degrade the contaminants. A successful bioremediation approach requires sufficient evidences for the non-toxicity of the contaminant. Current monitoring techniques require the disappearance of the contaminants. Different recombinant techniques also play great role to assess the pollution level. In the field and focus on interdisciplinary research, bioremediation technology will go a long way in cleaning our polluted environment in near future. Research on improved microbial strains and bioanalytical methods for measuring the level of contaminants should be strengthen.

INTRODUCTION

Intensification of agriculture and manufacturing industries has resulted in increased release of a wide range of xenobiotic compounds to the environment. Excess loading of hazardous waste has led to scarcity of clean water and disturbances of soil thus limiting crop production.

Microorganisms have an extensive but finite capacity to recycle synthetic organic molecules. One difficulty with microbial bioremediation is that not all components of chemical mixtures are degraded

DOI: 10.4018/978-1-5225-2325-3.ch020

with equal efficiency. By genetic engineering, the substrate range can be widened to include xenobiotics that are normally recalcitrant to degradation. Regulation of the deliberate release of genetically modified microorganisms requires, however, knowledge on the survival and effects of the inoculum in the environment (Gustafsson & Jansson, 1993). Government and the public become aware of the importance of the environment. To understand the changes in the environment, caused either by natural or mankind, large numbers of environmental parameters are considered. The range of environmental changes may from slow increase in global temperature to rapid accumulation of heavy metals and xenobiotics. Some changes are so slow that their determination requires careful monitoring for long time periods (Atlas, & Bartha, 1993). Pollution is defined by Holdgate in 1979 as "the introduction into the environment of substances or energy which causes hazards to human health, harm to living resources and ecological damage". The first major human influence on the environment was perhaps agriculture but the industrial revolution was also very significant. The main environment pollutant are inorganic compounds such as metals and nitrates, organic compounds, microorganisms including pathogens, gaseous compounds such as volatile, gases and particulates which can contaminate land, air and water (Table 1).

The environmental conditions and the properties of the pollutant affect the fate of pollutant in the environment. The properties of the compounds such as solubility, volatility and reactivity affect the environment. The contaminants are not restricted to their site of introduction but depending on conditions which can migrate and contaminate other parts of the environment such as lakes, rivers and sea. Always the pollutants are not single but can be complex mixtures of toxic compounds. National laws are normally associated with a number of acts covering many types of pollutants, industries and conditions (Figure 1).

In the USA the Environmental Protection Agency (EPA) has listed 12 persistent, bioaccumulative and toxic (PBT) chemicals and more extensive lists of primary standards for drinking water including microorganisms and inorganic and organic compounds. Many of these compounds have strict legal upper limits on land and water. These compounds are aldrin, pyrene, DDT, hexachloro benzene, alkyl lead, mercury, PCBs and dioxins etc. (Table 2).

Figure 1. Outline of the fate of compound release into the environment

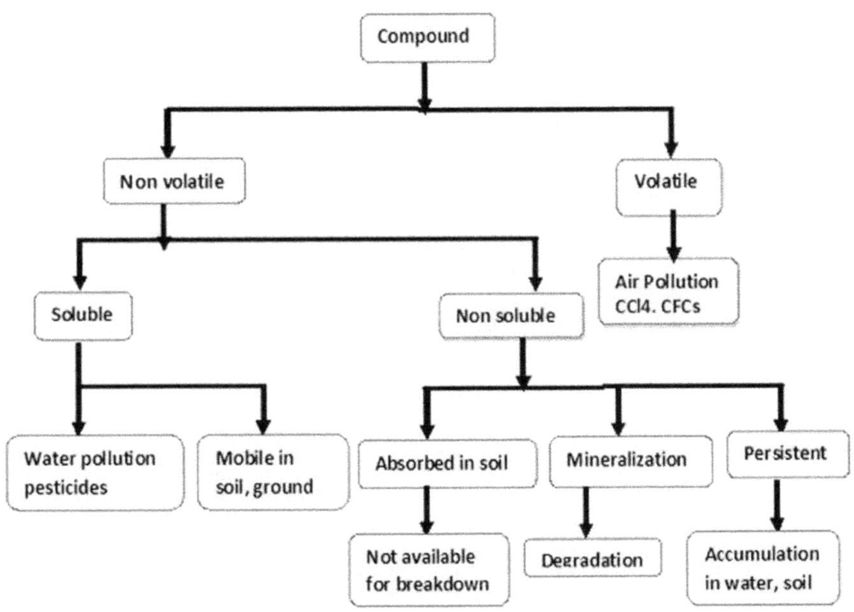

Table 1. Industry and their contaminants

Industry	Contaminants
Chemicals	Acid, alkali, metals, solvents, phenols, organic compounds, chlorophenols
Petrochemicals	Hydrocarbon, Phenols, Acids, alkali
Metals	Metals especially Fe, Cu, Ni, Cr, Zn, Cd
Energy	Phenols, Cyanide, Sulphur, Compounds, coal and coke dusts
Mineral Extraction	Metals Cu, Zn and Pb, gas leachate
Water supply and sewage	Microorganisms and methane

Reeve, 1994.

Table 2. US National Primary Standards for drinking water

Contaminants	Maximum Contaminant Level (mg/l)	Sources
	Microorganisms	
Cryptosporidum	99%	Faecal and human waste
Giardia lamlia	99.9%	Faecal and human waste
Total coliforms	5%	An indicator for the presence of harmful bacteria
Viruses	99.9%	Human and animal waste
	Disinfection by Products	
Bromate	0.01	Drinking water disinfection
Chlorine	1.0	Drinking water disinfection
	Disinfectants	
Chloramines	4.0	Water additives to control microbes
Chlorine	4.0	Water additives to control microbes
	Inorganic Chemicals	
Copper	1.3	Corrosion of plumbing system
Cyanide	0.2	Steel, metal and fertilizer industry
Cadmium	0.005	Corrosion pipes
Fluoride	4.0	Fertilizer, run off of sewage
	Organic Chemicals	
Atrazine	0.003	Herbicide used on crops
Benzene	0.005	Factory and landfill
Chloro benzene	0.1	Chemicals and agriculture factory
2,4 D	0.07	Herbicide run off
Dioxin	0.000003	Emission from waste incineration
Ethyl benzene	0.7	Discharge from refineries
Ethylene dibromide	0.0004	Refineries
Heptachlor	0.001	Banned termiticide
BHC	0.0005	Agriculture chemical factories
PCBs	0.0005	Run off from waste chemicals

www.epa.gov.

IMPORTANCE OF TESTING AND MONITORING IN BIOREMEDIATION

The environmental pollutants originate from contaminated sites such as gasworks, domestic, industrial, sewage and metal industries like electroplating and accidents or spillage as in cas of Seveso and the tanker Amoco Cadiz. The fate of pollutants in the environment depends on the properties of the compound and the environmental conditions. The properties of the compounds that affect their fate are their water solubility, volatility and reactivity. Often the contaminants are not restricted to their site of introduction but depending on the conditions can migrate and contaminate other parts of the environment such as lakes, rivers, reservoirs, estuaries and the sea. The pollutants are not always single compounds but can complex mixtures containing a number of toxic compounds. The possible important environmental pollutants are metals, radionucleotides, nitrates, nitrites, cyanides, petrochemicals, synthetic polymers, pathogens, volatiles compounds and gases.

Thus any monitoring must consider the type of contaminants present, their availability and the possibility of biomagnification and bioaccumulation in the environment. The environment must be monitored the presence of plant and animal pathogens. Thus environmental monitoring system must be very sensitive and accurate. In determination of the pollutant the present and their effect on the environment, the role of molecular techniques are increased day by day (Reeve, 1994). These approaches yield valuable information regarding effectiveness of remediation and help focus additional remediation strategies. Effective monitoring of physical, chemical and biological parameters is a necessary component of any remediation application. Each of the different parameters provides different insight into the strengths and deficiencies of a remediation process/system. Using this data, useful conclusions can be drawn that allow for modifications to be made that will maximise treatment efficiency.

TESTING OF SAMPLE

To analyse any component in the environment, sample must be collected and different features must be considered. These features include what part needs to be sampled, how many samples need to be covered, time of sampling and statistical analysis. The part of sample in the environment depends on the type of pollutant such as a chemical spill may contaminate the soil or it may also reach and contaminate the groundwater, lakes and rivers. In this case a broad spread of sampling is required from the surface as well subsurface area.

Soil Sampling

Contaminated sites often need some form of survey as all the information of the pollution present cannot be from the site history. Some directives (BSI, 1988) suggest that 17 samples should be taken from 5 hectare site at three different depths. The drawback is that a large number of samples have to be analysed which raise the cost of analysis. Therefore a site history can be of great use in directing the sampling process and reducing costs.

Water Sampling

The sampling of contaminated water is difficult due to their nature. To get accurate assessment, the samples will have to take over a time period and sections and different depths. Therefore various automatic methods can be used. Some industries discharge its wastewater into waterways intermittently which require and extend the time of sampling and whole waterway need to be assessed (Tebbut, 1998). When a specific organism is to be tested in the environment, the sample must be representative. One example is the presence of edible mussel *Mytilus edulis* indicates the metal pollution because it is metal accumulator.

Air Sampling

Air sampling also has the same problem as water but also influenced by wind direction and strength. Generally there are three methods are used as –pumped system, pre-concentration and grab samples. In the pumped system the sample is pumped directly from the air into analyser. In some cases where the concentration of the pollutant is present at very low, pre concentration is required before analysis. In grab sampling, the samples of air are captured in bottles, syringes, bellows and bags. If the collected samples cannot be analysed immediately, it can be stored but care is required because the pollutant can lost or change during storage.

ANALYSIS

The following parameters need to be known before analysis. These are as-

- The limit of detection of the compound.
- The variation with the method of analysis.
- The compound stability.

The analysis methods are divided into three categories as physical, chemical and biological analysis.

Physical Analysis

Physical methods are often used to determine the contaminants and are as follows:

- **Gravimetric Method:** It is used to determine suspended solid or, total or volatile solids.
- **pH:** It is used to determine very acid or alkaline conditions and measured with pH electrode.
- **Colorimetric:** Colour and turbidity are important in water quality and can be determined using comparison of tubes, colour discs and spectrophotometers.
- **Dissolved Oxygen:** It is very important in water quality and maintains aerobic biological organisms. It can be measured using an oxygen electrode.
- **Ion Specific Electrode:** These electrodes can be used to determine the levels of ammonia, nitrate, nitrite, calcium, sodium and other ions.

Table 3. Different methods for chemical analysis of environmental samples

Methods	Comments
Thin layer chromatography	Compounds move at various rates when placed in a solvent
Gas liquid chromatography	Volatile compounds separated on long column over which a mixture of gases is passed.
High performance liquid chromatography	The liquid phase is passed through the column at high pressure.
Mass spectrometry	The sample is bombarded in a vacuum with electrons and fragments produced get separated by their charge and mass
Nuclear magnetic resonance	The resonance frequency of atomic nuclei in a strong magnetic field will vary with their chemical environment
Infrared spectroscopy	The sample is placed in IR radiation. Particular bonds making up groups absorb radiation resulting in a spectrum characteristics for specific groups.
Atomic absorption spectrophotometer	It is used to determine metals. A solution is sprayed into a flame and the emission spectrum measured which is characteristic for specific metal.
GLC-MS	In this GLC is combined with MS to identify the peaks eluted.

Reeve, 1994.

Chemical Analysis

There are numerous standard chemical methods for the determination of environmental pollutants such as chloride, nitrate, nitrite and phosphate (Table 3).

Metals can be determined by chemical or by atomic absorption spectrophotometry. Organic compounds can be assayed by a number of techniques such as HPLC, GLC which can be combined to mass spectrometers in order to identify the compound separated. The oxygen demand in water can be estimated by measuring the COD which is determined by refluxing the sample with potassium dichromate in concentrated sulphuric acid. The sample is refluxed for 2 h and the potassium dichromate remaining is determined by titration against ferrous ammonium sulphide. It gives measure of the oxygen required for the oxidation of the contents in the sample. The most important limitation is that chemical measurements only provide information about the chemicals that are included in the analysis (Wernersson, 2012). One example that highlights the limitations with chemical monitoring is the recent interest in per fluoro octane sulfonic acid (PFOS), which was brought to attention in the early 2000s (Lehmler, 2005).

Biological Analysis

Different microbial techniques are used for the detection and estimation of microbial numbers. The following methods are given below-

- Dry weight produced by centrifugation or filtering.
- Cell numbers, haemocytometer, coulter counter, flow cytometer.
- Viable cell counts, plate counts, most probable number (MPN).
- Packed cell volume.
- Spectrophotometry, optical density at various wavelengths.

These microbiological have the limitations with environmental samples. The organisms get attached to sediments and difficult to quantify and results probably underestimate the cell numbers. The viable count methods depend on that each colony formed, has been derived from a single cell and by using selective media the number of specific bacteria as *Salmonella* and coliform can be estimated. In bioremediation processes, microorganisms use the contaminants as nutrient or energy sources (Hess, Zarda, Hahn, Hanner, & Stax, 1997; Agrawal, 1998; Tang, Criddle, & Leckie, 2007). The selective techniques favour the faster growing species and slower growing species can be missed. It has been estimated that 90-99% of the environment species are not culturable. Therefore, the growth characteristics of the organisms of interest need to be known before their estimation.

The Royal Commission on sewage disposal established biological oxygen demand (BOD) as important parameter in water quality. The BOD is a measure of the oxygen demand in a sample as a result of its biodegradable organic content. The oxygen demand of wastewater is determined by incubating the sample in sealed bottle in the dark at 20^0 C. After 5 days the oxygen consumed is measured and this is BOD_5 value. Sometime the sample is diluted with a volume of activated sludge. Activated sludge is a mixture of aerobic organisms produced by the treatment of waste. This test gives a reasonable result with normal sewage and affected by the time of incubation, temperature and microorganisms present. If the waste contains high levels of nitrogenous materials, longer incubation period is required. If the waste is too strong the oxygen will be depleted before oxidation is complete. Although BOD assay has a number of deficiencies:

- Slow method.
- Time consuming.
- Expensive method.
- Does not have fully natural conditions.
- Oxidation may not be complete in 5 days.
- Difficult to interpret with waste containing high levels of non-degradable organic material.
- Imprecise-particular with low levels of organic materials.

Genomic Analysis

Genomic analysis is the identification, measurement or comparison of genomic features such as DNA sequence, structural variation, gene expression or regulatory and functional element annotation at a genomic scale. Methods for genomic analysis typically require high-throughput sequencing or microarray hybridisation and bioinformatics. Different techniques are used molecular biology and recombinant DNA technology, have enabled the study of the environmental microbial population both in the lab and in situ which was not possible with conventional methods.

Some possible methods where DNA can be amplified by the polymerase chain reaction (PCR) from extracted DNA or RNA treated with reverse transcriptase. The most frequent primers used in the PCR technique are 16s ribosomal RNA (rRNA) sequences. Denaturing gradient gel electrophoresis (DGGE) or temperature gradient gel electrophoresis (TGGE) can separate the PCR product. Then the particular bands can be amplified by PCR and sequenced. Once the sequence obtained it can be compared with the available database of rDNA sequences for the identification (Figure 2).

Figure 2. Molecular biology techniques, used to determine the microbial population in environmental sample

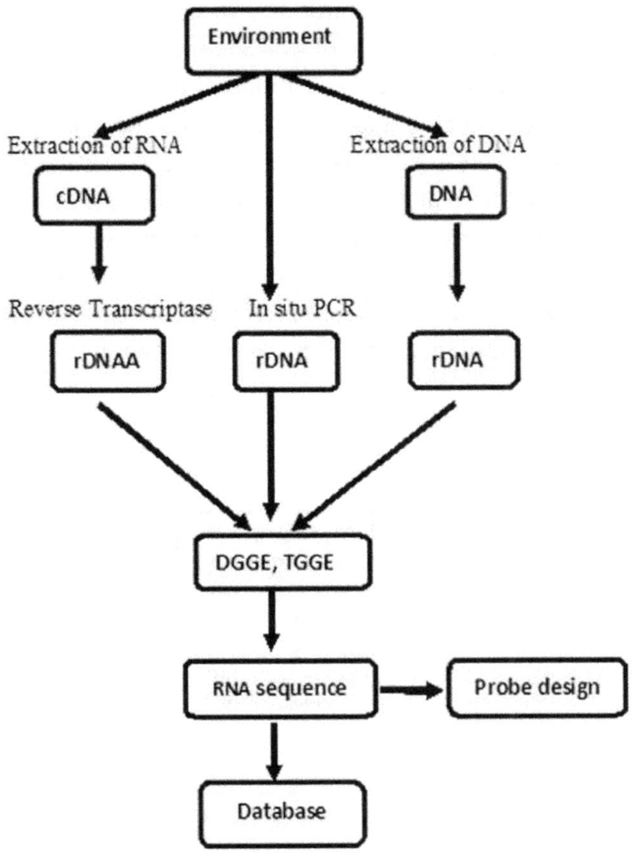

The sequences also allow the design of probes which become helpful in fluorescence in situ hybridisation (FISH). If specific primers are selected it help to amplify the low copy or single copy genes which cannot be otherwise (Hanson, & Larson, 2008). The following are the examples of the use of recombinant DNA technology in the study of microbial communities.

- The microbial ecology of soils and extreme habitats.
- Aerobic and anaerobic sewage system.
- Biodiversity.
- Microbial population in contaminated groundwater.
- Bioremediation.
- Release of genetically engineered microorganisms.

DNA technology also useful to follow the bacterial population in activated sludge. It has been clear that culture techniques are not capable of providing complete detail of the microbial population in activated sludge. RDT can be used to estimate the degree of genetic variation in environmental samples and used to detect bioleaching bacteria.

Proteomics Analysis

It is the systematic analysis of the protein population in a tissue, cell or subcellular compartment. The ability to separate and identify the proteins from cells allows the study at the protein level of the effects of pollutants. The protein analysis starts with 2D separation of proteins and then digestion of individual bands. The resulting peptides are extracted and their masses are determined by matrix-assisted laser desorption ionization (MALDI), electrospray ionization (ESI) and mass spectrometry (MS). Modern machines give accuracy, mass resolution and sensitivity which allow for the identification of picomolar and femtomolar amounts of proteins and peptides if matching genome sequences are available (Figure 3).

Proteomics has the potential to identify not only single proteins, but also to generate protein fingerprints that are specific for certain pollutants. Proteins are predicted to be less transient than gene expression patterns and more stable as potential biomarkers (Bourlat et al., 2012).

Figure 3. Proteomics methods, used for studying the protein expression in microorganisms

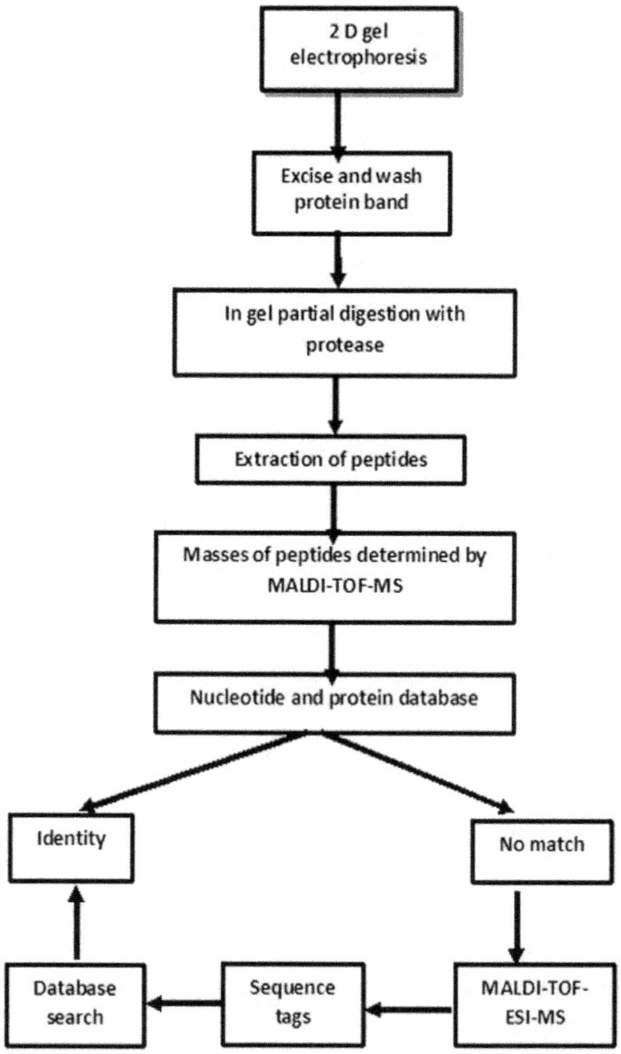

MONITORING POLLUTION

The determination of the concentration of the pollutant in any environmental conditions is the most common approach. Different chemical and other methods are available for this. However, this is not suitable to determine the effect of pollution on the organisms that make up the environment. Besides this, the use of biological system to measure the pollutant or effects *in situ* includes bioindicator, biomarkers and specific type organisms. Bioindicators and biomarkers have the advantage and measure the action of the pollutants on the environment.

Bioindicator

Bioindicator organisms are those that can be used to identify and quantify the effects of pollutants in the environment. Ideally bioindicators are:

- Distributed widely.
- Abundant and not mobile.
- Available all year.
- Easy to collect.
- High concentration of pollutant.
- Sensitive to the pollutant.

The type of changes which can be observed are ecological behavioural and physiological. Ecological factors involve changes in population density, key species and species diversity. Behavioural changes can be feeding activities, bacterial mobility and web spinning. The physiological changes can be the accumulation of heavy metals, carbon dioxide production, BOD and microbial activity. Lichens are a symbiotic relationship between an alga and fungus and have many of the properties required of a bioindicator. Both lichens and mosses are abundant, not mobile, easy to collect and particularly sensitive to air pollutant and widely used for air quality. Honey bees have been used as a bioindicator for the contamination of the atmosphere by heavy metals. Honey bees, honey, pollen and wax were assayed for the presence of the heavy metals.

Another immobile organism as vivalve molluscs is used as bioindicator. The filter feeding muscles are known to accumulate heavy metals and ideal for the monitoring of coastal waters. The measurements of chlorophyll content, biomass and ash free day weight are common indicators of pollution. This system is used to assess the acid mine drainage impact on streams. Marine pollution is monitored by observing the increase malformation in embryos of *Limanda limanda* in the North Sea. This increase malformation is related to high input of DDT and PCBs. Some behavioural, sublethal and lethal tests are applied to determine the concentration of TNT by using earthworm in contaminated soil. In Australian estuaries, *Heloecius cordiformis* is used as a good indicator of lead metal.

Cell Biology in Environmental Monitoring

Cell biology is the study of the structural and functional aspects of cells and the cellular organelles. It can be successfully used for the detection of environmental pollution, particularly mutagens and carcinogens. The aim of cell biological methods is to trace the harmful effects of pollutants on different cellular components like membranes, chloroplasts, mitochondria, chromosomes. In addition, the macromolecules namely nucleic acids (particularly DNA) and proteins. Further, cell biological methods also help in understanding the mechanisms of toxicity of pollutants. Some important cell biological methods used in environmental pollution monitoring are described.

Membrane Damage in Bioassay

Plasma membrane is an outer layer which surrounds the cell and protects from hostile environment. It is the first cellular component to be directly exposed to pollutants and damaged by the pollutants. The physical damage of cell membrane by the pollutants can be measured by light, phase contrast and electron microscopy. The semipermeable properties of the membranes may be disrupted by the pollutants. The disruption includes leakage of enzymes (e.g., lactate dehydrogenase), efflux of electrolytes or uptake of trypan blue. Lysosomes are also used as biomarkers for cell viability by neutral red retention test but the damaged lysosomes cannot retain this dye. Nowadays, animal and plant tissue culture techniques are also used for pollution monitoring by measuring cellular damages observed in cell cycles. An example is the use of human lymphocyte culture to monitor the persons exposed to toxic pollutants.

Cytogenic Bioassays

The genetic damages of the cells can be effectively used in bio-monitoring of pollution by measuring the changes in the chromosomes. In this method short life cycle for this purpose, animals (e.g. insect *Drosophila*) and plants (e.g. *Arabidiopsis*) with shorter life cycles are preferred. Other plants such as pea, maize and soy bean are also used in cytogenetic bioassays.

Chromosomal DAMAGE

Different genotoxicity tests are performed to show the effect of pollutant which causes changes in genetic material (Winegar, & Keith, 1993). The pollutants may cause several types of chromosomal damages-fragmentation, bridge formation, and disruption in cell division. These chromosomal damages or changes can be effectively used for pollution detection. It is cleared that the severity of chromosomal damage depends on the chemical nature of the pollutant.

Micronucleus Test

The severe damage to chromosomes may result in large scale fragmentation of chromosomes by the pollutants which results the micronuclei formation. The micronuclei development is directly related to the severity of the damage which is used for screening of mutagenic compounds by Micronucleus test (MNT).

BIOMARKERS

Biomarkers can be defined as quantified measures of changes in the biological system that responds to exposure to metals and xenobiotic substances that lead to biological effects (Huggett, Kimerle, Mehrle, & Bergman, 1992). Several biomarkers are available as tools for tagging bacteria. The choice of biomarker depends on the system studied and the questions to be addressed (Jansson, 1995). The use of biomarkers has its origin in human toxicology where they have proved to be very useful as measures of exposure to chemicals as well as to provide early warning signals for specific diseases (Timbrell, 1998). The advantages are as follows-

- The determination can be made over a time period.
- It can indicate the risk of exposure to a particular chemical.
- It provides the information on the toxicity of single compound.

Biochemical Indicator

These are based on the ability of the pollutant to generate a response at the gene level, inducing the specific enzyme involved in detoxification of the contaminant. Some hydrolytic enzymes such as esterase and amidase detoxify the xenobiotic compounds but the major parts of xenobiotics are catalysed by oxidation reaction of cytochrome P_{450}. Another process of detoxification found in both mammals and plants is the conjugation with glutathione catalysed by the enzyme glutathione-S-transferase (Figure 4).

The sequestration by metallothionenins in bacteria and phytochelatins in plants is one of the numerous mechanisms for the detoxification of heavy metals. There are several other enzymes are available for measuring the effects of pollutants are ethoxy resorufin-o-dethylas (EROD) and aryl hydrocarbon and poly aromatic hydrocarbons (PAHs).

Figure 4. Detoxification of xenobiotics and sequestration of heavy metals

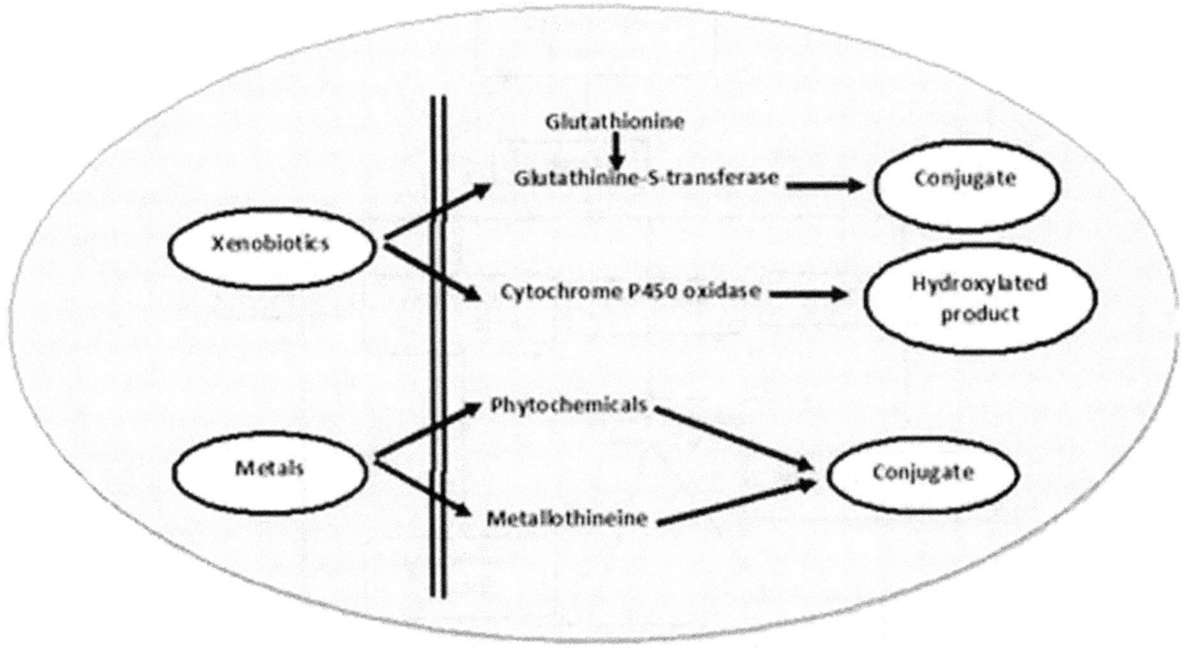

Genetic Indicator

Several molecular techniques are also used to determine the effect of pollutants in the environment and include the introduction of genetically engineered indicator organisms, antibiotics and heavy metal resistance genes. The most rapid and sensitive method for the gene expression is to fuse the relevant promoter sequence with product that is easily detected which is done by antibiotic and heavy metal resistance but the best one produce light. The *gus* A gene product cleaves the colourless substrate X *galt* to give a blue colour and this gene fused with promoter that respond to the pollutant.

Another example is the insertion of a plasmid into an *E. coli* strain containing *lac*Z gene, which code the enzyme β-galactosidase, linked to the *ars* R gene that codes for regulatory protein of the *ars* operon. The *ars* operon is involved in the removal of antimony and arsenic from the cell. Thus the cell induces enzyme β-galactosidase on exposure to antimony or arsenate which can be tested by the addition of p-aminophenyl β-galactopyranoside. The reaction product p-aminophenol can be determined electrochemically. Thus a bacterial sensing system can be developed for the detection of antimony and arsenate (Figure 5).

Figure 5. The use of gene coding for protein that can emit light in the detection of effects of pollution in the environment

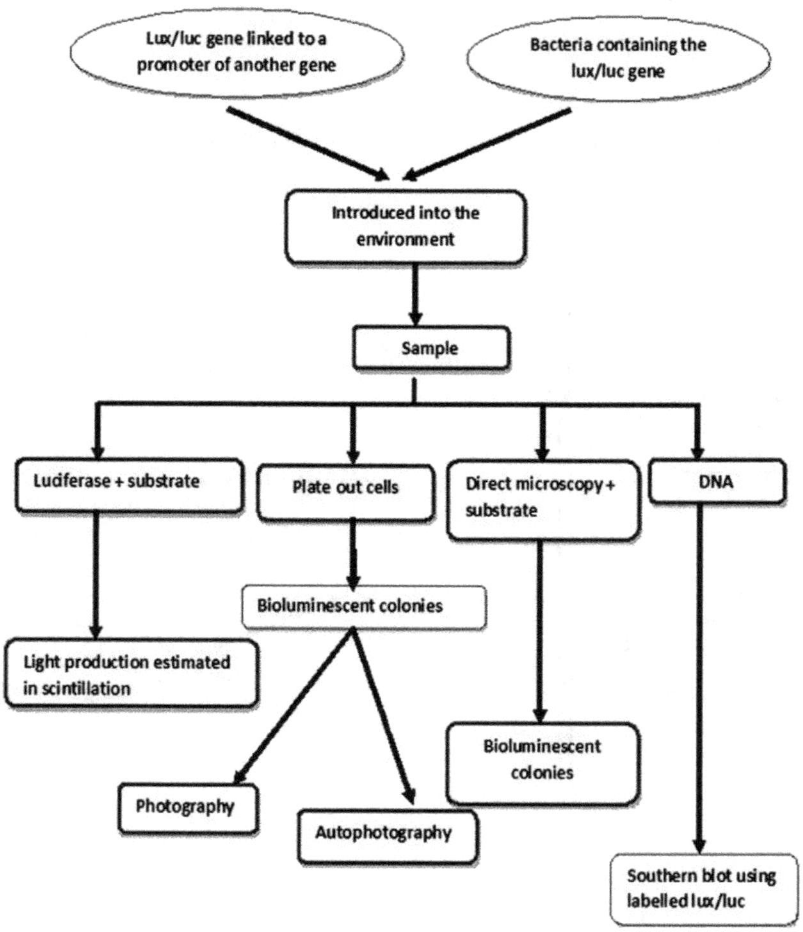

Another gene product does not require substrate because gene codes a light producing protein, the luciferase enzyme *lux* AB from prokaryotes and *luc* gene from eukaryote and a green fluorescent protein coded by the *gfp* gene from a jelly fish *Aequorea victoria.* These type marker gene places under control of promoter genes associated with a response to toxic compounds. Thus if engineered bacteria are exposed to a toxic compound, it trigger light and can be easily detected.

An *E. coli* strain has constructed which contained a promotor *alk* B from *Pseudomonas oleovorans* and *lux* AB genes of *Vibrio harveyi*. The cells showed light when exposed to alkanes and used to detect soil contamination by heating soil.

Bioluminescence was used to track genetically engineered *Xanthomonas compestris* after its release into the environment. The engineered strain of *X. compestris* contain lux gene which showed bioluminescence under specific conditions. This light emission detected by special cameras which provide real time measurement. Microbial cells are not only but recombinant mouse haptoma cell line containing luciferase gene under the control of cytochrome P_{450}, has been construct The *P. aeruginosa* strain was also tagged with *lac*ZY genes, encoding b-galactosidase and lactose permease, respectively, providing an additional phenotype; namely, blue colony formation on X-gal containing medium (Flemming, Leung, Lee, Trevors, & Greer, 1994). Similarly, a 2, 4-D degrading *Pseudomonas cepacia* strain, tagged with *lac*ZY and *lux*AB genes, was monitored by colony counting in 2, 4-D amended soil (Masson, Comeau, Brousseau, Samson, & Greer, 1993). The cytochrome P_{450} induced on the exposure to dioxins, PAHs and petrochemicals.

Immunoassays

The most popular immunochemical technique in environmental analyses today is immunoassay, which has been shown to detect and quantify many compounds of environmental interest such as pesticides, industrial chemicals, and products of xenobiotic metabolism. Basic immunoassays are performed by detection of a specific marker molecule immobilized either to antibody (Ab) or the antigen (Ag). Marker molecules may be in the form of fluorescent or chemi luminescent compounds, radioisotopes or enzymes. Enzyme-based immunoassay offer many advantages over other immune techniques, because of the great amount of product molecules, which results in signal amplification. The main enzymes used are horseradish peroxidase, alkaline phosphatase and β-galactosidase. A widely used immunoassay for environmental purposes is enzyme-linked immunosorbent assay (ELISA), which can be carried out according to different formats – direct competitive, indirect competitive or sandwich-type. Competitive assays are most common and can be performed in different ways. Analyte and the tracer (direct competitive ELISA) or analyte and the immobilised ligand (indirect ELISA) may compete for a limited number of binding sites. Sandwich-type ELISA is non-competitive assay, in which the analyte is recognised by two different antibodies immobilized Ab and marker Ab (Fare, & Barcelo, 2005; Harris, 1999).

Immunological techniques are useful for the detection of pollutants (pesticides, herbicides) and identification of pathogens that exhibit immunological properties. Different immunoassays are used for the detection of several pesticides e.g. aldrin, tri azines DDT, glyphosate. Immunoassays can be used to detect the metabolic products of certain bacteria. Antibodies have been the gold standard in molecular recognition for several decades and have been incorporated widely into biosensors and assays relating to food (Ricci, Volpe, Micheli, & Palleschi, 2007). For instance, assay systems have been developed for the detection of toxins of *Cholera* and *Salmonella*. In fact, assay techniques are available for detection of pesticide and herbicide contamination in water. *Salmonella* are Gram-negative bacteria naturally found in the gastrointestinal tract of warm blooded animals and humans (Nowak, Müffling, Chaunchom, & Hartung, 2007; Luo et al., 2009). Different microscopic fungi have been used for recombinant antibody production, including yeast species *Saccharomyces cerevisiae* and *Pichia pastoris* (Churchill, Sheedy, Yau, & Hall, 2002).

Toxicity Testing using Biological Materials

Despite, the toxic metal contaminants can cause an immeasurable damage in the environmental health especially affecting the vast array of microbial community. Microbes such as bacteria, protist, fungi, yeast, algae, etc. are omnipresent and play key geo active roles in the environment, particularly in element biotransformation and biogeochemical cycling, metal, mineral and nutrients transformations, decomposition, bio weathering, and soil and sediment formation processes (Gadd, 2010). In environment, the metal contaminants and its derivatives significantly interact with microorganisms and pose severe adverse impacts to the significant microbial communities under certain conditions (Giller, Witter, & McGrath, 1998; Bhakta, Ohnishi, Munekage, & Iwasaki, 2010; Bhakta, Munekage, Ohnishi, & Jana, 2012; Bhakta, Ohnishi, Munekage, Olaniran, 2010; Balgobind, & Pillay, 2013; Lenart-Boroń, & Boroń, 2014; Bhakta, Munekage, Ohnishi, Jana, & Balcazar, 2014; Kuperman, Siciliano, Römbke, & Oorts, 2014). Toxicity testing depends on measurements of the lethal effects of a compound on a specific biological system which is carried out with a single pollutant (Table 4).

Toxicity Testing Using Plants and Algae

The algae was introduced in 1914 for the toxicity testing by using *Selenastrum apricornutum* which is now popularly known as *Pesudokirchneriella subcaitaata* and now used to evaluate water quality. These testing methods can be represented by a number of trophic levels. The freshwater alga *Chlorella vulgaris, S. capricornatum* and *Senedesmus subspicatus* are used for trophic level 1. These tests involved the exposure of the algal culture to the pollutant for 3-4 days and the growth rate reduction calculates the EC_{50} value. The EC_{50} value is the concentration of the pollutant which reduces the growth about 50%. The EC_{50} value represents the level of toxicant which affects the environment. The results can be affected by the culture conditions such as test species, culture volume and light intensity.

In tropic level 2 the main organism is the water flea. The effect of pollutant is determined by the population of organism after 24 h of exposure. In trophic level 3 no test organisms exist due to their

Table 4. Molecular biology markers

Selection	Assay
Antibiotic resistance gene *npt II*	Selective plate, resistance to kanamycine
Heavy metal resistance gene *mer*	Selective plate, resistance to mercury
Chromogenic marker gene *lac* ZY	Plate assay with X –gal converted to a blue coloured product
Bacterial luciferase gene, *lux* AB	Bioluminescence with or without luciferin
Green luminescence gene *gfp*	Green fluorescence, flow cytometry

Huggett et al., 1992.

similarity in response to level 4 organisms. The final group, level 5 represents the top of the food chain and include the bird species.

Luminescent Organisms

Microtox TM, BiotoxTM and Tox Alert are system that uses the reduction in light emission by luminous bacteria such as *Photomicrobium fisceri* and *Photobacteriaum phosphoreum* as a measurement of toxicity of a compound over 15 minutes exposure period.

Ames Test

The test developed by Bruce Ames has found the widest acceptance (Ames, Mccann, & Yamasaki, 1975; Maron, & Ames, 1983). The test system proposed by Bruce Ames is commonly used for determining the genotoxicity of single chemical compounds and complex environmental samples (Adamiak, Jadczyk, & Kucharczyk, 1999; Berthi-Conti, Jacobi, Kleihauer, & Witte, 1998).

This test is used to determine the substance ability to cause mutation in bacteria. This test includes the treatment of *Salmonella typhimurium* mutant which requires amino acid histidine for growth (His^-) with the test compound along with an extract of a rat liver. The rat liver is added because many nontoxic compounds can be converted to mutagens by the enzyme present in mammalian liver which are not found in bacteria. If the compound is mutagen then reverse mutations occur and medium lacking histidine will form colonies. The number of colony gives a measure of the mutagenic potential.

Molecular Biology Biomarkers

The use of biological markers or biomarkers at the molecular and cellular level are considered as the first sentinels tools for sensitive effect in environmental quality or biotechnological processes assessment (Torre et al., 2005). Hence, a biomarker can be defined as the measure of the effect of pollutants and or the host response in biochemical, cellular and molecular terms (Bodin, Burgeot, Stanisie, Bocquene, Menard, 2004; Magni, Falco, Falugi, Franzoni, & Monteverde, 2005). Another development using genetic manipulation of biological material for toxicity determination is the generation of transgenic strain of the nematode *Caenorhabditis elegans.* The strain has *lac* Z gene from *E. coli* fused to the *hsp* 16 gene. The *lac* Z gene codes for the enzyme β galactosidase and the *hsp* 16 is an inducible gene, induced when the nematode undergoes in heat shock response. The enzyme can be detected by the addition of a substrate such as o-nitrophenyl- β-galactopyranoside (ONPG) which produces a blue colour when cleaved by the enzyme (Table 5).

Other example as biomarkers and bio-indicators are used to study the effect of pentachlorophenol. PCP is widely used as a wood preservative and because of its highly chlorinated nature is resistant to microbial attack. This compound is highly toxic thus require close monitoring.

Biosensors

A biosensor is defined as a self contained, integrated device, consisting of a biological recognition element interfaced to a physical signal transducer, that together reversibly respond to a chemical species

Table 5. Effect of pentachlorophenol on different biomarkers and bioindicators

System	Type	Indicator	Exposure Time (Hours)	EC_{50}
Chlorella vulgaris	Microalgae	Growth inhibition	24	59
Daphnia magna	Cladoceran	Immobilization	24	2.1
Allium cepa	Plant	Mitotic Index	24	10
RTC-2 Cell line	Rainbow trout	Neutral red uptake	24	90
Vero cell line	Monkey	Cell growth	24	34

Repetto et al., 2001.

in a concentration-dependant manner. A wider definition also includes some other forms of biological sensors, including genetically engineered microorganisms, which respond in observable ways to target analyte or group of related analytes (Rogers, 2006; Turner, Karube, & Wilson, 1987).

Biosensors essentially use the specific reaction of biological materials such as enzyme and antibodies to detect and quantify the molecules in complex mixtures. In biosensor the reaction is linked with a transducer that produces a signal proportional to the number of molecules present. Thus a biosensor can give online and continuous determination of the compound present. The biological material in biosensors may be enzyme, antibodies, hormone receptors, proteins, organelles, lectins, DNA and whole cells. The biological reaction can be converted into a signal by a very extensive range of transducers. In biosensor the biological material is immobilized and kept to close contact with the transducer (Figure 6). Different methods are used for immobilization which depends on the properties of materials to be immobilized. In the most of the biosensors described in the literature for environmental applications electrochemical transducers are used (Gong, & Dimerjian, 1997).

Figure 6. Basic components of a biosensor

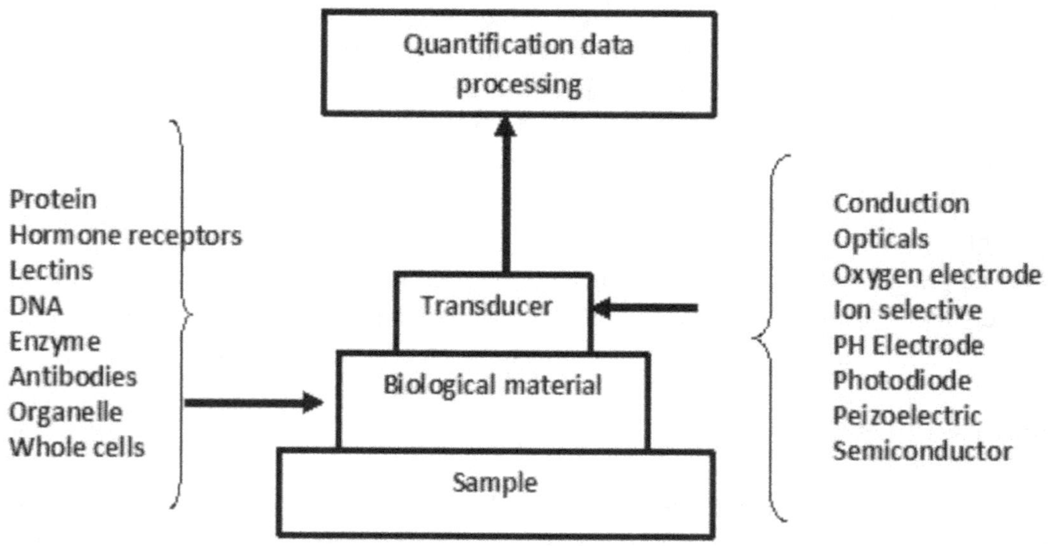

The advantage of biosensors for environmental analysis and monitoring include-

- Online sensing, real time monitoring.
- Fast response.
- Operation with complex mixtures.
- Very specific response.
- Response to a class of compounds, allowing the biosensors to monitor toxicity, mutagenicity and carcinogenicity.
- Highly sensitive.
- Lower cost and quicker operation than conventional assays.
- Portability in some cases.

The disadvantages of biosensors are that the biosensors cannot be sterilized and thus cannot be used in sterile bioreactor operation, the biosensor has a finite life and it can be poisoned by some compounds and metals (Hall, 1990). In beginning, biosensors are used to monitor the environment. The oxygen demand of wastewater has been determined traditionally by measuring the BOD5 value. This is standard method and requires only simple instrument and 5 days are required to get results. However, with BOD biosensor a fast online determination is possible.

Biosensors have been developed for the determination of pesticide concentration using the enzyme acetylcholinesterase and also combined with choline oxidase. Acetylcholinesterase converts acetylcholine to choline and choline oxidase converts the choline to betain and hydrogen peroxide. H_2O_2 formation can be monitored amperometrically or by an oxygen electrode. Pesticides inhibit the action of acetylcholinesterase and produce a reduction in peroxide formation, which can be used to estimate toxicity.

Metals such as lead and cadmium can be detected by their inactivation of oxidases and dehydrogenases linked again to an oxygen electrode. Thus biosensors can provide cheap, reliable and accurate monitoring of the environment which also gives real time analysis (Figure 7).

Gas Biosensors or Microbial Biosensors

Some microbial biosensors are used for the detection of gases as sulphur dioxide, methane and carbon dioxide. *Thiobacillus*-based biosensor can detect the pollutant SO_2, while methane (CH_4) can be detected by immobilized *Methalomonas*. For carbon dioxide monitoring, a particular strain of *Pseudomonas* is used. Recently, for example, a biosensor for nitrate monitoring has been constructed by transformation of *E. coli* with plasmid containing nitrate reductase operon fused to *gfp* reporter gene (Taylor, Bain, Richardson, Spiro, & Russell, 2004).

Recombinant DNA technology in biotechnology influence the environmental monitoring in the following areas-

- The technology allowed the study of microbial population in environmental samples and in situ without need of microbial cultures.
- RDT also influence the applications of bioindicators and biomarkers.
- Biomarkers using microorganisms with green fluorescent protein (gfp) gene inserted and used to monitor the pollution in the environment.
- RDT may be used to introduce the genetically modified organisms (GEM) into environment.
- Biosensors are able to determine the contaminant in the environment and perhaps provide real time analysis.

Figure 7. A simple biosensor based on a disposable print screen electrode

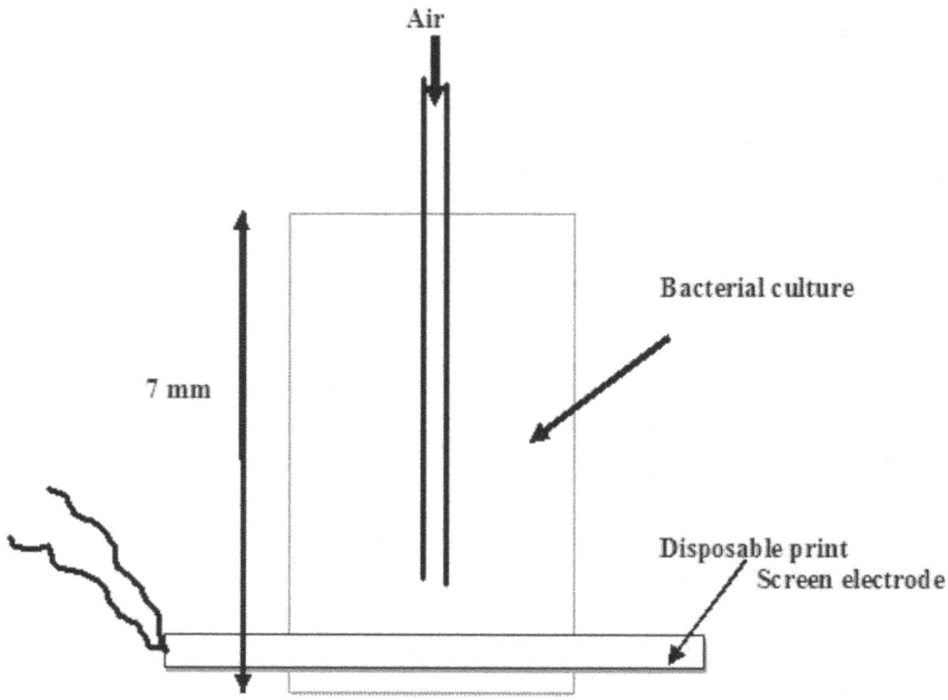

FUTURE PROSPECTS AND CONCLUSION

Bioremediation has the potential to restore contaminated environments inexpensively yet effectively. Lack of sufficient knowledge about the effect of various environmental factors on the rate and extent of biodegradation create a source of uncertainty. It is important to point out that many field tests have not been correctly designed, well controlled, or properly analyzed, leading to uncertainty when selecting response options. Hence, future field studies should invest serious efforts adopting scientifically legitimate approaches and acquiring the highest quality data possible. Although it is a useful component in determination of the effectiveness of remediation, it stops short of estimating potential health risk of the remediation products. Despite these limitations, several points can be drawn from these remediation assessment evaluations.

- First, it is important to evaluate the biological activity of products of remediation and also of stable intermediates produced during remediation.
- Another important point to be made is that a better understanding of the mechanisms of biological effects of contaminants will permit a more directed approach to evaluation of the activity of the remediation products. The selection of bioassays to be used as well as the specific details of experimental design can be based on known mechanisms of action of the parent compounds.
- Finally, knowledge of the spectrum of biological activities associated with remediated chemicals and their stable intermediates will provide the basis for more accurate risk assessment and guide remediation needs and approaches.

Future needs in this area of research include assays to evaluate biological effects under conditions of exposure that mimic those found in the environment and models to extrapolate effects to assess risk to people and wildlife.

REFERENCES

Adamiak, W., Jadczyk, P., & Kucharczyk, J. (1999). Application of *Salmonella* strains with altered nitroreductase and O-acetyl transferase activities to the evaluation of the mutagenicity of airborne particles. *Acta Microbiologica Polonica*, *48*(2), 131–140. PMID:10581669

Agarwal, S. K. (1998). Environmental Biotechnology. APH Publishing Corporation.

Ames, B. M., Mccann, J., & Yamasaki, E. (1975). Methods for detecting carcinogens and mutagens with the *Salmonella* / mammalian-microsome mutagenicity test. *Mutation Research*, *31*(6), 347–364. doi:10.1016/0165-1161(75)90046-1 PMID:768755

Atlas, R. M., & Bartha, R. (1993). *Microbial Ecology*. Harlow: Addison Wesley Longmann.

Benzie, J., Glockner, F., Rodríguez-Ezpeleta, N., Faith, D., Bean, T., & Obst, M. (2013). Genomics in marine monitoring: New opportunities for assessing marine health status. *Marine Pollution Bulletin*, *74*(1), 19–31. doi:10.1016/j.marpolbul.2013.05.042 PMID:23806673

Berthe-Conti, L., Jacobi, H., Kleihauer, S., & Witte, L. (1998). Cytotoxicity and mutagenicity of a 2,4,6 trinitrotoluene (TNT) and hexogen contaminated soil in *S. typhimurium* and mammalian cells. *Chemosphere*, *37*(2), 209–218. doi:10.1016/S0045-6535(98)00039-3 PMID:9650264

Bhakta, J. N., Munekage, Y., Ohnishi, K., & Jana, B. B. (2012). Isolation and identification of cadmium and lead resistant lactic acid bacteria for applying as metal removing probiotic. *International Journal of Environmental Science and Technology*, *9*(3), 433–440. doi:10.1007/s13762-012-0049-3

Bhakta, J. N., Munekage, Y., Ohnishi, K., Jana, B. B., & Balcazar, J. L. (2014). Isolation and characterization of cadmium and arsenic absorbing bacteria for bioremediation. *Water, Air, and Soil Pollution*, *225*(10), 2151. doi:10.1007/s11270-014-2151-2

Bhakta, J. N., Ohnishi, K., Munekage, Y., & Iwasaki, K. (2010). Isolation and probiotic characterization of arsenic-resistant lactic acid bacteria for uptaking arsenic. *International Journal of Chemical and Biological Engineering*, *3*(4), 167–174.

Bodin, N., Burgeot, T., Stanisie, J. Y., Bocquene, G., & Menard, D. (2004). Seasonal variations of a battery of biomarkers and physiological indices for the mussel *Mytilus galloprovincialis* transplanted into the northwest Mediterranean Sea. *Biochemistry Physiology and Toxicology Pharmacology*, *13*(4), 411–427. doi:10.1016/j.cca.2004.04.009 PMID:15536049

Bourlat, S., Borja, A., Taylor, M., Weisberg, S., Griffith, J., Lettieri, T., & Martyniuk, C. et al. (2012). Advancing the Omics in aquatic toxicology: SETAC North America 31st Annual Meeting. *Ecotoxicology and Environmental Safety*, *76*, 1–2. doi:10.1016/j.ecoenv.2011.11.001 PMID:22114953

Churchill, R. L. T., Sheedy, C., Yau, K. Y. F., & Hall, C. (2002). Micro-flow-injection analysis (MFIA) immunoassay of herbicide residue 2, 6-dichlorobenzamide – towards automated at-line monitoring using modular microfluidics. *Analytica Chimica Acta*, *468*, 185–194. doi:10.1016/S0003-2670(02)00093-4

Farre, M., Brix, R., & Barcelo, D. (2005). Screening water for pollutants using biological techniques under European Union funding during the last 10 years. *Trends in Analytical Chemistry*, *24*(6), 532–545. doi:10.1016/j.trac.2005.03.008

Flemming, C. A., Leung, K. T., Lee, H., Trevors, J. T., & Greer, C. W. (1994). Survival of lux-lac-marked biosurfactant-producing *Pseudomonas aeruginosa* UG2L in soil monitored by nonselective plating and PCR. *Applied and Environmental Microbiology*, *60*, 1606–1613. PMID:8017939

Gadd, G. M. (2010). Metals, minerals and microbes: Geomicrobiology and bioremediation. *Microbiology*, *156*(3), 609–643. doi:10.1099/mic.0.037143-0 PMID:20019082

Giller, K., Witter, E., & Mc. Grath, S. P. (1998). Toxicity of heavy metals to microorganisms and microbial processes in agricultural soils: A review. *Soil Biology & Biochemistry*, *30*(10–11), 1389–1414. doi:10.1016/S0038-0717(97)00270-8

Gong, K., & Demerjian, K. L. (1997). Measurement and analysis of hydrocarbons at White face Mountain, New York. *Geophysics Chemistry.*, *102*(D23), 28059–28069. doi:10.1029/97JD02703

Gustafsson, K., & Jansson, J. K. (1993). Ecological risk assessment of the deliberate release of genetically modified microorganisms. *Applied and Environmental Microbiology*, *22*, 236–242.

Hall, E. A. H. (1990). *Biosensors* (pp. 250–262). Milton Keynes, UK: Open University Press.

Hanson, N., & Larsson, A. (2008). Fixed wavelength fluorescence to detect PAH metabolites in fish bile: Increased statistical power with an alternative dilution method. *Environmental Technology*, *24*(2), 200–210. PMID:17899412

Harris, J. (1999). RAPDs in systematics-a useful methodology? In P. M. Hollingsworth, R. M. Bateman, & R. J. Gornall (Eds.), *Molecular systematics and plant evolution* (pp. 221–228). London: Taylor and Francis. doi:10.1201/9781439833278.ch11

Hess, A., Zarda, B., Hahn, D., Hanner, A., & Stax, D. (1997). In situ analysis of denitrifying toluene and mxylene degrading bacteria in a diesel fuel contaminated laboratory aquifer column. *Journal Applied Enviromental Microbiology*, *63*, 2136–2141. PMID:9172330

Huggett, R. J., Kimerle, R. A., Mehrle, P. M., & Bergman, H. J. (1992). Biomarkers. SETAC Special Publications Series, Lewis Publishers, Ann Arbor. K.R. Rogers. *Analytica Chimica Acta*, *568*, 222–226.

Jansson, J. K. (1995). Tracking genetically engineered microorganisms in nature. *Current Opinion in Biotechnology*, *6*(3), 275–283. doi:10.1016/0958-1669(95)80048-4 PMID:7780181

Kuperman, R. G., Siciliano, S. D., Römbke, J., & Oorts, K. (2014). Deriving Site-Specific Soil Clean-Up Values for Metals and Metalloids: Rationale for Including Protection of Soil Microbial Processes. *Integrated Environmental Assessment and Management*, *10*(3), 388–400. doi:10.1002/ieam.1513 PMID:24376192

Lenart-Boroń, A., & Boroń, P. (2014). The Effect of Industrial Heavy Metal Pollution on Microbial Abundance and Diversity in Soils-A Review. In M. C. Hernandez-Soriano (Ed.), *Environmental Risk Assessment of Soil Contamination* (pp. 759–783). doi:10.5772/57406

Luo, J., Liu, X., Tian, Q., Yue, W., Zeng, J., Chen, G., & Cai, X. (2009). Disposable marine monitoring: New opportunities for assessing marine health status. *Marine Pollution Monitoring and Assessment*, *144*, 221–228.

Magni, P., Falco, D., Falugi, G., Franzoni, C., & Monteverde, M. (2005). Genotoxicity biomarkers and acetyl V cholinesterase activity in natural populations of *Mytilus galloprovincialis* along a pollution gradient in the Gulf of Oristano (Sardinia, western Mediterranean). *Environmental Pollution*, *142*(1), 65–72. doi:10.1016/j.envpol.2005.09.018 PMID:16289776

Maron, D. M., & Ames, B. N. (1983). Methods for *Salmonella* mutagenicity tests. *Mutation Research*, *113*(3-4), 173–215. doi:10.1016/0165-1161(83)90010-9 PMID:6341825

Masson, L., Comeau, Y., Brousseau, R., Samson, R., & Greer, C. (1993). Construction and application of chromosomally integrated lac-lux gene markers to monitor the fate of a 2,4-dichlorophenoxyacetic acid-degrading bacterium in contaminated soils. *Microbial Releases*, *1*, 209–216. PMID:7506623

Nowak, B., Müffling, T., Chaunchom, S., & Hartung, J. (2007). *Salmonella* contamination in pigs at slaughter and on the farm: A field study using an antibody ELISA test and PCR techniques. *International Journal of Food Microbiology*, *115*(3), 259–267. doi:10.1016/j.ijfoodmicro.2006.10.045 PMID:17292500

Olaniran, A. O., Balgobind, A., & Pillay, B. (2013). Bioavailability of Heavy Metals in Soil: Impact on Microbial Biodegradation of Organic Compounds and Possible Improvement Strategies. *International Journal of Molecular Sciences*, *14*(5), 10197–10228. doi:10.3390/ijms140510197 PMID:23676353

Reeve, R. N. (1994). *Environmental Analysis* (3rd ed.). Chichester, UK: John Wiley.

Ricci, F., Volpe, G., Micheli, L., & Palleschi, G. (2007). A review on novel developments and applications of immunosensors in food analysis. *Annals in Chimmica Acta, 605*(2), 111-129.

Tang, C. Y., Criddle, Q. S., & Leckie, J. O. (2007). Effect of flux (transmembrane pressure) and membranes properties on fouling and rejection of reverse osmosis and nanofiltration membranes treating perfluorooctane sulfonate containing waste water. *Journal of Environmental Science & Technology*, *41*(6), 2008–2014. doi:10.1021/es062052f PMID:17410798

Taylor, C. J., Bain, L. A., Richardson, D. J., Spiro, S., & Russell, D. A. (2004). Construction of a whole-cell gene reporter for the fluorescent bioassay of nitrate. *Analytical Biochemistry*, *328*(1), 60–66. doi:10.1016/j.ab.2004.01.013 PMID:15081908

Tebbut, H. Y. (1998). *Principles of Water Quality Control* (5th ed.). Butterworth-Heineman Oxford.

Torre, F. R., Ferrari, L., & Salibia, N. A. (2005). Biomarkers of a native fish species (*Cnesterodon decemmaculatus*) application to the water toxicity assessment of a peri-urban polluted river of Argentina. *Chemosphere*, *59*(4), 577–583. doi:10.1016/j.chemosphere.2004.12.039 PMID:15788181

Turner, A. P. F., Karube, I., & Wilson, G. S. (1987). *Biosensors: Fundamentals and Applications*. Oxford, UK: Oxford University Press.

Wernersson, A. S. (2012). Swedish monitoring of hazardous substances in the aquatic environment current, required monitoring and potential developments. *Rapport*, *2*, 23.

Winegar, E. D., & Keith, L. H. (1993). *Sampling and analysis of air borne pollutants*. London: Lewis Publications Sciences, Boka-Raton –Ann Arbor.

KEY TERMS AND DEFINITIONS

Ames Test: A microbiological assay that is used to assess the mutagenic potential of chemical compounds.

Bio-Monitoring: Monitoring activities in human beings, using biomarkers, that focus on environmental exposures, diseases and/or disorders and genetic susceptibility, and their potential relationships.

Denaturing Gradient Gel Electrophoresis (DGGE): It is a kind of electrophoresis method for molecular fingerprinting that separates polymerase chain reaction (PCR)-generated DNA products on the basis of sequence differences that results in differential denaturing characteristics of the DNA.

Enzyme-linked immunosorbent assay (ELISA): It is a test that uses antibodies and colour change to identify a substance. After the antigen is immobilized, the detection antibody is added, forming a complex with the antigen.

Fluorescence in Situ Hybridisation (FISH): is a cytogenetic technique that uses fluorescent probes that bind to only those parts of the chromosome with a high degree of sequence complementarity.

Matrix-Assisted Laser Desorption Ionization (MALDI): It is a soft ionizing technique used in mass spectrometry, allowing the analysis of biomolecules (DNA, proteins, peptide) and large organic molecules (polymers and dendrimers) which tend to be fragile and fragment when ionized by more conventional ionizing methods.

Metallothioneins (MTs): It is a low molecular weight (6-7 kDa), cystine-rich proteins encoded by the Smt locus and essentially play key roles in metal detoxification process in many microorganisms first discovered in *Synechococcus* PCC 7942.

Microbial Metal-Sequestration: Microbe sequesters metals by various mechanisms such as, binding, crystallization, precipitation, metal and mineral nanoparticle deposition, complexion, chemical transformation, volatilization, etc., to tackle metal toxicity.

Xenobiotic: Xenobiotics are manmade foreign organic compounds that are found as pollutants in the natural environment.

Bio-Marker: Substances (molecules) used as indicators of a biologic state – i.e. as an indicator of normal or pathogenic processes or of biological response to an external agent.

Compilation of References

Abbasian, F., Palanisami, T., Megharaj, M., Naidu, R., Lockington, R., & Ramadass, K. (2016). Microbial diversity and hydrocarbon degrading gene capacity of a crude oil field soil as determined by metagenomics analysis. *Biotechnology Progress*, *32*(3), 638–648. doi:10.1002/btpr.2249 PMID:26914145

Abbasi, H., Hamedi, M. M., Lotfabad, T. B., Zahiri, H. S., Sharafi, H., Masoomi, F., & Noghabi, K. A. (2012). Biosurfactant-producing bacterium, *Pseudomonas aeruginosa* MA01 isolated from spoiled apples: Physicochemical and structural characteristics of isolated biosurfactant. *Journal of Bioscience and Bioengineering*, *113*(2), 211–219. doi:10.1016/j.jbiosc.2011.10.002 PMID:22036074

Abd El-Zaher, E. H. F., Mahmoud, Y. A.-G., & Magda, M. A. (2011). Effect of different concentrations of phenol on growth of some fungi isolated from contaminated soil. *African Journal of Biotechnology*, *8*, 1384–1392.

Abd-El-Haleem, D., Beshay, U., Abdelhamid Abdu, O., Moawad, H., & Zaki, S. (2003). Effects of mixed nitrogen sources on biodegradation of phenol by immobilized *Acinetobacter* sp. strain W-17. *African Journal of Biotechnology*, *2*(1), 8–12. doi:10.5897/AJB2003.000-1002

Abdelkafi, S., Chambkha, M., Casalot, L., Sayadi, S., & Labat, M. (2005). Isolation and characterization of a novel *Bacillus* sp., strain YAS1, capable of transforming tyrosol under hypersaline conditions. *FEMS Microbiology Letters*, *252*(1), 79–84. doi:10.1016/j.femsle.2005.08.032 PMID:16165329

Abdulrasaq, O. O., & Basiru, O. G. (2010). Removal of Copper (II), Iron (III) and Lead (II) ions from mono component stimulated waste effluent by adsorption on coconut husk. *African Journal of Environmental Science and Technology*, *4*(6), 382–387. doi:10.5897/AJEST09.224

Abed, R. M. M., & Köster, J. (2005). The direct role of aerobic heterotrophic bacteria associated with cyanobacteria in the degradation of oil compounds. *International Biodeterioration & Biodegradation*, *55*(1), 29–37. doi:10.1016/j.ibiod.2004.07.001

Abou-El-Souod, G., & El-Sheekh, M. M. (2016). Biodegradation of basic fuchsin and methyl red by the blue green algae *Hydrocoleum oligotrichum* and *Oscillatoria limnetica*. *Environmental Engineering and Management Journal*, *15*(2), 279–286.

Abou-shanab, R. A. I., El-Dalatony, M., El-Sheekh, M. M., Ji, M.-K., Salam, E., Kabra, A., & Jeon, B.-H. (2014). Cultivation of a new microalga, *Micractinium reisseri*, in municipal wastewater for nutrient removal, biomass, lipid, and fatty acid production. *Biotechnology Bioprocess Engineering*, *19*(3), 510–518. doi:10.1007/s12257-013-0485-z

Achal, V., & Pan, X. (2011). Characterization of urease and carbonic anhydrase producing bacteria and their role in calcite precipitation. *Current Microbiology*, *62*(3), 894–902. doi:10.1007/s00284-010-9801-4 PMID:21046391

Adamiak, W., Jadczyk, P., & Kucharczyk, J. (1999). Application of *Salmonella* strains with altered nitroreductase and O-acetyl transferase activities to the evaluation of the mutagenicity of airborne particles. *Acta Microbiologica Polonica, 48*(2), 131–140. PMID:10581669

Adams, G. O., Fufeyin, P. T., Okoro, S. E., & Ehinomen, I. (2015). Bioremediation, biostimulation and bioaugmention: A review. *International Journal of Environmental Bioremediation & Biodegradation, 3*(1), 28–39.

Adenipekun, C. O., & Lawal, R. (2012). Uses of mushrooms in bioremediation: A review. *Biotechnology and Molecular Biology Review, 7*(3), 62–68.

Adishkumar, S., Sivajothi, S., & Banu, J. R. (2012). Coupled solar photofenton process with aerobic sequential batch reactor for treatment of pharmaceutical wastewater. *Desalination and Water Treatment, 48*(1-3), 89–95. doi:10.1080/19443994.2012.698799

Adolfsson-Erici, M., Pettersson, M., Parkkonen, J., & Sturve, J. (2002). Triclosan, a commonly used bactericide found in human milk and in the aquatic environment in Sweden. *Chemosphere, 46*(9-10), 1485–1489. doi:10.1016/S0045-6535(01)00255-7 PMID:12002480

Adriano, D. C. (2003). *Trace elements in the terrestrial environment biogeochemistry, bioavailability and risks of heavy metals*. New York: Springer-Verlag.

Adu, J. K., & Oades, J. M. (1978). Utilization of organic materials in soil aggregates by bacteria and fungi. *Soil Biology & Biochemistry, 10*(2), 117–122. doi:10.1016/0038-0717(78)90081-0

Affar, E. B., Dufour, M., Poirier, G. G., & Nadeau, D. (1998). Isolation, purification and partial characterization of chloragocytes from the earthworm species *Lumbricus terrestris. Molecular and Cellular Biochemistry, 185*(1/2), 123–133. doi:10.1023/A:1006882207581 PMID:9746218

Agarry, S. E., Durojaiye, A. O., & Solomon, B. O. (2008). Microbial degradation of phenols: A review. *International Journal of Environment and Pollution, 32*(1), 12–28. doi:10.1504/IJEP.2008.016895

Agarry, S. E., & Solomon, B. O. (2008). Kinetics of batch microbial degradation of phenols by indigenous *Pseudomonas fluorescence. International Journal of Environmental Science and Technology, 5*(2), 223–232. doi:10.1007/BF03326016

Agarwal, S. K. (1998). Environmental Biotechnology. APH Publishing Corporation.

Agarwal, S. K. (1998). *Environmental Biotechnology*. New Delhi, India: APH Publishing Corporation.

Agdag, O. N., & Sponza, D. T. (2005). Anaerobic/aerobic treatment of municipal landfill leachate in sequential two-stage up-flow anaerobic sludge blanket reactor (UASB)/completely stirred tank reactor (CSTR) systems. *Process Biochemistry, 40*(2), 895–902. doi:10.1016/j.procbio.2004.02.021

Agency for Toxic Substances and Disease Registry. (2006). *Case studies in environmental medicine*. Academic Press.

Agger, J. W., Isaksen, T., Varnai, A., Vidal-Melgosa, S., Willats, W. G. T., Ludwig, R., & Westereng, B. et al. (2014). Discovery of LPMO activity on hemicelluloses shows the importance of oxidative processes in plant cell wall degradation. *Proceedings of the National Academy of Sciences of the United States of America, 111*(17), 6287–6292. doi:10.1073/pnas.1323629111 PMID:24733907

Aglave, B. A., Jagtap, K. U., Kalal, P. R., & Lokhande, M. O. (2009). Helping the ecosystem through mycoremediation. *Asian Sciences, 4*, 100–101.

Agnello, A., Bagard, M., van Hullebusch, E., Esposito, G., & Huguenot, D. (2016). Comparative bioremediation of heavy metals and petroleum hydrocarbons co-contaminated soil by natural attenuation, phytoremediation, bioaugmentation and bioaugmentation-assisted phytoremediation. *The Science of the Total Environment*, *563-564*, 693–703. doi:10.1016/j.scitotenv.2015.10.061 PMID:26524994

Agosin, E., & Odier, E. (1985). Solid-state fermentation, lignin degradation and resulting digestibility of wheat straw fermented by selected whiterot fungi. *Applied Microbiology and Biotechnology*, *21*(6), 397–403. doi:10.1007/BF00249988

Agrawal, S. P., Subramanian, R., Tripathi, R., Rautela, M. S., & Kumar, S. (2010). A case study of bioremediation of contaminated soil and waste water in upper Assam. Petrotech-2010, New Delhi, India.

Aguila, S., Vazquez-Duhalt, R., Tinoco, R., Rivera, M., Pecchi, G., & Alderete, J. B. (2008). Stereoselective oxidation of R-(+)-limonene by chloroperoxidase from *Caldariomyces fumago. Green Chemistry*, *10*(6), 647–653. doi:10.1039/b719992a

Aguilera, F., Méndez, J., Pásaro, E., & Laffon, B. (2010). Review on the effects of exposure to spilled oils on human health. *Journal of Applied Toxicology*, *30*(4), 291–301. PMID:20499335

Ahalya, N., Ramachandra, T. V., & Kanamadi, R. D. (2003). Biosorption of heavy metals. *Research Journal of Chemistry and Environment*, *7*(4), 71–79.

Ahemad, M. (2012). Implications of bacterial resistance against heavy metals in bioremediation: A review. *Institute of Integrative Omics and Applied Biotechnology Journal*, *3*, 39–46.

Ahemad, M., & Malik, A. (2011). *Bioaccumulation of heavy metals by Zinc resistant bacteria isolated from agricultural soils irrigated with waste water.* Bacteriology Journal; doi:10.3923/bj.2011

Ahire, K. C., Kapadnis, B. P., Kulkarni, G. J., Shouche, Y. S., & Deopurkar, R. L. (2012). Biodegradation of tributyl phosphate by novel bacteria isolated from enrichment cultures. *Biodegradation*, *23*(1), 165–176. doi:10.1007/s10532-011-9496-7 PMID:21755325

Ahluwalia, S. S., & Goyal, D. (2006). Microbial and plant derived biomass for removal of heavy metals from wastewater. *Bioresource Technology*, *98*(12), 2243–2257. doi:10.1016/j.biortech.2005.12.006 PMID:16427277

Ahluwalia, S. S., & Goyal, D. (2013). Microbial waste biomass for removal of chromium (VI) from chrome effluent. *Bioremediation Journal*, *17*(3), 190–199. doi:10.1080/10889868.2013.807770

Ahmed, F., Khan, A. U., & Yasser, A. (2013). Comparative phycoremediation of sewage water by various species of algae. *Proceeding of Pakistan Academy of Science*, *50*(2), 131–139.

Ahn, P., & Kim, B. (2016). Adsorption of heavy metals using chemically modified tea leaves. *International Journal of Biotechnology and Bioengineering*, *3*(1), 342.

Ahuja, S. K., Ferreira, G. M., & Moreira, A. R. (2004). Utilization of enzymes for environmental applications. *Critical Reviews in Biotechnology*, *24*(2-3), 125–154. doi:10.1080/07388550490493726 PMID:15493529

Aiking, H., Govers, H., & Van't Riet, J. (1985). Detoxification of mercury, cadmium, and lead in Klebsiella aerogenes NCTC 418 growing in continuous culture. *Applied and Environmental Microbiology*, *50*(5), 1262–1267. PMID:3911898

Ajao, A. T., Kannan, M., Yakubu, S. E., Umoh, V. J., & Ameh, J. B. (2012). Homology modeling, simulation and molecular docking studies of catechol-2, 3-Dioxygenase from *Burkholderia cepacia:* Involved in degradation of Petroleum hydrocarbons. *Bioinformation*, *8*(18), 848–854. doi:10.6026/97320630008848 PMID:23144539

Aken, B. V., Stahl, J. D., Naveau, H., Agathos, S. N., & Aust, S. D. (2000). Transformation of 2, 4, 6-trinitrotoluene (TNT) reduction products by lignin peroxidase (H8) from the white-rot basidiomycete Phanerochaete chrysosporium. *Bioremediation Journal*, *4*(2), 135–145. doi:10.1080/10889860091114194

Akhundova, E., & Atakishiyeva, Y. (2015). *Interaction Between Plants and Biosurfactant Producing Microorganisms in Petroleum Contaminated Absheron Soils. In Phytoremediation for Green Energy* (pp. 115–122). Springer.

Akin, D. E., Rigsby, L. L., Sethuraman, A., Morrison, W. H. III, Gamble, G. R., & Eriksson, K. E. L. (1995). Alterations in structure, chemistry, and biodegradability of grass lignocellulose treated with the white rot fungi *Ceriporiopsis subvermispora* and *Cyathus stercoreus*. *Applied and Environmental Microbiology*, *61*, 1591–1598. PMID:7747973

Akinola, O. O. (2014). Review of the role of plant in carbondioxide sequestration globally using chlorophy II or leaf index. *J. of Environ & Earth Sci*, *4*, 22–30.

Akunna, J. C., & Jefferies, C. (2000). Performance of family-size sequencing batch reactor and rotating biological contactor units treating sewage at various operating conditions. *Water Science and Technology*, *41*(1), 97–104.

Alarcón, A., & Zuberer, D. A. (2008). Arbuscular mycorrhiza and petroleum-degrading microorganisms enhance phytoremediation of petroleum-contaminated soil. *International Journal of Phytoremediation*, *10*(4), 251–263. doi:10.1080/15226510802096002 PMID:19260211

Al-Awadhi, H., Sulaiman, R. H., Mahmoud, H. M., & Radwan, S. S. (2007). Alkaliphilic and halophilic hydrocarbon-utilizing bacteria from Kuwaiti coasts of the Arabian Gulf. *Applied Microbiology and Biotechnology*, *77*(1), 183–186. doi:10.1007/s00253-007-1127-1 PMID:17710391

Al-Bahry, S. N., Al-Wahaibi, Y. M., Elshafie, A. E., Al-Bemani, A. S., Joshi, S. J., Al-Makhmari, H. S., & Al-Sulaimani, H. S. (2013). Biosurfactant production by *Bacillus subtilis* B20 using date molasses and its possible application in enhanced oil recovery. *International Biodeterioration & Biodegradation*, *81*, 141–146. doi:10.1016/j.ibiod.2012.01.006

Al-Baldawi, A., Abdullah, S. R. S., Anuar, N., Suja, F., & Mushrifah, I. (2015). Phytodegradation of total petroleum hydrocarbon (TPH) in diesel-contaminated water using Scirpus grossus. *Ecological Engineering*, *74*, 463–473. doi:10.1016/j.ecoleng.2014.11.007

Alban, A., David, S. O., Bjorkesten, L., Andersson, C., Sloge, E., Lewis, S., & Currie, I. (2003). A novel experimental design for comparative two-dimensional gel analysis: Two-dimensional difference gel electrophoresis incorporating a pooled internal standard. *Proteomics*, *3*(1), 36–44. doi:10.1002/pmic.200390006 PMID:12548632

Alberts, B., Bray, D., Hopkin, K., Johnson, A., Lewis, J., Raff, M., & Walter, P. et al. (2013). *Essential Cell Biology*. Garland Science.

Albrecht-Gary, A. M., & Crumbliss, A. L. (1998). Coordination chemistry of siderophores: thermodynamics and kinetics of iron chelation and release. *Metal Ions in Biological Systems, 35*(2), 39–327.

Alfawwaz, A. T., Jacob, A. J., & Al-Wahishe, T. E. (2016). Bioremoval capacity of phenol by green micro-algal and fungal species isolated from dry environment. *International Journal of Scientific & Technology Research*, *5*(2), 155–160.

Al-Harazin, I. M., Nakhla, G. F., & Farooq, S. (1991). Start-up of sequencing batch reactors for toxic wastewater treatment. *Journal of Environmental Science and Health. Part A, Environmental Science and Engineering & Toxic and Hazardous Substance Control*, *26*(5), 673–687. doi:10.1080/10934529109375662

Ali, S. M., Khalid, A. R., & Majid, R. M. (2014). The removal of Zinc, Chromium and Nickel from industrial waste water using corn cob. *Iraqi Journal of Science*, *55*(1), 123–131.

Al-Nasrawi, H. (2012). Biodegradation of crude oil by fungi isolated from Gulf of Mexico. *Journal of Bioremediation & Biodegradation.*

Altaf, M. S., & Ali, T. A. (2010). Wastewater treatment using sequential batch reactor and development of microbiological method for the analysis of relative toxicity. *Pakistan Journal of Nutrition, 9*(6), 574–576. doi:10.3923/pjn.2010.574.576

Alva, V., & Peyton, B. M. (2003). Phenol and catechol biodegradation by the haloalkaliphile *Halomonas campisalis*: Influence of pH and salinity. *Environmental Science & Technology, 37*(19), 4397–4402. doi:10.1021/es0341844 PMID:14572091

Amann, R. I., Ludwig, W., & Schleifer, K. H. (1995). Phylogenetic identification and in situ detection of individual microbial cells without cultivation. *Microbiological Reviews, 59*(1), 143–169. PMID:7535888

Ambure, P., & Roy, K. (2016). Scoring functions in docking experiments. In *Methods and Algorithms for Molecular Docking-Based Drug Design and Discovery* (pp. 54–98). IGI Global. doi:10.4018/978-1-5225-0115-2.ch003

Amellal, N., Portal, J.-M., Vogel, T., & Berthelin, J. (2001). Distribution and location of polycyclic aromatic hydrocarbons (PAHs) and PAH-degrading bacteria within polluted soil aggregates. *Biodegradation, 12*(1), 49–57. doi:10.1023/A:1011909107858 PMID:11693295

American Water Works Association Research and Technical Practice Committee on Organic Contaminant. (1981). An assessment of microbial activity on GAC. *Journal - American Water Works Association, 73*, 447–454.

Ames, B. M., Mccann, J., & Yamasaki, E. (1975). Methods for detecting carcinogens and mutagens with the *Salmonella* / mammalian-microsome mutagenicity test. *Mutation Research, 31*(6), 347–364. doi:10.1016/0165-1161(75)90046-1 PMID:768755

Amodu, O. S., Ntwampe, S. K., & Ojumu, T. V. (2014). Emulsification of Hydrocarbons by Biosurfactant: Exclusive Use of Agrowaste. *BioResources, 9*(2), 3508–3525. doi:10.15376/biores.9.2.3508-3525

Anand, A., & Satyanarayana, T. (2012). Applicability of carboxydotrophic bacterial Cmonoxide dehydrogenase (CODH) in Csequestration and bioenergy generation. *Journal of Scientific and Industrial Research, 71*, 381–384.

Anastasi, A., Tigini, V., & Varese, G. C. (2013). The bioremediation potential of different ecophysiological groups of fungi. In E. M. Goltapeh, Y. R. Danesh, & A. Varma (Eds.), Fungi as Bioremediators, Soil Biology 32 (pp. 29-49). Springer. doi:10.1007/978-3-642-33811-3_2

Anawar, H. M., Garcia-Sanchez, A., Kul Alam, M. T., & Rahman, M. M. (2008). Phytofiltration of water polluted with arsenic and heavy metals. *International Journal of Environment and Pollution, 33*(2-3), 292–312. doi:10.1504/IJEP.2008.019400

Andersen, H., Siegrist, H., Halling-Sorensen, B., & Ternes, T. A. (2003). Fate of estrogens in a municipal sewage treatment plant. *Environmental Science & Technology, 37*(18), 4021–4026. and doi:10.1021/es026192a

Anderson, T. R., & Tang, K. W. (2010). Carbon cycling and POC turn-over in the mesopelagic zone of the ocean: Insights from a simple model. *Deep-sea Research. Part II, Topical Studies in Oceanography, 57*(16), 1581–1592. doi:10.1016/j.dsr2.2010.02.024

Andreini, C., Bertini, I., & Rosato, A. (2004). A hint to search for metalloproteins in gene banks. *Bioinformatics (Oxford, England), 20*(9), 1373–1380. doi:10.1093/bioinformatics/bth095 PMID:14962940

Andrews, G. F., & Tien, C. (1981). Bacterial film growth in adsorbent surfaces. *AIChE Journal. American Institute of Chemical Engineers, 27*(3), 396–403. doi:10.1002/aic.690270309

Angayarkanni, J., Palaniswamy, M., & Swaminathan, K. (2003). Biotreatment of distillery effluent using *Aspergillus niveus*. *Bulletin of Environmental Contamination and Toxicology*, *70*(2), 268–277. and doi:10.1007/s00128-002-0187-2

Angelini, V. A., Orejas, J., Medina, M. I., & Agostini, E. (2011). Scale up of 2,4-dichlorophenol removal from aqueous solutions using *Brassica napus* hairy roots. *Journal of Hazardous Materials*, *85*(1), 269–274. and doi:10.1016/j.jhazmat.2010.09.028

An, J., Totrov, M., & Abagyan, R. (2005). Pocketome via comprehensive identification and classification of ligand binding envelopes. *Molecular & Cellular Proteomics*, *4*(6), 752–761. doi:10.1074/mcp.M400159-MCP200 PMID:15757999

Anjem, A., & Imlay, J. A. (2012). Mononuclear iron enzymes are primary targets of hydrogen peroxide stress. *The Journal of Biological Chemistry*, *287*(19), 15544–15556. doi:10.1074/jbc.M111.330365 PMID:22411989

Annachhatre, A. P., & Gheewala, S. H. (1996). Biodegradation of chlorinated phenolic compounds. *Biotechnology Advances*, *14*(1), 35–56. doi:10.1016/0734-9750(96)00002-X PMID:14536923

Annaduari, G., Rajesh, B. S., Mahesh, K. P. O., & Murugesan, T. (2000b). Absorption and biodegradation of phenol of Chitosan immobilised *Pseudomonas putida* (Mcm. 2174). *Bioprocess Engineering*, *22*(6), 493–501. doi:10.1007/s004499900092

Annadurai, G., Balan, M. S., & Murugesan, T. (2000a). Design of experiments in the biodegradation of phenol using immobilized *Pseudomonas pictorium* (NICM- 2077) on activated carbon. *Bioprocess Engineering*, *22*, 101–107.

Annemieke, V. (2013). A thready affair: Linking fungal diversity and community dynamics to terrestrial decomposition processes. *FEMS Microbiology Reviews*, *37*(4), 477–494. doi:10.1111/1574-6976.12001 PMID:22978352

Antizar-Ladislao, B., Spanova, K., Beck, A. J., & Russell, N. J. (2008). Microbial community structure changes during bioremediation of PAHs in an aged coal-tar contaminated soil by invessel composting. *International Biodeterioration & Biodegradation*, *61*(4), 357–364. doi:10.1016/j.ibiod.2007.10.002

Aparna, A., Srinikethan, G., & Hedge, S. (2011). Effect of addition of biosurfactant produced by *Pseudomonas* ssp. on biodegradation of crude oil. *International Proceedings of Chemical Biological & Environmental Engineering*, *6*, 71–75.

Aparna, A., Srinikethan, G., & Smitha, H. (2012). Production and characterization of biosurfactant produced by a novel Pseudomonas sp. 2B. *Colloids and Surfaces. B, Biointerfaces*, *95*, 23–29. doi:10.1016/j.colsurfb.2012.01.043 PMID:22445235

Aquino, N. B., Sevigny, M. B., Sabangan, J., & Louie, M. C. (2012). The role of cadmium and nickel in estrogen receptor signaling and breast cancer: Metalloestrogens or not? *Journal of Environmental Science and Health. Part C, Environmental Carcinogenesis & Ecotoxicology Reviews*, *30*(3), 189–224. doi:10.1080/10590501.2012.705159 PMID:22970719

Aracic, S., Manna, S., Petrovski, S., Wiltshire, J. L., Mann, G., & Franks, A. E. (2015). Innovative biological approaches for monitoring and improving water quality. *Frontiers in Microbiology*, *6*, 826. doi:10.3389/fmicb.2015.00826 PMID:26322034

Aracic, S., Semenec, L., & Franks, A. E. (2014). Investigating microbial activities of electrode-associated microorganisms in real-time. *Frontiers in Microbiology*, *5*, 663. doi:10.3389/fmicb.2014.00663 PMID:25506343

Arami, M., Limaee, N. Y., Mahmoodi, N. M., & Tabrizi, N. S. (2006). Equilibrium and kinetics studies for the adsorption of direct and acid dyes from aqueous solution by soy meal hull. *Journal of Hazardous Materials*, *135*(1-3), 171–179. doi:10.1016/j.jhazmat.2005.11.044 PMID:16442216

Arancon, R. A. D., Lin, C. S. K., Chan, K. M., Kwan, T. H., & Luque, R. (2013). Advances on waste valorization: New horizons for a more sustainable society. *Energy Science & Engineering*, *1*(2), 53–71. doi:10.1002/ese3.9

Arıca, M. Y., Bayramoglu, G., Yılmaz, M., Genc, O., & Bektas, S. (2004). Biosorption of Hg, Cd and Zn by Ca-alginate and immobilized wood rotting fungus *Funalia trogii*. *Journal of Hazardous Materials*, *109*(1-3), 191–199. doi:10.1016/j.jhazmat.2004.03.017 PMID:15177759

Aristidou, A., & Penttilä, M. (2000). Metabolic engineering applications to renewable resource utilization. *Current Opinion in Biotechnology*, *11*(2), 187–198. doi:10.1016/S0958-1669(00)00085-9 PMID:10753763

Arnold, C. W., Parfitt, D. G., & Kaltreider, M. (2007). Phytovolatilization of oxygenated gasoline-impacted groundwater at an underground storage tank site via conifers. *International Journal of Phytoremediation*, *9*(1), 53–69. doi:10.1080/15226510601139409 PMID:18246715

Arora, D. S., & Sharma, R. K. (2009). Enhancement in in vitro digestibility of wheat straw obtained from different geographical regions during solid state fermentation by white rot fungi. *BioResour*, *4*, 909–920.

Arun, A., Raja, P. P., Arthi, R., Ananthi, M., Kumar, K. S., & Eyini, M. (2008). Polycyclic aromatic hydrocarbons (PAHs) biodegradation by basidiomycetes fungi, *Pseudomonas* isolate, and their cocultures: Comparative *in vivo* and *in silico* approach. *Applied Biochemistry and Biotechnology*, *151*(2-3), 132–142. doi:10.1007/s12010-008-8160-0 PMID:18975143

Arunakumar, K. K. I. U., & Walpola, B. C. (2013). Banana peel: A green revolution for metal removal from contaminated waters. *Korean Journal of Environmental Agriculture*, *32*(2), 108–116. doi:10.5338/KJEA.2013.32.2.108

Arutchelvan, V., Kanakasabai, S., Nagarajan, S., & Muralikrishnan, V. (2005). Isolation and identification of novel high strength phenol degrading bacterial strains from phenol-formaldehyde resin manufacturing industrial wastewater. *Journal of Hazardous Materials*, *B127*(1-3), 238–243. doi:10.1016/j.jhazmat.2005.04.043 PMID:16154261

Arutchelvi, J., Sudhakar, M., Arkatkar, A., Doble, M., Bhaduri, S., & Uppara, P. V. (2008). Biodegradation of polyethylene and polypropylene. *Indian Journal of Biotechnology*, *7*(1), 9.

Asakura, K., Satoh, H., Chiba, M., Okamoto, M., Serizawa, K., Nakano, M., & Omae, K. (2009). Genotoxicity studies of heavy metals: Lead, bismuth, indium, silver and antimony. *Journal of Occupational Health*, *51*(6), 498–512. doi:10.1539/joh.L9080 PMID:19851040

Ashraf, H., & Husain, Q. (2010). Studies on bitter gourd peroxidase catalyzed removal of p-bromophenol from wastewater. *Desalination*, *262*(1-3), 267–272. doi:10.1016/j.desal.2010.05.044

Ashraf, M., Ozturk, M., & Ahmad, M. S. A. (2010). *Plant adaptation and phytoremediation*. New York: Springer. doi:10.1007/978-90-481-9370-7

Ashrafuzzaman, M., Hossen, F. A., Razi, I. M., Hoque, M. A., Islam, M. Z., Shahidullah, S. M., & Meon, S. (2009). Efficiency of plant growthpromoting rhizobacteria (PGPR) for the enhancement of rice growth. *African Journal of Biotechnology*, *8*, 1247–1252.

Asok, A. K., Fathima, P. A., & Jisha, M. S. (2015). Biodegradation of Linear Alkylbenzene Sulfonate (LAS) by Immobilized Pseudomonas sp. *Advances in Chemical Engineering and Science*, *5*(04), 465–475. doi:10.4236/aces.2015.54048

Atagana, H. I. (2011). Bioremediation of co-contamination of crude oil and heavy metals in soil by phytoremediation using Chromolaena odorata (L) King & HE Robinson. *Water, Air, and Soil Pollution*, *215*(1-4), 261–271. doi:10.1007/s11270-010-0476-z

Atkinson, R., Aschmann, S. M., & Winer, A. M. (1987). Kinetics of the reactions of NO3 radicals with a series of aromatic compounds. *Environmental Science & Technology*, *21*(11), 1123–1126. doi:10.1021/es00164a015

Atlas, R. M., & Bartha, R. (1993). *Microbial Ecology*. Harlow: Addison Wesley Longmann.

Atlas, R. M., & Pramer, D. (1990). Focus on bioremediation. *ASM News*, *56*, 352–354.

Aulenta, F., Canosa, A., Reale, P., Rossetti, S., Panero, S., & Majone, M. (2009). Microbial reductive dechlorination of trichloroethene to ethene with electrodes serving as electron donors without the external addition of redox mediators. *Biotechnology and Bioengineering*, *103*(1), 85–91. doi:10.1002/bit.22234 PMID:19160378

Aust, S. D., Swaner, P. R., & Stahl, J. D. (2004). Detoxification and metabolism of chemicals by white-rot fungi. In Pesticide decontamination and detoxification. Oxford.

Aust, S. D. (1990). Degradation of environmental pollutants by *Phanerochaete chyrosporium. Microbial Ecology*, *20*(1), 197–209. and doi:10.1007/BF02543877

Averill, C., & Finzi, A. (2011). Increasing plant use of organic nitrogen with elevation is reflected in nitrogen uptake rates and ecosystem δ15N. *Ecology*, *92*(4), 883–891. doi:10.1890/10-0746.1 PMID:21661551

Ayu, K. R., Tony, H., Tadashi, T., Yasuhiro, T., & Kazuhiro, M. (2011). Bioremediation of crude oil by white rot fungi *Polyporus* sp. S133. *Journal of Microbiology and Biotechnology*, *21*(9), 995–1000. doi:10.4014/jmb.1105.05047 PMID:21952378

Azam, F., & Malfatti, F. (2007). Microbial structuring of marine ecosystems. *Nature Reviews. Microbiology*, *5*(10), 782–791. doi:10.1038/nrmicro1747 PMID:17853906

Azar, S. K., & Azar, S. S. (2016). Waste Related Pollutions and Their Potential Effect on Cancer Incidences in Lebanon. *Journal of Environmental Protection*, *7*(06), 778–783. doi:10.4236/jep.2016.76070

Babel, S., & Kurniawan, T. A. (2003). Low-cost adsorbents for heavy metals uptake from contaminated water: A review. *Journal of Hazardous Materials*, *97*(1-3), 219–243. doi:10.1016/S0304-3894(02)00263-7 PMID:12573840

Badia-Fabregat, M., Lucas, D., Gros, M., Rodríguez-Mozaz, S., Barceló, D., Caminal, G., & Vicent, T. (2015). Identification of some factors affecting pharmaceutical active compounds (PhACs) removal in real wastewater. Case study of fungal treatment of reverse osmosis concentrate. *Journal of Hazardous Materials*, *283*, 663–671. and doi:10.1016/j.jhazmat.2014.10.007

Badia-Fabregat, M., Lucas, D., Pereira, M. A., Alves, M., Pennanen, T., Fritze, H., & Caminal, G. et al. (2016). Continuous fungal treatment of non-sterile veterinary hospital effluent: Pharmaceuticals removal and microbial community assessment. *Applied Microbial Biotechnology*, *100*(5), 2401–2415. and doi:10.1007/s00253-015-7105-0

Bafana, A., Kirishnamurthi, K., & Devi, S. & Chakrabarti T. (2008). Biological decolorization of C.I. Direct Black38 by *E. gallinarum. Journal of Hazardous Materials*, *157*(1), 183–193. doi:10.1016/j.jhazmat.2007.12.085 PMID:18280646

Bahadure, S., Kalia, R., & Chavan, R. (2013). Comparative study of bioremediation of hydrocarbon fuels. *International Journal of Biotechnology and Bioengineering Research.*, *4*(7), 677–686.

Baig, M. S., & Manickam, N. (2010). Homology modeling and docking studies of *Comamonas testosteroni* B-356 biphenyl-2, 3-dioxygenase involved in degradation of polychlorinated biphenyls. *International Journal of Biological Macromolecules*, *46*(1), 47–53. doi:10.1016/j.ijbiomac.2009.10.014 PMID:19879892

Bai, J., Wen, J. P., Li, H. M., & Jiang, Y. (2007). Kinetic modelling of growth and biodegradation of phenol and *m*-cresol using *Alcaligenes faecalis. Process Biochemistry*, *42*(4), 510–517. doi:10.1016/j.procbio.2006.10.004

Bailey, V. L., Peacock, A. D., Smith, J. L., & Bolton, H. Jr. (2002). Relationships between soil microbial biomass determined by chloroform fumigation–extraction, substrate-induced respiration and phospholipid fatty acid analysis. *Soil Biology & Biochemistry*, *34*(9), 1385–1389. doi:10.1016/S0038-0717(02)00070-6

Bai, S. R., & Abraham, T. E. (2001). Biosorption of Cr(VI) from aqueous solution by *Rhizopus nigricans. Bioresource Technology, 79*(1), 73–81. doi:10.1016/S0960-8524(00)00107-3 PMID:11396911

Bai, Y., Qi, W., Liang, J., & Qu, J. (2014). Using high-throughput sequencing to assess the impacts of treated and untreated wastewater discharge on prokaryotic communities in an urban river. *Applied Microbiology and Biotechnology, 98*(4), 1841–1851. doi:10.1007/s00253-013-5116-2 PMID:23912119

Bajguz, A. (2011). Suppression of *Chlorella vulgaris* growth by cadmium, lead, and copper stress and its restoration by endogenous brassinolide. *Archives of Environmental Contamination and Toxicology, 60*(3), 406–416. doi:10.1007/s00244-010-9551-0 PMID:20523975

Bajpai, P. (2012). *Biotechnology for pulp and paper processing*. New York: Springer. doi:10.1007/978-1-4614-1409-4

Baldrian, P. (2003). Interactions of heavy metals with white-rot fungi. *Enzyme and Microbial Technology, 32*(1), 78–91. doi:10.1016/S0141-0229(02)00245-4

Baldrian, P. (2008). Wood-inhabiting lignolytic basidiomycetes in soils: Ecology and constraints for application in bioremediation. *Fungal Ecology, 1*(1), 4–12. doi:10.1016/j.funeco.2008.02.001

Baldrian, P., Kolarˇık, M., & Sˇtursova´, M. (2012). Active and total microbial communities in forest soil are largely different and highly stratified during decomposition. *The ISME Journal, 6*(2), 248–258. doi:10.1038/ismej.2011.95 PMID:21776033

Baldrian, P., & Valášková, V. (2008). Degradation of cellulose by basidiomycetous fungi. *FEMS Microbiology Reviews, 32*(3), 501–521. doi:10.1111/j.1574-6976.2008.00106.x PMID:18371173

Baldrian, P., Vorˇisˇkova´, J., Dobiasˇova´, P., & Merhautova, V. (2011). Production of extracellular enzymes and degradation of biopolymers by saprotrophic microfungi from the upper layers of forest soil. *Plant and Soil, 338*(1-2), 111–125. doi:10.1007/s11104-010-0324-3

Balík, J., Pavlíková, D., Vanek, V., Kulhánek, M., & Kotková, B. (2007). The influence of long-term sewage sludge application on the activity of phosphatases in the rhizosphere of plants. *Plant, Soil and Environment, 53*(9), 375.

Banat, I. M., Satpute, S. K., Cameotra, S. S., Patil, R., & Nyayanit, N. V. (2014). Cost effective technologies and renewable substrates for biosurfactants' production. *Frontiers in Microbiology, 5*, 697.

Bancroft, K., Maloney, S. W., Mcelhaney, J., Suffet, I. H., & Pipes, W. O. (1983). Assessment of bacterial growth and total organic carbon removal on granular activated carbon contactors. *Applied and Environmental Microbiology, 46*(3), 683–688. PMID:6639023

Bapiri, A., Bååth, E., & Rousk, J. (2010). Drying-rewetting cycles affect fungal and bacterial growth differently in an arable soil. *Microbial Ecology, 60*(2), 419–428. doi:10.1007/s00248-010-9723-5 PMID:20635180

Baquero, F., Tedim, A. P., & Coque, T. M. (2013). Antibiotic resistance shaping multi-level population biology of bacteria. Frontiers in Microbiology, 4, 15.10.3389/fmicb.2013.00015

Barac, T., Taghavi, S., Borremans, B., Provoost, A., Oeyen, L., Colpaert, J. V., & van-der-Lelie, D. et al. (2004). Engineered endophytic bacteria improve phytoremediation of water-soluble, volatile, organic pollutants. *Nature Biotechnology, 22*(5), 583–588. doi:10.1038/nbt960 PMID:15077119

Bardgett, R. D., & van der Putten, W. H. (2014). Belowground biodiversity and ecosystem functioning. *Nature, 515*(7528), 505–511. doi:10.1038/nature13855 PMID:25428498

Baronti, C., Curini, R., DAscenzo, G., Di Corcia, A., Gentili, A., & Samperi, R. (2000). Monitoring natural and synthetic estrogens at activated sludge sewage treatment plants and in a receiving river water. *Environmental Science & Technology, 34*(24), 5059–5066. doi:10.1021/es001359q

Bartha, R., & Atlas, R. M. (1977). The microbiology of aquatic oil spills. *Advances in Applied Microbiology, 22*, 225–266. doi:10.1016/S0065-2164(08)70164-3 PMID:337769

Basharat, Z. (2015b). *In vitro and in silico analysis of enzymes from indigenous bacteria for remediation of azo dyes* (Unpublished M.Phil. dissertation). Fatima Jinnah Women University, Rawalpindi, Pakistan.

Basharat, Z., & Yasmin, A. (2016a). *Understanding properties of the master effector of phage shock operon in Mycobacterium tuberculosis via bioinformatics approach.* DOI:.10.1101/050047

Basharat, Z., & Yasmin, A. (2016b). Molecular docking of *Sulfobacillus acidophilus* barbiturase with s-triazine compounds. *PeerJ Preprints, 4*, e2070v1. DOI:10.7287/peerj.preprints.2070v1

Basharat, Z. (2015a). Computational sequence analysis and *in silico* modeling of a stripe rust resistance protein encoded by wheat TaHSC70 gene. *Journal of Computational Biology and Bioinformatics Research, 7*(1), 1–7. doi:10.5897/JCBBR2013.082

Basharat, Z., Messaoudi, A., Ruba, S., & Yasmin, A. (2016a). NQO1 rs1800566 polymorph is more prone to NOx induced lung injury: Endorsing deleterious functionality through informatics approach. *Gene, 591*(1), 14–20. doi:10.1016/j.gene.2016.06.048 PMID:27349566

Basharat, Z., & Yasmin, A. (2015a). *In silico* assessment of phosphorylation and O-β-GlcNAcylation sites in human NPC1 protein critical for Ebola virus entry. *Infection, Genetics and Evolution, 34*, 326–338. doi:10.1016/j.meegid.2015.06.001 PMID:26048414

Basharat, Z., & Yasmin, A. (2016c). Bioinformatic analysis of human Gαq Q209 missense mutations associated with *uveal melanoma*. In *13th International Bhurban Conference on Applied Sciences and Technology (IBCAST)* (pp. 106-111). IEEE. Doi:10.1109/IBCAST.2016.7429863

Basharat, Z., Zeb, S., & Yasmin, A. (2016b). *Computational study of some amoebicidal phytochemicals against heat shock protein of Naegleria fowleri.* Gene Reports; doi:10.1016/j.genrep.2016.09.003

Bastida, F., Jehmlich, N., Lima, K., Morris, B., Richnow, H., Hernández, T., & García, C. et al. (2016). The ecological and physiological responses of the microbial community from a semiarid soil to hydrocarbon contamination and its bioremediation using compost amendment. *Journal of Proteomics, 135*, 162–169. doi:10.1016/j.jprot.2015.07.023 PMID:26225916

Bastos, A. E. R., Moon, D. H., Rossi, A., Trevors, J. T., & Tsai, S. M. (2000b). Salt tolerant phenol-degrading microorganisms isolated from Amazon soil samples. *Archives of Microbiology, 174*(5), 346–352. doi:10.1007/s002030000216 PMID:11131025

Basu, A., Sen, P., & Jha, A. (2015). Environmental arsenic toxicity in West Bengal, India: A brief policy review. *Indian Journal of Public Health, 59*(4), 295–298. doi:10.4103/0019-557X.169659 PMID:26584169

Basumatary, B., Bordoloi, S., & Sarma, H. P. (2012). Crude oil-contaminated soil phytoremediation by using Cyperus brevifolius (Rottb.) Hassk. *Water, Air, and Soil Pollution, 223*(6), 3373–3383. doi:10.1007/s11270-012-1116-6

Basumatary, B., Saikia, R., & Bordoloi, S. (2012). Phytoremediation of crude oil contaminated soil using nut grass, Cyperus rotundus. *Journal of Environmental Biology, 33*(5), 891. PMID:23734455

Basu, S., Dastidar, S. G., Mukhopadhyay, S., & Gangopadhyay, A. (2012). Utilization of bulk drugs by a highly antibiotic resistant microorganism isolated from waste water of a bulk drug industry. *Journal of Pharmaceutical Science and Technology*, *2*(1), 41–46.

Batt, A. L., Kim, S., & Aga, D. S. (2006). Enhanced biodegradation of iopromide and trimethoprim in nitrifying activated sludge. *Environmental Science & Technology*, *40*(23), 7367–7373. and doi:10.1021/es060835v

Baudoin, E., Benizri, E., & Guckert, A. (2003). Impact of artificial root exudates on the bacterial community structure in bulk soil and maize rhizosphere. *Soil Biology & Biochemistry*, *35*(9), 1183–1192. doi:10.1016/S0038-0717(03)00179-2

Baumann, M., & Baxendale, I. R. (2016). Continuous photochemistry: The flow synthesis of ibuprofen via a photo-Favorskii rearrangement. *Reaction Chemistry & Engineering*, *1*(2), 147–150. doi:10.1039/C5RE00037H

Bazot, S., Blum, H., & Robin, C. (2008). Nitrogen rhizodeposition assessed by a 15NH3 shoot pulse-labelling of Lolium perenne L. grown on soil exposed to 9 years of CO2 enrichment. *Environmental and Experimental Botany*, *63*(1-3), 410–415. doi:10.1016/j.envexpbot.2007.10.020

BCAS. (2013). Special Reports[Chinese Academy of Sciences.]. *Bulletin of the Chinese Academy of Sciences*, *27*, 3.

BCC Research. (2015). *Trends in the global market for microbes and microbial products, 2010-2016 (in US$ Millions)*. Retrieved August 13, 2016, from http://www.bccresearch.com/report/microbial-productsmarkets-bio086a.html

Beare, M. H., Coleman, D. C., Crossley, D. A. Jr, Hendrix, P. F., & Odum, E. P. (1995). A hierarchical approach to evaluating the significance of soil biodiversity to biogeochemical cycling. *Plant and Soil*, *170*(1), 5–22. doi:10.1007/BF02183051

Becker, J., & Wittmann, C. (2015). Advanced Biotechnology: Metabolically Engineered Cells for the Bio-Based Production of Chemicals and Fuels, Materials, and Health-Care Products. *Angewandte Chemie International Edition*, *54*(11), 3328–3350. doi:10.1002/anie.201409033 PMID:25684732

Bedard, D. L., & Quensen III, J. F. (1995). Microbial reductive dechlorination of polychlorinated biphenyls. *Microbial Transformation and Degradation of Toxic Organic Chemicals*, 127-216.

Beh, M., Strauss, G., Huber, R., Stetter, K. O., & Fuchs, G. (1993). Enzymes of the reductive citric acid cycle in the autotrophic eubacterium Aquifex pyrophilus and in the archaebacterium Thermoproteus neutrophilus. *Archives of Microbiology*, *160*(4), 306–311. doi:10.1007/BF00292082

Beilby, J. P. (1983). Effects of inhibitors on early protein, RNA, and lipid synthesis in germinating vesicular-arbuscular mycorrhizal fungal spores of Glomus caledonium. *Canadian Journal of Botany*, *29*, 596–601. PMID:6883219

Beinert, H., & Kiley, P. J. (1999). Fe–S proteins in sensing and regulatory functions. *Current Opinion in Chemical Biology*, *3*(2), 152–157. doi:10.1016/S1367-5931(99)80027-1 PMID:10226040

Bellenger, J. P., Wichard, T., Kustka, A. B., & Kraepiel, A. M. L. (2008). Uptake of molybdenum and vanadium by a nitrogen-fixing soil bacterium using siderophores. *Nature Geoscience*, *1*(4), 243–246. doi:10.1038/ngeo161

Beltran, F. J., & Alvarez, P. (1996). Rate constant determination of ozone-organic fast reaction in water using agitated cell. *Environ. Sci. Health*, *31*, 1159–1178.

Bengtsson, G., Ek, H., Rundgren, S., & Bengtsson, G. (1992). Evolutionary response of earthworms to long-term metal exposure. *Oikos*, *63*(2), 289–297. doi:10.2307/3545390

Benzie, J., Glockner, F., Rodríguez-Ezpeleta, N., Faith, D., Bean, T., & Obst, M. (2013). Genomics in marine monitoring: New opportunities for assessing marine health status. *Marine Pollution Bulletin*, *74*(1), 19–31. doi:10.1016/j.marpolbul.2013.05.042 PMID:23806673

Bergauer, P., Fonteyne, P. A., Nolard, N., Schinner, F., & Margesin, R. (2005). Biodegradation of phenol and phenol-related compounds by psychrophilic and cold- tolerant alpine yeasts. *Chemosphere*, *59*(7), 909–918. doi:10.1016/j.chemosphere.2004.11.011 PMID:15823324

Berg, B., & Matzner, E. (1997). Effect of N deposition on decomposition of plant litter and soil organic matter in forest systems. *Environ Rev*, *5*(1), 1–25. doi:10.1139/a96-017

Berg, B., & McClaugherty, C. (2008). *Plant Litter: Decomposition, Humus Formation, Carbon Sequestration*. Heidelberg, Germany: Springer-Verlag. doi:10.1007/978-3-540-74923-3

Berg, G., Roskot, N., Steidle, A., Eberl, L., Zock, A., & Smalla, K. (2002). Plant-dependent genotypic and phenotypic diversity of antagonistic rhizobacteria isolated from different Verticillium host plants. *Applied and Environmental Microbiology*, *68*(7), 3328–3338. doi:10.1128/AEM.68.7.3328-3338.2002 PMID:12089011

Berthe-Conti, L., Jacobi, H., Kleihauer, S., & Witte, L. (1998). Cytotoxicity and mutagenicity of a 2,4,6 trinitrotoluene (TNT) and hexogen contaminated soil in *S. typhimurium* and mammalian cells. *Chemosphere*, *37*(2), 209–218. doi:10.1016/S0045-6535(98)00039-3 PMID:9650264

Bertoldo, C., Dock, C., & Antranikian, G. (2004). Thermoacidophilic microorganisms and their novel biocatalysts. *Engineering in Life Sciences*, *4*(4), 521–532. doi:10.1002/elsc.200402155

Beveridge, T. J. (1989). Role of cellular design in bacterial metal accumulation and mineralization. *Annual Review of Microbiology*, *43*(1), 147–171. doi:10.1146/annurev.mi.43.100189.001051 PMID:2679352

Beveridge, T. J., & Doyle, R. J. (1989). *Metal Ions and Bacteria*. New York: J Wiley.

Beyer, W. N. (1981). Metals and terrestrial earthworms (Annelida:Oligochaeta). In *Workshop on the role of earthworms in the stabilization of organic residues*. Beech Leaf Press.

Bezerra de Souza Sobrinho, H., de Luna, J. M., Rufino, R. D., Figueiredo Porto, A. L., & Sarubbo, L. A. (2013). Assessment of toxicity of a biosurfactant from *Candida sphaerica* UCP 0995 cultivated with industrial residues in a bioreactor. *Electronic Journal of Biotechnology*, *16*(4), 4–4.

Bezza, F. A., & Chirwa, E. M. N. (2016). Biosurfactant-enhanced bioremediation of aged polycyclic aromatic hydrocarbons (PAHs) in creosote contaminated soil. *Chemosphere*, *144*, 635–644. doi:10.1016/j.chemosphere.2015.08.027 PMID:26408261

Bhakta, J. N. (2016). Microbial response against metal toxicity. In A. K. Rathoure & V. K. Dhatwalia (Eds.), Toxicity and Waste Management Using Bioremediation (pp. 75-96). IGI Global. doi:10.4018/978-1-4666-9734-8.ch004

Bhakta, J. N., Lahiri, S., Pitman, J. K., & Jana, B. B. (2015). Carbon dioxide sequestration in wastewater by a consortium of elevated carbon dioxide-tolerant microalgae. *Journal of CO_2 Utilization, 10,* 105–112.

Bhakta, J. N., Son, L. T., & Munekage, Y. (2009). Distribution and potential impact of metal pollutants in the coastal environment: A case study with special reference to coastal aquaculture in red river delta of Viet Nam. *Electronic Journal of Biology, 5*(2), 22–27. Retrieved from http://www.ejbio.com

Bhakta, J. N. (2016). Microbial response against metal toxicity. In A. K. Rathoure & V. K. Dhatwalia (Eds.), *Toxicity and Waste Management Using Bioremediation* (pp. 75–96). IGI Global.

Bhakta, J. N., Jana, B. B., Lahiri, S., Panigrahi, A., Mandal, S. K., Rana, S., & Rokunuzzaman, M. et al. (2016). Ecological restoration: An emerging eco-technology for sustainable environmental conservation. *Int J Env Tech Sci*, *2*, 21–30.

Bhakta, J. N., & Munekage, Y. (2008). Role of ecosystem components in Cd removal process of aquatic ecosystem. *Ecological Engineering*, *32*(3), 274–280. doi:10.1016/j.ecoleng.2007.12.004

Bhakta, J. N., & Munekage, Y. (2010). Spatial distribution and contamination status of arsenic, cadmium and lead in some coastal shrimp (*Macrobrachium rosenbergii*) farming ponds of Viet Nam. *Pacific Journal of Science and Technology*, *11*(1), 606–615.

Bhakta, J. N., & Munekage, Y. (2013). Identification of potential soil adsorbent for the removal of hazardous metals from aqueous phase. *International Journal of Environmental Science and Technology*, *10*(2), 315–324. doi:10.1007/s13762-012-0116-9

Bhakta, J. N., Munekage, Y., Ohnishi, K., & Jana, B. B. (2012a). Isolation and identification of cadmium and lead resistant lactic acid bacteria for applying as metal removing probiotic. *International Journal of Environmental Science and Technology*, *9*(3), 433–440. doi:10.1007/s13762-012-0049-3

Bhakta, J. N., Munekage, Y., Ohnishi, K., Jana, B. B., & Balcazar, J. L. (2014). Isolation and characterization of cadmium and arsenic absorbing bacteria for bioremediation. *Water, Air, and Soil Pollution*, *225*(10), 2151. doi:10.1007/s11270-014-2151-2

Bhakta, J. N., Ohnishi, K., Munekage, Y., & Iwasaki, K. (2010). Isolation and probiotic characterization of arsenic-resistant lactic acid bacteria for uptaking arsenic. *International Journal of Chemical and Biological Engineering*, *3*(4), 167–174.

Bhakta, J. N., Ohnishi, K., Munekage, Y., Iwasaki, K., & Wei, M. (2012b). Characterization of lactic acid bacteria-based probiotics as heavy metals sorbents. *Journal of Applied Microbiology*, *112*(6), 1193–1206. doi:10.1111/j.1365-2672.2012.05284.x PMID:22404232

Bhakta, J. N., Rana, S., Lahiri, S., & Munekage, Y. (2014). *Sorption of mercury onto waste material derived low-cost activated carbon*. Applied Water Science. doi:10.1007/s13201-014-0236-0

Bhargava, C., Carmona, F. F., Bhargava, M., & Srivastava, S. (2012). Approaches for enhanced phytoextraction of heavy metals. *Journal of Environmental Management*, *105*, 103–120. doi:10.1016/j.jenvman.2012.04.002 PMID:22542973

Bhargava, H. N., & Leonard, P. A. (1996). Triclosan: Applications and safety. *American Journal of Infection Control*, *24*(3), 209–218. doi:10.1016/S0196-6553(96)90017-6 PMID:8807001

Bharti, R. K., Srivastava, S., & Thakur, I. S. (2014). Proteomic Analysis of CConcentrating Chemolithotrophic Bacteria Serratia sp. for Sequestration of CDioxide. *PLoS ONE*, *9*(3), e91300. doi:10.1371/journal.pone.0091300 PMID:24619032

Bhatnagar, S., & Kumari, R. (2013). Bioremediation: A sustainable tool for environmental management- A review. *Annual Review & Research in Biology*, *3*(4), 974–993.

Bhattacharya, H. N., & Chakrabarti, S. (2011). Incidence of fluoride in the groundwater of Purulia District, West Bengal: A geo-environmental appraisal Scientific Report. *Current Science*, *101*, 2.

Bhushan, B., Paquet, L., Halasz, A., Spain, J. C., & Hawari, J. (2003). Mechanism of xanthine oxidase catalyzed biotransformation of HMX under anaerobic conditions. *Biochemical and Biophysical Research Communications*, *306*(2), 509–515. doi:10.1016/S0006-291X(03)01001-5 PMID:12804594

Bietz, S., & Rarey, M. (2016). SIENA: Efficient compilation of selective protein binding site ensembles. *Journal of Chemical Information and Modeling*, *56*(1), 248–259. doi:10.1021/acs.jcim.5b00588 PMID:26759067

Bikadi, Z., & Hazai, E. (2009). Application of the PM6 semi-empirical method to modeling proteins enhances docking accuracy of AutoDock. *Journal of Cheminformatics*, *1*(1), 1. doi:10.1186/1758-2946-1-15 PMID:20150996

Bilej, M., Vetvicka, V., Tuckova, L., Trebichavsky, I., & Koukal,, M. M., & Ima, P. (1990). Phagocytosis of synthetic particles in earthworms: Effect of antigenic stimulation and opsonisation. *Folia Biologica*, *36*, 273–275. PMID:2279582

Billore, S. K., Singh, N., Ram, H. K., Sharma, J. K., Singh, V. P., Nelson, R. M., & Dass, P. (2001). Treatment of a molasses based distillery effluent in a constructed wetland in central India. *Water Science and Technology*, *44*(11–12), 441–448.

Binder, M., & Hibbett, D. S. (2002). Higher-level phylogenetic relationships of homo basidiomycetes (mushroom-forming fungi) inferred from four rDNA regions. *Molecular Phylogenetics and Evolution*, *22*(1), 76–90. doi:10.1006/mpev.2001.1043 PMID:11796031

Biostimulation. (n.d.). In *Wikipedia, the free encyclopedia*. Retrieved from http://en.wikipedia.org/wiki/

Bishnoi, N. R., & Garima. (2005). Fungus: An alternative for bioremediation of heavy metal containing wastewater: a review. *Journal of Scientific and Industrial Research*, *64*, 93–100.

Bisht, S., Pandey, P., Kaur, G., Aggarwal, H., Sood, A., Sharma, S., & Bisht, N. et al. (2014). Utilization of endophytic strain Bacillus sp. SBER3 for biodegradation of polyaromatic hydrocarbons (PAH) in soil model system. *European Journal of Soil Biology*, *60*, 67–76. doi:10.1016/j.ejsobi.2013.10.009

Bittsanszkya, A., Kfmivesb, T., Gullnerb, G., Gyulaia, G., Kissa, J., Heszkya, L., & Rennenberg, H. et al. (2005). Ability of transgenic poplars with elevated glutathione content to tolerate zinc (2^+) stress. *Environment International*, *31*(2), 251–254. doi:10.1016/j.envint.2004.10.001 PMID:15661291

Blakemore, R. J. (2010). Saga of Herr Hilgendorfs worms. *Zoology in the Middle East*, *49*(sup2), 7–22. doi:10.1080/09397140.2010.10638453

Blanchette, R. A., Burnes, T. A., Eerdmans, M. M., & Akhtar, M. (1992). Evaluating Isolates of *Phanerochaete chrysosporium* and *Ceriporiopsis subvermispora* for Use in Biological Pulping Processes. *Holzforschung*, *46*(2), 109–115. doi:10.1515/hfsg.1992.46.2.109

Blaudez, D., Botton, B., & Chalot, M. (2000). Cadmium uptake and subcellular compartmentation in the ectomycorrhizal fungus Paxillus involutus. *Microbiology*, *146*(5), 1109–1117. doi:10.1099/00221287-146-5-1109 PMID:10832638

Blyth, W., Shahsavari, E., Morrison, P. D., & Ball, A. S. (2015). Biosurfactant from red ash trees enhances the bioremediation of PAH contaminated soil at a former gasworks site. *Journal of Environmental Management*, *162*, 30–36. doi:10.1016/j.jenvman.2015.07.041 PMID:26217887

Boddy, L. (1999). Saprotrophic cord-forming fungi: Meeting the challenge of heterogeneous environments. *Mycologia*, *91*(1), 13–32. doi:10.2307/3761190

Boddy, L., Hynes, J., Bebber, D. P., & Fricker, M. D. (2009). Saprotrophic cord systems: Dispersal mechanisms in space and time. *Mycoscience*, *50*(1), 9–19. doi:10.1007/S10267-008-0450-4

Boden, A. B., & Marland, T. G. (2014). *Climate change 2014 – Impacts, adaptation and vulnerability: Global and sectoral aspect*. New York: Cambridge university press.

Bodin, N., Burgeot, T., Stanisie, J. Y., Bocquene, G., & Menard, D. (2004). Seasonal variations of a battery of biomarkers and physiological indices for the mussel *Mytilus galloprovincialis* transplanted into the northwest Mediterranean Sea. *Biochemistry Physiology and Toxicology Pharmacology*, *13*(4), 411–427. doi:10.1016/j.cca.2004.04.009 PMID:15536049

Boethling, R., Fenner, K., Howard, P., Klecka, G., Madsen, T., Snape, J. R., & Whelan, M. J. (2009). Environmental persis- tence of organic pollutants: Guidance for development and review of POP risk profiles. *Integrated Environmental Assessment and Management*, *5*(4), 39–556. doi:10.1897/IEAM_2008-090.1 PMID:19552498

Boldt, J. L., Martone, R., Samhouri, J., Perry, I., Itoh, S., Chung, I. K., & Yoshi, N. et al. (2014). Developing ecosystem indicators for responses to multiple stressors. *Oceanography (Washington, D.C.)*, *27*(4), 116–133. doi:10.5670/oceanog.2014.91

Bolen, N. S., Adriano, D. C., Natesan, R., & Koo, B. J. (2003). Effects of organic amendments on the phyoavalability of chromate in mineral soil. *Journal of Environmental Quality*, *32*(1), 120–128. doi:10.2134/jeq2003.1200 PMID:12549550

Boll, M., & Fuchs, G. (2005). Unusual reactions involved in anaerobic metabolism of phenolic compounds. *Biological Chemistry*, *386*(10), 989–997. doi:10.1515/BC.2005.115 PMID:16218871

Bond, P. L., Erhart, R., Wagner, M., Keller, J., & Blackall, L. L. (1999). Identification of some of the major groups of bacteria in efficient and non-efficient biological phosphorus removal activated sludge systems. *Applied and Environmental Microbiology*, *65*, 4077–4084. PMID:10473419

Bonfa, M. R. L., Grossman, M. J., Mellado, E., & Durrant, L. R. (2011). Biodegradation of aromatic hydrocarbons by Haloarchaea and their use for the reduction of the chemical oxygen demand of hypersaline petroleum produced water. *Chemosphere*, *84*(11), 1671–1676. doi:10.1016/j.chemosphere.2011.05.005 PMID:21621813

Boone, D. R., Johnson, R. L., Chen, D. C., Mathrani, I. M., & Mah, R. A. (1989). Methanogenesis and reductive dechlorinations in an alkaline, hypersaline sediment and groundwater. In M. S. Da Costa, J. C. Duarte, & R. A. D. Williams (Eds.), *Microbiology of Extreme Environments and its potential for Biotechnology* (pp. 205–215). London: Elsevier Applied Science.

Boonsaner, M., Borrirukwisitsak, S., & Boonsaner, A. (2011). Phytoremediation of BTEX contaminated soil by Canna× generalis. *Ecotoxicology and Environmental Safety*, *74*(6), 1700–1707. doi:10.1016/j.ecoenv.2011.04.011 PMID:21497398

Boopathy, R. (2000). Factors limiting bioremediation technologies. *Bioresource Technology*, *74*(1), 63–67. doi:10.1016/S0960-8524(99)00144-3

Borgeest, C., Greenfeld, C., Tomic, D., & Flaws, J. A. (2002). The effects of endocrine disrupting chemicals on the ovary. *Frontiers in Bioscience: A Journal and Virtual Library*, *7*, d1941-8.

Borja, R., Martin, A., Maestro, R., Alba, J., & Fiestas, J. A. (1992). Enhancement of the anaerobic digestion of olive mill wastewaters by removal of phenolic inhibitors. *Proc. Biochem.*, *27*(4), 231–237. doi:10.1016/0032-9592(92)80023-V

Borja, R., Sanchez, E., Marin, A., & Jimenez, A. M. (1996). Kinetic behaviour of waste tyre rubber as microorganism support in an anaerobic digester treating cane molasses distillery slops. *Bioprocess and Biosystems Engineering*, *16*(1), 17–23. doi:10.1007/s004490050282

Bossi, R., Seiden, P., Andersen, S. M., Jacobsen, C. S., & Streibig, J. C. (1999). Analysis of metsulfuron-methyl in soil by liquid chromatography/tandem mass spectrometry. Application to a field dissipation study. *Journal of Agricultural and Food Chemistry*, *47*(10), 4462–4468. doi:10.1021/jf981280t PMID:10552834

Botha, A. (2011). The importance and ecology of yeasts in soil. *Soil Biology & Biochemistry*, *43*(1), 1–8. doi:10.1016/j.soilbio.2010.10.001

Bourceret, A., Leyval, C., De Fouquet, C., & Cébron, A. (2015). Mapping the Centimeter-Scale Spatial Variability of PAHs and Microbial Populations in the Rhizosphere of Two Plants. *PLoS ONE*, *10*(11), 1–22. doi:10.1371/journal.pone.0142851 PMID:26599438

Bourlat, S., Borja, A., Taylor, M., Weisberg, S., Griffith, J., Lettieri, T., & Martyniuk, C. et al. (2012). Advancing the Omics in aquatic toxicology: SETAC North America 31st Annual Meeting. *Ecotoxicology and Environmental Safety*, *76*, 1–2. doi:10.1016/j.ecoenv.2011.11.001 PMID:22114953

Boyd, P. W., Jickells, T., Law, C. S., Blain, S., Boyle, E. A., Buesseler, K. O., & Watson, A. J. et al. (2007). Mesoscale iron enrichment experiments 1993–2005: Synthesis and future directions. *Science*, *315*(5812), 612–617. doi:10.1126/science.1131669 PMID:17272712

Brahmaiah, T., Spurthi, L., Chandrika, K., Ramanaiah, S., & Prasad, K. S. Sai. (2015). Kinetics of heavy metal (Cr & Ni) removal for the wastewater by using low cost adsorbent. *World Journal of Pharmacy and Pharmaceutical Sciences*, *4*(11), 1600–1610.

Brahmbhatt, N. H., Patel, R. V., & Jasrai, R. T. (2012). Bioremediation potential of *Spirogyra* sp. and *Oscillatoria* sp. for cadmium. *Asian Journal of Biochemical and Pharmaceutical Research*, *2*, 102–107.

Brandl, H., Bosshard, R., & Wegmann, M. (1999). Computer-munching microbes: Metal leaching from electronic scrap by bacteria and fungi. *Process Metallurgy*, *9B*, 569–576. doi:10.1016/S1572-4409(99)80146-1

Braud, A., Hoegy, F., Jezequel, K., Lebeau, T., & Schalk, I. J. (2009). New insights into the metal specificity of the *Pseudomonas aeruginosa* pyoverdine–iron uptake pathway. *Environmental Microbiology*, *11*(5), 1079–1091. doi:10.1111/j.1462-2920.2008.01838.x PMID:19207567

Breen, A., & Singleton, F. L. (1999). Fungi in lignocellulose breakdown and biopulping. *Current Opinion in Biotechnology*, *10*(3), 252–258. doi:10.1016/S0958-1669(99)80044-5 PMID:10361070

Brooijmans, N., & Kuntz, I. D. (2003). Molecular recognition and docking algorithms. *Annual Review of Biophysics and Biomolecular Structure*, *32*(1), 335–373. doi:10.1146/annurev.biophys.32.110601.142532 PMID:12574069

Brooker, R. J. (2011). *Genetics: Analysis and Principle*. New York, NY: McGraw-Hill Science.

Bruins, M. R., Kapil, S., & Oehme, F. W. (2000). Microbial resistance to metals in the environment. *Ecotoxicology and Environmental Safety*, *45*(3), 198–207. doi:10.1006/eesa.1999.1860 PMID:10702338

Brune, A. (2014). Symbiotic digestion of lignocellulose in termite guts. Detailed review of the complex role microbial endosymbionts in termite digestion. *Nature Reviews. Microbiology*, *12*(3), 168–180. doi:10.1038/nrmicro3182 PMID:24487819

Budavari, S. (1996). *The Merck Index: An encyclopedia of chemicals, drugs and biochemicals, Whitehouse station*. Merck.

Budin, K., Subhramanium, Y., Tair, R., & Ali, S. A. M. (2014). The ability of crab and cockle shell to adsorb lead and chromium from industrial effluent. *IOSR Journal of Environmental Science. Toxicology and Food Technology*, *8*(12), 4–6. doi:10.9790/2402-081210406

Buee, M., De Boer, W., Martin, F., van Overbeek, L., & Jurkevitch, E. (2009a). The rhizosphere zoo: An overview of plant associated communities of microorganisms, including phages, bacteria, archaea, and fungi, and some of their structuring factors. *Plant and Soil*, *321*(1-2), 189–212. doi:10.1007/s11104-009-9991-3

Buesseler, K. O., & Boyd, P. W. (2003). Will ocean fertilization work? *Science*, *300*(5616), 67–68. doi:10.1126/science.1082959 PMID:12677048

Bugg, T. D. H., Ahmad, M., Hardiman, E. M., & Singh, R. (2011). The emerging role for bacteria in lignin degradation and bioproduct formation. *Current Opinion in Biotechnology*, *22*(3), 394–400. doi:10.1016/j.copbio.2010.10.009 PMID:21071202

Buhner, S. H. (2002). *The Lost Language of Plants- The Ecological Importance of Plant Medicines to Life on Earth*. White River Junction, Vermont: Chelsea Green Publishing.

Buitenkamp, M., & Stintzing, R. A. (2008). *Europe's sanitation problem - 20 million Europeans need access to safe and affordable sanitation. Women in Europe for a Common Future*. Netherlands: WECF.

Bull, A. T. (2010). Actinobacteria of the extremobiosphere. In Extremophiles Handbook. Berlin: Springer-Verlag GmbH.

Burd, G. I., Dixon, D. G., & Glick, B. R. (2000). Plant growth-promoting bacteria that decrease heavy metal toxicity in plants. *Canadian Journal of Microbiology*, *46*(3), 237–245. doi:10.1139/w99-143 PMID:10749537

Burghout, P., Cron, L. E., Gradstedt, H., Quintero, B., Simonetti, E., Bijlsma, J. J., & Hermans, P. W. et al. (2010). Carbonic anhydrase is essential for Streptococcus pneumoniae growth in environmental ambient air. *Journal of Bacteriology*, *192*(15), 4054–4062. doi:10.1128/JB.00151-10 PMID:20525828

Burken, J. G., Shanks, J. V., & Thompson, P. L. (2000). Metabolism of Explosives and Nitroaromatic Compounds. *Biodegradation of Nitroaromatic Compounds and Explosives*, 239.

Buser, H. R., Müller, M. D., & Theobald, N. (1998b). Occurrence of the pharmaceutical drug clofibric acid and the herbicide mecoprop in various Swiss lakes and in the North Sea. *Environmental Science & Technology*, *32*(1), 188–192. doi:10.1021/es9705811

Buser, H. R., Poiger, T., & Müller, M. D. (1998a). Occurrence and face of the pharmaceutical drug diclofenac in surface waters: Rapid photodegradation in a lake. *Environmental Science & Technology*, *32*(22), 3449–3456. doi:10.1021/es980301x

Buser, H. R., Poiger, T., & Müller, M. D. (1999). Occurrence and environmental behaviour of chiral pharmaceutical drug ibuprofen in surface waters and in wastewater. *Environmental Science & Technology*, *33*(15), 2529–2535. doi:10.1021/es981014w

Busk, P. K., & Lange, L. (2015). Classification of fungal and bacterial lytic polysaccharide monooxygenases. *BMC Genomics*, *16*(1), 368. doi:10.1186/s12864-015-1601-6 PMID:25956378

Bustamante, M., Durán, N., & Diez, M. C. (2012). Biosurfactants are useful tools for the bioremediation of contaminated soil: a review. *Journal of Soil Science and Plant Nutrition, 12*(4), 667-687.

Cadet, C., Beteau, J. F., & Hernandez, S. C. (2004). Multicriteria control strategy for cost/quality compromise in wastewater treatment plants. *Control Engineering Practice*, *12*(3), 335–347. doi:10.1016/S0967-0661(03)00134-5

Cai, J. H., & Jia, C. Q. (2010). Mercury removal from aqueous solution using coke-derived sulfur impregnated activated carbons. *Industrial & Engineering Chemistry Research*, *49*(6), 2716–2721. doi:10.1021/ie901194r

Cairney, J. W. G. (2005). Basidiomycete mycelia in forest soils: Dimensions, dynamics and roles in nutrient distribution. *Mycological Research*, *109*(1), 7–20. doi:10.1017/S0953756204001753 PMID:15736859

Cai, Z., Zhou, Q., Peng, S., & Li, K. (2010). Promoted biodegradation and microbiological effects of petroleum hydrocarbons by Impatiens balsamina L. with strong endurance. *Journal of Hazardous Materials*, *183*(1), 731–737. doi:10.1016/j.jhazmat.2010.07.087 PMID:20724074

Camarero, S., Ibarra, D., Martínez, M. J., & Martínez, Á. T. (2005). Lignin-derived compounds as efficient laccase mediators for decolorization of different types of recalcitrant dyes. *Applied and Environmental Microbiology*, *71*(4), 1775–1784. doi:10.1128/AEM.71.4.1775-1784.2005 PMID:15812000

Cambria, M. T., Di Marino, D., Falconi, M., Garavaglia, S., & Cambria, A. (2010). Docking simulation and competitive experiments validate the interaction between the 2, 5-xylidine inhibitor and *Rigidoporus lignosus* laccase. *Journal of Biomolecular Structure & Dynamics*, *27*(4), 501–509. doi:10.1080/07391102.2010.10507334 PMID:19916571

Cameotra, S. S., & Makkar, R. S. (2010). Biosurfactant-enhanced bioremediation of hydrophobic pollutants. *Pure and Applied Chemistry*, *82*(1), 97–116. doi:10.1351/PAC-CON-09-02-10

Campos, J. M., Montenegro Stamford, T. L., Sarubbo, L. A., de Luna, J. M., Rufino, R. D., & Banat, I. M. (2013). Microbial biosurfactants as additives for food industries. *Biotechnology Progress*, *29*(5), 1097–1108. doi:10.1002/btpr.1796 PMID:23956227

Campos, V. M., Merino, I., Casado, R., Pacios, L. F., & Gómez, L. (2008). Phytoremediation of organic pollutants. *Spanish Journal of Agricultural Research*, *6*(S1), 38–47. doi:10.5424/sjar/200806S1-372

Canadell, J. G., Le Quéré, C., Raupach, M. R., Field, C. B., Buitenhuis, E. T., Ciais, P., & Marland, G. et al. (2007). Contributions to accelerating atmospheric CO2 growth from economic activity, Cintensity, and efficiency of natural sinks. *Proceedings of the National Academy of Sciences of the United States of America*, *104*(47), 18866–18870. doi:10.1073/pnas.0702737104 PMID:17962418

Canellas, L. P., Olivares, F. L., Okorokova-Façanha, A. L., & Façanha, A. R. (2002). Humic acids isolated from earthworm compost enhance root elongation, lateral root emergence, and plasma membrane H+-ATPase activity in maize roots. *Plant Physiology*, *130*(4), 1951–1957. doi:10.1104/pp.007088 PMID:12481077

Canfield, D. E., Stewart, F. J., Thamdrup, B., De Brabandere, L., Dalsgaard, T., Delong, E. F., & Ulloa, O. et al. (2010). A cryptic sulfur cycle in oxygen-minimum–zone waters off the Chilean coast. *Science*, *330*(6009), 1375–1378. doi:10.1126/science.1196889 PMID:21071631

Cao, X., Ying, X., & Janelle, L., (2013). The effects of micro-algae characteristics on the bioremediation rate of deepwater horizon crude oil. *Journal of Emerging Investigators,* 1-7.

Cao, W., Zhang, H., Wang, Y., & Pan, J. (2012). Bioremediation of polluted surface water by using biofilms on filamentous bamboo. *Ecological Engineering*, *42*, 146–149. doi:10.1016/j.ecoleng.2012.02.018

Carol, P. (2012). Caffeine in Wastewater Is a Tracer for Human Fecal Contamination. *Environmental Health Perspectives*, *120*(3), a108–a109. doi:10.1289/ehp.120-a108a PMID:22382130

Carpenter, S. R., Stanley, E. H., & Vander-Zanden, M. J. (2011). State of the worlds freshwater ecosystems: Physical, chemical and biological changes. *Annual Review of Environment and Resources*, *36*(1), 75–99. doi:10.1146/annurev-environ-021810-094524

Carredano, E., Karlsson, A., Kauppi, B., Choudhury, D., Parales, R. E., Parales, J. V., & Ramaswamy, S. et al. (2000). Substrate binding site of naphthalene 1, 2-dioxygenase: Functional implications of indole binding. *Journal of Molecular Biology*, *296*(2), 701–712. doi:10.1006/jmbi.1999.3462 PMID:10669618

Carreiro, M. M., Sinsabaugh, R. L., Repert, D. A., & Parkhurst, D. F. (2000). Microbial enzyme shifts explain litter decay responses to simulated nitrogen deposition. *Ecology*, *81*(9), 2359–02365. doi:10.1890/0012-9658(2000)081[2359:MESELD]2.0.CO;2

Carr, G. M., & Neary, J. P. (2008). *Water Quality for Ecosystem and Human Health* (2nd ed.). United Nations Environment Programme Global Environment Monitoring System.

Carvalho, P. N., Basto, M. C. P., Almeida, C. M. R., & Brix, H. (2014). A review of plant-pharmaceutical interactions: From uptake and effects in crop plants to phytoremediation in constructed wetlands. *Environmental Science and Pollution Research International*, *21*(20), 11729–11763. and doi:10.1007/s11356-014-2550-3

Cataldo, D. A., Bean, R. M., & Fellow, R. J. (1987). Uptake and fate of phenol aniline and quinoline in terrestrial plants. In R. H. Gray, . . . (Eds.), Health and environmental research on complex organic mixtures. Academic Press.

Cazes, de M., Abejón, R., Belleville, M. & Sanchez-Marcano, J. (2014). Membrane bioprocesses for pharmaceutical micropollutant removal from waters. *Membranes (Basel)*, *4*(4), 692–729.

Cébron, A., Louvel, B., Faure, P., France-Lanord, C., Chen, Y., Murrell, J. C., & Leyval, C. (2011). Root exudates modify bacterial diversity of phenanthrene degraders in PAH-polluted soil but not phenanthrene degradation rates. *Environmental Microbiology*, *13*(3), 722–736. doi:10.1111/j.1462-2920.2010.02376.x PMID:21087382

Celik, I., Gallicchio, L., Boyd, K., Lam, T. K., Matanoski, G., Tao, X., & Alberg, A. J. et al. (2008). Arsenic in drinking water and lung cancer: A systematic review. *Environmental Research*, *108*(1), 48–55. doi:10.1016/j.envres.2008.04.001 PMID:18511031

Celin, A., Sukla, L. B., & Misra, V. N. (2005). Biological elimination of sulphur from high sulphur coal by Aspergillus-like fungi. *Fuel*, *84*, 1597–1600.

Central Pollution Control Board (CPCB). (2008). Status of water quality in India 2007. New Delhi, India: Author.

Cernansky, S., Kolenck, M., Sevc, J., Unk, M., & Hiller, E. (2009). Fungal votalization of trivalent and pentavalent arsenic under laboratoryconditions. *Bioresource Technology*, *100*(30), 1037–1043. doi:10.1016/j.biortech.2008.07.030 PMID:18774290

Cerniglia, C. E. (1992). Biodegradatation of polycyclic aromatic hydrocarbons. *Biodegradation*, *3*(2-3), 351–368. doi:10.1007/BF00129093

Cerniglia, C. E., Gibson, D. T., & Van Baalen, C. (1980). Oxidation of naphthalene by cyanobacteria and microalgae. *Journal of General Microbiology*, *116*(2), 495–500. PMID:7003059

Chacko, J. T., & Subramaniam, K. (2011). Enzymatic degradation of azo dyes-A review. *International Journal of Environmental Sciences*, *1*(6), 1250.

Chakraborti, D., Das, B., Rahman, M. M., Chowdhury, U. K., Biswas, B., Goswami, A. B., & Das, D. et al. (2009). Status of groundwater arsenic contamination in the state of West Bengal, India: A 20-year study report. *Molecular Nutrition & Food Research*, *53*(5), 542–551. doi:10.1002/mnfr.200700517 PMID:19382148

Chamberlain, K., & Crawford, D. L. (2000). Thatch biodegradation and antifungal activities of two lignocellulolytic Streptomyces strains in laboratory cultures and in golf green turfgrass. *Canadian Journal of Microbiology*, *46*(6), 550–558. doi:10.1139/w00-025 PMID:10913977

Chandrasekaran, M., & Bahkali, A. H. (2013). Valorization of date palm (Phoenix dactylifera) fruit processing by-products and wastes using bioprocess technology –[Review]. *Saudi Journal of Biological Sciences*, *20*(2), 105–120. doi:10.1016/j.sjbs.2012.12.004 PMID:23961227

Chang, C. Y., Chang, J. S., Vigneswaran, S., & Kandasamy, J. (2008). Pharmaceutical wastewater treatment by membrane bioreactor process—a case study in southern Taiwan. *Desalination*, *234*(1-3), 393–401. doi:10.1016/j.desal.2007.09.109

Chang, F. C., Ko, C. H., Tsai, M. J., Wang, Y. N., & Chung, C. Y. (2014). Phytoremediation of heavy metal contaminated soil by *Jatrophacurcas*. *Ecotoxicology (London, England)*, *23*(10), 1969–1978. doi:10.1007/s10646-014-1343-2 PMID:25236867

Chatterjee, S., Gupta, D., Roy, P., Chatterjee, N. C., Saha, P., & Dutta, S. (2011). Study of a lead tolerant yeast strain BUSCY1 (MTCC9315). *African Journal of Microbiological Research*, *5*, 5362–5372.

Chatzipavlidis, I., Kefalogianni, I., Venieraki, A., & Holzapfel, W. (2013). Status and Trends of the Conservation and Sustainable Use of Microorganisms in Agroindustrial Processes. Commission on Genetic Resources for Food and Agriculture, Food and Agriculture Organization of the United Nations.

Chaudhry, Q., Blom-Zandstra, M., Gupta, S. K., & Joner, E. (2005). Utilising the synergy between plants and rhizosphere microorganisms to enhance breakdown of organic pollutants in the environment (15 pp). *Environmental Science and Pollution Research International*, *12*(1), 34–48. doi:10.1065/espr2004.08.213 PMID:15768739

Checkai, R. T., Corey, R. B., & Helmke, P. A. (1987). Effects of ionic and complexed metal concentrations on plant uptake of cadmium and micronutrient metals from solution. *Plant and Soil*, *99*(2-3), 335–345. doi:10.1007/BF02370879

Chekol, T., Vough, L. R., & Chaney, R. L. (2004). Phytoremediation of polychlorinated biphenyl-contaminated soils: The rhizosphere effect. *Environment International*, *30*(6), 799–804. doi:10.1016/j.envint.2004.01.008 PMID:15120198

Chekroun, K. B., & Mourad, B. (2013). The role of algae in phytoremediation of heavy metals: A review. *J Mater Environ Sci*, *4*(6), 873–880.

Chelliapan, S., & Golar, S. (2011). Pre-treatment of antibiotic wastewater using an anaerobic reactor. *Research Journal of Chemical Sciences*, *1*(6), 59–63.

Chelliapan, S., & Sallis, P. J. (2015). Anaerobic treatment of high sulphate containing pharmaceutical wastewater. *Journal of Scientific and Industrial Research*, *74*, 526–530.

Chelliapan, S., Wilby, T., & Sallis, P. J. (2006). Performance of an up-flow anaerobic stage reactor (UASR) in the treatment of pharmaceutical wastewater containing macrolide antibiotics. *Water Research*, *40*(3), 507–516. and doi:10.1016/j.watres.2005.11.020

Chelliapan, S., Yuzir, A., Din, M. F., & Sallis, P. J. (2011). Anaerobic pre-treatment of pharmaceutical wastewater using packed bed reactor. *International Journal of Chemical Engineering and Applications*, *2*(1), 32–37. doi:10.7763/IJCEA.2011.V2.71

Chen, F., Tan, M., Ma, J., Zhang, S., Li, G., & Qu, J. (2016). Efficient remediation of PAH-metal co-contaminated soil using microbial-plant combination: A greenhouse study. *Journal of Hazardous Materials*, *302*, 250–261. doi:10.1016/j.jhazmat.2015.09.068 PMID:26476312

Cheng, J., Huang, Y., Feng, J., Sun, J., Zhou, J., & Cen, K. (2013). Improving CO2 fixation efficiency by optimizing Chlorella PY-ZUI culture conditions in sequential bioreactors. *Bioresource Technology*, *144*, 321–327. doi:10.1016/j.biortech.2013.06.122 PMID:23891832

Chen, J., Fales, S. L., Varga, G. A., & Royse, D. J. (1995). Biodegradation of cell wall components of maize stover colonized by white-rot fungi and resulting impact on in-vitro digestibility. *Journal of the Science of Food and Agriculture*, *68*(1), 91–98. doi:10.1002/jsfa.2740680115

Chen, J., Sun, G. X., Wang, X. X., Lorenzo, V. C. D., Rosen, B. P., & Zhu, Y. G. (2014). Volatilization of arsenic from polluted soil by *Pseudomonas putida* engineered for expression of the arsM arsenic (III) S-adenosine methyltransferase gene. *Environmental Science & Technology*, *48*(17), 10337–10344. doi:10.1021/es502230b PMID:25122054

Chen, S., & Wilson, D. W. (2008). Genetic engineering of bacteria and their potential for Hg^{2+} bioremediation. *Bioremediation*, *8*(2), 97–103. PMID:9342882

Chen, W. C., Chang, N. B., & Chen, J. C. (2003). Rough set-based hybrid fuzzy-neural controller design for industrial wastewater treatment. *Water Research*, *37*(1), 95–107. doi:10.1016/S0043-1354(02)00255-5 PMID:12465791

Chen, W.-Y., Wu, J.-H., Lin, S.-C., & Chang, J.-E. (2016). Bioremediation of polychlorinated-p-dioxins/dibenzofurans contaminated soil using simulated compost-amended landfill reactors under hypoxic conditions. *Journal of Hazardous Materials*, *312*, 159–168. doi:10.1016/j.jhazmat.2016.03.060 PMID:27037469

Chen, X. J., Wu, X. H., Jian, Y., Yuan, H. Z., Zhou, P., Ge, T. D., & Wu, J. S. et al. (2015). Cdioxide assimilation potential, functional gene amount and RubisCO activity of autotrophic microorganisms in agricultural soils. *Huan Jing Ke Xue*, *35*, 1144–1150. PMID:24881409

Chen, X., Nielsen, J. L., Furgal, K., Liu, Y., Lolas, I. B. Y., & Bester, K. (2011). Biodegradation of triclosan and formation of methyl-triclosan in activated sludge under aerobic conditions. *Chemosphere*, *84*(4), 452–456. doi:10.1016/j.chemosphere.2011.03.042 PMID:21507452

Chen, Z., Ren, N., Wang, A., Zhang, Z., & Shi, Y. (2008). A novel application of TPAD–MBR system to the pilot treatment of chemical synthesis-based pharmaceutical wastewater. *Water Research*, *42*(13), 3385–3392. and doi:10.1016/j.watres.2008.04.020

Chen, Z., Wen, Q., Wang, J., & Li, F. (2006). Simultaneous removal of carbon and nitrogen from municipal-type synthetic wastewater using net-like rotating biological contactor (NRBC). *Process Biochemistry*, *41*(12), 2468–2472. doi:10.1016/j.procbio.2006.06.003

Cheraghi, M., Lorestani, B., & Yousefi, N. (2011). Introduction of hyper accumulator plants with phytoremediation potential of a lead-zinc mine in Iran. *International Journal of Environmental, Chemical, Ecological. Geological and Geophysical Engineering*, *5*(5), 289–294.

Cheung, P. Y., & Kinkle, B. K. (2001). Mycobacterium diversity and pyrene mineralization in petroleum-contaminated soils. *Applied and Environmental Microbiology*, *67*(5), 2222–2229. doi:10.1128/AEM.67.5.2222-2229.2001 PMID:11319104

Cheung, W. W. L., Lam, V. W. Y., Sarmiento, J. L., Kearney, K., Watson, R., Zeller, D., & Pauly, D. (2009). *Re-distribution offish catch by climate change: A summary of a new scientific analysis*. Global Change Biology, Pew Ocean Science Series.

Chien, M. F., Makita, R., Sugawara, K., & Inoue, C. (2015). Study on As Uptake and Rhizobacteria of Two as Hyperaccumulators Forward to As Phytoremediation. Paper presented at the *Advanced. Materials Research*, *1130*, 568–571.

Chijioke-Osuji, C. C., Ibegbulam-Njoku, P. N., & Belford, E. J. D. (2014). Biodegradation of crude oil polluted soil by co-composting with agricultural wastes and inorganic fertilizer. *Journal of Natural Sciences Research.*, *4*(6), 28–39.

Chikere, C. B., Okpokwasili, G. C., & Chikere, B. O. (2011). Monitoring of microbial hydrocarbon remediation in the soil. *3 Biotech, 1*(3), 117-138.

Chiong, T., Lau, S., Khor, E., & Danquah, M. (2014). Enzymatic approach to phenol removal from wastewater using peroxidases. *Biochemical Engineering & Bioprocess Engineering. OA Biotechnology*, *3*(1), 9.

Chitrani, B. D., Ghazani, A. A., & Chan, W. C. D. (2006). Determining the size and shape dependence of gold nanoparticles uptake into mammalian cells. *Nano Letters*, *6*(4), 662–668. doi:10.1021/nl052396o PMID:16608261

Chiu, S. Y., Kao, C. Y., Chen, C. H., Kuan, T. C., Ong, S. C., & Lin, C. S. (2008). Reduction of CO2 by a high density culture of Chlorella sp. in a semicontinuous photobioreactor. *Bioresource Technology*, *99*(9), 3389–3396. doi:10.1016/j.biortech.2007.08.013 PMID:17904359

Choudhary, V. R., & Mantri, K. (2000). Adsorption of aromatic hydrocarbons on highly siliceous MCM-41. *Langmuir*, *16*(17), 7031–7037. doi:10.1021/la991714u

Chowdhury, S., Bala, N. N., & Dhauria, P. (2012). Bioremediation- A Natural way for cleaner environment. *Int J Pharm Chem Biol Sci*, *2*(4), 600–611.

Christen, P., Davidson, S., Combet-Blanc, Y., & Auria, R. (2011). Phenol biodegradation by the thermoacidophilic archaeon *Sulfolobus solfataricus* 98/2 in a fed batch bioreactor. *Biodegradation*, *22*(3), 475–484. doi:10.1007/s10532-010-9420-6 PMID:20886261

Christen, P., Vega, A., Casalot, L., Simon, G., & Auria, R. (2012). Kinetics of aerobic phenol biodegradation by the acidophilic and hyperthermophilic archaeon *Sulfolobus solfataricus* 98/2. *Biochemical Engineering Journal*, *62*, 56–62. doi:10.1016/j.bej.2011.12.012

Chuan, M. C., Shu, G. Y., & Liu, J. C. (1996). Solubility of heavy metals in a contaminated soil: Effects of redox potential and pH. *Water, Air, and Soil Pollution*, *90*(3-4), 543–556. doi:10.1007/BF00282668

Churchill, R. L. T., Sheedy, C., Yau, K. Y. F., & Hall, C. (2002). Micro-flow-injection analysis (MFIA) immunoassay of herbicide residue 2, 6-dichlorobenzamide – towards automated at-line monitoring using modular microfluidics. *Analytica Chimica Acta*, *468*, 185–194. doi:10.1016/S0003-2670(02)00093-4

Chushak, Y. G., Chapleau, R. R., Frey, J. S., Mauzy, C. A., & Gearhart, J. M. (2015). Identifying potential protein targets for toluene using a molecular similarity search, *in silico* docking and *in vitro* validation. *Toxicological Reviews*, *4*(2), 519–526.

Ciriolo, M. R., Civitareale, P., Carri, M. T., De Martino, A., Galiazzo, F., & Rotilio, G. (1994). Purification and characterization of Ag, Zn-superoxide dismutase from *Saccharomyces cerevisiae* exposed to silver. *The Journal of Biological Chemistry*, *269*, 25783–25787. PMID:7929283

Cirja, M., Ivashechkin, P., Schäffer, A., & Corvini, P. (2008). Factors affecting the removal of organic micropollutants from wastewater in conventional treatment plants (CTP) and membrane bioreactors (MBR). *Reviews in Environmental Science and Biotechnology*, *7*(1), 61–78. doi:10.1007/s11157-007-9121-8

Clara, M., Kreuzinger, N., Strenn, B., Gans, O., & Kroiss, H. (2005). The solids retention time - a suitable design parameter to evaluate the capacity of wastewater treatment plants to remove micropollutants. *Water Research*, *39*(1), 97–106. and doi:10.1016/j.watres.2004.08.036

Clara, M., Strenn, B., Gans, O., Martinez, E., Kreuzinger, N., & Kroiss, H. (2005). Removal of selected pharmaceuticals, fragrances and endocrine disrupting compounds in a membrane bioreactor and conventional wastewater treatment plants. *Water Research*, *39*(19), 4797–4807. and doi:10.1016/j.watres.2005.09.015

Clark, T. P., & Piskin, R. (1977). Chemical quality and indicator parameters for monitoring leachate in Illinois. *Environ. Geol.*, *1*(6), 329–340. doi:10.1007/BF02380501

Claus, H., Gleixner, G., & Filip, Z. (1999). Formation of humic-like substances in mixed and pure cultures of aquatic microorganisms. *Acta Hydrochimica et Hydrobiologica*, *27*(4), 200–207. doi:10.1002/(SICI)1521-401X(199907)27:4<200::AID-AHEH200>3.0.CO;2-Q

Clemens, S., Kim, E. J., Neumann, D., & Schroeder, J. I. (1999). Tolerance to toxic metals by a gene family of phytochelatin synthases from plants and yeast. *The EMBO Journal*, *18*(12), 3325–3333. doi:10.1093/emboj/18.12.3325 PMID:10369673

Clemens, S., Palmgren, M. G., & Krämer, U. (2002). A long way ahead: Understanding and engineering plant metal accumulation. *Trends in Plant Science*, *7*(7), 309–315. doi:10.1016/S1360-1385(02)02295-1 PMID:12119168

Clemmensen, K., Bahr, A., Ovaskainen, O., Dahlberg, A., Ekblad, A., Wallander, H., & Lindahl, B. et al. (2013). Roots and Associated Fungi Drive Long-Term CSequestration in Boreal Forest. *Science*, *339*(6127), 1615–1618. doi:10.1126/science.1231923 PMID:23539604

Climate Change, Fisheries and Aquaculture (PaCFA). (2009). *Strategic framework for fisheries, aquaculture and climate change*. Author.

Cluis, C. (2004). Junk-greedy greens: Phytoremediation as a new option for soil decontamination. *BioTeach Journal*, *2*(6), 1-67.

Coelho, L. M., Rezende, H. C., de-Sousa, P. A. R., Melo, D. F. O., & Coelho, N. M. M. (2015). Bioremediation of Polluted Waters Using Microorganisms. In N. Shiomi (Ed.), *Advances in Bioremediation of Wastewater and Polluted Soil*. doi:10.5772/60770

Cohen, C. K., Fox, T. C., Garvin, D. F., & Kochian, L. V. (1998). The role of iron-deficiency stress responses in stimulating heavy-metal transport in plants. *Plant Physiology*, *116*(3), 1063–1072. doi:10.1104/pp.116.3.1063 PMID:9501139

Colberg, P. J. (1988). Anaerobic microbial degradation of cellulose, lignin, oligolignols, and monoaromatic lignin derivatives. In A. J. B. Zehnder (Ed.), *Biology of Anaerobic Microorganisms* (pp. 333–372). New York: John Wiley & Sons.

Coleman, J., Blake-Kalff, M., & Davies, E. (1997). Detoxification of xenobiotics by plants: Chemical modification and vacuolar compartmentation. *Trends in Plant Science*, *2*(4), 144–151. doi:10.1016/S1360-1385(97)01019-4

Collins, G., Woods, A., & Sharon, M. H. (2003). Microbial Community structure and Methanogenic Activity during Start-up of Psychrophilic Anaerobic Digesters Treating Synthetic Industrial Wastewaters. *FEMS Microbiology Ecology*, *46*(1), 159–170. doi:10.1016/S0168-6496(03)00217-4 PMID:19719569

Collins, Y. E., & Stotzky, G. (1989). Factors affecting the toxicity of heavy metals to microbes. In T. J. Beveridge & R. J. Doyle (Eds.), *Metal ions and bacteria* (pp. 31–90). New York, NY: John Wiley & Sons, Inc.

Coniglio, M. S., Busto, V. D., González, P. S., Medina, M. I., Milrad, S., & Agostini, E. (2008). Application of *Brassica napus* hairy root cultures for phenol removal from aqueous solutions. *Chemosphere*, *72*(7), 1035–1042. and doi:10.1016/j.chemosphere.2008.04.003

Connolly, M. L. (1983). Solvent-accessible surfaces of proteins and nucleic acids. *Science*, *221*(4612), 709–713. doi:10.1126/science.6879170 PMID:6879170

Coogan, M. A., Edziyie, R. E., La Point, T. W., & Venables, B. J. (2007). Algal bioaccumulation of triclocarban, triclosan, and methyl-triclosan in a North Texas wastewater treatment plant receiving stream. *Chemosphere*, *67*(10), 1911–1918. doi:10.1016/j.chemosphere.2006.12.027 PMID:17275881

Cook, R. L., & Hesterberg, D. (2013). Comparison of trees and grasses for rhizoremediation of petroleum hydrocarbons. *International Journal of Phytoremediation*, *15*(9), 844–860. doi:10.1080/15226514.2012.760518 PMID:23819280

Cooper, E. L., & Stein, E. A. (1981). *Invertebrate blood cells*. London: Academic Press.

Cornelissen, G., Gustafsson, Ö., Bucheli, T. D., Jonker, M. T., Koelmans, A. A., & van Noort, P. C. (2005). Extensive sorption of organic compounds to black carbon, coal, and kerogen in sediments and soils: Mechanisms and consequences for distribution, bioaccumulation, and biodegradation. *Environmental Science & Technology*, *39*(18), 6881–6895. doi:10.1021/es050191b PMID:16201609

Cornu, A., Besle, J. M., Mosoni, P., & Grent, E. (1994). Lignin-carbohydrate complexes in forages: Structure and consequences in the ruminal degradation of cell-wall carbohydrates. *Reproduction, Nutrition, Development*, *24*(5), 385–398. doi:10.1051/rnd:19940501 PMID:7802932

Cossarizza, A., Cooper, E. L., Suzuki, M. M., Salvioli, S., Capri, M., Gri, G., & Franceschi, C. et al. (1996). Earthworm leucocytes that are not phagocytic and cross-react with several human epitopes can kill human tumor cell lines. *Experimental Cell Research*, *224*(1), 174–182. doi:10.1006/excr.1996.0125 PMID:8612683

Cozzone, A. J. (1998). Post-translational modification of proteins by reversible phosphorylation in prokaryotes. *Biochimie*, *80*(1), 43–48. doi:10.1016/S0300-9084(98)80055-2 PMID:9587661

Crane, M., Watts, C., & Boucard, T. (2006). Chronic aquatic environmental risks from exposure to human pharmaceuticals. *The Science of the Total Environment*, *367*(1), 23–41. doi:10.1016/j.scitotenv.2006.04.010 PMID:16762401

Crawford, D. W., Bonnevie, N. C., & Wenning, R. J. (1995). Sources of pollution and sediment contamination in Newark Bay. *Exotoxicology and Environmental Safety, New Jersey*, *35*(1), 85–100. doi:10.1006/eesa.1995.1010 PMID:7540541

Crawford, R. L., & Crawford, D. L. (Eds.). (2005). *Bioremediation: Principles and applications*. Cambridge University Press.

Crecchio, C., Curci, M., Pizzigallo, M. D., Ricciuti, P., & Ruggiero, P. (2004). Effects of municipal solid waste compost amendments on soil enzyme activities and bacterial genetic diversity. *Soil Biology & Biochemistry*, *36*(10), 1595–1605. doi:10.1016/j.soilbio.2004.07.016

Cripps, C., Bumpus, J. A., & Aust, S. D. (1990). Biodegradation of azo and heterocyclic dyes by *Phanerochaete chyrosporium. Applied and Environmental Microbiology*, *56*, 1114–1118.

Crittenden, J. C., Reddy, P. S., Arora, H., Trynoski, J. H., Perram, D. L., & Summers, R. S. (1991). Predicting gas performance with rapid small-scale column tests. *Journal - American Water Works Association*, *83*, 77–87.

Cruz-Morató, C., Ferrando-Climent, L., Rodriguez-Mozaz, S., Barceló, D., Marco-Urrea, E., Vicent, T., & Sarrà, M. (2013). Degradation of pharmaceuticals in non-sterile urban wastewater by *Trametes versicolor* in a fluidized bed bioreactor. *Water Research*, *47*(14), 5200–5210. and doi:10.1016/j.watres.2013.06.007

Cruz-Morató, C., Lucas, D., Llorca, M., Rodriguez-Mozaz, S., Gorga, M., Petrovic, M., & Marco-Urrea, E. et al. (2014). Hospital wastewater treatment by fungal bioreactor: Removal efficiency for pharmaceuticals and endocrine disruptor compounds. *The Science of the Total Environment*, *493*, 365–376. and doi:10.1016/j.scitotenv.2014.05.117

Csutak, O., Stoica, I., Ghindea, R., Ana-Maria, T., & Vassu, T. (2010). Insights on yeast bioremediation processes. *Romanian Biotechnological Letters*, *15*(2), 5066–5071.

Cunningham, S. D., & Berti, W. R. (1993). Remediation of contaminated soils with green plants: An overview. *In Vitro Cellular & Developmental Biology. Plant*, *29*(4), 207–212. doi:10.1007/BF02632036

Cunningham, S. D., & Ow, D. W. (1996). Promises and prospects of phytoremediation. *Plant Physiology*, *110*(3), 715–719. doi:10.1104/pp.110.3.715 PMID:12226213

Cupples, A. M., Sanford, R. A., & Sims, G. K. (2005). Dehalogenation of Bromoxynil (3,5-Dibromo-4-Hydroxybenzonitrile) and Ioxynil (3,5-Diiodino-4-Hydroxybenzonitrile) by *Desulfitobacterium chlororespirans. Applied and Environmental Microbiology*, *71*(7), 3741–3746. doi:10.1128/AEM.71.7.3741-3746.2005 PMID:16000784

Curling, S. F., Clausen, C. A., & Winandy, J. E. (2002). Experimental method to quantify progressive stages of decay of wood by basidiomycete fungi. *International Biodeterioration & Biodegradation*, *49*(1), 13–19. doi:10.1016/S0964-8305(01)00101-9

Current Data for Atmospheric CO2. 2015. (http://co2now.org/,)

Dada, A. O., Ojediran, J. O., & Olalekan, A. P. (2013). Sorption of Pb2+ from aqueous solution unto modified rice husk: Isotherm studies. Advances in Physical Chemistry: Vol. 2013. (*Article ID: 842425,6*). Cairo, Egypt: Hindawi Publishing Corporation.

Daffonchio, D., Ferrer, M., Mapelli, F., Cherif, A., Lafraya, A., Malkawi, H. I., & Kalogerakis, N. et al. (2013). Bioremediation of Southern Mediterranean oil polluted sites comes of age. *New Biotechnology*, *30*(6), 743–748. doi:10.1016/j.nbt.2013.05.006 PMID:23727339

Daffonchio, D., Mapelli, F., Cherif, A., Malkawi, H. I., Yakimov, M. M., Abdel-Fattah, Y. R., & Boon, N. et al. (2012). ULIXES, unravelling and exploiting Mediterranean Sea microbial diversity and ecology for xenobiotics and pollutants cleanup. *Reviews in Environmental Science and Biotechnology*, *11*(3), 207–211. doi:10.1007/s11157-012-9283-x

Dagley, S. (1971). Catabolism of aromatic compounds by micro-organisms. *Advances in Microbial Physiology*, *6*(0), 1–46. doi:10.1016/S0065-2911(08)60066-1 PMID:4950664

Dagley, S., Chapman, P. J., Gibson, D. T., & Wood, J. M. (1964). Degradation of benzene nucleus by bacteria. *Nature*, *202*(4934), 775–778. doi:10.1038/202775a0 PMID:14187616

Dagley, S., Evans, W. C., & Ribbons, D. W. (1960). New pathways in the oxidative metabolism of aromatic compounds by microorganisms. *Nature*, *188*(4750), 560–566. doi:10.1038/188560a0 PMID:13719300

Dales, R. P., & Kalac, Y. (1992). Phagocytic defense by the earthworm *Eisenia fetida* against certain pathogenic bacteria. *Comparative Biochemistry and Physiology*, *101*(3), 487–490. doi:10.1016/0300-9629(92)90499-G

Dams, R. I., Paton, G. I., & Killham, K. (2007). Rhizo-remediation of pentachlorophenol by *Sphingobium chlorophenolicum*, ATCC 39723. *Chemosphere*, *68*(5), 864–870. doi:10.1016/j.chemosphere.2007.02.014 PMID:17376504

Danh, L. T., Truong, P., Mammucari, R., Tran, T., & Foster, N. (2009). Vetiver grass, Vetiveria zizanioides: A choice plant for phytoremediation of heavy metals and organic wastes. *International Journal of Phytoremediation*, *11*(8), 664–691. doi:10.1080/15226510902787302 PMID:19810597

Daniel, G. (2003). Microview of wood under degradation by bacteria and fungi. *Advances in Our Changing World*, *845*, 34–72. doi:10.1021/bk-2003-0845.ch004

Daniel, G., Asiegbu, F., & Johansson, M. (1998). The saprotrophic wood-degrading abilities of *Heterobasidium annosum* intersterility groups P and S. *Mycological Research*, *102*(8), 991–997. doi:10.1017/S0953756297005935

Das, N., & Chandran, P. (2011). Microbial Degradation of Petroleum Hydrocarbon Contaminants: An Overview. *Biotechnology Research International*. doi:10.4061/2011/941810

Dash, R. R., Gaur, A., & Balomajumder, C. (2009). Cyanide in industrial wastewaters and its removal: A review on biotreatment. *Journal of Hazardous Materials*, *163*(1), 1–11. doi:10.1016/j.jhazmat.2008.06.051 PMID:18657360

Dashtban, M., Schraft, H., & Qin, W. (2009). Fungal bioconversion of lignocellulosic residues; opportunities & perspectives. *International Journal of Biological Sciences*, *5*, 578–595. doi:10.7150/ijbs.5.578 PMID:19774110

Dashtban, M., Schraft, H., Syed, T. A., & Qin, W. (2010). Fungal biodegradation and enzymatic modification of lignin. *Int J Biochem Mol Biol*, *1*, 36–50. PMID:21968746

Das, M. P., Bashwant, M., Kumar, K., & Das, J. (2012). Control of pharmaceutical effluent parameters through bioremediation. *Journal of Chemical and Pharmaceutical Research*, *4*(2), 1061–1065.

Das, N. R., & Kumar, N. (2015). Assessment of phenol biodegradation capacity of indigenous bacteria isolated from sewage treatment plant. *African Journal of Microbiological Research*, *9*(16), 1150–1161. doi:10.5897/AJMR2013.5909

Das, N., & Chandran, P. (2010). Microbial degradation of petroleum hydrocarbon contaminants: An overview. *Biotechnology Research International*. PMID:21350672

Das, N., Vimala, R., & Karthika, P. (2008). Biosorption of heavy metals - An overview. *Indian Journal of Biotechnology*, *7*, 159–169.

Das, S., Dash, H. R., & Chakraborty, J. (2016). Genetic basis and importance of metal resistant genes in bacteria for bioremediation of contaminated environments with toxic metal pollutants. *Applied Microbiology and Biotechnology*, *100*(7), 2967–2984. doi:10.1007/s00253-016-7364-4 PMID:26860944

Datta, A. S., Chakrabortty, A., De-Dalal, S. S., & Lahiri, S. C. (2014). Fluoride contamination of ground water in West Bengal, India. *Research report. Fluoride*, *47*(3), 241–248.

Daughton, C. G., & Ternes, T. A. (1999). Pharmaceuticals and personal care products in the environment: Agents of subtle change? *Environmental Health Perspectives*, *107*(supplement 6), 907–942. doi:10.1289/ehp.99107s6907 PMID:10592150

Dave, D., & Ghaly, A. E. (2011). *Remediation technologies for marine oil spills: A critical review and comparative analysis*. Academic Press.

Davidson, E. A., & Janssens, I. A. (2006). Temperature sensitivity of soil Cdecomposition and feedbacks to climate change. *Nature*, *440*(7081), 165–173. doi:10.1038/nature04514 PMID:16525463

Davies, N. A., Hodson, M. E., & Black, S. (2002). Changes in toxicity and bioavailability of lead in contaminated soils to the earthworm *Eisenia fetida* (Savigny, 1826) after bone meal amendments to the soil. *Environmental Toxicology and Chemistry*, *21*(12), 2685–2691. doi:10.1002/etc.5620211222 PMID:12463565

De Angelis, K. M., Allgaier, M., Chavarria, Y., Fortney, J. L., Hugenholtz, P., Simmons, B., & Hazen, T. C. et al. (2011). Characterization of trapped lignin-degrading microbes in tropical forest soil. *PLoS ONE*, *6*(4), e19306. doi:10.1371/journal.pone.0019306 PMID:21559391

De Boer, W., Folman, L. B., Summerbell, R. C., & Boddy, L. (2005). Living in a fungal world: Impact of fungi on soil bacterial niche development. *FEMS Microbiology Reviews*, *29*(4), 795–811. doi:10.1016/j.femsre.2004.11.005 PMID:16102603

De Boer, W., Kowalchuk, G. A., & van Veen, J. A. (2006). Root-food and the rhizosphere microbial community composition. *The New Phytologist*, *170*(1), 3–6. doi:10.1111/j.1469-8137.2006.01674.x PMID:16539597

De Boer, W., & Vanderwal, A. (2008). Interactions between saprotrophic basidiomycetes and bacteria. In L. Boddy, J. C. Frankland, & P. Van West (Eds.), *Ecology of Saprotrophic Basidiomycetes* (pp. 142–15). Amsterdam: Academic Press. doi:10.1016/S0275-0287(08)80010-0

de Cássia, F. S., da Silva, R., Rufino, R. D., Luna, J. M., Farias, C. B., Filho, H. J., & Sarubbo, L. A. et al. (2013). Enhancement of biosurfactant production from *Pseudomonas cepacia* CCT6659 through Optimisation of Nutritional Parameters Using Response Surface Methodology. *Tenside Surfactants Detergents*, *50*(2), 137–142. doi:10.3139/113.110241

De Filipps, P., Di Palma, L., Petrucci, E., Scarsella, M., & Verdone, N. (2013). Production and characterization of adsorbent materials from sewage sludge by pyrolysis. *Chem Eng Trans*, *32*, 205–210.

De Fouquet, C. (2011). From exploratory data analysis to geostatistical estimation: Examples from the analysis of soil pollutants. *European Journal of Soil Science*, *62*(3), 454–466. doi:10.1111/j.1365-2389.2011.01374.x

De Graaff, M. A., Classen, A. T., Castro, H. F., & Schadt, C. W. (2010). Labile soil carbon inputs mediate the soil microbial community composition and plant residue decomposition rates. *The New Phytologist*, *188*(4), 1055–1064. doi:10.1111/j.1469-8137.2010.03427.x PMID:21058948 de la Cueva, S. C., Rodríguez, C. H., Cruz, N. O. S., Contreras, J. A. R., & Miranda, J. L. (2016). Changes in Bacterial Populations During Bioremediation of Soil Contaminated with Petroleum Hydrocarbons. *Water, Air, and Soil Pollution*, *227*(3), 1–12. doi:10.1007/s11270-016-2789-z

de Oliveira, D. W. F., França, Í. W. L., Félix, A. K. N., Martins, J. J. L., Giro, M. E. A., Melo, V. M. M., & Gonçalves, L. R. B. (2013). Kinetic study of biosurfactant production by *Bacillus subtilis* LAMI005 grown in clarified cashew apple juice. *Colloids and Surfaces. B, Biointerfaces*, *101*, 34–43. doi:10.1016/j.colsurfb.2012.06.011 PMID:22796769

De Souza, M. P., Pickering, I. J., Walla, M., & Terry, N. (2002). Selenium assimilation and volatilization from selenocyanate-treated Indian mustard and muskgrass. *Plant Physiology*, *128*(2), 625–633. doi:10.1104/pp.010686 PMID:11842165

De, D. N. (2000). *Plant cell vacuoles*. Academic Press.

DeDecker, B. S. (2000). Allosteric drugs: Thinking outside the active-site box. *Chemistry & Biology*, *7*(5), R103–R107. doi:10.1016/S1074-5521(00)00115-0 PMID:10801477

Deeb, R. A., & Cohen, L. (2000). Aerobic biotransformation of gasoline aromatics in multicomponent mixtures. *Bioremediation Journal*, *4*(1), 1–9.

Deegan, A. M., Shaik, B., Nolan, K., Urell, K., Oelgemo¨ller, M., Tobin, J., & Morrissey, A. (2011). Treatment options for wastewater effluents from pharmaceutical companies. *International Journal of Environmental Science and Technology*, *8*(3), 649–666. doi:10.1007/BF03326250

Dehghanian, F., Kay, M., & Kahrizi, D. (2015). A novel recombinant AzrC protein proposed by molecular docking and *in silico* analyses to improve azo dyes binding affinity. *Gene*, *569*(2), 233–238. doi:10.1016/j.gene.2015.05.063 PMID:26026905

Del Río, J. C., Gutiérrez, A., & Martínez, A.T. (2004). Identifying acetylated lignin units in non-wood fibers using pyrolysis-gas chromatography/ mass spectrometry. *Rapid Commun Mass Spectrom, 18,* 1181-1185.

Delaney, J. L., & Hughes, T. W. (1979). Source assessment: Manufacture of acetone and phenol from cumene. Mosanto Research Corp.

Delfino, J.J., & Dube, D.J. (1976). Persistent contamination of ground water by phenol. *Environmental Science and Health, A11*, 345-355.

Delhaize, E., Kataoka, T., Hebb, D. M., White, R. G., & Ryan, P. R. (2003). Genes encoding proteins of the cation diffusion facilitator family that confer manganese tolerance. *The Plant Cell*, *15*(5), 1131–1142. doi:10.1105/tpc.009134 PMID:12724539

DeLong, E. F., & Pace, N. R. (2001). Environmental diversity of bacteria and archaea. *Systematic Biology*, *50*(4), 470–478. doi:10.1080/106351501750435040 PMID:12116647

De-Lorenzo, M. E., Scott, G. I., & Ross, P. E. (1999). Effects of the agricultural pesticides atrazine, deethylatrazine, endosulfan, and chlorpyrifos on an estuarine microbial food web. *Environmental Toxicology and Chemistry*, *18*(12), 2824–2835. doi:10.1002/etc.5620181224

Den Boeft, J., Kruiswijk, F. J., & Schulting, F. L. (1984). Air pollution by combustion of solid fuels. The Hague: Ministry of Housing, Physical Planning and Environment (Publication Lucht No.37).

Deniza, F. (2015). Bioremediation of contaminated water with unnatural dye using blue-green alga *Spirulina Platensis. Environ Prog Sustain Energy*, *34*(5), 1414–1419. doi:10.1002/ep.12137

Deniz, O., Tozkoparan, E., Ìlvan, A., Çìftçì, F., Bozkanat, E., & Öztürk, S. (2000). A Case of Carbamazepine-Induced Lymphocytic Interstitial Pneumonitis. *Turkish Journal of Medical Science*, *30*, 509–510.

Derraik, J. G. B. (2002). The pollution of the marine environment by plastic debris: A review. *Marine Pollution Bulletin*, *44*(9), 842–852. doi:10.1016/S0025-326X(02)00220-5 PMID:12405208

Desbrow, C., Routledge, E. J., Brighty, C. G., Sumpter, J. P., & Waldock, M. (1998). Identification of estrogenic chemicals in STW effluent. 1. Chemical fractionation and in vitro biological screening. *Environmental Science & Technology*, *32*(11), 1549–1558. doi:10.1021/es9707973

Deshmukh, R., Khardenavis, A. A., & Purohit, H. J. (2016). Diverse Metabolic Capacities of Fungi for Bioremediation. *Indian Journal of Microbiology*, *56*(3), 247–264. doi:10.1007/s12088-016-0584-6 PMID:27407289

Deziel, N. (2014). Pharmaceuticals in wastewater treatment plant effluent waters. *Scholarly Horizons: University of Minnesota. Morris Undergraduate Journal*, *1*(2), 12.

Di Gregorio, S., Lampis, S., & Vallini, G. (2005). Selenite precipitation by a rhizospheric strain of Stenotrophomonas sp. isolated from the root system of Astragalus bisulcatus: A biotechnological perspective. *Environment International*, *31*(2), 233–241. doi:10.1016/j.envint.2004.09.021 PMID:15661289

Diamandis, E. P. (2004). Mass spectrometry as a diagnostic and a cancer biomarker discovery tool opportunities and potential limitations. *Molecular & Cellular Proteomics*, *3*(4), 367–378. doi:10.1074/mcp.R400007-MCP200 PMID:14990683

Diaz, E. (2004). Bacterial degradation of aromatic pollutants: A paradigm of metabolic versatility. *International Microbiology*, *7*, 173–180. PMID:15492931

Dibrov, P., Dzioba, J., Gosink, K. K., & Hase, C. C. (2002). Chemiosmotic mechanism of antimicrobial activity of Ag+ in *Vibrio cholerae. Antimicrob. Antimicrobial Agents and Chemotherapy*, *46*(8), 2668–2670. doi:10.1128/AAC.46.8.2668-2670.2002 PMID:12121953

Diez, M. C. (2010). Biological Aspects involved in the degradation of organic pollutants. *Journal of Soil Science and Plant Nutrition*, *10*(3), 244–267. doi:10.4067/S0718-95162010000100004

Di-Fiore, A., Alterio, V., Monti, S. M., DeSimone, G., & DAmbrosio, K. (2015). Thermostable Carbonic Anhydrases in Biotechnological Applications. *International Journal of Molecular Sciences*, *16*(7), 15456–15480. doi:10.3390/ijms160715456 PMID:26184158

Dighton, J. (2003). *The Role of Fungi in Ecosystem Processes*. New York, NY: Marcel Dekker. doi:10.1201/9780203911440

Dimitriu, P. A., Shukla, S. K., Conradt, J., Marquez, M. C., Ventosa, A., Maglia, A., & Mormile, M. R. et al. (2005). *Nitrincola lacisaponensis* gen. nov., sp. nov., a novel alkaliphilic bacterium isolated from an alkaline, saline lake. Int. J. Syst. *Evol. Micro.*, *55*(6), 2273–2278. doi:10.1099/ijs.0.63647-0 PMID:16280482

Dindal, D. L. (1990). *Soil biology guide*. New York: Wiley.

Distel, D. L., Amin, M., Burgoyne, A., Linton, E., Mamangkey, G., Morrill, W., & Yang, J. et al. (2011). Molecular phylogeny of Pholadoidea Lamarck, 1809 supports a single origin for xylotrophy (wood feeding) and xylotrophic bacterial endosymbiosis in Bivalvia. *Molecular Phylogenetics and Evolution*, *61*(2), 245–254. doi:10.1016/j.ympev.2011.05.019 PMID:21684342

Dixit, A., Dixit, S., & Goswami, C. S. (2015). Eco-friend alternatives for the removal of heavy metal using dry biomass of weeds and study the mechanism involved. *Journal of Bioremediation and Biodegradation*, *6*(3), 290–297.

Dixit, D., & Parmar, N. (2013). Treatment of pharmaceutical waste water by electro-coagulation and natural coagulation process[review]. *VSRD International Journal of Environmental Science and Technology*, *4*(5), 79–88.

Dixit, R., Malaviya, D., Pandiyan, K., Singh, U. B., Sahu, A., & Shukla, R. (2015). Bioremediation of heavy metals from soil and aquatic environment: An over view of principles and criteria of fundamental processes. *Sustainability*, *7*(2), 2189–2212. doi:10.3390/su7022189

Dizhbite, T., Zakis, G., Kizimia, A., Lazareva, E., Rossinskaya, G., Jurk-jane, V., & Viesturs, U. et al. (1999). Lignin – a useful bio-resource for the production of sorption-active materials. *Bioresource Technology*, *67*(3), 221–228. doi:10.1016/S0960-8524(98)80004-7

Doan, J., & Domingo, D. (2015). *Bioremediation: Microorganisms cleaning the environment*. Retrieved July 5, 2016, from: http://jdbioremed.wix.com/bioremediation#!timeline/c16s3

Dominguez, J., Velando, A., & Ferreiro, A. (2005). Are *Eisenia fetida* (Savigny, 1926) and *Eisenia andrei* Bouché (1972) (Oligochaeta, Lumbricidae) different biological species?[. *Pedobiologia*, *49*(1), 81–87. http://webs.uvigo.es/avelando/pdfs_archivos/Eisenias.pdf doi:10.1016/j.pedobi.2004.08.005

Dong, Y., Lin, J., Chen, Y., Fu, F., Chi, Y., & Chen, G. (2014). Graphene quantum dots, graphene oxide, carbon quantum dots and graphite nanocrystals in coals. *Nanscale*, *6*(13), 7410–7415. doi:10.1039/c4nr01482k PMID:24875280

Donmez, G. (2002). Bioaccumulation of the reactive textile dyes by *Candida tropicalis* growing in molasses medium. *Enzyme and Microbial Technology*, *20*(3), 363–366. doi:10.1016/S0141-0229(01)00511-7

Donzelli, D., De Michele, C., & Scholes, R. (2013). Competition between trees and grasses for both soil water and mineral nitrogen in dry savannas. *Journal of Theoretical Biology*, *332*, 181–190. doi:10.1016/j.jtbi.2013.04.003 PMID:23639405

Douds, J. D. D., Pfeffer, P. E., & Shachar-Hill, Y. (2000). Cpartitioning, cost, and metabolism of arbuscular mycorrhizas. In Arbuscular mycorrhizas: physiology and function. 107-129.

Doyle, R. J., Matthews, T. H., & Streips, U. N. (1980). Chemical basis for selectivity of metal ions by the Bacillus subtilis cell wall. *Journal of Bacteriology*, *143*, 471–480. PMID:6772632

Drewes, J. E. (2007). Removal of pharmaceutical residues during wastewater treatment. In M. Petrovic & D. Barcelo (Eds.), *Comprehensive analytical chemistry* (pp. 427–446). Oxford, UK: Elsevier.

Dridi, D., Bchir, F., Zouiten, A., Tahrani, L., & Mansour, H. B. (2013). Murine chronotoxicity to pharmaceutical wastewater. *Biological Rhythm Research*, *45*(2), 167–181. doi:10.1080/09291016.2013.787684

Drigo, B., Pijl, A. S., Duyts, H., Kielak, A. M., Gamper, H. A., Houtekamer, M. J., & vanVeen, J. A. et al. (2010). Shifting Cflow from roots into associated microbial communities in response to elevated atmospheric CO2. *Proceedings of the National Academy of Sciences of the United States of America*, *107*(24), 10938–10942. doi:10.1073/pnas.0912421107 PMID:20534474

Drillia, P., Dokianakis, S. N., Fountoulakis, M. S., Kornaros, M., Stamatelatou, K., & Lyberatos, G. (2005). On the occasional biodegradation of pharmaceuticals in the activated sludge process: The example of the antibiotic sulfamethoxazole. *Journal of Hazardous Materials*, *122*(3), 259–265. and doi:10.1016/j.jhazmat.2005.03.009

Druckenmiller, T., & Maroto-Valer, M. M. (2005). Csequestration using brine of adjusted pH to form mineral carbonates. *Matthew L. Fuel Processing Technology*, *86*(14-15), 1599–1614. doi:10.1016/j.fuproc.2005.01.007

Dubey, K.V., Charde, P.N., Meshram, S.U., Yadav, S.K., Singh, S., & Asha, A. J. (2012). Potential of New Microbial Isolates for Biosurfactant Production using Combinations of Distillery Waste with other Industrial Wastes. *J Petrol Environ Biotechnol S1:002*. doi:10.4172/2157-7463.S1-002

Ducklow, H. W., Steinberg, D. K., & Buesseler, K. O. (2001). Upper ocean carbon export and the biological pump. *Oceanography (Washington, D.C.)*, *14*(4), 50–58. doi:10.5670/oceanog.2001.06

Duffner, F. M., Kirchner, U., Bauer, M. P., & Muller, R. (2000). Phenol/cresol degradation by the thermophillic *Bacillus thermoglucosidasius* A7: Cloning and sequence analysis of five genes involved in the pathway. *Gene*, *256*(1-2), 215–221S. and doi:10.1016/S0378-1119(00)00352-8

Duffus, J. H. (2002). Heavy metals"—a meaningless term? *Pure and Applied Chemistry*, *74*(5), 793–807. doi:10.1351/pac200274050793

Dwivedi, A. K., & Rajput, D. P. S. (2014). Studies on adsorption removal of heavy metal (Cu and Cd) from aqueous solution by tea waste adsorbent. *Journal of Industrial Pollution Control*, *30*(1), 85–90.

Dwivedi, S. (2012). Bioremediation of heavy metal by algae: Current and future perspective. *Journal of Advanced Laboratory Research in Biology*, *3*(3), 195–199.

Eapen, S., Singh, S., & Dsouza, S. F. (2007). Advances in development of transgenic plants for remediation of xenobiotic pollutants. *Biotechnology Advances*, *25*(5), 442–451. doi:10.1016/j.biotechadv.2007.05.001 PMID:17553651

Eastmond, D. A., MacGregor, J. T., & Slesinski, R. S. (2008). Trivalent Chromium: Assessing the Genotoxic Risk of an Essential Trace Element and Widely Used Human and Animal Nutritional Supplement. *Critical Reviews in Toxicology*, *38*(3), 173–190. doi:10.1080/10408440701845401 PMID:18324515

Eastwood, D., Floudas, D., Binder, M., Majcherczyk, A., Schneider, P., Aerts, A., & Watkinson, S. C. et al. (2011). The plant cell wall-decomposing machinery underlies the functional diversity of forest fungi. *Science*, *333*(6043), 762–765. doi:10.1126/science.1205411 PMID:21764756

Eberhardt, M. (1976). Experience with the use of biologically effective activated carbon. In Reports on special problems in water technology (vol. 9, pp. 331-347). U.S. Environmental Protection Agency, Municipal Environmental Research Laboratory.

Eckmann, A., Felten, A., Mishchenko, A., Britnell, L., Krupke, R., Novoselov, K. S., & Casiraghi, C. (2012). Probing the nature of defects in graphene by Raman spectroscopy. *Nano Letters*, *12*(8), 3925–3930. doi:10.1021/nl300901a PMID:22764888

EDGAR. (2014). *Eemissions-report-93171*. Retrieved from http://edgar.jrc.ec.europa.eu/news_docs/jrc-2014-trends-in-global-co2-

Edwards, C. A., & Bohlen, P. J. (1996). *Biology and ecology of earthworms* (3rd ed.). London: Chapman & Hall.

Egland, P. G., Pelletier, D. A., Dispensa, M., Gibson, J., & Harwood, C. S. (1997). A cluster of bacterial genes for anaerobic benzene ring biodegradation. *Proceedings of the National Academy of Sciences of the United States of America*, *94*(12), 6484–6489. doi:10.1073/pnas.94.12.6484 PMID:9177244

Ehrhardt, H. M., & Rehm, H. J. (1985). Phenol degradation by microorganisms adsorbed on activated carbon. *Applied Microbiology and Biotechnology*, *21*, 32–36.

Ehrlich, G. G., Goelitz, D. F., & Godsy, E. M. (1982). Degradation of phenolic contaminants in ground water by anaerobic bacteria, St. Louis Park, MN. *Ground Water*, *20*(6), 703–710. doi:10.1111/j.1745-6584.1982.tb01390.x

Eickhorst, T., & Tippkötter, R. (2008). Improved detection of soil microorganisms using fluorescence in situ hybridization (FISH) and catalyzed reporter deposition (CARD-FISH). *Soil Biology & Biochemistry*, *40*(7), 1883–1891. doi:10.1016/j.soilbio.2008.03.024

Eisler, R. (1987). Polycyclic aromatic hydrocarbon hazards to fish, wildlife, and invertebrates: a synoptic review. *US Fish and Wildlife Service Biological Report, 85*(1.11), 1-55.

El Mahdi, A. M., Abdul Aziz, H., Amr, A., Salem, S., El-Gendy, N. S., & Nassar, H. (2015a). Performance of some New Bacterial Isolates on Biodegradation of Libyan Light Crude Oil Using Agro-Industrial Wastes as Co-Substrates. In Applied Mechanics and Materials (Vol. 802, pp. 496-500). Trans Tech Publications.

El Mahdi, A. M., Aziz, H. A., & Amr, A. (2015b). Performance of Isolated Kocuria sp. SAR1 in Light Crude Oil Biodegradation. *Journal of Bioremediation & Biodegradation, 6*(4), 1. doi:10.4172/2155-6199.1000303

El Mahdi, A. M., Aziz, H. A., El-Gendy, N. S., Amr, S. S. A., & Nassar, H. N. (2014). Optimization of Libyan Crude Oil Biodegradation by Using Solid Waste Date as a Natural Low-Cost Material. *J Bioremed Biodeg, 5*(7), 252.

El May, Y., Dorge, S., Jeguirim, M., Trouvé, G., & Said, R. (2012). Measurement of gaseous and particulate pollutants during combustion of date palm wastes for energy recovery. *Aerosol and Air Quality Research, 12*(5), 814–825.

El-Ashtoukhy, S. Z., El-Taweel, Y. A., Abdelwahab, O., & Nassef, E. M. (2013). Treatment of petrochemical wastewater containing phenolic compounds by electrocoagulation using a fixed bed electrochemical reactor. *International Journal of Electrochemical Science, 8*, 1534–1550.

Elba. (2014). Environmental hazards associated with pesticide import into Costa Rica, 1977 2009. *J. Environ. Biol,35.*

Elcey, C. D., & Manoj, B. (2010). Demineralization of coal by stepwise leaching: A study of sub- bituminous Indian coal by FTIR and SEM, *Journal of the University Chemical Technology and metallurgy, 45*(4), 385-390.

Elcey, C. D., & Manoj, B. (2013). Demineralization of sub-bituminous Coal by fungal leaching: structural characterization by X-ray and FTIR analysis. *Research Journal of Chemistry and Environment, 17*(8), 11-15.

Elcey, C. D., & Manoj, B. (2014). An investigation on bio-solubilization of coal by *Aspergillus niger. Chemical Science Transactions, 3*(3), 1200–1206.

Elcey, C. D., & Manoj, B. (2016). Graphitization of coal by Bio-solubilization: Structure probe by Raman spectroscopy. *Asian Journal of Chemistry, 28*(7), 1557–1560. doi:10.14233/ajchem.2016.19750

Elder, D. J., & Kelly, D. J. (1994). The bacterial degradation of benzoic acid and benzenoid compounds under anaerobic conditions: Unifying trends and new perspectives. *FEMS Microbiology Reviews, 13*(4), 441–468. doi:10.1111/j.1574-6976.1994.tb00061.x PMID:8011356

Elekwachi, C. O., Andresen, J., & Hodgman, T. C. (2014). Global use of bioremediation technologies for decontamination of ecosystems. *Journal of Bioremediation and Biodegradation, 5*, 1–9. doi:10.4172/2155-6199.1000225

El-Gendy, N. S., Ali, H. R., El-Nady, M. M., Deriase, S. F., Moustafa, Y. M., & Roushdy, M. I. (2014). Effect of different bioremediation techniques on petroleum biomarkers and asphaltene fraction in oil-polluted sea water. *Desalination and Water Treatment, 52*(40-42), 7484–7494. doi:10.1080/19443994.2013.831784

Elmolla, E. S., Ramdass, N., & Choudhari, M. (2012). Optimization of Sequencing Batch Reactor operating conditions for treatment of high strength pharmaceutical wastewater. *Journal of Environmental Science and Technology, 5*(6), 452–459. doi:10.3923/jest.2012.452.459

El-Naas, M., Al-Zuhair, S., & Makhlouf, S. (2008). Continuous biodegradation of phenol in a Spouted Bed Bioreactor (SBBR). *Chemical Engineering Journal, 160*(2), 565–570. doi:10.1016/j.cej.2010.03.068

El-Sayed, W. S., Ibrahim, M. K., Abu-Shady, M., El-Beih, F., Ohmura, N., Saiki, H., & Ando, A. (2003). Isolation and characterization of phenol-catabolizing bacteria from a coking plant. *Bioscience, Biotechnology, and Biochemistry*, *67*(9), 2026–2029. doi:10.1271/bbb.67.2026 PMID:14519997

El-Shahawi, M. S., Hamza, A., Bashammakhb, A. S., & Al-Saggaf, W. T. (2010). An overview on the accumulation, distribution, transformations, toxicity and analytical methods for the monitoring of persistent organic pollutants. *Talanta*, *80*(5), 1587–1597. doi:10.1016/j.talanta.2009.09.055 PMID:20152382

El-Sheekh, M. M., El-Shouny, W. A., Osman, M. E., & El-Gammal, E. W. (2005). Growth and heavy metals removal efficiency of *Nostoc muscorum* and *Anabaena subcylindrica* in sewage and industrial wastewater effluents. *Environmental Toxicology and Pharmacology*, *19*(2), 357–365. doi:10.1016/j.etap.2004.09.005 PMID:21783496

El-Sheekh, M. M., Farghl, A. A., Galal, H. R., & Bayoumi, H. S. (2016). Bioremediation of Different Types of Polluted Water Using Microalgae. *Rendconti Lincei-Scienze Fisiche e Naturali*, *27*(2), 401–410. doi:10.1007/s12210-015-0495-1

El-Sheekh, M. M., Gharieb, M. M., & Abou-El-Souod, G. W. (2009). Biodegradation of dyes by some green algae and cyanobacteria. *International Biodeterioration & Biodegradation*, *63*(6), 699–704. doi:10.1016/j.ibiod.2009.04.010

El-Sheekh, M. M., Gharieb, M. M., & Abou-El-Souod, G. W. (2012). Biodegradation of Phenolic and Polycyclic Aromatic Compounds by Some Algae and Cyanobacteria. *Bioremediation & Biodegradation*, *3*, 1–9.

El-Sheekh, M. M., & Hamouda, R. (2014). Biodegradation of crude oil by some cyanobacteria underheterotrophic conditions. *Desalination and Water Treatment*, *52*(7-9), 1448–1454. doi:10.1080/19443994.2013.794008

Emtiazi, G., & Hedayatkhah, A. (2015). Comparative modelling of 3D-structure of *Geobacter* sp. M21 (a metal reducing bacteria) Mn-Fe superoxide dismutase and its binding properties with bisphenol-A, aminotriazole and ethylene-diurea. *Iranian Journal of Environmental Technology*, *1*(1), 19–29.

Enright, A.-M., McHugh, S., Collins, G., & OFlaherty, V. (2005). Low-temperature anaerobic biological treatment of solvent containing pharmaceutical wastewater. Water Research, 39(19), 4587–4596.10.1016/j.watres.2005.08.037

Ensley, B. D. (2000). *Phytoremediation of toxic metals: using plants to clean up the environment*. John Wiley & s Sons, Incorporated.

Erakhrumen, A. (2015). Assessment of In-Situ Natural Dendroremediation Capability of Rhizophora racemosa in a Heavy Metal Polluted Mangrove Forest, Rivers State, Nigeria. *Journal of Applied Sciences and Environmental Management*, *19*(1), 21–27. doi:10.4314/jasem.v19i1.3

Eriksson, K. E. L., Blanchette, R. A., & Ander, P. (1990). *Microbial and enzymatic degradation of wood components*. Berlin: Springer-Verlag. doi:10.1007/978-3-642-46687-8

Ernst, W. H. O., Verkleij, J. A. C., & Schat, H. (1992). Metal tolerance in plants. *Acta Botanica Neerlandica*, *41*(3), 229–248. doi:10.1111/j.1438-8677.1992.tb01332.x

Erskine, P. T., Senior, N., Awan, S., Lambert, R., Lewis, G., Tickle, I. J., & Coope, J. B. (1997). X-ray structure of 5-aminolaevulinate dehydratase, a hybrid aldolase. *Nature Structural Biology*, *4*(12), 1025–1031. doi:10.1038/nsb1297-1025 PMID:9406553

e. Silva, N. M. P. R., Rufino, R. D., Luna, J. M., Santos, V. A., & Sarubbo, L. A. (2014). Screening of Pseudomonas species for biosurfactant production using low-cost substrates. *Biocatalysis and Agricultural Biotechnology*, *3*(2), 132–139. doi:10.1016/j.bcab.2013.09.005

Evanko, C. R., & Dzombak, D. A. (1997). Remediation of metals-contaminated soil and groundwater. *Environmental Sciences*, *412*, 1–45.

Evans, J. (2005). *Bioremediation mechanism discovered.* Royal Society of Chemistry. Retrieved October 1, 2016, from https://www.chemistryworld.com/news/6-july-2005-bioremediation-mechanism-discovered/1013110.article

Evans, C., & Hedger, J. (2001). Degradation of cell wall polymers. In G. Gadd (Ed.), *Fungi in Bioremediation.* doi:10.1017/CBO9780511541780.002

Fackler, K., Gradinger, C., Hinterstoisser, B., Messner, K., & Schwanninger, M. (2006). Lignin degradation by white rot fungi on spruce wood shavings during short-time solid-state fermentations monitored by near infrared spectroscopy. *Enzyme and Microbial Technology*, *39*(7), 1476–1483. doi:10.1016/j.enzmictec.2006.03.043

Fadeeva, M. S., Bertsova, Y. V., Euro, L., & Bogachev, A. V. (2011). Cys377 residue in NqrF subunit confers Ag^+ sensitivity of Na^+-translocating NADH:quinone oxidoreductase from *Vibrio harveyi. Biochemistry*, *76*, 186–195. PMID:21568851

Fahy, A., Ball, A. S., Lethbridge, G., Timmis, K. N., & McGenity, T. J. (2008). Isolation of alkali-tolerant benzene-degrading bacteria from a contaminated aquifer. *Letters in Applied Microbiology*, *47*(1), 60–66. doi:10.1111/j.1472-765X.2008.02386.x PMID:18544140

Falconer, R. E., Bown, J. L., White, N. A., & Crawford, J. W. (2007). Biomass recycling: A key to efficient foraging by fungal colonies. *Oikos*, *116*(9), 1558–1568. doi:10.1111/j.0030-1299.2007.15885.x

Fang, H. H. P., Chui, H. K., & Li, Y. Y. (1994). Microbial structure and activity of UASB granules treating different wastewaters. *Water Science and Technology*, *30*, 87–96.

FAO. (2014). *Food wastage: Key facts and figures.* Retrieved August 13, 2016 from "http://www.fao.org/news/story/en/item/196402/icode/

Farraji, H. (2014). Wastewater Treatment by Phytoremediation Methods. In M. A. H. A. Aziz (Ed.), *Wastewater Engineering: Types, Characteristics and Treatment Technologies* (pp. 205–218). Malaysia: IJSRPUB.

Farraji, H., Aziz, H. A., Tajuddin, R. M., & Mojiri, A. (2014). Optimization of Phytoremediation of Lead-Contaminated Soil by Spinach (Spinacia oleracea L). *International Journal of Scientific Research in Knowledge*, *2*(10), 480–486. doi:10.12983/ijsrk-2014-p0480-0486

Farraji, H., Zaman, N. Q., Tajuddin, R. M., & Faraji, H. (2016). Advantages and disadvantages of phytoremediation: A concise review. *Int J Env Tech Sci*, *2*, 69–75.

Farre, M., Brix, R., & Barcelo, D. (2005). Screening water for pollutants using biological techniques under European Union funding during the last 10 years. *Trends in Analytical Chemistry*, *24*(6), 532–545. doi:10.1016/j.trac.2005.03.008

Fathi, A. A., Azooz, M. M., & Al-Fredan, M. A. (2013). Phycoremediation and the potential of sustainable algal biofuel production using wastewater. *American Journal of Applied Sciences*, *10*(2), 189–194. doi:10.3844/ajassp.2013.189.194

Fatima, K., Afzal, M., Imran, A., & Khan, Q. M. (2015). Bacterial rhizosphere and endosphere populations associated with grasses and trees to be used for phytoremediation of crude oil contaminated soil. *Bulletin of Environmental Contamination and Toxicology*, *94*(3), 314–320. doi:10.1007/s00128-015-1489-5 PMID:25661008

Favero, N., Costa, P., & Massimino, M. L. (1991). *In vitro* uptake of cadmium by basidiomycete *Pleurotus ostreatus. Biotechnology Letters*, *10*(10), 701–704. doi:10.1007/BF01088172

Fayer, R., & Trout, J. M. (2005). Zoonotic protists in the marine environment. In S. Belkin & R. R. Colwell (Eds.), *Oceans and Health: Pathogens in the Marine Environment* (pp. 143–163). Springer. doi:10.1007/0-387-23709-7_7

Fazal, S., Zhang, B., Zhong, Z., Gao, L., & Lu, X. (2015). Membrane separation technology on pharmaceutical wastewater by using MBR (Membrane Bioreactor). *Journal of Environmental Protection*, *6*(04), 299–307. doi:10.4236/jep.2015.64030

Federle TW, Gasior SD, Nuck BA (1997) Extrapolating mineralization rates from the ready CO2 screening test to activated sludge, river water, and soil. *Environ. Toxicol. Chemist., 16*, 127–134. doi:10.1002 /etc.5620160205

Feig, M., Onufriev, A., Lee, M. S., Im, W., Case, D. A., & Brooks, C. L. (2004). Performance comparison of generalized born and Poisson methods in the calculation of electrostatic solvation energies for protein structures. *Journal of Computational Chemistry*, *25*(2), 265–284. doi:10.1002/jcc.10378 PMID:14648625

Feitkenhauer, H., Schnicke, S., Müller, R., & Märkl, H. (2001). Determination of the kinetic parameters of the phenol-degrading thermophile Bacillus thermoleovorans sp. A2. *Applied Microbiology and Biotechnology*, *57*, 744–750. doi:10.1007/s002530100823 PMID:11778888

Fengel, D., & Wegener, G. (1984). *Wood: Chemistry, ultrastructure, reactions*. Berlin: De Gruyter.

Feng, J., Kweon, O., Xu, H., Cerniglia, C. E., & Chen, H. (2012). Probing the NADH-and Methyl Red-binding site of a FMN-dependent azoreductase (AzoA) from *Enterococcus faecalis*. *Archives of Biochemistry and Biophysics*, *520*(2), 99–107. doi:10.1016/j.abb.2012.02.010 PMID:22387379

Feng, T., Lin, H., Guo, Q., & Feng, Y. (2014). Influence of an arsenate-reducing and polycyclic aromatic hydrocarbons-degrading Pseudomonas isolate on growth and arsenic accumulation in Pteris vittata L. and removal of phenanthrene. *International Biodeterioration & Biodegradation*, *94*, 12–18. doi:10.1016/j.ibiod.2014.06.005

Ferguson, J. F., & Pietari, J. M. H. (2000). Anaerobic transformations and bioremediation of chlorinated solvents. *Environmental Pollution*, *107*(2), 209–215. doi:10.1016/S0269-7491(99)00139-6 PMID:15092997

Fernández-Luqueño, F., Valenzuela-Encinas, C., Marsch, R., Martínez-Suárez, C., Vázquez-Núñez, E., & Dendooven, L. (2011). Microbial communities to mitigate contamination of PAHs in soil—possibilities and challenges: *A review. Environmental Science and Pollution Research International*, *18*(1), 12–30. doi:10.1007/s11356-010-0371-6 PMID:20623198

Fernández-Martínez, M., Vicca, S., Janssens, I. A., Sardans, J., Luyssaert, S., Campioli, M., & Peñuelas, J. et al. (2014). Nutrient availability as the key regulator of global forest Cbalance. *Nature Climate Change*, *4*(6), 471–476. doi:10.1038/nclimate2177

Ferrari, A. C., & Robertson, J. (2001). Resonant Raman spectroscopy of disordered, amorphous and diamond like carbon. *Phys Rev B., 64,* 075414-1–075414.

Ferrari, A. C., & Basko, M. (2013). Raman Spectroscopy as a versatile tool for studying the properties of graphene. *Nature Nanotechnology*, *8*(4), 235–246. doi:10.1038/nnano.2013.46 PMID:23552117

Finney, L. A., & OHalloran, T. V. (2003). Transition metal speciation in the cell: Insights from the chemistry of metal ion receptors. *Science*, *300*(5621), 931–936. doi:10.1126/science.1085049 PMID:12738850

Fischer, E., & Molnar, L. (1993). Environmental aspects of the chloragogenous tissue of earthworms. *Soil Biology & Biochemistry*, *24*(12), 1723–1727. doi:10.1016/0038-0717(92)90177-Y

FitzPatrick, M., Champagne, P., Cunningham, M. F., & Whitney, R. A. (2010). A biorefinery processing perspective: Treatment of lignocellulosic materials for the production of value-added products. *Bioresource Technology*, *101*(23), 8915–8922. doi:10.1016/j.biortech.2010.06.125 PMID:20667714

Flemming, C. A., Leung, K. T., Lee, H., Trevors, J. T., & Greer, C. W. (1994). Survival of lux-lac-marked biosurfactant-producing *Pseudomonas aeruginosa* UG2L in soil monitored by nonselective plating and PCR. *Applied and Environmental Microbiology*, *60*, 1606–1613. PMID:8017939

Floudas, D., Binder, M., Riley, R., Barry, K., Blanchette, R. A., Henrissat, B., & Yadav, J. S. et al. (2012). The Paleozoic origin of enzymatic lignin decomposition reconstructed from 31 fungal genomes. *Science*, *336*, 1715–1719. doi:10.1126/science.1221748 PMID:22745431

Folke, C., Carpenter, S., Walker, B., Scheffer, M., Elmqvist, T., Gunderson, L., & Holling, C. S. (2004). Regime shifts, resilience, and biodiversity in ecosystem management. *Annual Review of Ecology and Systematics*, *35*(1), 557–581. doi:10.1146/annurev.ecolsys.35.021103.105711

Ford, C. I., Walter, M., Northcott, G. L., Di, H. J., Cameron, K. C., & Trower, T. (2007). Fungal Inoculum Properties: Extracellular Enzyme Expression and Pentachlorophenol Removal by New Zealand Species in Contaminated Field Soils. *Journal of Environmental Quality*, *36*(6), 1749–1759. doi:10.2134/jeq2007.0007 PMID:17965377

Forney, L. J., Liu, W.-T., Guckert, J. B., Kumagai, Y., Namkung, E., Nishihara, T., & Larson, R. J. (2001). Structure of microbial communities in activated sludge: Potential implications for assessing the biodegradability of chemicals. *Ecotoxicology and Environmental Safety*, *49*(1), 40–53. doi:10.1006/eesa.2001.2034 PMID:11386714

Foucher, S., Battaglia-Brunet, F., Ignatiadis, I., & Morin, D. (2001). Treatment by sulfate-reducing bacteria of Chessy acid-mine drainage and metals recovery. *Chemical Engineering Science*, *56*(4), 1639–1645. doi:10.1016/S0009-2509(00)00392-4

Franchi, E., Agazzi, G., Rolli, E., Borin, S., Marasco, R., Chiaberge, S., & Barbafieri, M. (2016). Exploiting hydrocarbon-degrader indigenous bacteria for bioremediation and phytoremediation of a multi-contaminated soil. *Chemical Engineering & Technology*, *39*(9), 1676–1684. doi:10.1002/ceat.201500573

Frankenberger, D. G., & Fotopoulou, A. (2005). Non-specific biodegradation of the organophosphorus pesticides, cadusafos and ethoprophos, by two bacterial isolates. *FEMS Microbiology Ecology*, *53*(3), 369–378. doi:10.1016/j.femsec.2005.01.012 PMID:16329956

Franzetti, A., Tamburini, E., & Banat, I. M. (2010). Applications of biological surface active compounds in remediation technologies. In *Biosurfactants* (pp. 121–134). Springer New York. doi:10.1007/978-1-4419-5979-9_9

Free, C. M., Jensen, O. P., Mason, S. A., Eriksen, M., Williamson, N. J., & Boldgiv, B. (2014). High-levels of microplastic pollution in a large, remote, mountain lake. *Marine Pollution Bulletin*, *85*(1), 156–163. doi:10.1016/j.marpolbul.2014.06.001 PMID:24973278

Freer, N., & Detroy, R. W. (1982). Biological delignification of C-labeled lignocelluloses by basidiomycetes: Degradation and solubilization of the lignin and cellulose components. *Mycologia*, *74*(6), 943–951. doi:10.2307/3792724

Freitag, D., Lay, J. P., & Korte, F. (1984). Environmental hazard- the results as related to structures & transplantation into the environment. In QSAR Environ. Toxicol. Proc. Work. Quant. Struct. Act. Relat. D. Reidel Publishing Co.

Frey, S. D., Elliott, E. T., Paustian, K., & Peterson, G. A. (2000). Fungal translocation as a mechanism for soil nitrogen inputs to surface residue decomposition in a no-tillage agroecosystem. *Soil Biology & Biochemistry*, *32*(5), 689–698. doi:10.1016/S0038-0717(99)00205-9

Frey, S. D., Gupta, V. V. S. R., Elliott, E. T., & Paustian, K. (2001). Protozoan grazing affects estimates of Cutilization efficiency of the soil microbial community. *Soil Biology & Biochemistry*, *33*(12-13), 1759–1768. doi:10.1016/S0038-0717(01)00101-8

Fritsche, W., & Holrichter, M. (2005). Aerobic degradation of recalcitrant organic compounds by Microorganisms. In J. Joderning & J. Winter (Eds.), *Environmental biotechnology, Concepts and applications. Wiley-VCH Verlag GmbH & Co. KGaA*. doi:10.1002/3527604286.ch7

Fuchs, G., Boll, M., & Heider, J. (2011). Microbial degradation of aromatic compounds — from one strategy to four. *Nature Reviews. Microbiology*, *9*(11), 803–816. doi:10.1038/nrmicro2652 PMID:21963803

Fuentes, S., Barra, B., Caporaso, J. G., & Seeger, M. (2016). From rare to dominant: A fine-tuned soil bacterial bloom during petroleum hydrocarbon bioremediation. *Applied and Environmental Microbiology*, *82*(3), 888–896. doi:10.1128/AEM.02625-15 PMID:26590285

Fu, F., & Wang, Q. (2011). Removal of heavy metal ions from wastewaters: A review. *Journal of Environmental Management*, *92*(3), 407–418. doi:10.1016/j.jenvman.2010.11.011 PMID:21138785

Fulekar, M. H. (2009). Bioremediation of fenvalerate by *Pseudomonas aeruginosa* in a scale up bioreactor. *Romanian Biotechnological Letters*, *14*(6), 4900–4905.

Fulekar, M. H., Singh, A., & Bhaduri, A. M. (2009). Genetic engineering strategies for enhancing phytoremediation of heavy metals. *African Journal of Biotechnology*, *8*(4).

Fulke, B. A., Mudliar, S. N., Yadav, R., Shekh, A., Srinivasan, N., Ramanan, R., & Chakrabarti, T. et al. (2010). Bio-mitigation of CO2, calcite formation and simultaneous biodiesel precursors production using Chlorella species. *Bioresource Technology*, *101*(21), 8473–8476. doi:10.1016/j.biortech.2010.06.012 PMID:20580227

Gadd, G. M. (1990). Heavy metal accumulation by bacteria and other microorganisms. *Experientia*, *46*(8), 834–840. doi:10.1007/BF01935534

Gadd, G. M. (2004). Microbial influence on metal mobility and application for bioremediation. *Geoderma*, *122*(2), 109–119. doi:10.1016/j.geoderma.2004.01.002

Gadd, G. M. (2010). Metals, minerals and microbes: Geomicrobiology and bioremediation. *Microbiology*, *156*(3), 609–643. doi:10.1099/mic.0.037143-0 PMID:20019082

Gadd, G. M., & Griffiths, A. J. (1978). Microorganisms and heavy metal toxicity. *Microbial Ecology*, *4*(4), 303–317. doi:10.1007/BF02013274 PMID:24232222

Gadd, G. M., & Pan, X. (2016). Biomineralization, Bioremediation and Biorecovery of Toxic Metals and Radionuclides. *Geomicrobiology Journal*, *33*(3-4), 175–178. doi:10.1080/01490451.2015.1087603

Gahlawat, S., Makhijani, M., Chauhan, K., Valsangkar, S., & Gauba, P. Accessing the phytoremediation potential of *Cicer arietinum* for aspirin. *International Journal of Genetic Engineering and Biotechnology, 5*(2), 161-168.

Gahlawat, S., & Gauba, P. (2016). Phytoremediation of aspirin and tetracycline by *Brassica juncea. International Journal of Phytoremediation*, *18*(9), 929–935. and doi:10.1080/15226514.2015.1131230

Gaillard, V., Chenu, C., & Recous, S. (2003). Cmineralisation in soil adjacent to plant residues of contrasting biochemical quality. *Soil Biology & Biochemistry*, *35*(1), 93–99. doi:10.1016/S0038-0717(02)00241-9

Galassi, D. M. P., Lombardo, P., Fiasca, B., Di-Cioccio, A., Di-Lorenzo, T., Petitta, M., & Di-Carlo, P. (2014). Earthquakes trigger the loss of groundwater biodiversity. *Scientific Reports*, *4*, 6273. doi:10.1038/srep06273 PMID:25182013

Gali, V. S., Kumar, P., & Mehrotra, I. (2006). Biodegradation of phenol with wastewater as a cosubstrate in upflow anaerobic sludge blanket. *Journal of Environmental Engineering*, *132*(11), 1539–1542. doi:10.1061/(ASCE)0733-9372(2006)132:11(1539)

Galletti, G. C., Piccaglia, R., & Concialini, V. (1990). Optimization of electrochemical detection in the high-performance liquid chromatography of lignin phenolics from lignocellulosic byproducts. *Journal of Chromatography. A*, *507*, 439–450. doi:10.1016/S0021-9673(01)84223-0 PMID:2380306

Gambhir, R. S., Kapoor, V., Nirola, A., Sohi, R., & Bansal, V. (2012). Water Pollution: Impact of Pollutants and New Promising Techniques in Purification Process. *Journal of Human Ecology (Delhi, India)*, *37*(2), 103–109.

Ganesan, V. (2008). Rhizoremediation of cadmium soil using a cadmium-resistant plant growth-promoting rhizopseudomonad. *Current Microbiology*, *56*(4), 403–407. doi:10.1007/s00284-008-9099-7 PMID:18239844

Gan, S., Lau, E. V., & Ng, H. K. (2009). Remediation of soils contaminated with polycyclic aromatic hydrocarbons (PAHs). *Journal of Hazardous Materials*, *172*(2-3), 532–549. doi:10.1016/j.jhazmat.2009.07.118 PMID:19700241

Gans, J., Wolinsky, M., & Dunbar, J. (2005). Computational improvements reveal great bacterial diversity and high metal toxicity in soil. *Science*, *309*(5739), 1387–1390. doi:10.1126/science.1112665 PMID:16123304

Gao, Y., Guo, S., Wang, J., Li, D., Wang, H., & Zeng, D.-H. (2014). Effects of different remediation treatments on crude oil contaminated saline soil. *Chemosphere*, *117*, 486–493. doi:10.1016/j.chemosphere.2014.08.070 PMID:25240723

Garbeva, P., Hol, W. H. G., Termorshuizen, A. J., Kowalchuk, G. A., & de Boer, W. (2011). Fungistasis and general soil biostasis – a new synthesis. *Soil Biology & Biochemistry*, *43*(3), 469–477. doi:10.1016/j.soilbio.2010.11.020

Garcia, A., Rivas, H. M., Figueroa, J. L., & Monroe, A. L. (1995). Case history: Pharmaceutical wastewater treatment plant upgrade, Smith Kline Beecham Pharmaceuticals Company. *Desalination*, *102*(1–3), 255–263. doi:10.1016/0011-9164(95)00061-6

Garcia, E., Mellado, J. C. O., & Ventosa, A. (2004). *Halomonas organivorans* sp. nov., a moderate halophile able to degrade aromatic compounds. *International Journal of Systematic and Evolutionary Microbiology*, *54*(5), 1723–1728. doi:10.1099/ijs.0.63114-0 PMID:15388735

Garcia, M. T., Ventosa, A., & Mellado, E. (2005). Catabolic versatility of aromatic compound-degrading halophilic bacteria. *FEMS Microbiology Ecology*, *54*, 97–109. doi:10.1016/j.femsec.2005.03.009 PMID:16329976

Gardea-Torresdey, J. L., De La Rosa, G., & Peralta-Videa, J. R. (2004). Use of phytofiltration technologies in the removal of heavy metals: A review. *Pure and Applied Chemistry*, *76*(4), 801–813. doi:10.1351/pac200476040801

Garg, S. N., Baranwal, R. M., Mishra, S. C., Chaudhuri, T. K., & Bisaria, V. S. (2008). Laccase of *Cyathusbulleri*: Structural, catalytic characterization and expression in *Escherichia coli. Biochimica et Biophysica Acta*, *1784*(2), 259–268. doi:10.1016/j.bbapap.2007.11.006 PMID:18083129

Gartiser, S., Elke, U., Radka, A., & Kümmerer, K. (2007). Ultimate biodegradation and elimination of antibiotics in inherent tests. *Chemosphere*, *67*(3), 604–613. and doi:10.1016/j.chemosphere.2006.08.038

Gasch, A. P. (2007). Comparative genomics of the environmental stress response in ascomycete fungi. *Yeast (Chichester, England)*, *24*(11), 961–976. doi:10.1002/yea.1512 PMID:17605132

Gasic, K., & Korban, S. S. (2007). Expression of Arabidopsis phytochelatin synthase in Indian mustard (Brassica juncea) plants enhances tolerance for Cd and Zn. *Planta*, *225*(5), 1277–1285. doi:10.1007/s00425-006-0421-y PMID:17086401

Gaur, A., & Adholeya, A. (2004). Prospects of arbuscularmycorrhizal fungi in phytoremediation of heavy metal contaminated soils. *Current Science*, *86*(4), 528–534.

Gavis, J., & Ferguson, J. F. (1972). The cycling of mercury through the environment. *Water Research*, *6*(9), 989–1008. doi:10.1016/0043-1354(72)90053-X

Geesey, G., & Jang, L. (1990). Extracellular polymers for metal binding. In H. L. Ehrlich & C. L. Brierley (Eds.), *Microbial Mineral Recovery* (pp. 223–247). New York: Mc-GrawHill.

Gekeler, W., Grill, E., Winnacker, E. L., & Zenk, M. H. (1989). Survey of the plant kingdom for the ability to bind heavy metals through phytochelatins. *Zeitschrift für Naturforschung C*, *44*(5-6), 361–369.

Gerhardt, K. E., Huang, X.-D., Glick, B. R., & Greenberg, B. M. (2009). Phytoremediation and rhizoremediation of organic soil contaminants: Potential and challenges. *Plant Science*, *176*(1), 20–30. doi:10.1016/j.plantsci.2008.09.014

Gerrity, D., Gamage, S., Holady, J. C., Mawhinney, D. B., Quinones, O., Trenholm, R. A., & Snyder, S. A. (2011). Pilot-scale evaluation of ozone and biological activated carbon for trace organic contaminant mitigation and disinfection. *Water Research*, *45*(5), 2155–2165. doi:10.1016/j.watres.2010.12.031 PMID:21288550

Geslin, C., Llanos, J., Prieur, D., & Jeanthon, C. (2001). The manganese and iron superoxide dismutases protect *Escherichia coli* from heavy metal toxicity. *Research in Microbiology*, *152*(10), 901–905. doi:10.1016/S0923-2508(01)01273-6 PMID:11766965

Geys, R., Soetaert, W., & Van Bogaert, I. (2014). Biotechnological opportunities in biosurfactant production. *Current Opinion in Biotechnology*, *30*, 66–72. doi:10.1016/j.copbio.2014.06.002 PMID:24995572

Ghani, B., Takai, M., Hishani, N. Z., Kishimoto, N., Ismail, A. K., Tano, T., & Sugio, T. (1993). Isolation and Characterisation of a Mo^{6+}-reducing bacterium. *Applied and Environmental Microbiology*, *59*, 1176–1180. PMID:16348915

Ghorbani, Y., Oliazadeh, M., & Shavedi, A. (2008). Aluminum solubilization from red mud by some indigenous fungi in Iran. *Journal of Applied Biosciences*, *7*, 207–213.

Ghorbani, Y., Oliazadeh, M., Shavedi, A., Roohi, R., & Pirayehgar, A. (2007). Use of some isolated fungi in biological Leaching of Aluminium from low grade bauxite. *African Journal of Biotechnology*, *6*(11), 284–1288.

Ghosh, M., & Singh, S. (2005). A review on phytoremediation of heavy metals and utilization of it's by products. *Asian J Energy Environ*, *6*(4), 214–231.

Ghosh, M., Verma, S. C., Mengoni, A., & Tripathi, A. K. (2004). Enrichment and identification of bacteria capable of reducing chemical oxygen demand of anaerobically treated spent wash. *Journal of Applied Microbiology*, *6*, 241–278.

Ghosh, P., & Gangopadyay, R. (2000). Photofunctionalization of cellulose and lignocellulose fibers using photoactive organic acids. *European Polymer Journal*, *36*(3), 625–634. doi:10.1016/S0014-3057(99)00093-2

Gifford, S., Dunstan, R. H., O'Connor, W., Koller, C. E., & MacFarlan, G. R. (2006). Aquatic zooremediation: Deploying animals to remediate contaminated aquatic environments. *Trends Biotechnol*, *25*(2), 0-65.

Gilbert, N. (2012). Drug-pollution law all washed up. Nature. *International Weekly Journal of Science.*, *491*(7425), 503–504. and doi:10.1038/491503a

Giller, K., Witter, E., & McGrath, S. P. (1998). Toxicity of heavy metals to microorganisms and microbial processes in agricultural soils: A review. *Soil Biology & Biochemistry*, *30*(10–11), 1389–1414. doi:10.1016/S0038-0717(97)00270-8

Girotti, S., Ferri, E. N., Fumo, M. G., & Maiolini, E. (2008). Monitoring of environmental pollutants by bioluminescent bacteria. *Analytica Chimica Acta*, *608*(1), 2–29. doi:10.1016/j.aca.2007.12.008 PMID:18206990

Girotto, F., Alibardi, L., & Cossu, R. (2015). Food waste generation and industrial uses: *A review. Waste Management (New York, N.Y.)*, *45*, 32–41. doi:10.1016/j.wasman.2015.06.008 PMID:26130171

Gisbert, C., Ros, R., De Haro, A., Walker, D. J., Bernal, M. P., Serrano, R., & Navarro-Aviñó, J. (2003). A plant genetically modified that accumulates Pb is especially promising for phytoremediation. *Biochemical and Biophysical Research Communications*, *303*(2), 440–445. doi:10.1016/S0006-291X(03)00349-8 PMID:12659836

Gkorezis, P., Daghio, M., Franzetti, A., Van Hamme, J. D., Sillen, W., & Vangronsveld, J. (2016). The Interaction between Plants and Bacteria in the Remediation of Petroleum Hydrocarbons: An Environmental Perspective. *Frontiers in Microbiology,7*.

Godjevargova, T., Ivanova, D., Aleksieva, Z., & Dimova, N. (2003). Biodegradation of toxic organic components from industrial phenol production wastewaters by free and immobilized *Trichosporon cutaneum* R57. *Proc. Biochem.*, *38*(6), 915–920. doi:10.1016/S0032-9592(02)00170-X

Goerlitz, D. F., Troutman, D. E., & Gody, E. M. (1985). Migration of wood processing chemical in contaminated ground water in a sand aquifer at Pensacola, Florida. *Environmental Science & Technology*, *19*(10), 955–961. doi:10.1021/es00140a012

Gogate, P. R., & Pandit, A. R. (2004). A Review of imperative technologies for wastewater treatment I: Oxidation technologies at ambient conditions. *Advances in Environmental Research*, *8*(3-4), 501–551. doi:10.1016/S1093-0191(03)00032-7

Gohlke, H., Hendlich, M., & Klebe, G. (2000). Knowledge-based scoring function to predict protein-ligand interactions. *Journal of Molecular Biology*, *295*(2), 337–356. doi:10.1006/jmbi.1999.3371 PMID:10623530

Gold, M. H., Youngs, H. L., & Gelpke, M. D. (2000). Manganese peroxidase. *Metal Ions in Biological Systems*, *37*, 559–586. PMID:10693145

Gomes, H. I., Dias-Ferreira, C., & Ribeiro, A. B. (2013). Overview of in situ and ex situ remediation technologies for PCB-contaminated soils and sediments and obstacles for full-scale application. *The Science of the Total Environment*, *445*, 237–260. doi:10.1016/j.scitotenv.2012.11.098 PMID:23334318

Gomes, N. C. M., Rosa, C. A., Pimentel, P. F., & Mendonça-Hagler, L. C. S. (2002). Uptake of free and complexed silver ions by different strains of *Rhodotorula mucilaginosa*. *Brazilian Journal of Microbiology*, *33*(1), 62–66. doi:10.1590/S1517-83822002000100013

Gómez, J. L., Bódalo, A., Gómez, E., Hidalgo, A. M., Gómez, M., & Murcia, M. D. (2008). A transient design model of a continuous tank reactor for removing phenol with immobilized soybean peroxidase and hydrogen peroxide. *Chemical Engineering Journal*, *145*(1), 142–148. doi:10.1016/j.cej.2008.07.050

Gönen, F., & Serin, D. S. (2012). Adsorption study on orange peel: Removal of Ni (II) ions from aqueous solution. *African Journal of Biotechnology*, *11*(5), 1250–1258.

Gönen, F., & Serin, D. S. (2012). Adsorption study on orange peel: Removal of Ni(II) ions from aqueous solution. *African Journal of Biotechnology*, *11*(5), 1250–1258.

Gong, K., & Demerjian, K. L. (1997). Measurement and analysis of hydrocarbons at White face Mountain, New York. *Geophysics Chemistry.*, *102*(D23), 28059–28069. doi:10.1029/97JD02703

Gonick, H. C. (2008). Nephrotoxicity of cadmium & lead. *The Indian Journal of Medical Research*, *128*, 335–352. PMID:19106433

Gonzalez, G., Herrera, M. G., Garcia, M. T., & Peña, M. (2001). Biodegradation of phenol in a continuous process: Comparative study of stirred tank and fluidized-bed bioreactors. *Bioresource Technology*, *76*(3), 245–251. doi:10.1016/S0960-8524(00)00092-4 PMID:11198177

Gordon, O., Slenters, T. V., Brunetto, P. S., Villaruz, A. E., Sturdevant, D. E., Otto, M., & Fromm, K. M. et al. (2010). Silver coordination polymers for prevention of implant infection: Thiol interaction, impact on respiratory chain enzymes, and hydroxyl radical induction. *Antimicrobial Agents and Chemotherapy*, *54*(10), 4208–4218. doi:10.1128/AAC.01830-09 PMID:20660682

Gosselin, P. (2013). *Most of the rise in CO2 Likely Comes From Natural Sources.* Retrieved from http://notrickszone.com/2013/03/02/most-of-the-rise-in-co2-likely-comes-from-natural-sources/# sthash.p7H9e59p.dpbs

Goswami, R. C. D., Kalita, N., & Kalita, M. C. (2012). A study on growth and CO2 mitigation by microalgae Selenastrum sp.: Its growth behaviour under different nutrient environments and lipid production. *Ann Biol Res*, *3*, 499–510.

Gothalwal, R., & Chillara, S. (2012). *Cyanoremediation: A Green Clean Technology. In Microorganisms in Environmental Management* (pp. 767–786). Springer. doi:10.1007/978-94-007-2229-3_34

Gottschalk, F., Sonderer, T., Scholz, R. W., & Nowack, B. (2009). Modeled environmental concentrations of engineered nanomaterials (TiO2, ZnO, Ag, CNT, Fullerenes) for different regions. *Environmental Science & Technology*, *43*(24), 9216–9222. doi:10.1021/es9015553 PMID:20000512

Govindaraj, S., Srimurali, S., & Kumar, S. K. (2014). Non-steroidal anti-inflammatory drugs in Indian rivers. *Environmental Science and Pollution Research International*, *21*(2), 921–931. doi:10.1007/s11356-013-1957-6 PMID:23832803

Graedel, T. E. (1978). *Chemical compounds in the atmosphere.* New York: Academic Press.

Graham, J. H., Hodge, N. C., & Morton, J. B. (1995). Fatty Acid methyl ester profiles for characterization of glomalean fungi and their endomycorrhizae. *Applied and Environmental Microbiology*, *61*(1), 58–64. PMID:16534923

Grand View Research. (2014). *Biosurfactants market by product (rhamnolipids, sophorolipids, MES, APG, sorbitan esters, sucrose esters) expected to reach USD 2308.8 Million by 2020.* Retrieved August 13, 2016 from http://www.grandviewresearch.com/industryanalysis/biosurfactants-industry

Gray, C. J., & Engel, A. S. (2013). Microbial diversity and impact on carbonate geochemistry across a changing geochemical gradient in a karst aquifer. *International Society for Microbial Ecology*, *7*, 325–337. PMID:23151637

Green, M. H., Muriel, W. J., & Bridges, B. A. (1976). Use of a simplified fluctuation test to detect low levels of mutagens. *Mutation Research*, *38*(1), 33–42. doi:10.1016/0165-1161(76)90077-7 PMID:768758

Gregory, K. B., & Lovley, D. R. (2005). Remediation and recovery of uranium from contaminated subsurface environments with electrodes. *Environmental Science & Technology*, *39*(22), 8943–8947. doi:10.1021/es050457e PMID:16323798

Grill, E., Winnacker, E. L., & Zenk, M. H. (1987). Phytochelatins, a class of heavy-metal-binding peptides from plants, are functionally analogous to metallothioneins. *Proceedings of the National Academy of Sciences of the United States of America*, *84*(2), 439–443. doi:10.1073/pnas.84.2.439 PMID:16593801

Grizzetti, B., Bouraoui, F., & Aloe, A. (2012). Changes of nitrogen and phosphorus loads to European seas. *Global Change Biology*, *18*(2), 769–782. doi:10.1111/j.1365-2486.2011.02576.x

Groenen, P. J. (1978). Components of tobacco smoke. In Nature and quantity; potential influence on health. CIVO-TNO Institute (Rep. No. R/5787).

Grosdidier, A., Zoete, V., & Michielin, O. (2007). EADock: Docking of small molecules into protein active sites with a multiobjective evolutionary optimization. *Proteins: Structure, Function, and Bioinformatics*, *67*(4), 1010–1025. doi:10.1002/prot.21367 PMID:17380512

Grosdidier, A., Zoete, V., & Michielin, O. (2011). SwissDock, a protein-small molecule docking web service based on EADock DSS. *Nucleic Acids Research*, *39*(suppl 2), W270–W277. doi:10.1093/nar/gkr366 PMID:21624888

Grossi, V., Cravo-Laureau, C., Rontani, J. F., Cros, M., & Hirschler-Réa, A. (2011). Anaerobic oxidation of n-alkenes by sulphate-reducing bacteria from the genus Desulfatiferula: N-ketones as potential metabolites. *Research in Microbiology*, *162*(9), 915–922. doi:10.1016/j.resmic.2011.07.004 PMID:21810468

Gu, C., Laverman, A. M., & Pallud, C. (2012). Environmental controls on nitrogen and sulfur cycles in surficial aquatic sediments. *Frontiers in Microbiology*, *3*, 45. doi:10.3389/fmicb.2012.00045 PMID:22375137

Gudiña, E. J., Rodrigues, A. I., Alves, E., Domingues, M. R., Teixeira, J. A., & Rodrigues, L. R. (2015). Bioconversion of agro-industrial by-products in rhamnolipids toward applications in enhanced oil recovery and bioremediation. *Bioresource Technology*, *177*, 87–93. doi:10.1016/j.biortech.2014.11.069 PMID:25479398

Guerra, A., Mendonca, R., Ferraz, A., Lu, F., & Ralph, J. (2004). Structural characterization of lignin during Pinus taeda wood treatment with *Ceriporiopsis subvermispora*. *Applied and Environmental Microbiology*, *70*(7), 4073–4078. doi:10.1128/AEM.70.7.4073-4078.2004 PMID:15240285

Guggenberger, G., Frey, S. D., Six, J., Paustian, K., & Elliot, E. T. (1999). Bacterial and fungal cell-wall residues in conventional and no-tillage agroecosystems. *Soil Science Society of America Journal*, *63*(5), 1188–1198. doi:10.2136/sssaj1999.6351188x

Guillén, F., Martínez, A. T., & Martínez, M. J. (1992). Substrate specificity and properties of the aryl-alcohol oxidase from the ligninolytic fungus *Pleurotus eryngii*. *European Journal of Biochemistry*, *209*(2), 603–611. doi:10.1111/j.1432-1033.1992.tb17326.x PMID:1425667

Guillén, F., Martínez, M. J., Muñoz, C., & Martínez, A. T. (1997). Quinone redox cycling in the ligninolytic fungus *Pleurotus eryngii* leading to extracellular production of superoxide anion radical. *Archives of Biochemistry and Biophysics*, *339*(1), 190–199. doi:10.1006/abbi.1996.9834 PMID:9056249

Guleria, A. (2014). Biodegradation of pharmaceutical waste, 4-nitrophenol by *Pseudumonas putida*. *Life Sciences*, *3*(5), 127–130.

Gunasekera, T. S., Striebich, R. C., Mueller, S. S., Strobel, E. M., & Ruiz, O. N. (2013). Transcriptional profiling suggests that multiple metabolic adaptations are required for effective proliferation of Pseudomonas aeruginosa in jet fuel. *Environmental Science & Technology*, *47*(23), 13449–13458. doi:10.1021/es403163k PMID:24164330

Guo, H., Luoa, S., Chena, L., Xiaoa, X., Xi, Q., Wei, W., & Yejuan, H. et al. (2010). Bioremediation of heavy metals by growing hyper-accumulator endophytic bacterium *Bacillus* spp. L14. *Bioresource Technology*, *101*(22), 8599–8605. doi:10.1016/j.biortech.2010.06.085 PMID:20637605

Gupta, A., Barrow, C. J., & Puri, M. (2012). Omega-3 biotechnology: Thraustochytrids as a novel source of omega-3 oils. *Biotechnology Advances*, *30*(6), 1733–1745. doi:10.1016/j.biotechadv.2012.02.014 PMID:22406165

Gupta, S. K., & Singh, J. (2011). Evaluation of mollusc as sensitive indicator of heavy metal pollution in aquatic system: A review. *IIOAB-India*, *2*(1), 49–57.

Gupta, S., & Kushwah, T. (2015). Exposure of ZnO-NPs enhanced gut-associated microbial activity in *Eisenia foetida*. *Journal of Toxicology and Environmental Health Sciences*, *7*(2), 9–17. doi:10.5897/JTEHS2015.0330

Gupta, S., Kushwah, T., & Yadav, S. (2014). Earthworm coelomocytes as a nanoscavanger to ZnO-NPs. *Nanoscale Research Letters*, *9*(1), 259. doi:10.1186/1556-276X-9-259 PMID:24959107

Gupta, V. S. S. R., & Ponge, J. F. (1998). Regulation of microbial activity and organic matter dynamics by macroorganisms in the detritusphere.*Proc World Cong Soil Sci,*20–26.

Gurujeyalakshmi, G., & Oreil, P. (1988). Isolation of phenol degrading *Bacillus stearothermophilus* and partial characterization of the phenol hydroxylase. *Applied and Environmental Microbiology*, *55*, 500–502. PMID:2719481

Gustafsson, K., & Jansson, J. K. (1993). Ecological risk assessment of the deliberate release of genetically modified microorganisms. *Applied and Environmental Microbiology*, *22*, 236–242.

Gutiérrez, A., Caramelo, L., Prieto, A., Martínez, M. J., & Martínez, A. T. (1994). Anisaldehyde production and aryl-alcohol oxidase and dehydrogenase activities in ligninolytic fungi of the genus *Pleurotus*. *Applied and Environmental Microbiology*, *60*, 1783–1788. PMID:8031078

Haas, K. L., & Franz, K. J. (2009). Application of metal coordination chemistry to explore and manipulate cell biology. *Chemical Reviews*, *109*(10), 4921–4960. doi:10.1021/cr900134a PMID:19715312

Haddadi, A., & Shavandi, M. (2013). Biodegradation of phenol in hypersaline conditions by *Halomonas* sp. Strain PH2–2 isolated from saline soil. *International Biodeterioration & Biodegradation*, *85*, 29–34. doi:10.1016/j.ibiod.2013.06.005

Haghshenas, H., Kay, M., Dehghanian, F., & Tavakol, H. (2016). Molecular dynamics study of biodegradation of azo dyes via their interactions with AzrC azoreductase. *Journal of Biomolecular Structure & Dynamics*, *34*(3), 453–462. doi:10.1080/07391102.2015.1039585 PMID:26325128

Haider, K., Jagnow, G., Kohnen, R., & Lim, S. U. (1974). Degradation of chlorinated benzenes, phenols and chlorinated derivatives by benzene phenol utilizing bacteria under aerobic conditions. *Archives of Microbiology*, *96*(1), 183–200. doi:10.1007/BF00590175 PMID:4134769

Haigh, P. M. (1994). A review of agronomic factors influencing grass silage effluent production in England and Wales. *Journal of Agricultural Engineering Research*, *57*(2), 73–87. doi:10.1006/jaer.1994.1007

Haimi, J. (2000). Decomposer animals and bioremediation of soils. *Environmental Pollution*, *107*(2), 233–238. doi:10.1016/S0269-7491(99)00142-6 PMID:15093000

Haimi, J., & Siira-Pietikainen, A. (1996). Decomposer animal communities in forest soil along heavy metal pollution gradient. *Fresenius' Journal of Analytical Chemistry*, *354*, 672–675. PMID:15067469

Haki, G. D., & Rakshit, S. K. (2003). Developments in industrial important thermostable enzymes: A review. *Bioresource Technology*, *89*(1), 17–34. doi:10.1016/S0960-8524(03)00033-6 PMID:12676497

Haldar, A., & Ray, S. K. (2014). Arsenic and Fluoride Problems of Groundwater in West Bengal and Available Technologies for Remediation. *International Journal of Innovative Research in Science*, *3*(6).

Hall, E. A. H. (1990). *Biosensors* (pp. 250–262). Milton Keynes, UK: Open University Press.

Halling-Sørensen, B., Nielsen, S. N., Lanzky, P. F., & Ingerslev, F. (1998). Occurrence, fate and effects of pharmaceutical substances in the environment—A review. *Chemosphere*, *36*(2), 357–393. doi:10.1016/S0045-6535(97)00354-8 PMID:9569937

Hall, J. L. (2002). Cellular mechanisms for heavy metal detoxification and tolerance. *Journal of Experimental Botany*, *53*(366), 1–11. doi:10.1093/jexbot/53.366.1 PMID:11741035

Hamdi, M. (1992). Toxicity and biodegradability of olive mill wastewaters in batch anaerobic digestion. *Applied Biochemistry and Biotechnology*, *37*(2), 155–163. doi:10.1007/BF02921667

Hamdy, M., & Noyes, O. (1975). Formation of methyl mercury by bacteria. *Applied Microbiology*, *30*(3), 424–432. PMID:1180551

Hamed, S. S., Kauchke, E., & Cooper, E. L. (2002). Cytochemical properties of earthworm coelomocytes enriched by Percoll. In New model for analyzing antimicrobial properties with biomedical applications (pp. 29-37). IOS Press.

Hamid, A. A. A., Hamid, T. A., Haziyamin, T., Wahab, R. A., & Huyop, F. (2015). Identification of functional residues essential for dehalogenation by the non-stereospecific α-haloalkanoic acid dehalogenase from Rhizobium sp. RC1. *Journal of Basic Microbiology*, *55*(3), 324–330. doi:10.1002/jobm.201300526 PMID:25727054

Han, M. J., Choi, H. T., & Song, H. G. (2004). Degradation of phenanthrene by Trametes versicolor and its laccase. *JOURNAL OF MICROBIOLOGY-SEOUL*, *42*(2), 94–98. PMID:15357301

Hannula, S. E., Boschker, H. T. S., De Boer, W., & Veen, J. A. (2012a). 13C pulse-labeling assessment of the community structure of active fungi in the rhizosphere of a genetically starch-modified potato culivar and its parental isoline. *New Phytol, 194*, 784–799.

Hanson, N., & Larsson, A. (2008). Fixed wavelength fluorescence to detect PAH metabolites in fish bile: Increased statistical power with an alternative dilution method. *Environmental Technology*, *24*(2), 200–210. PMID:17899412

Hara, E., Kurihara, M., Nomura, N., Nakajima, T., & Uchiyama, H. (2013). Bioremediation field trial of oil-contaminated soil with food-waste compost. *Journal of JSCE*, *1*(1), 125–132. doi:10.2208/journalofjsce.1.1_125

Haritash, A. K., & Kaushik, C. P. (2009). Biodegradation aspects of polycyclic aromatic hydrocarbons (PAHs): A review. *Journal of Hazardous Materials*, *169*(1-3), 1–15. doi:10.1016/j.jhazmat.2009.03.137 PMID:19442441

Harris, J. (1999). RAPDs in systematics-a useful methodology? In P. M. Hollingsworth, R. M. Bateman, & R. J. Gornall (Eds.), *Molecular systematics and plant evolution* (pp. 221–228). London: Taylor and Francis. doi:10.1201/9781439833278.ch11

Harrison, J. J., Tremaroli, V., Stan, M. A., Chan, C. S., Vacchi-Suzzi, C., Heyne, B. J., & Turner, R. J. et al. (2009). Chromosomal antioxidant genes have metal ion-specifc roles as determinants of bacterial metal tolerance. *Environmental Microbiology*, *11*(10), 2491–2509. doi:10.1111/j.1462-2920.2009.01973.x PMID:19555372

Harter, T. (2003). *Groundwater Quality and Groundwater Pollution*. University of California, Division of Agriculture and Natural Resources, Publication 8084.

Harvell, C. D., Kim, K., Burkholder, J. M., Colwell, R. R., Epstein, P. R., Grimes, D. J., & Vasta, G. R. et al. (1999). Emerging marine diseases: Climate links and anthropogenic factors. *Science*, *285*(5433), 1505–1510. doi:10.1126/science.285.5433.1505 PMID:10498537

Harwood, C. S., & Parales, R. E. (1996). The β-ketoadipate pathway and the biology of self-identity. *Annual Review of Microbiology*, *50*(1), 553–590. doi:10.1146/annurev.micro.50.1.553 PMID:8905091

Hashimi, M. A. I., & Abbas, A. H. (2007). Sequential anaerobic/aerobic treatment of pharmaceutical wastewater. *Tikrit Journal of Eng. Sciences*, *14*(2), 1–31.

Hashimi, M. A. I., Al-Ali, M. I., & Abbas, A. H. (2011). Anaerobic treatment of pharmaceutical wastewater using packed bed reactor. *Journal of Environmental Studies*, *7*, 33–40.

Hasin, A. A., Gurman, S. J., Murphy, L. M., Perry, A., Smith, T. J., & Gardiner, P. E. (2010). Remediation of chromium (VI) by a methane-oxidizing bacterium. *Environmental Science & Technology*, *44*(1), 400–405. doi:10.1021/es901723c PMID:20039753

Hassanshahian, M., & Cappello, S. (2013). Crude oil biodegradation in the marine environments. *Chamy R. Biodegradation-Engineering and Technology,* 101-135.

Haszeldine, R. S. (2009). Carbon capture and storage: How green can black be? *Science*, *325*(5948), 1647–1652. doi:10.1126/science.1172246 PMID:19779187

Hawley, G. G. (1981). The condensed chemical dictionary (10th ed.). New York: Van Nostrand Reinh8old Co.

Hawthorne, S. B., & Sievers, R. E. (1984). Emission of organic air pollutants from shale oil wastewaters. *Environmental Science & Technology*, *18*(6), 483–490. doi:10.1021/es00124a016 PMID:22247953

Hayaishi, O. (2008). From oxygenase to sleep. *The Journal of Biological Chemistry*, *283*(28), 19165–19175. doi:10.1074/jbc.X800002200 PMID:18390546

Hayes, M. H. B. (1983). Darwin's "vegetable mould" and some modern concepts of humus structure and soil aggregation. In J. E. Satchell (Ed.), *Earthworm Ecology: From Darwin to Vermiculture*. London: Chapman & Hall. doi:10.1007/978-94-009-5965-1_3

Hayes, V. E., Ternan, N. G., & McMillan, G. (2000). Organophosphonate metabolism by a moderately halophilic bacterial isolate. *FEMS Microbiol.*, *186*(2), 171–175. doi:10.1111/j.1574-6968.2000.tb09099.x PMID:10802166

Hazra, C., Kundu, D., & Chaudhari, A. (2012). Biosurfactant-assisted bioaugmentation in bioremediation. In *Microorganisms in Environmental Management* (pp. 631–664). Springer Netherlands. doi:10.1007/978-94-007-2229-3_28

Head, I. M., Jones, D. M., & Roling, W. F. M. (2006). Marine microorganisms make a meal of oil. *Nature Reviews. Microbiology*, *4*(3), 173–182. doi:10.1038/nrmicro1348 PMID:16489346

Head, M. A., & Oleszkiewicz, J. A. (2004). Bio-augmentation for nitrification at cold temperatures. *Water Research*, *38*(3), 523–530. doi:10.1016/j.watres.2003.11.003 PMID:14723920

Heberer, T., Fuhrmann, B., Schmidt-Ba¨umler, K., Tsipi, D. V., Kout souba, & Hiskia. (2001). Occurrence of pharmaceutical residues in sewage, river, ground and drinking water in Greece and Ger many. In *Pharmaceuticals and personal care products in the environment: Scientific and regulatory issues* (pp. 70–83). American Chemical Society.

Heberer, T. (2002). Occurrence, fate, and removal of pharmaceutical residues in the aquatic environment: A review of recent research data. *Toxicology Letters*, *131*(1-2), 5–17. doi:10.1016/S0378-4274(02)00041-3 PMID:11988354

Hedges, J. I., & Oades, J. M. (1997). Comparative organic geochemistries of soils and marine sediments. *Organic Geochemistry*, *27*(7-8), 319–361. doi:10.1016/S0146-6380(97)00056-9

Hegazy, A. K., & Emam, M. H. (2011). Accumulation and soil-to-plant transfer of radionuclides in the Nile Delta coastal black sand habitats. *International Journal of Phytoremediation*, *13*(2), 140–155. doi:10.1080/15226511003753961 PMID:21598782

Hegde, R. S., & Fletcher, J. S. (1996). Influence of plant growth stage and season on the release of root phenolics by mulberry as related to development of phytoremediation technology. *Chemosphere*, *32*(12), 2471–2479. doi:10.1016/0045-6535(96)00144-0

Heider, J., & Fuchs, G. (1996). Anaerobic metabolism of aromatic compounds. *European Journal of Biochemistry*, *243*(3), 577–596. doi:10.1111/j.1432-1033.1997.00577.x PMID:9057820

Heidler, J., & Halden, R. U. (2008). Meta-analysis of mass balances examining chemical fate during wastewater treatment. *Environmental Science & Technology*, *42*(17), 6324–6332. doi:10.1021/es703008y PMID:18800497

Heilbuth, N. M., Linardi, V. R., & Santos, V. L. (2003). Phenol biodegradation by free and immobilized cells of *Acinetobacter johnsoni* (M.Sc. Thesis). Instituto De Ciencias Biologicas/Pos-Graduacao Em. Microbiologia.

Heinaru, E., Merimaa, M., Viggor, S., Lehiste, I., Truu, J., & Heinaru, A. (2005). Biodegradation efficiency of functionally important populations selected for bioaugmentation in phenol- and oil-polluted area. *FEMS Microbiology Ecology*, *51*(3), 363–373. doi:10.1016/j.femsec.2004.09.009 PMID:16329884

Heinfling, A., Ruiz-Dueñas, F. J., Martínez, M. J., Bergbauer, M., Szewzyk, U., & Martínez, A. T. (1998). A study on reducing substrates of manganese-oxidizing peroxidases from *Pleurotus eryngii* and *Bjerkandera adusta*. *FEBS Letters*, *428*(3), 141–146. doi:10.1016/S0014-5793(98)00512-2 PMID:9654123

Heip, C., Hummel, H., van-Avesaath, P., Appeltans, W., Arvanitidis, C., Aspden, R., & Nash, R. et al. (2009). *Marine Biodiversity and Ecosystem Functioning*. Dublin, Ireland: Printbase.

Helbig, K., Grosse, C., & Nies, D. H. (2008). Cadmium toxicity in glutathione mutants of *Escherichia coli. Journal of Bacteriology*, *190*(15), 5439–5454. doi:10.1128/JB.00272-08 PMID:18539742

Helmke, P.A., Robarge, W.P., Korotev, R.L., & Schomber, P.J. (1979).Effects of soil-applied sewage sludge on concentrations of elements in eartworms. *J.Environ. Qual., 8,* 322-327.

Helt, C. D., Weber, K. P., Legge, R. L., & Slawson, R. M. (2012). Antibiotic resistance profiles of representative wetland bacteria and faecal indicators following ciprofloxacin exposure in lab-scale constructed mesocosms. *Ecological Engineering*, *39*, 113–122. doi:10.1016/j.ecoleng.2011.11.007

Henkel, M., Müller, M. M., Kügler, J. H., Lovaglio, R. B., Contiero, J., Syldatk, C., & Hausmann, R. (2012). Rhamnolipids as biosurfactants from renewable resources: Concepts for next-generation rhamnolipid production. *Process Biochemistry*, *47*(8), 1207–1219. doi:10.1016/j.procbio.2012.04.018

Herman, M. A. B., Nault, B. A., & Smart, C. D. (2008). Effects of plant growth promoting rhizobacteria on bell pepper production and green peach aphid infestations in New York. *Crop Protection (Guildford, Surrey)*, *27*(6), 996–1002. doi:10.1016/j.cropro.2007.12.004

Hermann, J. C., Ghanem, E., Li, Y., Raushel, F. M., Irwin, J. J., & Shoichet, B. K. (2006). Predicting substrates by docking high-energy intermediates to enzyme structures. *Journal of the American Chemical Society*, *128*(49), 15882–15891. doi:10.1021/ja065860f PMID:17147401

Hess, A., Zarda, B., & Hahn, D. (1997). In situ analysis of denitrifying toluene and m-xylene degrading bacteria in a diesel fuel contaminated laboratory aquifer column. Applied and Environmental Microbiology. *J. App. Enviro. Micro*, *63*, 21362141.

Hess, A., Zarda, B., Hahn, D., Hanner, A., & Stax, D. (1997). In situ analysis of denitrifying toluene and mxylene degrading bacteria in a diesel fuel contaminated laboratory aquifer column. *Journal Applied Enviromental Microbiology*, *63*, 2136–2141. PMID:9172330

Higuchi, T. (1997). *Biochemistry and molecular biology of wood*. London: Springer Verlag. doi:10.1007/978-3-642-60469-0

Hilden, K., Hakala, T. K., Maijala, P., Lundell, T. K., & Hatakka, A. (2007). Novel thermotolerant laccases produced by the white-rot fungus *Physisporinus rivulosus. Applied Microbiology and Biotechnology*, *77*(2), 301–309. doi:10.1007/s00253-007-1155-x PMID:17805527

Hileman, B. (2005). Common Herbicide Kills Tadpoles. *Chemical and Engineering News*, *83*(15), 11.

Hill, G.A., & Robinson, C.W. (1972). Substrate inhibition kinetics, phenol degradation by *P.putida. Biotechnol. Bioeng., 17,* 1599-1615.

Hill, G. T., Mitkowski, N. A., Aldrich-Wolfe, L., Emele, L. R., Jurkonie, D. D., Ficke, A., & Nelsona, E. B. et al. (2000). Methods for assessing the composition and diversity of soil microbial communities. *Applied Soil Ecology*, *15*(1), 25–36. doi:10.1016/S0929-1393(00)00069-X

Hillyer, J. F., & Albrecht, R. M. (2001). Gastrointestinal persorption and tissue distribution of differently sized colloidal gold nanoparticles. *Journal of Pharmaceutical Sciences*, *90*(12), 1927–1936. doi:10.1002/jps.1143 PMID:11745751

Hinteregger, C., & Streichsbier, F. (1997). *Halomonas* sp., a moderately halophilic strain, for biotreatment of saline phenolic wastewater. *Biotechnology Letters*, *19*(11), 1099–1102. doi:10.1023/A:1018488410102

Hiras, D. N., Manariotis, I. D., & Grigoropoulos, S. G. (2004). Organic and nitrogen removal in a two-stage rotating biological contactor treating municipal wastewater. *Bioresource Technology*, *93*(1), 91–98. and doi:10.1016/j.biortech.2003.06.005

Hivrale, Pawar, Rane, & Govindwar. (2016). Application of Genomics and Proteomics in Bioremediation. *Toxicity and Waste Management Using Bioremediation*. IGI Global. doi:10.4018/978-1-4666-9734-8.ch005

Hnaien, M., Lagarde, F., Bausells, J., Errachid, A., & Jaffrezic-Renault, N. (2011). A new bacterial biosensor for trichloroethylene detection based on a three dimensional carbon nanotubes bio-architecture. *Analytical and Bioanalytical Chemistry*, *400*(4), 1083–1092. doi:10.1007/s00216-010-4336-x PMID:21052645

Hobbelen, P. H. F., Koolhaas, J. E., & Van Gestel, C. A. M. (2004). Risk assessment of heavy metals for detritivores in floodplain soils in the Biesbosch, the Netherlands, taking bioavailability into account. *EnvironmentPollution*, *129*, 409–419. PMID:15016462

Hobraud, M., Viande, E., & de Theix, A. (1999). Differential protein expression by Pseudomonas fragi submitted to various stresses. *Electrophoresis*, *20*(11), 2204–2213. doi:10.1002/(SICI)1522-2683(19990801)20:11<2204::AID-ELPS2204>3.0.CO;2-I PMID:10493125

Hoffland, E., Kuyper, T. W., Wallander, H., Plassard, C., Gorbushina, A. A., Haselwandter, K., & van Breemen, N. et al. (2004). The role of fungi in weathering. *Frontiers in Ecology and the Environment*, *2*(5), 258–264. doi:10.1890/1540-9295(2004)002[0258:TROFIW]2.0.CO;2

Hogan, C. M. (2010). Abiotic factor in Encyclopedia of Earth. Washington, DC: National Council for Science and Environment.

Ho, I., & Trappe, J. M. (1973). Translocation of 14C from Festuca plants to their endomycorrhizal fungi. *Nature*, *224*, 30–31.

Hoinkis, J., Deowan, S. A., Panten, V., Figoli, A., Huang, R. R., & Drioli, E. (2012). Membrane bioreactor (MBR) technology – a promising approach for industrial water reuse. Procedia Engineering, 33, 234–241. 10.1016/j.proeng.2012.01.1199

Holben, W. E., Noto, H., Sumino, T., & Suwa, Y. (1998). Molecular analysis of bacterial communities in a three-compartment granular activated sludge system indicates community-level control by incompatible nitrification processes. *Applied and Environmental Microbiology*, *64*, 2528–2532. PMID:9647825

Hong, H.-B., Chang, Y.-S., Nam, I.-H., Fortnagel, P., & Schmidt, S. (2002). Biotransformation of 2,7-dichloro- and 1,2,3, 4-tetrachlorodibenzo-p-dioxin by *Sphingomonas wittichii* RW1. *Applied and Environmental Microbiology*, *68*(5), 2584–2588. doi:10.1128/AEM.68.5.2584-2588.2002 PMID:11976140

Hong, R., Kang, T. Y., Michels, C. A., & Gadura, N. (2012). Membrane lipid peroxidation in copper alloy-mediated contact killing of *Escherichia coli. Applied and Environmental Microbiology*, *78*(6), 1776–1784. doi:10.1128/AEM.07068-11 PMID:22247141

Hong, S. H., Ryu, H., Kim, J., & Cho, K.-S. (2011). Rhizoremediation of diesel-contaminated soil using the plant growth-promoting rhizobacterium Gordonia sp. S2RP-17. *Biodegradation*, *22*(3), 593–601. doi:10.1007/s10532-010-9432-2 PMID:21082332

Hopkin, S. P. (1989). *Ecophysiology of Metals in Terrestrial Invertebrates*. London: Elsevier.

Hornung, U. (1997). Soil Venting. Environmental Biotechnology, CRC Press, Jul 19, 2010 -. *Nature*.

Horwat, M., Tice, M., & Kjellerup, B. V. (2015). Biofilms at work: Bio-, phyto-and rhizoremediation approaches for soils contaminated with polychlorinated biphenyls. *AIMS Bioengineering*, *2*(4), 224–234. doi:10.3934/bioeng.2015.4.324

Hoshino, M., Akimoto, H., & Okuda, M. (1978). Photochemical oxidation of benzene, toluene and ethylbenzene initiated by OH radicals in the gas phase. *Bulletin of the Chemical Society of Japan*, *51*(3), 718–724. doi:10.1246/bcsj.51.718

Hosseinzadeh, R., Moosavi-Movahedi, A. A., & Ghourchian, H. (2014). Electrochemistry and molecular modeling of the hemoglobin–benzene interaction with a nanocrystalline mixed metal oxide. *RSC Advances*, *4*(90), 49128–49136. doi:10.1039/C4RA08018A

Houghton, J. E., & Shanley, M. S. (1994). Catabolic potential of pseudomonads: a regulatory perspective. In R. G. Chaudhry (Ed.), *Biological Degradation and Bioremediation of Toxic Chemicals* (pp. 11–32). London: Chapman & Hall.

Howard, P. H. (1989). Handbook of environmental fate and exposure data for organic chemicals. Lewis Publishers.

Hsieh, J.-L., Chen, C.-Y., Chiu, M.-H., Chein, M., Chang, J.-S., Endo, G., & Huang, C.-C. (2009). Expressing a bacterial mercuric ion binding protein in plant for phytoremediation of heavy metals. *Journal of Hazardous Materials*, *161*(2), 920–925. doi:10.1016/j.jhazmat.2008.04.079 PMID:18538925

Hsu, J. L., & Chen, S. H. (2005). Recent progress in quantitative proteomics using stable isotope labeling, multidimensional liquid chromatography and mass spectrometry. *Current Proteomics*, *2*(4), 287–302. doi:10.2174/157016405775201766

Huang, B. (2009). MetaPocket: A meta approach to improve protein ligand binding site prediction. *OMICS: A Journal of Integrative Biology*, *13*(4), 325–330. doi:10.1089/omi.2009.0045 PMID:19645590

Huang, S. Y., Grinter, S. Z., & Zou, X. (2010). Scoring functions and their evaluation methods for protein–ligand docking: Recent advances and future directions. *Physical Chemistry Chemical Physics*, *12*(40), 12899–12908. doi:10.1039/c0cp00151a PMID:20730182

Huang, T., Guo, L., Zhang, H., Su, J., Wen, G., & Zhang, K. (2015). Nitrogen-removal efficiency of a novel aerobic denitrifying bacterium, *Pseudomonas stutzeri* strain ZF31, isolated from a drinking-water reservoir. *Bioresource Technology*, *196*, 209–216. doi:10.1016/j.biortech.2015.07.059 PMID:26241840

Huang, X.-D., El-Alawi, Y., Gurska, J., Glick, B. R., & Greenberg, B. M. (2005). A multi-process phytoremediation system for decontamination of persistent total petroleum hydrocarbons (TPHs) from soils. *Microchemical Journal*, *81*(1), 139–147. doi:10.1016/j.microc.2005.01.009

Huang, X.-D., El-Alawi, Y., Penrose, D. M., Glick, B. R., & Greenberg, B. M. (2004). A multi-process phytoremediation system for removal of polycyclic aromatic hydrocarbons from contaminated soils. *Environmental Pollution*, *130*(3), 465–476. doi:10.1016/j.envpol.2003.09.031 PMID:15182977

Hubble, B. R., Stetter, J. R., Gebert, E., Harkness, J. B. L., & Flotard, R. D. (1981). Experimental measurements from residential wood-burning stoves. *Proc. Int. Conf. Resid. Solid Fuels: Environ. Impacts Solut.*

Huerta-Fontela, M., Galceran, M. T., & Ventura, F. (2011). Occurrence and removal of pharmaceuticals and hormones through drinking water treatment. *Water Research*, *45*(3), 1432–1442. doi:10.1016/j.watres.2010.10.036 PMID:21122885

Huey, R., Morris, G. M., Olson, A. J., & Goodsell, D. S. (2007). A semiempirical free energy force field with charge-based desolvation. *Journal of Computational Chemistry*, *28*(6), 1145–1152. doi:10.1002/jcc.20634 PMID:17274016

Huggett, R. J., Kimerle, R. A., Mehrle, P. M., & Bergman, H. J. (1992). Biomarkers. SETAC Special Publications Series, Lewis Publishers, Ann Arbor. K.R. Rogers. *Analytica Chimica Acta*, *568*, 222–226.

Hughes, S. M., & Cooper, D. G. (1996). Biodegradation of phenol using the selfcycling fermentation (SCF) process. *Biotechnology and Bioengineering*, *51*(1), 112–119. doi:10.1002/(SICI)1097-0290(19960705)51:1<112::AID-BIT13>3.0.CO;2-S PMID:18627094

Hu, L., Xu, S., Pan, C., Yuan, C., Zou, H., & Jiang, G. (2005). Matrix-assisted laser desorption/ionization time-of-flight mass spectrometry with a matrix of carbon nanotubes for the analysis of low-mass compounds in environmental samples. *Environmental Science & Technology*, *39*(21), 8442–8447. doi:10.1021/es0508572 PMID:16294885

Hussain, A., Kumar, P., & Mehrotra, I. (2010). Nitrogen biotransformation in anaerobic treatment of phenolic wastewater. *Desalination*, *250*(1), 35–41. doi:10.1016/j.desal.2009.09.018

Hussein, H. S., Ruiz, O. N., Terry, N., & Daniell, H. (2007). Phytoremediation of mercury and organomercurials in chloroplast transgenic plants: Enhanced root uptake, translocation to shoots, and volatilization. *Environmental Science & Technology*, *41*(24), 8439–8446. doi:10.1021/es070908q PMID:18200876

Hussein, K. A., & Joo, J. H. (2013). Heavy metal resistance of bacteria and its impact on the production of antioxidant enzymes. *African Journal of Microbiological Research*, *7*(20), 2288–2296. doi:10.5897/AJMR12.1764

Hutchinson, D. H., & Robinson, C. W. (1988). Kinetics of the simultaneous batch degradation of *p*-Cresol and Phenol by *Pseudomonas putida*. *Applied Microbiology and Biotechnology*, *29*(6), 599–604. doi:10.1007/BF00260991

Huttermann, A., Orikiriza, L. J. B., & Agaba, H. (2009). Application of superabsorbent polymers for improving the ecological chemistry of degraded or polluted lands. *Clean Soil Air Water*, *37*(7), 517–526. doi:10.1002/clen.200900048

Hu, W. J., Harding, S. A., Lung, J., Popko, J. L., Ralph, J., Stokke, D. D., & Chiang, V. L. et al. (1999). Repression of lignin biosynthesis promotes cellulose accumulation and growth in transgenic trees. *Nature Biotechnology*, *17*(8), 808–812. doi:10.1038/11758 PMID:10429249

Hwang, S., & Cutright, T. J. (2002). Biodegradability of Aged Pyrene and Phenenthrene in a Natural Soil. *Chemosphere*, *47*(9), 891–899. doi:10.1016/S0045-6535(02)00016-4 PMID:12108695

Hybholt, T. K., Aamand, J., & Johnsen, A. R. (2011). Quantification of centimeter-scale spatial variation in PAH, glucose and benzoic acid mineralization and soil organic matter in road-side soil. *Environmental Pollution*, *159*(5), 1085–1091. doi:10.1016/j.envpol.2011.02.028 PMID:21396755

Iacurci, J. (2013). *Plastic Particles Harm Freshwater Organisms Too. Nature World News*. New Orleans, LA: Americal Chemical Society.

Ideriah, T. J. K., David, O. D., & Ogbonna, D. N. (2012). Removal of heavy metal ions in aqueous solutions using palm fruit fibre as adsorbent. *Journal of Environmental Chemistry and Exotoxicology*, *4*(4), 82–90.

Ileri, R., Sengil, I. A., Kulac, S., & Damar, Y. (2003). Treatment of mixed pharmaceutical industry and domestic wastewater by sequencing batch reactor. *Journal of Environmental Science and Health. Part A, Toxic/Hazardous Substances & Environmental Engineering*, *38*(10), 2101–2111. and doi:10.1081/ESE-120023336

Ilyina, A., Castillo Sanchez, M. I., Villarreal Sanchez, J. A., Ramirez Esquivel, G., & Candelas Ramirez, J. (2003). Isolation of soil bacteria for bioremediation of hydrocarbon contamination. *Bulletin of Moscow University*, *44*(1), 88–91.

Imlay, J. A. (2003). Pathways of oxidative damage. *Annual Review of Microbiology*, *57*(1), 395–418. doi:10.1146/annurev.micro.57.030502.090938 PMID:14527285

Imlay, J. A., Chin, S. M., & Linn, S. (1988). Toxic DNA damage by hydrogen peroxide through the Fenton reaction *in vivo* and *in vitro*. *Science*, *240*(4852), 640–642. doi:10.1126/science.2834821 PMID:2834821

Inouhe, M. (2005). Phytochelatins. *Brazilian Journal of Plant Physiology*, *17*(1), 65–78. doi:10.1590/S1677-04202005000100006

Intergovernmental Panel on Climate Change (IPCC). (2013). The Physical Science Basis. In T. F. Stocker, D. Qin, G. Plattner, M. Tignor, S. K. Allen, J. Boschung, & P. M. Midgley et al. (Eds.), *Working Group I Contribution to the Fifth Assessment Report of the Intergovernmental Panel on Climate Change, 1535*. Cambridge, UK: Cambridge University Press.

IPCC. (2013). Fourth assessment report: climate change (AR5). Stockholm, Sweden: IPCC.

Ireland, M. P. (1976). Excretion of lead, zinc and calcium by the earthworms by the earthworm *Dendrobaena rubida* living in soil contaminated with zinc and lead. *Soil Biology & Biochemistry*, *8*(5), 347–350. doi:10.1016/0038-0717(76)90031-6

Ireland, M. P. (1978). Heavy metal binding properties of earthworm chloragosomes. *Acta Biol.Sci.Hung*, *29*, 385–394. PMID:757448

Ireland, M. P. (1979). Metal accumulation by the earthworm *Lumbricus rubellus, Dendrobaena veneta* and *Eiseniella tetraedra* living in heavy metal polluted sites. *Environmental Pollution*, *19*(3), 201–206. doi:10.1016/0013-9327(79)90041-7

Irvine, R., Earley, J., Kehrberger, G. J., & Tod Delaney, B. (1993). Bioremediation of soils contaminated with bis-(2-ethylhexyl) phthalate (BEHP) in a soil slurry-sequencing batch reactor. *Environment and Progress*, *12*(1), 39–44. doi:10.1002/ep.670120108

Irving, H., & Williams, R. J. P. (1953). The stability of transition-metal complexes. *Journal of the Chemical Society*, *1953*, 3192–3210. doi:10.1039/jr9530003192

Isa, M. H., Kutty, S. M., & Bin Abd Rahim, M. F. (2010). *Anaerobic treatment of pharmaceutical wastewater under mesophilic and thermophilic conditions.International Conference on Environment.*

Isik, M., & Sponza, D. T. (2008). Anaerobic/aerobic treatment of a simulated textile wastewater. *Separation and Purification Technology*, *60*(1), 64–72. doi:10.1016/j.seppur.2007.07.043

Islam, M. (2014). *Phytofiltration of arsenic and cadmium from the water environment using Micranthemum umbrosum.* Academic Press.

Issoufi, I., Rhykerd, R., & Smiciklas, K. (2006). Seedling growth of agronomic crops in crude oil contaminated soil. *Journal Agronomy & Crop Science*, *192*(4), 310–317. doi:10.1111/j.1439-037X.2006.00212.x

Itoh, M., Nakamura, M., Suzuki, T., Kawai, K., Horitsu, H., & Takamizawa, K. (1995). Mechanism of chromium(VI) toxicity in *Escherichia coli*: Is hydrogen peroxide essential in Cr(VI) toxicity? *Journal of Biochemistry*, *117*, 780–786. PMID:7592539

Ivanković, T., & Hrenović, J. (2010). Surfactants in the environment. *Arhiv za Higijenu Rada i Toksikologiju*, *61*(1), 95–109. doi:10.2478/10004-1254-61-2010-1943 PMID:20338873

Izzo, V., Notomista, E., Picardi, A., Pennacchio, F., & Di Donato, A. (2005). The thermophilic archaeon Sulfolobus solfataricus is able to grow on phenol. *Research in Microbiology*, *156*(5-6), 677–689. doi:10.1016/j.resmic.2005.04.001 PMID:15921893

Jacob-Lopes, E., Lacarda, L. M. C. F., & Franco, T. T. (2008). Biomass production and Cdioxide fixation by Aphanothecemicroscopica Nageli in a bubble column photobioreactor. *Biochemical Engineering Journal*, *40*(1), 27–34. doi:10.1016/j.bej.2007.11.013

Jacobs, I. J., & Menon, U. (2004). Progress and challenges in screening for early detection of ovarian cancer. *Molecular & Cellular Proteomics*, *3*(4), 355–366. doi:10.1074/mcp.R400006-MCP200 PMID:14764655

Jadhav, D. N., & Vanjara, A. K. (2004). Removal of phenol from wastewater using sawdust, polymerized sawdust and sawdust carbon. *Indian J Chem Technol*, *11*, 35–41.

Jain, R. K., Kapur, M., Labana, S., Lal, B., Sarma, P. M., Bhattacharya, D., & Thakur, I. S. (2005). Microbial diversity: Application of micro-organisms for the biodegradation of xenobiotics. *Current Science*, *89*(1), 101–112.

Jaklitsch, W. M., Kubicek, C. P., & Scrutten, M. C. (1991). Intracellular location of enzymes involved in citrate production by *Aspergillus niger. Canadian Journal of Microbiology*, *37*(11), 823–827. doi:10.1139/m91-142 PMID:1777859

Jalali, M., & Aboulghazi, F. (2013). Sunflower stalk, an agricultural waste, as an adsorbent for the removal of lead and cadmium from aqueous solutions. *Journal of Material Cycles and Waste Management*, *15*(4), 548–555. doi:10.1007/s10163-012-0096-3

Janin, J., Henrick, K., Moult, J., Eyck, L. T., Sternberg, M. J., Vajda, S., & Wodak, S. J. et al. (2003). CAPRI: A critical assessment of predicted interactions. *Proteins: Structure, Function, and Bioinformatics*, *52*(1), 2–9. doi:10.1002/prot.10381 PMID:12784359

Janssen, C. (1998). Alternative assays for routine toxicity assessments: a review. In G. Schüürmann & B. Markert (Eds.), *Ecotoxicology: Ecological Fundamentals, Chemical Exposure, and Biological Effects* (pp. 813–839). New York: John Wiley & Sons, Inc. and Spektrum AkademischerVerlag.

Jansson, J. K. (1995). Tracking genetically engineered microorganisms in nature. *Current Opinion in Biotechnology*, *6*(3), 275–283. doi:10.1016/0958-1669(95)80048-4 PMID:7780181

Jansson, J. K., Björklöf, K., Elvang, A. M., & Jørgensen, K. S. (2000). Biomarkers for monitoring efficacy of bioremediation by microbial inoculants. *Environmental Pollution*, *107*(2), 217–223. doi:10.1016/S0269-7491(99)00140-2 PMID:15092998

Javaid, M. K., Ashiq, M., & Tahir, M. (2016). Potential of Biological Agents in Decontamination of Agricultural Soil. *Scientifica, 2016.*

Jayasekara, R., Harding, I., Bowater, I., & Lonergan, G. (2005). Biodegradability of a selected range of polymers and polymer blends and standard methods for assessment of biodegradation. *Journal of Polymers and the Environment*, *13*(3), 231–251. doi:10.1007/s10924-005-4758-2

Jayashree, R., Nithya, S. E., Rajesh, P. P., & Krishnaraju, M. (2012). Biodegradation capability of bacterial species isolated from oil contaminated soil. *Journal of Academia Industrial Research*, *1*(3), 140–143.

Jay, K., & Steiglitz, L. (1995). Identification and quantification of volatile organic components in emissions of waste incineration plants. *Chemosphere*, *30*(7), 1249–1260. doi:10.1016/0045-6535(95)00021-Y

Jeevitha, P., Keserkar, J. V., & Hemapriya, J. (2015). Economically Feasible-Bacterial Mediated Biosurfactant Production using Agro Wastes as Carbon Source. *International Journal of Scientific Engineering & Technology Research*, *04*(9), 5797–5803.

Jennings, D. H. (1995). *The Physiology of Fungal Nutrition*. Cambridge, UK: Cambridge University Press. doi:10.1017/CBO9780511525421

Jensen, H. L. (1957). Decomposition of chloro-substituted aliphatic acids by soil bacteria. *Canadian Journal of Microbiology*, *3*(2), 151–164. doi:10.1139/m57-019

Jensen, K. A. Jr, Bao, W., Kawai, S., Srebotnik, E., & Hammel, K. E. (1996). Manganese-dependent cleavage of nonphenolic lignin structures by *Ceriporiopsis subvermispora* in the absence of lignin peroxidase. *Applied and Environmental Microbiology*, *62*, 3679–3686. PMID:16535418

Jeppesen, E., Brucet, S., Naselli-Flores, L., Papastergiadou, E., Stefanidis, K., Nõges, T., & Beklioğlu, M. et al. (2015). Ecological impacts of global warming and water abstraction on lakes and reservoirs due to changes in water level and related changes in salinity. *Hydrobiologia*, *750*(1), 201–227. doi:10.1007/s10750-014-2169-x

Jeswani, H., & Mukherji, S. (2012). Degradation of phenolics, nitrogen-heterocyclics and polynuclear aromatic hydrocarbons in a rotating biological contactor. *Bioresource Technology*, *111*, 12–20. doi:10.1016/j.biortech.2012.01.157 PMID:22382298

Ji, Y., Mao, G., Wang, Y., & Bartlam, M. (2013). Structural insights into diversity and n-alkane biodegradation mechanisms of alkane hydroxylases. *Frontiers in Microbiology*, *4*, 58.

Jiang, F., & Kim, S. H. (1991). Soft docking: Matching of molecular surface cubes. *Journal of Molecular Biology*, *219*(1), 79–102. doi:10.1016/0022-2836(91)90859-5 PMID:2023263

Jian-guang, Y., Jun-yuan, L., Jian-ying, Y., Xu-liang, Z., & Zi-xiang, D. (2014). Hydrothermal processing of arsenic containing bioremediation biomass: Pteris vittata. *Journal of Environmental Chemical Engineering*, *2*(3), 1358–1364. doi:10.1016/j.jece.2014.04.011

Jiang, Y., Yang, K., Wang, H., Shang, Y., & Yang, X. (2015). Characteristics of phenol degradation in saline conditions of a halophilic strain JS3 isolated from industrial activated sludge. *Marine Pollution Bulletin*, *99*(1-2), 1–2, 230–234. doi:10.1016/j.marpolbul.2015.07.021 PMID:26187399

Jiang, Z. J., Fang, J. G., Mao, Y. Z., & Wang, W. (2010). Eutrophication assessment and bioremediation strategy in a marine fish cage culture area in Nansha Bay, China. *Journal of Applied Phycology*, *22*(4), 421–426. doi:10.1007/s10811-009-9474-1

Jianlong, W., Xiangchun, Q., Libo, W., Yi, Q., & Hegemann, W. (2002). Bioaugmentation as a Tool to Enhance the Removal of Refractory Compound in Coke Plant Wastewater. *Process Biochem, Oxford, UK*, *38*(5), 777–781. doi:10.1016/S0032-9592(02)00227-3

Jiao, N. (2013). Microbial Carbon Sequestration in the Ocean: The Big Role of Small Bugs in Mitigating Global Climate Change. Bulletin of the Chinese Academy of Sciences. *Special Reports*, *27*, 3.

Jiao, N., Robinson, C., Azam, F., Thomas, H., Baltar, F., Dang, H., & Zhang, R. et al. (2014). Mechanisms of microbial carbon sequestration in the ocean – future research directions. *Biogeosciences*, *11*(19), 5285–5306. doi:10.5194/bg-11-5285-2014

Jillian, E. G., Robert, S., & Rajakaruna, N. (2015). Transfer of heavy metals through terrestrial food webs: A review. *Environmental Monitoring and Assessment*, *187*(4). doi:10.1007/s10661-015-4436-3 PMID:25800370

Jin, P., Jin, X., Wang, X., Feng, Y., & Wang, X. C. (2013). Biological Activated Carbon Treatment Process for Advanced Water and Wastewater Treatment. In Biomass Now-Cultivation and Utilization (pp. 153-192). InTech. doi:10.5772/52021

Jin, J. N., Yao, J., Zhang, Q. Y., Yu, C., Chen, P., Liu, W. J., & Choi, M. M. et al. (2015). An integrated approach of bioassay and molecular docking to study the dihydroxylation mechanism of pyrene by naphthalene dioxygenase in *Rhodococcus* sp. ustb-1. *Chemosphere*, *128*, 307–313. doi:10.1016/j.chemosphere.2015.02.012 PMID:25747183

Ji, X., Ripp, S. A., Layton, A. C., Sayler, G. S., & DeBruyn, J. M. (2013). Assessing long term effects of bioremediation: Soil bacterial communities 14 years after polycyclic aromatic hydrocarbon contamination and introduction of a genetically engineered microorganism. *Journal of Bioremediation & Biodegradation*, *4*(8), 1–8.

Jjemna, P. K. (2004). *Environmental microbiology: Principles and applications*. Enfield, NH: Science Publishers. Inc.

Joergensen, R. G., & Wichern, F. (2008). Quantitative assessment of the fungal contribution to microbial tissue in soil. *Soil Biology & Biochemistry*, *40*(12), 2977–2991. doi:10.1016/j.soilbio.2008.08.017

Johal, M. S. (2011). *Understanding nanomaterials*. CRC Press.

John, E. A., William, A. A., Brady, D. L., David, N. T., Jeffrey, A. L., Deborah, T. N., & Vicki, S. et al. (2016). Degradation of phenolic compounds by the lignocelluloses deconstructing thermoacidophilic bacterium *Alicyclobacillus acidocaldarius*. *Journal of Industrial Microbiology & Biotechnology*, *43*(1), 13–23. doi:10.1007/s10295-015-1700-z PMID:26542284

John, S. G., Ruggiero, C. E., Hersman, L. E., Tung, C.-S., & Neu, M. P. (2001). Siderophore mediated plutonium accumulation by *Microbacterium flavescens* (JG-9). *Environmental Science & Technology*, *35*(14), 2942–2948. doi:10.1021/es010590g PMID:11478246

Johnson, A. C., Belfroid, A., & Di Corcia, A. (2000). Estimating steroid oestrogen inputs into activated sludge treatment works and observations on their removal from the effluent. *The Science of the Total Environment*, *256*(2-3), 163–173. doi:10.1016/S0048-9697(00)00481-2 PMID:10902843

Johnson, D., Anderson, D., & McGrath, S. (2005). Soil microbial response during the phytoremediation of a PAH contaminated soil. *Soil Biology & Biochemistry*, *37*(12), 2334–2336. doi:10.1016/j.soilbio.2005.04.001

Johnson, D., Maguire, K., Anderson, D., & McGrath, S. (2004). Enhanced dissipation of chrysene in planted soil: The impact of a rhizobial inoculum. *Soil Biology & Biochemistry*, *36*(1), 33–38. doi:10.1016/j.soilbio.2003.07.004

Joint, I., Doney, S. C., & Karl, D. M. (2010). Will ocean acidification affect marine microbes? *International Society for Microbial Ecology*, *5*, 1–7. PMID:20535222

Joner, E. J., & Leyval, C. (2003). Rhizosphere gradients of polycyclic aromatic hydrocarbon (PAH) dissipation in two industrial soils and the impact of arbuscular mycorrhiza. *Environmental Science & Technology*, *37*(11), 2371–2375. doi:10.1021/es020196y PMID:12831019

Joner, E., Corgie, S., Amellal, N., & Leyval, C. (2002). Nutritional constraints to degradation of polycyclic aromatic hydrocarbons in a simulated rhizosphere. *Soil Biology & Biochemistry*, *34*(6), 859–864. doi:10.1016/S0038-0717(02)00018-4

Jones, O. A., Lester, J. N., & Voulvoulis, N. (2005). Pharamceuticals: A threat to drinking water? *Trends in Biology*, *23*(4), 163–167. doi:10.1016/j.tibtech.2005.02.001 PMID:15780706

Jørgensen, N., & Stepanauskas, R. (2009). Biomass of pelagic fungi in Baltic rivers. *Hydrobiologia*, *623*(1), 105–112. doi:10.1007/s10750-008-9651-2

Joshi, S. J., & Desai, A. J. (2010). Biosurfactant's role in bioremediation of NAPL and fermentative production. In Biosurfactants (pp. 222-235). Springer New York. doi:10.1007/978-1-4419-5979-9_17

Joss, A., Keller, E., Alder, A. C., Gobel, A., McArdell, C. S., Ternes, T. A., & Siegrist, H. (2005). Removal of pharmaceuticals and fragrances in biological wastewater treatment. *Water Research*, *39*(14), 3139–3152. and doi:10.1016/j.watres.2005.05.031

Joutey, N. T., Bahafid, W., Sayel, H., & El Ghachtouli, N. (2013). *Biodegradation: involved microorganisms and genetically engineered microorganisms. In Biodegradation-life of science* (pp. 289–320). Rijeka: InTech.

Joutey, N. T., Bahafid, W., Sayel, H., & Ghachtouli, N. E. (2013). Biodegradation: Involved microorganisms and genetically engineered microorganisms. In R. Chamy & F. Rosenkranz (Eds.), *Biodegradation-Life of Science* (pp. 289–318). InTech.

Jové, P., Olivella, M. À., Camarero, S., Caixach, J., Planas, C., Cano, L., & Heras, F. X. De L. (2016). Fungal biodegradation of anthracene-polluted cork: A comparative study. *J Environ Sci Health A Tox Hazard Subst Environ Eng, 51*(1), 70-77.

Joy, S., Nair, P. S., Hariharan, R., & Pillai, M. R. (2006). Detailed comparison of the protein-ligand docking efficiencies of GOLD, a commercial package and ArgusLab, a licensable freeware. *In Silico Biology*, *6*(6), 601–605. PMID:17518767

Juang, R.S., & Tsai, S.Y. (2006). Growth kinetics of *Pseudomonas putida* in the Biodegradation of single and mixed phenol and sodium salicylate. *Biochem. Eng., 2*(1), 133-140.

Judson, R. S., Martin, M. T., Reif, D. M., Houck, K. A., Knudsen, T. B., Rotroff, D. M., & Dix, D. J. et al. (2010). Analysis of eight oil spill dispersants using rapid, in vitro tests for endocrine and other biological activity. *Environmental Science & Technology*, *44*(15), 5979–5985. doi:10.1021/es102150z PMID:20602530

Juhasz, A. L., & Naidu, R. (2000). Bioremediation of high molecular weight polycyclic aromatic hydrocarbons: A review of the microbial degradation of benzo[a]pyrene. *International Biodeterioration & Biodegradation*, *45*(1-2), 57–88. doi:10.1016/S0964-8305(00)00052-4

Jungclaus, G. A., Lopez-Avila, V., & Hites, R. A. (1978). Organic compounds in an industrial wastewater: A case study of their environmental impact. *Environmental Science & Technology*, *12*(1), 88–96. doi:10.1021/es60137a015

Kadhim, H., Graham, C., Baratt, P., Evane, C. S., & Rastall, R. A. (1999). Removal of phenolic compounds by *Coriolus versicolor* grown on wheat bran. *Enzyme and Microbial Technology*, *24*(5-6), 303–307. doi:10.1016/S0141-0229(98)00123-9

Kahru, A., Reiman, R., & Rätseep, A. (1998). The efficiency of different degrading bacteria and activated sludges in detoxification of phenolic leachates. *Chemosphere*, *37*(2), 301–318. doi:10.1016/S0045-6535(98)00048-4 PMID:9650267

Kamaludeen, S., & Ramasamy, K. (2008). Rhizoremediation of metals: Harnessing microbial communities. *Indian Journal of Microbiology*, *48*(1), 80–88. doi:10.1007/s12088-008-0008-3 PMID:23100702

Kamika, I., & Momba, M. N. B. (2015). Effect of nickel on nutrient removal by selected indigenous protozoan species in wastewater systems. *Saudi Journal of Biological Sciences*, *22*(2), 147–156. doi:10.1016/j.sjbs.2014.09.010 PMID:25737645

Kanekar, P. P., Sarnaik, S. S., & Kelkar, A. S. (1999). Bioremediation of phenol by alkaline bacteria isolated from alkaline lake of Lonar. *Journal of Applied Microbiology*, *85*(S1), 128S–133S. doi:10.1111/j.1365-2672.1998.tb05291.x PMID:21182701

Kang, M. H., & Park, J. M. (1997). Sequential degradation of phenol and cyanide by a commensal interaction between two microorganisms. *Chem. Technol. Biotechnol.*, *69*(2), 226–230. doi:10.1002/(SICI)1097-4660(199706)69:2<226::AID-JCTB677>3.0.CO;2-N

Kang, S.-W., Kim, Y.-B., Shin, J.-D., & Kim, E.-K. (2010). Enhanced biodegradation of hydrocarbons in soil by microbial biosurfactant, sophorolipid. *Applied Biochemistry and Biotechnology*, *160*(3), 780–790. doi:10.1007/s12010-009-8580-5 PMID:19253005

Kannan, V., Vijayasanthi, M., & Chinnasamy, M. (2012). Bioremediation of chromium in tannery effluent by filamentous Cyanobacteria *Anabaena flos-aquae* West. *International Journal of Environmental Sciences*, *2*(4), 2360–2366.

Kapadia, S. G., & Yagnik, B. N. (2013). Current trend and potential for microbial biosurfactants. Asian J. Exp. *Biol. Sci*, *4*(1), 1–8.

Kapley, A., Prasad, S., & Purohit, H. J. (2007). Changes in microbial diversity in fed-batch reactor operation with wastewater containing nitroaromatic residues. *Bioresource Technology*, *98*(13), 2479–2484. doi:10.1016/j.biortech.2006.09.012 PMID:17070036

Kapley, A., & Purohit, H. J. (2009). Genomic tools in bioremediation. *Indian Journal of Microbiology*, *49*(2), 108–113. doi:10.1007/s12088-009-0012-2 PMID:23100758

Kapoor, M., & Rajagopal, R. (2011). Enzymatic bioremediation of organophosphorus insecticides by recombinant organophosphorous hydrolase. *International Biodeterioration & Biodegradation*, *65*(6), 896–901. doi:10.1016/j.ibiod.2010.12.017

Kargi, F., & Dinçer, A. R. (1996). Enhancement of biological treatment performance of saline wastewater by halophilic bacteria. *Bioprocess Engineering*, *15*, 1. doi:10.1007/BF00435529

Kargi, F., & Uygur, A. (1996). Biological treatment of saline wastewater in an aerated percolator unit utilizing halophilic bacteria. *Environmental Technology*, *17*(3), 325–330. doi:10.1080/09593331708616391

Kargupta, W. (2012). *CSequestration by Microalgae* (M.Tech. thesis). IIT Bombay, Mumbai, India.

Kargupta, W., Ganesh, A., & Mukherji, S. (2015). Estimation of Cdioxide sequestration potential of microalgae grown in a batch photobioreactor. *Bioresource Technology*, *180*, 370–375. doi:10.1016/j.biortech.2015.01.017 PMID:25616748

Karigar, C. S., & Rao, S. S. (2011). Role of microbial enzymes in the bioremediation of pollutants: A review. *Enzyme Research*, *2011*, 1–11. doi:10.4061/2011/805187 PMID:21912739

Karin, F. (2012). How spectroscopy and microspectroscopy of degraded wood contribute to understand fungal wood decay. *Applied Microbiology and Biotechnology*, *96*(3), 587–599. doi:10.1007/s00253-012-4369-5 PMID:22983562

Karlson, U., & Frankenberger, W. T. (1993). Biological alkylation of selenium and tellurium. In H. Sigel & A. Sigel (Eds.), Metal ions in Biological Systems: Biological Properties of metal Alkyl derivatives, (pp. 185-227). New York: Academic Press.

Karlsson, J., Clarke, A. K., Chen, Z. Y., Hugghins, S. Y., Park, Y. I., Husic, H. D., & Samuelsson, G. et al. (1998). A novel alpha-type carbonic anhydrase associated with the thylakoid membrane in Chlamydomonas reinhardtii is required for growth at ambient CO2. *The EMBO Journal*, *17*(5), 1208–1216. doi:10.1093/emboj/17.5.1208 PMID:9482718

Kar, S., Swaminathan, T., & Baradarajan, A. (1996). Studies on biodegradation of a mixture of toxic and non-toxic pollutant using *Arthrobacter* species. *Bioprocess Engineering*, *15*(4), 195–199. doi:10.1007/BF00369481

Kashem, M. A., & Singh, B. R. (2001). Metal availability in contaminated soils: I. Effects of flooding and organic matter on changes in Eh, pH and solubility of Cd, Ni and Zn. *Nutrient Cycling in Agroecosystems*, *61*(3), 247–255. doi:10.1023/A:1013762204510

Kästner, M., & Miltner, A. (2016). Application of compost for effective bioremediation of organic contaminants and pollutants in soil. *Applied Microbiology and Biotechnology*, *100*(8), 3433–3449. doi:10.1007/s00253-016-7378-y PMID:26921182

Katzung, B. G. (2001). *Basic and clinical pharmacology* (8th ed.). New York: McGraw- Hill.

Kaur, G., & Hundal, S. S. (2015). Bioremediation of heavy metal contaminated soil using earthworm *Eisenia fetida*. *Journal of Environment*, *4*(2), 25–29.

Kavitha, G. V., & Beebi, S. K. (2003). Biodegradation of phenol in a packed bed reactor using peat media. *Asian Journal of Microbiology, Biotechnology &. Environmental Sciences*, *5*(2), 157–159.

Kavitha, V., & Palanivelu, K. (2004). The role of ferrous ion in Fenton and photo-Fenton process for the degradation of phenol. *Chemosphere*, *55*(9), 1235–1243. doi:10.1016/j.chemosphere.2003.12.022 PMID:15081764

Keisham, S., & Paul, B. (2015). A Review on the Recent Scenario of Municipal Solid Waste Management in India. *International Journal of Engineering Research and General Science*.

Keith, L. H. (1976). Identification of organic compounds in unbleached treated Kraft paper mill wastewaters. *Environmental Science & Technology*, *10*(6), 555–564. doi:10.1021/es60117a009

Kersten, P. J. (1990). Glyoxal oxidase of *Phanerochaete chrysosporium*: Its characterization and activation by lignin peroxidase. *Proceedings of the National Academy of Sciences of the United States of America*, *87*(8), 2936–2940. doi:10.1073/pnas.87.8.2936 PMID:11607073

Keyeo, F., Aishah, N., & Amir, H. G. (2011). The effect of nitrogen fixing activity and phytohormone production of diazotroph in promoting growth of rice seedlings. *Biotechnol*, *10*(3), 267–273. doi:10.3923/biotech.2011.267.273

Khalfaoui, A., & Meniai, A. H. (2012). Application of chemically modified orange peels for removal of Copper (II) from aqueous solution. *Theoretical Foundations of Chemical Engineering*, *46*(6), 732–739. doi:10.1134/S0040579512060103

Khalifeh, A., Roozbehani, B., Moradi, A. M., Imani Moqadam, S., & Mirdrikvand, M. (2013). Isolation of Crude Oil from Polluted Waters Using Biosurfactants Pseudomonas Bacteria: Assessment of Bacteria Concentration Effects. Engineering, *Technology & Applied Science Research, 3*(2).

Khan, A. A., Wang, R. F., Cao, W. W., Doerge, D. R., Wennerstrom, D., & Cerniglia, C. E. (2001). Molecular cloning, nucleotide sequence, and expression of genes encoding a polycyclic aromatic ring dioxygenase from *Mycobacterium sp.* strain PYR-1. *Applied and Environmental Microbiology*, *67*(8), 3577–3585. doi:10.1128/AEM.67.8.3577-3585.2001 PMID:11472934

Khan, A. G. (2005). Role of soil microbes in the rhizospheres of plants growing on trace metal contaminated soils in phytoremediation. *Journal of Trace Elements in Medicine and Biology*, *18*(4), 355–364. doi:10.1016/j.jtemb.2005.02.006 PMID:16028497

Khan, M. N., & Mohammad, F. (2014). Eutrophication: Challenges and Solutions. In A. A. Ansari & S. S. Gill (Eds.), *Eutrophication: Causes, Consequences and Control.* Springer Science & Business, Media Dordrecht. doi:10.1007/978-94-007-7814-6_1

Khan, M. S., Zaidi, A., Wani, P. A., & Oves, M. (2009). Role of plant growth promoting rhizobacteria in the remediation of metal contaminated soils. *Environmental Chemistry Letters*, *7*(1), 1–19. doi:10.1007/s10311-008-0155-0

Khan, M. Z. H., & Mostafa, M. G. (2011). Aerobic treatment of pharmaceutical wastewater in a biological reactor. *International Journal of Environmental Sciences*, *1*(7), 1797–1805.

Khardenavis, A. A., Kapley, A., & Purohit, H. J. (2007). Simultaneous nitrification and denitrification by diverse Diaphorobacter sp. *Applied Microbiology and Biotechnology*, *77*(2), 403–409. doi:10.1007/s00253-007-1176-5 PMID:17876578

Khatisashvili, G., Matchavariani, L., & Gakhokidze, R. (2015). Improving Phytoremediation of Soil Polluted with Oil Hydrocarbons in Georgia. In K. R. Hakim & M. Sabir et al. (Eds.), *Soil Remediation and Plants* (pp. 547–570). doi:10.1016/B978-0-12-799937-1.00019-X

Kidd, K., Blanchfield, P., Mills, K., Palace, V., Evans, R., Lazorchak, J., & Flick, R. (2007). Collapse of a fish population after exposure to a synthetic estrogen. *Proceedings of the National Academy of Sciences of the United States of America*, *104*(21), 8897–8901. doi:10.1073/pnas.0609568104 PMID:17517636

Kim, D. Y., Bovet, L., Maeshima, M., Martinoia, E., & Lee, Y. (2007). The ABC transporter AtPDR8 is a cadmium extrusion pump conferring heavy metal resistance. *The Plant Journal*, *50*(2), 207–218. doi:10.1111/j.1365-313X.2007.03044.x PMID:17355438

Kim, E. H., Mason, R. P., Porter, E. T., & Soulen, H. L. (2004). The effect of resuspension on the fate of total mercury and methyl mercury in a shallow estuarine ecosystem: A mesocosm study. *Marine Chemistry*, *86*(3-4), 121–137. doi:10.1016/j.marchem.2003.12.004

Kim, J., Kang, S.-H., Min, K.-A., Cho, K.-S., & Lee, I.-S. (2006). Rhizosphere microbial activity during phytoremediation of diesel-contaminated soil. *Journal of Environmental Science and Health Part A*, *41*(11), 2503–2516. doi:10.1080/10934520600927658 PMID:17000542

Kim, J., Min, K., Cho, K., & Lee, I. (2007). Enhanced bioremediation and modified bacterial community structure by barnyard grass in diesel-contaminated soil. *Environmental engineering research-seoul*, *12*(2), 37–45. doi:10.4491/eer.2007.12.2.037

Kim, K.-H., Jahan, S. A., Kabir, E., & Brown, R. J. (2013). A review of airborne polycyclic aromatic hydrocarbons (PAHs) and their human health effects. *Environment International*, *60*, 71–80. doi:10.1016/j.envint.2013.07.019 PMID:24013021

Kim, S. I. I., Kim, S. J., Nam, M. H., Kim, S., Ha, K. S., Oh, K. H., & Park, Y. M. et al. (2002). Proteome analysis of aniline-induced proteins in *Acinetobacter lwoffi* K24. *Current Microbiology*, *44*(1), 61–66. doi:10.1007/s00284-001-0075-8 PMID:11727043

Kim, S. J., Jones, R. C., Cha, C. J., Kweon, O., Edmondson, R. D., & Cerniglia, C. E. (2004). Identification of proteins induced by polycyclic aromatic hydrocarbon in *Mycobacterium vanbaalenii* PYR-1 using two-dimensional polyacrylamide gel electrophoresis and de novo sequencing methods. *Proteomics*, *4*(12), 3899–3908. doi:10.1002/pmic.200400872 PMID:15540208

Kimura, K., Hara, H., & Watanabe, Y. (2007). Elimination of selected acidic pharmaceuticals from municipal wastewater by an activated sludge system and membrane bioreactors. *Environmental Science & Technology*, *41*(10), 3708–3714. and doi:10.1021/es061684z

Kim, Y. H., Cho, K., Yun, S. H., Kim, J. Y., Kwon, K. H., Yoo, J. S., & Kim, S. I. (2006). Analysis of aromatic catabolic pathways in Pseudomonas putida KT 2440 using a combined proteomic approach: 2-DE/MS and cleavable isotope-coded affinity tag analysis. *Proteomics*, *6*(4), 1301–1318. doi:10.1002/pmic.200500329 PMID:16470664

King, G. M. (2011). Enhancing soil Cstorage for Cremediation: Potential contributions and constraints by microbes. *Trends in Microbiology*, *19*(2), 75–84. doi:10.1016/j.tim.2010.11.006 PMID:21167717

Kinney, J. E. (1960). Evaluating the taste and odour control problems. *Journal - American Water Works Association*, *52*, 505–516.

Kirby, R. (2005). Actinomycetes and lignin degradation. *Advances in Applied Microbiology*, *58*, 125–168. doi:10.1016/S0065-2164(05)58004-3 PMID:16543032

Kirchgessner, M. (1978). *Trace element metabolism in man and animals*. Freising, Weihenstephan: Technische Universitat Munchen.

Kirchmair, J., Williamson, M. J., Tyzack, J. D., Tan, L., Bond, P. J., Bender, A., & Glen, R. C. (2012). Computational prediction of metabolism: Sites, products, SAR, P450 enzyme dynamics, and mechanisms. *Journal of Chemical Information and Modeling*, *52*(3), 617–648. doi:10.1021/ci200542m PMID:22339582

Kirk, J. L., Beaudette, L. A., Hart, M., Moutoglis, P., Klironomos, J. N., Lee, H., & Trevors, J. T. (2004). Methods of studying soil microbial diversity. *Journal of Microbiological Methods*, *58*(2), 169–188. doi:10.1016/j.mimet.2004.04.006 PMID:15234515

Kirkpatrick, W. D., White, P. Jr, Wolf, D., Thoma, G., & Reynolds, C. (2006). Selecting plants and nitrogen rates to vegetate crude-oil–contaminated soil. *International Journal of Phytoremediation*, *8*(4), 285–297. doi:10.1080/15226510600992840 PMID:17305303

Kiyohara, H., Matsumoto, T., & Yamada, H. (2000). Lignin-carbohydrate complexes: Intestinal immune system modulating ingredients in kampo (Japanese herbal) medicine, juzen-taiho-to. *Planta Medica*, *66*, 20–24. doi:10.1055/s-2000-11116 PMID:10705728

Klein, D. A., & Paschke, M. W. (2004). Filamentous fungi: The indeterminate lifestyle and microbial ecology. *Microbial Ecology*, *47*(3), 224–235. doi:10.1007/s00248-003-1037-4 PMID:15037964

Klekner, V., & Kosaric, N. (1992). Degradation of phenols by algae. *Environmental Technology*, *13*(5), 493–502. doi:10.1080/09593339209385176

Klibanov, A. M. (1983). Enzymatic removal of hazardous pollutants from industrial aqueous effluents. *Enzy. Eng.*, *6*, 319–323.

Klibanov, A. M., Tu, T. M., & Scott, K. P. (1983). Peroxidase-catalyzed removal of phenols from coal-conversion waste waters. *Science*, *221*(4607), 259–260. doi:10.1126/science.221.4607.259-a PMID:17815191

Klironomos, J. N., Rillig, M. C., & Allen, M. F. (1996). Below-ground microbial and microfaunal responses to Artemisia tridentata grown under elevated atmospheric CO2. *Functional Ecology*, *10*(4), 527–534. doi:10.2307/2389946

Klironomos, J. N., Rillig, M. C., Allen, M. F., Zak, D. R., Kubiske, M., & Pregitzer, K. S. (1997). Soil fungal–arthropod responses to Populus tremuloides grown under enriched atmospheric CO# under field conditions. *Global Change Biology*, *3*(6), 473–478. doi:10.1046/j.1365-2486.1997.00085.x

Klironomos, J. N., Ursic, M., Rillig, M. C., & Allen, M. F. (1998). Interspecific differences in the response of arbuscular mycorrhizal fungi to Artemisia tridentata grown under elevated atmospheric CO2. *The New Phytologist*, *138*(4), 599–605. doi:10.1046/j.1469-8137.1998.00141.x

Knappe, D. R. U., Snoeyink, V. L., Roche, P., Prados, M. J., & Bourbigot, M. (1999). Atrazine removal by preloaded GAC. *Journal - American Water Works Association*, (10): 97–109.

Knigge, T., Monsinjon, T., & Andersen, O. K. (2004). Surface-enhanced laser desorption/ionization-time of flight-mass spectrometry approach to biomarker discovery in blue mussels (Mytilus edulis) exposed to polyaromatic hydrocarbons and heavy metals under field conditions. *Proteomics*, *4*(9), 2722–2727. doi:10.1002/pmic.200300828 PMID:15352246

Knupp, G., Rücker, G., Ramos-Cormenzana, A., Garrido Hoyos, S., Neugebauer, M., & Ossenkop, T. (1996). Problems of identifying phenolic compounds during the microbial degradation of olive oil mill wastewater, Inter. *Biodeterio. Biodeg.*, *38*, 277–282.

Kobayashi, T., Hasinaga, T., Mikami, E., & Suzuki, T. (1989). Methanogenic degradation of phenol and benzoate in acclimated sludges. *Water Science and Technology*, *21*, 55–56.

Kolpin, D. W., Furlong, E. T., Meyer, M. T., Thurman, E. M., Zaugg, S. D., Barber, L. B., & Buxton, H. T. (2002). Pharmaceuticals, Hormones and other organic wastewater contaminants in U.S. streams, 19992000: A national reconnaissance. *Environmental Science & Technology*, *36*(6), 1202–1211. doi:10.1021/es011055j PMID:11944670

Kōmura, I., & Izaki, K. (1971). Mechanism of Mercuric Chloride Resistance in Microorganisms: I. Vaporization of a Mercury Compound from Mercuric Chloride by Multiple Drug Resistant Strains of Escherichia coli. *Journal of Biochemistry*, *70*(6), 885–893. PMID:4947308

Kōmura, I., Funaba, T., & Izaki, K. (1971). Mechanism of Mercuric Chloride Resistance in Microorganisms: II. NADPH-Dependent Reduction of Mercuric Chloride and Vaporization of Mercury from Mercuric Chloride by a Multiple Drug Resistant Strain of Escherichia coli. *Journal of Biochemistry*, *70*(6), 895–901. PMID:4401341

Kong, W., Ren, J., Wang, S., & Chen, Q. (2014). Removal of heavy metals from aqueous solution using acrylic-modified sugar cane bagasse – based adsorbents: Equilibrium and kinetic studies. *BioResources*, *9*(2), 3184–3196. doi:10.15376/biores.9.2.3184-3196

Koo, S.-Y., & Cho, K. S. (2009). Isolation and characterization of a plant growth-promoting rhizobacterium, Serratia sp. SY5. *Journal of Microbiology and Biotechnology*, *19*(11), 1431–1438. PMID:19996698

Koschorreck, K., Schmid, R. D., & Urlacher, V. B. (2009). Improving the functional expression of a *Bacillus licheniformis* laccase by random and site-directed mutagenesis. *BMC Biotechnology*, *9*(1), 1. doi:10.1186/1472-6750-9-12 PMID:19236694

Kostal, J. R. Y., Wu, C. H., Mulchandani, A., & Chen, W. (2004). Enhanced arsenic accumulation in engineered bacterial cells expressing ArsR. *Applied and Environmental Microbiology*, *70*(8), 4582–4587. doi:10.1128/AEM.70.8.4582-4587.2004 PMID:15294789

Kostka, J. E., Teske, A. P., Joye, S. B., & Head, I. M. (2014). The metabolic pathways and environmental controls of hydrocarbon biodegradation in marine ecosystems. *Frontiers in Microbiology, 5*, 471. doi: 10.3389/fmicb.2014.00471

Kot, A., & Namiesńik, J. (2000). The role of speciation in analytical chemistry. *Trends in Analytical Chemistry*, *19*(2-3), 69–79. doi:10.1016/S0165-9936(99)00195-8

Kothari, V., Panchal, M., & Srivastava, N. (2013). *Microbial Degradation of Hydrocarbons*. Ahmedabad, Gujarat, India: Institute of Science: Nirma University.

Kotz, J. C. (2012). Chemistry and Chemical reactivity (8th ed.). Brookes/Cole, Cengage Learning.

Koukkou, A. I. (Ed.). (2011). *Microbial Bioremediation of Non-metals: Current Research.* Caister Academic Press.

Kozlov, M. V., Haukioja, E. E., & Yarmishko, V. T. (1993). Aerial pollution in kola peninsula. *International Workshop*, St.Petersburg, Russia.

Krallish, I., Svetlana, G., Ludmila, S., Phillip, B., & Margesin, R. (2006). Phenol degradation by immobilized cold-adapted yeast strains of *Cryptococcus terreus* and *Rhodotorula creatinivora*. *Extremophiles*, *10*(5), 441–449. doi:10.1007/s00792-006-0517-0 PMID:16601913

Kramer, B., Rarey, M., & Lengauer, T. (1999). Evaluation of the FLEXX incremental construction algorithm for protein–ligand docking. *Proteins: Structure, Function, and Bioinformatics*, *37*(2), 228–241. doi:10.1002/(SICI)1097-0134(19991101)37:2<228::AID-PROT8>3.0.CO;2-8 PMID:10584068

Krämer, U. (2010). Metal hyperaccumulation in plants. *Annual Review of Plant Biology*, *61*(1), 517–534. doi:10.1146/annurev-arplant-042809-112156 PMID:20192749

Krämer, U., & Chardonnens, A. (2001). The use of transgenic plants in the bioremediation of soils contaminated with trace elements. *Applied Microbiology and Biotechnology*, *55*(6), 661–672. doi:10.1007/s002530100631 PMID:11525612

Krebs, R., Gupta, S. K., Furrer, G., & Schulin, R. (1998). Solubility and plant uptake of metals with and without liming of sludge-amended soils. *Journal of Environmental Quality*, *27*(1), 18–23. doi:10.2134/jeq1998.00472425002700010004x

Krishnakumar, A. M. (2007). *Structural studies of enzymes involved in propylene and acetone metabolism in Xanthobacter autotrophicus.* Academic Press.

Krishnan, A. K., & Anirudhan, T. S. (2002). Removal of mercury (II) from aqueous solutions and chlor-alkali industry effluent by steam activated and sulphurised activated carbons prepared from bagasse pith: Kinetics and equilibrium studies. *Journal of Hazardous Materials*, *92*(2), 161–183. doi:10.1016/S0304-3894(02)00014-6 PMID:11992701

Krishnani, K. K. (2016). Lignocellulostic wheat straw – derived ion exchange adsorbent for heavy metals removal. *Applied Biochemistry and Biotechnology*, *178*(4), 670–686. doi:10.1007/s12010-015-1901-y PMID:26494139

Kristanti, R. A., Toyama, T., Hadibarata, T., Tanaka, Y., & Mori, K. (2014). Sustainable Removal of Nitrophenols by Rhizoremediation Using Four Strains of Bacteria and Giant Duckweed (Spirodela polyrhiza). *Water, Air, and Soil Pollution*, *225*(4), 1–10. doi:10.1007/s11270-014-1928-7

Kristiansson, E., Fick, J., Janzon, A., Grabic, R., Rutgersson, C., Weidegård, B., & Larsson, D. G. J. et al. (2011). Pyrosequencing of antibiotic-contaminated river sediments reveals high levels of resistance and gene transfer elements. *PLoS ONE*, *6*(2), e17038. and doi:10.1371/journal.pone.0017038

Krivobok, S., Kuony, S., Meyer, C., Louwagie, M., Willison, J. C., & Jouanneau, Y. (2003). Identification of pyrene-induced proteins in *Mycobacterium sp.* strain 6PY1: Evidence for two ring-hydroxylating dioxygenases. *Journal of Bacteriology*, *185*(13), 3828–3841. doi:10.1128/JB.185.13.3828-3841.2003 PMID:12813077

Krug, M., Ziegler, H., & Straube, G. (1985). Degradation of phenolic compounds by the yeast Candida tropicalis HP15.I. Physiology of growth and substrate utilization. *Journal of Basic Microbiology*, *25*(2), 103–110. and doi:10.1002/jobm.3620250206

Ksheminska, H., Jaglarz, A., Fedorovych, D., Babyak, L., Yanovych, D., Kaszycki, P., & Koloczek, H. (2003). Bioremediation of chromium by the yeast *Pichia guilliermondii*: Toxicity and accumulation of Cr (III) and Cr (VI) and the influence of riboflavin on Cr tolerance. *Microbiological Research*, *158*(1), 59–67. doi:10.1078/0944-5013-00178 PMID:12608581

Kubicek, C. P., Scherferl, K. G., & Wöhres, W. (1988). Evidence for a cytoplasmic pathway of oxalate biosynthesis in *Aspergillus niger. Applied and Environmental Microbiology*, *54*(3), 633–637. PMID:3132096

Kudanga, T., & Le Roes-Hill, M. (2014). Laccase applications in biofuels production: Current status and future prospects. *Applied Microbiology and Biotechnology*, *98*(15), 6525–6542. doi:10.1007/s00253-014-5810-8 PMID:24841120

Kufareva, I., Ilatovskiy, A. V., & Abagyan, R. (2012). Pocketome: An encyclopedia of small-molecule binding sites in 4D. *Nucleic Acids Research*, *40*(D1), D535–D540. doi:10.1093/nar/gkr825 PMID:22080553

Kühner, S., Wöhlbrand, L., Fritz, I., Wruck, W., Hultschig, C., Hufnagel, P., & Rabus, R. et al. (2005). Substrate-dependent regulation of anaerobic degradation pathways for toluene and ethylbenzene in a denitrifying bacterium, strain EbN1. *Journal of Bacteriology*, *187*(4), 1493–1503. doi:10.1128/JB.187.4.1493-1503.2005 PMID:15687214

Kuiper, I., Lagendijk, E. L., Bloemberg, G. V., & Lugtenberg, B. J. (2004). Rhizoremediation: A beneficial plant-microbe interaction. *Molecular Plant-Microbe Interactions*, *17*(1), 6–15. doi:10.1094/MPMI.2004.17.1.6 PMID:14714863

Kujan, P., Prell, A., Šafa, H., Sobotka, M., Ezanka, T., & Holler, P. (2005). Removal of copper ions from dilute solutions by Streptomyces noursei mycelium. Comparison with yeast biomass. *Folia Microbiologica*, *50*(4), 309–314. doi:10.1007/BF02931411 PMID:16408849

Kujan, P., Prell, A., Safar, H., Sobotka, M., Rezanka, T., & Holler, P. P. (2006). Use of the industrial yeast *Candida utilis* for cadmium sorption. *Folia Microbiologica*, *51*(4), 257–260. doi:10.1007/BF02931807 PMID:17007420

Kulshreshtha, S. (2013). Genetically engineered microorganisms: A problem solving approach for bioremediation. *Bioremediation & Biodegradation*, *4*(4), 1–2.

Kumar, A. P., Janardhan, A., Viswanath, B., Monika, K., Jung, J. Y., & Narasimha, G. (2016). Evaluation of orange peel for biosurfactant production by *Bacillus licheniformis* and their ability to degrade naphthalene and crude oil. *3 Biotech, 6*(1), 1-10.

Kumar, A., Bisht, B. S., Joshi, V. D., & Dhewa, T. (2011). Review on bioremediation of polluted environment: A management tool. *International Journal of Environmental Sciences., 1*(6), 1079–1093.

Kumar, A., Kumar, S., & Kumar, S. (2005). Biodegradation kinetics of phenol and catechol using *Pseudomonas putida* MTCC 1194. *Biochemical Engineering Journal, 22*(2), 151–159. doi:10.1016/j.bej.2004.09.006

Kumar, A., & Sahu, O. (2013). Sugar industry wastes removal of toxic metals from waste water. *World Journal of Chemical Education, 1*(1), 17–20.

Kumaran, P. (1986). Development of biotechnology of toxic waste treatment. *Proc. Symp. Environmental Biology of Coastal Ecosystems, 11.*

Kumaran, P., & Paruchuri, Y. L. (1997). Kinetics of phenol biotransformation. *Water Research, 31*(1), 11–22. doi:10.1016/S0043-1354(99)80001-3

Kumar, M., Chaudhary, P., Dwivedi, M., Kumar, R., Paul, D., Jain, R. K., & Kumar, A. et al. (2005). Enhanced biodegradation of β-and δ-hexachlorocyclohexane in the presence of α-and γ-isomers in contaminated soils. *Environmental Science & Technology, 39*(11), 4005–4011. doi:10.1021/es048497q PMID:15984776

Kumar, M., & Philip, L. (2006). Enrichment and Isolation of a Mixed Bacterial Culture for Complete Mineralization of Endosulfan. *J Environ Sci Health, 41*(1), 81–96. doi:10.1080/03601230500234935 PMID:16393897

Kumar, P. S., Ramlingam, S., Abhinaya, R. V., Kiruphe, S. D., Murugesan, M., & Sevanesan, S. (2012). Adsorption of metal ions onto the chemically modified agricultural waste. *Clean-Soil, Air. Water (Basel), 40*(2), 188–197.

Kumar, R., Pandey, S., & Pandey, A. (2006). Plant roots and carbon sequestration. *Current Science, 91*(7), 885–890.

Kumar, S. R., Arumugam, T., Anandakumar, C. R., Balakrishnan, S., & Rajavel, D. S. (2013). Use of plant species in controlling environmental pollution-A review. *Bulletin of Environment. Pharmacology and Life Sciences, 2*(2), 52–63.

Kumar, S. S., Shantkriti, S., Muruganandham, T., Murugesh, E., Rane, N., & Govindwar, S. P. (2016). Bioinformatics aided microbial approach for bioremediation of wastewater containing textile dyes. *Ecological Informatics, 31*, 112–121. doi:10.1016/j.ecoinf.2015.12.001

Kummerer, K. (2001). Drugs in the Environment: Emission of Drugs, Diagnostic Aids and Disinfectants into Wastewater by Hospitals in Relation to Other Sources- a Review. *Chemosphere, 45*(6-7), 957–969. doi:10.1016/S0045-6535(01)00144-8 PMID:11695619

Kundu, D., Mondal, S., Dutta, D., Haque, S., & Ghosh, A. R. (2016). Accumulation and contamination of lead in different trophic levels of food chain in sewage-fed East Kolkata Wetland, West Bengal, India. *Int J Env Tech Sci, 2*, 61–68.

Kuntz, I. D., Blaney, J. M., Oatley, S. J., Langridge, R., & Ferrin, T. E. (1982). A geometric approach to macromolecule-ligand interactions. *Journal of Molecular Biology, 161*(2), 269–288. doi:10.1016/0022-2836(82)90153-X PMID:7154081

Kuo, H.-C., Juang, D.-F., Yang, L., Kuo, W.-C., & Wu, Y.-M. (2014). Phytoremediation of soil contaminated by heavy oil with plants colonized by mycorrhizal fungi. *International Journal of Environmental Science and Technology, 11*(6), 1661–1668. doi:10.1007/s13762-013-0353-6

Kuperman, R. G., Siciliano, S. D., Römbke, J., & Oorts, K. (2014). Deriving Site-Specific Soil Clean-Up Values for Metals and Metalloids: Rationale for Including Protection of Soil Microbial Processes. *Integrated Environmental Assessment and Management, 10*(3), 388–400. doi:10.1002/ieam.1513 PMID:24376192

Kurzawova, V., Stursa, P., Uhlik, O., Norkova, K., Strohalm, M., Lipov, J., & Mackova, M. et al. (2012). Plant–microorganism interactions in bioremediation of polychlorinated biphenyl-contaminated soil. *New Biotechnology, 30*(1), 15–22. doi:10.1016/j.nbt.2012.06.004 PMID:22728721

Kusian, B., Sultemeyer, D., & Bowien, B. (2002). Carbonic anhydrase is essential for growth of Ralstonia eutropha at ambient CO2concentrations. *Journal of Bacteriology, 184*(18), 5018–5026. doi:10.1128/JB.184.18.5018-5026.2002 PMID:12193617

Kutty, R., Purohit, H. J., & Khanna, P. (2000). Isolation and characterization of a Pseudomonas sp. strain PH1 utilizing meta-aminophenol. *Canadian Journal of Microbiology, 46*(3), 211–217. doi:10.1139/w99-132 PMID:10749534

Kuwata, K., Uebori, M., & Yamazaki, Y. (1980). Determination of phenol in polluted air as p-nitrobenzene azophenol derivatives by reversed phase high performance liquid chromatography. *Analytical Chemistry, 52*(6), 857–860. doi:10.1021/ac50056a020

Kuyucak, N., & Volesky, B. (1989). Accumulation of cobalt by marine alga. *Biotechnology and Bioengineering, 33*(7), 809–814. doi:10.1002/bit.260330703 PMID:18587987

Kvesitadze, G., Khatisashvili, G., Sadunishvili, T., & Ramsden, J. J. (2006). *Biochemical mechanisms of detoxification in higher plants: basis of phytoremediation.* Springer Science & Business Media.

La Guardia, M. J., Hale, R. C., Harvey, E., & Mainor, T. M. (2001). Alkylphenol ethoxylate degradation products in land-applied sew age sludge (biosolids). *Environmental Science & Technology, 35*(24), 4798–4804. doi:10.1021/es0109040 PMID:11775155

LAbbée, J. B., Tu, Y., Barriault, D., & Sylvestre, M. (2011). Insight into the metabolism of 1, 1, 1-trichloro-2, 2-bis (4-chlorophenyl) ethane (DDT) by biphenyl dioxygenases. *Archives of Biochemistry and Biophysics, 516*(1), 35–44. doi:10.1016/j.abb.2011.09.016 PMID:22001737

Lahiri (Ganguly), S., Sarkar (Paria), D., & Jana, B. B. (2015). Evaluating the Ecological Resilient Driven Performance of a Tropical Waste Stabilization Pond System using Ecological Signature of Biological Integrity. *Journal of Ecological Engineering, 16*(3), 97–107.

Lai, L., Lin, C., Xu, Z. Q., Han, X. L., Tian, F. F., Mei, P., & Liu, Y. et al. (2012). Spectroscopic studies on the interactions between CdTe quantum dots coated with different ligands and human serum albumin. *Spectrochimica Acta. Part A: Molecular and Biomolecular Spectroscopy, 97*, 366–376. doi:10.1016/j.saa.2012.06.025 PMID:22797377

Lakshmi, C., & Sridevi, V. (2009). A review on biodegradation of phenol from Industrial effluents. *Jr. of Industrial Pollution Control, 25*(1), 13–27.

Lal, R. (2014). Abating climate change and feeding the world through soil Csequestration. In Soil as world heritage. Springer.

Lal, R. (2008). Carbon sequestration. *Philosophical Transactions of the Royal Society of London. Series B, Biological Sciences, 363*(1492), 815–830. doi:10.1098/rstb.2007.2185 PMID:17761468

Lal, R. (2014). Soil Csequestration to mitigate climate change. *Geoderma, 123*(1-2), 1–22. doi:10.1016/j.geoderma.2004.01.032

Lal, R., Kimble, J. M., Follett, R. F., & Cole, C. V. (1998). *The potential of US cropland to sequester Cand mitigate the greenhouse effect*. Boca Raton, FL: CRC Press, LLC.

Landis, W. G., & Yu, M. H. (2003). *Introduction to environmental toxicology- Impacts of chemicals upon ecological systems* (3rd ed.). New York: Lewis Publishers.

Langdon, C. J., Piearce, T. G., Meharg, A. A., & Emple, K. T. (2001). Survival and behavior of behavior of the earthworms *Lumbricus rubellus* and *Dendrodrilus rubidus* from arsenate-contaminated and non-contaminated sites. *Soil Biology & Biochemistry*, *33*(9), 1239–1244. doi:10.1016/S0038-0717(01)00029-3

Langley, J. A., & Megonigal, J. P. (2010). Ecosystem response to elevated CO2 levels limited by nitrogen-induced plant species shift. *Nature*, *466*(7302), 96–99. doi:10.1038/nature09176 PMID:20596018

Lapara, T. M., Nakatsu, C. H., Pantea, L. M., & Alleman, J. E. (2001). Aerobic biological treatment of a pharmaceutical wastewater: Effect of temperature on COD removal and bacterial community development. *Water Research*, *35*(18), 4417–4425. and doi:10.1016/S0043-1354(01)00178-6

Larsen, M. H., Blackburn, N., Larsen, J. L., & Olsen, J. E. (2004). Influences of temperature, salinity and starvation on the motility and chemotactic response of Vibrio anguillarum. *Microbiology*, *150*(5), 1283–1290. doi:10.1099/mic.0.26379-0 PMID:15133091

Larsson, D. G. J. (2014). Antibiotics in the environment. *Upsala Journal of Medical Sciences*, *119*(2), 108–112. and do i:10.3109/03009734.2014.896438

Larsson, D. G. J., Adolfsson-Erici, M., Parkkonen, J., Pettersson, M., Berg, A. H., Olsson, P. E., & Forlin, L. (1999). Ethinyloestradiol - an undesired fish contraceptive? *Aquatic Toxicology (Amsterdam, Netherlands)*, *45*(2-3), 2–3, 91–97. doi:10.1016/S0166-445X(98)00112-X

Lasat, M. M. (2000). Phytoextraction of metals from contaminated soil: A review of plant/soil/metal interaction and assessment of pertinent agronomic issues. *Journal of Hazardous Substance Research*, *2*(5), 1–25.

Lashkenari, M. S., Davodo, B., & Eisazadeh, H. (2011). Removal of Arsenic from aqueous solution using poly aniline rice husk nano composite. *Korean Journal of Chemical Engineering*, *28*(7), 1532–1538. doi:10.1007/s11814-011-0014-8

Laskin, J., & Futrell, J. H. (2005). Activation of large lons in FT-ICR mass spectrometry. *Mass Spectrometry Reviews*, *24*(2), 135–167. doi:10.1002/mas.20012 PMID:15389858

Lavelle, P. (1983). The structure of earthworm communities. In J. E. Satchell (Ed.), *Earthworm Ecology*. London: Chapman & Hall. doi:10.1007/978-94-009-5965-1_39

Laville, N., Ait-Aissa, S., Gomez, E., Casellas, C., & Porcher, J. M. (2004). Effects of humanpharmaceuticals on cytotoxicity, EROD activity and ROS production in fish hepatocytes. *Toxicology*, *196*(1-2), 41–55. doi:10.1016/j.tox.2003.11.002 PMID:15036755

Law, G. T., Geissler, A., Lloyd, J. R., Livens, F. R., Boothman, C., Begg, J. D., & Morris, K. et al. (2010). Geomicrobiological redox cycling of the transuranic element neptunium. *Environmental Science & Technology*, *44*(23), 8924–8929. doi:10.1021/es101911v PMID:21047117

Lay, J. O. (2001). MALDI-TOF mass spectrometry of bacteria. *Mass Spectrometry Reviews*, *20*(4), 172–194. doi:10.1002/mas.10003 PMID:11835305

Leach, A. R., Shoichet, B. K., & Peishoff, C. E. (2006). Prediction of protein-ligand interactions. Docking and scoring: Successes and gaps. *Journal of Medicinal Chemistry*, *49*(20), 5851–5855. doi:10.1021/jm060999m PMID:17004700

Leahy, J. G., & Colwell, R. R. (1990). Microbial degradation of hydrocarbons in the environment. *Microbiological Reviews*, *54*, 305–315. PMID:2215423

Lee, C. M., Lu, C. J., & Chuang, M. S. (1994). Effects of immobilized cells on the biodegradation of chlorinated phenols. *Water Science and Technology*, *30*, 87–90.

Lee, H., Yun, S. Y., Jang, S., Kim, G. H., & Kim, J. J. (2015). Bioremediation of polycyclic aromatic hydrocarbons in creosote contaminated soil by *Peniophora incarnata* KUC8836. *Bioremediation Journal*, *19*(1), 1–8. doi:10.1080/10889868.2014.939136

Lee, K. E. (1985). *Earthworms - Their Ecology and Relationships with Soils and Land Use*. Sydney: Academic Press.

Lee, R. F., & Ryan, C. (1979). Microbial degradation of oraganochlorine compounds in estuarine waters and sediments. *Proc. Workshop: Microbe degradation Pollut Mar Environ.*

Lee, S.-H., Lee, W.-S., Lee, C.-H., & Kim, J.-G. (2008). Degradation of phenanthrene and pyrene in rhizosphere of grasses and legumes. *Journal of Hazardous Materials*, *153*(1), 892–898. doi:10.1016/j.jhazmat.2007.09.041 PMID:17959304

Lee, S., Moon, J. S., Ko, T. S., Petros, D., Goldsbrough, P. B., & Korban, S. S. (2003). Overexpression of Arabidopsis phytochelatin synthase paradoxically leads to hypersensitivity to cadmium stress. *Plant Physiology*, *131*(2), 656–663. doi:10.1104/pp.014118 PMID:12586889

Lee, S.-W., Glickmann, E., & Cooksey, D. A. (2001). Chromosomal locus for cadmium resistance in Pseudomonas putida consisting of a cadmium-transporting ATPase and a MerR family response regulator. *Applied and Environmental Microbiology*, *67*(4), 1437–1444. doi:10.1128/AEM.67.4.1437-1444.2001 PMID:11282588

Lefebvre, O., Habouzit, F., Bru, V., Delgenes, J. P., Godon, J. J., & Moletta, R. (2004). Treatment of hypersaline industrial wastewater by a microbial consortium in a sequencing batch reactor. *Environmental Technology*, *25*(5), 543–553. doi:10.1080/09593330.2004.9619345 PMID:15242230

Lefebvre, O., Vasudevan, N., Thanasekaran, K., Moletta, R., & Godon, J. J. (2006). Microbial diversity in hypersaline wastewater: The example of tanneries. *Extremophiles*, *10*(6), 505–513. doi:10.1007/s00792-006-0524-1 PMID:16738814

Leigh, M. B., Prouzová, P., Macková, M., Macek, T., Nagle, D. P., & Fletcher, J. S. (2006). Polychlorinated biphenyl (PCB)-degrading bacteria associated with trees in a PCB-contaminated site. *Applied and Environmental Microbiology*, *72*(4), 2331–2342. doi:10.1128/AEM.72.4.2331-2342.2006 PMID:16597927

Lemire, J. A., Harrison, J. J., & Turner, R. J. (2013). Antimicrobial activity of metals: Mechanisms, molecular targets and applications. *Nature Reviews. Microbiology*, *11*(6), 371–384. doi:10.1038/nrmicro3028 PMID:23669886

Lenart-Boroń, A., & Boroń, P. (2014). The Effect of Industrial Heavy Metal Pollution on Microbial Abundance and Diversity in Soils-A Review. In M. C. Hernandez-Soriano (Ed.), *Environmental Risk Assessment of Soil Contamination* (pp. 759–783). doi:10.5772/57406

Lengauer, T., & Rarey, M. (1996). Computational methods for biomolecular docking. *Current Opinion in Structural Biology*, *6*(3), 402–406. doi:10.1016/S0959-440X(96)80061-3 PMID:8804827

LeQuéré, C., Raupach, M. R., Canadell, J. G., Marland, G., Bopp, L., Ciais, P., & Foster, P. et al. (2009). Trends in the sources and sinks of Cdioxide. *Nature Geoscience*, *2*(12), 831–836. doi:10.1038/ngeo689

Lettinga, G., van Velsen, A. F., Hohma, S. M., de Zeeuw, W., & Klapwijk, A. (1980). Use of upflow sludge blanket (UASB) reactor concept for biological wastewater treatment. *Biotechnology and Bioengineering*, *22*, 699–734. doi:10.1002/bit.260220402

Leung, D. Y., Caramanna, G., & Maroto-Valer, M. M. (2014). An overview of current status of Cdioxide capture and storage technologies. *Renewable & Sustainable Energy Reviews*, *39*, 426–443. doi:10.1016/j.rser.2014.07.093

Leven, L., Nyberg, K., Korkea-aho, L., & Schnurer, A. (2006). Phenols in anaerobic digestion processes and inhibition of ammonia oxidising bacteria (AOB) in soil. *The Science of the Total Environment*, *364*(1-3), 229–238. doi:10.1016/j.scitotenv.2005.06.003 PMID:16125214

Levin, L., Viale, A., & Forchiassin, A. (2003). Degradation of organic pollutants by the white rotbasidiomycete *Trametes trogii*. *International Biodeterioration & Biodegradation*, *52*(1), 1–5. doi:10.1016/S0964-8305(02)00091-4

Levin, R., Brown, M. J., Kashtock, M. E., Jacobs, D. E., Whelan, E. A., Rodman, J., & Sinks, T. et al. (2008). Lead Exposures in U.S. Children, 2008: Implications for Prevention. *Environmental Health Perspectives*, 116. PMID:18941567

Levin, S. A. (1992). The Problem of Pattern and Scale in Ecology: The Robert H. Mac Arthur Award. *Ecology*, *73*(6), 1943–1967. doi:10.2307/1941447

Liang, C., & Balser, T. C. (2011). Microbial production of recalcitrant organic matter in global soils: Implications for productivity and climate policy. *Nature Reviews. Microbiology*, *9*(1), 75. doi:10.1038/nrmicro2386-c1 PMID:21113179

Liang, Y. J., Chi, L. Y., Min, X. B., Tang, C. J., & Zhang, H.-J. (2012). Hydrothermal sulfidation and floatationtreatment of heavy-metal-containing sludge for recovery and stabilization. *Journal of Hazardous Materials*, *217*, 307–310. doi:10.1016/j.jhazmat.2012.03.025 PMID:22487137

Liang, Y., Meggo, R., Hu, D., Schnoor, J. L., & Mattes, T. E. (2015). Microbial community analysis of switchgrass planted and unplanted soil microcosms displaying PCB dechlorination. *Applied Microbiology and Biotechnology*, *99*(15), 6515–6526. doi:10.1007/s00253-015-6545-x PMID:25820643

Liau, S. Y., Read, D. C., Pugh, W. J., Furr, J. R., & Russell, A. D. (1997). Interaction of silver nitrate with readily identifiable groups: Relationship to the antibacterial action of silver ions. *Letters in Applied Microbiology*, *25*(4), 279–283. doi:10.1046/j.1472-765X.1997.00219.x PMID:9351278

Librando, V., & Forte, S. (2005). Computer evaluation of protein segments removal effects from naphthalene 1, 2-dioxygenase enzyme on polycyclic aromatic hydrocarbons interaction. *Biochemical Engineering Journal*, *27*(2), 161–166. doi:10.1016/j.bej.2005.08.011

Librando, V., & Pappalardo, M. (2011). Computational study on the interaction of a ring-hydroxylating dioxygenase from *Sphingomonas* CHY-1 with PAHs. *Journal of Molecular Graphics & Modelling*, *29*(7), 915–919. doi:10.1016/j.jmgm.2011.03.001 PMID:21441049

Lichtarge, O., Bourne, H. R., & Cohen, F. E. (1996). An evolutionary trace method defines binding surfaces common to protein families. *Journal of Molecular Biology*, *257*(2), 342–358. doi:10.1006/jmbi.1996.0167 PMID:8609628

Lide, D. R. (1993). *CRC handbook of chemistry and physics*. CRC Press.

Lika, K., & Papadakis, I. A. (2009). Modeling the biodegradation of phenolic compounds by microalgae. *Journal of Sea Research*, *62*(2-3), 135–146. doi:10.1016/j.seares.2009.02.005

Lim Ai, P., & Aris, A. Z. (2014). A review on economically adsorbents on heavy metals removal in water and wastewater. *Reviews in Environmental Science and Biotechnology*, *13*(2), 163–181. doi:10.1007/s11157-013-9330-2

Lin, C. C., & Lin, H. L. (2005). Remediation of soil contaminated with the heavy metal (Cd^{2+}). *Journal of Hazardous Materials*, *122*(1-2), 7–15. doi:10.1016/j.jhazmat.2005.02.017 PMID:15943924

Lin, C., & Olson, B. H. (1995). Occurrence of cop-like resistance genes among bacteria isolated from a water distribution system. *Canadian Journal of Microbiology*, *41*(7), 642–646. doi:10.1139/m95-087

Lin, D., Zhou, Q., Xie, X., & Liu, Y. (2010). Potential biochemical and genetic toxicity of triclosan as an emerging pollutant on earthworms (Eisenia fetida). *Chemosphere*, *81*(10), 1328–1333. doi:10.1016/j.chemosphere.2010.08.027 PMID:20825966

Lindahl, B. D., Ihrmark, K., Boberg, J., Trumbore, S. E., Ho¨gberg, P., Stenlid, J., & Finlay, R. D. (2007). Spatial separation of litter decomposition and mycorrhizal nitrogen uptake in a boreal forest. *The New Phytologist*, *173*(3), 611–620. doi:10.1111/j.1469-8137.2006.01936.x PMID:17244056

Linley, E., Denyer, S. P., McDonnell, G., Simons, C., & Maillard, J. Y. (2012). Use of hydrogen peroxide as a biocide: New consideration of its mechanisms of biocidal action. *The Journal of Antimicrobial Chemotherapy*, *67*(7), 1589–1596. doi:10.1093/jac/dks129 PMID:22532463

Lin, T. F., Huang, H. I., Shen, F. T., & Young, C.-C. (2006). The protons of gluconic acid are the major factor responsible for the dissolution of tricalcium phosphate by *Burkholderia cepacia* CC-A174. *Bioresource Technology*, *97*(7), 957–960. doi:10.1016/j.biortech.2005.02.017 PMID:16356711

Lin, Y. H., Wu, C. L., Hsu, C. H., & Li, H. L. (2009). Biodegradation of phenol with chromium(VI) reduction in an anaerobic fixed-biofilm process—kinetic model and reactor performance. *Journal of Hazardous Materials*, *172*(2-3), 1394–1401. doi:10.1016/j.jhazmat.2009.08.005 PMID:19726129

Lin, Z., Zhen, Z., Wu, Z., Yang, J., Zhong, L., Hu, H., & Zhang, D. et al. (2016). The impact on the soil microbial community and enzyme activity of two earthworm species during the bioremediation of pentachlorophenol-contaminated soils. *Journal of Hazardous Materials*, *301*, 35–45. doi:10.1016/j.jhazmat.2015.08.034 PMID:26342149

Lin, Z., Zhou, C., Wu, J., Zhou, J., & Wang, L. (2005). A further insight into the mechanism of Ag^+ biosorption by *Lactobacillus* spp. strain A09. *Spectrochimica Acta. Part A: Molecular and Biomolecular Spectroscopy*, *61*(6), 1195–1200. doi:10.1016/j.saa.2004.06.041 PMID:15741121

Lippard, S. J., & Berg, J. M. (1994). *Principles of Bioinorganic Chemistry*. Mill Valley, CA: University Science Books.

Lipscomb, J. D. (2008). Mechanism of extradiol aromatic ring- cleaving dioxygenases. *Current Opinion in Structural Biology*, *18*(6), 644–649. doi:10.1016/j.sbi.2008.11.001 PMID:19007887

Liu, H., Guo, S., Jiao, K., Hou, J., Xie, H., & Xu, H. (2015). Bioremediation of soils co-contaminated with heavy metals and 2, 4, 5-trichlorophenol by fruiting body of Clitocybe maxima. *Journal of Hazardous Materials*, *294*, 121–127. doi:10.1016/j.jhazmat.2015.04.004 PMID:25863026

Liu, L., Jiang, C. Y., Liu, X. Y., Wu, J. F., Han, J. G., & Liu, S. J. (2007). Plant–microbe association for rhizoremediation of chloronitroaromatic pollutants with Comamonas sp. strain CNB-1. *Environmental Microbiology*, *9*(2), 465–473. doi:10.1111/j.1462-2920.2006.01163.x PMID:17222144

Liu, X., Liu, J., Zhu, T., Zhang, L., He, X., & Zhang, J. Z. (2016). PBSA_E: A PBSA based free energy estimator for protein-ligand binding affinity. *Journal of Chemical Information and Modeling*, *56*(5), 854–861. doi:10.1021/acs.jcim.6b00001 PMID:27088302

Liu, Y., Chen, M., Jiang, L., & Song, L. (2014). New insight into molecular interaction of heavy metal pollutant—cadmium (II) with human serum albumin. *Environmental Science and Pollution Research International*, *21*(11), 6994–7005. doi:10.1007/s11356-014-2610-8 PMID:24522399

Li, W., & Wong, M. (2012). Interaction of Cd/Zn hyperaccumulating plant (Sedum alfredii) and rhizosphere bacteria on metal uptake and removal of phenanthrene. *Journal of Hazardous Materials, 209*, 421–433. doi:10.1016/j.jhazmat.2012.01.055 PMID:22309655

Li, X., Schuler, M. A., & Berenbaum, M. R. (2007). Molecular mechanisms of metabolic resistance to synthetic and natural xenobiotics. *Annual Review of Entomology, 52*(1), 231–253. doi:10.1146/annurev.ento.51.110104.151104 PMID:16925478

Li, Y., Dhankher, O. P., Carreira, L., Lee, D., Chen, A., Schroeder, J. I., & Meagher, R. B. (2004). Overexpression of phytochelatin synthase in Arabidopsis leads to enhanced arsenic tolerance and cadmium hypersensitivity. *Plant & Cell Physiology, 45*(12), 1787–1797. doi:10.1093/pcp/pch202 PMID:15653797

Li, Y., Yang, H., Liu, N., Luo, J., Wang, Q., & Wang, L. (2015). Cadmium accumulation and metallothionein biosynthesis in cadmium-treated freshwater mussel *Anodonta woodiana. PLoS ONE, 10*(2), e0117037. doi:10.1371/journal.pone.0117037 PMID:25647043

Li, Y., Zhu, G., Ng, J. W., & Tan, S. K. (2014). A review on removing pharmaceutical contaminants from wastewater by constructed wetlands: Design, performance and mechanism. *The Science of the Total Environment, 468-469*, 908–932. and doi:10.1016/j.scitotenv.2013.09.018

Li, Z. S., Lu, Y. P., Zhen, R. G., Szczypka, M., Thiele, D. J., & Rea, P. A. (1997). A new pathway for vacuolar cadmium sequestration in Saccharomyces cerevisiae: YCF1-catalyzed transport of bis (glutathionato) cadmium. *Proceedings of the National Academy of Sciences of the United States of America, 94*(1), 42–47. doi:10.1073/pnas.94.1.42 PMID:8990158

Lobersli, E. M., & Steinnes, E. (1988). Metal uptake in plants from a birch forest area near a copper smelter in Norway. *Water Air Pollution, 37*, 25–39.

Loehr, R. C., & Krishnamoorthy, R. (1988). Terrestrial bioaccumulation potential of phenolic compounds. *Hazardous Waste & Hazardous Materials, 5*(2), 109–119. doi:10.1089/hwm.1988.5.109

Loh, K. C., & Cao, B. (2008). Paradigm in biodegradation using *Pseudomonas putida:* A review of proteomics studies. *Enzyme and Microbial Technology, 43*(1), 1–12. doi:10.1016/j.enzmictec.2008.03.004

Lokeshwari, H., & Chandrappa, G. T. (2007). Effects of heavy metal contamination from anthropogenic sources on Dasarahalli tank, India. *Lakes and Reservoirs: Research and Management, 12*(3), 121–128. doi:10.1111/j.1440-1770.2007.00337.x

Lokhande, R. S., Singare, P. U., & Pimple, D. S. (2011). Toxicity Study of Heavy Metals Pollutants in Waste Water Effluent Samples Collected from Taloja Industrial Estate of Mumbai, India. *Resources and Environment, 1*(1), 13–19. doi:10.5923/j.re.20110101.02

Lokke, H., & Van Gestel, C. A. M. (1998). *Handbook of soil invertebrate toxicity tests*. Chichester, UK: John Wiley & Sons.

Lonsane, B. K., Ghildyal, N. P., Budiatman, S., & Ramakrishna, S. V. (1985). Engineering aspects of solid state fermentation. *Enzyme and Microbial Technology, 7*(6), 258–265. doi:10.1016/0141-0229(85)90083-3

Lonsane, B. K., Saucedo-Castaneda, G., Raimbault, M., Roussos, S., Viniegra-Gonzalez, G., Ghildyal, N. P., & Krishnaiah, M. M. et al. (1992). Scale-up strategies for solid state fermentation systems. *Process Biochemistry (Barking, London, England), 27*(5), 259–273. doi:10.1016/0032-9592(92)85011-P

Lopez, A., Lazaro, N., Morales, S., & Margues, A. M. (2002). Nickel biosorption by free and immobilized cells of *Pseudomonas fluorescens* 4F39: A comparative study. *Water, Air, and Soil Pollution, 135*(1/4), 157–172. doi:10.1023/A:1014706827124

López-Fernández, R., Martínez, L., & Villaverde, S. (2012). Membrane bioreactor for the treatment of pharmaceutical wastewater containing corticosteroids. *Desalination, 300*, 19–23. doi:10.1016/j.desal.2012.05.032

Loredano, P. (2015). Lignin-degrading enzymes. *The FEBS Journal, 282*(7), 1190–1213. doi:10.1111/febs.13224 PMID:25649492

Lorimor, J., Fulhage, C., Zhang, R., Funk, T., Sheffield, R., Sheppard, C., & Newton, G. L. (2001). *Manure Management Strategies/ Technologies.* White Paper on Animal Agriculture and the Environment for National Centre for Manure and Animal Waste Management.

Loureiro, S., Soares, A. M. V. M., & Nogueira, A. J. A. (2005). Terrestrial avoidance behaviour tests as screening tool to assess soil contamination. *Environmental Pollution, 138*(1), 121–131. doi:10.1016/j.envpol.2005.02.013 PMID:15885863

Lovely, D. R. (2002). Dissimulatory metal reduction: From early life to bioremediation. *ASM News, 68*(5), 231–237.

Lovley, D. R., & Lloyd, J. R. (2000). Microbes with a metal for bioremediation. *Nature Biotechnology, 18*(6), 600–601. doi:10.1038/76433 PMID:10835594

Ludzack, F. J., & Ettinger, M. B. (1960). Chemical structures resistant to aerobic biochemical stabilization. *Wat. Poll. Cont. Fed., 32*, 1173–1200.

Luna, J. M., Rufino, R. D., Albuquerque, C. D., Sarubbo, L. A., & Campos-Takaki, G. M. (2011). Economic optimized medium for tensio-active agent production by *Candida sphaerica* UCP0995 and application in the removal of hydrophobic contaminant from sand. *International Journal of Molecular Sciences, 12*(4), 2463–2476. doi:10.3390/ijms12042463 PMID:21731452

Luo, J., Liu, X., Tian, Q., Yue, W., Zeng, J., Chen, G., & Cai, X. (2009). Disposable marine monitoring: New opportunities for assessing marine health status. *Marine Pollution Monitoring and Assessment, 144*, 221–228.

Luo, Y., Guo, W., Ngo, H. H., Nghiem, L. D., Hai, F. I., Zhang, J., & Wang, X. C. et al. (2014). A review on the occurrence of micropollutants in the aquatic environment and their fate and removal during wastewater treatment. *The Science of the Total Environment, 473-474*, 619–641. and doi:10.1016/j.scitotenv.2013.12.065

Luo, Y., Jiang, X., Wu, L., Song, J., Wu, S., Lu, R., & Christie, P. (2003). Accumulation and chemical fractionation of Cu in a paddy soil irrigated with Cu-enriched organic waste materials. *Ecological Bulletins, 39*, 53–56.

Lussenhop, J., Treonis, A., Curtis, P. S., Teeri, J. A., & Vogel, C. S. (1998). Response of soil biota to elevated atmospheric CO! in poplar model systems. *Oecologia, 113*(2), 247–251. doi:10.1007/s004420050375

Luther, M., & Shaaban, M. M. (1990). Algal degradation of aromatic compounds. *Forum Mikrobiology, 1*, 42.

Lu, Y., & Conrad, R. (2005). In situ stable isotope probing of methanogenic archaea in the rice rhizosphere. *Science, 309*(5737), 1088–1090. doi:10.1126/science.1113435 PMID:16099988

Lu, Y.-F., Lu, M., Peng, F., Wan, Y., & Liao, M.-H. (2014). Remediation of polychlorinated biphenyl-contaminated soil by using a combination of ryegrass, arbuscular mycorrhizal fungi and earthworms. *Chemosphere, 106*, 44–50. doi:10.1016/j.chemosphere.2013.12.089 PMID:24457052

Lu, Y., Lueders, T., Friedrich, M. W., & Conrad, R. (2005). Detecting active methanogenic populations on rice roots using stable isotope probing. *Environmental Microbiology, 7*(3), 326–336. doi:10.1111/j.1462-2920.2005.00697.x PMID:15683393

Lu, Z. Y., & Xiao-Jue, G. (2015). High-Throughput Screening for a Moderately Halophilic Phenol-Degrading Strain and Its Salt Tolerance Response. *International Journal of Molecular Sciences, 16*(6), 11834–11848. doi:10.3390/ijms160611834 PMID:26020478

Lyon, D. Y., & Vogel, T. M. (2013). *Bioaugmentation for Groundwater Remediation: An Overview. In Bioaugmentation for Groundwater Remediation* (pp. 1–37). Springer. doi:10.1007/978-1-4614-4115-1_1

Mabire, F., Audebert, R., & Quivoron, C. (1984). Flocculation properties of some water-soluble cationic copolymers toward silica suspensions: A semi-quantative interpretation of the role of molecular weight and cationicity through a patchwork model. *Journal of Colloid and Interface Science*, *97*(1), 120–136. doi:10.1016/0021-9797(84)90280-7

Macaulay, B. M. (2015). Understanding the behaviour of oil-degrading micro-organisms to enhance the microbial remediation of spilled petroleum. *Applied Ecology and Environmental Research*, *13*(1), 247–262.

Macaulay, B. M., & Rees, D. (2014). Bioremediation of Oil Spills: A Review of Challenges for Research Advancement. *Annals of Environmental Science*, *8*(1), 2.

Mackay, D., & Webster, E. (2005). Environmental persistence of chemicals. *Environmental Science and Pollution Research International*, *13*(1), 43–49. doi:10.1065/espr2006.01.008 PMID:16417131

Mackova, M., Dowling, D., & Macek, T. (2006). *Phytoremediation and rhizoremediation* (Vol. 9). Springer Science & Business Media. doi:10.1007/978-1-4020-4999-4

Mackova, M., Prouzova, P., Stursa, P., Ryslava, E., Uhlik, O., Beranova, K., & Macek, T. et al. (2009). Phyto/rhizoremediation studies using long-term PCB-contaminated soil. *Environmental Science and Pollution Research International*, *16*(7), 817–829. doi:10.1007/s11356-009-0240-3 PMID:19823887

Macomber, L., Elsey, S. P., & Hausinger, R. P. (2011). Fructose-1,6-bisphosphate aldolase (class II) is the primary site of nickel toxicity in *Escherichia coli*. *Molecular Microbiology*, *82*(5), 1291–1300. doi:10.1111/j.1365-2958.2011.07891.x PMID:22014167

Macomber, L., & Imlay, J. A. (2009). The iron-sulfur clusters of dehydratases are primary intracellular targets of copper toxicity. *Proceedings of the National Academy of Sciences of the United States of America*, *106*(20), 8344–8349. doi:10.1073/pnas.0812808106 PMID:19416816

Maddipati, P., Atiyeh, H. K., Bellmer, D. D., & Huhnke, R. L. (2011). Ethanol production from syngas by *Clostridium* strain P11 using corn steep liquor as a nutrient replacement to yeast extract. *Bioresource Technology*, *102*(11), 6494–6501. doi:10.1016/j.biortech.2011.03.047 PMID:21474306

Madigan, M. T., Martinko, J. M., & Parker, J. (1997). Brock biology of microorganisms (Vol. 514). Upper Saddle River, NJ: Prentice Hall.

Madukasi, E. I., Dai, X., He, C., & Zhou, J. (2010). Potentials of phototrophic bacteria in treating pharmaceutical wastewater. *International Journal of Environmental Science and Technology*, *7*(1), 165–174. doi:10.1007/BF03326128

Magellan, K., Barral-Fraga, L., Rovira, M., Srean, P., Urrea, G., García-Berthou, E., & Guasch, H. (2014). Behavioural and physical effects of arsenic exposure in fish are aggravated by aquatic algae. *Aquatic Toxicology (Amsterdam, Netherlands)*, *156*, 116–124. doi:10.1016/j.aquatox.2014.08.006 PMID:25190483

Magni, P., Falco, D., Falugi, G., Franzoni, C., & Monteverde, M. (2005). Genotoxicity biomarkers and acetyl V cholinesterase activity in natural populations of *Mytilus galloprovincialis* along a pollution gradient in the Gulf of Oristano (Sardinia, western Mediterranean). *Environmental Pollution*, *142*(1), 65–72. doi:10.1016/j.envpol.2005.09.018 PMID:16289776

Mah, H., Boyd, D., Gadaginamath, G., Wood, A., Buening, M., Chang, R., . . . Conney, A. (2013). *Polycyclic aromatic hydrocarbons*. Paper presented at the Toxicology: Proceedings of the 7th International Congress of Pharmacology, Paris, France.

Mahammed, R. R. (2012). Removal of heavy metals from wastewater using black tea waste. *Arabian Journal for Science and Engineering*, *37*(6), 1505–1520. doi:10.1007/s13369-012-0264-8

Maheshwari, R., Singh, U., Singh, P., Singh, N., Lal, J., & Rani, B. (2014). To decontaminate wastewater employing bioremediation technologies. *J Adv Sci Res*, *5*(2), 7–15.

Mahmoud, Y. A.-G. (2004). Uptake of radionuclides by some fungi. *Mycobiology*, *32*(3), 110–114. doi:10.4489/MYCO.2004.32.3.110

Mai, C., Kues, U., & Militz, H. (2004). Biotechnology in the wood industry. *Applied Microbiology and Biotechnology*, *63*, 477–494. doi:10.1007/s00253-003-1411-7 PMID:12937955

Mai, C., Majcherczyk, A., & Huttermann, A. (2000). Chemo-enzymatic synthesis and characterization of graft copolymers from lignin and acrylic compounds. *Enzyme and Microbial Technology*, *27*(1-2), 167–175. doi:10.1016/S0141-0229(00)00214-3 PMID:10862917

Maier, R. M. (2000). Microorganisms and organic pollutants. In R. M. Maier, I. L. Pepper, & C. P. Gerba (Eds.), *Environmental Microbiology* (pp. 363–402). San Diego, CA: Academic Press.

Maila, M. P., Randima, P., & Cloete, T. E. (2005). Multispecies and monoculture rhizoremediation of polycyclic aromatic hydrocarbons (PAHs) from the soil. *International Journal of Phytoremediation*, *7*(2), 87–98. doi:10.1080/16226510590950397 PMID:16128441

Majeau, J. A., Satinder, K. B., & Tyagi, R. D. (2010). Laccases for removal of recalcitrant and emerging pollutants. *BiresourTechnol*, *101*(7), 2331–2350. doi:10.1016/j.biortech.2009.10.087 PMID:19948398

Major, D. N., & Wang, H. (2012). How public health impact is addressed: A retrospective view on three different oil spills. *Toxicological and Environmental Chemistry*, *94*(3), 442–467. doi:10.1080/02772248.2012.654633

Makhijani, M., Gahlawat, S., Chauhan, K., Valsangkar, S., & Gauba, P. (2014). Phytoremediation potential of *Cicer arietinum* for Tetracycline. *International Journal of Genetic Engineering and Biotechnology*, *5*(2), 153–160.

Makkar, R. S., Cameotra, S. S., & Banat, I. M. (2011). Advances in utilization of renewable substrates for biosurfactant production. *AMB Express*, *1*(5), 1–19. PMID:21906330

Malik, A. (2006). Bioremediation. In *Environmental Microbiology*. National Science Digital Library.

Malosso, E., English, L., Hopkins, D. W., & ODonnell, A. G. (2004). Use of 13C-labelled plant materials and ergosterol, PLFA and NLFA analyses to investigate organic matter decomposition in Antarctic soil. *Soil Biology & Biochemistry*, *36*(1), 165–175. doi:10.1016/j.soilbio.2003.09.004

Maltseva, O., & Oriel, P. (1997). Monitoring of an Alkaline 2,4, 6-Trichlorophenol-Degrading Enrichment Culture by DNA Fingerprinting Methods and Isolation of the Responsible Organism, Haloalkaliphilic *Nocardioides* sp. Strain M6. *Applied and Environmental Microbiology*, *63*, 4145–4149. PMID:16535723

Mandal, A. K., Sarma, P. M., Jeyaseelan, C. P., Channashettar, V. A., Singh, B., Lal, B., & Datta, J. (2012). Large-scale bioremediation of petroleum hydrocarbon contaminated waste at Indian oil refineries: Case studies. *Int. J. Life Sci. Pharm. Res*, *2*(4), L114–L128.

Mandal, A. K., Sarma, P. M., & Jeyaseelan, C. Paul., Channashettar, V. A., Singh, B., Lal, B., & Datta, J. (2012). Large scale bioremediation of petroleum hydrocarbon contaminated waste at Indian oil refineries: Case studies. *Int J Life Sci Pharma Res*, *2*(4), L-114–L-128.

Mandal, A. K., Sarma, P. M., Singh, B., & Jeyaseelan, C. P. (2012). Bioremediation: An environment friendly sustainable biotechnological solution for remediation of petroleum hydrocarbon contaminated waste. *ARPN Journal of Science and Technology*, *2*(Special Issue), 1–12.

Mandal, A. K., Sarma, P. M., Singh, B., Jeyaseelan, C. P., Channashettar, V. A., Lal, B., & Datta, J. (2011). Bioremediation: A sustainable eco-friendly biotechnological solution for environmentalpollution in oil industries. *Journal of Sustainable Develeopment and Environmental Protection*, *1*(3), 5–23.

Mandal, A. K., Sarma, P. M., Singh, B., Jeyaseelan, C. P., Channashettar, V. A., Lal, B., & Datta, J. (2012). Bioremediation: An environment friendly sustainable biotechnological solution for remediation of petroleum hydrocarbon contaminated waste. *Journal of Science and Technology*, *2*, 1–12.

Mandels, M., & Sternberg, D. (1976). Recent advances in cellulase technology. *Ferment. Technol*, *54*, 267–286.

Mani, D., & Kumar, C. (2014). Biotechnological advances in bioremediation of heavy metals contaminated ecosystems: An overview with special reference to phytoremediation. *International Journal of Environmental Science and Technology*, *11*(3), 843–872. doi:10.1007/s13762-013-0299-8

Mann, M., & Jensen, O. N. (2003). Proteomic analysis of post-translational modifications. *Nature Biotechnology*, *21*(3), 255–261. doi:10.1038/nbt0303-255 PMID:12610572

Manoj, B. (2012). Chemical demineralization of high volatile Indian bituminous coal by carboxylic acid and characterization of the products by SEM/EDS. *Journal of Environmental Research and Development*, *6*(3A), 653-659.

Manoj, B. (2014). Investigation of nano-crystalline structure in selected carbonaceous materials. *International Journal of Minerals Materials and Metallurgy*, *21*(9), 940-946.

Manoj, B. (2013). Bio-demineralization of Indian bituminous coal by *Aspergillus niger* and characterization of the products. *Research Journal of Biotechnology*, *8*(3), 49–54.

Manoj, B. (2016). A Comprehensive analysis of various structural parameters of Indian coals with the aid of advanced analytical tools. *International Journal of Coal Science and Technology*, *3*(2), 123–132. doi:10.1007/s40789-016-0134-1

Manoj, B., & Kunjomana, A. G. (2014). Systematic Investigations of graphene layers in sub-bituminous coal. *Russian Journal of Applied Chemistry*, *87*(11), 1726–1733. doi:10.1134/S1070427214110251

Mansour, H. B., Dellai, A., & Ayed, Y. (2012). Toxicities effects of pharmaceutical, olive mill and textile wastewaters before and after degradation by *Pseudomonas putida* mt-2. *Cancer Cell International*, *12*(1), 4. and doi:10.1186/1475-2867-12-4

Mansour, H. B., Mosrati, R., Barillier, D., Ghedira, K., & Chekir-Ghedira, L. (2012). Bioremediation of industrial pharmaceutical drugs. *Drug and Chemical Toxicology*, *35*(3), 235–240. and doi:10.3109/01480545.2011.591799

Mao, X., Jiang, R., Xiao, W., & Yu, J. (2015). Use of surfactants for the remediation of contaminated soils: A review. *Journal of Hazardous Materials*, *285*, 419–435. doi:10.1016/j.jhazmat.2014.12.009 PMID:25528485

Margesin, R. P., Bergauer, S. G., & Gander, S. (2004). Degradation of phenol and toxicity of phenolic compounds: A comparison of cold-tolerant *Arthrobacter* sp. and mesophilic *Pseudomonas putida*. *Extremophiles*, *8*(3), 201–207. doi:10.1007/s00792-004-0378-3 PMID:14872323

Margesin, R., & Schinner, F. (2001). Biodegradation and bioremediation of hydrocarbons in extreme environments. *Applied Microbiology and Biotechnology*, *56*(5-6), 650–663. doi:10.1007/s002530100701 PMID:11601610

Maron, D. M., & Ames, B. N. (1983). Methods for *Salmonella* mutagenicity tests. *Mutation Research*, *113*(3-4), 173–215. doi:10.1016/0165-1161(83)90010-9 PMID:6341825

Marouga, R., David, S., & Hawkins, E. (2005). The development of the DIGE system: 2D fluorescence difference gel analysis technology. *Analytical and Bioanalytical Chemistry*, *382*(3), 669–678. doi:10.1007/s00216-005-3126-3 PMID:15900442

Marques, A. P., Rangel, A. O., & Castro, P. M. (2009). Remediation of heavy metal contaminated soils: Phytoremediation as a potentially promising clean-up technology. *Critical Reviews in Environmental Science and Technology*, *39*(8), 622–654. doi:10.1080/10643380701798272

Marrs, K. A., Alfenito, M. R., Lloyd, A. M., & Walbot, V. (1995). *A glutathione S-transferase involved in vacuolar transfer encoded by the maize gene Bronze-2*. Academic Press.

Marschner, H. (2012). *Marschner's mineral nutrition of higher plants* (Vol. 89). Academic press.

Marschner, P., Yang, C. H., Lieberei, R., & Crowley, D. E. (2001). Soil and plant specific effects on bacteria community composition in the rhizosphere. *Soil Biology & Biochemistry*, *33*(11), 1437–1445. doi:10.1016/S0038-0717(01)00052-9

Martínez, Speranza, Ruiz-Dueñas, Ferreira, Camarero, Guillén, … Del Río. (2005). Biodegradation of lignocellulosics: Microbial, chemical, and enzymatic aspects of the fungal attack of lignin. *International Microbiology*, *8*, 195–204. PMID:16200498

Martínez, A. T. (2002). Molecular biology and structure-function of lignin degrading heme peroxidases. *Enzyme and Microbial Technology*, *30*(4), 425–444. doi:10.1016/S0141-0229(01)00521-X

Martinez, A. T., Rencoret, J., Nieto, L., Jimenez-Barbero, J., Gutierrez, A., & del Rıo, J. C. (2011). Selective lignin and polysaccharide removal in fungal decay as evidenced by in situ structural analyses. *Environmental Microbiology*, *13*(1), 96–107. doi:10.1111/j.1462-2920.2010.02312.x PMID:21199251

Martínez, M. J., Ruiz-Dueñas, F. J., Guillén, F., & Martínez, A. T. (1996). Purification and catalytic properties of two manganese-peroxidase isoenzymes from *Pleurotus eryngii*. *European Journal of Biochemistry*, *237*(2), 424–432. doi:10.1111/j.1432-1033.1996.0424k.x PMID:8647081

Martinez-Neito, L., Ramos Cormenzana, A., Garcia Pareja, M. P., & Garrido Hoyos, S. E. (1992). Phenolic compounds biodegradation of olive mill wastewater with *Aspergillus terreus*. *Grasas y Aceites*, *43*, 75–81.

Martínez-Sotres, C., Rutiaga-Quiñones, J. G., Herrera-Bucio, R., Gallo, M., & López-Albarrán, P. (2015). Molecular docking insights into the inhibition of laccase activity by medicarpin. *Wood Science and Technology*, *49*(4), 857–868. doi:10.1007/s00226-015-0734-8

Martín, M., Gibello, A., Lobo, C., Nande, M., Garbi, C., Fajardo, C., & Martínez-Iñigo, M. J. (2008). Application of fluorescence in situ hybridization technique to detect simazine-degrading bacteria in soil samples. *Chemosphere*, *71*(4), 703–710. doi:10.1016/j.chemosphere.2007.10.071 PMID:18082866

Martinoia, E., Maeshima, M., & Neuhaus, H. E. (2007). Vacuolar transporters and their essential role in plant metabolism. *Journal of Experimental Botany*, *58*(1), 83–102. doi:10.1093/jxb/erl183 PMID:17110589

Martin, S., & Griswold, W. (2009). Human health effects of heavy metals. *Environmental Science and Technology Briefs for Citizens*, *15*, 1–6.

Martiny, J. B. H., Bohannan, B. J. M., Brown, J. H., Colwell, R. K., Fuhrman, J. A., Green, J. L., & Staley, J. T. et al. (2006). Microbial biogeography: Putting microorganisms on the map. *Nature Reviews. Microbiology*, *4*(2), 102–112. doi:10.1038/nrmicro1341 PMID:16415926

Marungruenga, K., & Pavasant, P. (2006). Removal of basic dye (Astrazon blue FGRL) using macroalgae *Caulerpa lentillifera. Journal of Environmental Management, 78*(3), 268–274. doi:10.1016/j.jenvman.2005.04.022 PMID:16112338

Mary Agbogidi, O., Dickens Dolor, E., & Mercy Okechukwu, E. (2007). Evaluation of Tectona grandis (Linn.) and Gmelina arborea (Roxb.) for phytoremediation in crude oil contaminated soils.[ACS]. *Agriculturae Conspectus Scientificus, 72*(2), 149–152.

Masood, F., & Malik, A. (2013). *Current Aspects of Metal Resistant Bacteria in Bioremediation: From Genes to Ecosystem. In Management of Microbial Resources in the Environment* (pp. 289–311). Springer.

Masson, L., Comeau, Y., Brousseau, R., Samson, R., & Greer, C. (1993). Construction and application of chromosomally integrated lac-lux gene markers to monitor the fate of a 2,4-dichlorophenoxyacetic acid-degrading bacterium in contaminated soils. *Microbial Releases, 1*, 209–216. PMID:7506623

Matelich, G. (2014). *Freshwater: The new frontier in micro-plastics research.* AMB Express.

Matsumura, Y., Yamabe, K., & Takahashi, H. (1985). The effects of hydrophilic structures of active-carbon on the adsorption of benzene and methanol vapors. *Carbon, 23*(3), 263–271. doi:10.1016/0008-6223(85)90111-3

Mattson, D. J., French, M. G., & French, S. P. (2002). Consumption of earthworms by yellow stone grizzly bears. *Ursus, 13*, 153–158.

Mauna Loa. (2015). *Hawaii observatory, Earth System Research Laboratory, Global monitoring division.* Retrieved from http://www.esrl.noaa.gov/gmd/obop/mlo/

Ma, W. C. (1988). Toxicity of copper to lumbricid earthworms in sandy agricultural soils amended with Cu-enriched organic waste material. *Ecological Bulletins, 39*, 53–56.

Ma, X., Ding, N., Peterson, E. C., & Daugulis, A. J. (2016). Heavy metals species affect fungal-bacterial synergism during the bioremediation of fluoranthene. *Applied Microbiology and Biotechnology, 100*(17), 7741–7750. doi:10.1007/s00253-016-7595-4 PMID:27178182

Ma, Y., Prasad, M. N., Rajkumar, M., & Freitas, H. (2011). Plant growth promoting rhizobacteria and endophytes accelerate phytoremediation of metalliferous soils. *Biotechnology Advances, 29*(2), 248–258. doi:10.1016/j.biotechadv.2010.12.001 PMID:21147211

Mayabhate, S. P., Gupta, S. K., & Joshi, S. G. (1988). Biological treatment of pharmaceutical wastewater. *Water, Air, and Soil Pollution, 38*(1–2), 189–197.

Mayer, A. M., & Staples, R. C. (2002). Laccase: New functions for an old enzyme. *Phytochemistry, 60*(6), 551–565. doi:10.1016/S0031-9422(02)00171-1 PMID:12126701

Ma, Z., Jacobsen, F. E., & Giedroc, D. P. (2009). Coordination chemistry of bacterial metal transport and sensing. *Chemical Reviews, 109*(10), 4644–4681. doi:10.1021/cr900077w PMID:19788177

Mc Alinden, J. (2008). *The use of liquid dye penetrants containing the azo compound CI Solvent Red 164 in the detection of flaws or cracks in metal components occupational hygiene, Bootle (VPN 523 4545).2008.* Retrieved from http://www.britishmetalforming.com/uploads/file/pdf/Azo%20dyes%20-%20Nov%2008.pdf

McAllister, K. A., Lee, H., & Trevors, J. T. (1996). Microbial degradation of pentachlorophenol. *Biodegradation, 7*(1), 1–40. doi:10.1007/BF00056556

McCutcheon, S., & Schnoor, J. (2003). Overview of phytotransformation and control of wastes. In Phytoremediation: Transformation and control of contaminants (pp. 3-58). doi:10.1002/047127304X.ch1

McDonald, P. (1981). The biochemistry of silage. John Wiley & Sons.

McGenity, T. J. (2014). Hydrocarbon biodegradation in intertidal wetland sediments. *Current Opinion in Biotechnology*, *27*, 46–54. doi:10.1016/j.copbio.2013.10.010 PMID:24863896

McGrath, S. P., & Zhao, F. J. (2003). Phytoextraction of metals and metalloids from contaminated soils. *Current Opinion in Biotechnology*, *14*(3), 277–282. doi:10.1016/S0958-1669(03)00060-0 PMID:12849780

McIntyre, T., & Lewis, G. M. (1997). Advancement of phytoremediation as an innovative environmental technology for stabilization, remediation, or restoration of contaminated sites in Canada: A discussion paper. *Journal of Soil Contamination*, *6*(3), 227–224. doi:10.1080/15320389709383562

McKay, B., & Mulvaney, K. (2001). A Review of Marine Major Ecological Disturbances (MMEDs). *Endangered Species Update*, *18*(1), 14–24.

McLaughlin, M. C., Borch, T., & Blotevogel, J. (2016). Spills of hydraulic fracturing chemicals on agricultural topsoil: Biodegradation, sorption, and co-contaminant interactions. *Environmental Science & Technology*, *50*(11), 6071–6078. doi:10.1021/acs.est.6b00240 PMID:27171137

McLean, K., Sabri, M., Marshall, K., Lawson, R., Lewis, D., Clift, D., & McVey, J. et al. (2005). Biodiversity of cytochrome P450 redox systems. *Biochemical Society Transactions*, *33*(4), 796–801. doi:10.1042/BST0330796 PMID:16042601

Meador, J., Yeh, A., Young, G., & Gallagher, E. P. (2016). Contaminants of emerging concern in a large temperate estuary. *Environmental Pollution*, *213*, 254–267. doi:10.1016/j.envpol.2016.01.088 PMID:26907702

Meena, S. K. (2013). Bioterrorism is a dangerous to environment-A review. *Innovare Journal of Agricultural Sciences*, 5-7.

Meggo, R. E., Schnoor, J. L., & Hu, D. (2013). Dechlorination of PCBs in the rhizosphere of switchgrass and poplar. *Environmental Pollution*, *178*, 312–321. doi:10.1016/j.envpol.2013.02.035 PMID:23603468

Meharg, A. A., & Cairney, J. W. G. (2000). Ecto-mycorrhizae -Extending the capabilities of rhizosphere remediation. *Soil Biology & Biochemistry*, *32*(11-12), 1475–1484. doi:10.1016/S0038-0717(00)00076-6

Mehmannavaz, R., Prasher, S. O., & Ahmad, D. (2002). Rhizospheric effects of alfalfa on biotransformation of polychlorinated biphenyls in a contaminated soil augmented with Sinorhizobium meliloti. *Process Biochemistry*, *37*(9), 955–963. doi:10.1016/S0032-9592(01)00305-3

Mehnaz, S., Mirza, M. S., Haurat, J., Bally, R., Normand, P., Bano, A., & Malik, K. A. (2001). Isolation and 16S rRNA sequence analysis of the beneficial bacteria from the rhizosphere of rice. *Canadian Journal of Microbiology*, *47*(2), 110–117. doi:10.1139/w00-132 PMID:11261489

Mehta, S. K., & Gaur, J. P. (2005). Use of algae for removing heavy metal ions from wastewater: Progress and prospects. *Critical Reviews in Biotechnology*, *25*(3), 113–152. doi:10.1080/07388550500248571 PMID:16294830

Meissner, P., & Falkinham, J. (1984). Plasmid-encoded mercuric reductase in Mycobacterium scrofulaceum. *Journal of Bacteriology*, *157*(2), 669–672. PMID:6693354

Melling, F. B. Jr. (1993). *Soil microbial ecology: Applications in agricultural and environmental management.* New York: Marcel Dekker.

Melo, J. S., Kholi, S., Patwardhan, A. W., & DSouza, S. F. (2005). Effect of oxygen transfer limitations in phenol biodegradation. *Process Biochemistry (Barking, London, England)*, *40*(2), 625–628. doi:10.1016/j.procbio.2004.01.049

Mena-Benitez, G. L., Gandia-Herrero, F., Graham, S., Larson, T. R., McQueen-Mason, S. J., French, C. E., & Bruce, N. C. et al. (2008). Engineering a catabolic pathway in plants for the degradation of 1, 2-dichloroethane. *Plant Physiology*, *147*(3), 1192–1198. doi:10.1104/pp.108.119008 PMID:18467461

Méndez, R., Leplae, R., Lensink, M. F., & Wodak, S. J. (2005). Assessment of CAPRI predictions in rounds 3–5 shows progress in docking procedures. *Proteins: Structure, Function, and Bioinformatics*, *60*(2), 150–169. doi:10.1002/prot.20551 PMID:15981261

Meng, E. C., Shoichet, B. K., & Kuntz, I. D. (1992). Automated docking with grid-based energy evaluation. *Journal of Computational Chemistry*, *13*(4), 505–524. doi:10.1002/jcc.540130412

Menn, F.-M., James, P. E., & Gary, S. S. (2008). Genetically engineered microorganisms and bioremediation. In H. J. Rehmand, & G. Reed (Eds.), Bio technology (2nd ed.; pp. 441-463). New York: VCH Publishers Inc.

Menon, B., Dou, Z., Heinhorst, S., & Cannon, G. C. (2008). Halothiobacillus neapolitanus Carboxysomes Sequester Heterologous and Chimeric RubisCO Species. *PLoS ONE*, *3*(10), e3570. doi:10.1371/journal.pone.0003570 PMID:18974784

Meorin, E. C. (1989). Urban run-off treatment in a fresh/brakish water marsh in Fremont, California. In D. A. Hammer (Ed.), Constructed Wetlands for Wastewater Treatment, (pp. 667-685). Industrial and Agricultural Lewis Publishers.

Merkl, N., Schultze-Kraft, R., & Infante, C. (2004). Phytoremediation in the tropics—the effect of crude oil on the growth of tropical plants. *Bioremediation Journal*, *8*(3-4), 177–184. doi:10.1080/10889860490887527

Metcalf & Eddy. (2005). *Wastewater Engineering-Treatment, disposal and reuse* (4th ed.). Tata McGraw Hill.

Metcalfe, C. D., Miao, X. S., Koenig, B. G., & Struger, J. (2003). Distribution of acidic and neutral drugs in surface waters near sewage treatment plants in the lower Great Lakes, Canada. *Environmental Toxicology and Chemistry*, *22*(12), 2881–2889. doi:10.1897/02-627 PMID:14713027

Methé, B. A., Nelson, K. E., Deming, J. W., Momen, B., Melamud, E., Zhang, X., & Brinkac, L. M. et al. (2005). The psychrophilic lifestyle as revealed by the genome sequence of Colwellia psychrerythraea 34H through genomic and proteomic analyses. *Proceedings of the National Academy of Sciences of the United States of America*, *102*(31), 10913–10918. doi:10.1073/pnas.0504766102 PMID:16043709

Mi, J., Orbea, A., Syme, N., Ahmed, M., Cajaraville, M. P., & Cristóbal, S. (2005). Peroxisomal proteomics, a new tool for risk assessment of peroxisome proliferating pollutants in the marine environment. *Proteomics*, *5*(15), 3954–3965. doi:10.1002/pmic.200401243 PMID:16130170

Mikola, J., Bardgett, R. D., & Hedlund, K. (2002). Biodiversity, ecosystem functioning and soil decomposer food webs. In M. Loreau, S. Naeem, & P. Inchausti (Eds.), *Biodiversity andecosystem functioning–synthesis and perspectives* (pp. 169–180). Oxford, UK: Oxford University Press.

Mill, T., & Mabey, W. (1985). Photochemical transformations. In Environmental exposure from chemicals. Boca Raton, FL: CRC Press.

Millard, P., & Singh, B. K. (2010). Does grassland vegetation drive soil microbial diversity? *Nutrient Cycling in Agroecosystems*, *88*(2), 147–158. doi:10.1007/s10705-009-9314-3

Millennium Ecosystem Assessment (MA). (2005). *Ecosystems and Human Well-being: Synthesis*. Washington, DC: Island Press.

Miller, R. M., & Jastrow, J. D. (1990). Hierarchy of root and mycorrhizal fungal interactions with soil aggregation. *Soil Biology & Biochemistry*, *22*(5), 579–584. doi:10.1016/0038-0717(90)90001-G

Miller, R. M., & Jastrow, J. D. (2000). Mycorrhizal fungi influence soil structure. In Y. Kapulnik & D. D. Douds Jr., (Eds.), *Arbuscular mycorrhizas: physiology and function* (pp. 3–18). Dordrecht, The Netherlands: Kluwer. doi:10.1007/978-94-017-0776-3_1

Miller, S., Pallan, S., Gangji, A. S., Lukic, D., & Clase, C. M. (2013). Mercury-associated nephrotic syndrome: A case report and systematic review of the literature. *American Journal of Kidney Diseases*, *62*(1), 135–138. doi:10.1053/j.ajkd.2013.02.372 PMID:23602193

Milochau, A., Lassegues, M., & Valembois, P. (1997). Purificaton,characterization and activities of two hemolytic and antibacterial proteins from coelomic fluid of the annelid *Eisenia fetida andrei. Biochimica et Biophysica Acta*, *1337*(1), 123–132. doi:10.1016/S0167-4838(96)00160-4 PMID:9003444

Milstein, O., Vered, Y., Gressel, J., & Flowers, H. M. (1981). Biodegradation of wheat straw lignocarbohydrate complexes (LCC) II. Fungal growth on aqueous hydrolysate liquors and particulate residues of wheat straw. *Eur. J. Appl. Microbiol. Biotechnol*, *13*(2), 117–127. doi:10.1007/BF00499699

Miransari, M. (2011). Hyperaccumulators, arbuscular mycorrhizal fungi and stress of heavy metals. *Biotechnology Advances*, *29*(6), 645–653. doi:10.1016/j.biotechadv.2011.04.006 PMID:21557996

Misal, S. A., Lingojwar, D. P., Shinde, R. M., & Gawai, K. R. (2011). Purification and characterization of azoreductase from alkaliphilic strains *Bacillus badius. Process Biochemistry*, *46*(6), 264–269. doi:10.1016/j.procbio.2011.02.013

Miyabe, S., Izawa, S., & Inoue, Y. (2000). Expression of ZRC1 coding for suppressor of zinc toxicity is induced by zinc-starvation stress in Zap1-dependent fashion in Saccharomyces cerevisiae. *Biochemical and Biophysical Research Communications*, *276*(3), 879–884. doi:10.1006/bbrc.2000.3580 PMID:11027563

Mohan Krishna, V., Hrushikesh, N., & Sreehari, K., Aravind Kumar, T., Vidyavathi, N., & Pallavi, A. (2011). Studies on bioremediation of phenol by earthworm. *International Journal of Environmental Sciences*, *1*(6), 1268–1273.

Mohan, S. V., Prakasham, R. S., Satyavathi, B., Annapurna, J., & Ramakrishna, S. V. (2001). Biotreatability studies of pharmaceutical wastewater using an anaerobic suspended film contact reactor. *Water Science and Technology*, *43*(2), 271–276.

Mohanty, S., & Mukherji, S. (2013). Surfactant aided biodegradation of NAPLs by Burkholderiamultivorans: Comparison between Triton X-100 and rhamnolipid JBR-515. *Colloids and Surfaces. B, Biointerfaces*, *102*, 644–652. doi:10.1016/j.colsurfb.2012.08.064 PMID:23104033

Mohapatra, H., & Gupta, R. (2005). Concurrent sorption of Zn(II), Cu(II) and Co(II) by *Oscillatoria angustissima* as a function of pH in binary and ternary metal solutions. *Bioresource Technology*, *96*(12), 1387–1398. doi:10.1016/j.biortech.2004.11.004 PMID:15792587

Moharikar, A., Purohit, H. J., & Kumar, R. (2005). Microbial population dynamics at effluent treatment plants. *Journal of Environmental Monitoring*, *7*(6), 552–558. doi:10.1039/b406576j PMID:15931414

Mohsenzadeh, F., Nasseri, S., Mesdaghinia, A., Nabizadeh, R., Zafari, D., Khodakaramian, G., & Chehregani, A. (2010). Phytoremediation of petroleum-polluted soils: Application of Polygonum aviculare and its root-associated (penetrated) fungal strains for bioremediation of petroleum-polluted soils. *Ecotoxicology and Environmental Safety*, *73*(4), 613–619. doi:10.1016/j.ecoenv.2009.08.020 PMID:19932506

Mojiri, A., Ziyang, L., Tajuddin, R. M., Farraji, H., & Alifar, N. (2016). Co-treatment of landfill leachate and municipal wastewater using the ZELIAC/zeolite constructed wetland system. *Journal of Environmental Management*, *166*, 124–130. doi:10.1016/j.jenvman.2015.10.020 PMID:26496842

Money, N. P. (2007). Biomechanics of invasive hyphal growth. In Biology of the Fungal Cell Wall. Springer-Verlag.

Mons M.N., Hoogenboom A.C., & Noij T.H.M., (2003). *Pharmaceuticals and drinking water supply in the Netherlands.* Nieuwegein, Kiwa Water Research (Kiwa Report No. BTO 2003.040).

Montero-Rodríguez, D., Andrade, R. F. S., Ribeiro, D. L. R., Lima, R. A., Araujo, H. W. C., & Campos-Takaki, G. M. (2014). Ability of Serratia marcescens UCP/WFCC 1549 for biosurfactant production using industrial wastes and fuels biodegradation In Industrial, medical and environmental applications of microorganisms: Current status and trends, (pp. 211-216). Academic Press.

Monzon, A. M., Juritz, E., Fornasari, M. S., & Parisi, G. (2013). CoDNaS: A database of conformational diversity in the native state of proteins. *Bioinformatics (Oxford, England)*, *29*(19), 2512–2514. doi:10.1093/bioinformatics/btt405 PMID:23846747

Moore-Kucera, J., & Dick, R. P. (2008). Application of 13C-labeled litter and root materials for in situ decomposition studies using phospholipid fatty acids. *Soil Biology & Biochemistry*, *40*(10), 2485–2493. doi:10.1016/j.soilbio.2008.06.002

Moragan, J. E., & Morgan, A. J. (1999). The accumulation of metals (Cd,Cu,Pb,Zn and Ca) by two ecologically contrasting earthworm species (*Lumbricus rubellus* and *Aporrectodea caliginosa*): Implications for ecotoxicological testing. *Applied Soil Ecology*, *13*(1), 9–20. doi:10.1016/S0929-1393(99)00012-8

Moredo, N., Lorenzo, M., Domínguez, A., Moldes, D., Cameselle, C., & Sanroman, A. (2003). Enhanced ligninolytic enzyme production and degrading capability of Phanerochaete chrysosporium and Trametes versicolor. *World Journal of Microbiology & Biotechnology*, *19*(7), 665–669. doi:10.1023/A:1025198917474

Morel, M., Crouzet, J., Gravot, A., Auroy, P., Leonhardt, N., Vavasseur, A., & Richaud, P. (2009). AtHMA3, a P1B-ATPase allowing Cd/Zn/Co/Pb vacuolar storage in Arabidopsis. *Plant Physiology*, *149*(2), 894–904. doi:10.1104/pp.108.130294 PMID:19036834

Moreno-Pirajan, J. C., & Giraldo, L. (2012). Heavy metal ions adsorption from wastewater using activated carbon from orange peel. *E-Journal of Chemistry*, *9*(2), 926–937. doi:10.1155/2012/383742

Morgan, A. J., & O'Reilly, M. (2002). The manganese relationships of epigeic and endogeic earthworm species living in soils associated with disused manganese mines.*Proceedings Sixth International Earthworm Ecology Symposium.*

Morgan, J. E., & Morgan, A. J. (1993). Seasonal changes in the tissue-metal (Cd, Zn and Pb) concentrations in two eco-physiologically dissimilar earthworm species: Pollution-monitoring implications. *Environmental Pollution*, *82*(1), 1–17. doi:10.1016/0269-7491(93)90155-H PMID:15091792

Moubasher, H., Hegazy, A., Mohamed, N., Moustafa, Y., Kabiel, H., & Hamad, A. (2015). Phytoremediation of soils polluted with crude petroleum oil using Bassia scoparia and its associated rhizosphere microorganisms. *International Biodeterioration & Biodegradation*, *98*, 113–120. doi:10.1016/j.ibiod.2014.11.019

Mouhamadou, B., Faure, M., Sage, L., Marcais, J., Souard, F., & Geremia, R. A. (2013). Potential of autochthonous fungal strains isolated from contaminated soils for degradation of polychlorinated biphenyls. *Fungal Biology*, *117*(4), 268–274. doi:10.1016/j.funbio.2013.02.004 PMID:23622721

Moulis, J.-M., Davasse, V., Golinelli, M.-P., Meyer, J., & Quinkal, I. (1996). The coordination sphere of iron-sulfur clusters: Lessons from site-directed mutagenesis experiments. *Journal of Biological Inorganic Chemistry*, *1*(1), 2–14. doi:10.1007/s007750050017

Moutaouakkil, A., Zeroual, Y., Dzayri, F. Y., & Talbi, M., Lee, K., & Blaghen, M. (2003). Bacterial decolorization of the azo dye methyl red by *Enterobacter agglomerans. Annals of Microbiology*, *53*, 161–169.

Mrozik, A., Cycon, M., & Piotrowska-Seget, Z. (2010). Changes of FAME profiles as a marker of phenol degradation in different soils inoculated with *Pseudomonas* sp. CF600. *International Biodeterioration & Biodegradation*, *64*(1), 86–96. doi:10.1016/j.ibiod.2009.11.002

Muda, K., Aris, A., Salim, M. R., Ibrahim, Z., & Yahya, A. (2012). Textile wastewater treatment using biogranules under intermittent anaerobic/aerobic reaction phase. *Journal of Water and Environment Technology*, *10*(3), 303–315. doi:10.2965/jwet.2012.303

Mudgett, R. E. (1986). Solid-state fermentations. In A. L. Demain & N. A. Solomon (Eds.), *Manual of Industrial Microbiology and Biotechnology* (pp. 66–83). Washington, DC: American Society of Microbiology.

Mukherjee, A. K., & Bordoloi, N. K. (2011). Bioremediation and reclamation of soil contaminated with petroleum oil hydrocarbons by exogenously seeded bacterial consortium: A pilot-scale study. *Environl Sci Pollut Res*, *18*(3), 471–478. doi:10.1007/s11356-010-0391-2 PMID:20835890

Mukherji, S., & Jeswani, H. (2011). Treatment of biomass gasifier wastewater in a rotating biological contactor. *Conference proceedings,IWA Biofilm Conference, Processes in Biofilms*, 215-216.

Muley, P., Dhumal, M., & Vora, D. (2014). Sequestration of atmospheric Cdioxide by microbial carbonic anhydrase. *IOSR J Environ Sci, Toxicol &. Food Technology*, *8*, 45–48.

Müller, M. M., & Hausmann, R. (2011). Regulatory and metabolic network of rhamnolipid biosynthesis: Traditional and advanced engineering towards biotechnological production. *Applied Microbiology and Biotechnology*, *91*(2), 251–264. doi:10.1007/s00253-011-3368-2 PMID:21667084

Mulligan, C. N., Yong, R. N., & Gibbs, B. F. (2001). Remediation alternative treatment option for heavy metal bearing wastewaters: A review. *Bioresource Technology*, *53*, 195–206.

Munoz, J. A., Perez-Esteban, B., Esteban, M., de la Escalera, S., Gomez, M. A., Martinez-Toledo, M. V., & Gonzalez-Lopez, J. (2001). Growth of moderately halophilic bacteria isolated from sea water using phenol as the sole carbon source. *Folia Microbiologica*, *46*(4), 297–302. doi:10.1007/BF02815617 PMID:11830940

Muraleedharan, T. R., & Venkobachar, C. (1990). Mechanism of biosorption of copper (II) by *Ganoderma lucidum. Biotechnol Bioenginering*, *5*(3), 320–325. doi:10.1002/bit.260350314

Murdoch, R. W., & Hay, A. G. (2013). Genetic and chemical characterization of ibuprofen degradation by *Sphingomonas* Ibu-2. *Microbiology*, *159*(3), 621–632. doi:10.1099/mic.0.062273-0 PMID:23329679

Mutamim, N. S. A., Noor, Z. Z., Hassan, M. A. A., & Olsson, G. (2012). Application of membrane bioreactor technology in treating high strength industrial wastewater: A performance review. *Desalination*, *305*, 1–11. doi:10.1016/j.desal.2012.07.033

Mutzel, A., Reinscheid, U. M., Antranikian, G., & Müller, R. (1996). Isolation and characterization of a thermophilic bacillus strain that degrades phenol and cresols as sole carbon source at 70 oC. *Applied Microbiology and Biotechnology*, *46*(5-6), 593–596. doi:10.1007/s002530050866

Muyzer, G. (1999). DGGE/TGGE a method for identifying genes from natural ecosystems. *Current Opinion in Microbiology*, *2*(3), 317–322. doi:10.1016/S1369-5274(99)80055-1 PMID:10383868

Mwangi, I.W., & Ngile, J.C. (2012). Removal of heavy metals from contaminated water using ethylene diamine modified green sea weed (*Caulerpa serrulata*). *Physics and Chemistry of the Earth, 50 – 52*, 111 – 120.

Naidu, R., Bolan, N., Megharaj, M., Juhanz, A., Lombi, E., & Smith, E. (2010).Human and ecological risk assessment of contaminated sites-Key knowledge gaps. *19th Congress of Soil Science. Soil Solutions for a changing World.*

Naik, M. G., & Duraphe, M. D. (2012). Review paper on-Parameters affecting bioremediation. *Int J Life Sci Pharma Res*, *2*(3), L-77–L-80.

Nair, I. C., & Shashidhar, S. (2004). Microbial degradation of phenol by a species of *Alcaligenes* isolated from a tropical soil. *Soil Science*, *T-5*, 3, 4, 47–51.

Nair, I., Jayachandran, C. K., & Shankar, S. (2008). Biodegradation of phenol. *African Journal of Biotechnology*, *7*(25), 4951–4958.

Nakada, N., Shinohara, H., Murata, A., Kiri, K., Managaki, S., Sato, N., & Takada, H. (2007). Removal of selected pharmaceuticals and personal care products (PPCPs) and endocrine-disrupting chemicals (EDCs) during sand filtration and ozonation at a municipal sewage treatment plant. *Water Research*, *41*(19), 4373–4382. doi:10.1016/j.watres.2007.06.038 PMID:17632207

Nakajima, A., & Tsuruta, T. (2004). Competitive biosorption of thorium and uranium by Micrococcus luteus. *Journal of Radioanalytical and Nuclear Chemistry*, *260*(1), 13–18. doi:10.1023/B:JRNC.0000027055.16768.1e

Nakamura, Y., & Sawada, T. (2000). Biodegradation of phenol in presence of heavy metals. *Journal of Chemical Technology and Biotechnology (Oxford, Oxfordshire)*, *75*(2), 137–142. doi:10.1002/(SICI)1097-4660(200002)75:2<137::AID-JCTB194>3.0.CO;2-0

Nakashima, T., Okada, T., Asahi, J., Yamashita, A., Kawa, K., Kasai, H., & Hirano, T. et al. (2008). 8-Hydroxydeoxyguanosine generated in the earthworm *Eisenia fetida* grown in metal-containing soil. *Mutation Research*, *654*(2), 138–144. doi:10.1016/j.mrgentox.2008.05.011 PMID:18585474

Namasivayam, C., & Periasamy, K. (1993). Bicarbonate treated peanut hull carbon for Hg (II) removal from aqueous solution. *Water Resources*, *27*, 1663–1668.

Namkoong, W., Hwang, E.-Y., Park, J.-S., & Choi, J.-Y. (2002). Bioremediation of diesel-contaminated soil with composting. *Environmental Pollution*, *119*(1), 23–31. doi:10.1016/S0269-7491(01)00328-1 PMID:12125726

Nanayakkara, S., Patti, A. F., & Saito, K. (2014). Chemical depolymerization of lignin involving the redistribution mechanism with phenols and repolymerization of depolymerized products. *Green Chemistry*, *16*(4), 1897–1903. doi:10.1039/C3GC41708E

Nandy, T., & Kaul, S. N. (2001). Anaerobic pre-treatment of herbal-based wastewater using fixed-film reactor with recourse to energy recovery. *Water Research*, *35*(2), 351–362. and doi:10.1016/S0043-1354(00)00263-3

Nascimento, C. W. A. D., & Xing, B. (2006). Phytoextraction: a review on enhanced metal availability and plant accumulation. *Scientia Agricola, 63*(3), 299-311.

Nassef, M., Matsumoto, S., Seki, M., Khalil, F., Kang, I. J., Shimasaki, Y., & Honjo, T. et al. (2010). Acute effects of triclosan, diclofenac and carbamazepine on feeding performance of Japanese medaka fish (Oryzias latipes). *Chemosphere*, *80*(9), 1095–1100. doi:10.1016/j.chemosphere.2010.04.073 PMID:20537681

National Research Council (NRC). (1992). *Restoration of aquatic ecosystems: Science, technology and public policy*. Washington, DC: National Academy Press.

Neelima, G., Garg, K. L., & Mukerji, K. G. (1994). Microbial degradation of environmental pollutants. In Recent Advances in Bio-deterioration and biodegradation (vol. 2, pp. 329-362). Naya Prokash.

Nehl, D., Allen, S., & Brown, J. (1997). Deleterious rhizosphere bacteria: An integrating perspective. *Applied Soil Ecology*, *5*(1), 1–20. doi:10.1016/S0929-1393(96)00124-2

Newman, L. A., & Reynolds, C. M. (2004). Phytodegradation of organic compounds. *Current Opinion in Biotechnology*, *15*(3), 225–230. doi:10.1016/j.copbio.2004.04.006 PMID:15193330

Ngah, W. S. W., & Hanafiah, M. A. K. M. (2008). Removal of heavy metal ions from wastewater by chemically modified plant wastes as adsorbent: A review. *Bioresource Technology*, *99*(10), 3935–3948. doi:10.1016/j.biortech.2007.06.011 PMID:17681755

Nicolaou, S. A., Gaida, S. M., & Papoutsakis, E. T. (2010). A comparative view of metabolite and substrate stress and tolerance in microbial bioprocessing: From biofuels and chemicals, to biocatalysis and bioremediation. *Metabolic Engineering*, *12*(4), 307–331. doi:10.1016/j.ymben.2010.03.004 PMID:20346409

Nieboer, E., & Richardson, D. H. S. (1980). The replacement of the nondescript term 'heavy metals' by a biologically and chemically significant classification of metal ions. *Environmental Pollution, 1*(B), 3-26.

Nieboer, E., & Richardson, D. H. S. (1980). The replacement of the nondescript term "heavy metals" by a biologically and chemically significant classification of metal ions. *Environmental Pollution*, 3–26.

Niehaus, F., Bertoldo, C., Kähler, M., & Antranikian, G. (1999). Extremophiles as a source of novel enzymes for industrial application. *Applied Microbiology and Biotechnology*, *51*(6), 711–729. doi:10.1007/s002530051456 PMID:10422220

Nies, D. H. (1995). The cobalt, zinc, and cadmium efflux system CzcABC from Alcaligenes eutrophus functions as a cation-proton antiporter in Escherichia coli. *Journal of Bacteriology*, *177*(10), 2707–2712. doi:10.1128/jb.177.10.2707-2712.1995 PMID:7751279

Nies, D. H. (1999). Microbial heavy-metal resistance. *Applied Microbiology and Biotechnology*, *51*(6), 730–750. doi:10.1007/s002530051457 PMID:10422221

Nikolopoulou, M., Eickenbusch, P., Pasadakis, N., Venieri, D., & Kalogerakis, N. (2013 a). Microcosm evaluation of autochthonous bioaugmentation to combat marine oil spills. *New Biotechnology*, *30*(6), 734–742. doi:10.1016/j.nbt.2013.06.005 PMID:23835403

Nikolopoulou, M., & Kalogerakis, N. (2011). Petroleum spill control with biological means. *Comprehensive Biotechnology*, *6*, 263–274. doi:10.1016/B978-0-08-088504-9.00392-5

Nikolopoulou, M., Pasadakis, N., & Kalogerakis, N. (2013 b). Evaluation of autochthonous bioaugmentation and biostimulation during microcosm-simulated oil spills. *Marine Pollution Bulletin*, *72*(1), 165–173. doi:10.1016/j.marpolbul.2013.04.007 PMID:23660443

Nimick, D. A., Caldwell, R. R., Skaar, D. R., & Selch, T. M. (2013). Fate of geothermal mercury from Yellowstone National Park in the Madison and Missouri Rivers, USA. *The Science of the Total Environment*, *443*, 40–54. doi:10.1016/j.scitotenv.2012.10.080 PMID:23178889

Nishioka, H. (1975). Mutagenic activities of metal compounds in bacteria. *Mutation Research*, *31*(3), 185–189. doi:10.1016/0165-1161(75)90088-6 PMID:805366

Niu, G. L., Zhang, J. J., Zhao, S., Liu, H., Boon, N., & Zhou, N. Y. (2009). Bioaugmentation of a 4-chloronitrobenzene contaminated soil with Pseudomonas putida ZWL73. *Environmental Pollution*, *57*(3), 763–771. doi:10.1016/j.envpol.2008.11.024 PMID:19108939

Niu, M., Wei, S., Bai, J., Wang, S., & Ji, D. (2015). Remediation and safe production of Cd contaminated soil via multiple cropping hyperaccumulator Solanum nigrum L. and low accumulation Chinese cabbage. *International Journal of Phytoremediation*, *17*(7), 657–661. doi:10.1080/15226514.2014.955168 PMID:25976879

Nnenna, F.-P., Lekiah, P., & Obemeata, O. (2011). Degradation of antibiotics by bacteria and fungi from the aquatic environment. *Journal of Toxicology and Environmental Health Sciences*, *3*(10), 275–285.

Noori, A., et al. (2015). Leucanthemum vulgare Lam. crude oil phytoremediation. *International Journal of Phytoremediation*.

Noori, A., Zare Maivan, H., & Alaie, E. (2014). Leucanthemum vulgare Lam. germination, growth and mycorrhizal symbiosis under crude oil contamination. *International Journal of Phytoremediation*, *16*(9), 962–970. doi:10.1080/15226514.2013.810577 PMID:24933896

Noparat, P., Maneerat, S., & Saimmai, A. (2014). Application of biosurfactant from *Sphingobacterium spiritivorum* AS43 in the biodegradation of used lubricating oil. *Applied Biochemistry and Biotechnology*, *172*(8), 3949–3963. doi:10.1007/s12010-014-0829-y PMID:24590892

Norby, R. J., & Cotrufo, M. F. (1998). Global change: A question of litter quality. *Nature*, *396*(6706), 17–18. doi:10.1038/23812

Norel, R., Lin, S. L., Wolfson, H. J., & Nussinov, R. (1994). Shape complementarity at protein–protein interfaces. *Biopolymers*, *34*(7), 933–940. doi:10.1002/bip.360340711 PMID:8054472

Notomista, E., Cafaro, V., Bozza, G., & Di Donato, A. (2009). Molecular determinants of the regioselectivity of toluene/o-xylene monooxygenase from *Pseudomonas* sp. strain OX1. *Applied and Environmental Microbiology*, *75*(3), 823–836. doi:10.1128/AEM.01951-08 PMID:19074607

Nowak, B., Müffling, T., Chaunchom, S., & Hartung, J. (2007). *Salmonella* contamination in pigs at slaughter and on the farm: A field study using an antibody ELISA test and PCR techniques. *International Journal of Food Microbiology*, *115*(3), 259–267. doi:10.1016/j.ijfoodmicro.2006.10.045 PMID:17292500

Nozari, M., Samaei, M. R., & Dehghani, M. (2014). Investigation of the Effect of Co-Metabolism on Removal of Dodecane by Microbial Consortium from Soil in a Slurry Sequencing Bioreactor. *Journal of Bioremediation and Biodegradation*, *5*, 253.

Nunoshiba, T., Obata, F., Boss, A. C., Oikawa, S., Mori, T., Kawanishi, S., & Yamamoto, K. (1999). Role of iron and superoxide for generation of hydroxyl radical, oxidative DNA lesions, and mutagenesis in *Escherichia coli*. *The Journal of Biological Chemistry*, *274*(49), 34832–34837. doi:10.1074/jbc.274.49.34832 PMID:10574955

Nwoko, C. O. (2010). Trends in phytoremediation of toxic elemental and organic pollutants. *African Journal of Biotechnology*, *9*(37), 6010–6016.

Nyer, E. K. (1992). *Treatment for organic contaminants: Biological methods. In Groundwater Treatment Technology* (pp. 129–214). New York: Van Nostrand Reinhold.

Nyholm, N. (1991). The European system of standardized legal tests for assessing the biodegradability of chemicals. *Environmental Toxicology and Chemistry*, *10*(10), 1237–1246. doi:10.1002/etc.5620101002

Oakland, H. (2013). *Bioremediation mariculture in Zanzibar, Tanzania: A viability assessment of using bath sponge and pearl oyster farms to filter highly polluted waters in the Zanzibar channel.* Independent Study Project (ISP) Collection. Paper 1526. Retrieved October 5, 2016, from http://digitalcollections.sit.edu/isp_collection/1526

Oaks J.L., Gilbert M., Virani M.Z., Watson R.T., Meteyer C.U., Rideout B.A., … Khan A.A. (2004). Diclofenac residues as the cause of vulture population decline in Pakistan. *Nature*, *427*(6975), 630-633.

Obasi, N. A., Eze, E., Anyanwu, D. I., & Okorie, U. C. (2013). Effects of organic manures on the physicochemical properties of crude oil polluted soils. *Afr J Biochem Res*, *7*(6), 67–75.

Obayori, O. S., Ilori, M. O., Adebusoye, S. A., Oyetibo, G. O., Omotayo, A. E., & Amund, O. O. (2009). Degradation of hydrocarbons and biosurfactant production by *Pseudomonas* sp. strain LP1. *World Journal of Microbiology & Biotechnology*, *25*(9), 1615–1623. doi:10.1007/s11274-009-0053-z

Obayori, O. S., Salam, L. B., & Ogunwumi, O. S. (2014). Biodegradation of Fresh and Used Engine Oils by *Pseudomonas aeruginosa* LP5. *J Bioremed Biodeg*, *5*(1), 213. doi:10.4172/2155-6199.1000213

Oberdorster, G., Oberdorster, E., & Oberdorster, J. (2005). Nanotoxicology: An emerging discipline evolving from studies of ultrafine particles. *Environmental Health Perspectives*, *113*(7), 825–839. doi:10.1289/ehp.7339 PMID:16002369

Obiakalaije, U. M., Makinde, O. A., & Amakoromo, E. R. (2015). Bioremediation of crude oil polluted soil using animal waste. *International Journal of Environmental Bioremediation & Biodegradation*, *3*(3), 79–85.

OECD. (1992). *Test no. 301: ready biodegradability*. Paris: Organisation for Economic Co-operation and Development.

Ogunseitan, O. A., Yang, S., & Ericson, J. (2000). Microbial d-aminolevulinate dehydratase as a biosensor of lead bioavailability in contaminated environments. *Soil Biology & Biochemistry*, *32*(13), 1899–1906. doi:10.1016/S0038-0717(00)00164-4

Ohtsubo, Y., Kudo, T., Tsuda, M., & Nagata, Y. (2004). Strategies for bioremediation of polychlorinated biphenyls. *Applied Microbiology and Biotechnology*, *65*(3), 250–258. doi:10.1007/s00253-004-1654-y PMID:15248039

Okeke, J. J., Isreal, O. I., & Akubukor, F. C. (2014). Role of invertebrates as bioremediators: A Review. *Am J Life Sci Res*, *3*(1), 13–20.

Okpokwasili, G. C. (2007). Biotechnology and clean environment. In *Proceedings of the 20th Annual conference of the Biotechnology Society of Nigeria (BSN)*. Ebonyi State University.

Oksman-Caldentey, K. M., & Barz, W. H. (Eds.). (2002). *Plant biotechnology and transgenic plants* (Vol. 92). CRC press. doi:10.1201/9780203910849

Oktem, Y. A., Ince, O., Sallis, P., Donnelly, T., & Ince, B. K. (2007). Anaerobic treatment of a chemical synthesis-based pharmaceutical wastewater in a hybrid upflow anaerobic sludge blanket reactor. *Bioresource Technology*, *99*(5), 1089–1096. and doi:10.1016/j.biortech.2007.02.036

Olaniran, A. O., Balgobind, A., & Pillay, B. (2013). Bioavailability of heavy metals in soil impact on microbial biodegradati on of organic compounds and possible improvement strategies. *International Journal of Molecular Sciences*, *14*(5), 10197–102228. doi:10.3390/ijms140510197 PMID:23676353

Olgui, E. J. (2003). Phycoremediation: Key issues for cost-effective nutrient removal processes. *Biotechnology Advances*, *22*(1-2), 81–91. doi:10.1016/S0734-9750(03)00130-7 PMID:14623045

Oliveira, J. G. D., & Garcia-Cruz, C. H. (2013). Properties of a biosurfactant produced by Bacillus pumilus using vinasse and waste frying oil as alternative carbon sources. *Brazilian Archives of Biology and Technology*, *56*(1), 155–160. doi:10.1590/S1516-89132013000100020

Ollis, D. J., Ewart-Smith, J. L., Day, J. A., Job, N. M., Macfarlane, D. M., Snaddon, C. D., & Mbona, N. et al. (2015). The development of a classification system for inland aquatic ecosystems in South Africa. *Water Science and Technology*, *41*, 5.

Olson, G. J., Porter, F., Rubinstein, J., & Silver, S. (1982). Mercuric reductase enzyme from a mercury-volatilizing strain of Thiobacillus ferrooxidans. *Journal of Bacteriology*, *151*(3), 1230–1236. PMID:6286594

ONeill, E. G. (1994). Responses of soil biota to elevated atmospheric Cdioxide. *Plant and Soil*, *165*(1), 55–65. doi:10.1007/BF00009962

ONeill, E. G., Luxmoore, R. J., & Norby, R. J. (1987). Increases in mycorrhizal colonization and seedling growth in Pinus echinata and Quercus alba in an enriched CO! atmosphere. *Canadian Journal of Forest Research, 17*(8), 878–883. doi:10.1139/x87-139

Onodera, S. M., Saito, M., Sawano, T., Hosono, M., Taniguchi, J., Shimada, Y., & Buapeng, R. D. et al. (2008). Effects of intensive urbanization on the intrusion of shallow groundwater into deep groundwater: Examples from Bangkok and Jakarta. *The Science of the Total Environment, 404*(2-3), 401–410. doi:10.1016/j.scitotenv.2008.08.003 PMID:18804843

Ontañon, O. M., González, P. S., Ambrosio, L. F., Paisio, C. E., & Agostini, E. (2014). Rhizoremediation of phenol and chromium by the synergistic combination of a native bacterial strain and Brassica napus hairy roots. *International Biodeterioration & Biodegradation, 88*, 192–198. doi:10.1016/j.ibiod.2013.10.017

Oren, A. (2002). Diversity of halophilic microorganisms: Environments, phylogeny, physiology, and applications. *Journal of Industrial Microbiology & Biotechnology, 28*(1), 56–63. doi:10.1038/sj/jim/7000176 PMID:11938472

Ornston, L. N., & Stanier, R. Y. (1966). The conversion of catechol and protocatechuate to β-ketoadipate by Pseudomonas putida. *The Journal of Biological Chemistry, 241*, 3776–3786. PMID:5916391

Orsod, M., Joseph, M., & Huyop, F. (2012). Characterization of exopolysaccharides produced by *Bacillus cereus* and *Brachybacterium* spp. isolated from Asian Sea Bass (*Lates calcarifer*). *Malaysian Journal of Microbiology, 8*(3), 170–174.

Ortiz, D. F., Kreppel, L., Speiser, D. M., Scheel, G., McDonald, G., & Ow, D. (1992). Heavy metal tolerance in the fission yeast requires an ATP-binding cassette-type vacuolar membrane transporter. *The EMBO Journal, 11*(10), 3491. PMID:1396551

Ortiz, D. F., Ruscitti, T., McCue, K. F., & Ow, D. W. (1995). Transport of metal-binding peptides by HMT1, fission yeast ABC-type vacuolar membrane protein. *The Journal of Biological Chemistry, 270*(9), 4721–4728. doi:10.1074/jbc.270.9.4721 PMID:7876244

Ortiz-Hernández, M. L., Sánchez-Salinas, E., Dantán-González, E., & Castrejón-Godínez, M. L. (2013). Pesticide Biodegradation: Mechanisms, Genetics and Strategies to Enhance the Process. In R. Chamy & F. Rosenkranz (Eds.), Biodegradation - Life of Science. Academic Press.

Osono, T. (2007). Ecology of ligninolytic fungi associated with leaf litter decomposition. *Ecological Research, 22*(6), 955–974. doi:10.1007/s11284-007-0390-z

Osono, T., Hagiwara, Y., & Masuya, H. (2011a). Effects of temperature and litter type on fungal growth and decomposition of leaf litter. *Mycoscience, 52*(5), 327–332. doi:10.1007/S10267-011-0112-9

Osono, T., To-Anun, C., Hagiwara, Y., & Hirose, D. (2011b). Decomposition of wood, petiole and leaf litter by Xylaria species from northern Thailand. *Fungal Ecology, 4*(3), 210–218. doi:10.1016/j.funeco.2010.11.003

Otjen, L., & Blanchette, R. A. (1986). A discussion of microstructural changes in wood during decomposition by white rot basidiomycetes. *Canadian Journal of Botany, 64*(5), 905–911. doi:10.1139/b86-121

Oves, M., Khan, M.S., & Zaidi, A. (2013). Biosorption of heavy metals by Bacillus thuringiensis strain OSM29 originating from industrial effluent contaminated north Indian soil. *Saudi Journal of Biological Sciences, 20,* 121-129.

Oviya, M., Giri, S.S., Sukumaran, V., Natarajan, P., (2012). Immobilization of carbonic anhydrase enzyme purified from Bacillus subtilis VSG-4 and its application as CO2 sequesterer. *Prep Biochem & Iotechnol, 42,* 462–475.

Ozdemir, S., Cirik, K., Akman, D., Sahinkaya, E., & Cinar, O. (2013). Treatment of azo-dye containing synthetic textile dye effluent using sulfidogenic anaerobic baffled reactor. *Bioresource Technology, 146*, 135–143. and doi:10.1016/j.biortech.2013.07.066

Pacwa-Płociniczak, M., Płaza, G. A., Piotrowska-Seget, Z., & Cameotra, S. S. (2011). Environmental applications of biosurfactants: Recent advances. *International Journal of Molecular Sciences*, *12*(1), 633–654. doi:10.3390/ijms12010633 PMID:21340005

Pacwa-Płociniczak, M., Płaza, G. A., Poliwoda, A., & Piotrowska-Seget, Z. (2014). Characterization of hydrocarbon-degrading and biosurfactant-producing *Pseudomonas* sp. P-1 strain as a potential tool for bioremediation of petroleum-contaminated soil. *Environmental Science and Pollution Research International*, *21*(15), 9385–9395. doi:10.1007/s11356-014-2872-1 PMID:24743958

Paine, R. T., Ruesink, J. L., Sun, A., Soulanille, E. L., Wonham, M. J., Harley, C. D. G., & Secord, D. L. et al. (1996). Trouble on oiled waters: Lessons from the Exxon Valdez oil spill. *Annual Review of Ecology and Systematics*, *27*(1), 197–235. doi:10.1146/annurev.ecolsys.27.1.197

Palmroth, M. R., Pichtel, J., & Puhakka, J. A. (2002). Phytoremediation of subarctic soil contaminated with diesel fuel. *Bioresource Technology*, *84*(3), 221–228. doi:10.1016/S0960-8524(02)00055-X PMID:12118697

Pal, S., Patra, A. K., Reza, S. K., Wildi, W., & Poté, J. (2010). Use of bio-resources for remediation of soil pollution. *Nat Resour*, *1*, 110–125.

Pandey, A., & Gundevia, H. S. (2008). Role of the fungus—*Periconiella* sp. in destruction of biomedical waste. *Journal of Environmental Science and Engineering*, *50*(3), 239–240.

Pandey, B., & Fulekar, M. H. (2012). Bioremediation technology: A new horizon for environmental clean-up. *Biologie Medicale*, *4*(1), 51–59.

Pandey, J., Sood, S. S., & Jain, R. K. (2007). Terminal restriction fragment length polymorphism (T-RFLP) analysis: Characterizing the unseen. *Indian Journal of Microbiology*, *47*(1), 90–91. doi:10.1007/s12088-007-0017-7 PMID:23100647

Pandey, M. P., & Kim, C. S. (2011). Lignin depolymerization and conversion: A review of thermochemical methods. *Chemical Engineering & Technology*, *34*(1), 29–41. doi:10.1002/ceat.201000270

Pandey, P. K., Kass, P. H., Soupir, M. L., Biswas, S., & Singh, V. P. (2015). Contamination of water resources by pathogenic bacteria. *Pathogens*, *4*(2), 307–334. PMID:26011827

Pandey, S. S., & Bagga, D. (2013). Phytoremediation an alternative. *International Journal of Environmental Engineering and Management*, *4*(5), 483–488.

Panesar, R., Panesar, P. S., & Bera, M. B. (2011). Development of low cost medium for the production of biosurfactants. *Asian J. Biotechnol*, *3*(4), 388–396. doi:10.3923/ajbkr.2011.388.396

Panetta, L. E. (2003). *America's living oceans: charting a course for sea change*. Pew Oceans Commission.

Pan-Hou, H. S., & Imura, N. (1982). Involvement of mercury methylation in microbial mercury detoxication. *Archives of Microbiology*, *131*(2), 176–177. doi:10.1007/BF01054003 PMID:6462125

Pan-Hou, H. S., Nishimoto, M., & Imura, N. (1981). Possible role of membrane proteins in mercury resistance of Enterobacter aerogenes. *Archives of Microbiology*, *130*(2), 93–95. doi:10.1007/BF00411057 PMID:6459062

Paniagua-Michel, J., & Rosales, A. (2015). Marine Bioremediation - A Sustainable Biotechnology of Petroleum Hydrocarbons Biodegradation in Coastal and Marine Environments. *J Bioremed Biodeg*, *6*, 273. doi:10.4172/2155-6199.1000273

Panigrahi, S., Hussain, J., Ramesh, T., & Mohanty, A. K. (2011). Natural disaster compels biota to adapt. *Current Science*.

Pan, X., Lin, D., Zheng, Y., Zhang, Q., Yin, Y., Cai, L., & Yu, Y. (2016). Biodegradation of DDT by Stenotrophomonas sp. DDT-1: Characterization and genome functional analysis. *Scientific Reports*, *6*. PMID:26888254

Paoletti, A. C., Zybailov, B., & Washburn, M. P. (2004). Principles and applications of multidimensional protein identification technology. *Expert Review of Proteomics*, *1*(3), 275–282. doi:10.1586/14789450.1.3.275 PMID:15966824

Parales, R. E., Bruce, N. C., Schmid, A., & Wackett, L. P. (2002). Biodegradation, biotransformation, and biocatalysis (B3). *Applied and Environmental Microbiology*, *68*(10), 2699–4709. doi:10.1128/AEM.68.10.4699-4709.2002 PMID:12324310

Park, D., Yun, Y. S., Jo, J. H., & Park, J. M. (2005). Mechanism of hexavalent chromium removal by dead fungal biomass of *Aspergillus niger*. *Water Research*, *39*(4), 533–540. doi:10.1016/j.watres.2004.11.002 PMID:15707625

Park, J. D., & Zheng, W. (2012). Human exposure and health effects of inorganic and elemental mercury. *Journal of Preventive Medicine and Public Health = Yebang Uihakhoe Chi*, *45*(6), 344–352. doi:10.3961/jpmph.2012.45.6.344 PMID:23230464

Park, J. H., Lamb, D., Paneerselvam, P., Choppala, G., Bolan, N., & Chung, J.-W. (2011). Role of organic amendments on enhanced bioremediation of heavy metal (loid) contaminated soils. *Journal of Hazardous Materials*, *185*(2), 549–574. doi:10.1016/j.jhazmat.2010.09.082 PMID:20974519

Parry, M. J. (1975). Evidence of mitotic divisions of coelomocytes in the normal, wounded and grafted earthworm *Eisenia fetida*. *Experimentia*, *32*(4), 449–451. doi:10.1007/BF01920788

Passatore, L., Rossetti, S., Juwarkar, A. A., & Massacci, A. (2014). Phytoremediation and bioremediation of polychlorinated biphenyls (PCBs): State of knowledge and research perspectives. *Journal of Hazardous Materials*, *278*, 189–202. doi:10.1016/j.jhazmat.2014.05.051 PMID:24976127

Pathirana, S., Connell, D., & Vowles, P. (1994). Distribution of polycylic aromatic hydrocarbons (PAHs) in an urban roadway system. *Ecotoxicology and Environmental Safety*, *28*(3), 256–269. doi:10.1006/eesa.1994.1051 PMID:7525221

Patil, P. G., Kulkarni, G. S., Kore, S. V., & Kore, V. S. (2013). Aerobic sequencing batch reactor for wastewater treatment: A review. *International Journal of Engineering Research and Technology*, *2*(10), 534–550.

Patneedi, C. B., & Prasadu, K. D. (2015). Impact of pharmaceutical wastes on human life and environment. *Rasayan Journal of Chemistry*, *8*(1), 67–70.

Patowary, K., Saikia, R. R., Kalita, M. C., & Deka, S. (2014). Degradation of polyaromatic hydrocarbons employing biosurfactant-producing Bacillus pumilusKS2. *Annals of Microbiology*, 1–10.

Patowary, R., Patowary, K., Kalita, M. C., & Deka, S. (2016). Utilization of Paneer Whey Waste for Cost-Effective Production of Rhamnolipid Biosurfactant. *Applied Biochemistry and Biotechnology*, 1–17. PMID:27142272

Pattanasupong, A., Nagase, H., Inoue, M., Hirata, K., Tani, K., Nasu, M., & Miyamoto, K. (2004). Ability of a microbial consortium to remove pesticide, carbendazim and 2,4-dichlorophenoxyacetic acid. *World Journal of Microbiology & Biotechnology*, *20*(5), 517–522. doi:10.1023/B:WIBI.0000040405.56291.52

Patterson, J. W. (1985). *Industrial wastewater treatment technology*. Boston: Butterworths.

Pattus, F., & Abdallah, M. (2000). Siderophores and iron-transport in microorganisms. *Review Journal of Chinese Chemical Society*, *47*(1), 1–20. doi:10.1002/jccs.200000001

Patwa, A., Thiery, A., Fabien, L., & Martin, L. K.S., Boisset Claire, Bramard, Jean-Francois, Bottero Jean-Yves, &Barthelemy Philippe. (2015). Accumulation of nanoparticles in jellyfish mucus: A bio-inspired route to decontamination of nano-waste. *Scientific Reports*. doi:10.1038/screp1138

Payne, W. J. (1970). Energy yields and growth of heterotrophs. *Annual Review of Microbiology*, *24*(1), 17–52. doi:10.1146/annurev.mi.24.100170.000313 PMID:4927132

Paz-Alberto, A. M., & Sigua, G. C. (2013). Phytoremediation: A green technology to remove environmental pollutants. *American Journal of Climate Change*, *2*(01), 71–86. doi:10.4236/ajcc.2013.21008

Pearce, F. (1993). What turns an oil spill into a disaster? *New Scientist*, *1858*, 11–13.

Pearson, R. G. (1963). Hard and soft acids and bases. *Journal of the American Chemical Society*, *85*(22), 3533–3539. doi:10.1021/ja00905a001

Peiter, E., Montanini, B., Gobert, A., Pedas, P., Husted, S., Maathuis, F. J., & Sanders, D. (2007). A secretory pathway-localized cation diffusion facilitator confers plant manganese tolerance. *Proceedings of the National Academy of Sciences of the United States of America*, *104*(20), 8532–8537. doi:10.1073/pnas.0609507104 PMID:17494768

Pekari, K., Vainotalo, S., Heikkila, P., Palotie, A., Luotamo, M., & Riihimaki, V. (1992). Biological monitoring of occupational exposure to low levels of benzene. *Scandinavian Journal of Work, Environment & Health*, *18*(5), 317–322. doi:10.5271/sjweh.1570 PMID:1439659

Pelczar, M. J., Chan, E. C. S., & Kreig, N. K. (1993). *Microbiology concepts and application*. McGraw Hill, Inc.

Pelz, O., Cifuentes, L. A., Hammer, B. T., Kelley, C. A., & Coffin, R. B. (1998). Tracing the assimilation of organic compounds using δ13C analysis of unique amino acids in the bacterial peptidoglycan cell wall. *FEMS Microbiology Ecology*, *25*, 229–240.

Peng, S., Zhou, Q., Cai, Z., & Zhang, Z. (2009). Phytoremediation of petroleum contaminated soils by Mirabilis Jalapa L. in a greenhouse plot experiment. *Journal of Hazardous Materials*, *168*(2), 1490–1496. doi:10.1016/j.jhazmat.2009.03.036 PMID:19346069

Peralta-Zamora, P., Gimenes, L. F., Cordi, L., Reyes, J., Alves, O. L., & Durán, N. (1998). Remediation of effluents from pulp and paper industry using horseradish peroxidase immobilized on ceramic materials. *Proc. 3rd Lat. Am. Biodegradation and Biodeterioration Symposium*.

Pereira, Y., Lagniel, G., Godat, E., Baudouin-Cornu, P., Junot, C., & Labarre, J. (2008). Chromate causes sulfur starvation in yeast. *Toxicological Sciences*, *106*(2), 400–412. doi:10.1093/toxsci/kfn193 PMID:18794233

Perelo, L. W. (2010). Review: In situ and bioremediation of organic pollutants in aquatic sediments. *Journal of Hazardous Materials*, *177*(1-3), 81–89. doi:10.1016/j.jhazmat.2009.12.090 PMID:20138425

Perez Marín, A. B., Aguilar, M. I., Meseguer, V. F., Ortũno, J. F., Sáez, J., & Lloréns, M. (2008). Biosorption of chromium (III) by orange (Citrus cinensis) waste: Batch and continuous studies. *Chemical Engineering Journal*, *155*(1-2), 199–206. doi:10.1016/j.cej.2009.07.034

Perez, J. M., Calderón, I. L., Arenas, F. A., Fuentes, D. E., Pradenas, G. A., Fuentes, E. L., & Vásquez, C. C. et al. (2007). Bacterial toxicity of potassium tellurite: Unveiling an ancient enigma. *PLoS ONE*, *2*(2), e211. doi:10.1371/journal.pone.0000211 PMID:17299591

Perez, M., Romero, L. I., Nebot, E., & Sales, D. (1997). Colonisation of a porous sintered glass support in anaerobic thermophilic bioreactors. *Bioresource Technology*, *59*(2-3), 177–183. doi:10.1016/S0960-8524(96)00154-X

Perez, M., Torrades, F., Garcia, H. J. A., Domennech, X., & Peral, J. (2002). Removal of organic contaminants in paper pulp treatment effluents under Fenton and photofenton conditions. *App. Cat. B. Environment*, *36*, 63–74.

Perez, S., Eichhorn, P., & Aga, D. S. (2005). Evaluating the biodegradability of sulfamethazine, sulfamethoxazole, sulfathiazole, and trimethoprim at different stages of sewage treatment. *Environmental Toxicology and Chemistry*, *24*(6), 1361–1367. and doi:10.1897/04-211R.1

Perfumo, A., Smyth, T. J. P., Marchant, R., & Banat, I. M. (2010). Production and roles of biosurfactants and bioemulsifiers in accessing hydrophobic substrates. In *Handbook of hydrocarbon and lipid microbiology* (pp. 1501–1512). Springer Berlin Heidelberg. doi:10.1007/978-3-540-77587-4_103

Perhar, G., & Arhonditsis, G. B. (2014). Aquatic ecosystem dynamics following petroleum hydrocarbon perturbations: A review of the current state of knowledge. *Journal of Great Lakes Research*, *40*(3), 56–72. doi:10.1016/j.jglr.2014.05.013

Perpetuo, E. A., Souza, C. B., & Nascimento, C. A. O. (2011). Engineering bacteria for bioremediation. In A. Carpi (Ed.), *Progress in Molecular and Environmental Bioengineering. From Analysis and Modeling to Technology Applications* (pp. 605–632). Rijeka: InTech.

Perry, J. J. (1984). Perry's chemical engineer handbook (6th ed.). McGraw Hill.

Pessione, E., Giuffrida, M. G., Prunotto, L., Barello, C., Mazzoli, R., Fortunato, D., & Giunta, C. (2003). Membrane proteome of Acinetobacter radioresistens S13 during aromatic exposure. *Proteomics*, *3*(6), 1070–1076. doi:10.1002/pmic.200300425 PMID:12833532

Peters, M., Heinaru, E., Talpsep, H., Ward, H., Stottmeister, U., Heinaru, A., & Nurk, A. (1997). Acquisition of a deliberately introduced phenol degradation operon, *pheAB*, by different indigenous *Pseudomonas* species. *Applied and Environmental Microbiology*, *63*, 4899–4906. PMID:9406411

Peterson, C. H., Rice, S. D., Short, J. W., Esler, D., Bodkin, J. L., Ballachey, B. E., & Irons, D. B. (2003). Long-term ecosystem response to the Exxon Valdez oil spill. *Science*, *302*(5653), 2082–2086. doi:10.1126/science.1084282 PMID:14684812

Peyton, B. M., Wilson, T., & Yonge, D. R. (2002). Kinetics of phenol biodegradation in high salt solutions. *Water Research*, *36*(19), 4811–4820. doi:10.1016/S0043-1354(02)00200-2 PMID:12448524

Pfeiffer, F. M. (1979). The 1977 screening survey for measurement of organic priority pollutants in petroleum refinery wastewaters. *ASTM Special Technical Publication*, 181–190.

Piccoli, S., Musiani, F., & Giorgetti, A. (2014). Dynamic characterization and substrate binding of cis-2, 3-dihydrobiphenyl-2, 3-diol dehydrogenase—an enzyme used in bioremediation. *Journal of Molecular Modeling*, *20*(12), 1–6. doi:10.1007/s00894-014-2531-y PMID:25433599

Piearce, T. G. (1978). The calcium relations of selected Lumbricidae. *Journal of Animal Ecology*, *41*(1), 167–188. doi:10.2307/3511

Pieper, D. H., dos Santos, V. A. M., & Golyshin, P. N. (2004). Genomic and mechanistic insights into the biodegradation of organic pollutants. *Current Opinion in Biotechnology*, *15*(3), 215–224. doi:10.1016/j.copbio.2004.03.008 PMID:15193329

Pieper, D. H., & Reineke, W. (2000). Engineering bacteria for bioremediation. *Current Opinion in Biotechnology*, *11*(3), 262–270. doi:10.1016/S0958-1669(00)00094-X PMID:10851148

Pimentel, D., Wilson, C., McCullum, C., Huang, R., Dwen, P., Flack, J., & Cliff, B. et al. (1997). Economic and Environmental Benefits of Biodiversity. *Bioscience*, *47*(11), 747–757. doi:10.2307/1313097

Pinckney, J. L., Paerl, H. W., & Harrington, M. B. (1999). Responses of the phytoplankton community growth rate to nutrient pulses in variable estuarine environments. *Journal of Phycology*, *35*(6), 1455–1463. doi:10.1046/j.1529-8817.1999.3561455.x

Pino, N. J., Muñera, L. M., & Peñuela, G. A. (2016). Root exudates and plant secondary metabolites of different plants enhance polychlorinated biphenyl degradation by rhizobacteria. *Bioremediation Journal*, *20*(2), 108–116. doi:10.1080/10889868.2015.1124065

Pinto, G., Pollio, A., Previtera, L., & Temussi, F. (2002). Degradation of phenols by microalgae. *Biotechnology Letters*, *24*(24), 2047–2051. doi:10.1023/A:1021367304315

Piromyou, P., Buranabanyat, B., Tantasawat, P., Tittabutr, P., Boonkerd, N., & Teaumroong, N. (2011). Effect of plant growth promoting rhizobacteria (PGPR) inoculation on microbial community structure in rhizosphere of forage corn cultivated in Thailand. *European Journal of Soil Biology*, *11*(47), 44–54. doi:10.1016/j.ejsobi.2010.11.004

Pizarro-Tobías, P., Fernández, M., Niqui, J. L., Solano, J., Duque, E., Ramos, J. L., & Roca, A. (2015). Restoration of a Mediterranean forest after a fire: Bioremediation and rhizoremediation field-scale trial. *Microbial Biotechnology*, *8*(1), 77–92. doi:10.1111/1751-7915.12138 PMID:25079309

Pizarro-Tobías, P., Niqui, J. L., Roca, A., Solano, J., Fernández, M., Bastida, F., & Ramos, J. L. et al. (2015). Field trial on removal of petroleum-hydrocarbon pollutants using a microbial consortium for bioremediation and rhizoremediation. *Environmental Microbiology Reports*, *7*(1), 85–94. doi:10.1111/1758-2229.12174 PMID:25870876

Plangklang, P., & Reungsang, A. (2010). Bio-augmentation of carbo furan by *Burkholderia cepacia* PCL_3 in a bio slurry phase sequencing batch reactor. *Process Biochemistry*, *45*(2), 230–238. doi:10.1016/j.procbio.2009.09.013

Plumb, R. H. (1987). A comparison of ground water monitoring data from CERCLA and RCRA sites. *Ground Water Monitoring Review*, *7*(4), 94–100. doi:10.1111/j.1745-6592.1987.tb00968.x

Pointing, S. B. (2001). Feasibility of bioremediation by white-rot fungi. *Applied Microbiology and Biotechnology*, *57*(1-2), 20–33. doi:10.1007/s002530100745 PMID:11693920

Polymenakou, P. N., & Stephanou, E. G. (2005). Effect of temperature and additional carbon sources on phenol degradation by an indigenous soil *Pseudomonas*. *Biodegradation*, *16*(5), 403–413. doi:10.1007/s10532-004-3333-1 PMID:15865154

Pomati, F., Castiglioni, S., Zuccato, E., Fanelli, R., Vigetti, D., Rossetti, C., & Calamari, D. (2006). Effects of a complex mixture of therapeutic drugs at environmental levels on human embryonic cells. *Environmental Science & Technology*, *40*(7), 2442–2447. and doi:10.1021/es051715a

Poorter, H. (1993). Interspecific variation in the growth response of plants to an elevated ambient CO2 concentration. *Vegetatio*, *140*(1), 77–97. doi:10.1007/BF00048146

Pophali, G. R., Khan, R., Dhodapkar, R. S., Nandy, T., & Devotta, S. (2007). Anaerobic–aerobic treatment of purified terephthalic acid (PTA) effluent: A technoeconomic alternative to two-stage aerobic process. *Journal of Environmental Management*, *85*(4), 1024–1033. and doi:10.1016/j.jenvman.2006.11.016

Powlowski, J., & Shingler, V. (1994). Genetics and biochemistry of phenol degradation by *Pseudomonas* sp. CF600. *Biodegradation*, *5*(3-4), 219–236. doi:10.1007/BF00696461 PMID:7765834

Prabha, Y., Soni, S. K., Gupta, S., & Sonal, . (2016). Potential of Algae in Bioremediation of Wastewater: Current Research. *International Journal of Current Microbiology and Applied Sciences*, *5*(2), 693–700. doi:10.20546/ijcmas.2016.502.076

Prakash, O., & Lal, R. (2006). Description of Sphingobium fuliginis sp. nov., a phenanthrene-degrading bacterium from a fly ash dumping site, and reclassification of Sphingomonas cloacae as Sphingobium cloacae comb. nov. *International Journal of Systematic and Evolutionary Microbiology*, *56*(9), 2147–2152. doi:10.1099/ijs.0.64080-0 PMID:16957112

Prasad, M. (1995). Cadmium toxicity and tolerance in vascular plants. *Environmental and Experimental Botany*, *35*(4), 525–545. doi:10.1016/0098-8472(95)00024-0

Prasad, M. N. V. (1999). Metallothioneins and metal binding complexes in plants. In *Heavy metal stress in plants* (pp. 51–72). Springer Berlin Heidelberg. doi:10.1007/978-3-662-07745-0_3

Prasad, M. N. V. (2011). *A state-of-the-art report on bioremediation, its applications to contaminated sites in India*. Ministry of Environment & Forests, Government of India.

Prasad, N. K., Vindal, V., Narayana, S. L., Ramakrishna, V., Kunal, S. P., & Srinivas, M. (2012). *In silico* analysis of *Pycnoporus cinnabarinus* laccase active site with toxic industrial dyes. *Journal of Molecular Modeling*, *18*(5), 2013–2019. doi:10.1007/s00894-011-1215-0 PMID:21877154

Prasanna, R., Jaiswal, P., Singh, Y. V., & Singh, P. K. (2008). Influence of biofertilizers and organic amendments on nitrogenase activity and phototrophic biomass of soil under wheat. *Acta Agron Hung*, *56*(2), 149–159. doi:10.1556/AAgr.56.2008.2.4

Prasanna, R., Joshi, M., Rana, A., & Nain, L. (2010). Modulation of IAA production in cyanobacteria by tryptophan and light. *Polish Journal of Microbiology / Polskie Towarzystwo Mikrobiologow = the Polish Society of Microbiologists*, *59*, 99–105. PMID:20734754

Prasanna, R., Joshi, M., Rana, A., Shivay, Y. S., & Nain, L. (2012). Influence of co-inoculation of bacteria-cyanobacteria on crop yield and C–N sequestration in soil under rice crop. *World Journal of Microbiology & Biotechnology*, *28*(3), 1223–1235. doi:10.1007/s11274-011-0926-9 PMID:22805842

Predki, P. F. (2004). Functional protein microarrays: Ripe for discovery. *Current Opinion in Chemical Biology*, *8*(1), 8–13. doi:10.1016/j.cbpa.2003.12.005 PMID:15036150

Preis, S., Terentyeva, Y., & Rozkov, A. (1997). Photocatalytic oxidation of phenolic compounds in wastewater from oil shale treatment. *Water Science and Technology*, *35*(4), 165–174. doi:10.1016/S0273-1223(97)88394-1

Prentice, I. C., Farquhar, G. D., Fasham, M. J. R., Goulden, M. I., Heinmann, M., Jaramillo, V. J., … Wallace, D.W. (2001). The Ccycle and atmospheric Cdioxide, in Climate Change. The Scientific Basis. Contribution of Working Group I to the Third Assessment Report of the Intergovernmental Panel on Climate Change. Cambridge University Press.

Prento, P. (1979). Metal and phosphate in the chloragosomes of *Lumbricus terrestris* and their possible physiological significance. *Cell and Tissue Research*, *196*, 123–134. doi:10.1007/BF00236353 PMID:421244

Prescott, L. M., Harley, J. P., & Klein, D. A. (2002). *Microbiology* (5th ed.). New York: McGrawHill.

Prevost-Boure, N. C., Christen, R., Dequiedt, S., Mougel, C., Lelievre, M., Jolivet, C., & Ranjard, L. et al. (2011). Validation and application of a PCR primer set to quantify fungal communities in the soil environment by real-time quantitative PCR. *PLoS ONE*, *6*(9), e24166. doi:10.1371/journal.pone.0024166 PMID:21931659

Priyadarshani, I., Sahu, D., & Rath, B. (2011). Microalgal bioremediation: Current practices and perspectives. *J Biochem Technol*, *3*(3), 299–304.

Priya, E. S., & Selvan, P. S. (2014). *Water hyacinth (Eichhornia crassipes) – An efficient and economic adsorbent for textile effluent treatment – A review*. Arabian Journal of Chemistry; doi:10.1016/j.arabjc.2014.03.002

Public Health Engineering Department (PHED). (2013). *A Note on Fluoride Contamination of Ground Water in West Bengal*. Author.

Purohit, H. J. (2003). Biosensors as molecular tools for use in bioremediation. *Journal of Cleaner Production*, *11*(3), 293–301. doi:10.1016/S0959-6526(02)00072-0

Purohit, H. J., Raje, D. V., Kapley, A., Padmanabhan, P., & Singh, R. N. (2003). Peer reviewed: Genomics tools in environmental impact assessment. *Environmental Science & Technology*, *37*(19), 356A–363A. doi:10.1021/es032594m PMID:14572077

Puškárová, A., Bučková, M., Chovanová, K., Harichová, J., Karelová, E., Godočíková, J., & Pangallo, D. et al. (2013). Diversity and PAH growth abilities of bacterial strains isolated from a contaminated soil in Slovakia. *Biologia*, *68*(4), 587–591. doi:10.2478/s11756-013-0193-3

Puvaneswari, N., Muthukrishnan, J., & Gunasekaran, P. (2006). Toxicity assessment and microbial degradation of azo dyes. *Indian Journal of Experimental Biology*, *44*, 618–626. PMID:16924831

Qari, H. A., & Hassan, I. A. (2014). Removal of pollutants from waste water using *Dunaliella* algae. Biomedical & Pharmacology Journal, 7(1), 147–151. 10.13005/bpj/465

Quan, X., Shi, H., Liu, H., Wang, J., & Qian, Y. (2004). Removal of 2,4-dichlorophenol in a conventional activated sludge system through bioaugmentation. *Process Biochem, Oxford, UK*, *39*(11), 1701–1707. doi:10.1016/S0032-9592(03)00307-8

Qureshi, A., Prabu, S. K., & Purohit, H. J. (2001). Isolation and Characterization of Pseudomonas Strain for Degradation of 4-nitrophenol. *Microbes and Environments*, *16*(1), 49–52. doi:10.1264/jsme2.2001.49

Qureshi, A., Verma, V., Kapley, A., & Purohit, H. J. (2007). Degradation of 4-nitroaniline by Stenotrophomonas strain HPC 135. *International Biodeterioration & Biodegradation*, *60*(4), 215–218. doi:10.1016/j.ibiod.2007.03.004

Qu, Y. Y., Zhang, R., Fang, M., Zhou, J., & Yan, B. (1926). Bioaugmentation with a novel alkali-tolerant Pseudomonas strain for alkaline phenol wastewater treatment in sequencing batch Reactor. *World Journal of Microbiology & Biotechnology*, *27*(8), 1919–1926. doi:10.1007/s11274-011-0653-2

Qu, Y. Y., Zhou, J. T., Wang, J., Song, Z. Y., Xing, L. L., & Fu, X. (2006). Bioaugmentation of bromoamine acid degradation with *Sphingomonas xenophaga* QYY & DNA fingerprint analysis of augmented systems. *Biodegradation*, *17*(1), 83–91. doi:10.1007/s10532-005-3544-0 PMID:16453174

Radajewski, S., Ineson, P., Parekh, N. R., & Murrell, J. C. (2000). Stable-isotope probing aș a tool in microbial ecology. *Nature*, *403*(6770), 646–649. doi:10.1038/35001054 PMID:10688198

Radjenović, J., Petrovic, M., & Barcelo, D. (2009). Fate and distribution of pharmaceuticals in wastewater and sewage sludge of the conventional activated sludge (CAS) and advanced membrane bioreactor (MBR) treatment. *Water Research*, *43*(3), 831–841. and doi:10.1016/j.watres.2008.11.043

Radosevich, S. R., Holt, J. S., & Ghersa, C. M. (2007). Ecology of Weeds and Invasive Plants: Relationship to Agriculture and Natural Resource Management. Wiley-Interscience. doi:10.1002/9780470168943

Rahman, M. A., Soumya, K. K., Tripathi, A., Sundaram, S., Singh, S., & Gupta, A. (2011). Evaluation and sensitivity of cyanobacteria, *Nostoc muscorum* and *Synechococcus* PCC 7942 for heavy metals stress - a step toward biosensor. *Toxicological and Environmental Chemistry*, *93*(10), 1982–1990. doi:10.1080/02772248.2011.606110

Rajesh, Y., Pujari, M., & Uppaluri, R. (2014). Equilibrium and kinetic studies of Ni(II) adsorption using pine apple and bamboo stem based adsorbents. *Separation Science and Technology*, *49*(4), 533–544. doi:10.1080/01496395.2013.845426

Rajeswari, R., & Kanmani, S. (2009). A study on degradation of pesticide wastewater by TiO_2 catalysis. *Journal of Scientific and Industrial Research*, *68*, 1063–1067.

Rajkumar, M., Ma, Y., & Freitas, H. (2013). Improvement of Ni phytostabilization by inoculation of Ni resistant Bacillus megaterium SR28C. *Journal of Environmental Management*, *128*, 973–980. doi:10.1016/j.jenvman.2013.07.001 PMID:23895909

Rajput, M. S., Sharma, A. K., Sharma, S., & Verma, S. (2015). Removal of Pb (II) from aqueous solution by orange peel. *International Journal of Applied Research*, *1*(9), 411–413.

Rakshit, P. C. (2004). *Physical Chemistry* (7th ed.). Kolkata, India: Sarat Book Distributors.

Ramachandran, S., Pierre, F., & Pandey, A. (2006). Gluconic acid: Properties, applications and microbial production. *Food Technology and Biotechnology*, *44*(2), 185–195.

Ramanathan, K., Shanthi, V., & Sethumadhavan, R. (2009). *In silico* identification of catalytic residues in azobenzene reductase from *Bacillus subtilis* and its docking studies with azo dyes. *Interdisciplinary Sciences: Computational Life Sciences*, *1*(4), 290–297. PMID:20640807

Ramaprabhu, S. A., & Kaniyoor, A. (2012). Raman spectroscopic investigation of graphite oxide derived graphene. *AIP Advances*, *2*(3), 032183, 032183–13. doi:10.1063/1.4756995

Ramírez-Castillo, F. Y., Loera-Muro, A., Jacques, M., Garneau, P., Avelar-González, F. J., Harel, J., & Guerrero-Barrera, A. L. (2015). Waterborne Pathogens: Detection Methods and Challenges. *Pathogens*, *4*(2), 307–334. doi:10.3390/pathogens4020307 PMID:26011827

Ramos, D. T., Maranho, L. T., Godoi, A. F. L., da Silva Carvalho Filho, M. A., Lacerda, L. G., & de Vasconcelos, E. C. (2009). Petroleum hydrocarbons rhizodegradation by Sebastiania commersoniana (BAILL.) LB SM. & Downs. *Water Air and Soil Pollution Focus*, *9*(3-4), 293–302. doi:10.1007/s11267-009-9208-z

Rampelotto, P. H. (2013). Extremophiles and Extreme Environments. *Life (Basel)*, *3*(3), 482–485. PMID:25369817

Rana, R. S., Singh, P., Kandari, V., Singh, R., Dobhal, R., & Gupta, S. (2014). A review on characterization and bioremediation of pharmaceutical industries' wastewater: An Indian perspective. *Applied Water Science*, 1-12. 10.1007/s13201-014-0225-3

Randhawa, G. K., & Kullar, J. K. (2011). Bioremediation of pharmaceuticals, pesticides, and petrochemicals with gomeya/cow dung. *ISRN Pharmacology*, *2011*, 362459. and doi:10.5402/2011/362459

Rane, N. R., Chandanshive, V. V., Watharkar, A. D., Khandare, R. V., Patil, T. S., Pawar, P. K., & Govindwar, S. P. (2015). Phytoremediation of sulfonated remazol red dye and textile effluents by *Alternanthera philoxeroides*: An anatomical, enzymatic and pilot scale study. *Water Research*, *83*, 271–281. doi:10.1016/j.watres.2015.06.046 PMID:26164661

Rani, A., & Goel, R. (2009). Strategies for crop improvement in contaminated soils using metal-tolerant bioinoculants. In M. S. Khan, A. Zaidi & J. Musarrat, (Eds.), Microbial strategies for crop improvement, (pp. 105–132). Springer. doi:10.1007/978-3-642-01979-1_5

Rani, M. J., Hemambika, B., Hemapriya, J., & Kannan, V. R. (2010). Comparative assessment of heavy metal removal by immobilized and dead bacterial cells: A biosorption approach. *African Journal of Environmental Science and Technology*, *4*(2), 77-83.

Rani, A., Porwal, S., Sharma, R., Kapley, A., Purohit, H. J., & Kalia, V. C. (2008). Assessment of microbial diversity in effluent treatment plants by culture dependent and culture independent approaches. *Bioresource Technology*, *99*(15), 7098–7107. doi:10.1016/j.biortech.2008.01.003 PMID:18280146

Ranjan, S., Dasgupta, N., Chinnappan, S., Ramalingam, C., & Kumar, A. (2015). A novel approach to evaluate titanium dioxide nanoparticle–protein interaction through docking: An insight into mechanism of action. *Proceedings of the National Academy of Sciences, India Section B: Biological Sciences*, 1-7.

Ranjard, L., Dequiedt, S., Chemidlin Prévost-Bouré, N., Thioulouse, J., Saby, N. P. A., Lelievre, M., & Lemanceau, P. et al. (2013). Turnover of soil bacterial diversity driven by wide-scale environ- mental heterogeneity. *Nature Communications*, *4*, 1434. doi:10.1038/ncomms2431 PMID:23385579

Ranzelli-Cain, R., & Kaloustian, K. V. (1995). Evidence for the involvement of opioid peptides in phagocytosis, conformation, granulation and aggregation of immunocompetent *Lumbricus terrestris* amoebocytes. *Comparative Biochemistry and Physiology*, *111*, 205–211. PMID:8521243

Rao, A. G., Naidu, G. V., Prasad, K. K., Rao, N. C., Mohan, S. V., Jetty, A., & Sharma, P. N. (2004). Anaerobic treatment of wastewater with high suspended solids from a bulk drug industry using fixed film reactor (AFFR). *Bioresource Technology*, *93*(3), 241–247. and doi:10.1016/j.biortech.2003.11.004

Rao, P. N., Habiulla, S., Prasad, K. V., & Anand, Ch. (2014). A treatment study of pharmaceutical industrial wastewater using UASB. *Environmental Sciences*, *3*(10), 195–197.

Rarey, M., Kramer, B., Lengauer, T., & Klebe, G. (1996). A fast flexible docking method using an incremental construction algorithm. *Journal of Molecular Biology*, *261*(3), 470–489. doi:10.1006/jmbi.1996.0477 PMID:8780787

Rascio, N., & Navari-Izzo, F. (2011). Heavy metal hyperaccumulating plants: How and why do they do it? And what makes them so interesting? *Plant Science*, *180*(2), 169–181. doi:10.1016/j.plantsci.2010.08.016 PMID:21421358

Ray, S., & Banerjee, A. (2015). Molecular level biodegradation of phenol and its derivatives through dmp operon of *Pseudomonas putida:* A bio-molecular modeling and docking analysis. *Journal of Environmental Sciences (China)*, *36*, 144–151. doi:10.1016/j.jes.2015.03.035 PMID:26456616

Read, D. J. (1998b). The role of mycorrhiza in determining the response of Plantago lanceolata to CO! enrichment. *The New Phytologist*, *139*(2), 367–373. doi:10.1046/j.1469-8137.1998.00200.x

Rea, P. A., Li, Z. S., Lu, Y. P., Drozdowicz, Y. M., & Martinoia, E. (1998). From vacuolar GS-X pumps to multispecific ABC transporters. *Annual Review of Plant Biology*, *49*(1), 727–760. doi:10.1146/annurev.arplant.49.1.727 PMID:15012252

Reddersen, K., Heberer, T., & Dünnbier, U. (2002). Identification and significance of phenazone drugs and their metabolites in ground- and drinking water. *Chemosphere*, *49*(6), 539–544. doi:10.1016/S0045-6535(02)00387-9 PMID:12430641

Reeve, R. N. (1994). *Environmental Analysis* (3rd ed.). Chichester, UK: John Wiley.

Reeves, R. D., & Baker, A. J. (2000). *Metal-accumulating plants. Phytoremediation of toxic metals: using plants to clean up the environment*. New York: Wiley.

Reicosky, D. (2003). Conservation Agriculture: Global Environmental Benefits of Soil CManagement. In L. García-Torres, J. Benites, A. Martínez-Vilela, & A. Holgado-Cabrera (Eds.), *Conservation agriculture* (pp. 291–300). Springer. doi:10.1007/978-94-017-1143-2_35

Reincke, A. J., Reinecke, S. A., & Maboeta, M. S. (2001). Cocoon production and viability as endpoints in toxicity testing of heavy metals with three earthworm species. *Pedobiologia*, *45*(1), 61–68. doi:10.1078/0031-4056-00068

Reis, R. S., Pacheco, G. J., Pereira, A. G., & Freire, D. M. G. (2013). Biosurfactants: production and applications. *Biodegradation: Life of science,* 978-953.

Relya, R. A. (2005). The lethal impact of Roundup® on aquatic and terrestrial amphibians. *Ecological Applications*, *15*, 618.

Reungoat, J., Escher, B. I., Macova, M., & Keller, J. (2011). Biofiltration of wastewater treatment plant effluent: Effective removal of pharmaceuticals and personal care products and reduction of toxicity. *Water Research*, *45*(9), 2751–2762. doi:10.1016/j.watres.2011.02.013 PMID:21450327

Revathi, S., & Venugopal, S. (2013). Physiological and biochemical mechanisms of heavy metal tolerance. *International Journal of Environmental Sciences*, *3*(5), 1339.

Ricci, F., Volpe, G., Micheli, L., & Palleschi, G. (2007). A review on novel developments and applications of immunosensors in food analysis. *Annals in Chimmica Acta, 605*(2), 111-129.

Richardson, C. J. (2008). *The Everglades Experiments: Lessons for Ecosystem Restoration*. New York: Springer-Verlag. doi:10.1007/978-0-387-68923-4

Richardson, N., Gordon, A. K., Muller, W. J., Pletschke, B. I., & Whitfield, A. K. (2010). The use of liver histopathology, lipid peroxidation and acetylcholine esterase assays as biomarkers of contaminant induced stress in the Cape stumpnose (*Rhabdosargus holubi*) (Teleostei: Sparidae), from selected South African estuaries. *Water S.A., 36*(4), 407–415. doi:10.4314/wsa.v36i4.58409

Richins, R. D., Kaneva, I., Mulchandani, A., & Chen, W. (1997). Biodegradation of organophosphorus pesticides by surface-expressed. *Nature Biotechnology*, 15. PMID:9335050

Rillig, M. C. (2004). Arbuscular mycorrhizae, glomalin, and soil aggregation. *Canadian Journal of Soil Science, 84*(4), 355–363. doi:10.4141/S04-003

Rillig, M. C., & Allen, M. F. (1998). Arbuscular mycorrhizae of Gutierrezia sarothrae and elevated Cdioxide: Evidence for shifts in C allocation to and within the mycobiont. *Soil Biology & Biochemistry, 30*(14), 2001–2008. doi:10.1016/S0038-0717(98)00073-X

Rillig, M. C., & Allen, M. F. (1999). What is the role of arbuscular mycorrhizal fungi in plant-to-ecosystem responses to elevated atmospheric CO2? *Mycorrhiza, 9*(1), 1–8. doi:10.1007/s005720050257

Rillig, M. C., Field, C. B., & Allen, M. F. (1999a). Soil biota responses to long-term atmospheric CO! enrichment in two California annual grasslands. *Oecologia, 119*(4), 572–577. doi:10.1007/s004420050821

Rimkus, G. G., Gatermann, R., & Hühnerfuss, H. (1999). Musk xylene and musk ketone amino metabolites in the aquatic environment. *Toxicology Letters, 111*(1-2), 5–15. doi:10.1016/S0378-4274(99)00190-3 PMID:10630699

Riser-Roberts, E. (1998). *Remediation of Petroleum Contaminated Soil: Biological, Physical, and Chemical Processes*. Boca Raton, FL: Lewis Publishers. doi:10.1201/9781420050578

Rittmann, B. E., & McCarty, P. L. (2001). *Environmental biotechnology: Principles and applications*. New York: McGraw-Hill.

Rivera-Hoyos, C. M., Morales-Álvarez, E. D., Poveda-Cuevas, S. A., Reyes-Guzmán, E. A., Poutou-Piñales, R. A., Reyes-Montaño, E. A., & Cardozo-Bernal, Á. M. et al. (2015). Computational analysis and low-scale constitutive expression of laccases synthetic genes GlLCC1 from *Ganoderma lucidum* and POXA 1B from *Pleurotus ostreatus* in *Pichia pastoris*. *PLoS ONE, 10*(1), e0116524. doi:10.1371/journal.pone.0116524 PMID:25611746

RIVM. (1986). Bilthoven, The Netherlands: National Institute of Public Health and Environment Protection Document No. 738513002.

Robertson, L. W., & Hansen, L. G. (2015). *PCBs: recent advances in environmental toxicology and health effects*. University Press of Kentucky.

Robinson, B., Bolan, N., Mahimairaja, S., & Clothier, B. (2005). Solubility, mobility, and bioaccumulation of trace elements: Abiotic processes in the rhizosphere. In *Trace elements in the environment: Biogeochemistry, Biotechnology and Bioremediation* (pp. 97–110). Boca Raton, FL: CRC Press. doi:10.1201/9781420032048.sec2

Robinson, J. B., & Tuovinen, O. H. (1984). Mechanisms of microbial resistance and detoxification of mercury and organomercury compounds: Physiological, biochemical, and genetic analyses. *Microbiological Reviews, 48*(2), 95. PMID:6377034

Robinson, N. J., Tommey, A. M., Kuske, C., & Jackson, P. J. (1993). Plant metallothioneins. *The Biochemical Journal, 295*(Pt 1), 1–10. doi:10.1042/bj2950001 PMID:8216201

Robles-Gonzalez, I. V., Fava, F., & Poggi-Varaldo, H. M. (2008). A review on slurry bioreactors for bioremediation of soils and sediments. *Microbial Cell Factories, 7*(1), 1–16. doi:10.1186/1475-2859-7-5 PMID:18312630

Rodrigues, A. I., Gudiña, E. J., Teixeira, J. A., & Rodrigues, L. R. (2015). Development of low-cost culture media for effective biosurfactant production. *SINAFERM 2015 XX Simpósio Nacional de Bioprocessos*, 1-5.

Rodriguez, E., Campinas, M., Acero, J. L., & Rosa, M. J. (2016). Investigating PPCP Removal from Wastewater by Powdered Activated Carbon/ Ultrafiltration. *Water, Air, and Soil Pollution, 227*(6), 177. doi:10.1007/s11270-016-2870-7

Rogers, H. H., Prior, S. A., Runion, G. B., & Mitchell, R. J. (1996). Root to shoot ratio of crops as influenced by CO2. *Plant and Soil, 187*(2), 229–248. doi:10.1007/BF00017090

Rohrbacher, F., & St-Arnaud, M. (2016). Root Exudation: The Ecological Driver of Hydrocarbon Rhizoremediation. *Agronomy, 6*(1), 19. doi:10.3390/agronomy6010019

Rojas-Avelizapa, N. G., Roldan-Carrillo, T., Zegarra-Martınez, H., Muñoz-Colunga, A. M., & Fernández-Linares, L. C. (2007). A field trial for an ex-situ bioremediation of a drilling mud-polluted site. *Chemosphere, 66*(9), 1595–1600. doi:10.1016/j.chemosphere.2006.08.011 PMID:16997351

Röling, W. F., & van Bodegom, P. M. (2014). Toward quantitative understanding on microbial community structure and functioning: A modeling-centered approach using degradation of marine oil spills as example. *Front. Microbiol, 5*, 125. doi:10.3389/fmicb.2014.00125 PMID:24723922

Rombouts, I., Beaugrand, G., Artigas, L. F., Dauvin, J.-C., Gevaert, F., Goberville, E., & Kirby, R. R. et al. (2013). Evaluating marine ecosystem health: Case studies of indicators using direct observations and modelling methods. *Ecological Indicators, 24*, 353–365. doi:10.1016/j.ecolind.2012.07.001

Rosario-Ortiz, F. L., Wert, E. C., & Snyder, S. A. (2010). Evaluation of UV/H_2O_2 treatment for the oxidation of pharmaceuticals in wastewater. *Water Research, 44*(5), 1440–1448. doi:10.1016/j.watres.2009.10.031 PMID:19931113

Rosselló-Mora, R., & Amann, R. (2001). The species concept for prokaryotes. *FEMS Microbiology Reviews, 25*(1), 39–67. doi:10.1111/j.1574-6976.2001.tb00571.x PMID:11152940

Ross, G. (2004). The public health implications of polychlorinated biphenyls (PCBs) in the environment. *Ecotoxicology and Environmental Safety, 59*(3), 275–291. doi:10.1016/j.ecoenv.2004.06.003 PMID:15388267

Rossi, F., Olguin, E., Diels, L., & De Philippis, R. (2015). Microbial fixation of CO_2 in water bodies and in dry lands to combat climate change, soil loss and desertification. *New Biotechnology, 32*(1), 109–120. doi:10.1016/j.nbt.2013.12.002 PMID:24355428

Rousk, J., Brookes, P. C., & Baath, E. (2010a). Investigating the mechanisms for opposing pH relationships of fungal and bacterial growth in soil. *Soil Biology & Biochemistry, 42*(6), 926–934. doi:10.1016/j.soilbio.2010.02.009

Roy, A. S., Baruah, R., Borah, M., Singh, A. K., Boruah, H. P. D., Saikia, N., & Bora, T. C. et al. (2014). Bioremediation potential of native hydrocarbon degrading bacterial strains in crude oil contaminated soil under microcosm study. *International Biodeterioration & Biodegradation, 94*, 79–89. doi:10.1016/j.ibiod.2014.03.024

Rubin, H. E., & Alexander, M. (1983). Effects of nutrients on the rates of mineralization of trace concentrations of phenol and *p*-nitrophenol. *Environmental Science & Technology, 17*(2), 104–107. doi:10.1021/es00108a008 PMID:22295962

Ruelle, V., Moualij, B. E., Zorzi, W., Ledent, P., & Pauw, E. D. (2004). Rapid identification of environmental bacterial strains by matrix-assisted laser desorption/ionization time-of-flight mass spectrometry. *Rapid Communications in Mass Spectrometry*, *18*(18), 2013–2019. doi:10.1002/rcm.1584 PMID:15378711

Ruiz-Dueñas, F. J., Martínez, M. J., & Martínez, A. T. (1999). Molecular characterization of a novel peroxidase isolated from the ligninolytic fungus *Pleurotus eryngii. Molecular Microbiology*, *31*(1), 223–235. doi:10.1046/j.1365-2958.1999.01164.x PMID:9987124

Ruquan, Y. C., Xiang, J., Lin, Z., & Peng, K., Huang, & Torr. (2013). Coal as an abundant source of graphene quantum dots. *Nature Communications*, *4*, 2943–2947. PMID:24309588

Ruzicka, F. J., & Beinert, H. (1978). The soluble "high potential" type iron-sulfur protein from mitochondria is aconitase. *The Journal of Biological Chemistry*, *253*(8), 2514–2517. PMID:204652

Sadtler, N. A. (1960). *Sadtler standard spectra*. Sadtler Research Lab.

Saharan, B. S., Sahu, R. K., & Sharma, D. (2011). A review on biosurfactants: fermentation, current developments and perspectives. *Genetic Engineering and Biotechnology Journal, 2011*(1), 1-14.

Sahin, I., Keskin, S. Y., & Keski, C. S. (2013). Biosorption of cadmium, manganese, nickel, lead, and zinc ions by *Aspergillus tamari. Desalination and Water Treatment*, 1–6.

Sahoo, N. K., Ramesh, A., & Pakshirajan, K. (2012). *Bacterial degradation of aromatic xenobiotic compounds: an overview on metabolic pathways and molecular approaches. In Microorganisms in Environmental Management* (pp. 201–220). Springer Netherlands.

Said, M., Ahmad, A., & Abdul Wahab, M. (2013). Removal of phenol during ultrafiltration of Palm oil mill effluent (POME): Effect of pH, ionic strength, pressure and temperature. *Der Pharma Chemical*, *5*(3), 190–196.

Saikia, R. R., Deka, S., Deka, M., & Banat, I. M. (2012). Isolation of biosurfactant-producing *Pseudomonas aeruginosa* RS29 from oil-contaminated soil and evaluation of different nitrogen sources in biosurfactant production. *Annals of Microbiology*, *62*(2), 753–763. doi:10.1007/s13213-011-0315-5

Saimmai, A., Sobhon, V., & Maneerat, S. (2011). Molasses as a whole medium for biosurfactants production by Bacillus strains and their application. *Applied Biochemistry and Biotechnology*, *165*(1), 315–335. doi:10.1007/s12010-011-9253-8 PMID:21509601

Sakagami, H., Satoh, K., Ida, Y., Koyama, N., Premanathan, M., Arakaki, R., & Yoshida, T. et al. (1999). Induction of apoptosis and anti-HIV activity by tannin- and lignin-related substances. *Basic Life Sciences*, *66*, 595–611. PMID:10800465

Saleem, M. (2007). Pharmaceutical wastewater treatment: A physicochemical study. *Journal of Research (Science)*, *18*(2), 125–134.

Sallis, P. J., Christian, V., Shrivastava, R., Shukla, D., Modi, H. A., & Vyas, B. R. M. (2005). Degradation of xenobiotic compounds by lignin-degradibg white-rot fungi: Enzymology and mechanisma involved. *Indian Journal of Experimental Biology*, *43*, 301–312.

Salmén, L., & Burgert, I. (2009). Cell wall features with regard to mechanical performance. A review. *Holzforschung*, *63*, 121–129.

Sanchez, C. (2009). Lignocellulosic residues: Biodegradation and bioconversion by fungi. *Biotechnology Advances*, *27*(2), 185–194. doi:10.1016/j.biotechadv.2008.11.001 PMID:19100826

Sancinetti, G. P., Sader, L. T., Varesche, M. B. A., Amorim, E. L. C., Omena, S. P. F., & Silva, E. L. (2012). Phenol degradation in an anaerobic fluidized bed reactor packed with low density support materials. *Brazilian Journal of Chemical Engineering*, *29*(1), 87–98. doi:10.1590/S0104-66322012000100010

Sandermann, H. Jr. (1994). Higher plant metabolism of xenobiotics: Thegreen liverconcept. *Pharmacogenetics and Genomics*, *4*(5), 225–241. doi:10.1097/00008571-199410000-00001 PMID:7894495

Sandhya, S., Padmavathy, S., Swaminathan, K., Subrahmanyam, Y. V., & Kaul, S. N. (2005). Micro aerophilic- aerobic sequential batch reactor for treatment of azo dyes containing simulated wastewater. *Process Biochemistry*, *40*(2), 885–890. doi:10.1016/j.procbio.2004.02.015

Sangamithirai, K. M., Jayapriya, J., Hema, J., & Manoj, R. (2015). Evaluation of in-vessel co-composting of yard waste and development of kinetic models for co-composting. *International Journal of Recycling of Organic Waste in Agriculture*, *4*(3), 157–165. doi:10.1007/s40093-015-0095-1

Sangster, J. (1997). *Octanol-water partition coefficients: Fundamentals and physical chemistry.* John Wiley and Sons.

Santos, A. C., Bezerra, M. S., dos Santos Pereira, H., Dos Santos, E. S., & De Macedo, G. R. (2010). Production and recovery of rhamnolipids using sugar cane molasses as carbon source. *Journal of Chemistry and Chemical Engineering*, *4*(11), 27–33.

Santos, D. K. F., Rufino, R. D., Luna, J. M., Santos, V. A., & Sarubbo, L. A. (2016). Biosurfactants: Multifunctional Biomolecules of the 21st Century. *International Journal of Molecular Sciences*, *17*(3), 401. doi:10.3390/ijms17030401 PMID:26999123

Santos, D. K., Rufino, R. D., Luna, J. M., Santos, V. A., Salgueiro, A. A., & Sarubbo, L. A. (2013). Synthesis and evaluation of biosurfactant produced by *Candida lipolytica* using animal fat and corn steep liquor. *Journal of Petroleum Science Engineering*, *105*, 43–50. doi:10.1016/j.petrol.2013.03.028

Santos, P. M., Benndorf, D., & Sá-Correia, I. (2004). Insights into Pseudomonas putida KT2440 response to phenol-induced stress by quantitative proteomics. *Proteomics*, *4*(9), 2640–2652. doi:10.1002/pmic.200300793 PMID:15352239

Saravanane, R., Murthy, D. V., & Krishnaiah, K. (2001a). Bioaugmentation and anaerobic treatment of pharmaceutical effluent in fluidized bed reactor. *Journal of Environmental Science and Health. Part A, Toxic/Hazardous Substances & Environmental Engineering*, *36*(5), 779–791. and doi:10.1081/ESE-100103760

Saravanane, R., Murthy, D. V., & Krishnaiah, K. (2001b). Bioaugmentation and treatment of cephalexin drug-based pharmaceutical effluent in an upflow anaerobic fluidized bed system. *Bioresource Technology*, *76*(3), 279–281. and doi:10.1016/S0960-8524(00)00121-8

Saravanane, R., Murthy, D. V., & Krishnaiah, K. (2001c). Treatment of anti-osmotic drug based pharmaceutical effluent in an upflow anaerobic fluidized bed system. *Waste Management (New York, N.Y.)*, *21*(6), 563–568. and doi:10.1016/S0956-053X(00)00109-4

Saravanan, P., Pakshirajan, K., & Saha, P. (2008). Growth kinetics of an indigenous mixed microbial consortium during phenol degradation in a batch reactor. *Bioresource Technology*, *99*(1), 205–209. doi:10.1016/j.biortech.2006.11.045 PMID:17236761

Saravanan, V., & Subramaniyan, V. (2014). Production of biosurfactant by *Pseudomonas aeruginosa* PB3A using agro-industrial wastes as a carbon source. *Malays. J. Microbiol.*, *10*, 57–62.

Sarikaya, A., & Ladisch, M. R. (1997). Mechanism and potential applications of bio-ligninolytic systems in a CELSS. *Applied Biochemistry and Biotechnology*, *62*(2-3), 131–149. doi:10.1007/BF02787990 PMID:9170251

Sarkar, A., Ravindran, G., & Krishnamurthy, V. (2013). A brief review on the effect of cadmium toxicity: From cellular to organ level. *International Journal of Bio-Technology and Research*, *3*(1), 17–36.

Sarkar, S., Witham, S., Zhang, J., Zhenirovskyy, M., Rocchia, W., & Alexov, E. (2013). DelPhi Web Server: A comprehensive online suite for electrostatic calculations of biological macromolecules and their complexes. *Communications in Computational Physics*, *13*(01), 269–284. doi:10.4208/cicp.300611.201011s PMID:24683424

Sarma, H. (2011). Metal hyperaccumulation in plants: A review focusing on phytoremediation technology. *Journal of Environmental Science and Technology*, *4*(2), 118–138. doi:10.3923/jest.2011.118.138

Sarma, P. M., Bhattacharya, D., Krishnan, S., & Lal, B. (2004). Assessment of intra-species diversity among strains of *Acinetobacter baumannii* isolated from sites contaminated with petroleum hydrocarbons. *Canadian Journal of Microbiology*, *50*(6), 405–414. doi:10.1139/w04-018 PMID:15284886

Sarwade, V.D., & Gawai, K.R. (2014). Biodegradation of phenol by alkaliphilic *Bacillus badius* D1IOSR. *Journal of Environmental Science, Toxicology and Food Technology*, *8*(2), 28-35.

Satpute, S. K., Banat, I. M., Dhakephalkar, P. K., Banpurkar, A. G., & Chopade, B. A. (2010). Biosurfactants, bioemulsifiers and exopolysaccharides from marine microorganisms. *Biotechnology Advances*, *28*(4), 436–450. doi:10.1016/j.biotechadv.2010.02.006 PMID:20172021

Satyanarayana, T., Raghukumar, C., & Shivaji, S. (2005). Extremophilic microbes: Diversity and perspectives. *Current Science*, *89*, 78–90.

Satyanarayan, S., Karambe, A., & Vanerkar, A. P. (2009). Herbal pharmaceutical wastewater treatment by a pilot scale upflow anaerobic sludge blanket (UASB) reactor. *Water Science and Technology*, *59*(11), 2265–2272. and doi:10.2166/wst.2009.273

Sauer, S., Lange, B. M., Gobom, J., Nyarsik, L., Seitz, H., & Lehrach, H. (2005). Miniaturization in functional genomics and proteomics. *Nature Reviews. Genetics*, *6*(6), 465–476. doi:10.1038/nrg1618 PMID:15931170

Savage, D. F., Afonso, B., Chen, A. H., & Silver, P. A. (2010). Spatially ordered dynamics of the bacterial carbon fixation machinery. *Science*, *327*(5970), 1258–1261. doi:10.1126/science.1186090 PMID:20203050

Savin, M. C., Wolf, D. C., Davis, K. J., Gbur, E. E., & Thoma, G. J. (2015). Nematodes as Bioindicators of Ecosystem Recovery During Phytoremediation of Crude Oil Contaminated Soil. *International Journal of Phytoremediation*, *17*(2), 182–190. doi:10.1080/15226514.2013.876964 PMID:25361231

Saxena, P. K. KrishnaRaj, S., Dan, T., Perras, M. R., & Vettakkorumakankav, N. N. (1999). Phytoremediation of heavy metal contaminated and polluted soils. In Heavy metal stress in plants (pp. 305-329). Springer Berlin Heidelberg.

Sayad, S., & Ellouz, R. (1995). Roles of lignin peroxidase and manganese peroxidase from phanerochaete chrysosporium in the decolorization of olive mill wastewaters. *Applied and Environmental Microbiology*, *61*(3), 1098–1103.

Sayler, G. S., & Ripp, S. (2000). Field applications of genetically engineered microorganisms for bioremediation processes. *Current Opinion in Biotechnology*, *11*(3), 286–289. doi:10.1016/S0958-1669(00)00097-5 PMID:10851144

Schechter, I. (2005). Mapping of the active site of proteases in the 1960s and rational design of inhibitors/drugs in the 1990s. *Current Protein & Peptide Science*, *6*(6), 501–512. doi:10.2174/138920305774933286 PMID:16381600

Scheibenbogen, K., Zybner, R. G., Lee, H., & Trevors, J. J. (1994). Enhanced removal of selected hydrocarbons from soil by *Pseudomonas aeruginosa* UG 2 biosurfactants and some chemical surfactants. *Journal of Chemical Technology and Biotechnology (Oxford, Oxfordshire)*, *59*(1), 55–59. doi:10.1002/jctb.280590109

Schleppi, P., Wallin, I. B., Hagedorn, F., & Korner, C. (2012). Increased nitrate availability in the soil of a mixed mature temperate forest subjected to elevated CO2 concentration (canopy FACE). *Global Change Biology*, *18*(2), 757–768. doi:10.1111/j.1365-2486.2011.02559.x

Schmidt, M. W. I., Torn, M. S., Abiven, S., Dittmar, T., Guggenberger, G., Janssens, I. A., & Trumbore, S. E. et al. (2011). Persistence of soil organic matter as an ecosystem property. *Nature*, *478*(7367), 49–56. doi:10.1038/nature10386 PMID:21979045

Schmidt, O. (2006). *Wood and Tree Fungi. Biology, Damage, Protection, and Use*. Heidelberg, Germany: Springer.

Schnecke, V., & Kuhn, L. A. (2000). Virtual screening with solvation and ligand-induced complementarity. *Perspectives in Drug Discovery and Design*, *20*(1), 171–190. doi:10.1023/A:1008737207775

Schneider, T., Keiblinger, K. M., Schmid, E., Streflinger-Gleixner, K., Ellersdorfer, G., Roschitzki, B., & Riedel, K. et al. (2012). Who is who in litter decomposition? Metaproteomics reveals major microbial players and their biogeochemical functions. *The ISME Journal*, *6*(9), 1749–1762. doi:10.1038/ismej.2012.11 PMID:22402400

Schneider, T., Schellenberg, M., Meyer, S., Keller, F., Gehrig, P., Riedel, K., & Martinoia, E. (2009). Quantitative detection of changes in the leaf-mesophyll tonoplast proteome in dependency of a cadmium exposure of barley (Hordeum vulgare L.) plants. *Proteomics*, *9*(10), 2668–2677. doi:10.1002/pmic.200800806 PMID:19391183

Schneidman-Duhovny, D., Inbar, Y., Nussinov, R., & Wolfson, H. J. (2005). PatchDock and SymmDock: Servers for rigid and symmetric docking. *Nucleic Acids Research*, *33*(Web Serversuppl 2), W363–W367. doi:10.1093/nar/gki481 PMID:15980490

Schöftner, P., Watzinger, A., Holzknecht, P., Wimmer, B., & Reichenauer, T. G. (2015). Transpiration and metabolisation of TCE by willow plants–a pot experiment. *International Journal of Phytoremediation*.

Scholz, M., & Martin, R. J. (1997). Ecological Equilibrium on Biological Activated Carbon. *Water Research*, *31*(12), 2959–2968. doi:10.1016/S0043-1354(97)00155-3

Schröder, H. F., Tambosi, J. L., Sena, R. F., Moreira, R. F., José, H. J., & Pinnekamp, J. (2012). The removal and degradation of pharmaceutical compounds during membrane bioreactor treatment. *Water Science and Technology*, *65*(5), 833–839. and doi:10.2166/wst.2012.828

Schwarze, F. W. M. R., Engels, J., & Mattheck, C. (2000). *Fungal strategies of decay in trees*. Berlin: Springer. doi:10.1007/978-3-642-57302-6

Schwarzenbach, R. P., Gschwend, P. M., & Imoden, D. M. (2004). Environmental Organic Chemistry (2nd ed.). J. Wiley and Sons.

Scoparo, C. H. G., Queiroz, M. I., & Franco, T. T. (2010). Biotransformations of Cdioxide in photobioreactors. *Energy Conversion and Management*, *51*(5), 894–900. doi:10.1016/j.enconman.2009.11.027

Scott, C., Lewis, S. E., Milla, R., Taylor, M. C., Rodgers, A. J., Dumsday, G., & Russell, R. J. et al. (2010). A free-enzyme catalyst for the bioremediation of environmental atrazine contamination. *Journal of Environmental Management*, *91*(10), 2075–2078. doi:10.1016/j.jenvman.2010.05.007 PMID:20570036

Scott, C., Pandey, G., Hartley, C. J., Jackson, C. J., Cheesman, M. J., Taylor, M. C., & Oakeshott, J. G. et al. (2008). The enzymatic basis for pesticide bioremediation. *Indian Journal of Microbiology*, *48*(1), 65–79. doi:10.1007/s12088-008-0007-4 PMID:23100701

Scott, G. M., Akhtar, M., Lentz, M. J., Kirk, T. K., & Swaney, R. (1998). New technology for paper making: Commercializing biopulping. *Tappi J*, *81*, 220–225.

Scott, H. D., Wolf, D. C., & Lavy, T. L. (1982). Adsorption and degradation of phenol at low concentration in soil. *Environ. Qual.*, *11*, 107–111. doi:10.2134/jeq1982.00472425001100010025x

Secretariat of the Convention on Biological Diversity (SCBD). (2009). Scienti- c Synthesis of the Impacts of Ocean Fertilization on Marine Biodiversity. Technical Series No. 45. Author.

Sedlak, D. L., Gray, J. L., & Pinkston, K. E. (2000). Understanding microcontaminants in recycled water. *Environmental Science & Technology*, *34*(23), 509A–515A. doi:10.1021/es003513e PMID:21662317

Seghezzo, L., Zeeman, G., van Lier, J. B., Hamelers, H. V. M., & Lettinga, G. (1998). A review: The anaerobic treatment of sewage in UASB and EGSB reactors. *Bioresource Technology*, *65*(3), 175–190. doi:10.1016/S0960-8524(98)00046-7

Segura, A., & Ramos, J. L. (2013). Plant–bacteria interactions in the removal of pollutants. *Current Opinion in Biotechnology*, *24*(3), 467–473. doi:10.1016/j.copbio.2012.09.011 PMID:23098915

Semple, K. T., & Cain, R. B. (1996). Biodegradation of Phenols by the Alga *Ochromonas danica. Applied and Environmental Microbiology*, *62*(4), 1265–1273. PMID:8919787

Sengupta, J. (2002). Recycling of Agro- industrial wastes for manufacturing of building materials and components in India: An overview. *Civil Engineering & Construction Review*, *15*(2), 23–33.

Senthil, K. R., & Parthiban, R. (2011). Removal of zinc (Zn++) ions with crab shell particles from aqueous solution. *International Journal of Environmental Sciences*, *1*(7), 1965–1977.

Seo, J. S., Keum, Y. S., & Li, Q. X. (2009). Bacterial degradation of aromatic compounds. *International Journal of Environmental Research and Public Health*, *6*(1), 278–309. doi:10.3390/ijerph6010278 PMID:19440284

Seufferheld, M. J., & Curzi, M. J. (2010). Recent discoveries on the roles of polyphosphates in plants. *Plant Molecular Biology Reporter*, *28*(4), 549–559. doi:10.1007/s11105-010-0187-z

Shailubhai, K. (1986). *Treatment of Petroleum industry oil sludge in soil, Tibtech August 1986*. Amsterdam: Elsevier Science Publishers B.V.

Shamsuddoha, A. S. M., Bulbul, A., & Huq, S. M. I. (2006). Accumulation of arsenic in green algae and its subsequent transfer to the soil-plant system. *Bangladesh Journal of Medical Microbiology*, *22*(2), 148–151.

Shankar, K. S., Devi, Ch. A., & Rao, C. S. V. R. (2011). Phytoremediation of heavy metals with *Amaranthus dubius* in semi-arid soils of Patancheru, Andhra Pradesh. *Indian Journal of Dryland Agricultural Research and Development*, *26*(2), 71–76.

Shao, J., Chen, J., Li, T., & Zhao, X. (2014). Spectroscopic and molecular docking studies of the *in vitro* interaction between puerarin and cytochrome P450. *Molecules (Basel, Switzerland)*, *19*(4), 4760–4769. doi:10.3390/molecules19044760 PMID:24743933

Sharholy, M., Ahmad, K., Mahmood, G., & Trivedi, R. C. (2008). Municipal solid waste management in Indian cities. A review. *Journal of Waste Management*, *28*(2), 459–467. doi:10.1016/j.wasman.2007.02.008 PMID:17433664

Shariati, F. P., Mehrnia, M. R., Salmasi, B. M., Heran, M., Wisniewski, C., & Sarrafzadeh, M. H. (2010). Membrane bioreactor for treatment of pharmaceutical wastewater containing acetaminophen. *Desalination*, *250*(2), 798–800. doi:10.1016/j.desal.2008.11.044

Sharma, A., Mishra, M., Sheet, S., & Thite, M. (2013). Role of microbes as cleaning degrading industrial wastes for environmental sustainability- a review. *Recent Res. Sci. Technol*, *5*, 21–25.

Sharma, D. K., & Wadhwa, G. (1997). Demineralization of coal by stepwise bioleaching: A comparative study of three Indian coals by Fourier transform infrared and X-ray diffraction techniques. *World Journal of Microbiology & Biotechnology*, *13*(1), 29–36. doi:10.1007/BF02770804

Sharma, H., Barber, J. T., Ensley, H. E., & Polito, M. A. (1997). A comparison of the toxicity of phenol and chlorinated phenols by *Lemna gibba*, with reference to 2,4,5-trichlorophenol. *Environmental Toxicology and Chemistry*, *16*, 346–350. doi:10.1002/etc.5620160233

Sharma, R. K., & Arora, D. S. (2015). Fungal degradation of lignocellulosic residues: An aspect of improved nutritive quality. *Critical Reviews in Microbiology*, *41*(1), 52–60. doi:10.3109/1040841X.2013.791247 PMID:23855359

Shashirekha, S., Uma, L., & Subramaniam, G. (1997). Phenol degradation by the marine cyanobacterium *Phormidium valderianum* BDU 30501, Ind. *Microbial Biotechnology*, *19*(2), 130–133. doi:10.1038/sj.jim.2900438

Sheeja, R. Y., & Murugesan, T. (2002). Mass transfer studies on the biodegradation of phenols in up-flow packed bed reactors. *Journal of Hazardous Materials*, *B89*(2-3), 287–301. doi:10.1016/S0304-3894(01)00319-3 PMID:11744212

Shen, X., Liu, Z., & Wang, B. (2004). Isolation, identification of phenol degrading *Rhadococcus* sp. strain PNAN5 and characterization of its ring cleavage dioxygenases. *Acta Scientiae Circumstantiae*, *24*, 482–486.

Shen, Y., West, C., & Hutchins, S. R. (2000). In vitro cytotoxicity of aromatic aerobic biotransformation products in bluegill sunfish BF-2 cells. *Ecotoxicology and Environmental Safety*, *45*(1), 27–32. doi:10.1006/eesa.1999.1843 PMID:10677264

Sheppard, S. (1997).Toxicity testing using microcosms. In Soil Toxicology (pp. 345-374). Lewis Publishers.

Shibata, A., Inoue, Y., & Katayama, A. (2006). Aerobic and anaerobic biodegradation of phenol derivatives in various paddy soils. *The Science of the Total Environment*, *367*(2-3), 979–987. doi:10.1016/j.scitotenv.2006.01.031 PMID:16530810

Shimada, T., Kim, D., Murayama, N., Tanaka, K., Takenaka, S., Nagy, L. D., & Guengerich, F. P. et al. (2013). Binding of diverse environmental chemicals with human cytochromes P450 2A13, 2A6, and 1B1 and enzyme inhibition. *Chemical Research in Toxicology*, *26*(4), 517–528. doi:10.1021/tx300492j PMID:23432429

Shimao, M. (2001). Biodegradation of plastics. *Current Opinion in Biotechnology*, *12*(3), 242–247. doi:10.1016/S0958-1669(00)00206-8 PMID:11404101

Shimp, R. J., & Pfaender, F. K. (1982). Effects of surface area and flow rate on marinal bacterial growth in activated carbon columns. *Applied and Environmental Microbiology*, *44*, 471–477. PMID:16346080

Shinoda, T., Sakai, Y., Ue, M., Hiraishi, A., & Kato, N. (2000). Isolation and characterization of a new denitrifying *spirillum* capable of anaerobic degradation of phenol. *Applied and Environmental Microbiology*, *66*(4), 1286–1291. doi:10.1128/AEM.66.4.1286-1291.2000 PMID:10742201

Shivaraman, N., & Pandey, R. A. (2000). Characterization and biodegradation of phenolic wastewater, J. Ind. Assos. *Environmental Management*, *27*, 12.

Shi, X., Lefebvre, O., Ng, K. K., & Ng, H. W. (2014). Sequential anaerobic–aerobic treatment of pharmaceutical wastewater with high salinity. *Bioresource Technology*, *153*, 79–86. and doi:10.1016/j.biortech.2013.11.045

Shrestha, B., Kour, J., Homagai, P. L., Pokhrel, M. R., & Ghimire, K. N. (2013). Surface modification of the biowaste for purification of wastewater contaminated with toxic heavy metals – lead and cadmium. *Advances in Chemical Engineering and Science*, *3*(03), 178–184. doi:10.4236/aces.2013.33022

Shukla, K. P., Sharma, S., Singh, N. K., Singh, V., Bisht, S., & Kumar, V. (2013). Rhizoremediation: A Promising Rhizosphere Technology. In Y. B. Patil & P. Rao (Eds.), *Applied Bioremediation-active and passive approaches, 331*. doi:10.5772/56905

Siddiquee, S., Rovina, K., Azad, S. A., Naher, L., Suryani, S., & Chaikaew, P. (2015). Heavy metal contaminants removal from wastewater using the potential filamentous fungi biomass: A review. *Journal of Microbial and Biochemical Technology*, *7*(6), 384–393. doi:10.4172/1948-5948.1000243

Siddique, T., Okeke, B. C., Arshad, M., & Frankenberger, W. T. (2003). Enrichment and isolation of endosulfan-degrading microorganisms. *Journal of Environmental Quality*, *32*(1), 47–54. doi:10.2134/jeq2003.4700 PMID:12549541

Siegenthaler, U., & Sarmiento, J. (1993). Atmospheric Cdioxide and the ocean. *Nature*, *365*(6442), 119–125. doi:10.1038/365119a0

Sierra-Alvarez, R., Field, J. A., Kortekaas, S., & Lettinga, G. (1994). Overview of the anaerobic toxicity caused by organic forest industry wastewater pollutants. *Water Science and Technology*, *29*, 353–363.

Silva-Castro, G. A., Rodelas, B., Perucha, C., Laguna, J., González-López, J., & Calvo, C. (2013). Bioremediation of diesel-polluted soil using biostimulation as post-treatment after oxidation with Fenton-like reagents: Assays in a pilot plant. *The Science of the Total Environment*, *445*, 347–355. doi:10.1016/j.scitotenv.2012.12.081 PMID:23354375

Silva, E. J., Silva, N. M. P. R., Rufino, R. D., Luna, J. M., Silva, R. O., & Sarubbo, L. A. (2014). Characterization of a biosurfactant produced by *Pseudomonas cepacia* CCT6659 in the presence of industrial wastes and its application in the biodegradation of hydrophobic compounds in soil. *Colloids and Surfaces. B, Biointerfaces*, *117*, 36–41. doi:10.1016/j.colsurfb.2014.02.012 PMID:24613853

Silva, R. D. C. F., Almeida, D. G., Rufino, R. D., Luna, J. M., Santos, V. A., & Sarubbo, L. A. (2014). Applications of biosurfactants in the petroleum industry and the remediation of oil spills. *International Journal of Molecular Sciences*, *15*(7), 12523–12542. doi:10.3390/ijms150712523 PMID:25029542

Simon, M. (2015). Lignocellulose degradation mechanisms across the Tree of Life. *Current Opinion in Chemical Biology*, *29*, 108–119. doi:10.1016/j.cbpa.2015.10.018 PMID:26583519

Sims, R. W., & Gerard, B. M. (1999). *Earthworms: Notes for the identification of British species* (4th ed.). Shrewsbury, UK: Published for The Linnean Society of London and The Estuarine and Coastal Sciences Association by Field Studies Council, Montford Bridge.

Singer, A. C., Smith, D., Jury, W. A., Hathuc, K., & Crowley, D. E. (2003). Impact of the plant rhizosphere and augmentation on remediation of polychlorinated biphenyl contaminated soil. *Environmental Toxicology and Chemistry*, *22*(9), 1998–2004. doi:10.1897/02-471 PMID:12959523

Singer, A., Gilbert, E., Luepromchai, E., & Crowley, D. (2000). Bioremediation of polychlorinated biphenyl-contaminated soil using carvone and surfactant-grown bacteria. *Applied Microbiology and Biotechnology*, *54*(6), 838–843. doi:10.1007/s002530000472 PMID:11152078

Singh, G., Singh, A. K., & Bhatt, K. (n.d.). Biodegradation of polythenes by bacteria isolated from soil. *International Journal of Research and Development in Pharmacy and Life Sciences*, *5*(2), 2056-2062.

Singh, A., & Gauba, P. (2014). Mycoremediation: A Treatment for Heavy Metal Pollution of Soil. *Journal of Civil Engineering and Environment Technology*, *1*(4), 59–61.

Singh, A., Kuhad, R. C., & Ward, O. P. (2009). Biological remediation of soil: an overview of global market and available technologies. In A. Singh & C. Ramesh et al. (Eds.), *Advances in applied bioremediation* (pp. 1–19). Springer. doi:10.1007/978-3-540-89621-0_1

Singh, B. K., Dawson, L. A., Macdonald, C. A., & Buckland, S. M. (2009a). Impact of biotic and abiotic interaction on soil microbial communities and functions: A field study. *Applied Soil Ecology*, *41*(3), 239–248. doi:10.1016/j.apsoil.2008.10.003

Singh, B., Singh, Y., & Sekhon, G. S. (1995). Fertilizer-N use efficiency and nitrate pollution of groundwater in developing countries, Integrated Nitrogen Management in Relation to Leaching and Groundwater Quality. *Journal of Contaminant Hydrology*, *20*(3–4), 167–184. doi:10.1016/0169-7722(95)00067-4

Singh, D. (2008). Biodegradation and bioremediation of pesticide in soil: Concept, method and recent developments. *Indian Journal of Microbiology*, *48*(1), 35–40. doi:10.1007/s12088-008-0004-7 PMID:23100698

Singh, D., & Chen, S. (2008). The white-rot fungus Phanerochaete chrysosporium: Conditions for the production of lignin-degrading enzymes. *Applied Microbiology and Biotechnology*, *81*(3), 399–417. doi:10.1007/s00253-008-1706-9 PMID:18810426

Singh, D., & Fulekar, M. H. (2009). Bioremediation of phenol by a novel partitioning bioreactor using cow dung microbial consortia. *Biotechnology Journal*, *4*(3), 423–431. and doi:10.1002/biot.200800204

Singh, D., & Fulekar, M. H. (2010). Benzene bioremediation using cow dung microflora in two phase partitioning bioreactor. *Journal of Hazardous Materials*, *175*(1-3), 336–343. and doi:10.1016/j.jhazmat.2009.10.008

Singh, D., & Fulekar, M. H. (2010). Biodegradation of petroleum hydrocarbons by Pseudomonas putida strain MHF 7109 isolated from cow dung microbial consortium. *Clean (Weinh)*, *38*(8), 781–786.

Singh, D., Rawat, S., Waseem, M., Gupta, S., Lynn, A., Nitin, M., & Sharma, K. K. et al. (2016). Molecular modeling and simulation studies of recombinant laccase from *Yersinia enterocolitica* suggests significant role in the biotransformation of non-steroidal anti-inflammatory drugs. *Biochemical and Biophysical Research Communications*, *469*(2), 306–312. doi:10.1016/j.bbrc.2015.11.096 PMID:26631965

Singh, D., Sharma, K. K., Jacob, S., & Gakhar, S. K. (2014). Molecular docking of laccase protein from *Bacillus Safensis* DSKK5 Isolated from earthworm gut: A novel method to study dye decolorization potential. *Water, Air, and Soil Pollution*, *225*(11), 1–12. doi:10.1007/s11270-014-2175-7

Singh, O. M., Nagaraj, V., & Nagathihalli, S. (2006). Transcriptomics, proteomics and interactomics: Unique approaches to track the insights of bioremediation. *Briefings in Functional Genomics & Proteomics*, *4*(4), 355–362. doi:10.1093/bfgp/eli006 PMID:17202126

Singh, O. V. (2006). Proteomics and metabolomics: The molecular make-up of toxic aromatic pollutant bioremediation. *Proteomics*, *6*(20), 5481–5492. doi:10.1002/pmic.200600200 PMID:16972298

Singh, O. V., Labana, S., Pandey, G., Budhiraja, R., & Jain, R. K. (2003). Phytoremediation: An overview of metallic ion decontamination from soil. *Applied Microbiology and Biotechnology*, *61*(5-6), 405–412. doi:10.1007/s00253-003-1244-4 PMID:12764555

Singh, R., Paul, D., & Jain, R. K. (2006). Biofilms: Implications in bioremediation. *Trends in Microbiology*, *14*(9), 389–397. doi:10.1016/j.tim.2006.07.001 PMID:16857359

Singh, R., Singh, P., & Sharma, R. (2014). Microorganism as a tool of bioremediation technology for cleaning environment. *RE:view*, *4*(1), 1–6.

Singh, R., Singh, P., & Sharma, R. (2014). Microorganism as a tool of bioremediation technology for cleaning environment: A review. *Proceedings of the International Academy of Ecology and Environmental Sciences*, *4*(1), 1–6.

Singh, R., Singh, P., Sharma, R., Selvalakshmi, S., Jayakumar, S., & Ramachandran, V. S. (2014). Microorganism as a tool of bioremediation technology for cleaning environment: A review. *Proceedings of* International Academy *of* Ecology *and* Environmental. *The Sciences*, *4*, 1–6.

Singh, S., & Gupta, V. K. (2016). Biodegradation and bioremediation of pollutants: Perspectives strategies and applications. *International Journal of Pharmacology and Biological Sciences*, *10*(1), 53–65.

Singh, V., Mani, I., & Chaudhary, D. K. (2012). Analysis of the multicopper oxidase gene regulatory network of *Aeromonas hydrophila. Systems and Synthetic Biology*, *6*(3-4), 51–59. doi:10.1007/s11693-012-9096-9 PMID:24294339

Sinha, J., Reyes, S. J., & Gallivan, J. P. (2010). Reprogramming bacteria to seek and destroy an herbicide. *Nature Chemical Biology*, *6*(6), 464–470. doi:10.1038/nchembio.369 PMID:20453864

Sinha, R. K., Agarwal, S., Chauhan, K., & Valani, D. (2010). The wonders of earthworms &its vermicompost to farm production:Charles Darwins friends of farmers with potential to replace destructive chemical fertilizers. *Agricultural Sciences*, *1*(02), 76–94. doi:10.4236/as.2010.12011

Sinha, R. K., Valani, D., Sinha, S., Singh, S., & Heart, S. (2009). Bioremediation of contaminated sites: A low-cost nature's biotechnology for environmental clean up by versatile microbes, plants & earthworms. In T. Faerber & J. Herzog (Eds.), *Solid Waste Management and Environmental Remediation* (pp. 1–72). Nova Science.

Sinsabaugh, R. L., & Shah, J. J. (2010). Integrating resource utilization and temperature in metabolic scaling of riverine bacterial production. *Ecology*, *91*(5), 1455–1465. doi:10.1890/08-2192.1 PMID:20503877

Sirtori, C., Zapata, A., Oller, I., Gernjak, W., Aguera, A., & Malato, S. (2009). Decontamination industrial pharmaceutical wastewater by combining solar photo-Fenton and biological treatment. *Water Research*, *43*(3), 661–668. and doi:10.1016/j.watres.2008.11.013

Sivapathasekaran, C., Mukherjee, S., Ray, A., Gupta, A., & Sen, R. (2010). Artificial neural network modeling and genetic algorithm based medium optimization for the improved production of marine biosurfactant. *Bioresource Technology*, *101*(8), 2884–2887. doi:10.1016/j.biortech.2009.09.093 PMID:19914826

Sivasubramanian, V. (2011). Microalgae for potential phycoremediation of Industrial effluent - A case study. *ENVIS CENTRE Newsletter, 9*(4).

Sivasubramanian, V., Subramanian, V. V., & Muthukumaran, M. (2010). Bioremediation of chrome-sludge from an electroplating industry using the micro alga *Desmococcus olivaceus* - A pilot study. *J Algal Biomass Util*, *1*(3), 104–128.

Siwela, A. H., Nyathi, C. B., & Naik, Y. S. (2009). Metal accumulation and antioxidant enzyme activity in *C. gariepinus*, Catfish, and *O. mossambicus*, tilapia, collected from lower Mguza and Wright Dams, Zimbabwe. *Bulletin of Environmental Contamination and Toxicology*, *83*(5), 648–651. doi:10.1007/s00128-009-9861-y PMID:19768358

Six, J., Frey, S. D., Thiet, R. K., & Batten, K. M. (2006). Bacterial and Fungal Contributions to CSequestration in Agroecosystems. *Soil Science Society of America Journal*, *70*(2), 555–569. doi:10.2136/sssaj2004.0347

Slater, H., Gouin, T., & Leigh, M. B. (2011). Assessing the potential for rhizoremediation of PCB contaminated soils in northern regions using native tree species. *Chemosphere*, *84*(2), 199–206. doi:10.1016/j.chemosphere.2011.04.058 PMID:21596420

Sliwoski, G., Kothiwale, S., Meiler, J., & Lowe, E. W. (2014). Computational methods in drug discovery. *Pharmacological Reviews*, *66*(1), 334–395. doi:10.1124/pr.112.007336 PMID:24381236

Smalla, K., Wieland, G., Buchner, A., Zock, A., Parzy, J., Kaiser, S., & Berg, G. et al. (2001). Bulk and rhizosphere soil bacterial communities studied by denaturing gradient gel electrophoresis: Plant-dependent enrichment and seasonal shifts revealed. *Applied and Environmental Microbiology*, *67*(10), 4742–4751. doi:10.1128/AEM.67.10.4742-4751.2001 PMID:11571180

Smith, J. L., & Paul, E. A. (1990). The significance of soil microbial biomass estimations. In J. Bollag & G. Stotsky (Eds.), *Soil biochemistry* (Vol. 8, pp. 357–393). Mercel Dekker, New York.

Smith, K. S., Jakubzick, C., Whittam, T. S., & Ferry, J. G. (1999). Carbonic anhydrase is an ancient enzyme widespread in prokaryotes. *Proceedings of the National Academy of Sciences of the United States of America*, *96*(26), 15184–15189. doi:10.1073/pnas.96.26.15184 PMID:10611359

Smith, N., Witham, S., Sarkar, S., Zhang, J., Li, L., Li, C., & Alexov, E. (2012). DelPhi web server v2: Incorporating atomic-style geometrical figures into the computational protocol. *Bioinformatics (Oxford, England)*, *28*(12), 1655–1657. doi:10.1093/bioinformatics/bts200 PMID:22531215

Smith, R. D., Damm-Ganamet, K. L., Dunbar, J. B. Jr, Ahmed, A., Chinnaswamy, K., Delproposto, J. E., & Doyle, L. et al. (2016). CSAR benchmark exercise 2013: Evaluation of results from a combined computational protein design, docking, and scoring/ranking challenge. *Journal of Chemical Information and Modeling*, *56*(6), 1022–1031. doi:10.1021/acs.jcim.5b00387 PMID:26419257

Smith, S. E., & Read, D. J. (1997). *Mycorrhizal symbiosis*. San Diego, CA: Academic Press.

Snyder, S. A., Adham, S., Redding, A. M., Cannon, F. S., De Carolis, J., Oppenheimer, J., & Yoon, Y. et al. (2007). Role of membranes and activated carbon in the removal of endocrine disruptors and pharmaceuticals. *Desalination*, *202*(1-3), 156–181. doi:10.1016/j.desal.2005.12.052

Snyder, S. A., Westerhoff, P., Yoon, Y., & Sedlak, D. L. (2003). Pharmaceuticals, personal care products and endocrine disruptors in water:implications for the water industry. *Environmental Engineering Science*, *20*(5), 449–469. doi:10.1089/109287503768335931

Soberón-Chávez, G., & Maier, R. M. (2011). Biosurfactants: a general overview. In Biosurfactants (pp. 1-11). Springer Berlin Heidelberg. doi:10.1007/978-3-642-14490-5_1

Sobrinho, H. B., Luna, J. M., Rufino, R. D., Porto, A. L. F., & Sarubbo, L. A. (2013). Biosurfactants: Classification, properties and environmental applications. *Recent Developments in Biotechnology*, *11*, 1–29.

Solovchenko, A., Pogosyan, S., Chivkunova, O., Selyakh, I., Semenova, L., Voronova, E., & Lobakova, E. et al. (2014). Phycoremediation of alcohol distillery wastewater with a novel *chlorella sorokiniana* strain cultivated in a photobioreactor monitored on-line via chlorophyll fluorescence. *Algal Res*, *6*, 234–241. doi:10.1016/j.algal.2014.01.002

Somiari, R. I., Somiari, S., Russell, S., & Shriver, C. D. (2005). Proteomics of breast carcinoma. *Journal of Chromatography. B, Analytical Technologies in the Biomedical and Life Sciences*, *815*(1), 215–225. doi:10.1016/j.jchromb.2004.11.012 PMID:15652811

Sommer, R., & Bossio, D. (2014). Dynamics and climate change mitigation potential of soil organic Csequestration. *Journal of Environmental Management*, *144*, 83–87. doi:10.1016/j.jenvman.2014.05.017 PMID:24929498

Somsamak, P., Cowan, R. M., & Haggblom, M. M. (2001). Anaerobic biotransformation of fuel oxygenates under sulfate-reducing conditions. *FEMS Microbiology Ecology*, *37*, 259–264. doi:10.1111/j.1574-6941.2001.tb00873.x

Sonawdekar, S. (2012). Bioremediation: A boon to hydrocarbon degradation. *International Journal of Environmental Sciences*, *2*(4), 2408–2424.

Sönmez, İ., Kaplan, M., & Sönmez, S. (2007). An investigation of seasonal changes in nitrate contents of soils and irrigation waters in greenhouses located in antalya-demre region. *Asian Journal of Chemistry*, *19*(7), 5639–5646.

Sosa Alderete, L. G., Ibáñez, S. G., Agostini, E., & Medina, M. I. (2012). Phytoremediation of phenol at pilot scale by tobacco hairy roots. *International Journal of Environmental Sciences*, *3*(1), 398–407.

Sousa, M., Melo, V. M. M., Rodrigues, S., Santana, H. B., & Gonçalves, L. R. B. (2012). Screening of biosurfactant-producing Bacillus strains using glycerol from the biodiesel synthesis as main carbon source. *Bioprocess and Biosystems Engineering*, *35*(6), 897–906. doi:10.1007/s00449-011-0674-0 PMID:22218992

Souza, E. C., Vessoni-Penna, T. C., & de Souza Oliveira, R. P. (2014). Biosurfactant-enhanced hydrocarbon bioremediation: An overview. *International Biodeterioration & Biodegradation*, *89*, 88–94. doi:10.1016/j.ibiod.2014.01.007

Spain, A., & Alm, E. (2003). Implications of microbial heavy metal tolerance in the environment. *Review of Undergraduate Research*, *2*, 1–6.

Spina, F., Anastasi, A., Prigione, V., Tigini, V., & Varese, G. C. (2012). Biological treatment of industrial wastewaters: A fungal approach. *Chemical Engineering Transactions*, *27*, 175–180.

Spurgeon, D. J., & Hopkin, S. P. (1999). Tolerance to zinc in populations of earthworm Lumbricusrubellus from uncontaminated and metal- contaminated ecosystem. *Archives of Environmental Contamination and Toxicology*, *37*(3), 332–337. doi:10.1007/s002449900522 PMID:10473789

Srebotnik, E., Messner, K., & Foisner, R. (1988). Penetrability of white rot-degraded pine wood by the lignin peroxidase of *Phanerochaete chrysosporium. Applied and Environmental Microbiology*, *54*, 2608–2614. PMID:16347765

Sreekanth, D., Sivaramakrishna, D., Himabindu, V., & Anjaneyulu, Y. (2009). Thermophilic treatment of bulk drug pharmaceutical industrial wastewaters by using hybrid up flow anaerobic sludge blanket reactor. *Bioresource Technology*, *100*(9), 2534–2539. and doi:10.1016/j.biortech.2008.11.028

Staddon, P. L., Fitter, A. H., & Graves, J. D. (1999). Effect of elevated atmospheric CO! on mycorrhizal colonization, external mycorrhizal hyphal production and phosphorus inflow in Plantago lanceolata and Trifolium repens in association with the arbuscular mycorrhizal fungus Glomus mosseae. *Global Change Biology*, *5*(3), 347–358. doi:10.1046/j.1365-2486.1999.00230.x

Stadtman, E. R. (1993). Oxidation of free amino acids and amino acid residues in proteins by radiolysis and by metal-catalyzed reactions. *Annual Review of Biochemistry*, *62*(1), 797–821. doi:10.1146/annurev.bi.62.070193.004053 PMID:8352601

Stadtman, E. R., & Levine, R. L. (2003). Free radical-mediated oxidation of free amino acids and amino acid residues in proteins. *Amino Acids*, *25*(3-4), 207–218. doi:10.1007/s00726-003-0011-2 PMID:14661084

Stanier, R. Y., & Ornston, L. N. (1973). The β-ketoadipate pathway. *Advances in Microbial Physiology*, *9*, 89–151. doi:10.1016/S0065-2911(08)60377-X PMID:4599397

Steinberg, P. D., & Rillig, M. C. (2003). Differential decomposition of arbuscular mycorrhizal fungal hyphae and glomalin. *Soil Biology & Biochemistry*, *35*(1), 191–194. doi:10.1016/S0038-0717(02)00249-3

Stein, E. A., & Cooper, E. L. (1981). Cytochemical observations of coelomocytes the earthworm *Lumbricusterrestris. Developmental and Comparative Immunology*, *5*, 15–25. PMID:7009243

Stephenson, G. L., Kaushik, A., Kaushik, N. K., Solmon, K. R., Steele, T., & Scroggins, R. F. (1998). Use of an avoidance-response test to assess the toxicity of contaminated soils to earthworms. In Advances in earthworm toxicology (pp. 67-81). Setac Press.

Stewart, E. J., Aslund, F., & Beckwith, J. (1998). Disulfide bond formation in the *Escherichia coli* cytoplasm: An *in vivo* role reversal for the thioredoxins. *The EMBO Journal*, *17*(19), 5543–5550. doi:10.1093/emboj/17.19.5543 PMID:9755155

Stohs, S. J., & Bagchi, D. (1995). Oxidative mechanisms in the toxicity of metal ions. *Free Radical Biology & Medicine*, *18*(2), 321–336. doi:10.1016/0891-5849(94)00159-H PMID:7744317

Stoilova, I., Krastanov, A., Yanakieva, I., Kratchanova, M., & Yemendjiev, H. (2007). Biodegradation of mixed phenolic compounds by *Aspergillus awamori* NRRL 3112. *International Biodeterioration & Biodegradation*, *60*(4), 342–346. doi:10.1016/j.ibiod.2007.05.011

Storey, S., Ashaari, M., McCabe, G., Harty, M., Dempsey, R., Doyle, O., & Doyle, E. et al. (2014). Microbial community structure during fluoranthene degradation in the presence of plants. *Journal of Applied Microbiology*, *117*(1), 74–84. doi:10.1111/jam.12518 PMID:24712542

Strickland, M. S., & Rousk, J. (2010). Considering fungal: Bacterial dominance in soils – methods, controls, and ecosystem implications. *Soil Biology & Biochemistry*, *42*(9), 1385–1395. doi:10.1016/j.soilbio.2010.05.007

Strlič, M., Kolar, J., Šelih, V.-S., Kočar, D., & Pihlar, B. (2003). A comparative study of several transition metals in Fenton-like reaction systems at circum-neutral pH. *Acta Chimica Slovenica*, *50*, 619–632.

Stuart, S. L., & Woods, S. L. (1998). Kinetic evidence for pentachlorophenol-dependent growth of a dehalogenating population in a pentachlorophenol- and acetate-fed methanogenic culture. *Biotechnology and Bioengineering*, *57*(4), 420–429. doi:10.1002/(SICI)1097-0290(19980220)57:4<420::AID-BIT5>3.0.CO;2-O PMID:10099218

Stucki, G., & Thuer, M. (1994). Increased removal capacity for 1,2- dichloromethane by biological modification of the granular modification of the granular activated carbon process. *Applied Microbiology and Biotechnology*, *42*(1), 167–172. doi:10.1007/BF00170241 PMID:7765815

Stumpf, M., Ternes, T. A., Wilken, R. D., & Rodrigues, S. V. (1999). Polar drug residues in sewage and natural waters in state of Rio de Janeiro, Brazil. *The Science of the Total Environment*, *225*(1-2), 135–141. doi:10.1016/S0048-9697(98)00339-8 PMID:10028710

Suarez, S., Lema, J. M., & Omil, F. (2010). Removal of pharmaceutical and personal care products (PPCPs) under nitrifying and denitrifying conditions. *Water Research*, *44*(10), 3214–3224. and doi:10.1016/j.watres.2010.02.040

Subashchandrabose, S. R., Ramakrishnan, B., Megharaj, M., Venkateswarlu, K., & Naidu, R. R. (2013). Mixotrophic cyanobacteria and microalgae as distinctive biological agents for organic pollutant degradation. *Environment International*, *51*, 59–72. doi:10.1016/j.envint.2012.10.007 PMID:23201778

Suberkropp, K., & Weyers, H. (1996). Application of fungal and bacterial production methodologies to decomposing leaves in streams. *Applied and Environmental Microbiology*, *62*, 1610–1615. PMID:16535312

Sukumaran, D. (2013). Phytoremediation of heavy metals from industrial effluent using constructed wetland technology. *Applied Ecology and Environmental Sciences*, *1*(5), 92–97. doi:10.12691/aees-1-5-4

Summers, A. O., & Silver, S. (1978). Microbial transformations of metals. *Annual Review of Microbiology*, *32*(1), 637–672. doi:10.1146/annurev.mi.32.100178.003225 PMID:360977

Sumner, E. R., Shanmuganathan, A., Sideri, T. C., Willetts, S. A., Houghton, J. E., & Avery, S. V. (2005). Oxidative protein damage causes chromium toxicity in yeast. *Microbiology*, *151*(6), 1939–1948. doi:10.1099/mic.0.27945-0 PMID:15942001

Sun, B., Ling, W., & Wang, Y. (2013). Can root exudate components influence the availability of pyrene in soil? *Journal of Soils and Sediments*, *13*(7), 1161–1169. doi:10.1007/s11368-013-0712-4

Su, R., Zhang, G., Wang, P., Li, S., Ravenelle, R. M., & Crittenden, J. C. (2015). Treatment of antibiotic pharmaceutical wastewater using a rotating biological contactor. *Journal of Chemistry*, *2015*, 705275. doi:10.1155/2015/705275

Suresh, B., & Ravishankar, G. A. (2004). Phytoremediationa novel and promising approach for environmental clean-up. *Critical Reviews in Biotechnology*, *24*(2-3), 97–124. and doi:10.1080/07388550490493627

Suresh, P. S., Kumar, A., Kumar, R., & Singh, V. P. (2008). An *Insilco* approach to bioremediation: Laccase as a case study. *Journal of Molecular Graphics & Modelling*, *26*(5), 845–849. doi:10.1016/j.jmgm.2007.05.005 PMID:17606396

Sutherland, T. D., Horne, I., Lacey, M. J., Harcourt, R. L., Russell, R. J., & Oakeshott, J. G. (2000). Enrichment of an endosulfan-degrading mixed bacterial culture. *Applied and Environmental Microbiology*, *66*(7), 2822–2828. doi:10.1128/AEM.66.7.2822-2828.2000 PMID:10877774

Sweetly, D. J., Sangeetha, K., & Suganthi, B. (2014). Biosorption of heavy metal lead from aqueous solution by non-living biomass of S*argassum myriocystum. International Journal of Application and Innovation in Engineering and Management*, *3*(4), 39–45.

Syldatk, C., & Hausmann, R. (2010). Microbial biosurfactants. *European Journal of Lipid Science and Technology*, *112*(6), 615–616. doi:10.1002/ejlt.201000294

Sylvestre, M., Macek, T., & Mackova, M. (2009). Transgenic plants to improve rhizoremediation of polychlorinated biphenyls (PCBs). *Current Opinion in Biotechnology*, *20*(2), 242–247. doi:10.1016/j.copbio.2009.01.006 PMID:19250817

Szczypka, M. S., Wemmie, J. A., Moye-Rowley, W. S., & Thiele, D. J. (1994). A yeast metal resistance protein similar to human cystic fibrosis transmembrane conductance regulator (CFTR) and multidrug resistance-associated protein. *The Journal of Biological Chemistry*, *269*(36), 22853–22857. PMID:7521334

Tabak, H. H., Lens, P., van Hullebusch, E. D., & Dejonghe, W. (2005). Developments in bioremediation of soils and sediments polluted with metals and radionuclides – 1. Microbial processes and mechanisms affecting bioremediation of metal contamination and influencing metal toxicity and transport. *Reviews in Environmental Science and Biotechnology*, *4*(3), 115–156. doi:10.1007/s11157-005-2169-4

Tabib, A., Haddadi, M., Shavandi, M. S., & Azizinikoo, P. (2012). Biodegradation of Phenol by Newly Isolated Phenol-degrading Bacterium *Ralstonia* sp. Strain PH-S1. *Journal of Petroleum Science and Technology*, *2*, 58–62.

Tabita, F. R. (1999). Microbial ribulose-1,5-bisphosphate carboxylase/oxygenase: A different perspective. *Photosynthesis Research*, *60*(1), 1–28. doi:10.1023/A:1006211417981

Tack, F. M. G., Vanhaesebroeck, T., Verloo, M. G., VanRompaey, K., & VanRanst, E. (2005). Mercury baseline levels in Flemish soils (Belgium). *Environmental Pollution*, *134*(1), 173–179. doi:10.1016/j.envpol.2004.05.031 PMID:15572235

Tagermina, H., Ebadi, T., Nasernejad, B., & Ghafori, M. (2014). Arsenic removal from water by sugar cane bagasse: An application of response surface methodology (RSM). *Water, Air, and Soil Pollution*, *225*(7), 2028–2050. doi:10.1007/s11270-014-2028-4

Taghreed, A., & Muftah, H. (2012). Aerobic Biodegradation of Phenols: A Comprehensive Review. *Critical Reviews in Environmental Science and Technology*, *42*(16), 1631–1690. doi:10.1080/10643389.2011.569872

Taha, M. F., Keat, C. F., Shaharum, M. S., & Raml, A. (2011). *Removal of Ni(II), Zn(II) and Pb (II) ions from single metal aqueous solution using activated carbon prepared from rice husk.* World Academy of Science, Engineering and Technology.

Tahrani, L., Soufi, L., Mehri, I., Najjari, A., Hassan, A., Van Loco, J., & Mansour, H. B. et al. (2015). Isolation and characterization of antibiotic-resistant bacteria from pharmaceutical industrial wastewaters. *Microbial Pathogenesis*, *89*, 54–61. and doi:10.1016/j.micpath.2015.09.001

Taiz, L. (1992). The plant vacuole. *The Journal of Experimental Biology*, *172*(1), 113–122. PMID:9874729

Takeuchi, Y., Suzuki, Y., Mochidzuki, K., Yagishita, Y., Fukuda, T., Amakusa, H., & Abe, H. (1996). Biological activated carbon treatment wastewater containing organics and heavy metal ions. In M. D. LeVan (Ed.), *Proceedings of the Fifth International Conference on Fundamentals of Adsorption* (pp. 929-236). Kluwer Academic Publishers. doi:10.1007/978-1-4613-1375-5_116

Tambosi, J. L., de Sena, R. F., Favier, M., Gebhardt, W., Joséa, H. J., Schröderb, H. F., & Moreiraa, R. P. M. (2010). Removal of pharmaceutical compounds in membrane bioreactors (MBR) applying submerged membrane. *Desalination*, *261*(1-2), 148–156. doi:10.1016/j.desal.2010.05.014

Tanaka, H., Itakura, S., & Enoki, A. (1999). Hydroxyl radical generation by an extracellular low-molecular -weight substance and phenol oxidase activity during wood degradation by the white-rot basidiomycete Trametes versicolor. *Journal of Biotechnology*, *75*(1), 57–70. doi:10.1016/S0168-1656(99)00138-8 PMID:10704993

Tanase, A. M., Vassu, T., Csutak, O., Pelinescu, D., Robertina, I., & Stoica, I. (2012). Phylogenetic analysis of oil polluted soil microbial strains. *Rom. Biotechnol. Lett*, *17*, 7093–7103.

Tandon, S. A., Kumar, R., Bajwa, R. K., & Yadav, S. A. (2013). Phytoremediation of fluoroquinolone group of antibiotics from waste water. Natural Science, 5(1212a), 21–27. 10.4236/ns.2013.512A004

Tangahu, B. V., Sheikh Abdullah, S. R., Basri, H., Idris, M., Anuar, N., & Mukhlisin, M. (2011). A review on heavy metals (As, Pb, and Hg) uptake by plants through phytoremediation. *International Journal of Chemical Engineering*, *2011*, 1–31. doi:10.1155/2011/939161

Tang, C. Y., Criddle, Q. S., Fu, C. S., & Leckie, J. O. (2007). Effect of flux and technique. *Biologie Medicale*, *1*(3), 1–6.

Tang, C. Y., Criddle, Q. S., & Leckie, J. O. (2007). Effect of flux (transmembrane pressure) and membranes properties on fouling and rejection of reverse osmosis and nanofiltration membranes treating perfluorooctane sulfonate containing waste water. *Journal of Environmental Science & Technology*, *41*(6), 2008–2014. doi:10.1021/es062052f PMID:17410798

Tanghe, T., & Verstraete, W. (2001). Adsorption of nonylphenol onto granular activated carbon. *Water, Air, and Soil Pollution*, *131*(1/4), 61–72. doi:10.1023/A:1011966914827

Tartakovsky, B., Manuel, M. F., Beaumier, D., Greer, C. W., & Guiot, S. R. (2001). Enhanced selection of an anaerobic pentachlorophenol- degrading consortium. *Biotechnology and Bioengineering*, *73*(6), 476–483. doi:10.1002/bit.1082 PMID:11344452

Taseli, B. K., G¨okc¸ay, C. F., & G¨urol, A. (2008). Influence of nickel (II) and chromium (VI) on the laboratory scale rotating biological contactor. *Journal of Industrial Microbiology & Biotechnology*, *35*(9), 1033–1039. and doi:10.1007/s10295-008-0379-9

Taub, F. B. (2004). *Fish 430 lectures: Biological Impacts of Pollutants on Aquatic Organisms*. Seattle, WA: University of Washington College of Ocean and Fishery Sciences.

Tawfik, A., Klapwijk, A., El-Gohary, F., & Lettinga, G. (2005). Potentials of using a rotating biological contactor (RBC) for posttreatment of anaerobically pre-treated domestic wastewater. *Biochemical Engineering Journal*, *25*(1), 89–98. doi:10.1016/j.bej.2005.03.013

Taylor, C. J., Bain, L. A., Richardson, D. J., Spiro, S., & Russell, D. A. (2004). Construction of a whole-cell gene reporter for the fluorescent bioassay of nitrate. *Analytical Biochemistry*, *328*(1), 60–66. doi:10.1016/j.ab.2004.01.013 PMID:15081908

Tchounwon, P. B., Ayensu, W. K., & Ninashvili, N. (2003). Environmental exposure to mercury and its toxico-pathologic implications for public health. *Environmental Toxicology*, *18*(3), 149–175. doi:10.1002/tox.10116 PMID:12740802

Tebbut, H. Y. (1998). *Principles of Water Quality Control* (5th ed.). Butterworth-Heineman Oxford.

Techer, D., Martinez-Chois, C., Laval-Gilly, P., Henry, S., Bennasroune, A., DInnocenzo, M., & Falla, J. (2012). Assessment of Miscanthus× giganteus for rhizoremediation of long term PAH contaminated soils. *Applied Soil Ecology*, *62*, 42–49. doi:10.1016/j.apsoil.2012.07.009

Teh, C. Y., Wu, T. Y., & Juan, J. C. (2014). Optimization of agro-industrial wastewater treatment using unmodified rice starch as a natural coagulant. *Industrial Crops and Products*, *56*, 17–26. doi:10.1016/j.indcrop.2014.02.018

Tekolo, O. M., Mckenzie, J., Botha, A., & Prior, B. A. (2010). The osmotic stress tolerance of basidiomycetous yeasts. *FEMS Yeast Research*, *10*(4), 482–491. doi:10.1111/j.1567-1364.2010.00612.x PMID:20214685

Teng, Y., Li, X., Chen, T., Zhang, M., Wang, X., Li, Z., & Luo, Y. (2016). Isolation of the PCB-degrading bacteria Mesorhizobium sp. ZY1 and its combined remediation with Astragalus sinicus L. for contaminated soil. *International Journal of Phytoremediation*, *18*(2), 141–149. doi:10.1080/15226514.2015.1073667 PMID:26292091

Teng, Y., Wang, X., Zou, L., Huang, M., & Du, X. (2015). Experimental and theoretical study on the binding of 2-mercaptothiazoline to bovine serum albumin. *Journal of Luminescence*, *161*, 14–19. doi:10.1016/j.jlumin.2014.12.046

Ten, H. (2001). Oxidative mechanisms involved in lignin degradation by white-rot fungi. *Chemical Reviews*, *101*(11), 3397–3414. doi:10.1021/cr000115l PMID:11749405

Ternes, T. A. (1998). Occurrence of drugs in German sewage treatment plants and rivers. *Water Research*, *32*(11), 3245–3260. doi:10.1016/S0043-1354(98)00099-2

Ternes, T. A., Joss, A., & Siegrist, H. (2004). Scrutinizing pharmaceuticals and personal care products in wastewater treatment. *Environmental Science & Technology*, *38*(20), 392A–399A. and doi:10.1021/es040639t

Tesar, M., Reichenauer, T. G., & Sessitsch, A. (2002). Bacterial rhizosphere populations of black poplar and herbal plants to be used for phytoremediation of diesel fuel. *Soil Biology & Biochemistry*, *34*(12), 1883–1892. doi:10.1016/S0038-0717(02)00202-X

Tezel, U., Guven, E., Erguder, T. H., & Demirer, G. N. (2001). Sequential (anaerobic/aerobic) biological treatment of dalaman SEKA pulp and paper industry effluent. *Waste Management (New York, N.Y.)*, *21*(8), 717–724. and doi:10.1016/S0956-053X(01)00013-7

Thai, P. K., Lai, F. Y., Bruno, R., van Dyken, E., Hall, W., OBrien, J., & Mueller, J. F. et al. (2016). Refining the excretion factors of methadone and codeine for wastewater analysis- Combining data from pharmacokinetic and wastewater studies. *Environment International*, *94*, 307–314. doi:10.1016/j.envint.2016.05.033 PMID:27295047

Thakuria, B., Jungai, N., & Adhikari, S. (2015). Catalytic Site Prediction of azoreductase enzyme of *E. coli* with potentially important industrial dyes using molecular docking tools. *International Journal of Bioscience. Biochemistry and Bioinformatics*, *5*(2), 91.

Thakuria, B., Singha, C. J., Maisnam, P., & Adhikari, S. (2015). Functional and catalytic active sites prediction and docking analysis of azoreductase enzyme in *Pseudomonas putida* with a variety of commercially available azodyes. *African Journal of Biotechnology*, *14*(26), 2162–2169. doi:10.5897/AJB2015.14699

Thapa, B., Kc, A. K., & Ghimire, A. (2012). A review on bioremediation of petroleum hydrocarbon contaminants in soil. *Kathmandu University Journal of Science, Engineering and Technology, 8*(1), 164-170.

Thavasi, R., Jayalakshmi, S., & Banat, I. M. (2011). Application of biosurfactant produced from peanut oil cake by *Lactobacillus delbrueckii* in biodegradation of crude oil. *Bioresource Technology, 102*(3), 3366–3372. doi:10.1016/j.biortech.2010.11.071 PMID:21144745

Thavasi, R., Nambaru, V. S., Jayalakshmi, S., Balasubramanian, T., & Banat, I. M. (2011). Biosurfactant production by *Pseudomonas aeruginosa* from renewable resources. *Indian Journal of Microbiology, 51*(1), 30–36. doi:10.1007/s12088-011-0076-7 PMID:22282625

Thiel, V. (2011). Extreme environments. In Encyclopedia of Geobiology (pp. 362–366). Academic Press.

Thieta, R. K., Freya, S. D., & Six, J. (2006). Do growth yield efficiencies differ between soil microbial communities differing in fungal:bacterial ratios? Reality check and methodological issues. *Soil Biology & Biochemistry, 38*(4), 837–844. doi:10.1016/j.soilbio.2005.07.010

Thomas, J. C., Davies, E. C., Malick, F. K., Endreszl, C., Williams, C. R., Abbas, M., & Ostenkowski, P. et al. (2003). Yeast metallothionein in transgenic tobacco promotes copper uptake from contaminated soils. *Biotechnology Progress, 19*(2), 273–280. doi:10.1021/bp025623q PMID:12675559

Thomsen, R., & Christensen, M. H. (2006). MolDock: A new technique for high-accuracy molecular docking. *Journal of Medicinal Chemistry, 49*(11), 3315–3321. doi:10.1021/jm051197e PMID:16722650

Thorpe, C. L., Law, G. T. W., Boothman, C., Lloyd, J. R., Burke, I. T., & Morris, K. (2012). The synergistic effects of high nitrate concentrations on sediment bioreduction. *Geomicrobiology Journal, 29*(5), 484–493. doi:10.1080/01490451.2011.581332

Thorpe, K. L., Cummings, R. I., Tutchinson, T. H., Scholz, M., Brighty, G., Sumpter, J. P., & Tyler, C. R. (2003). Relative potencies and combination effects of steroidal estrogens in fish. *Environmental Science & Technology, 37*(6), 1142–1149. doi:10.1021/es0201348 PMID:12680667

Thurman, E. M. (1985). Phenol. In Kirk-Othmer encyclopedia of chemical technology (3rd ed.). New York, NY: John Wiley and Sons.

Tian, Q., & Chen, J. (2006). Application of bioactivated carbon (BAC) Process in Water and Wastewater Treatment.[In Chinese]. *Environmental Engineering, 24*(1), 84–86.

Tibbles, B. J., & Baecker, A. A. W. (1989). Effect of pH and inoculum size on phenol degradation by bacteria isolated from from landfill waste. *Environmental Pollution, 59*(3), 227–239. doi:10.1016/0269-7491(89)90228-5 PMID:15092404

Tiedje, J. M., Asuming-Brempong, S., Nusslein, K., Marsh, T. L., & Flynn, S. J. (1999). Opening the black box of soil microbial diversity. *Applied Soil Ecology, 13*(2), 109–122. doi:10.1016/S0929-1393(99)00026-8

Tiirola, M.A., Wang, H., Paulin, L., & Kulomaa, M.S. (2002). Evidence for natural horizontal transfer of the pcpB gene in the evolution of polychlorophenol-degrading sphingomonads. *Appl. Environ. Microbiol., 68*, 4495–4501.

Tirado-Rives, J., & Jorgensen, W. L. (2006). Contribution of conformer focusing to the uncertainty in predicting free energies for protein-ligand binding. *Journal of Medicinal Chemistry, 49*(20), 5880–5884. doi:10.1021/jm060763i PMID:17004703

To'th, A. J., Gergely, F., & Mizsey, P. (2011). Physicochemical treatment of pharmaceutical process wastewater: Distillation and membrane processes. *Chemical Engineering (Albany, N.Y.), 55*(2), 59–67.

Todorcius, T., Bulgariu, L., & Popa, V. I. (2015). Adsorption of Cu (II) from aqueous solution on wheat straw lignin: Equilibrium and kinetic studies. *Cellulose Chemistry and Technology*, *49*(5-6), 439–447.

Toledo-Cervantes, A., Morales, M., Novelo, E., & Revah, S. (2013). Cdioxide fixation and lipid storage by Scenedesmus obtusiusculus. *Bioresource Technology*, *130*, 652–658. doi:10.1016/j.biortech.2012.12.081 PMID:23334023

Tomás-Gallardo, L., Canosa, I., Santero, E., Camafeita, E., Calvo, E., López, J. A., & Floriano, B. (2006). Proteomic and transcriptional characterization of aromatic degradation pathways in Rhodoccocus sp. strain TFB. *Proteomics*, *6*(S1), S119–S132. doi:10.1002/pmic.200500422 PMID:16544280

Tong, Y. P., Kneer, R., & Zhu, Y. G. (2004). Vacuolar compartmentalization: A second-generation approach to engineering plants for phytoremediation. *Trends in Plant Science*, *9*(1), 7–9. doi:10.1016/j.tplants.2003.11.009 PMID:14729212

Torkian, A., Yazdani, O., & Alinejad, K. (2003). Treatability evaluation of municipal wastewater and anaerobically-treated industrial effluent in a rotating biological contactor. *Industrial Journal of Engineering*, *16*(6), 16–22.

Torre, F. R., Ferrari, L., & Salibia, N. A. (2005). Biomarkers of a native fish species (*Cnesterodon decemmaculatus*) application to the water toxicity assessment of a peri-urban polluted river of Argentina. *Chemosphere*, *59*(4), 577–583. doi:10.1016/j.chemosphere.2004.12.039 PMID:15788181

Torsvik, V., & Øvreås, L. (2002). Microbial diversity and function in soil: From genes to ecosystems. *Current Opinion in Microbiology*, *5*(3), 240–245. doi:10.1016/S1369-5274(02)00324-7 PMID:12057676

Touati, D., Jacques, M., Tardat, B., Bouchard, L., & Despied, S. (1995). Lethal oxidative damage and mutagenesis are generated by iron in Δ*fur* mutants of *Escherichia coli*: Protective role of superoxide dismutase. *Journal of Bacteriology*, *177*(9), 2305–2314. doi:10.1128/jb.177.9.2305-2314.1995 PMID:7730258

Trama, B., Fernandes, J., Labuto, G., Oliveira, J., Viana-Niero, C., Pascon, R., & Vallim, M. (2014). The Evaluation of Bioremediation Potential of a Yeast Collection Isolated from Composting. *Advances in Microbiology*, *4*(12), 796–807. doi:10.4236/aim.2014.412088

Tran, N. H., Urase, T., & Kusakabe, O. (2009). The characteristics of enriched nitrifier culture in the degradation of selected pharmaceutically active compounds. *Journal of Hazardous Materials*, *171*(1-3), 1051–1057. and doi:10.1016/j.jhazmat.2009.06.114

Treseder, K. K., & Allen, M. F. (2000). Mycorrhizal fungi have a potential role in soil Cstorage under elevated CO2 and nitrogen deposition. *The New Phytologist*, *147*(1), 189–200. doi:10.1046/j.1469-8137.2000.00690.x

TRI. (1998). *Toxic Chemical Release Inventory*. National Library of Medicine, National Toxicology Information Program.

TRI05 TRI explorer: Providing access to EPA's toxics release inventory data. (2007). Retrieved from http://www.epa.gov/triexplorer/

Trinci, A. P. J., Davies, D. R., Gull, K., Lawrence, M. I., Bonde Nielsen, B., Rickers, A., & Theodorou, M. K. (1994). Anaerobic fungi in herbivorous animals. *Mycological Research*, *98*(2), 129–152. doi:10.1016/S0953-7562(09)80178-0

Trindade, P. V. O., Sobral, L. G., Rizzo, A. C. L., Leite, S. G. F., & Soriano, A. U. (2005). Bioremediation of a weathered and recently oil-contaminated soil from Brazil: A comparison study. *Chemosphere*, *58*(4), 515–522. doi:10.1016/j.chemosphere.2004.09.021 PMID:15620743

Tripathi, A. K., Harsh, N. S. K., & Gupta, N. (2007). Fungal treatment of industrial effluents: A mini-review. *Life Science Journal*, *4*(2), 78–81.

Tripathi, M. K., Yasir, M., Gurjar, V. S., Bose, P., Dubey, A., & Shrivastava, R. (2015). Insights from the molecular docking of hydrolytic products of methyl isocyanate (MIC) to inhibition of human immune proteins. *Interdisciplinary Sciences: Computational Life Sciences*, *7*(3), 287–294. PMID:26297312

Tu, C., Foster, L., Alvarado, A., McKenna, R., Silverman, D. N., & Frost, S. C. (2012). Role of Zinc in Catalytic Activity of Carbonic Anhydrase IX. *Archives of Biochemistry and Biophysics*, *521*(1-2), 90–94. doi:10.1016/j.abb.2012.03.017 PMID:22465027

Turabekova, M., Rasulev, B., Theodore, M., Jackman, J., Leszczynska, D., & Leszczynski, J. (2014). Immunotoxicity of nanoparticles: A computational study suggests that CNTs and C 60 fullerenes might be recognized as pathogens by Toll-like receptors. *Nanoscale*, *6*(7), 3488–3495. doi:10.1039/c3nr05772k PMID:24548972

Turner, A. P. F., Karube, I., & Wilson, G. S. (1987). *Biosensors: Fundamentals and Applications*. Oxford, UK: Oxford University Press.

Twiggy, L. E., & Socha, L. V. (2001). Defluorination of sodium mono fluoroacetate by soil microorganisms from central Australia. *Soil Biology & Biochemistry*, *33*(2), 227–234. doi:10.1016/S0038-0717(00)00134-6

Ucun, H., Yildiz, E., & Nuhoglu, A. (2010). Phenol biodegradation in a batch jet loop bioreactor (JLB): Kinetics study and pH variation. *Bioresource Technology*, *101*(9), 2965–2971. doi:10.1016/j.biortech.2009.12.005 PMID:20053552

UE. (2006). Regulation (EC) No 1907/2006 of the European Parliament and of the Council of 18 December 2006 concerning the Registration, Evaluation, Authorisation and Restriction of Chemicals (REACH), establishing a European Chemicals Agency, amending Directive 1999/45/EC and repealing Council Regulation (EEC) No 793/93 and Commission Regulation (EC) No 1488/94 as well as Council Directive 76/769/EEC and Commission Directives 91/155/EEC, 93/67/EEC, 93/105/EC and 2000/21/EC. *Official Journal of the European Union, L, 136*, 1–280.

Ullah, A., Heng, S., Munis, M. F. H., Fahad, S., & Yang, X. (2015). Phytoremediation of heavy metals assisted by plant growth promoting (PGP) bacteria: A review. *Environmental and Experimental Botany*, *117*, 28–40. doi:10.1016/j.envexpbot.2015.05.001

Ulrich, S., Seijo, M., Laguecir, A., & Stoll, S. (2006). Nanoparticle adsorption on a week polyelectrolyte. Stiffness, pH, charge mobility and ionic concentration effects investigated by Monte Carlo simulations. *Journal of Physical Chemistry*, *110*(42), 20954–20964. doi:10.1021/jp063671d PMID:17048913

Umar, A. F., Tahir, F., Larkin, M., Oyawoye, O. M., & Musa, B. (2012). In-Situ biostimulatory effect of selected organic wastes on bacterial atrazine biodegradation. *Adv Microbiol*, *2*(04), 587–592. doi:10.4236/aim.2012.24076

Umrania, V. V. (2006). Bioremediation of toxic heavy metals using acidothermophilic autotrophes. *Bioresource Technology*, *97*(10), 1237–1242. doi:10.1016/j.biortech.2005.04.048 PMID:16324838

UNFCCC. (2007). *Report of the conference of parties on its thirteenth session*. Bali, Indonesia: Author.

United Nations Environmental Programme (UNEP). (1998). Environmental effects of ozone depletion: 1998 Assessment. Nairobi, Kenya: UNEP.

United Nations World Water Assessment Programme (UNWWAP). (2009). *The World Water Development Report 3: Water in a Changing World. United Nations World Water Assessment Programme*. Paris, France: UNESCO.

Unlu, M., Morgan, M. E., & Minden, J. S. (1997). Difference gel electrophoresis. A single gel method for detecting changes in protein extracts. *Electrophoresis*, *18*(11), 2071–2077. doi:10.1002/elps.1150181133 PMID:9420172

Unrine, J. M., Hunyadi, S. E., Tsyusko, O. V., Rao, W., Shoults-Wilson, W. A., & Bertsch, P. M. (2010). Evidence for bioavailability of an nanoparticles from soil and biodistribution with in earthworms (*Eisenia fetida*). *Environmental Science & Technology*, *44*(21), 8308–8313. doi:10.1021/es101885w PMID:20879765

Uqab, B., Mudasir, S., Sheikh, A. Q., & Nazir, R. (2016). Bioremediation: A management tool. *J Bioremediat Biodegrad*, *7*(2), 1–5.

US Agriculture Centre. (2004). *A Farmer's Guide to Agriculture and Water Quality Issues*. Author.

US EPA Toxics release inventory. (1989). Washington, DC: US Environmental Protection Agency, Office of Pesticides and Toxic Substances.

Uwadiae, S. E., Yerima, Y., & Azike, R. U. (2011). Enzymatic biodegradation of pharmaceutical wastewater. *International Journal of Energy and Environment*, *2*(4), 683–690.

Vader, J. S., van Ginkel, C. G., Sperling, F., de Jong, J., de Boer, W., de Graaf, J. S., & Stokman, P. G. W. et al. (2000). Degradation of ethinyl estradiol by nitrifying activated sludge. *Chemosphere*, *41*(8), 1239–1243. and doi:10.1016/S0045-6535(99)00556-1

Valembois, P., Roch, P., & Lassegues, M. (1988). Evidence of plasma clotting system in earthworms. *Journal of Invertebrate Pathology*, *51*(3), 221–228. doi:10.1016/0022-2011(88)90029-8

Valembois, P., Roch, P., Lassegues, M., & Cassand, P. (1982). Antibacterial activity of the hemolytic system from the earthworm *Eisenia fetida andrei*. *Journal of Invertebrate Pathology*, *40*(1), 21–27. doi:10.1016/0022-2011(82)90032-5

Valko, M., Morris, H., & Cronin, M. T. D. (2005). Metals, toxicity and oxidative stress. *Current Medicinal Chemistry*, *12*(10), 1161–1208. doi:10.2174/0929867053764635 PMID:15892631

Van Aken, B. (2008). Transgenic plants for phytoremediation: Helping nature to clean up environmental pollution. *Trends in Biotechnology*, *26*(5), 225–227. doi:10.1016/j.tibtech.2008.02.001 PMID:18353473

van der Lelie, D., Springael, D., Römling, U., Ahmed, N., & Mergeay, M. (1999). Identification of a gene cluster, czr, involved in cadmium and zinc resistance in Pseudomonas aeruginosa. *Gene*, *238*(2), 417–425. doi:10.1016/S0378-1119(99)00349-2 PMID:10570969

Van der Wal, A., De Boer, W., Klein, Gunnewiek, P. J. A., & Van Veen, J. A. (2009b). Possible mechanism for spontaneous establishment of Calluna vulgaris in a recently abandoned agricultural field. *Restor Ecol, 17,* 308–313.

Van der Wal, A., Van Veen, J. A., Pijl, A. S., Summerbell, R. C., & De Boer, W. (2006a). Constraints on development of fungal biomass and decomposition processes during restoration of arable sandy soils. *Soil Biol Biochem, 38,* 2890–2902.

Van Eerd, L. L., Hoagland, R. E., Zablotowicz, R. M., & Hall, J. C. (2003). Pesticide metabolism in plants and microorganisms. *Weed Science*, *51*(4), 472–495. doi:10.1614/0043-1745(2003)051[0472:PMIPAM]2.0.CO;2

Van Gestel, C. A. M., Leon, C. D., & Van Straalen, N. M. (1997). Evaluation of soil fauna ecotoxicity tests regarding their use in risk assessment. In J. Tarradellas, G. Bitton, & D. Rossel (Eds.), *Soil Ecotoxicology* (pp. 291–318). New York: Lewis Publishers, New Publishers.

Vančura, V., & Hovadik, A. (1965). Root exudates of plants: II. Composition of root exudates of some vegetables. *Plant and Soil*, *22*(1), 21–32. doi:10.1007/BF01377686

Vanerkar, A. P., Satyanarayan, S., & Dharmadhikari, D. M. (2013). Full scale treatment of herbal pharmaceutical industry wastewater. *International Journal of Chemical & Physical Sciences*, *2*, 52–62.

Vara Prasad, M. N., & de Oliveira Freitas, H. M. (2003). Metal hyperaccumulation in plants: Biodiversity prospecting for phytoremediation technology. *Electronic Journal of Biotechnology*, *6*(3), 285–321.

Varsha, Y. M., & Chenna, S. (2012). An emphasis on xenobiotic degradation in environmental cleanup. *Journal of Bioremediation & Biodegradation*.

Vasiliadou, I. A., Molina, R., Martínez, F., & Melero, J. A. (2014). Experimental and modeling study on removal of pharmaceutically active compounds in rotating biological contactors. *Journal of Hazardous Materials*, *274*, 473–482. and doi:10.1016/j.jhazmat.2014.04.034

Vasiliadou, I. A., Sánchez-Vázquez, R., Molina, R., Martínez, F., Melero, J. A., Bautista, L. F., & Morales, G. et al. (2016). Biological removal of pharmaceutical compounds using white-rot fungi with concomitant FAME production of the residual biomass. *Journal of Environmental Management*, *180*, 228–237. and doi:10.1016/j.jenvman.2016.05.035

Vasilyeva, G., & Strijakova, E. (2007). Bioremediation of soils and sediments contaminated by polychlorinated biphenyls. *Microbiology*, *76*(6), 639–653. doi:10.1134/S002626170706001X PMID:18069324

Vaughan, N. E., Lenton, T. M., & Shepherd, J. G. (2009). Climate change mitigation: Trade-offs between delay and strength of action required. *Climatic Change*, *96*(1-2), 29–43. doi:10.1007/s10584-009-9573-7

Vàzquez, G., Mosquera, O., Freire, M. S., Antorrena, G., Gouzàalez, A., & Levrez, J. (2012). Alkaline pre-treatment of waste chest nut shell from a food industry to enhance Cadmium, Copper, Lead and Zinc ions removal. *Chemical Engineering Journal*, *184*, 147–155. doi:10.1016/j.cej.2012.01.019

Veena Gayathri, K., & Vasudevan, N. (2010). Aerobic degradation of Catechol and Resorcinol by Bacterial strains and isolated from wastewater. *Int. J. Biotechnol. Res.*, *3*(2), 35–38.

Veena Gayathri, K., & Vasudevan, N. (2010). Effect of pH, Nitrogen sources and salts on the degradation of Phenol by the Bacterial Consortium under Saline Conditions. *Int. J. Biotech. Biochem*, *6*, 783–791.

Veena Gayathri, K., & Vasudevan, N. (2010). Enrichment and Isolation of Phenol degrading Moderately Halophilic Bacterial Consortium from Saline Environment *J. Biorem. Biodeg.*, *1*(1), 104.

Veena Gayathri, K., & Vasudevan, N. (2011). Ortho and meta cleavage dioxygenases detected during the degradation of phenolic compounds by a moderately halophilic bacterial consortium. *Int. Res. J. Microbiol.*, *2*(10), 406–414.

Veena Gayathri, K., & Vasudevan, N. (2013). Degradation of 4-Chlorophenol by a Moderately Halophilic Bacterial Consortium under Saline Conditions. *British Microbiology Research Journal*, *3*(4), 513–524. doi:10.9734/BMRJ/2013/3171

Veena Gayathri, K., & Vasudevan, N. (2015). Degradation of Chlorophenolic compounds by a moderately halotolerant bacterial consortium isolated from saline environment. *International Journal of Bioassays*, *4*(7), 4102–4108.

Veena Gayathri, K., & Vasudevan, N. (2015). Degradation of dual phenolics by a moderately halophilic bacterial consortium and its degradation products. International. *Journal of Current Microbiology and Applied Sciences.*, *4*(5), 1083–1095.

Veeresh, G. S., Kumar, P., & Mehrotra, I. (2004). Treatment of phenol and cresols in up-flow anaerobic sludge blanket (UASB) process: A review. *Water Research*, *39*(1), 154–170. doi:10.1016/j.watres.2004.07.028 PMID:15607175

Verburg, P. H., Kok, K., Pontius, R. G., & Veldkamp, A. J. (2004). Modeling Land-Use and Land-Cover Change. Part of the series Global Change - The IGBP Series.

Verdonk, M. L., Chessari, G., Cole, J. C., Hartshorn, M. J., Murray, C. W., Nissink, J. W. M., & Taylor, R. et al. (2005). Modeling water molecules in protein-ligand docking using GOLD. *Journal of Medicinal Chemistry*, *48*(20), 6504–6515. doi:10.1021/jm050543p PMID:16190776

Verdonk, M. L., Cole, J. C., Hartshorn, M. J., Murray, C. W., & Taylor, R. D. (2003). Improved protein–ligand docking using GOLD. *Proteins: Structure, Function, and Bioinformatics*, *52*(4), 609–623. doi:10.1002/prot.10465 PMID:12910460

Verdonk, M. L., Mortenson, P. N., Hall, R. J., Hartshorn, M. J., & Murray, C. W. (2008). Protein– ligand docking against non-native protein conformers. *Journal of Chemical Information and Modeling*, *48*(11), 2214–2225. doi:10.1021/ci8002254 PMID:18954138

Vered, Y., Milstein, O., Flowers, H. M., & Gressel, J. (1981). Biodegradation of wheat straw lignocarbohydrate complexes (LCC). I. Dynamics of liberation of hot aqueous LCCs from wheat straw and partial characterization of the products. *Eur. J. Applied Microbiology and Biotechnology*, *12*(3), 183–188. doi:10.1007/BF01008341

Verschueren, K. (1983). *Handbook of Environmental Data on Organic Chemicals* (2nd ed.). New York: Van Nostrand Reinhold Co.

Vidali, M. (2001). Bioremediation: An overview. *Pure and Applied Chemistry*, *73*(7), 1163–1172. doi:10.1351/pac200173071163

Vidal, N., Bañares-Alcántara, R., Rodríguez-Roda, I., & Poch, M. (2002). Design of wastewater treatment plants using a conceptual design methodology. *Industrial & Engineering Chemistry Research*, *41*(20), 4993–5005. doi:10.1021/ie010652b

Vidhyalakshmi, R., Vallinachiyar, C., & Radhika, R. (2012). Production of Xanthan from Agro-Industrial Waste. *Journal of Advanced Scientific Research*, *3*(2).

Vieira, A. C., Marschalk, C., Biavatti, D. C., Lorscheider, C. A., Peralta, R. M., & Seixas, F. A. V. (2015). Modeling based structural insights into biodegradation of the herbicide diuron by laccase-1 from *Ceriporiopsis subvermispora*. *Bioinformation*, *11*(5), 224–228. doi:10.6026/97320630011224 PMID:26124565

Vijayakumar, S., & Saravanan, V. (2015). Biosurfactants-Types, Sources and Applications. Research. *Journal of Microbiology (Seoul, Korea)*, *10*(5), 181.

Villanneau, E. J., Saby, N. P., Orton, T. G., Jolivet, C. C., Boulonne, L., Caria, G., & Arrouays, D. et al. (2013). First evidence of large-scale PAH trends in French soils. *Environmental Chemistry Letters*, *11*(1), 99–104. doi:10.1007/s10311-013-0401-y

Villas-Boas, S. G., & Bruheim, P. (2007). The potential of metabolomics tools in bioremediation studies. *Omics: A Journal of Integrative Biology, 11*(3), 305-313.

Ville, P., Roch, P., Cooper, E., Masson, P., & Narrbomnne, J. (1995). PCBs increase molecular-related activities (lysosome, antibacterial, hemolysis,proteases) but inhibit macrophage-related functions (phagocytosis, wound healing) in earthworms. *Journal of Invertebrate Pathology*, *65*(3), 217–224. doi:10.1006/jipa.1995.1033 PMID:7745278

Vilmi, A., Karjalainen, S. M., Kuoppala, M., Tolonen, K. T., & Heino, J. (2016). Taxonomic distinctness along nutrient gradients: More diverse, less diverse or not different from random? *Ecological Indicators*, *61*(2), 1033–1041. doi:10.1016/j.ecolind.2015.10.061

Vink, J. P. M., & Van Der Zee, S. E. A. T. M. (1997). Pesticide biotransformation in surface waters: Multivariate analyses of environmental factors at field sites. *Water Research*, *31*(11), 2858–2868. doi:10.1016/S0043-1354(97)00139-5

Vitousek, P. M. (2004). *Nutrient cycling and limitation: Hawai'i as a model system.* Princeton University Press.

Volesky, B. (2001). Detoxification of metal-bearing effluents: Biosorption for the next century. *Hydrometallurgy*, *59*(2-3), 203–216. doi:10.1016/S0304-386X(00)00160-2

Volesky, B. (2004). *Sorption and biosorption*. Quebec: BV-Sorbex, Inc.

Volesky, B. (2007). Biosorption and me. *Water Research*, *41*(18), 4017–4029. doi:10.1016/j.watres.2007.05.062 PMID:17632204

Voltolina, D., Gómez-Villa, H., & Correa, G. (2005). Nitrogen removal and recycling by *Scenedesmus obliquus* in semi continuous cultures using artificial wastewater and simulated light and temperature cycle. *Bioresource Technology*, *96*(3), 359–362. doi:10.1016/j.biortech.2004.04.004 PMID:15474938

Vonk, J., & Sijpesteijn, A. K. (1973). Studies on the methylation of mercuric chloride by pure cultures of bacteria and fungi. *Antonie van Leeuwenhoek*, *39*(1), 505–513. doi:10.1007/BF02578894 PMID:4201588

Vymazal, J. (1984). Short term uptake of heavy metals by periphyton algae. *Hydrobiologia*, *119*(3), 171–179. doi:10.1007/BF00015208

Wadhwa, M., Bakshi, M. P., & Makkar, H. P. (2015). Waste to worth: Fruit wastes and by-products as animal feed. *CAB Reviews*, *10*(031), 1–26.

Waldron, K. J., & Robinson, N. J. (2009). How do bacterial cells ensure that metalloproteins get the correct metal? *Nature Reviews. Microbiology*, *7*(1), 25–35. doi:10.1038/nrmicro2057 PMID:19079350

Waldrop, M. P., Zak, D. R., Sinsabaugh, R. L., Gallo, M., & Lauber, C. (2004). Nitrogen deposition modifies soil Cstorage through changes in microbial enzymatic activity. *Ecological Applications*, *14*(4), 1172–1177. doi:10.1890/03-5120

Walker, J. D., Colwell, R. R., Vaituzis, Z., & Meyer, S. A. (1975). Petroleum degrading achlorophyllous alga *Prototheca zopfii. Nature*, *254*(5499), 423–424. doi:10.1038/254423a0 PMID:1118032

Walters, E., McClellan, K., & Halden, R. U. (2010). Occurrence and loss over three years of 72 pharmaceuticals and personal care products from biosolids-soil mixtures in outdoor mesocosms. *Water Research*, *44*(20), 6011–6020. and doi:10.1016/j.watres.2010.07.051

Walton, B. T., & Anderson, T. A. (1990). Microbial degradation of trichloroethylene in the rhizosphere: Potential application to biological remediation of waste sites. *Applied and Environmental Microbiology*, *56*(4), 1012–1016. PMID:2339867

Wang, B., Li, Y., Wu, N., & Lan, C. Q. (2008). CO2 bio-mitigation using microalgae. *Applied Microbiology and Biotechnology*, *79*(5), 707–718. doi:10.1007/s00253-008-1518-y PMID:18483734

Wang, C. C., Lee, C. M., & Kuan, C. H. (2000). Removal of dichlorophenol by suspended and immobilized *Bacillus insolitus. Chemosphere*, *41*(3), 447–452. doi:10.1016/S0045-6535(99)00263-5 PMID:11057607

Wang, J., & Chen, C. (2006). Biosorption of heavy metals by *Saccharomyces cerevisiac*: A review. *Biotechnology Advances*, *4*, 417–451.

Wang, J., Feng, X., Anderson, C. W., Xing, Y., & Shang, L. (2012). Remediation of mercury contaminated sites – a review. *Journal of Hazardous Materials*, *221–222*, 1–18. PMID:22579459

Wang, J., & Katz, S. E. (1994). The persistence of fecal-borne antibiotics in soil. *Journal - Association of Official Analytical Chemists*, *77*, 481–484.

Wang, K., Liu, S., Zhang, Q., & He, Y. (2009). Pharmaceutical wastewater treatment by internal micro-electrolysis-coagulation, biological treatment and activated carbon adsorption. *Environmental Technology*, *30*(13), 1469–1474. and doi:10.1080/09593330903229164

Wang, L., Li, Y., Yu, P., Xie, Z., Luo, Y., & Lin, Y. (2010). Biodegradation of phenol at high concentration by a novel fungal strain *Paecilomyces variotii* JH6. *Journal of Hazardous Materials*, *183*(1-3), 366–371. doi:10.1016/j.jhazmat.2010.07.033 PMID:20685040

Wang, L., & Luo, Q. (2003). Study on Degradation of Immobilized Microorganism on Endocrine Disruptor Di-n-butyl Phthalate. *Chinese Journal of Public Health*, *19*(11), 1302–1303.

Wang, L., Yan, W., Chen, J., Huang, F., & Gao, P. (2008). Function of the iron-binding chelator produced by Coriolus versicolor in lignin biodegradation. *Science in China. Series C, Life Sciences*, *51*(3), 214–221. doi:10.1007/s11427-008-0033-9 PMID:18246309

Wang, Q.-Q., Bradford, S. A., Zheng, W., & Yates, S. R. (2006). Sulfadimethoxine degradation kinetics in manure as affected by initial concentration, moisture, and temperature. *Journal of Environmental Quality*, *35*(6), 2162–2169. and doi:10.2134/jeq2006.0178

Wang, R. F., Wennerstrom, D., Cao, W. W., Khan, A. A., & Cerniglia, C. E. (2000). Cloning, expression, and characterization of the katG gene, encoding catalase-peroxidase, from the polycyclic aromatic hydro- carbon-degrading bacterium *Mycobacterium sp.* strain PYR-1. *Applied and Environmental Microbiology*, *66*(10), 4300–4304. doi:10.1128/AEM.66.10.4300-4304.2000 PMID:11010873

Wang, S., Wang, X., Poon, K., Wang, Y., Li, S., Liu, H., & Cai, Z. et al. (2013). Removal and reductive dechlorination of triclosan by *Chlorella pyrenoidosa*. *Chemosphere*, *92*(11), 1498–1505. doi:10.1016/j.chemosphere.2013.03.067 PMID:23648333

Wang, S., Wei, S., Ji, D., & Bai, J. (2015). Co-Planting Cd Contaminated Field Using Hyperaccumulator Solanum Nigrum L. Through Interplant with Low Accumulation Welsh Onion. *International Journal of Phytoremediation*, *17*(9), 879–884. doi:10.1080/15226514.2014.981247 PMID:25581317

Wang, Y. Q., Zhang, H. M., Cao, J., & Tang, B. P. (2014). Binding of a new bisphenol analogue, bisphenol S to bovine serum albumin and calf thymus DNA. *Journal of Photochemistry and Photobiology. B, Biology*, *138*, 182–190. doi:10.1016/j.jphotobiol.2014.06.002 PMID:24972352

Wang, Y., Rawlings, M., Gibson, D. T., Labbé, D., Bergeron, H., Brousseau, R., & Lau, P. C. (1995). Identification of a membrane protein and a truncated LysR-type regulator associated with the toluene degradation pathway in Pseudomonas putida F1. *Molecular and General Genetics MGG*, *246*(5), 570–579. doi:10.1007/BF00298963 PMID:7535376

Wang, Y., Tian, Y., Han, B., Zhaw, H. B., Bi, J. N., & Cai, B. L. (2007). Biodegradation of phenol by free and immobilized *Acinetobacter* sp. Strain PD12. *Journal of Environmental Sciences (China)*, *19*(2), 222–225. doi:10.1016/S1001-0742(07)60036-9 PMID:17915733

Warnes, S. L., Caves, V., & Keevil, C. W. (2012). Mechanism of copper surface toxicity in *Escherichia coli* O157:H7 and *Salmonella* involves immediate membrane depolarization followed by slower rate of DNA destruction which differs from that observed for Gram-positive bacteria. *Environmental Microbiology*, *14*(7), 1730–1743. doi:10.1111/j.1462-2920.2011.02677.x PMID:22176893

Wasilkowski, D., Swędzioł, Ż., & Mrozik, A. (2012). The applicability of genetically modified microorganisms in bioremediation of contaminated environments. *CHEMIK nauka-technika-rynek*, *1*(66), 817-826.

Watanabe, K. (2001). Microorganisms relevant to bioremediation. *Current Opinion in Biotechnology*, *12*(3), 237–241. doi:10.1016/S0958-1669(00)00205-6 PMID:11404100

Watanabe, K., Hino, S., & Takahashi, N. (1996). Effects of exogenous phenol degrading bacteria performance on ecosystem of activated sludge. *Ferm. Bioeng.*, *82*(3), 291–298. doi:10.1016/0922-338X(96)88822-2

Watanabe, K., Kodoma, Y., Stutsubo, K., & Harayama, S. (2001). Molecular characterization of bacterial populations in petroleum contaminated ground water discharge from undergoing crude oil storage cavities. *Applied and Environmental Microbiology*, *66*(11), 4803–4809. doi:10.1128/AEM.66.11.4803-4809.2000 PMID:11055927

Watanabe, M. E. (1997). Phytoremediation on the brink of commercialization. *Environmental Science & Technology*, *31*(4), 182A–186A. and doi:10.1021/es972219s

Watanabe, R. A., Sales, P. T. F., Campos, L. C., Garcia, T. A., Valadares, M. C., Schimidt, F., & Santiago, M. F. (2012). Evaluation of the use of *Pycnoporus sanguineus* fungus for phenolics and genotoxicity decay of a pharmaceutical effluent treatment. *Ambi-Agua. Taubaté*, *7*(3), 41–50.

Weber, C. F., & King, G. M. (2007). Physiological ecological, and phylogenetic characterization of stappia, a Marine CO-Oxidizing Bacterial Genus. *Applied and Environmental Microbiology*, *73*(4), 1266–1276. doi:10.1128/AEM.01724-06 PMID:17142374

Weis, A., Katebzadeh, K., Söderhjelm, P., Nilsson, I., & Ryde, U. (2006). Ligand affinities predicted with the MM/PBSA method: Dependence on the simulation method and the force field. *Journal of Medicinal Chemistry*, *49*(22), 6596–6606. doi:10.1021/jm0608210 PMID:17064078

Wei, X., Li, B., Zhao, S., Wang, L., Zhang, H., Li, C., & Wang, S. (2012). Mixed pharmaceutical wastewater treatment by integrated membraneaerated biofilm reactor (MABR) system—a pilot-scale study. *Bioresource Technology*, *122*, 189–195. and doi:10.1016/j.biortech.2012.06.041

Wernersson, A. S. (2012). Swedish monitoring of hazardous substances in the aquatic environment current, required monitoring and potential developments. *Rapport*, *2*, 23.

Werner, T., Morris, M. B., Dastmalchi, S., & Church, W. B. (2012). Structural modelling and dynamics of proteins for insights into drug interactions. *Advanced Drug Delivery Reviews*, *64*(4), 323–343. doi:10.1016/j.addr.2011.11.011 PMID:22155026

Wesenberg, D., Kyriakides, I., & Agathos, S. N. (2003). White-rot fungi and their enzymes for the treatment of industrial dye effluents. *Biotechnology Advances*, *22*(1-2), 161–187. doi:10.1016/j.biotechadv.2003.08.011 PMID:14623049

Westerhoff, P., Yoon, Y., Snyder, S. A., & Wert, E. (2005). Fate of endocrine-disruptor, pharmaceutical, and personal care product chemicals during simulated drinking water treatment processes. *Environmental Science & Technology*, *39*(17), 6649–6663. doi:10.1021/es0484799 PMID:16190224

Westhead, D. R., Clark, D. E., & Murray, C. W. (1997). A comparison of heuristic search algorithms for molecular docking. *Journal of Computer-Aided Molecular Design*, *11*(3), 209–228. doi:10.1023/A:1007934310264 PMID:9263849

Wetzstein, H. G., Schmeer, N., & Karl, W. (1997). Degradation of the fluoroquinolone enrofloxacin by the brown rot fungus Gloeophyllum striatum: Identification of metabolites. *Applied and Environmental Microbiology*, *63*(11), 4272–4281.

Whipps, J. M. (2001). Microbial interactions and biocontrol in the rhizosphere. *Journal of Experimental Botany*, *52*(suppl 1), 487–511. doi:10.1093/jexbot/52.suppl_1.487 PMID:11326055

White, M. C., Baker, F. D., Chaney, R. L., & Decker, A. M. (1981). Metal complexation in xylem fluid II. Theoretical equilibrium model and computational computer program. *Plant Physiology*, *67*(2), 301–310. doi:10.1104/pp.67.2.301 PMID:16661665

White, P. M. Jr, Wolf, D. C., Thoma, G. J., & Reynolds, C. M. (2006). Phytoremediation of alkylated polycyclic aromatic hydrocarbons in a crude oil-contaminated soil. *Water, Air, and Soil Pollution*, *169*(1-4), 207–220. doi:10.1007/s11270-006-2194-0

White, P., Dranke, M., & Hindle, P. (2012). *Integrated solid waste management: A lifecycle inventory*. Springer Science & Business Media.

Whiting, S. N., Broadley, M. R., & White, P. J. (2003). Applying a solute transfer model to phytoextraction: Zinc acquisition by Thlaspi caerulescens. *Plant and Soil*, *249*(1), 45–56. doi:10.1023/A:1022542725880

Whitton, B. A., & Potts, M. (2007). *The ecology of cyanobacteria: their diversity in time and space*. Kluwer Academic Publishers.

WHO. (1989). *Chlorophenols other than Pentachlorophenol*. Geneva: World Health Organization.

Wicklow, D. T. (1992). The coprophilous fungal community: an experimental system. In G. C. Carrol & D. T. Wicklow (Eds.), *The Fungal Community. Its Organisation and Role in the Ecosystem* (2nd ed., pp. 715–728). New York: Marcel Dekker.

Wiedemeier, T. H. (1999). *Natural attenuation of fuels and chlorinated solvents in the subsurface*. John Wiley & Sons. doi:10.1002/9780470172964

Wielgus-Serafinska, E., & Kawa, E. (1976). Accumulation and localization of lead in *Eisenia fetida* (Oligochaeta) tissues. *Folia Histochemica et Cytochemica*, *14*, 315–320. PMID:1030682

Wignarajah, K., Pisharody, S., & Fisher, J. W. (2000). Can incineration technology convert CELSS wastes to resources for crop production? A working hypothesis and some preliminary findings. *Advances in Space Research*, *26*(2), 327–333. doi:10.1016/S0273-1177(99)01072-8 PMID:11543171

Wilkins, J. C., Homer, K. A., & Beighton, D. (2001). Altered protein expression of *Streptococcus oralis* cultured at low pH revealed by two-dimensional gel electrophoresis. *Applied and Environmental Microbiology*, *67*(8), 3396–3405. doi:10.1128/AEM.67.8.3396-3405.2001 PMID:11472910

Wilkinson, C. (2002). Status of Coral Reefs of the World: 2002. Global Coral Reef Monitoring Network and Australian Institute of Marine Science.

Wilson, J. R., & Mertens, D. R. (1995). Cell wall accessibility and cell structure limitations to microbial digestion of forage. *Crop Science*, *35*(1), 251–259. doi:10.2135/cropsci1995.0011183X003500010046x

Winegar, E. D., & Keith, L. H. (1993). *Sampling and analysis of air borne pollutants*. London: Lewis Publications Sciences, Boka-Raton –Ann Arbor.

Wojtera-Kwiczor, J., Żukowska, W., Graj, W., Małecka, A., Piechalak, A., Ciszewska, L., & Barałkiewicz, D. et al. (2014). Rhizoremediation of Diesel-Contaminated Soil with Two Rapeseed Varieties and Petroleum degraders Reveals Different Responses of the Plant Defense Mechanisms. *International Journal of Phytoremediation*, *16*(7-8), 770–789. doi:10.1080/15226514.2013.856848 PMID:24933884

Wolfaardt, J. F., Jacobs, A., Rabie, C. J., Smit, S., Taljaardt, J. L., & Wing-field, M. J. (1999). Factors affecting colonization of freshly chipped softwood by white-rot fungi. Poster presented at an International Energy Agency Bioenergy Workshop on the Bioconversion of Lignocellulose, Itala Game Reserve, South Africa.

Wolf, J., Prüss-Ustün, A., Cumming, O., Bartram, J., Bonjour, S., Cairncross, S., & Higgins, J. P. T. et al. (2014). Systematic review: Assessing the impact of drinking water and sanitation on diarrhoeal disease in low- and middle-income settings: systematic review and meta-regression. *Tropical Medicine & International Health*, *19*(8), 928–942. doi:10.1111/tmi.12331 PMID:24811732

Wolf, L., Nick, A., & Cronin, A. (2015). *How to keep your groundwater drinkable: Safer siting of sanitation systems – Working Group 11 Publication*. Sustainable Sanitation Alliance.

Wolka, K., & Melaku, B. (2015). Exploring selected plant nutrient in compost prepared from food waste and cattle manure and its effect on soil properties and maize yield at Wondo Genet, Ethiopia. *Environmental Systems Research*, *4*(1), 1–7. doi:10.1186/s40068-015-0035-0

Wolterbeek, H. T. (1987). Cation exchange in isolated xylem cell walls of tomato. I. Cd2+ and Rb+ exchange in adsorption experiments. *Plant, Cell & Environment*, *10*(1), 39–44. doi:10.1111/j.1365-3040.1987.tb02077.x

Wolt, J. (1994). *Soil solution chemistry*. New York, NY: John Wiley.

Wong, D. W. S. (2009). Structure and action mechanism of ligninolytic enzymes. *Applied Biochemistry and Biotechnology*, *157*(2), 174–209. doi:10.1007/s12010-008-8279-z PMID:18581264

Wood, T. K. (2008). Molecular approaches in bioremediation. *Current Opinion in Biotechnology*, *19*(6), 572–578. doi:10.1016/j.copbio.2008.10.003 PMID:19000765

Woodward, K. B., Fellows, C. S., Conway, C. L., & Hunter, H. M. (2009). Nitrate removal, denitrification and nitrous oxide production in the riparian zone of an ephemeral stream. *Soil Biology & Biochemistry*, *41*(4), 671–680. doi:10.1016/j.soilbio.2009.01.002

Woolard, C.R., & Irvine, R.L. (1995). Treatment of hypersaline wastewater in the sequencing batch reactor. *Wat. Res.*, *29*(4), 1159-68.

Woolard, C. R., & Irvine, R. L. (1994). Biological treatment of hypersaline wastewater by a biofilm of halophilic bacteria. *Water Environment Research*, *66*(3), 230–235. doi:10.2175/WER.66.3.8

World Health Organization (WHO). (2003). *Guidelines for Safe Recreational Water Environments, 1*. Geneva, Switzerland: Coastal and Freshwaters.

World Health Organization (WHO). (2006). Guidelines for drinking-water quality, 1, Recommendations. Geneva, Switzerland: WHO.

World Health Organization (WHO). (2008). Guidelines for Drinking-water Quality, 1, Incorporating 1st and 2nd Addenda, Recommendations (3rd ed.). Geneva, Switzerland: WHO.

Worrall, J. J., Anagnost, S. E., & Zabel, R. A. (1997). Comparison of wood decay among diverse lignicolous fungi. *Mycologia*, *89*(2), 199–219. doi:10.2307/3761073

Wuana, A., Raymond, O., & Felix, E. (2014). Mixed interactions in soil: Implications for bioavailability,risk assessment and remediation. *African J. Environmental Science & Technology*, *8*(12), 691–708.

Wu, C. H., Wood, T. K., Mulchandani, A., & Chen, W. (2006). Engineering plant-microbe symbiosis for rhizoremediation of heavy metals. *Applied and Environmental Microbiology*, *72*(2), 1129–1134. doi:10.1128/AEM.72.2.1129-1134.2006 PMID:16461658

Wu, J., & Yu, H. Q. (2007). Biosorption of 2,4-dichlorophenol by immobilized white-rot fungus *Phanerochaete chrysosporiyum* from aqueous solutions. *Bioresource Technology*, *98*(2), 253–259. doi:10.1016/j.biortech.2006.01.018 PMID:16517157

Wu, M. Z., Reynolds, D. A., Fourie, A., & Thomas, D. G. (2013). Optimal field approaches for electrokinetic *in situ* oxidation remediation. *Ground Water Monitoring and Remediation*, *33*(1), 62–74. doi:10.1111/j.1745-6592.2012.01410.x

Wurzbacher, C. M., Ba¨rlocher, F., & Grossart, H. P. (2010). Fungi in lake ecosystems. *Aquatic Microbial Ecology*, *59*, 125–149. doi:10.3354/ame01385

Wu, T., & Crapper, M. (2009). Simulation of biopile processes using a hydraulics approach. *Journal of Hazardous Materials*, *171*(1), 1103–1111. doi:10.1016/j.jhazmat.2009.06.131 PMID:19631454

Wu, Z. G., Wang, Y. M., Xing, Z. H., Wu, X. W., & Song, S. S. (2005). Study on degrading phenol by immobilized *Ralstonia metallidurans* CH34. *Microbiology*, *32*, 31–36.

Wu, Z., & Steel, K. M. (2007). Demineralization of a UK bituminous coal using HF and ferric ions. *Fuel*, *86*(14), 2194–2200. doi:10.1016/j.fuel.2007.02.038

Xiao, J., Liu, P., & Yang, W. (2015). Nanodiamonds from coal under ambient conditions. *Nanoscale*, *7*(14), 6114–6125. doi:10.1039/C4NR06186A PMID:25774000

Xiao, W., Wang, H., Li, T., Zhu, Z., Zhang, J., He, Z., & Yang, X. (2013). Bioremediation of Cd and carbendazim co-contaminated soil by Cd-hyperaccumulator Sedum alfredii associated with carbendazim-degrading bacterial strains. *Environmental Science and Pollution Research International*, *20*(1), 380–389. doi:10.1007/s11356-012-0902-4 PMID:22529002

Xia, S., Frigo-Vaz, B., Zhao, X., Kim, J., & Wang, P. (2014). Biocatalytic Ccapture via reversible reaction cycle catalyzed by isocitrate dehydrogenase. *Biochemical and Biophysical Research Communications*, *452*(1), 147–150. doi:10.1016/j.bbrc.2014.08.058 PMID:25152403

Xia, S., Zhao, X., Frigo-Vaz, B., Zheng, W., Kim, J., & Wang, P. (2015). Cascade enzymatic reactions for efficient Csequestration. *Bioresource Technology*, *182*, 368–372. doi:10.1016/j.biortech.2015.01.093 PMID:25708541

Xie, X., Lü, W., & Chen, X. (2013). Binding of the endocrine disruptors 4-tert-octylphenol and 4-nonylphenol to human serum albumin. *Journal of Hazardous Materials*, *248*, 347–354. doi:10.1016/j.jhazmat.2013.01.036 PMID:23416871

Xing, B., McGill, W. B., Dudas, M. J., Maham, Y., & Hepler, L. (1994). Sorption of phenol by selected polymers: Isotherms, energetics, and polarity. *Environmental Science & Technology*, *28*(3), 466–473. doi:10.1021/es00052a019 PMID:22165882

Xu, C., Ma, F., & Zhang, X. (2009). Lignocellulose degradation and enzyme production by Irpex lacteus CD2 during solid-state fermentation of cornstover. *Journal of Bioscience and Bioengineering*, *108*(5), 372–375. doi:10.1016/j.jbiosc.2009.04.023 PMID:19804859

Xueheng, Z., Hardin, I. R., & Huey-Min, H. (2006). Biodegradation of a model azo disperse dye bythe white rot fungus *Pleurotus ostreatus. International Biodeterioration & Biodegradation*, *57*(1), 1–6. doi:10.1016/j.ibiod.2005.10.008

Xu, F. F., & Imlay, J. A. (2012). Silver(I), mercury(II), cadmium(II), and zinc(II) target exposed enzymic iron-sulfur clusters when they toxify *Escherichia coli. Applied and Environmental Microbiology*, *78*(10), 3614–3621. doi:10.1128/AEM.07368-11 PMID:22344668

Xu, L., Teng, Y., Li, Z.-G., Norton, J. M., & Luo, Y.-M. (2010). Enhanced removal of polychlorinated biphenyls from alfalfa rhizosphere soil in a field study: The impact of a rhizobial inoculum. *The Science of the Total Environment*, *408*(5), 1007–1013. doi:10.1016/j.scitotenv.2009.11.031 PMID:19995667

Xu, M., Zhang, Q., Xia, C., Zhong, Y., Sun, G., Guo Jun, Y., & Zhi, L. et al. (2014). Elevated nitrate enriches microbial functional genes for potential bioremediation of complexly contaminated sediments. *ISME*, *8*(9), 1932–1944. doi:10.1038/ismej.2014.42 PMID:24671084

Xun, L. Y., Topp, E., & Orser, C. S. (1992). Confirmation of oxidative dehalogenation of pentachlorophenol by a *Flavobacterium* pentachlorophenol hydroxylase. *Journal of Bacteriology*, *174*(17), 5745–5747. doi:10.1128/jb.174.17.5745-5747.1992 PMID:1512208

Xu, Q., Nakajima, M., Liu, Z., & Shiina, T. (2011). Biosurfactants for microbubble preparation and application. *International Journal of Molecular Sciences*, *12*(1), 462–475. doi:10.3390/ijms12010462 PMID:21339998

Yadav, S. (2014). Glycoprotein of coelomic fluid triggers formation of aggregates/hetero-aggregates of ZnO-NPs in Earthworms. *Int. J. Curr. Microbiol. App. Sci.*, *3*(11), 1040–1045.

Yadav, S. (2016a). *Earthworm Biology: Recent Advances in techniques to study earthworms*. Lambert Academic Publishing.

Yadav, S. (2016b). Screening of Immuno-competent coelomic cells in earthworms. *International J of Sciences*, *5*(4), 43–51. doi:10.18483/ijSci.999

Yang, B., Hao, F., Li, J., Wei, K., Wang, W., & Liu, R. (2014). Characterization of the binding of chrysoidine, an illegal food additive to bovine serum albumin. *Food and Chemical Toxicology*, *65*, 227–232. doi:10.1016/j.fct.2013.12.047 PMID:24394486

Yang, M., Jawitz, J. W., & Lee, M. (2015). Uranium and cesium accumulation in bean (Phaseolus vulgaris L. var. vulgaris) and its potential for uranium rhizofiltration. *Journal of Environmental Radioactivity*, *140*, 42–49. doi:10.1016/j.jenvrad.2014.10.015 PMID:25461514

Yang, S. H., Cho, J. K., Lee, S. Y., Abanto, O. D., Kim, S. K., Ghosh, C., & Hwang, S. G. et al. (2013). Isolation and Characterization of Novel Denitrifying Bacterium *Geobacillus* spp. SG-01 Strain from Wood Chips Composted with Swine Manure. *Asian-Australasian Journal of Animal Sciences*, *26*(11), 1651–1658. doi:10.5713/ajas.2013.13184 PMID:25049754

Yang, S., Liang, S., Yi, L., Xu, B., Cao, J., Guo, Y., & Zhou, Y. (2014). Heavy metal accumulation and phytostabilization potential of dominant plant species growing on manganese mine tailings. *Frontiers of Environmental Science & Engineering*, *8*(3), 394–404. doi:10.1007/s11783-013-0602-4

Yan, J., Jianping, W., Hongmei, L., Suliang, Y., & Zongding, H. (2005). The biodegradation of phenol at high initial concentration by yeast *Candida tropicalis*. *Biochem. Eng.*, *24*(3), 243–247. doi:10.1016/j.bej.2005.02.016

Yanni, Y. G., & Dazzo, F. B. (2010). Enhancement of rice production using endophytic strains of Rhizobium leguminosarum bv. trifolii in extensive field inoculation trials within the Egypt Nile delta. *Plant and Soil*, *336*(1-2), 129–142. doi:10.1007/s11104-010-0454-7

Yan, Z., & Wang, J. (2016). Scoring Functions of Protein-Ligand Interactions. In *Methods and Algorithms for Molecular Docking-Based Drug Design and Discovery* (pp. 220–245). IGI Global. doi:10.4018/978-1-5225-0115-2.ch009

Yaron, B., Calvet, R., & Prost, R. (1996). *Soil pollution*. Berlin: Springer-Verlag. doi:10.1007/978-3-642-61147-6

Yasuda, M., Ozaki, K., Nohata, Y., Ohshiro, A., & Sonda, T. (2010). Removal of heavy metal ions from water by used green tea leaves. *Journal of Home Economics of Japan*, *61*(6), 349–354.

Yee, D. C., Maynard, J. A., & Wood, T. K. (1998). Rhizoremediation of Trichloroethylene by a Recombinant, Root-Colonizing Pseudomonas fluorescensStrain Expressing Toluene ortho-Monooxygenase Constitutively. *Applied and Environmental Microbiology*, *64*(1), 112–118. PMID:9435067

Ye, H., Zhang, L., Zhang, B., Wu, G., & Du, D. (2012). Adsorption removal of Cu (II) from aqueous solution using modified rice husk. *International Journal of Engineering Research and Applications*, *2*(2), 855–863.

Yelle, D. J., Ralph, J., Lu, F., & Hammel, K. E. (2008). Evidence for cleavage of lignin by a brown rot basidiomycete. *Environmental Microbiology*, *10*(7), 1844–1849. doi:10.1111/j.1462-2920.2008.01605.x PMID:18363712

Yenn, R., Borah, M., Boruah, H. D., Roy, A. S., Baruah, R., Saikia, N., & Tamuli, A. et al. (2014). Phytoremediation of abandoned crude oil contaminated drill sites of Assam with the aid of a hydrocarbon-degrading bacterial formulation. *International Journal of Phytoremediation*, *16*(9), 909–925. doi:10.1080/15226514.2013.810573 PMID:24933892

Yergeau, E., Sanschagrin, S., & Maynard, C. S. (2014). Microbial expression profiles in the rhizosphere of willows depend on soil contamination. *ISME*, *8*(2), 344-3458.

Yergeau, E., Sanschagrin, S., Beaumier, D., & Greer, C. W. (2012). Metagenomic analysis of the bioremediation of diesel-contaminated Canadian high arctic soils. *PLoS ONE*, *7*(1), e30058. doi:10.1371/journal.pone.0030058 PMID:22253877

Yoshitomi, K., & Shann, J. (2001). Corn (Zea mays L.) root exudates and their impact on 14 C-pyrene mineralization. *Soil Biology & Biochemistry*, *33*(12), 1769–1776. doi:10.1016/S0038-0717(01)00102-X

Yoshizawa, S., Tanaka, M., & Shekdar, A. V. (2004). Global trends in waste generation. In Recycling waste treatment and clean technology. TMS Mineral, Metals and Materials Publishers.

Younis, H. M., Abo-El-Saad, M. M., Abdel-Razik, R. K., & Abo-Seda, S. A. (2002). Resolving the DDT target protein in insects as a subunit of the ATP synthase. *Biotechnology and Applied Biochemistry*, *35*, 9–17. PMID:11834125

Youssef, A. M., Ghazy, T. M., & Elnabarawy, T. (1982). Moisture sorption by modified-activated carbons. *Carbon*, *20*(2), 113–116. doi:10.1016/0008-6223(82)90415-8

Yuan, H., Ge, T., Chen, C., ODonnell, A. G., & Wu, J. (2012). Significant role for microbial autotrophy in the sequestration of soil carbon. *Applied and Environmental Microbiology*, *78*(7), 2328–2336. doi:10.1128/AEM.06881-11 PMID:22286999

Yu, K., & Zhang, T. (2012). Metagenomic and metatranscriptomic analysis of microbial community structure and gene expression of activated sludge. *PLoS ONE*, *7*(5), e38183. doi:10.1371/journal.pone.0038183 PMID:22666477

Yuste, J. C., Penuelas, J., Estiarte, M., Garcia-Mas, J., Mattana, S., Ogaya, R., & Sardans, J. et al. (2011). Drought-resistant fungi control soil organic matter decomposition and its response to temperature. *Global Change Biology*, *17*(3), 1475–1486. doi:10.1111/j.1365-2486.2010.02300.x

Zabłudowska, E., Kowalska, J., Wojas, S., Skłodowska, A., & Antosiewicz, D. (2009). Search for a plant for phytoremediation–What can we learn from field and hydroponic studies? *Chemosphere*, *77*(3), 301–307. doi:10.1016/j.chemosphere.2009.07.064 PMID:19733893

Zadrazil, F., & Isikhuemhen, O. (1997). Solid state fermentation of lignocellulosics into animal feed with white rot fungi. In S. Roussos, B. K. Lonsane, M. Raimbault, & G. Viniegra-Gonzalez (Eds.), *Advances in Solid State Fermentation* (pp. 23–38). Dordrecht, The Netherlands: Kluwer Academic Publishers. doi:10.1007/978-94-017-0661-2_22

Zaidi, A., Ahemad, M., Oves, M., Ahmad, E., & Khan, M. S. (2010). Role of phosphate solubilizing bacteria in legume improvement.Microbes for Legume Improvement, 273-292. doi:10.1007/978-3-211-99753-6_11

Zamanpour, G., & Mehrabani-Zeinabad, A. (2014). An experimental study on bioremediation and photolysis of enrofloxacin. *Water Science and Technology*, *70*(5), 932–938. and doi:10.2166/wst.2014.320

Zanaroli, G., Perez-Jimenez, J., Young, L., Marchetti, L., & Fava, F. (2006). Microbial reductive dechlorination of weathered and exogenous co-planar polychlorinated biphenyls (PCBs) in an anaerobic sediment of Venice Lagoon. *Biodegradation*, *17*(2), 19–27. doi:10.1007/s10532-005-3752-7 PMID:16477348

Zelles, L., Bai, Q. Y., Beck, T., & Beese, F. (1992). Signature fatty acids in phospholipids and lipopolysaccharides as indicators of microbial biomass and community structure in agricultural soils. *Soil Biology & Biochemistry*, *24*(4), 317–323. doi:10.1016/0038-0717(92)90191-Y

Zelzer, M., Todd, S. J., Hirst, A. R., McDonald, T. O., & Ulijn, R. V. (2013). Enzyme responsive materials: Design strategies and future developments. *Biomaterials Science*, *1*(1), 11–39. doi:10.1039/C2BM00041E

Zeng, Z., Patel, J., Lee, S.-H., McCallum, M., Tyagi, A., Yan, M., & Shea, K. J. (2015). Synthetic polymer nanoparticle-polysaccharide interactions: A systematic study. *Journal of the American Chemical Society*, *134*(5), 2681–2690. doi:10.1021/ja209959t PMID:22229911

Zhan, D., Zhou, Z., Guan, S., & Han, W. (2013). The Effect of conformational variability of phosphotriesterase upon N-acyl-L-homoserine lactone and paraoxon binding: Insights from molecular dynamics studies. *Molecules (Basel, Switzerland)*, *18*(12), 15501–15518. doi:10.3390/molecules181215501 PMID:24352010

Zhang, L., Pemberton, J. E., & Maier, R. M. (2014). Effect of fatty acid substrate chain length on *Pseudomonas aeruginosa* ATCC 9027 monorhamnolipid yield and congener distribution. *Process Biochemistry*, *49*(6), 989–995. doi:10.1016/j.procbio.2014.03.003

Zhang, P., Sheng, G. Y., Feng, Y. C., & Miller, D. M. (2005). Role of wheat residue-derived char in the biodegradation of benzonitrile in soil: Nutritional stimulation versus adsorptive inhibition. *Environmental Science & Technology*, *39*(14), 5442–5448. doi:10.1021/es0480670 PMID:16082978

Zhang, T., Li, S., Sun, X., Zhang, Y., Gong, X., Fu, Y., & Jia, L. (2015). The earthworm *Eisenia fetida* can help desalinate a coastal saline soil in Tianjin, North China. *PLoS ONE*, *10*(12), e0144709. doi:10.1371/journal.pone.0144709 PMID:26699869

Zhang, Y., Zeng, Z., Zeng, G., Liu, X., Liu, Z., Chen, M., & Xie, G. et al. (2012). Effect of Triton X-100 on the removal of aqueous phenol by laccase analyzed with a combined approach of experiments and molecular docking. *Colloids and Surfaces. B, Biointerfaces*, *97*, 7–12. doi:10.1016/j.colsurfb.2012.04.001 PMID:22580478

Zhang, Z., Hou, Z., Yang, C., Ma, C., Tao, F., & Xu, P. (2011). Degradation of n-alkanes and polycyclic aromatic hydrocarbons in petroleum by a newly isolated *Pseudomonas aeruginosa* DQ8. *Bioresource Technology*, *102*(5), 4111–4116. doi:10.1016/j.biortech.2010.12.064 PMID:21227683

Zhang, Z., Lian, B., Hou, W., Chen, M., Li, X., & Li, Y. (2011). Bacillus mucilaginosus can capture atmospheric CO2 by carbonic anhydrase. *African Journal of Microbiological Research*, *5*, 106–112.

Zhao, H. P., Ilhan, Z. E., Ontiveros-Valencia, A., Tang, Y., Rittmann, B. E., & Krajmalnik-Brown, R. (2013). Effects of multiple electron acceptors on microbial interactions in a hydrogen-based biofilm. *Environmental Science & Technology*, *47*(13), 7396–7403. doi:10.1021/es401310j PMID:23721373

Zhao, X. W., Zhou, M. H., Li, Q. B., Lu, Y. H., He, N., Sun, D. H., & Deng, X. (2005). Simultaneous mercury bioaccumulation and cell propagation by genetically engineered *Escherichia coli*. *Process Biochemistry*, *40*(5), 1611–1616. doi:10.1016/j.procbio.2004.06.014

Zhao, Y. S., Zheng, Q. C., Zhang, H. X., Chu, H. Y., & Sun, C. C. (2009). Interaction between 1, 2, 3-trichloropropane and haloalkane dehalogenase LinB. *Chemical Research in Chinese Universities*, *25*(4), 556–559.

Zhou, C. H., Xia, X., Lin, C. X., Tong, D. S., & Beltramini, J. (2011). Catalytic conversion of lignocellulosic biomass to fine chemicals and fuels. *Chemical Society Reviews*, *40*(11), 5588–5617. doi:10.1039/c1cs15124j PMID:21863197

Zhou, M., Tang, Y., & Wu, Y. (2013). Plant hairy roots for remediation of aqueous pollutants. *Plant Molecular Biology Reporter*, *31*(1), 1–8. doi:10.1007/s11105-012-0465-z

Zhou, P., Su, C., Li, B., & Qian, Y. (2006). Treatment of high-strength pharmaceutical wastewater and removal of antibiotics in anaerobic and aerobic biological treatment processes. *Journal of Environmental Engineering*, *132*(1), 129–136. doi:10.1061/(ASCE)0733-9372(2006)132:1(129)

Zhou, S., & Ingram, L. O. (2000). Synergistic hydrolysis of carboxymethyl cellulose and acid-swollen cellulose by two endoglucanases (CelZ and CelY) from *Erwinia chrysanthemi*. *Journal of Bacteriology*, *182*(20), 5676–5682. doi:10.1128/JB.182.20.5676-5682.2000 PMID:11004164

Zhou, X., Li, X., & Chen, X. (2013a). Binding mechanism of Orange G to human serum albumin: Saturation transfer difference-NMR, spectroscopic and computational techniques. *Dyes and Pigments*, *98*(2), 212–220. doi:10.1016/j.dyepig.2013.01.023

Zhou, X., Wang, Y., Hu, T., Or, P. M., Wong, J., Kwan, Y. W., & Yeung, J. H. et al. (2013b). Enzyme kinetic and molecular docking studies for the inhibitions of miltirone on major human cytochrome P450 isozymes. *Phytomedicine*, *20*(3), 367–374. doi:10.1016/j.phymed.2012.09.021 PMID:23102508

Zhuang, X., Chen, J., Shim, H., & Bai, Z. (2007). New advances in plant growth-promoting rhizobacteria for bioremediation. *Environment International*, *33*(3), 406–413. doi:10.1016/j.envint.2006.12.005 PMID:17275086

Zhu, Y. G., & Miller, R. M. (2003). Ccycling by arbuscular mycorrhizal fungi in soil–plant systems. *Trends in Plant Science*, *8*(9), 407–409. doi:10.1016/S1360-1385(03)00184-5 PMID:13678905

Zhu, Z., Yang, X.-e., Wang, K., Huang, H., Zhang, X., Fang, H., & He, Z. et al. (2012). Bioremediation of Cd-DDT co-contaminated soil using the Cd-hyperaccumulator Sedum alfredii and DDT-degrading microbes. *Journal of Hazardous Materials*, *235*, 144–151. doi:10.1016/j.jhazmat.2012.07.033 PMID:22868749

Zielinski, M., Kahl, S., Hecht, H. J., & Hofer, B. (2003). Pinpointing biphenyl dioxygenase residues that are crucial for substrate interaction. *Journal of Bacteriology*, *185*(23), 6976–6980. doi:10.1128/JB.185.23.6976-6980.2003 PMID:14617661

Zuccato, E., Calamari, D., Natangelo, M., & Fanelli, R. (2000). Presence of therapeutic drugs in the environment. *Lancet*, *355*(9217), 1789–1790. and doi:10.1016/S0140-6736(00)02270-4

Zühlke, S., Dunnbier, U., & Heberer, T. (2004). Detection and identification of phenazone-type drugs and their microbial metabolites in ground and drinking water applying solid-phase extraction and gas chromatography with mass spectrometric detection. *Journal of Chromatography. A*, *1050*(2), 201–209. doi:10.1016/S0021-9673(04)01401-3 PMID:15508313

About the Contributors

Vasudha Agnihotri is working as a scientist in G. B. Pant National Institute of Himalayan Environment and Sustainable Development, Kosi- Katarmal, Almora (HQ). Focus area of her research is adsorption/ separation based studies including chromatographic and bulk chemistry studies of medicinal plants, water, and soil of Indian Himalayan region. She has masters in Chemical Sciences from Indian Institute of Technology, Roorkee and a doctorate in Chemical Sciences from Indian Institute of Petroleum, Mohkampur, Dehradun after receiving CSIR-NET fellowship. Her work in Indian Institute of Petroleum, Mohkampur, Dehradun involved the development of ion- exchanged zeolites for removing sulfur from gasoline, obtained from various refineries, up to the level of Bharat stage III emission norms. She is a member of Chemical Research Society of India and Indian National Science Congress. She has participated in several workshops and conferences to share knowledge and improve her research skills. She is engaged in drying isotherm studies of the medicinal plant of IHR (post harvest processing) and studies on biomass-based adsorbents for removal of pollutants.

Mohammad Ali Zahed obtained his PhD in Environmental Engineering in 2010. He was a post-doctoral fellow at Auburn University, AL, United States, where his research was on the monitoring of petroleum in the Gulf of Mexico after the Deepwater horizon oil spill. He also worked in the environmental projects of United Nation Compensation Commission (UNCC) for three years. Currently, he is an assistant professor in the faculty of biological sciences, Kharazmi University, Tehran, Iran. His research is focused on bioremediation and phytoremediation of organic pollutants.

Hamidi Abdul Aziz is a professor in environmental engineering in the School of Civil Engineering and the Head of the Solid Waste Management Cluster, Universiti Sains Malaysia. Professor Aziz received his PhD degree in civil engineering (environmental engineering) from the University of Strathclyde in Scotland in 1992. To date, he has published over 200 refereed articles in international journals and proceedings. He is the Editor-in-chief of the International Journal of Scientific Research in Environmental Sciences (IJSRES), and the Managing Editor of the International Journal of Environment and Waste Management (IJEWM) and International Journal of Environment and Waste Management (IJEE). Professor Aziz's research focuses on alleviating problems associated with water pollution issues from industrial wastewater discharges and from solid waste management via landfilling, such as landfill leachate. Advanced oxidation processes is one of his research focuses. He also has strong interest in biodegradation and bioremediation of oil spills and also in noise pollution. The details of his published works are available through Researcher ID: F-6836-2010 and SCOPUS ID: 7005960760.

Manoj B is Associate Professor in Physics, Christ University, Bengaluru. Prof. Manoj has received his PhD degree in Physics from Bharathiar University, Tamilnadu. To date, he has published over 45 refereed articles in international journals and is in the editorial board of many journals. His research group focuses on the synthesis of novel carbon materials including graphene quantum dots, nanocarbon and other nano carbon composites. He also has interest in biodemineralization, bioleaching and Energy studies. He has guided more than 10 MPhil and PhD scholars.

Alka Bali, M. Pharm. Ph.D., is working as an Associate Professor in Pharmaceutical Chemistry at University Institute of Pharmaceutical Sciences, UGC Centre of Advanced Study, Panjab University, Chandigarh, India. She has more than 20 years of teaching and research experience. Her areas of research interest include computer-aided drug design, synthetic medicinal chemistry and pharmaceutical analysis. She has been working on the design and synthesis of novel potential atypical antipsychotic agents and cyclooxygenase-2 inhibitors.

Zarrin Basharat graduated in bioinformatics from International Islamic University, Islamabad in 2011. Later, she enrolled in Fatima Jinnah Women University, Rawalpindi to pursue a PhD. with project at the intersection of microbiology, biotechnology and bioinformatics. Since then, she has been trying to apply statistical and computational methods to obtain new insights and propose solutions to health and environmental related problems. She has worked in collaboration with several national and international scientists outside her lab. In addition to research work, she is currently running 𝒝ioin𝓕orge βωβ (bio-informatics without borders) project to flourish knowledge sharing and bridging gaps in bioinformatics.

Monazza Bibi is a PhD scholar at Fatima Jiinnah Women University. She has worked on laccase enzymes for bioremediation.

Abdullah M. El Mahdi has a bachelor's degree in Environmental Engineering with specialization in Water Engineering from the University of Sebha, Libya. He obtained a Master's degree in Environmental Science and Engineering from the Post Graduate Academy of Benghazi at Libya and Doctorate degree in Environmental Engineering with specialization in Water and Waste Water Engineering from the University of Sains Malaysia (USM) recently. He has over five Publications with impact factors at ISI and Scope.He also has a strong interest in corrosion and its environment and also in corrosion control. The details of his published works are available through Researcher ID: A-5752-2015 and ORCID ID 0000-0002-2065-5552.

Hossein Farraji is a Ph.D candidate in Civil (Environmental) Engineering at Universiti Sains Malaysia (USM) from 2012.Author completed his Master of (Horticulture) Engineering at Azad University (AU), central unit of science and researches in 1999.Author has completed his Bachelor of (Horticulture) Engineering in 1997.He has 11 years governmental job experiences in Institute of Standard and Industrial Researches of Iran (ISIRI). His research area is application green technology (phytoremediation and constructed wetland) for wastewater treatment. Designing vertical, horizontal and combined constructed wetlands for higher decontamination efficiency especially for high strength agroindustrial wastewater following traditional treatment methods is major of his studies.

Reena Gupta is a Professor in Department of Biotechnology, Himachal Pradesh University, Shimla.

Veena Gayathri Krishnaswamy received her Ph.D from Center for Environmental Studies, Anna University.She received Fellowship -UGC-JRF for Meritorious students in Science during her Ph.D research work. Currently she is working as an Assistant Professor, Department of Biotechnology, Stella Maris College, Chennai. Her research work is on Biodegradation of tannins, tannic acid, Reative azo dyes decolorisation, Chlorophenols degradation. Vermicomposting for the treatment of solid wastes and chromium removal. Most of her work is specialized on degradation of pollutants using extremophilic bacteria. She has published her findings in about 25 international journals.

Ajay Kumar has secured his Ph.D. degree from department of Botany (spl-Environmental Microbiology) Dayalbagh Educational Institute under Prof. J.N. Shrivastava (Head) Dept. Botany, in 2008. He has guided three Ph.D. students and completed two research projects. He is also actively engaged in various academic and research bodies. He has more than eight year teaching experience for UG and PG classes in various institutes. He has published several research papers in reputed Journals and he has presented many research papers in National and international Conferences.

Debajyoti Kundu has completed his M.sc from The Unviersity of Burdwan and M.Tech from University of Kalyani. Till now he has published 3 research papers in international journal.

Susmita Lahiri is serving as Assistant Professor in the International Centre for Ecological Engineering (ICEE) and Department of Ecological Studies, University of Kalyani, India. She has successfully worked as Co-PI in the completed project entitled "Assessment of Impact of Climate Change on Inland Aquaculture in selected districts of West Bengal" from 2010-2013 and ongoing project "Safeguarding Water resources in India with Green and Sustainable technologies". She is carrying out research for more than 15 years on aquatic microbiology, wastewater management, ecological engineering and aquaculture and received Ph. D. from Department of Zoology, University of Kalyani, India. Other research areas include carbon sequestration using microbes as eco-signature, adsorption, reclamation of aquatic environment (wastewater & eutrophic wetlands). She has experience of teaching on environment management and has organized and participated in several international and national seminars, conferences etc. She has published more than 20 papers in peer reviewed international and national journals.

Geetanjali Manchanda's research focus is on the application of plant physiology, molecular biology and microbiology for rhizoremediation of soils affected by contaminants. She has worked on the interactions of plants, bacteria and mycorrhizal fungi in relation to mycorrhizoremediation polluted soils.

Sunil Kumar Narwal is a Research Scholar in the Department of Biotechnology, Himachal Pradesh University, Shimla.

Nastaein Qamaruz Zaman obtained PhD in Civil (Environmental) Engineering at University of Canterbury, New Zealand in 2010. She joined Universiti Sains Malaysia in 2010 and is now focusing her teaching and research passion in waste management and odor pollution. Dr. Nastaein has authored several peer reviewed journal and book chapters. Currently sits in the Expert Committee on Odour Regulation with the Department of Environment Malaysia and actively involved with consultancy projects in odor pollution in Malaysia. She has also received several research grants from the Ministry of Higher Education Malaysia (ERGS and FRGS grants), the USM RUI grant and from MPOB.

Sukanta Rana received Ph. D. from International Centre for Ecological Engineering, University of Kalyani. Dr. Rana has research experience in ecological engineering, Aquaponics, Hydroponics, Ecological sanitation, urine-fed- aquaculture, Fertilizer synthesis from human urine etc. He published about 20 research papers in his field.

Raghvendra Pratap Singh has wide array of work on proteomics and Genomics. His current research profile is to comprehensive mining of proteome of predator and their Ecology and interaction study with other microflora.

Azra Yasmin has a knack for interdisciplinary work and her core areas of interest include microbiology, micro and macro environment dynamics, cancer genetics and bioinformatics. More than fifty peer reviewed publications in national and international journals of repute are to her credit. She has been awarded various scholarships during her studies and is the recipient of HEC best teacher award (2009). Apart from her administrative duties as a dean and chairperson, she has worked as a consultant for microbial assisted bioremediation projects, designed undergrad and graduate curriculum and held several honorary positions, including being on board of directors of Rawalpindi waste management company. She is also member of several learned societies and volunteers for women organization 'Soroptimist International' apart from her entrepreneurship ventures.

Index

R

S

T

V

W

X

Y

Z

www.ingramcontent.com/pod-product-compliance
Lightning Source LLC
LaVergne TN
LVHW080617300926
849162LV00005B/1194